APPLET CORRELATION

W9-AGF-555

Applet	Concept Illustrated	Description	Applet Exercise Number and Page
Sample from a population	This applet illustrates the concept that a sample should be representative of the population from which it is drawn. It may be used to assess how well a sample represents the population and the role that sample size plays in this process.	This applet produces a random sample from a population. The user may specify the sample size and the shape of the population distribution. The applet reports the mean, median, and standard deviation of the sample and creates a plot of the sample.	**4.4**, p. 194; **5.1**, p. 231; **5.3**, p. 247 **4.6**, p. 211
Sampling distributions	This applet illustrates the concept that the mean and standard deviation of the distribution of sample means are unbiased estimators of the mean and standard deviation of the population distribution. It may be used to compare the means and standard deviations of the distributions and to assess the effect of sample size. It can also be used to study sample proportions.	This applet simulates repeatedly choosing samples of a fixed size n from a population. The user may specify the size of the sample, the number of samples to be chosen, and the shape of the population distribution. The applet reports the means, medians, and standard deviations of both the sample means and the sample medians and also creates plots for both.	**4.7**, p. 240; **4.8**, p. 240
Random numbers	This applet illustrates the concept of randomness. It may be used to simulate selecting a random sample from a population by first assigning a unique integer to each experimental unit and then using the random numbers generated by the applet to determine the experimental units that will be included in the sample.	This applet generates random numbers from a range of integers specified by the user.	**1.1**, p. 18; **1.2**, p. 19; **3.6**, p. 159; **4.1**, p. 180; **5.2**, p. 231
Long run probability demonstrations: Simulating the probability of rolling a 6	This applet illustrates the concept that theoretical probabilities are long run experimental probabilities. It can be used to investigate the relationship between the theoretical and experimental probabilities of rolling a 6 as the number of times the die is rolled increases.	This applet simulates rolling a fair die. The user specifies the number of rolls. The applet reports the outcome of each roll and creates a frequency histogram for the outcomes. It also calculates and plots the proportion of 6s rolled during the simulation.	**3.1**, p. 124; **3.3**, p. 135; **3.4**, p. 135; **3.5**, p. 149
Simulating the probability of rolling a 3 or 4	This applet illustrates the concept that theoretical probabilities are long run experimental probabilities. It can be used to investigate the relationship between the theoretical and experimental probabilities of rolling a 3 or a 4 as the number of times the die is rolled increases.	This applet simulates rolling a fair die. The user specifies the number of rolls. The applet reports the outcome of each roll and creates a frequency histogram for the outcomes. It also calculates and plots the proportion of 3s and 4s rolled during the simulation.	**3.3**, p. 135; **3.4**, p. 135
Simulating the probability of a head with a fair coin	This applet illustrates the concept that theoretical probabilities are long run experimental probabilities. It can be used to investigate the relationship between the theoretical and experimental probabilities of getting heads as the number of times the coin is flipped increases.	This applet simulates flipping a fair coin. The user specifies the number of flips. The applet reports the outcome of each flip and creates a bar graph for the outcomes. It also calculates and plots the proportion of heads rolled during the simulation.	**3.2**, p. 124; **4.2**, p. 180
Simulating the probability of a head with an unfair coin $(P(H) = .02)$	This applet illustrates the concept that theoretical probabilities are long run experimental probabilities. It can be used to investigate the relationship between the theoretical and experimental probabilities of getting heads as the number of times an unfair coin is flipped increases.	This applet simulates flipping a coin where heads is less likely to occur than tails. The user specifies the number of flips. The applet reports the outcome of each flip and creates a bar graph for the outcomes. It also calculates and plots the proportion of heads rolled during the simulation.	**4.3**, p. 194
Simulating the probability of a head with an unfair coin $(P(H) = .08)$	This applet illustrates the concept that theoretical probabilities are long run experimental probabilities. It can be used to investigate the relationship between the theoretical and experimental probabilities of getting heads as the number of times an unfair coin is flipped increases.	This applet simulates flipping a coin where heads is more likely to occur than tails. The user specifies the number of flips. The applet reports the outcome of each flip and creates a bar graph for the outcomes. It also calculates and plots the proportion of heads rolled during the simulation.	**4.3**, p. 194
Simulating the stock market	This applet illustrates the concept that theoretical probabilities are long run experimental probabilities.	This applet simulates fluctuation in the stock market, where on any given day going up is equally likely as going down. The user specifies the number of days. The applet reports whether the stock market goes up or down each day and creates a bar graph for the outcomes. It also calculates and plots the proportion of days that the stock market goes up during the simulation.	**4.5**, p. 194

Applet	Concept Illustrated	Description	Applet Exercise Number and Page
Mean versus median	This applet illustrates the concept that the mean and the median of a data set respond differently to changes in the data. It can be used to investigate how skewness and outliers affect measures of central tendency.	This applet is designed to allow the user to visualize the relationship between the mean and median of a data set. The user may easily add and delete data points. The applet automatically updates the mean and median for each change in the data.	**2.1**, p. 58; **2.2**, p. 58; **2.3**, p. 58
Standard deviation	This applet illustrates the concept that standard deviation measures the spread of a data set. It can be used to investigate how the shape and spread of a distribution affect the standard deviation.	This applet is designed to allow the user to visualize the relationship between the mean and standard deviation of a data set. The user may easily add and delete data points. The applet automatically updates the mean and standard deviation for each change in the data.	**2.4**, p. 66; **2.5**, p. 66; **2.6**, p. 66; **2.7**, p. 87
Confidence intervals for a proportion	This applet illustrates the concept that not all confidence intervals contain the population proportion. It can be used to investigate the meaning of 95% and 99% confidence.	This applet generates confidence intervals for a population proportion. The user specifies the population proportion and the sample size. The applet simulates selecting 100 random samples from the population and finds the 95% and 99% confidence intervals for each sample. The confidence intervals are plotted and the number and proportion containing the true proportion are reported.	**5.5**, p. 281; **5.6**, p. 281
Confidence intervals for a mean (the impact of confidence level)	This applet illustrates the concept that not all confidence intervals contain the population mean. It can be used to investigate the meaning of 95% and 99% confidence.	This applet generates confidence intervals for a population mean. The user specifies the sample size, the shape of the distribution, the population mean, and the population standard deviation. The applet simulates selecting 100 random samples from the population and finds the 95% and 99% confidence intervals for each sample. The confidence intervals are plotted and the number and proportion containing the true mean are reported.	**5.1**, p. 263; **5.2**, p. 263
Confidence intervals for a mean (the impact of not knowing the standard deviation)	This applet illustrates the concept that confidence intervals obtained using the sample standard deviation are different from those obtained using the population standard deviation. It can be used to investigate the effect of not knowing the population standard deviation.	This applet generates confidence intervals for a population mean. The user specifies the sample size, the shape of the distribution, the population mean, and the population standard deviation. The applet simulates selecting 100 random samples from the population and finds the 95% z-interval and 95% t-interval for each sample. The confidence intervals are plotted and the number and proportion containing the true mean are reported.	**5.3**, p. 273; **5.4**, p. 273
Hypotheses tests for a proportion	This applet illustrates the concept that not all tests of hypotheses lead correctly to either rejecting or failing to reject the null hypothesis. It can be used to investigate the relationship between the level of confidence and the probabilities of making Type I and Type II errors.	This applet performs hypotheses tests for a population proportion. The user specifies the population proportion, the sample size, and the null and alternative hypotheses. The applet simulates selecting 100 random samples from the population and calculates and plots the z statistic and p-value for each sample. The applet reports the number and proportion of times the null hypotheses is rejected at both the .05 level and the .01 level.	**6.5**, p. 334; **6.6**, p. 334
Hypotheses tests for a mean	This applet illustrates the concept that not all tests of hypotheses lead correctly to either rejecting or failing to reject the null hypothesis. It can be used to investigate the relationship between the level of confidence and the probabilities of making Type I and Type II errors.	This applet performs hypotheses tests for a population mean. The user specifies the shape of the population distribution, the population mean and standard deviation, the sample size, and the null and alternative hypotheses. The applet simulates selecting 100 random samples from the population and calculates and plots the t statistic and p-value for each sample. The applet reports the number and proportion of times the null hypotheses is rejected at both the .05 level and the .01 level.	**6.1**, p. 307; **6.2**, p. 313; **6.3**, p. 313; **6.4**, p. 313
Correlation by eye	This applet illustrates the concept that the correlation coefficient measures the strength of a linear relationship between two variables. It helps the user learn to assess the strength of a linear relationship from a scattergram.	This applet computes the correlation coefficient r for a set of bivariate data plotted on a scattergram. The user can easily add or delete points and guess the value of r. The applet then compares the guess to its calculated value.	**9.2**, p. 523
Regression by eye	This applet illustrates the concept that the least squares regression line has a smaller SSE than any other line that might approximate a set of bivariate data. It helps the user learn to approximate the location of a regression line on a scattergram.	This applet computes the least squares regression line for a set of bivariate data plotted on a scattergram. The user can easily add or delete points and guess the location of the regression line by manipulating a line provided on the scattergram. The applet will then plot the least squares line. It displays the equations and the SSEs for both lines.	**9.1**, p. 497

A First Course in Statistics

A First Course in Statistics

TENTH EDITION

James T. McClave

Info Tech, Inc.
University of Florida

Terry Sincich

University of South Florida

PEARSON

Prentice
Hall

Upper Saddle River, New Jersey 07458

McClave, James T.
  A first course in statistics / James T. McClave. — 10th ed.
     p. cm.
Includes index.
ISBN-13: 978-0-13-615259-0
1. Statistics. I. Title.
QA276.M378 2009
519.5—dc22   2007044390

Vice President and Editorial Director, Mathematics: *Christine Hoag*
Editor-in-Chief, Mathematics & Statistics: *Deirdre Lynch*
Sponsoring Editor: *Dawn Murrin*
Editorial Assistant/Print Supplements Editor: *Joanne Wendelken*
Project Manager, Production: *Raegan Keida Heerema*
Associate Managing Editor: *Bayani Mendoza deLeon*
Senior Managing Editor: *Linda Mihatov Behrens*
Senior Operations Supervisor: *Diane Peirano*
Media Project Manager: *Richard Bretan*
Marketing Manager: *Wayne Parkins*
Marketing Assistant: *Kathleen deChavez*
Senior Art Director: *Juan R. López*
Interior Designer: *Kristine Carney*
Cover Designer: *Michael J. Fruhbeis*
AV Project Manager: *Thomas Benfatti*
Manager, Cover Visual Research & Permissions: *Karen Sanatar*
Director, Image Resource Center: *Melinda Patelli*
Manager, Rights and Permissions: *Zina Arabia*
Manager, Visual Research: *Beth Brenzel*
Image Permission Coordinator: *Fran Toepfer*
Art Studio: *GEX*
Compositor: *S4Carlisle Publishing Services*

COVER IMAGE CREDITS
Beach scene   Purestock / *Purestock*
Little Girl   Reportage / *Scott Nelson*
Children   Blend Images / *Andersen Ross*

 © 2009 by Pearson Education, Inc.
Pearson Prentice Hall
Pearson Education, Inc.
Upper Saddle River, NJ 07458

Printed in the United States of America

10 9 8 7 6 5 4 3 2 1

ISBN-13: 978-0-13-615259-0
ISBN-10:   0-13-615259-7

Pearson Education, Ltd., *London*
Pearson Education Australia PTY. Limited, *Sydney*
Pearson Education Singapore, Pte., Ltd
Pearson Education North Asia Ltd, *Hong Kong*
Pearson Education Canada, Ltd., *Toronto*
Pearson Educación de Mexico, S.A. de C.V.
Pearson Education – Japan, *Tokyo*
Pearson Education Malaysia, Pte. Ltd

Contents

Preface   xi

Applications Index   xvii

CHAPTER 1   Statistics, Data, and Statistical Thinking   2

1.1   The Science of Statistics   4

1.2   Types of Statistical Applications   4

1.3   Fundamental Elements of Statistics   6

1.4   Types of Data   11

1.5   Collecting Data   13

1.6   The Role of Statistics in Critical Thinking   15

Statistics in Action: *USA Weekend* Teen Surveys – Are Boys Really from Mars and Girls from Venus?   3

Using Technology: Creating and Listing Data with MINITAB   23

CHAPTER 2   Methods for Describing Sets of Data   26

2.1   Describing Qualitative Data   28

2.2   Graphical Methods for Describing Quantitative Data   38

2.3   Summation Notation   49

2.4   Numerical Measures of Central Tendency   51

2.5   Numerical Measures of Variability   62

2.6   Interpreting the Standard Deviation   67

2.7   Numerical Measures of Relative Standing   75

2.8   Methods for Detecting Outliers (Optional)   79

2.9   Graphing Bivariate Relationships (Optional)   89

2.10   Distorting the Truth with Descriptive Techniques   94

Statistics In Action: The "Eye Cue" Test: Does Experience Improve Performance?   27

Using Technology: Describing Data with MINITAB   109

CHAPTER 3   Probability   102

3.1   Events, Sample Spaces, and Probability   114

3.2   Unions and Intersections   127

3.3   Complementary Events   130

3.4   The Additive Rule and Mutually Exclusive Events   132

3.5   Conditional Probability   138

3.6   The Multiplicative Rule and Independent Events   141

3.7   Random Sampling   154

Statistics In Action: Lotto Buster! – Can You Improve Your Chances of Winning the Lottery?   113

Using Technology: Generating a Random Sample with MINITAB   166

CHAPTER 4   Random Variables and Probability Distributions   168

4.1   Two Types of Random Variables   170

4.2   Probability Distributions for Discrete Random Variables   173

4.3   The Binomial Distribution   183

4.4   Probability Distributions for Continuous Random Variables   196

4.5   The Normal Distribution   197

4.6   Descriptive Methods for Assessing Normality   213

4.7   Approximating a Binomial Distribution with a Normal Distribution (Optional)   221

4.8   Sampling Distributions   227

4.9   The Central Limit Theorem   234

Statistics in Action: Super Weapons Development – Is the Hit Ratio Optimized?   169

Using Technology: Binomial Probabilities, Normal Probabilities, and Normal Probability Plots with MINITAB   249

CHAPTER 5   Inferences Based on a Single Sample: Estimation with Confidence Intervals   254

5.1   Identifying the Target Parameter   255

5.2   Large-Sample Confidence Interval for a Population Mean   256

5.3   Small-Sample Confidence Interval for a Population Mean   266

5.4   Large-Sample Confidence Interval for a Population Proportion   276

5.5   Determining the Sample Size   283

Statistics in Action: Speed – Can a High School Football Player Improve His Sprint Time?   255

Using Technology: Confidence Intervals with MINITAB   297

CHAPTER 6   Inferences Based on a Single Sample: Tests of Hypothesis   300

6.1   The Elements of a Test of Hypothesis   301

6.2   Large-Sample Test of Hypothesis About a Population Mean   308

6.3   Observed Significance Levels: p-Values   315

6.4   Small-Sample Test of Hypothesis About a Population Mean   322

6.5   Large-Sample Test of Hypothesis About a Population Proportion   329

6.6   A Nonparametric Test about a Population Median (Optional)   336

Statistics in Action: Diary of a Kleenex User – How Many Tissues in a Box?   301

Using Technology: Tests of Hypothesis with MINITAB   348

CHAPTER 7   **Comparing Population Means   350**

7.1   Comparing Two Population Means: Independent Sampling   352

7.2   Comparing Two Population Means: Paired Difference Experiments   370

7.3   Determining the Sample Size   383

7.4   A Nonparametric Test for Comparing Two Population: Independent Sampling (Optional)   386

7.5   A Nonparametric Test for Comparing Two Populations: Paired Difference Experiments (Optional)   395

7.6   Comparing Three or More Population Means: Analysis of Variance (Optional)   403

Statistics in Action: On the Trail of the Cockroach: Do Roaches Travel at Random?   351

Using Technology: Comparing Means with MINITAB   428

CHAPTER 8   **Comparing Population Proportions   432**

8.1   Comparing Two Population Proportions: Independent Sampling   434

8.2   Determining the Sample Size   442

8.3   Testing Categorical Probabilities: Multinomial Experiment   444

8.4   Testing Categorical Probabilities: Two-Way (Contingency) Table   454

Statistics in Action: College Students and Alcohol – Is Amount Consumed Related to Drinking Frequency?   433

Using Technology: Chi-Square Analyses with MINITAB   478

CHAPTER 9   **Simple Linear Regression   482**

9.1   Probabilistic Models   484

9.2   Fitting the Model: The Least Squares Approach   487

9.3   Model Assumptions   502

9.4   Assessing the Utility of the Model: Making Inferences About the Slope β_1 507

9.5   The Coefficients of Correlation and Determination   515

9.6   Using the Model for Estimation and Prediction   525

9.7   A Complete Example   533

9.8   A Nonparametric Test for Correlation (Optional)   536

Statistics in Action: Can "Dowsers" Really Detect Water?   483

Using Technology: Simple Linear Regression with MINITAB   557

APPENDICES

Appendix A   TABLES

Table I	Random Numbers 560
Table II	Binomial Probabilities 563
Table III	Normal Curve Areas 567
Table IV	Critical Values of t 568
Table V	Critical Values of T_L and T_U for the Wilcoxon Rank Sum Test: Independent Samples 569
Table VI	Critical Values of T_0 in the Wilcoxon Paired Difference Signed Rank Test 570
Table VII	Percentage Points of the F Distribution, $\alpha = .10$ 571
Table VIII	Percentage Points of the F Distribution, $\alpha = .05$ 573
Table IX	Percentage Points of the F Distribution, $\alpha = .025$ 575
Table X	Percentage Points of the F Distribution, $\alpha = .01$ 577
Table XI	Critical Values of χ^2 579
Table XII	Critical Values of Spearman's Rank Correlation Coefficient 581

Appendix B   CALCULATION FORMULAS FOR ANALYSIS OF VARIANCE (INDEPENDENT SAMPLING)   582

Short Answers to Selected Odd-Numbered Exercises   583

Index   589

Preface

This 10th edition of *A First Course in Statistics* is an introductory text emphasizing inference, with extensive coverage of data collection and analysis as needed to evaluate the reported results of statistical studies and make good decisions. As in earlier editions, the text stresses the development of statistical thinking, the assessment of credibility, and the value of the inferences made from data, both by those who consume and those who produce them. It assumes a mathematical background of basic algebra.

The text incorporates the following features developed from the American Statistical Association's Guidelines for Assessment and Instruction in Statistics Education (GAISE) Project:

- Emphasize statistical literacy and develop statistical thinking.
- Use real data in applications.
- Use technology for developing a conceptual understanding and for analyzing data.
- Foster active learning in the classroom.
- Stress conceptual understanding rather than mere knowledge of procedures.

NEW IN THE 10TH EDITION

- ***Over 1,000 Exercises, with Revisions and Updates to 30%*** Many new and updated exercises, based on contemporary studies and real data, have been added. Most of these exercises foster and promote critical thinking skills. In addition to "Learning the Mechanics" exercises, "Applied Exercises" are categorized into "Basic", "Intermediate", and "Advanced" at the end of each section.

- ***New Visual End-of-Chapter Summaries*** Flow graphs for selecting the appropriate statistical method, as well as boxed notes with key words, formulas, definitions, lists, and key concepts, are now provided at the end of each chapter. This graphical presentation is especially helpful to those students who are visual learners. It aids students by summarizing and reinforcing the important points from the chapter.

- ***"Hands-On" Activities for Students*** In each chapter, students are provided with several opportunities to participate in hands-on classroom activities, ranging from real data collection to formal statistical analysis. These powerful, optional activities, based on the key concepts and procedures covered in the chapter, can be performed by students individually or as a class.

- ***Applet Exercises*** The text is accompanied by a CD containing applets (short JAVA computer programs). These point-and-click applets allow students to easily run simulations that demonstrate some of the more difficult statistical concepts (e.g., sampling distributions and confidence levels). Each chapter contains several optional applet exercises in the exercise sets. They are denoted with the following Applet icon: APPLET .

- ***New Statistics in Action Case*** A new *Statistics in Action* case on the relationship between how often and how much college students drink alcohol (Chapter 9) has been added. Each *Statistics in Action* case centers on a contemporary controversial or high-profile issue and the accompanying data.

- ***Data files and Applets*** The CD that accompanies the text contains files for all of the data sets marked with a CD icon. These include data sets for text examples, exercises, and *Statistics in Action* cases. All data files are saved in four different formats: MINITAB, SAS, SPSS, and ASCII (for easy importing into other statistical software packages). The CD also contains the applets that are used to illustrate statistical concepts.

Content-Specific Changes to This Edition

- *Chapter 9 ("Simple Linear Regression").* Several sections from the previous edition have been combined and streamlined to shorten the chapter. The section on estimating σ^2 is now included in Section 9.3 (Model Assumptions), while the sections on the coefficients of correlation and determination are combined into a single section (Section 9.5).

TRADITIONAL STRENGTHS

We have maintained the pedagogical features of *A First Course in Statistics* that we believe make it unique among introductory statistics texts. These features, which assist the student in achieving an overview of statistics and an understanding of its relevance to everyday life, are as follows:

- *Use Examples as a Teaching Device* Almost all new ideas are introduced and illustrated by a collection of over 100 data-based applications and examples. We believe that students understand definitions, generalizations, and theoretical concepts better after seeing an application. All examples have three components: (1) "Problem"; (2) "Solution"; and (3) "Look Back". This step-by-step process provides students with a defined structure by which to approach problems and enhances their problem-solving skills. The "Look Back" feature often gives helpful hints toward solving the problem and/or provides a further reflection or insight into the concept or procedure that is covered.

- *Now Work*– A "Now Work" exercise follows each example. "Now Work" directs the student to an end-of-section exercise that is similar in style and concept to the example in the text. This gives the student the opportunity to test and confirm his or her understanding of the concept taught.

- *Statistics in Action* Each chapter begins with a case study of an actual contemporary controversial or high-profile issue. Relevant research questions and data from the study are presented, and the proper analysis is demonstrated in short "Statistics in Action Revisited" sections throughout the chapter. These sections motivate students to critically evaluate the findings and think through the statistical issues involved

- *Real Data Exercises*– The text includes more than 1,000 exercises based on applications in a variety of disciplines and research areas. All of the applied exercises employ the use of current real data extracted from a wide variety of current publications (e.g., newspapers, magazines, journals, and the Internet). Still, some students have difficulty learning the mechanics of statistical techniques when all problems are couched in terms of realistic applications. For this reason, all exercise sections are divided into five parts:

  Understanding the Principles Short-answer exercises deal with definitions, concepts, and assumptions.

  Learning the Mechanics Designed as straightforward applications of new concepts, these exercises allow students to test their ability to comprehend a mathematical concept or a definition.

  Applying the Concepts—Basic Based on applications taken from a wide variety of journals, newspapers, and other sources, these short exercises help the student to begin developing the skills necessary to diagnose and analyze real-world problems.

  Applying the Concepts—Intermediate Based on more detailed real-world applications, these exercises require the student to apply critical thinking and knowledge of the technique presented in the section

  Applying the Concepts—Advanced These more challenging real-data exercises require the student to utilize critical thinking skills.

- *Critical Thinking Challenges* At the end of the "Chapter Supplementary Exercises," students are asked to apply their critical thinking skills to solve one or two challenging real-life problems. These exercises expose students to real-world problems with solutions that are derived from careful, logical thought and the selection of the appropriate statistical analysis tool

- *Exploring Data with Statistical Computer Software and the Graphing Calculator* Each statistical analysis method presented is demonstrated with the use of output from three leading Windows-based statistical software packages: SPSS, MINITAB, and SAS. In addition, output and keystroke instructions for the TI-83 Graphing Calculator are covered in optional boxes that are easy to locate throughout the text.

- *Statistical Software Printouts* These appear throughout the text in examples and exercises, and include MINITAB, as well as SPSS and SAS, printouts. Students are exposed to the computer printouts they will encounter in the hi-tech world.

- *"Using Technology" Tutorials* MINITAB Software Tutorials appear at the end of each chapter. They include point-and-click instructions (with screen shots) for MINITAB. These tutorials are easily located and provide students with useful information on how to best use and maximize MINITAB statistical software.

- *Profiles of Statisticians in History (Biography)* Boxes featuring famous statisticians give brief descriptions of their achievements. With these profiles, students will develop an appreciation of statisticians efforts and the discipline of statistics as a whole.

ACKNOWLEDGMENTS

This book reflects the efforts of a great many people over a number of years. First, we would like to thank the following professors, whose reviews and comments on this and previous editions have contributed to the 10th edition:

Reviewers Involved with the Current Edition of *A First Course in Statistics*

Dwight Galster, South Dakota State University
Geoffrey Exoo, Indiana State University
Yvonne Chueh, Central Washington University
Georgiana Baker, University of South Carolina
Mohammad Kazemi, University of North Carolina, Charlotte
John Holcomb, Cleveland State University
Mary Ehlers, Seattle University
Barbara Wainwright, Salisbury University
Rasul Khan, Cleveland State University
Jan Case, Jacksonville State University

Reviewers of Previous Editions

Bill Adamson, South Dakota State; Ibrahim Ahmad, Northern Illinois University; Roddy Akbari, Guilford Technical Community College; David Atkinson, Olivet Nazarene University; Mary Sue Beersman, Northeast Missouri State University; William H. Beyer, University of Akron; Marvin Bishop, Manhattan College; Patricia M. Buchanan, Pennsylvania State University; Dean S. Burbank, Gulf Coast Community College; Ann Cascarelle, St. Petersburg College; Kathryn Chaloner, University of Minnesota; Hanfeng Chen, Bowling Green State University; Gerardo Chin-Leo, The Everygreen State College; Linda Brant Collins, Iowa State University; Brant Deppa, Winona State University; John Dirkse, California State University—Bakersfield; N. B. Ebrahimi, Northern Illinois University; John Egenolf, University of Alaska—Anchorage; Dale Everson, University of Idaho; Christine Franklin, University of Georgia; Khadiga Gamgoum, Northern Virginia CC; Rudy Gideon, University of Montana; Victoria Marie Gribshaw, Seton Hill

awaking sleepers early, 275–276
baby weight, cigarette smoking vs., 539–541
blood loss, drug designed to reduce, 31–32
body fat, in men, 227
brain-specimen research, 48, 88, 275
bulimia, 367, 386
burn patients, blood loss in, 21
Caesarian births, 195, 226
cancer:
    brain, 149
    and cigar smoking, 151
    lung, CT scanning for, 20
    skin, 177–178
    and smoking, 139–140
    types treated, 101
carbon monoxide poisoning, 160–161
CDC health survey, 292
Chinese herbal drugs, 105
cholesterol levels in psychiatric patients, 246
clinical trial, 156–157
dementia, linking leisure activities and, 381–382
Depo-Provera, 424
depression, treating with St. John's wort, 440
dust mite allergies, 181
epilepsy, 242
eye refractive study, 61, 78
fertility rates, 555
fitness of heart patients, 212
gestation for pregnant women, length of, 212–213
hand washing vs. hand rubbing, 73–74, 242
head trauma, 329
health hazards of housework, 386
hearing impaired, conversing with, 553
hearing loss, 293
    in senior citizens, 106
heart rate during laughter, 314
heart patients:
    animal-assisted therapy for heart patients, 6, 74–75, 380, 418–419
    healing with music/imagery/touch/ prayer, 465
herbal medicines, 20, 150
herbal therapy, 307–308
HIV vaccine, 469
hospital admission study, 132
infant's listening time, 307–308
inflamed ear lobes, 386
jaw dysfunction, 452
latex allergy in health-care workers, 264, 283, 314, 321
lumbar disease risk factor, 473
lung cancer, CT scanning for, 20
melanoma deaths, 226
multiple-sclerosis drug, 475
neurological impairment of POWs, 403
organ transplant, 164
pain empathy and brain activity, 513–514, 544
pain tolerance, 525
patient medical instruction sheets, 163

physical activity of obese young adults, 240, 523
physical fitness exam, passing, 185–187
physical fitness problem, 188–190
placebo effect and pain, 380–381
post-op nausea study, 126–127
pregnancy test accuracy, 164
public perceptions of health risks, 545
quit-smoking program, 245
relation of eye and head movements, 514
salmonella in ice-cream bars, 295
short-term memory, 93
sickle-cell anemia, 246
sleep deprivation, 344
treating psoriasis with the Doctorfish of Kangal, 87, 401
visual acuity of children, 394
walking study, 92
weight loss, 352–353
West Nile virus cases, 47
women's height range, 212
Housing/building/real estate applications:
land purchase decision, 107
pipe, surface roughness of, 247
real estate sales, 57
sale prices of homes, 549–550
wood roof, bending strength of, 273–274, 289
Internet applications, *See* Electronics/computer/Internet applications
Legal/law applications:
federal civil trial appeals, 137, 182, 335
jury trial outcomes, 308
patent infringement case, 367–368
polygraph test error rates, 346
Manufacturing applications:
calculators, 223–224
contaminated gun cartridges, 182
defect rate of two machines, 442
dye discharge in paint, 247–248
industrial filling process, 213
manufacturer's claim, testing, 238
material safety data sheets, 295
steel sheet thickness, 231–232
Marine/marine life applications, *See also* Biology/life science applications
fish:
    brill, tapeworms in, 472–473
    feeding behavior of, 92
    feeding habits, 551
great white shark length, 329
scallops, sampling and the law, 345–346
hull failures of oil tankers, 106–107
whistling dolphins, 104–105
Medical/hospital/alternative medicine applications, *See also* Health-care applications
dosing errors at hospitals, 473
drug reaction, 504–505, 520–521, 527–528
drug response time, 310–312, 317, 389–390
FDA mandatory new-drug testing, 344
LASIK surgery complications, 226
mean hospital length of stay, 318–319

medical test errors, 345
placebo effect, 336
scopolamine, effect on memory, 417–418
teaching nursing skills, 424–425
waiting time at doctor's office, 227
Miscellaneous applications:
active nuclear power plants, 61, 67
ages of TV news viewers, 8
air threat classification with heuristics, 467
Al Qaeda attacks on the United States, 46, 273, 327, 341
aluminum cans contaminated by fire, 289
aluminum smelter pot, extending the life of, 499–500, 506, 542
animal-assisted therapy for heart patients, 6, 74–75, 380, 418–419
anthrax, mail rooms contaminated with, 181–182
Army Corps of Engineers Study of a contaminated river, 12
Bank of America *Keep the Change* program, 12, 65, 148, 216, 239, 326, 438, 495–496, 518–519
battle simulation trials, 476
Beanie Babies, 104, 552
Benford's law of numbers, 38, 181
biometric recognition methods, 543
bottle bursting strength, 393
box office receipts, 377
buyers of TVs, 443–444
cable-TV home shoppers, 443
    ages of, 322
census sampling, 157
chemical insect attractant, 161
children's recall of TV ads, 366, 393
cigarette smokers, 436–437
cigarette smoking:
    babies' weights vs., 539–541
    kicking the habit, 476
cockroach growth, 266
coin-tossing experiment, 119–120, 173–174
consumer complaint, 145
coupon usage, 477
creating menus to influence others, 466
cutting tools, life tests of, 507, 533
dance/movement therapy, 525
dates of pennies, 34
defective batteries, 331–332
die-tossing experiment, 119, 127–128, 144–145
executives who cheat at golf, 139
five-star hotels, rating services at, 367
freckling of superalloy ingots, 105, 341–342
free press, 247
Galileo's passe-dix game, 137–138
Gallup poll of teenagers, 20
gas turbines, cooling method for, 314–315
geography journals, 92, 501, 507, 513
goodness-of-fit test, 477
Holocaust, 21
homeless in the United States, 293
Hot Tamale caper, 346–347

hotel guest rooms, 118
households from 100,000, 154–155
hull failures of oil tankers, 106–107
human earlobes, 183
Index of Biotic Integrity (IBI), 366
Internet addiction study, 14
intrusion detection systems, 151, 308
Iraq War poll, 19, 475
land cover, remote-sensing data to identity, 314
laughter among deaf signers, 380, 386
Let's Make a Deal, 164–165
levelness of concrete slabs, 241
library book checkouts, 88
lie detector test, 152
life expectancy of Oscar winners, 380
load on frame structures, 213
"Made in the USA" survey, 36, 293, 334, 452–453
male nannies, 125
matching socks, 126
math programming, 294
measuring instruments, 384–385
media coverage of the 9–11 attacks and public opinion, 543
Monty Hall Dilemma, 469–470
mosquito repellent testing, 328
most likely coin-toss sequence, 165
most powerful women in America, 59, 67, 73, 88, 211, 219, 416–417
mother's race vs. maternal age, 128–130
music, 504–505
    quantitative models of, 497, 506
National Bridge Inventory, 20
National Firearms Survey, 34, 150, 282
natural-gas pipeline accident risk, 153
new-book reviews, 103, 246
new Hite Report, 107–108
No Child Left Behind Act, 108
object recall, 21
Odd Man Out, 164
"one-shot" device reliability, 247
organic chemistry, 552
portable grill displays, 126, 181, 346
power equipment quality, monitoring of, 152–153
psychic's ability, testing, 150
psychic's ESP, testing, 195
random-digit dial, 157
recall of TV commercials, 415–416
rigged school milk prices, detection of, 369, 441
rotated objects, view of, 524
sample selection, 160
scanning errors at Wal-Mart, 135, 288–289, 334
ShowCase Showdown, 182–183
sick at home, 282
skin cream effectiveness, 336
songs with violent lyrics, effect of, 157
spall damage in bricks, 555
speech listening, 49, 88
sports news on local TV broadcasts, 523
spreading rate of spilled liquid, 94, 502, 533
sterile couples in Jordan, 160
Subway shop customers, 172

susceptibility to hypnosis, 15, 241, 308, 468
tanker oil spills, 294
tax-exempt charities, 292–293
term insurance policy, 175–176
testing normality, 477
thematic atlas topics, 402–403
TNT, detecting traces of, 152
tongue twisters, reading, 381
training zoo animals, 60, 289
TV program on marijuana, effectiveness of, 448–449
"20/20" survey exposés, 22
urban counties, factors identifying, 161, 246
verbs and double-object datives, 335
visual search and memory, 381
warehouse clubs, 189
welfare workers, 142–143
Winchester bullets, velocity of, 74
wind turbine blade stress, 497–498, 523
"winner's curse" in auction bidding, 474
zinc phosphide in pest control, 106
Nutrition/food applications, *See also* **Beverage applications**
barbecue sauce, 55
binge eaters, 426
brown-bag lunches at work, 293–294
calories in school lunches, 307
eating disorders, 48, 240–241, 393
    in females, 275
food cravings, 442
high-fiber food and cancer, 473–474
McDonald's lunch, 345
oven cooking, 274, 289
red snapper, 150–151, 282
sugarcane, rat damage in, 444
Political applications:
economy, proportion optimistic about, 278–279
exit polls, 142, 461
political poll, 443
political representation of religious groups, 454
politics and religion, 466–467
rigged election, 477
voting for mayor, 191–193
Psychology applications, *See also* **Behavioral study applications**
aggressive behavior:
    and birth order, 162
    and personality, 264
alcohol, threats, and electric shocks, 212
child bipolar disorders, 544–545
cholesterol levels in psychiatric patients, 246
control-of-life perception, 104
depression, treating with St. John's wort, 440
developmental delays, 246
exam question order, impact of, 423
facial expressions, 425
famous psychological experiment, 142
IQ and mental retardation, 465–466
language impairment in children, 164
loneliness in families, 553–554

mental illness, public image of, 103
mental patients, social interaction of, 315
mental performance and sleep, 441
name game, 418, 501–502, 514, 525, 533, 543–544
obese adolescents, 346
participation and satisfaction, 294
personalities of cocaine abusers, 369–370
post-traumatic stress of POWs, 314, 322
role importance for the elderly, 35
"tip-of-the-tongue" phenomenon, 441
twins, attention time given to, 265–266
violence and stress, 246
Religion/beliefs/faith applications:
afterlife, belief in, 195
Bible, belief in, 36, 263, 453
marital status and religion, 459–461
political representation of religious groups, 454
religion and politics, 466–467
Science applications, *See* **Biology/life science applications; Earth science applications**
Sexuality/gender applications:
feminizing human faces, 322
masculinizing human faces, 345
new Hite Report, 107–108
sex composition patterns of children in families, 125–126
Sociological applications:
boys in family, 196
divorced couples, 120–121, 147
ethnicity and pain perception, 370
married-women study, 245–246
mother's race vs. maternal age, 128–130
planning habits, 440
single-parent families, 335
social play of children, 476
sociology field work methods, 36, 125, 453
Sports/fitness/exercise applications:
aerobic exercise, 425
baseball:
    batting averages, 221
    batting averages versus wins, 92
    batting averages vs. wins, 550–551
    elevation and hitting performance, 514–515
    no-hitters in, 245
basketball:
    NCAA March Madness, 346
    shooting free throws, 162
bowler's hot hand, 403
boxing, massage vs. rest in, 20–21, 513, 525, 542–543
climbers, effect of altitude on, 422–423
college tennis recruiting with a team website, 415
dart-throwing errors, 220
exercise workout dropouts, 263
football:
    mean inflation pressure, 285–286
    point spreads of NFL games, 314
    scouting an NFL free agent, 386

Go (game), 153
golf:
    executives who cheat at, 139
    golf ball brands, 408–410
    USGA golf ball specifications, 196
    USGA golf ball tests, 289
horse race, odds of winning,
        163–164
long-jump "takeoff error," 554–555
major sports venues, location of, 452
marathon winning times, 498, 506
NBA:
    draft, 153
    stacking in, 137, 152

NFL, successful running plays, 182
Olympic athlete drug test
        effectiveness, 441
professional athlete salaries, 107
professional golfers, ranking driving
        performance of, 59, 93, 221, 500,
        512–513, 531
shuffleboard tournament, 152
soccer, "headers" and IQ, 21
sports participation, 532–533
sprint speed training, 19
student gambling on sports, 195
walking, 92, 551–552
World Series winners, 150

Travel applications:
    luggage inspection at Newark
        airport, 227
    sanitation inspection of cruise ships,
        46–47, 73, 78, 88, 221
    travel habits of retirees, 474
    travel professional salaries, 246
Weather applications:
    rainfall:
        chance of, 125
        in Colorado, 394
        and desert ants, 274
        estimating, 500–501, 506–507, 513
    temperature, concrete-pavement
        response to, 402

A First Course in Statistics

CHAPTER 1

Statistics, Data, and Statistical Thinking

CONTENTS

1.1 The Science of Statistics

1.2 Types of Statistical Applications

1.3 Fundamental Elements of Statistics

1.4 Types of Data

1.5 Collecting Data

1.6 The Role of Statistics in Critical Thinking

STATISTICS IN ACTION

USA WEEKEND Teen Surveys: Are Boys Really from Mars and Girls from Venus?

USING TECHNOLOGY

Creating and Listing Data Using MINITAB

WHERE WE'RE GOING

■ Introduce the field of statistics.

■ Demonstrate how statistics applies to real-world problems.

■ Establish the link between statistics and data.

■ Identify the different types of data and data collection methods.

■ Differentiate between population and sample data.

■ Differentiate between descriptive and inferential statistics.

STATISTICS IN ACTION

USA WEEKEND Teen Surveys: Are Boys Really from Mars and Girls from Venus?

"Welcome to USA WEEK-END, the magazine that makes a difference."

—www.usaweekend. com

USA WEEKEND is a magazine supplement in approximately 600 U.S. newspapers with an estimated 50 million readers. Each year, the magazine conducts a "Teen Survey" in which America's teenagers give opinions on a wide variety of topics, including their views of the opposite sex (2003 survey) and their fascination with celebrities (2006 sur-

vey). The surveys are made available in print in the magazine supplement, at the www. usaweekend. com website and via the mail through newspaper subscriber lists.

Each survey is typically sponsored by a special-interest group. For example, YouthNoise.com and *Extra* magazine were sponsors of the 2006 survey, while *YM Magazine* and John Gray (author of the best-selling book *Men Are from Mars, Women Are from Venus*) sponsored the 2003 survey. The number of teens responding to the survey varies as well (37,000 teenagers in 2003 and 65,000 in 2006).

Some of the questions (and corresponding results) for the 2003 and 2006 Teen Surveys are shown here:

2006 Survey: Teens and Celebrities

1. When it comes to media coverage of celebrities, do you think there is _____ coverage?

Too much:	56%
Too little:	8%
Just enough:	36%

2. On a scale of 0 to 5 (with 0 being "no influence" and 5 being "extremely influential"), how much influence do celebrities have about important issues like war and politics?

Average = 1.77

3. Have you ever dieted because you want to look more like a celebrity?

	All	Girls	Boys
Yes:	13%	16%	6%
No:	87%	84%	94%

4. Do you think celebrities get paid too much?

Yes:	67%
No:	33%

2003 Survey: Teens and the Opposite Sex

1. Which quality first catches your attention in the opposite sex?

	Boys	Girls
Looks:	58%	40%
Personality:	27%	36%
Athleticism:	1%	2%
Confidence:	2%	3%
Sense of Humor:	9%	16%
Intelligence:	2%	2%
Popularity:	1%	1%

2. Do you think it's OK for girls in your grade to call guys?

	Boys	Girls
Yes:	96%	95%
No:	4%	5%

3. Do you think it's OK for girls in your grade to ask guys out?

	Boys	Girls
Yes:	90%	80%
No:	10%	20%

4. Is it possible to have a good friend of the opposite sex?

	Boys	Girls
No:	6%	3%
Yes, but only if part of a group:	8%	7%
Yes, anytime:	86%	90%

In the following "Statistics in Action Revisited" sections, we discuss several key statistical concepts covered in this chapter that are relevant to the *USA WEEKEND* Annual Teen Surveys:

"Statistics in Action Revisited" for Chapter 1

- Identifying the Population, Sample, and Inference, for the *USA WEEKEND* Teen Survey (p. 10)

- Identifying the Data Collection Method and Data Type (p. 15)

- Critically Assessing the Ethics of a Statistical Study (p. 17)

1.1 The Science of Statistics

What does statistics mean to you? Does it bring to mind batting averages, Gallup polls, unemployment figures, or numerical distortions of facts (lying with statistics!)? Or is it simply a college requirement you have to complete? We hope to persuade you that statistics is a meaningful, useful science whose broad scope of applications to business, government, and the physical and social sciences is almost limitless. We also want to show that statistics can lie only when they are misapplied. Finally, we wish to demonstrate the key role statistics plays in critical thinking—whether in the classroom, on the job, or in everyday life. Our objective is to leave you with the impression that the time you spend studying this subject will repay you in many ways.

The *Random House College Dictionary* defines *statistics* as "the science that deals with the collection, classification, analysis, and interpretation of information or data." Thus, a statistician isn't just someone who calculates batting averages at baseball games or tabulates the results of a Gallup Poll. Professional statisticians are trained in *statistical science*. That is, they are trained in collecting numerical information in the form of **data**, evaluating the information, and drawing conclusions from it. Furthermore, statisticians determine what information is relevant in a given problem and whether the conclusions drawn from a study are to be trusted.

> **Definition 1.1**
>
> **Statistics** is the science of data. This involves collecting, classifying, summarizing, organizing, analyzing, and interpreting numerical information.

In the next section, you'll see several real-life examples of statistical applications that involve making decisions and drawing conclusions.

1.2 Types of Statistical Applications

"Statistics" means "numerical descriptions" to most people. Monthly housing starts, the failure rate of liver transplants, and the proportion of African-Americans who feel brutalized by local police all represent statistical descriptions of large sets of data collected on some phenomenon. Often the data are selected from some larger set of data whose characteristics we wish to estimate. We call this selection process *sampling*. For example, you might collect the ages of a sample of customers at a video store to estimate the average age of *all* customers of the store. Then you could use your estimate to target the store's advertisements to the appropriate age group. Notice that statistics involves two different processes: (1) describing sets of data and (2) drawing conclusions (making estimates, decisions, predictions, etc.) about the sets of data on the basis of sampling. So, the applications of statistics can be divided into two broad areas: *descriptive statistics* and *inferential statistics*.

Biography

**FLORENCE NIGHTINGALE (1820–1910)—
The Passionate Statistician**

In Victorian England, the "Lady of the Lamp" had a mission to improve the squalid field hospital conditions of the British army during the Crimean War. Today, most historians consider Florence Nightingale to be the founder of the nursing profession. To convince members of the British Parliament of the need for supplying nursing and medical care to soldiers in the field, Nightingale compiled massive amounts of data from army files. Through a remarkable series of graphs (which included the first pie chart), she demonstrated that most of the deaths in the war either were due to illnesses contracted outside the battlefield or occurred long after battle action from wounds that went untreated. Florence Nightingale's compassion and self-sacrificing nature, coupled with her ability to collect, arrange, and present large amounts of data, led some to call her the Passionate Statistician.

> **Definition 1.2**
> **Descriptive statistics** utilizes numerical and graphical methods to look for patterns in a data set, to summarize the information revealed in a data set, and to present that information in a convenient form.

> **Definition 1.3**
> **Inferential statistics** utilizes sample data to make estimates, decisions, predictions, or other generalizations about a larger set of data.

Although we'll discuss both descriptive and inferential statistics in the chapters that follow, the primary theme of the text is **inference**.

Let's begin by examining some studies that illustrate applications of statistics.

STUDY 1.1: *"Best-Selling Girl Scout Cookies"*

Since 1917, the Girl Scouts of America have been selling boxes of cookies. Currently, there are eight varieties for sale: Thin Mints, Samoas, Caramel DeLites, Tagalongs, Peanut Butter Patties, Do-si-dos, Peanut Butter Sandwiches, and Trefoils. Each of the approximately 150 million boxes of girl scout cookies sold in 2006 was classified by variety. The results are summarized in Figure 1.1. From the graph, you can clearly see that the best-selling variety is Thin Mints (25%), followed by Samoas (19%) and Tagalongs (13%). Since the figure describes the various categories of boxes of girl scout cookies sold, the graphic is an example of descriptive statistics.

(*USA Today*, Feb. 15, 2007)

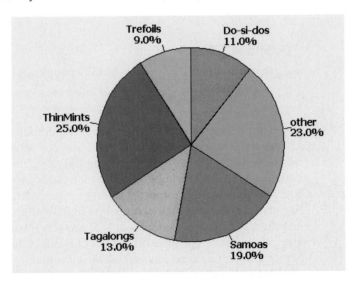

FIGURE 1.1
Best-selling girl scout cookies

STUDY 1.2: *"Does Height Influence Progression through Primary School Grades?"*

Researchers from the University of Melbourne (Australia) analyzed the heights of over 2,800 students in primary school (grades 1 through 6). After dividing the students into three equal groups based on age (youngest third, middle third, and oldest third) within each grade level, the researchers found that the oldest group of students was the shortest in height, on average—and that this phenomenon was due chiefly to the 133 children (mostly boys) who had been held back a grade level. From the analysis, the researchers inferred "that older boys within grades were relatively shorter than their younger peers" and that "when boys experienced school difficulties, height [was] one factor influencing the final decision to [hold back the student a grade level]." Thus, inferential statistics was applied to arrive at this conclusion.

(*Archives of Disease in Childhood*, April 2000)

STUDY 1.3: *"Animal Assisted Therapy . . . [for] Hospitalized Heart Failure Patients"*

A team from the UCLA Medical Center and School of Nursing, led by RN Kathie Cole, conducted a study to gauge whether animal-assisted therapy can improve the physiological responses of heart failure patients. Cole and her colleagues studied 76 heart patients, randomly divided into three groups. Each person in one group of patients was visited by a human volunteer accompanied by a trained dog, each person in another group was visited by a volunteer only, and the third group was not visited at all. The researchers measured patients' physiological responses (levels of anxiety, stress, and blood pressure) before and after the vis-

its. An analysis of the data revealed that those patients with animal-assisted therapy had significantly greater drops in levels of anxiety, stress, and blood pressure. Thus, the researchers concluded that "pet therapy has the potential to be an effective treatment . . . for patients hospitalized with heart failure."

Like Study 1.2, this study is an example of the use of inferential statistics. The medical researchers used data from 76 patients to make inferences about the effectiveness of animal-assisted therapy for all heart failure patients.

(*American Heart Association Conference*, November, 2005)

These studies provide three real-life examples of the uses of statistics. Notice that each involves an analysis of data, either for the purpose of describing the data set (Study 1.1) or for making inferences about a data set (Studies 1.2 and 1.3).

1.3 Fundamental Elements of Statistics

Statistical methods are particularly useful for studying, analyzing, and learning about *populations* of *experimental units*.

Definition 1.4

An **experimental unit** is an object (e.g., person, thing, transaction, or event) about which we collect data.

Definition 1.5

A **population** is a set of units (usually people, objects, transactions, or events) that we are interested in studying.

For example, populations may include (1) *all* employed workers in the United States, (2) *all* registered voters in California, (3) *everyone* who is afflicted with AIDS, (4) *all* the cars produced last year by a particular assembly line, (5) the *entire* stock of spare parts available at United Airlines' maintenance facility, (6) *all* sales made at the drive-in window of a McDonald's restaurant during a given year, or (7) the set of *all* accidents occurring on a particular stretch of interstate highway during a holiday period. Notice that the first three population examples (1 – 3) are sets (groups) of people, the next two (4 – 5) are sets of objects, the next (6) is a set of transactions, and the last (7) is a set of events. Notice also that *each set includes all the units in the population.*

In studying a population, we focus on one or more characteristics or properties of the units in the population. We call such characteristics *variables.* For example, we may be interested in the variables age, gender, and number of years of education of the people currently unemployed in the United States.

Definition 1.6

A **variable** is a characteristic or property of an individual population unit.

The name *variable* is derived from the fact that any particular characteristic may vary among the units in a population.

In studying a particular variable, it is helpful to be able to obtain a numerical representation for it. Often, however, numerical representations are not readily available, so measurement plays an important supporting role in statistical studies. **Measurement** is the process we use to assign numbers to variables of individual population units. We might, for instance, measure the performance of the president by asking a registered voter to rate it on a scale from 1 to 10. Or we might measure the age of the U.S. workforce simply by asking each worker, "How old are you?" In other cases, measurement involves the use of instruments such as stopwatches, scales, and calipers.

If the population you wish to study is small, it is possible to measure a variable for every unit in the population. For example, if you are measuring the GPA for all incoming first-year students at your university, it is at least feasible to obtain every GPA. When we measure a variable for every unit of a population, it is called a **census** of the population. Typically, however, the populations of interest in most applications are much larger, involving perhaps many thousands, or even an infinite number, of units. Examples of large populations are those following Definition 1.5, as well as all graduates of your university or college, all potential buyers of a new fax machine, and all pieces of first-class mail handled by the U.S. Post Office. For such populations, conducting a census would be prohibitively time consuming or costly. A reasonable alternative would be to select and study a *subset* (or portion) of the units in the population.

> **Definition 1.7**
> A **sample** is a subset of the units of a population.

For example, instead of polling all 140 million registered voters in the United States during a presidential election year, a pollster might select and question a sample of just 1,500 voters. (See Figure 1.2.) If he is interested in the variable "presidential preference," he would record (measure) the preference of each vote sampled.

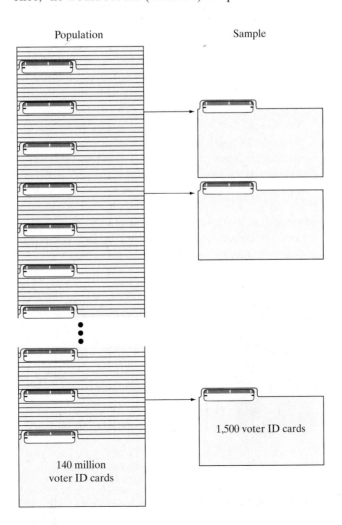

FIGURE 1.2

A sample of voter registration cards for all registered voters

After the variables of interest for every unit in the sample (or population) are measured, the data are analyzed, either by descriptive or inferential statistical methods. The pollster, for example, may be interested only in *describing* the voting patterns of the sample of 1,500 voters. More likely, however, he will want to use the information in the sample to make inferences about the population of all 120 million voters.

> **Definition 1.8**
>
> A **statistical inference** is an estimate, prediction, or some other generalization about a population based on information contained in a sample.

That is, *we use the information contained in the smaller sample to learn about the larger population.** Thus, from the sample of 1,500 voters, the pollster may estimate the percentage of all the voters who would vote for each presidential candidate if the election were held on the day the poll was conducted, or he might use the results to predict the outcome on election day.

EXAMPLE 1.1
KEY ELEMENTS OF A STATISTICAL PROBLEM— Ages of TV News Viewers

Problem According to *The State of the News Media*, 2006, the average age of viewers of *ABC World News Tonight* is 59 years. Suppose a rival network executive hypothesizes that the average age of ABC news viewers is less than 59. To test her hypothesis, she samples 500 viewers of *ABC World News Tonight* and determines the age of each.

a. Describe the population.

b. Describe the variable of interest.

c. Describe the sample.

d. Describe the inference.

Solution

a. The population is the set of units of interest to the TV executive, which is the set of all viewers of *ABC World News Tonight*.

b. The age (in years) of each viewer is the variable of interest.

c. The sample must be a subset of the population. In this case, it is the 500 viewers of *ABC World News Tonight* selected by the executive.

d. The inference of interest involves the *generalization* of the information contained in the sample of 500 viewers to the population of all viewers of *ABC World News Tonight*. In particular, the executive wants to *estimate* the average age of the viewers in order to determine whether it is less than 59 years. She might accomplish this by calculating the average age of the sample and using that average to estimate the average age of the population.

Look Back A key to diagnosing a statistical problem is to identify the data set collected (in this example, the ages of the 500 viewers of *ABC World News Tonight*) as a population or a sample.

EXAMPLE 1.2
KEY ELEMENTS OF A STATISTICAL PROBLEM— Pepsi vs. Coca-Cola

Problem "Cola wars" is the popular term for the intense competition between Coca-Cola and Pepsi displayed in their marketing campaigns, which have featured movie and television stars, rock videos, athletic endorsements, and claims of consumer preference based on taste tests. Suppose, as part of a Pepsi marketing campaign, 1,000 cola consumers are given a blind taste test (i.e., a taste test in which the two brand names are disguised). Each consumer is asked to state a preference for brand A or brand B.

a. Describe the population.

b. Describe the variable of interest.

---

*The terms *population* and *sample* are often used to refer to the sets of measurements themselves, as well as to the units on which the measurements are made. When a single variable of interest is being measured, this usage causes little confusion. But when the terminology is ambiguous, we'll refer to the measurements as *population data sets* and *sample data sets*, respectively.

1.4 Types of Data

You have learned that statistics is the science of data and that data are obtained by measuring the values of one or more variables on the units in the sample (or population). All data (and hence the variables we measure) can be classified as one of two general types: *quantitative data* and *qualitative data.*

Quantitative data are data that are measured on a naturally occurring numerical scale.* The following are examples of quantitative data:

1. The temperature (in degrees Celsius) at which each piece in a sample of 20 pieces of heat-resistant plastic begins to melt

2. The current unemployment rate (measured as a percentage) in each of the 50 states

3. The scores of a sample of 150 law school applicants on the LSAT, a standardized law school entrance exam administered nationwide

4. The number of convicted murderers who receive the death penalty each year over a 10-year period

> **Definition 1.10**
>
> **Quantitative data** are measurements that are recorded on a naturally occurring numerical scale.

In contrast, qualitative data cannot be measured on a natural numerical scale; they can only be classified into categories.† Examples of qualitative data include the following:

1. The political party affiliation (Democrat, Republican, or Independent) in a sample of 50 voters

2. The defective status (defective or not) of each of 100 computer chips manufactured by Intel

3. The size of a car (subcompact, compact, midsize, or full size) rented by each of a sample of 30 business travelers

4. A taste tester's ranking (best, worst, etc.) of four brands of barbecue sauce for a panel of 10 testers

Often, we assign arbitrary numerical values to qualitative data for ease of computer entry and analysis. But these assigned numerical values are simply codes: They cannot be meaningfully added, subtracted, multiplied, or divided. For example, we might code Democrat = 1, Republican = 2, and Independent = 3. Similarly, a taste tester might rank the barbecue sauces from 1 (best) to 4 (worst). These are simply arbitrarily selected numerical codes for the categories and have no utility beyond that.

> **Definition 1.11**
>
> **Qualitative data** are measurements that cannot be measured on a natural numerical scale; they can only be classified into one of a group of categories.

---

*Quantitative data can be subclassified as either *interval data* or *ratio data.* For ratio data, the origin (i.e., the value 0) is a meaningful number. But the origin has no meaning with interval data. Consequently, we can add and subtract interval data, but we can't multiply and divide them. Of the four quantitative data sets listed as examples, (1) and (3) are interval data while (2) and (4) are ratio data.
†Qualitative data can be subclassified as either *nominal data* or *ordinal data.* The categories of an ordinal data set can be ranked or meaningfully ordered, but the categories of a nominal data set can't be ordered. Of the four qualitative data sets listed as examples, (1) and (2) are nominal and (3) and (4) are ordinal.

EXAMPLE 1.4

DATA TYPES—
Army Corps of
Engineers Study of a
Contaminated River

Problem Chemical and manufacturing plants often discharge toxic-waste materials such as DDT into nearby rivers and streams. These toxins can adversely affect the plants and animals inhabiting the river and the riverbank. The U.S. Army Corps of Engineers conducted a study of fish in the Tennessee River (in Alabama) and its three tributary creeks: Flint Creek, Limestone Creek, and Spring Creek. A total of 144 fish were captured, and the following variables were measured for each:

1. River/creek where each fish was captured
2. Species (channel catfish, largemouth bass, or smallmouth buffalo fish)
3. Length (centimeters)
4. Weight (grams)
5. DDT concentration (parts per million)

Classify each of the five variables measured as quantitative or qualitative.

Solution The variables length, weight, and DDT concentration are quantitative because each is measured on a numerical scale: length in centimeters, weight in grams, and DDT in parts per million. In contrast, river/creek and species cannot be measured quantitatively: They can only be classified into categories (e.g., channel catfish, largemouth bass, and smallmouth buffalo fish for species). Consequently, data on river/creek and species are qualitative.

Look Back It is essential that you understand whether the data you are interested in are quantitative or qualitative, since the statistical method appropriate for describing, reporting, and analyzing the data depends on the data type (quantitative or qualitative).

Now Work Exercise 1.12

We demonstrate many useful methods for analyzing quantitative and qualitative data in the remaining chapters of the text. But first, we discuss some important ideas on data collection in the next section.

ACTIVITY 1.1: *Keep the Change:* Collecting Data

Recently, Bank of America introduced a savings program called *Keep the Change*. Each time a customer who is enrolled in the program uses his debit card to make a purchase, the difference between the purchase amount and the next higher dollar amount is transferred from the customer's checking account to a savings account. For example, if you were enrolled in the program and used your debit card to purchase a latte for $3.75, then $0.25 would be transferred from your checking to your savings account. For the first 90 days that a customer is enrolled in the program, Bank of America matches the amounts transferred, up to $250. In this and subsequent activities, we will investigate the potential benefit to the customer and the cost to the bank.

1. Simulate the program by keeping track of all purchases you make during one week that could be made with a debit card, even if you actually use a different form of payment. For each purchase, record both the purchase amount and the amount that would be transferred from checking to savings with the *Keep the Change* program.

2. You now have two sets of data: *Purchase Amounts* and *Amounts Transferred*. Both sets contain quantitative data. For each data set, identify the corresponding naturally occurring numerical scale. Explain why each set has an obvious lower bound but only one set has a definite upper bound.

3. Find the total of the amounts transferred for the one-week period. Since 90 days is approximately 13 weeks, multiply the total by 13 to estimate how much the bank would have to match during the first 90 days. Form a third data set, *Bank Matching*, by collecting the 90-day estimates of all the students in your class. Identify the naturally occurring scale, including bounds, for this set of data.

Keep the data sets from this activity for use in other activities. We suggest you use statistical software (e.g., MINITAB) or a graphing calculator to save the data.

1.5 Collecting Data

Once you decide on the type of data—quantitative or qualitative—appropriate for the problem at hand, you'll need to collect the data. Generally, you can obtain data in four different ways:

1. From a *published source*
2. From a *designed experiment*
3. From a *survey*
4. From an *observational study*

Sometimes, the data set of interest has already been collected for you and is available in a **published source**, such as a book, journal, or newspaper. For example, you may want to examine and summarize the divorce rates (i.e., number of divorces per 1,000 population) in the 50 states of the United States. You can find this data set (as well as numerous other data sets) at your library in the *Statistical Abstract of the United States*, published annually by the U.S. government. Similarly, someone who is interested in monthly mortgage applications for new home construction would find this data set in the *Survey of Current Business*, another government publication. Other examples of published data sources include *The Wall Street Journal* (financial data) and *The Sporting News* (sports information). The Internet (World Wide Web) now provides a medium by which data from published sources are readily obtained.*

A second method of collecting data involves conducting a **designed experiment**, in which the researcher exerts strict control over the units (people, objects, or things) in the study. For example, a recent medical study investigated the potential of aspirin in preventing heart attacks. Volunteer physicians were divided into two groups: the *treatment* group and the *control* group. Each physician in the treatment group took one aspirin tablet a day for one year, while each physician in the control group took an aspirin-free placebo made to look like an aspirin tablet. The researchers—not the physicians under study—controlled who received the aspirin (the treatment) and who received the placebo. As you'll learn in Chapter 10, a properly designed experiment allows you to extract more information from the data than is possible with an uncontrolled study.

Surveys are a third source of data. With a **survey**, the researcher samples a group of people, asks one or more questions, and records the responses. Probably the most familiar type of survey is the political poll, conducted by any one of a number of organizations (e.g., Harris, Gallup, Roper, and CNN) and designed to predict the outcome of a political election. Another familiar survey is the Nielsen survey, which provides the major networks with information on the most-watched programs on television. Surveys can be conducted through the mail, with telephone interviews, or with in-person interviews. Although in-person surveys are more expensive than mail or telephone surveys, they may be necessary when complex information is to be collected.

Finally, observational studies can be employed to collect data. In an **observational study**, the researcher observes the experimental units in their natural setting and records the variable(s) of interest. For example, a child psychologist might observe and record the level of aggressive behavior of a sample of fifth graders playing on a school playground. Similarly, a zoologist may observe and measure the weights of newborn elephants born in captivity. Unlike a designed experiment, an observational study is a study in which the researcher makes no attempt to control any aspect of the experimental units.

Regardless of which data collection method is employed, it is likely that the data will be a sample from some population. And if we wish to apply inferential statistics, we must obtain a *representative sample*.

> **Definition 1.12**
>
> A **representative sample** exhibits characteristics typical of those possessed by the target population.

---

*With published data, we often make a distinction between the *primary source* and a *secondary source*. If the publisher is the original collector of the data, the source is primary. Otherwise, the data is secondary-source data.

For example, consider a political poll conducted during a presidential election year. Assume that the pollster wants to estimate the percentage of all 140 million registered voters in the United States who favor the incumbent president. The pollster would be unwise to base the estimate on survey data collected for a sample of voters from the incumbent's own state. Such an estimate would almost certainly be *biased* high; consequently, it would not be very reliable.

The most common way to satisfy the representative sample requirement is to select a random sample. A *random sample* ensures that every subset of fixed size in the population has the same chance of being included in the sample. If the pollster samples 1,500 of the 140 million voters in the population so that every subset of 1,500 voters has an equal chance of being selected, she has devised a random sample. The procedure for selecting a random sample is discussed in Chapter 3. Here, we look at two examples involving actual sampling studies.

> **Definition 1.13**
>
> A **random sample** of *n* experimental units is a sample selected from the population in such a way that every different sample of size *n* has an equal chance of selection.

EXAMPLE 1.5
METHOD OF DATA COLLECTION— Internet Addiction Study

Problem What percentage of Web users are addicted to the Internet? To find out, a psychologist designed a series of 10 questions based on a widely used set of criteria for gambling addiction and distributed them through the website *ABCNews.com*. (A sample question: "Do you use the Internet to escape problems?") A total of 17,251 Web users responded to the questionnaire. If participants answered "yes" to at least half of the questions, they were viewed as addicted. The findings, released at a recent annual meeting of the American Psychological Association, revealed that 990 respondents, or 5.7%, are addicted to the Internet.

a. Identify the data collection method.

b. Identify the target population.

c. Are the sample data representative of the population?

Solution

a. The data collection method is a survey: 17,251 Internet users responded to the questions posed at the *ABCNews.com* website.

b. Since the website can be accessed by anyone surfing the Internet, presumably the target population is *all* Internet users.

c. Because the 17,251 respondents clearly make up a subset of the target population, they do form a sample. Whether or not the sample is representative is unclear, since we are given no information on the 17,251 respondents. However, a survey like this one in which the respondents are *self-selected* (i.e., each Internet user who saw the survey chose whether or not to respond to it) often suffers from *nonresponse bias*. It is possible that many Internet users who chose not to respond (or who never saw the survey) would have answered the questions differently, leading to a higher (or lower) percentage of affirmative answers.

Look Back Any inferences based on survey samples that employ self-selection are suspect due to potential nonresponse bias.

EXAMPLE 1.6
METHOD OF DATA COLLECTION—
Study of Susceptibility to Hypnosis

Problem Psychologists at the University of Tennessee carried out a study of the susceptibility of people to hypnosis (*Psychological Assessment*, Mar. 1995). In a random sample of 130 undergraduate psychology students at the university, each experienced both traditional hypnosis and computer-assisted hypnosis. Approximately half were randomly assigned to undergo the traditional procedure first, followed by the computer-assisted procedure. The other half were randomly assigned to experience computer-assisted hypnosis first, then traditional hypnosis. Following the hypnosis episodes, all students filled out questionnaires designed to measure a student's susceptibility to hypnosis. The susceptibility scores of the two groups of students were compared.

a. Identify the data collection method.

b. Is the sample data representative of the target population?

Solution

a. Here, the experimental units are the psychology students. Since the researchers controlled which type of hypnosis—traditional or computer assisted—the students experienced first (through random assignment), a designed experiment was used to collect the data.

b. The sample of 130 psychology students was randomly selected from all psychology students at the University of Tennessee. If the target population is *all University of Tennessee psychology students*, it is likely that the sample is representative. However, the researchers warn that the sample data should not be used to make inferences about other, more general, populations.

Look Back By using randomization in a designed experiment, the researcher is attempting to eliminate different types of bias, including self-selection bias.

Now Work Exercise 1.15

STATISTICS IN ACTION REVISITED

Identifying the Data Collection Method and Data Type for the *USA WEEKEND* Teen Survey

In the *USA WEEKEND* Annual Teen Survey, American teenagers are asked to respond to a variety of questions on topics such as newspaper usage (2004 survey) and the opposite sex (2003 survey). The title of the study implies that the data are collected through a survey of teenagers. The *USA WEEKEND* surveys were made available in printed form in the magazine, in an online survey form, and through the mail using newspaper subscriber lists.

Both quantitative and qualitative data were collected in the surveys. For example, the survey question "Do you think it's OK for girls in your grade to call guys?" is phrased to elicit a "yes" or "no" response. Since the responses produced for this question are categorical in nature, the data are qualitative. However, the question "How old were you when you went out on your first date?" will give meaningful numerical responses, such as 15, 16 or 17 years old. Thus, these data are quantitative.

1.6 The Role of Statistics in Critical Thinking

According to H. G. Wells, author of such science-fiction classics as *The War of the Worlds* and *The Time Machine,* "Statistical thinking will one day be as necessary for efficient citizenship as the ability to read and write." Written more than a hundred years ago, Wells's prediction is proving true today.

The growth in data collection associated with scientific phenomena, business operations, and government activities (quality control, statistical auditing, forecasting, etc.) has been remarkable in the past several decades. Every day the media present us with published results of political, economic, and social surveys. In increasing government emphasis on drug and product testing, for example, we see vivid evidence of the need to

Biography

H. G. WELLS (1866–1946)—
Writer and Novelist

English-born Herbert George Wells published his first novel, *The Time Machine*, in 1895 as a parody of the English class division and as a satirical warning that human progress is inevitable. Although most famous as a science-fiction novelist, Wells was a prolific writer as a journalist, sociologist, historian, and philosopher. Wells's prediction about statistical thinking (see Definition 1.14) is just one of a plethora of observations he made about life on this world. Here are a few more of H. G. Wells's more famous quotes:

"Advertising is legalized lying."
"Crude classification and false generalizations are the curse of organized life."
"The crisis of today is the joke of tomorrow."
"Fools make researchers and wise men exploit them."
"The only true measure of success is the ratio between what we might have done and what we might have been on the one hand, and the thing we have made and the things we have made of ourselves on the other."

be able to evaluate data sets intelligently. Consequently, each of us has to develop a discerning sense—an ability to use rational thought to interpret the meaning of data. This ability can help you make intelligent decisions, inferences, and generalizations; that is, it helps you *think critically* using statistics.

Definition 1.14

Statistical thinking involves applying rational thought and the science of statistics to critically assess data and inferences.

To gain some insight into the role statistics plays in critical thinking, let's look at a *New York Times* (June 17, 1995) article titled "The Case for No Helmets." The article, written by the editor of a magazine for Harley-Davidson bikers, lists several reasons motorcyclists should not be required by law to wear helmets. First, he argues that helmets may actually kill, since, in collisions at speeds greater than 15 miles an hour, the heavy helmet may protect the head but snap the spine. Second, the editor cites a "study" which claims that "nine states without helmet laws had a lower fatality rate (3.05 deaths per 10,000 motorcycles) than those that mandated helmets (3.38)." Finally, he reports that "in a survey of 2,500 [at a rally], 98% of the respondents opposed such laws."

You can use "statistical thinking" to help you critically evaluate the arguments presented in this article. For example, before you accept the 98% estimate, you would want to know how the data were collected for the survey cited by the editor of the biker magazine. It's possible that the 2,500 bikers in the sample were not selected at random from the target population of all bikers, but rather were "self-selected." (Remember, they were all attending a rally—maybe even a rally for bikers who oppose the law.) If the respondents were likely to have strong opinions regarding the helmet law (e.g., to strongly oppose the law), the resulting estimate is probably biased high. Further, if the biased sample was intentional, with the sole purpose to mislead the public, the researcher would be guilty of **unethical statistical practice**.

You'd also want more information about the study comparing the motorcycle fatality rate of the nine states without a helmet law with those states which mandate helmets: Were the data obtained from a published source? Were all 50 states included in the study? That is, are you seeing sample data or population data? Furthermore, do the helmet laws vary among states? If so, can you really compare the fatality rates?

These questions actually led to the discovery of two scientific and statistically sound studies on helmets. The first, a UCLA study of nonfatal injuries, disputed the charge that helmets shift injuries to the spine. The second study reported a dramatic *decline* in motorcycle crash deaths after California passed its helmet law.

As in the motorcycle helmet study, many statistical studies are based on survey data. Most of the problems with these surveys result from the use of *nonrandom* samples. These samples are subject to errors such as *selection bias, nonresponse bias* (recall Example 1.5), and *measurement error*. Researchers who are aware of these problems yet continue to use the sample data to make inferences are practicing unethical statistics.

> **Definition 1.15**
>
> **Selection bias** results when a subset of the experimental units in the population is excluded so that these units have no chance of being selected in the sample.

> **Definition 1.16**
>
> **Nonresponse bias** results when the researchers conducting a survey or study are unable to obtain data on all experimental units selected for the sample.

> **Definition 1.17**
>
> **Measurement error** refers to inaccuracies in the values of the data recorded. In surveys, this kind of error may be due to ambiguous or leading questions and the interviewer's effect on the respondent.

STATISTICS IN ACTION REVISITED

Critically Assessing the Results of the *USA WEEKEND* Teen Surveys

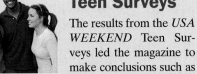

The results from the *USA WEEKEND* Teen Surveys led the magazine to make conclusions such as "Boys are from Mars, girls are from Venus" in their relations with the opposite sex (2003 survey) and "Teens want to look and act like famous people, and . . . they're taking more drastic steps [e.g., dieting] to do so" (2006 survey). Both surveys were based on a large sample of teen respondents: 37,000 in 2003 and 65,000 in 2006. However, several problems lead a critical thinker to cast doubt on the validity of the inferences.

First, are the samples representative of all American teens? *USA WEEKEND* admits that its annual teen survey is "unscientific," meaning that the teens are not randomly selected from all teens in the United States. In fact, the teens who responded are "self-selected," choosing to respond to the survey via the printed copy in the magazine, the Internet version, or the copy mailed to their home. Therefore, there is a strong selection bias in the survey results. Second, because not all American teens receive the magazine, were aware of the Internet survey, or were on the magazine subscriber mailing list, the potential for nonresponse bias is extremely high.

Finally, the wording in several of the survey questions is ambiguous or leading. This can occur when the survey is designed to support a specific point of view or organization (such as a celebrity teen magazine). For example, the question "Have you ever dieted because you want to look more like a celebrity?" may be answered "yes" for a variety of reasons. Consider a teenager who is dieting for health reasons and who answers "yes" because a by-product of dieting may be better looks. Ambiguities such as these will likely result in measurement error in the data.

ACTIVITY 1.2: *Identifying Misleading Statistics*

In a study found to have false or misleading statistics, the Food Research and Action Center claimed that *"One in four American children under age 12 is hungry or at risk of hunger."* The study was an attempt to persuade the public or the government to donate or allocate more money to charitable groups that feed the poor. In another flawed study, a researcher claimed that a relationship exists between two seemingly unrelated quantities: a CEO's golf handicap and the company's stock performance.

1. Look for an article in a newspaper or on the Internet in which a large proportion or percentage of a population is purported to be "at risk" of some calamity, as in the example of childhood hunger. Does the article cite a source or provide any information to support the proportion or percentage reported? Is the goal of the article to persuade some individual or group to take some action? If so, what action is being requested? Do you believe that the writer of the article may have some motive for exaggerating the problem? If so, give some possible motives.

2. Look for another article in which a relationship between two seemingly unrelated quantities is purported to exist, as in the study of the CEO's golf handicap and the performance of stocks. Select an article that contains some information on how the data were collected. Identify the target population and the data collection method. Based on what is presented in the article, do you believe that the data are representative of the population? Explain. Is the purported relationship of any practical interest? Explain.

In the remaining chapters of the text, you'll become familiar with the tools essential for building a firm foundation in statistics and statistical thinking.

KEY TERMS

Census	7	Measurement error	17	Representative sample	13
Data	4	Nonresponse bias	17	Sample	7
Descriptive statistics	5	Observational study	13	Selection bias	17
Designed experiment	13	Population	6	Statistical inference	8
Experimental unit	6	Published source	13	Statistical thinking	16
Inference	5	Qualitative data	11	Statistics	4
Inferential statistics	5	Quantitative data	11	Survey	13
Measure of reliability	9	Random sample	14	Unethical statistical practice	16
Measurement	7	Reliability	9	Variable	6

CHAPTER NOTES

Types of Statistical Applications

Descriptive

1. Identify **population** or **sample** (collection of experimental units)
2. Identify **variable(s)**
3. Collect **data**
4. **Describe** data

Inferential

1. Identify **population** (collection of *all* **experimental units**)
2. Identify **variable(s)**
3. Collect **sample** data (*subset* of population)
4. **Inference** about population based on sample
5. **Measure of reliability** of inference

Types of Data

1. **Quantitative** (numerical in nature)
2. **Qualitative** (categorical in nature)

Data Collection Methods

1. **Observational**
2. **Published source**
3. **Survey**
4. **Designed experiment**

Problems with Nonrandom Samples

1. **Selection bias**
2. **Nonresponse bias**
3. **Measurement error**

EXERCISES 1.1–1.33

Understanding the Principles

1.1 What is statistics?

1.2 Explain the difference between descriptive and inferential statistics.

1.3 List and define the five elements of an inferential statistical analysis.

1.4 List the four major methods of collecting data, and explain their differences.

1.5 Explain the difference between quantitative and qualitative data.

1.6 Explain how populations and variables differ.

1.7 Explain how populations and samples differ.

1.8 What is a representative sample? What is its value?

1.9 Why would a statistician consider an inference incomplete without an accompanying measure of its reliability?

1.10 Define statistical thinking.

1.11 Suppose you're given a data set that classifies each sample unit into one of four categories: A, B, C, or D. You plan to create a computer database consisting of these data, and you decide to code the data as A = 1, B = 2, C = 3, and D = 4. Are the data consisting of the classifications A, B, C, and D qualitative or quantitative? After the data are input as 1, 2, 3, or 4, are they qualitative or quantitative? Explain your answers.

APPLET **Applet Exercise 1.1**

The *Random Numbers* applet generates a list of *n* random numbers from 1 to *N*, where *n* is the size of the sample and *N* is the size of the population. The list generated often contains repetitions of one or more numbers.

a. Using the applet *Random Numbers*, enter 1 for the minimum value, 10 for the maximum value, and 10 for the number of samples. Then click on *Sample*. Look at the results, and list any numbers that are repeated and the number of times each of these numbers occurs.

b. Repeat part (a), changing the maximum value to 20 and keeping the size of the sample fixed at 10. If you still have repetitions, repeat the process, increasing the maximum value by 10 each time but keeping the size of the sample fixed. What is the smallest maximum value for which you had no repetitions?

c. Describe the relationship between the population size (maximum value) and the number of repetitions in the list of random numbers as the population size increases and the sample size remains the same. What can you conclude about using a random number generator to choose a relatively small sample from a large population?

APPLET **Applet Exercise 1.2**

The *Random Numbers* applet can be used to select a random sample from a population, but can it be used to simulate data? In parts (a) and (b), you will use the applet to create data sets. Then you will explore whether those data sets are realistic.

a. In Activity 1.1: *Keep the Change* (p. 12), a data set called *Amounts Transferred* is described. Use the *Random Numbers* applet to simulate this data set by setting the minimum value equal to 0, the maximum value equal to 99, and the sample size equal to 30. Explain what the numbers in the list produced by the applet represent in the context of the activity. (You may need to read the activity.) Do the numbers produced by the applet seem reasonable? Explain.

b. Use the *Random Numbers* applet to simulate grades on a statistics test by setting the minimum value equal to 0, the maximum value equal to 100, and the sample size equal to 30. Explain what the numbers in the list produced by the applet represent in this context. Do the numbers produced by the applet seem reasonable? Explain.

c. Referring to parts (a) and (b), why do the randomly generated data seem more reasonable in one situation than in the other? Comment on the usefulness of using a random-number generator to produce data.

Applying the Concepts—Basic

1.12 College application. Colleges and universities are requir-
NW ing an increasing amount of information about applicants before making acceptance and financial aid decisions. Classify each of the following types of data required on a college application as quantitative or qualitative.
a. High school GPA
b. High school class rank
c. Applicant's score on the SAT or ACT
d. Gender of applicant
e. Parents' income
f. Age of applicant

1.13 Ground motion of earthquakes. In the *Journal of Earthquake Engineering* (Nov. 2004), a team of civil and environmental engineers studied the ground motion characteristics of 15 earthquakes that occurred around the world between 1940 and 1995. Three (of many) variables measured on each earthquake were the type of ground motion (short, long, or forward directive), the magnitude of the earthquake (on the Richter scale), and peak ground acceleration (feet per second). One of the goals of the study was to estimate the inelastic spectra of any ground motion cycle.
a. Identify the experimental units for this study.
b. Do the data for the 15 earthquakes represent a population or a sample? Explain.
c. Identify the variables measured as quantitative or qualitative.

1.14 Sprint speed training. *The Sport Journal* (Winter 2004) reported on a study of a speed-training program for high school football players. Each participant was timed in a 40-yard sprint both before and after training. The researchers measured two variables: (1) the difference between the before and after sprint times (in seconds), and (2) the category of improvement ("improved," "no change," and "worse") for each player.
a. Identify the type (quantitative or qualitative) of each variable measured.
NW **b.** A total of 14 high school football players participated in the speed-training program. Does the data set collected represent a population or a sample? Explain.

1.15 Iraq War poll. A poll is to be conducted in which 2,000
NW individuals are asked whether the United States was justified in invading Iraq. The 2,000 individuals are selected by random-digit telephone dialing and asked the question over the phone.
a. What is the relevant population?
b. What is the variable of interest? Is it quantitative or qualitative?
c. What is the sample?
d. What is the inference of interest to the pollster?
e. What method of data collection is employed?
f. How likely is the sample to be representative?

1.16 Student GPAs. Consider the set of all students enrolled in your statistics course this term. Suppose you're interested in learning about the current grade point averages (GPAs) of this group.
a. Define the population and variable of interest.
b. Is the variable qualitative or quantitative?
c. Suppose you determine the GPA of every member of the class. Would this determination represent a census or a sample?
d. Suppose you determine the GPA of 10 members of the class. Would this determination represent a census or a sample?
e. If you determine the GPA of every member of the class and then calculate the average, how much reliability does your calculation have as an "estimate" of the class average GPA?
f. If you determine the GPA of 10 members of the class and then calculate the average, will the number you get necessarily be the same as the average GPA for the whole class? On what factors would you expect the reliability of the estimate to depend?
g. What must be true in order for the sample of 10 students you select from your class to be considered a random sample?

CEOPAY05

1.17 The Executive Compensation Scoreboard. Each year, *Forbes* publishes its "Executive Compensation Scoreboard." For the 2005 scoreboard, data were collected on chief executive officers at the 500 largest U.S. companies and the following variables were measured for each CEO: (1) the type of industry of the CEO's company (e.g., banking, retailing), (2) the CEO's total compensation ($ millions) for the year, (3) the CEO's total compensation ($ millions) over the previous five years, (4) the number of company stock shares (millions) held, (5) the CEO's age (years), and (6) the CEO's efficiency rating.
a. Do the data for the 500 CEOs in the 2005 "Executive Compensation Scoreboard" represent a population or sample? Explain.
b. Identify the type (quantitative or qualitative) of each variable measured.

1.18 Extinct birds. Biologists at the University of California (Riverside) are studying the patterns of extinction in the New Zealand bird population. (*Evolutionary Ecology Research*, July 2003.) At the time of the Maori colonization of New Zealand (prior to European contact), the following variables were measured for each bird species:
a. Flight capability (volant or flightless)
b. Type of habitat (aquatic, ground terrestrial, or aerial terrestrial)
c. Nesting site (ground, cavity within ground, tree, cavity above ground)
d. Nest density (high or low)
e. Diet (fish, vertebrates, vegetables, or invertebrates)
f. Body mass (grams)
g. Egg length (millimeters)
h. Extinct status (extinct, absent from island, present)

Identify each variable as quantitative or qualitative.

1.19 Study of quality of drinking water. *Disasters* (Vol. 28, 2004) published a study of the effects of a tropical cyclone on the quality of drinking water on a remote Pacific island. Water samples (size 500 milliliters) were collected approximately four weeks after Cyclone Ami hit the island. The following variables were recorded for each water sample:
a. Town where sample was collected
b. Type of water supply (river intake, stream, or borehole)
c. Acidic level (pH scale, 1 to 14)
d. Turbidity level (nephalometric turbidity units = NTUs)
e. Temperature (degrees centigrade)
f. Number of fecal coliforms per 100 milliliters
g. Free-chlorine residual (milligrams per liter)
h. Presence of hydrogen sulphide (yes or no)

Identify each variable as quantitative or qualitative.

Applying the Concepts—Intermediate

1.20 Herbal medicines. *The American Association of Nurse Anesthetists Journal* (Feb. 2000) published the results of a study on the use of herbal medicines before surgery. Of 500 surgical patients randomly selected for the study, 51% used herbal or alternative medicines (e.g., garlic, ginkgo, kava, fish oil) against their doctor's advice prior to surgery.
a. Do the 500 surgical patients represent a population or a sample? Explain.
b. If your answer was "sample" in part **a,** is the sample representative of the population? If you answered "population" in part **a,** explain how to obtain a representative sample from the population.
c. For each patient, what variable is measured? Are the data collected quantitative or qualitative?

1.21 National Bridge Inventory. All highway bridges in the United States are inspected periodically for structural deficiency by the Federal Highway Administration (FHWA). Data from the FHWA inspections are compiled into the National Bridge Inventory (NBI). Several of the nearly 100 variables maintained by the NBI are listed next. Classify each variable as quantitative or qualitative.
a. Length of maximum span (feet)
b. Number of vehicle lanes
c. Toll bridge (yes or no)
d. Average daily traffic
e. Condition of deck (good, fair, or poor)
f. Bypass or detour length (miles)
g. Type of route (interstate, U.S., state, county, or city)

1.22 Annual survey of computer crimes. The Computer Security Institute (CSI) conducts an annual survey of computer crime committed at U.S. businesses. CSI sends survey questionnaires to computer security personnel at all U.S. corporations and government agencies. In 2006, 616 organizations responded to the CSI survey. Fifty-two percent of the respondents admitted unauthorized use of computer systems at their firms during the year. (*Computer Security Issues & Trends*, Spring 2006.)
a. Identify the population of interest to CSI.
b. Identify the data collection method used by CSI. Are there any potential biases in the method used?
c. Describe the variable measured in the CSI survey. Is it quantitative or qualitative?
d. What inference can be made from the result of the study?

1.23 CT scanning for lung cancer. According to the American Lung Association, lung cancer accounts for 28% of all cancer deaths in the United States. A new type of screening for lung cancer, computed tomography (CT), has been developed. Medical researchers believe that CT scans are more sensitive than regular X-rays in pinpointing small tumors. The H. Lee Moffitt Cancer Center at the University of South Florida is currently conducting a clinical trial of 50,000 smokers nationwide to compare the effectiveness of CT scans with X-rays for detecting lung cancer. (*Todays' Tomorrows*, Fall 2002.) Each participating smoker is randomly assigned to one of two screening methods, CT or chest X-ray, and their progress tracked over time. The age at which the scanning method first detects a tumor is the variable of interest.
a. Identify the data collection method used by the cancer researchers.
b. Identify the experimental units of the study.
c. Identify the type (quantitative or qualitative) of the variable measured.
d. Identify the population and sample.
e. What is the inference that will ultimately be drawn from the clinical trial?

1.24 Gallup poll of teenagers. A Gallup Youth Poll was conducted to determine the topics that teenagers most want to discuss with their parents. The findings show that 46% would like more discussion about the family's financial situation, 37% would like to talk about school, and 30% would like to talk about religion. The survey was based on a national sampling of 505 teenagers, selected at random from all U.S. teenagers.
a. Describe the sample.
b. Describe the population from which the sample was selected.
c. Is the sample representative of the population?
d. What is the variable of interest?
e. How is the inference expressed?
f. Newspaper accounts of most polls usually give a *margin of error* (e.g., plus or minus 3%) for the survey result. What is the purpose of the margin of error and what is its interpretation?

1.25 Massage vs. rest in boxing. Does a massage enable the muscles of tired athletes to recover from exertion faster than usual? To answer this question, researchers recruited eight amateur boxers to participate in an experiment. (*British Journal of Sports Medicine*, April 2000.) After a 10-minute workout in which each boxer threw 400 punches, half the boxers were given a 20-minute massage and half just rested for 20 minutes. Before they returned to the ring for a second workout, the heart rate (beats per minute) and blood lactate level (micromoles) were recorded for each boxer. The researchers found no difference in the means of the two groups of boxers for either variable.

a. Identify the data collection method used by the researchers.

b. Identify the experimental units of the study.

c. Identify the variables measured and their type (quantitative or qualitative).

d. What is the inference drawn from the analysis?

e. Comment on whether this inference can be made about all athletes.

1.26 Insomnia and education. Is insomnia related to education status? Researchers at the Universities of Memphis, Alabama at Birmingham, and Tennessee investigated this question in the *Journal of Abnormal Psychology* (Feb. 2005). Adults living in Tennessee were selected to participate in the study, which used a random-digit telephone dialing procedure. Two of the many variables measured for each of the 575 study participants were number of years of education and insomnia status (normal sleeper or chronic insomniac). The researchers discovered that the fewer the years of education, the more likely the person was to have chronic insomnia.

a. Identify the population and sample of interest to the researchers.

b. Identify the data collection method. Are there any potential biases in the method used?

c. Describe the variables measured in the study as quantitative or qualitative.

d. What inference did the researchers make?

1.27 Blood loss in burn patients. A group of University of South Florida surgery researchers believes that the drug aprotinin is effective in reducing the blood loss of burn patients who undergo skin replacement surgery. (*USF Office of Research Annual Report, 1997–1998.*) In a clinical trial of 14 burn patients, half were randomly assigned to receive the drug and half a placebo (no drug). A preliminary analysis of the patients' blood loss revealed that the group with the drug had a significant reduction in bleeding.

a. Identify the data collection method used for this study.

b. Does the study involve descriptive or inferential statistics? Explain.

c. What is the population (or sample) of interest to the researchers?

Applying the Concepts — Advanced

1.28 Object recall study. Are men or women more adept at remembering where they leave misplaced items (such as car keys)? According to University of Florida psychology professor Robin West, women show greater competence in actually finding these objects (*Explore*, Fall 1998). Approximately 300 men and women from Gainesville, Florida, participated in a study in which each person placed 20 common objects in a 12-room "virtual" house represented on a computer screen. Thirty minutes later, the subjects were asked to recall where they put each of the objects. For each object, a recall variable was measured as "yes" or "no."

a. Identify the population of interest to the psychology professor.

b. Identify the sample.

c. Does the study involve descriptive or inferential statistics? Explain.

d. Are the variables measured in the study quantitative or qualitative?

1.29 Soccer "headers" and IQ. *USA Today* (Aug. 14, 1995) reported on a study which suggests that "frequently 'heading' the ball in soccer lowers players' IQs." A psychologist tested 60 male soccer players, ages 14–29, who played up to five times a week. Players who averaged 10 or more headers a game had an average IQ of 103, while players who headed one or fewer times per game had an average IQ of 112.

a. Describe the population of interest to the psychologist.

b. Identify the variables of interest.

c. Identify the type (qualitative or quantitative) of the variables in part **b**.

d. Describe the sample.

e. What is the inference made by the psychologist?

f. Discuss some reasons the inference in part **e** may be misleading.

SWREUSE

1.30 Success/failure of software reuse. The PROMISE Software Engineering Repository, hosted by the University of Ottawa, is a collection of publicly available data sets to serve researchers in building prediction software models. A PROMISE data set on software reuse, saved in the **SWREUSE** file, provides information on the success or failure of reusing previously developed software for each in a sample of 24 new software development projects. (*Data source: IEEE Transactions on Software Engineering*, Vol. 28, 2002.) Of the 24 projects, 9 were judged failures and 15 were successfully implemented.

a. Identify the experimental units for this study.

b. Describe the population from which the sample is selected.

c. What is the variable of interest in the study? Is it quantitative or qualitative?

d. Critically evaluate the statement "Since 15/24 = .625, it follows that 62.5% of all new software development projects will be successfully implemented."

Critical Thinking Challenges

1.31 Roper poll on the Holocaust. The Holocaust—the Nazi Germany extermination of the Jews and others during World War II—has been well documented through film, books, and interviews with the concentration camp survivors. But a recent Roper poll found that one in five U.S. residents doubted that the Holocaust really occurred. A national sample of more than 1,000 adults and high school students was asked, "Does it seem possible or does it seem impossible to you that the Nazi extermination of the Jews never happened?" Twenty-two percent of the adults and 20% of the high school students responded "Yes." (*New York Times*, May 16, 1994.) Discuss reasons for doubting the reliability of the inference.

1.32 Poll on alien spacecraft. "Have you ever seen anything that you believe was a spacecraft from another planet?" This was the question put to 1,500 American adults in a national poll conducted by ABC News and *The Washington Post*. The pollsters used random-digit telephone dialing to contact adult Americans until 1,500 responded. Ten percent (i.e., 150) of the respondents answered that they had, in fact, seen an alien spacecraft. (*Chance*, Summer 1997.) No information was provided on how many adults were called and, for one reason or another, did not answer the question.

a. Identify the data collection method.

b. Identify the target population.

c. Comment on the validity of the survey results.

1.33 **"20/20" survey exposés.** The popular prime-time ABC television program "20/20" presented several misleading (and possibly unethical) surveys in a segment titled "Facts or Fiction? Exposés of So-Called Surveys" (March 31, 1995).The information reported from four of these surveys, conducted by businesses or special-interest groups with specific objectives in mind, are given. (Actual survey facts are provided in parentheses.)

Quaker Oats Study: Eating oat bran is a cheap and easy way to reduce your cholesterol count. (*Fact*: Diet must consist of nothing but oat bran to achieve a slightly lower cholesterol count.)

March of Dimes Report: Domestic violence causes more birth defects than all medical issues combined. (*Fact*: No study—false report.)

American Association of University Women (AAUW) study: Only 29% of high school girls are happy with themselves, compared with 66% of elementary school girls. (*Fact*: Of 3,000 high school girls, 29% responded "Always true" to the statement, "I am happy the way I am." Most answered, "Sort of true" and "Sometimes true.")

Food Research and Action Center study: One in four American children under age 12 is hungry or at risk of hunger. (*Fact*: Survey results are based on responses to the questions "Do you ever cut the size of meals?", "Do you ever eat less than you feel you should?" and "Did you ever rely on limited numbers of foods to feed your children because you were running out of money to buy food for a meal?")

a. Refer to the Quaker Oats study relating oat bran to cholesterol levels. Discuss why it is unethical to report the results as stated.

b. Consider the false March of Dimes report on domestic violence and birth defects. Discuss the type of data required to investigate the impact of domestic violence on birth defects. What data collection method would you recommend?

c. Refer to the AAUW study of the self-esteem of high school girls. Explain why the results of the study are likely to be misleading. What data might be appropriate for assessing the self-esteem of high school girls?

d. Refer to the Food Research and Action Center study of hunger in America. Explain why the results of the study are likely to be misleading. What data would provide insight into the proportion of hungry American children?

SUMMARY ACTIVITY: Data in the News

Scan your daily newspaper or weekly news magazine, or search the Internet for articles that contain data. The data might be a summary of the results of a public opinion poll, the results of a vote by the U.S. Senate, or a list of crime rates, birth or death rates, etc. For each article you find, answer the following questions:

a. Do the data constitute a sample or an entire population? If a sample has been taken, clearly identify both the sample and the population; otherwise, identify the population.

b. What type of data (quantitative or qualitative) has been collected?

c. What is the source of the data?

d. If a sample has been observed, is it likely to be representative of the population?

e. If a sample has been observed, does the article present an explicit (or implied) inference about the population of interest? If so, state the inference made in the article.

f. If an inference has been made, has a measure of reliability been included? What is it?

g. Use your answers to questions d–f to critically evaluate the article.

REFERENCES

Brochures about Survey Research, Section on Survey Research Methods, American Statistical Association, 2004. (*www.amstat.org*)

Careers in Statistics, American Statistical Association, Biometric Society, Institute of Mathematical Statistics and Statistical Society of Canada, 2004. (*www.amstat.org*)

Cochran, W. G. *Sampling Techniques*, 3d ed. New York: Wiley, 1977.

Deming, W. E. *Sample Design in Business Research*. New York: Wiley, 1963.

Ethical Guidelines for Statistical Practice. American Statistical Association, 1995.

Hansen, M. H., Hurwitz, W. N., and Madow, W. G. *Sample Survey Methods and Theory*, Vol. 1. New York: Wiley, 1953.

Kirk, R. E., ed. *Statistical Issues: A Reader for the Behavioral Sciences*. Monterey, CA: Brooks/Cole, 1972.

Kish, L. *Survey Sampling*. New York: Wiley, 1965.

Scheaffer, R., Mendenhall, W., and Ott, R. L. *Elementary Survey Sampling*, 2d ed. Boston: Duxbury, 1979.

Tanur, J. M., Mosteller, F., Kruskal, W. H., Link, R. E., Pieters, R. S., and Rising, G. R. *Statistics: A Guide to the Unknown* (E. L. Lehmann, special editor). San Francisco: Holden-Day, 1989.

Yamane, T. *Elementary Sampling Theory*, 3d ed. Englewood Cliffs, NJ: Prentice-Hall, 1967.

Using Technology

Creating and Listing Data with MINITAB

Upon entering into a MINITAB session, you will see a screen similar to Figure 1.M.1. The bottom portion of the screen is an empty spreadsheet—called a MINITAB worksheet—with columns representing variables and rows representing observations (or cases). The very top of the screen is the MINITAB main menu bar, with buttons for the different functions and procedures available in MINITAB. Once you have entered data into the spreadsheet, you can analyze the data by clicking the appropriate menu buttons. The results will appear in the Session window.

You can create a MINITAB data file by entering data directly into the worksheet. Figure 1.M.2 shows data entered for a variable called "Ratio". The variables (columns) can be named by typing in the name of each variable in the box below the column number.

If the data are saved in an external data file, you can access it by using the options available in MINITAB. Click the "File" button on the menu bar, and then click "Open Worksheet", as shown in Figure 1.M.3. The dialog box shown in Figure 1.M.4 will appear.

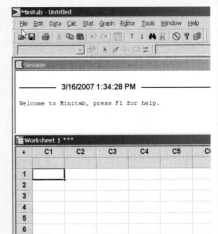

FIGURE 1.M.1
Initial screen viewed by MINITAB user

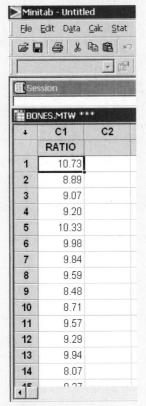

FIGURE 1.M.2
Data entered into the MINITAB worksheet

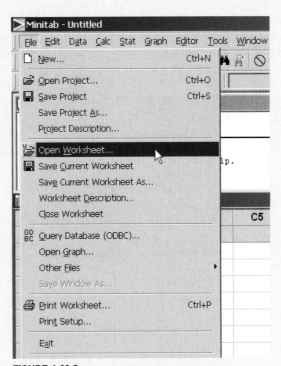

FIGURE 1.M.3
MINITAB options for reading data from an external file

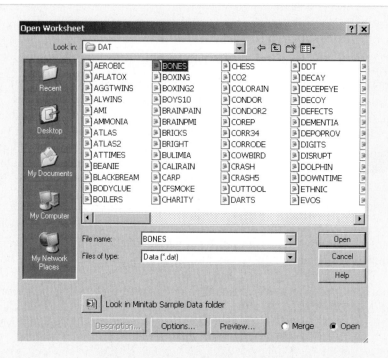

FIGURE 1.M.4
Selecting the external data file

Specify the disk drive and folder that contain the external data file and the file type, and then click on the file name, as shown in Figure 1.M.4. If the data set contains qualitative data or data with special characters, click on the "Options" button as shown in Figure 1.M.4. The Options dialog box, shown in Figure 1.M.5, will appear. Specify the appropriate options for the data set, and then click "OK" to return to the "Open Worksheet" dialog box (Figure 1.M.4). Click "Open" and the MINITAB worksheet will appear with the data from the external data file, as shown in Figure 1.M.6.

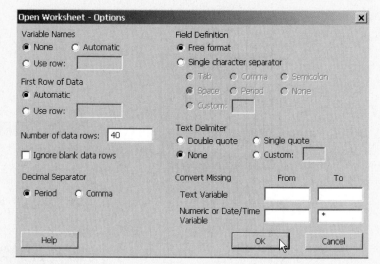

FIGURE 1.M.5
Selecting the data input options

	C1	C2-T	C3	C4-T	C5	C6
1	1	Ally	52	R	55	
2	2	Batty	12	C	12	
3	3	Bongo	28	R	40	
4	4	Blackie	52	C	10	
5	5	Bucky	40	R	45	
6	6	Bumble	28	R	600	
7	7	Crunch	21	C	10	
8	8	Congo	28	C	10	
9	9	Derby	28	R	30	
10	10	Digger	40	R	150	
11	11	Echo	17	R	20	
12	12	Fetch	5	C	15	
13	13	Early	5	C	20	
14	14	Flip	28	R	40	
15	15	Garcia	28	R	200	
16	16	Happy	52	R	20	
17	17	Grunt	28	R	175	
18	18	Gigi	5	C	15	
19	19	Goldie	52	R	45	

BEANIE.DAT ***

FIGURE 1.M.6
The MINITAB worksheet with the imported data

Reminder: The variables (columns) can be named by typing in the name of each variable in the box under the column number.

To obtain a listing (printout) of your data, click on the "Data" button on the MINITAB main menu bar, and then click on "Display Data". (See Figure 1.M.7.) The resulting menu, or dialog box, appears as in Figure 1.M.8. Enter the names of the variables you want to print in the "Columns, constants, and matrices to display" box (you can do this simply by clicking on the variables), and then click "OK". The printout will show up on your MINITAB session screen.

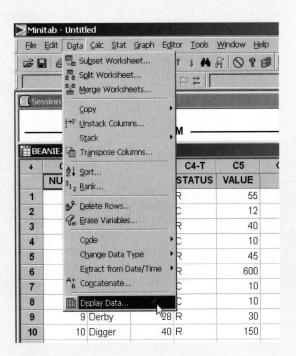

FIGURE 1.M.7
MINITAB menu options for obtaining a list of your data

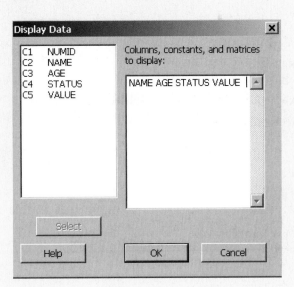

FIGURE 1.M.8
Display data dialog box

CHAPTER 2

Methods for Describing Sets of Data

CONTENTS

2.1 Describing Qualitative Data

2.2 Graphical Methods for Describing Quantitative Data

2.3 Summation Notation

2.4 Numerical Measures of Central Tendency

2.5 Numerical Measures of Variability

2.6 Interpreting the Standard Deviation

2.7 Numerical Measures of Relative Standing

2.8 Methods for Detecting Outliers (Optional)

2.9 Graphing Bivariate Relationships (Optional)

2.10 Distorting the Truth with Descriptive Techniques

STATISTICS IN ACTION
The "Eye Cue" Test: Does Experience Improve Performance?

USING TECHNOLOGY
Describing Data Using MINITAB

WHERE WE'VE BEEN
- Examined the difference between inferential and descriptive statistics
- Described the key elements of a statistical problem
- Learned about the two types of data: quantitative and qualitative
- Discussed the role of statistical thinking in managerial decision making

WHERE WE'RE GOING
- Describe data by using graphs
- Describe data by using numerical measures

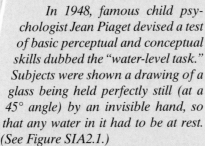

The "Eye Cue" Test:
Does Experience Improve Performance?

In 1948, famous child psychologist Jean Piaget devised a test of basic perceptual and conceptual skills dubbed the "water-level task." Subjects were shown a drawing of a glass being held perfectly still (at a 45° angle) by an invisible hand, so that any water in it had to be at rest. (See Figure SIA2.1.)

The task for the subject was to draw a line representing the surface of the water—a line that would touch the black dot pictured on the right side of the glass. Piaget found that young children typically failed the test. Fifty years later, research psychologists still use the water-level task to test the perception of both adults and children. Surprisingly, about 40% of the adult population also fail. In addition, males tend to do better than females, and younger adults tend to do better than older adults.

Will people with experience handling liquid-filled containers perform better on this "eye cue" test? This question was the focus of research conducted by psychologists Heiko Hecht (NASA) and Dennis R. Proffitt (University of Virginia) and published in *Psychological Science* (Mar. 1995). The researchers presented the task to each of six different groups: (1) male college students, (2) female college

students, (3) professional waitresses, (4) housewives, (5) male bartenders, and (6) male bus drivers. A total of 120 subjects (20 per group) participated in the study. Two of the groups, waitresses and bartenders, were assumed to have considerable experience handling liquid-filled glasses.

After each subject completed the drawing task, the researchers recorded the deviation* (in angle degrees) of the judged line from the true line. If the deviation was within 5° of the true water surface angle, the answer was considered correct. Deviations of more than 5° in either direction were considered incorrect answers.

The data for the water-level task (simulated on the basis of summary results presented in the journal article) are provided in the **EYECUE** file. For each of the 120 subjects in the experiment, the following variables (in the order they appear on the data file) were measured:

 EYECUE

Variables

GENDER (F or M)
GROUP (Student, Waitress, Wife, Bartender, or Bus driver)
DEVIATION (angle, in degrees, of the judged line from the true line)
JUDGE (Within5, More5Above, More5Below)

The researchers were interested in testing several theories concerning performance on the water-level task, including the following: (1) Males perform better than females, (2) younger adults perform better than older adults, and (3) experience improves task performance.

In the *Statistics in Action Revisited* sections that follow, we apply the graphical and numerical descriptive techniques of this chapter to the **EYECUE** data to answer some of the researchers' questions.

"Statistics in Action Revisited" for Chapter 2

- Interpreting Pie Charts (p. 33)
- Interpreting Histograms (p. 45)
- Interpreting Descriptive Statistics (p. 72)
- Detecting Outliers (p. 86)

*The true surface line is perfectly parallel to the tabletop.

FIGURE SIA2.1
Drawing of the water-level task

Suppose you wish to evaluate the mathematical capabilities of a class of 1,000 first-year college students, based on their quantitative Scholastic Aptitude Test (SAT) scores. How would you describe these 1,000 measurements? Characteristics of interest include the typical, or most frequent, SAT score; the variability in the scores; the highest and lowest scores; the "shape" of the data; and whether the data set contains any unusual scores. Extracting this information isn't easy. The 1,000 scores provide too many bits of information for our minds to comprehend. Clearly, we need some method for summarizing and characterizing the information in such a data set. Methods for describing data sets are also essential for statistical inference. Most populations make for large data sets. Consequently, we need methods for describing a data set that let us make descriptive statements (inferences) about a population on the basis of information contained in a sample.

Two methods for describing data are presented in this chapter, one *graphical* and the other *numerical*. Both play an important role in statistics. Section 2.1 presents both graphical and numerical methods for describing qualitative data. Graphical methods for describing quantitative data are illustrated in Sections 2.2, 2.8, and 2.9; numerical descriptive methods for quantitative data are presented in Sections 2.3–2.7. We end the chapter with a section on the *misuse* of descriptive techniques.

2.1 Describing Qualitative Data

Consider a study of aphasia published in the *Journal of Communication Disorders* (Mar. 1995). Aphasia is the "impairment or loss of the faculty of using or understanding spoken or written language." Three types of aphasia have been identified by researchers: Broca's, conduction, and anomic. The researchers wanted to determine whether one type of aphasia occurs more often than any other and, if so, how often. Consequently, they measured the type of aphasia for a sample of 22 adult aphasics. Table 2.1 gives the type of aphasia diagnosed for each aphasic in the sample.

For this study, the variable of interest, type of aphasia, is qualitative in nature. Qualitative data are nonnumerical in nature; thus, the value of a qualitative variable can only be classified into categories called *classes*. The possible types of aphasia—Broca's, conduction, and anomic—represent the classes for this qualitative variable. We can summarize such data numerically in two ways: (1) by computing the *class frequency*—the number of observations in the data set that fall into each class—or (2) by computing the *class relative frequency*—the proportion of the total number of observations falling into each class.

Definition 2.1

A **class** is one of the categories into which qualitative data can be classified.

APHASIA

TABLE 2.1 Data on 22 Adult Aphasias

Subject	Type of Aphasia	Subject	Type of Aphasia
1	Broca's	12	Broca's
2	Anomic	13	Anomic
3	Anomic	14	Broca's
4	Conduction	15	Anomic
5	Broca's	16	Anomic
6	Conduction	17	Anomic
7	Conduction	18	Conduction
8	Anomic	19	Broca's
9	Conduction	20	Anomic
10	Anomic	21	Conduction
11	Conduction	22	Anomic

Source: Li, E. C., Williams, S. E., and Volpe, R. D., "The effects of topic and listener familiarity of discourse variables in procedural and narrative discourse tasks." *Journal of Communication Disorders,* Vol. 28, No. 1, Mar. 1995, p. 44 (Table 1).

> ### Definition 2.2
> The **class frequency** is the number of observations in the data set that fall into a particular class.

> ### Definition 2.3
> The **class relative frequency** is the class frequency divided by the total number of observations in the data set; that is,
> $$\text{class relative frequency} = \frac{\text{class frequency}}{n}$$

> ### Definition 2.4
> The **class percentage** is the class relative frequency multiplied by 100; that is,
> $$\text{class percentage} = (\text{class relative frequency}) \times 100$$

Examining Table 2.1, we observe that 5 aphasics in the study were diagnosed as suffering from Broca's aphasia, 7 from conduction aphasia, and 10 from anomic aphasia. These numbers—5, 7, and 10—represent the class frequencies for the three classes and are shown in the summary table, Figure 2.1, produced with SPSS.

Figure 2.1 also gives the relative frequency of each of the three aphasia classes. From Definition 2.3, we calculate the relative frequency by dividing the class frequency by the total number of observations in the data set. Thus, the relative frequencies for the three types of aphasia are

$$\text{Broca's:} \quad \frac{5}{22} = .227$$

$$\text{Conduction:} \quad \frac{7}{22} = .318$$

$$\text{Anomic:} \quad \frac{10}{22} = .455$$

These values, expressed as a percent, are shown in the SPSS summary table of Figure 2.1. From these relative frequencies, we observe that nearly half (45.5%) of the 22 subjects in the study are suffering from anomic aphasia.

Although the summary table of Figure 2.1 adequately describes the data of Table 2.1, we often want a graphical presentation as well. Figures 2.2 and 2.3 show two of the most widely used graphical methods for describing qualitative data: bar graphs and pie charts. Figure 2.2 shows the frequencies of the three types of aphasia in a **bar graph** produced with SAS. Note that the height of the rectangle, or "bar," over each class is equal to the class frequency. (Optionally, the bar heights can be proportional to class relative frequencies.)

TYPE

		Frequency	Percent	Valid Percent	Cumulative Percent
Valid	Anomic	10	45.5	45.5	45.5
	Brocas	5	22.7	22.7	68.2
	Conduction	7	31.8	31.8	100.0
	Total	22	100.0	100.0	

FIGURE 2.1

SPSS summary table for types of aphasia

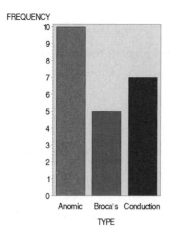

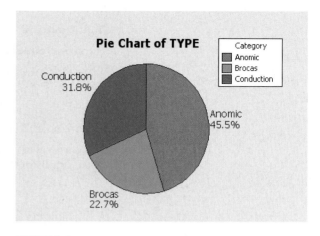

FIGURE 2.2

SAS bar graph for type of aphasia

FIGURE 2.3

MINITAB pie chart for type of aphasia

In contrast, Figure 2.3 shows the relative frequencies of the three types of aphasia in a **pie chart** generated with MINITAB. Note that the pie is a circle (spanning 360°) and the size (angle) of the "pie slice" assigned to each class is proportional to the class relative frequency. For example, the slice assigned to anomic aphasia is 45.5% of 360°, or $(.455)(360°) = 163.8°$.

Before leaving the data set in Table 2.1, consider the bar graph shown in Figure 2.4, produced with SPSS. Note that the bars for the types of aphasia are arranged in descending order of height, from left to right across the horizontal axis. That is, the tallest bar (Anomic) is positioned at the far left and the shortest bar (Broca's) is at the far right. This rearrangement of the bars in a bar graph is called a **Pareto diagram**. One goal of a Pareto diagram (named for the Italian economist Vilfredo Pareto) is to make it easy to locate the "most important" categories—those with the largest frequencies.

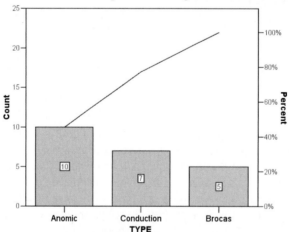

FIGURE 2.4

SPSS Pareto diagram for type of aphasia

Biography

Vilfredo Pareto (1843–1923)— The Pareto Principle

Born in Paris to an Italian aristocratic family, Vilfredo Pareto was educated at the University of Turin, where he studied engineering and mathematics. After the death of his parents, Pareto quit his job as an engineer and began writing and lecturing on the evils of the economic policies of the Italian government. While at the University of Lausanne in Switzerland in 1896, he published his first paper, *Cours d'économie politique*. In the paper, Pareto derived a complicated mathematical formula to prove that the distribution of income and wealth in society is not random, but that a consistent pattern appears throughout history in all societies. Essentially, Pareto showed that approximately 80% of the total wealth in a society lies with only 20% of the families. This famous law about the "vital few and the trivial many" is widely known as the Pareto principle in economics.

Summary of Graphical Descriptive Methods for Qualitative Data

Bar Graph: The categories (classes) of the qualitative variable are represented by bars, where the height of each bar is either the class frequency, class relative frequency, or class percentage.

Pie Chart: The categories (classes) of the qualitative variable are represented by slices of a pie (circle). The size of each slice is proportional to the class relative frequency.

Pareto Diagram: A bar graph with the categories (classes) of the qualitative variable (i.e., the bars) arranged by height in descending order from left to right.

Now Work Exercise 2.6

Let's look at a practical example that requires interpretation of the graphical results.

EXAMPLE 2.1
GRAPHING AND SUMMARIZING QUALITATIVE DATA—Drug Designed to Reduce Blood Loss

 **BLOODLOSS**

Problem

A group of cardiac physicians in southwest Florida has been studying a new drug designed to reduce blood loss in coronary bypass operations. Blood loss data for 114 coronary bypass patients (some who received a dosage of the drug and others who did not) are saved in the **BLOODLOSS** file. Although the drug shows promise in reducing blood loss, the physicians are concerned about possible side effects and complications. So their data set includes not only the qualitative variable DRUG, which indicates whether or not the patient received the drug, but also the qualitative variable COMP, which specifies the type (if any) of complication experienced by the patient. The four values of COMP are (1) redo surgery, (2) post-op infection, (3) both, or (4) none.

a. Figure 2.5, generated by SAS, shows summary tables for the two qualitative variables, DRUG and COMP. Interpret the results.
b. Interpret the MINITAB and SPSS printouts shown in Figures 2.6 and 2.7, respectively.

Solution

a. The top table in Figure 2.5 is a summary frequency table for DRUG. Note that exactly half (57) of the 114 coronary bypass patients received the drug and half did not. The bottom table in Figure 2.5 is a summary frequency table for COMP. The

The FREQ Procedure

DRUG	Frequency	Percent	Cumulative Frequency	Cumulative Percent
NO	57	50.00	57	50.00
YES	57	50.00	114	100.00

COMP	Frequency	Percent	Cumulative Frequency	Cumulative Percent
BOTH	6	5.26	6	5.26
INFECT	15	13.16	21	18.42
NONE	79	69.30	100	87.72
REDO	14	12.28	114	100.00

FIGURE 2.5

SAS summary tables for DRUG and COMP

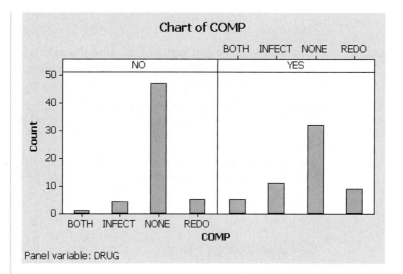

FIGURE 2.6

MINITAB side-by-side bar graphs for COMP by value of DRUG

Panel variable: DRUG

COMP

DRUG			Frequency	Percent	Valid Percent	Cumulative Percent
NO	Valid	BOTH	1	1.8	1.8	1.8
		INFECT	4	7.0	7.0	8.8
		NONE	47	82.5	82.5	91.2
		REDO	5	8.8	8.8	100.0
		Total	57	100.0	100.0	
YES	Valid	BOTH	5	8.8	8.8	8.8
		INFECT	11	19.3	19.3	28.1
		NONE	32	56.1	56.1	84.2
		REDO	9	15.8	15.8	100.0
		Total	57	100.0	100.0	

FIGURE 2.7

SPSS summary tables for COMP by value of drug

class percentages are given in the Percent column. We see that about 69% of the 114 patients had no complications, leaving about 31% who experienced either a redo surgery, a post-op infection, or both.

b. Figure 2.6 is a MINITAB side-by-side bar graph of the data. The four bars in the left-side graph represent the frequencies of COMP for the 57 patients who did not receive the drug; the four bars in the right-side graph represent the frequencies of COMP for the 57 patients who did receive a dosage of the drug. The graph clearly shows that patients who did not receive the drug suffered fewer complications. The exact percentages are displayed in the SPSS summary tables of Figure 2.7. Over 56% of the patients who got the drug had no complications, compared with about 83% for the patients who got no drug.

Look Back Although these results show that the drug may be effective in reducing blood loss, Figures 2.6 and 2.7 imply that patients on the drug may have a higher risk of incurring complications. But before using this information to make a decision about the drug, the physicians will need to provide a measure of reliability for the inference. That is, the physicians will want to know whether the difference between the percentages of patients with complications observed in this sample of 114 patients is generalizable to the population of all coronary bypass patients.

Now Work Exercise 2.18

STATISTICS IN ACTION REVISITED

Interpreting Pie Charts for the "Eye Cue" Test Data

In the *Psychological Science* "water-level task" experiment, the researchers measured three qualitative variables: *Gender* (F or M), *Subject Group* (student, waitress, wife, bartender, or bus driver), and *Judged Task Performance* (within 5° of the line, more than 5° above the line, or more than 5° below the line). Pie charts and bar graphs can be used to summarize and describe the responses in these variables and categories. Recall that the data are saved in the **EYECUE** file. The variables are named GENDER, GROUP, and JUDGE in the data file. We used MINITAB to create pie charts for these variables.

Figure SIA2.2 is a pie chart for the JUDGE variable. Clearly, the large slice for "Within5" indicates that a majority of subjects (52.5%) were judged to be within 5° of the correct line. The researchers want to know if the water-level task performance will vary across gender. Figure SIA2.3 shows side-by-side pie charts of the JUDGE variable for each level of GENDER. You can see that 65% of the male subjects (right-side chart) were judged to be within 5° of the line, compared with only 40% for female subjects (left-side chart). These graphs support the prevailing theory that men perform better than women on the "water-level" task.

We produced a similar set of side-by-side pie charts in Figure SIA2.4 to compare the performances of the different groups of subjects. These charts again show that bus drivers (75% judged to be within 5° of the correct line) and college students (72.5%) perform the best, while bartenders (40%), housewives (30%), and (surprisingly) waitresses (25%) perform the worst. These graphs do not seem to support the researchers' theory that experience improves task performance.

FIGURE SIA2.2

MINITAB pie chart for judged task performance

Pie Chart of Judged Line

FIGURE SIA2.3

MINITAB pie charts for judged task performance: females versus males

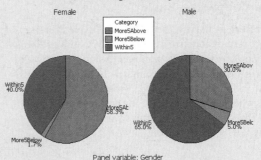

FIGURE SIA2.4

MINITAB pie charts for judged task performance: group comparisons

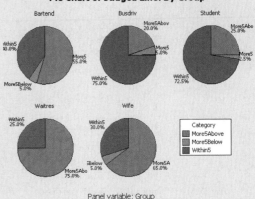

Exercises 2.1–2.20

Understanding the Principles

2.1 Explain the difference between class frequency, class relative frequency, and class percentage for a qualitative variable.

2.2 Explain the difference between a bar graph and a pie chart.

2.3 Explain the difference between a bar graph and a Pareto diagram.

Learning the Mechanics

2.4 Complete the following table:

Grade on Statistics Exam	Frequency	Relative Frequency
A: 90–100		.08
B: 80–89	36	
C: 65–79	90	
D: 50–64	30	
F: Below 50	28	
Total	**200**	**1.00**

2.5 A qualitative variable with three classes (X, Y, and Z) is measured for each of 20 units randomly sampled from a target population. The data (observed class for each unit) are as follows:

Y X X Z X Y Y Y Y X X Z X
Y Y X Z Y Y Y X

a. Compute the frequency for each of the three classes.
b. Compute the relative frequency for each of the three classes.
c. Display the results from part **a** in a frequency bar graph.
d. Display the results from part **b** in a pie chart.

Applying the Concepts—Basic

2.6 **National Firearms Survey.** In the journal *Injury Prevention* (Jan. 2007), researchers from the Harvard School of Public Health reported on the size and composition of privately held firearm stock in the United States. In a representative household telephone survey of 2,770 adults, 26% reported that they owned at least one gun. The accompanying graphic summarizes the types of firearms owned.
a. What type of graph is shown?

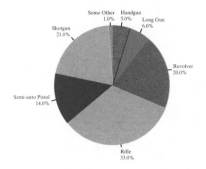

Source: Hepburn, M., Miller, D. A., and Hemenway, D. "The US gun stock: Results from the 2004 national firearms survey," *Injury Prevention*, Vol. 13, No. 1, Jan. 2007 (Figure 1).

b. Identify the qualitative variable described in the graph.

c. From the graph, identify the most common type of firearms.

d. Convert the graph into a Pareto diagram. Interpret the results.

2.7 **Japanese reading levels.** University of Hawaii language professors C. Hitosugi and R. Day incorporated a 10-week extensive reading program into a second-semester Japanese language course in an effort to improve students' Japanese reading comprehension. (*Reading in a Foreign Language*, Apr. 2004.) The professors collected 266 books originally written for Japanese children and required their students to read at least 40 of them as part of the grade in the course. The books were categorized into reading levels (color coded for easy selection) according to length and complexity. The reading levels for the 266 books are summarized in the following table:

Reading Level	Number
Level 1 (Red)	39
Level 2 (Blue)	76
Level 3 (Yellow)	50
Level 4 (Pink)	87
Level 5 (Orange)	11
Level 6 (Green)	3
Total	**266**

Source: Hitosugi, C. I., and Day, R. R. "Extensive reading in Japanese," *Reading in a Foreign Language*, Vol. 16, No. 1. Apr. 2004 (Table 2).

a. Calculate the proportion of books at reading level 1 (red).
b. Repeat part **a** for each of the remaining reading levels.
c. Verify that the proportions in parts **a** and **b** sum to 1.
d. Use the previous results to form a bar graph for the reading levels.
e. Construct a Pareto diagram for the data. Use the diagram to identify the reading level that occurs most often.

2.8 **Dates of pennies.** *Chance* (Spring 2000) reported on a study to estimate the number of pennies required to fill a coin collector's album. The data used in the study were obtained by noting the mint date on each of a sample of 2,000 pennies. The distribution of mint dates is summarized in the following table:

Mint Date	Number
Pre-1960	18
1960s	125
1970s	330
1980s	727
1990s	800

Source: Lu, S., and Skiena, S. "Filling a penny album," *Chance*, Vol. 13, No. 2, Spring 2000, p. 26 (Table 1).

a. Identify the experimental unit for the study.
b. Identify the variable measured.
c. What proportion of pennies in the sample have mint dates in the 1960s?
d. Construct a pie chart to describe the distribution of mint dates for the 2,000 sampled pennies.

2.9 **Estimating the rhino population.** The International Rhino Federation estimates that there are 17,800 rhinoceroses living in the wild in Africa and Asia. A breakdown of the number of rhinos of each species is reported in the accompanying table:

Rhino Species	Population Estimate
African Black	3,610
African White	11,330
(Asian) Sumatran	300
(Asian) Javan	60
(Asian) Indian	2,500
Total	17,800

Source: International Rhino Federation, Mar. 2007.

a. Construct a relative frequency table for the data.
b. Display the relative frequencies in a bar graph.
c. What proportion of the 17,800 rhinos are African rhinos? Asian?

2.10 **Role importance for the elderly.** In *Psychology and Aging* (Dec. 2000), University of Michigan School of Public Health researchers studied the roles that elderly people feel are the most important to them in late life. The following table summarizes the most salient roles identified by each respondent in a national sample of 1,102 adults 65 years or older.

Most Salient Role	Number
Spouse	424
Parent	269
Grandparent	148
Other relative	59
Friend	73
Homemaker	59
Provider	34
Volunteer, club, church member	36
Total	1,102

Source: Krause, N., and Shaw, B. A. "Role-specific feelings of control and mortality," *Psychology and Aging*, Vol. 15, No. 4, Dec. 2000 (Table 2).

a. Describe the qualitative variable summarized in the table. Give the categories associated with the variable.
b. Are the numbers in the table frequencies or relative frequencies?
c. Display the information in the table in a bar graph.

d. Which role is identified by the highest percentage of elderly adults? Interpret the relative frequency associated with this role.

PONDICE

2.11 **Characteristics of ice melt ponds.** The National Snow and Ice Data Center (NSIDC) collects data on the albedo, depth, and physical characteristics of ice melt ponds in the Canadian Arctic. Environmental engineers at the University of Colorado are using these data to study how climate affects the sea ice. Data on 504 ice melt ponds located in the Barrow Strait in the Canadian Arctic are saved in the **PONDICE** file. One variable of interest is the type of ice observed for each pond, classified as first-year ice, multiyear ice, or landfast ice. A SAS summary table and a horizontal bar graph that describe the types of ice of the 504 melt ponds are shown at the bottom of the page.
a. Of the 504 melt ponds, what proportion had landfast ice?
b. The University of Colorado researchers estimated that about 17% of melt ponds in the Canadian Arctic have first-year ice. Do you agree?
c. Convert the horizontal bar graph into a Pareto diagram. Interpret the graph.

Applying the Concepts—Intermediate

2.12 **Excavating ancient pottery.** Archaeologists excavating the ancient Greek settlement at Phylakopi classified the pottery found in trenches. (*Chance*, Fall 2000.) The accompanying table describes the collection of 837 pottery pieces uncovered in a particular layer at the excavation site. Construct and interpret a graph that will aid the archaeologists in understanding the distribution of the types of pottery found at the site.

Pot Category	Number Found
Burnished	133
Monochrome	460
Slipped	55
Painted in curvilinear decoration	14
Painted in geometric decoration	165
Painted in naturalistic decoration	4
Cycladic white clay	4
Conical cup clay	2
Total	837

Source: Berg, I., and Bliedon, S. "The Pots of Phylakopi: Applying Statistical Techniques to Archaeology," *Chance*, Vol. 13, No. 4, Fall 2000.

SAS output for Exercise 2.11

The FREQ Procedure

ICETYPE	Frequency	Percent	Cumulative Frequency	Cumulative Percent
First-year	88	17.46	88	17.46
Landfast	196	38.89	284	56.35
Multi-year	220	43.65	504	100.00

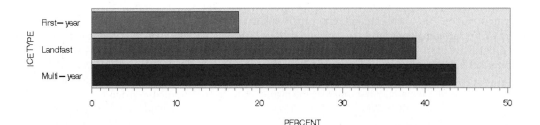

2.13 **"Made in the USA" survey.** "Made in the USA" is a claim stated in many product advertisements or on product labels. Advertisers want consumers to believe that their product is manufactured with 100% U.S. labor and materials—which is often not the case. What does "Made in the USA" mean to the typical consumer? To answer this question, a group of marketing professors conducted an experiment at a shopping mall in Muncie, Indiana. (*Journal of Global Business*, Spring 2002.) They asked every fourth adult entrant to the mall to participate in the study. A total of 106 shoppers agreed to answer the question, "'Made in the USA' means what percentage of U.S. labor and materials?" The responses of the 106 shoppers are summarized in the following table:

Response to "Made in the USA"	Number of Shoppers
100%	64
75 to 99%	20
50 to 74%	18
Less than 50%	4

Source: "'Made in the USA': Consumer perceptions, deception and policy alternatives," *Journal of Global Business*, Vol. 13, No. 24, Spring 2002 (Table 3).

a. What type of data collection method was used?

b. What type of variable, quantitative or qualitative, was measured?

c. Present the data in the table in graphical form. Use the graph to make a statement about the percentage of consumers who believe that "Made in the USA" means 100% U.S. labor and materials.

2.14 **Sociology field work methods.** University of New Mexico professor Jane Hood investigated the fieldwork methods used by qualitative sociologists. (*Teaching Sociology*, July 2006.) Searching for all published journal articles, dissertations, and conference proceedings over the previous seven years in the *Sociological Abstracts* database, she discovered that fieldwork methods could be categorized as follows: Interview, Observation plus Participation, Observation Only, and Grounded Theory. The accompanying table shows the number of papers in each category. Use an appropriate graph to portray the results of the study. Interpret the graph.

⊙ **FIELDWORK**

Fieldwork Method	Number of Papers
Interview	5,079
Observation + Participation	1,042
Observation Only	848
Grounded Theory	537

Source: Hood, J.C. "Teaching against the text: The case of qualitative methods," *Teaching Sociology*, Vol. 34, July 2006 (Exhibit 2).

2.15 **Best-paid CEOs.** *Forbes* magazine conducts an annual survey of the salaries of chief executive officers. In addition to salary information, *Forbes* collects and reports personal data on the CEOs, including their level of education and age. Do most CEOs have advanced degrees, such as master's degrees or doctorates? The data in the table on page 37 represent the highest degree earned by each of the top 40 best-paid CEOs of 2006. Use a graphical method to summarize the highest degree earned by these CEOs. What is your opinion about whether most CEOs have advanced degrees?

2.16 **Frog census in South Australia.** Periodically, the Environmental Protection Authority conducts a census of frogs in South Australia in order to monitor the health of aquatic environments. In the latest (2002) census, data were collected on 15 of the 28 frog species indigent to South Australia. Since every species of frog has a distinctive mating call, field researchers can identify the abundance of the species at a particular location simply by listening for the mating calls. In 2002, there were 152 recordings of the eastern banjo frog. For each recording, the field researcher estimated the number of frogs at the site. The accompanying table summarizes these data.

⊙ **FROGS**

Abundance (number of frogs)	Number of Recordings
One	33
Few (2–9)	75
Many (10–50)	27
Lots (>50)	17
Total	152

Source: *Frog Census 2002*, Australian Environmental Protection Authority.

a. Find the relative frequency associated with each frog abundance category.

b. Portray the data in the table with a relative frequency bar graph.

c. Use the bar graph to identify the abundance category with the greatest relative frequency.

⊙ **BIBLE**

2.17 **Do you believe in the Bible?** In its annual General Social Survey (GSS), the National Opinion Research Center (NORC) elicits the opinions of Americans on a wide variety of social topics. One question in the survey asked about a person's belief in the Bible. Data for the approximately 2,800 Americans who participated in the 2004 GSS are saved in the **BIBLE** file. Each respondent selected from one of the following answers: (1) The Bible is the actual word of God and is to be taken literally; (2) the Bible is the inspired word of God, but not everything is to be taken literally; (3) the Bible is an ancient book of fables; and (4) the Bible has some other origin, but is recorded by men. The variable "Bible1" contains the responses. (Note: A response value of 8 represents "Don't Know"; a value of 9 represents a missing value.)

a. Summarize the responses to the Bible question in the form of a relative frequency table.

b. Summarize the responses to the Bible question in a pie chart.

c. Write a few sentences that give a practical interpretation of the results shown in the summary table and graph.

Applying the Concepts—Advanced

⊙ **NZBIRDS**

2.18 **Extinct New Zealand birds.** Refer to the *Evolutionary Ecology Research* (July 2003) study of the patterns of extinction in the New Zealand bird population, Exercise 1.18 (p. 20). Data on flight capability (volant or flightless), habitat (aquatic, ground terrestrial, or aerial terrestrial), nesting site (ground, cavity within ground, tree, or cavity above ground), nest density (high or low), diet (fish, vertebrates, vegetables,

FORBES40

CEO	COMPANY	SALARY ($millions)	AGE (years)	DEGREE
Richard D. Fairbank	Capital One Financial	249.42	55	MBA
Terry S. Semel	Yahoo	230.55	63	MBA
Henry R. Silverman	Cendant	139.96	65	Law
Bruce Karatz	KB Home	135.53	60	Law
Richard S. Fuld, Jr.	Lehman Bros	122.67	60	MBA
Ray R. Irani	Occidental Petroleum	80.73	71	PhD
Lawrence J. Ellison	Oracle	75.33	61	None
John W. Thompson	Symantec	71.84	57	Masters
Edwin M. Crawford	Caremark Rx	69.66	57	Bachelor's
Angelo R. Mozilo	Countrywide Financial	68.96	67	Bachelor's
John T. Chambers	Cisco Systems	62.99	56	MBA
R. Chad Dreier	Ryland Group	56.47	58	Bachelor's
Lew Frankfort	Coach	55.99	60	MBA
Ara K. Hovnanian	Hovnanian Enterprises	47.83	48	MBA
John G. Drosdick	Sunoco	46.19	62	Master's
Robert I. Toll	Toll Brothers	41.31	65	Law
Robert J. Ulrich	Target	39.64	63	Bachelor's
Kevin B. Rollins	Dell	39.32	53	MBA
Clarence P. Cazalot, Jr.	Marathon Oil	37.48	55	Bachelor's
David C. Novak	Yurn Brands	37.42	53	Bachelor's
Mark G. Papa	EOG Resources	36.54	59	MBA
Henri A. Termeer	Genzyme	36.38	60	MBA
Richard C. Adkerson	Freeport Copper	35.41	59	MBA
Kevin W. Sharer	Amgen	34.49	58	Master's
Jay Sugarman	IStar Financial	32.94	43	MBA
George David	United Technologies	32.73	64	MBA
Bob R. Simpson	XTO Energy	32.19	57	MBA
J. Terrence Lanni	MGM Mirage	31.54	63	MBA
Paul F. Jacobs	Qualcomm	31.44	64	PhD
Stephen F. Bollenbach	Hilton Hotels	31.43	63	MBA
James J. Mulva	ConocoPhillips	31.34	59	MBA
John J. Mack	Morgan Stanley	31.23	61	Bachelor's
Ronald A. Williams	Aetna	30.87	57	Master's
David J. Lesar	Halliburton	29.36	53	MBA
H. Edward Hanway	Cigna	28.82	54	MBA
James E. Cayne	Bear Stearns Companies	28.40	72	None
Daniel P. Amos	Aflac	27.97	54	Bachelor's
Kent J. Thiry	DaVita	27.89	50	MBA
John W. Rowe	Exelon	26.90	60	Law
James M. Cornelius	Guidant	25.18	62	MBA

Source: Forbes, May 8, 2006.

or invertebrates), body mass (grams), egg length (millimeters), and extinct status (extinct, absent from island, or present) for 132 bird species that existed at the time of the Maori colonization of New Zealand are saved in the **NZBIRDS** file. Use a graphical method to investigate the theory that extinct status is related to flight capability, habitat, and nest density.

MTBE

2.19 Groundwater contamination in wells. In New Hampshire, about half the counties mandate the use of reformulated gasoline. This has led to an increase in the contamination of groundwater with methyl *tert*-butyl ether (MTBE). *Environmental Science & Technology* (Jan. 2005) reported on the factors related to MTBE contamination in public and private New Hamsphire wells. Data were collected on a sample of 223 wells. These data are saved in the **MTBE** file. Three of the variables are qualitative in nature: well class (public or private), aquifer (bedrock or unconsolidated), and detectable level of MTBE (below limit or detect). [Note: A detectable level of MTBE occurs if the MTBE value exceeds .2 microgram per liter.] The data on 11 selected wells are shown in the accompanying table.

MTBE (11 selected observations from 223)

Well Class	Aquifer	Detect MTBE?
Private	Bedrock	Below Limit
Private	Bedrock	Below Limit
Public	Unconsolidated	Detect
Public	Unconsolidated	Below Limit
Public	Unconsolidated	Below Limit
Public	Unconsolidated	Below Limit
Public	Unconsolidated	Detect
Public	Unconsolidated	Below Limit
Public	Unconsolidated	Below Limit
Public	Bedrock	Detect
Public	Bedrock	Detect

Source: Ayotte, J.D., Argue, D.M., and McGarry, F.J. "Methyl *tert*-Butyl Ether Occurrence and Related Factors in Public and Private Wells in Southeast New Hampshire," *Environmental Science & Technology*, Vol. 39, No. 1, Jan. 2005.

a. Use graphical methods to describe each of the three qualitative variables for all 223 wells.

b. Use side-by-side bar charts to compare the proportions of contaminated wells for private and public well classes.

c. Use side-by-side bar charts to compare the proportions of contaminated wells for bedrock and unconsolidated aquifers.

d. What inferences can be made from the bar charts of parts a–c?

2.20 Benford's Law of Numbers. According to *Benford's law*, certain digits (1, 2, 3, . . . , 9) are more likely to occur as the first significant digit in a randomly selected number than are other digits. For example, the law predicts that the number "1" is the most likely to occur (30% of the time) as the first digit. In a study reported in the *American Scientist* (July–Aug. 1998) to test Benford's law, 743 first-year college students were asked to write down a six-digit number at random. The first significant digit of each number was recorded and its distribution summarized in the following table:

 DIGITS

First Digit	Number of Occurrences
1	109
2	75
3	77
4	99
5	72
6	117
7	89
8	62
9	43
Total	743

Source: Hill, T. P. "The First Digit Phenomenon," *American Scientist,* Vol. 86, No. 4, July–Aug. 1998, p. 363 (Figure 5).

a. Describe the first digit of the "random guess" data with an appropriate graph.

b. Does the graph support Benford's law? Explain.

2.2 Graphical Methods for Describing Quantitative Data

Recall from Section 1.4 that quantitative data sets consist of data that are recorded on a meaningful numerical scale. To describe, summarize, and detect patterns in such data, we can use three graphical methods: *dot plots, stem-and-leaf displays,* and *histograms*. Since most statistical software packages can be used to construct these displays, we'll focus here on their interpretation rather than their construction.

For example, the Environmental Protection Agency (EPA) performs extensive tests on all new car models to determine their mileage ratings. Suppose that the 100 measurements in Table 2.2 represent the results of such tests on a certain new car model. How can we summarize the information in this rather large sample?

EPAGAS

TABLE 2.2 EPA Mileage Ratings on 100 Cars									
36.3	41.0	36.9	37.1	44.9	36.8	30.0	37.2	42.1	36.7
32.7	37.3	41.2	36.6	32.9	36.5	33.2	37.4	37.5	33.6
40.5	36.5	37.6	33.9	40.2	36.4	37.7	37.7	40.0	34.2
36.2	37.9	36.0	37.9	35.9	38.2	38.3	35.7	35.6	35.1
38.5	39.0	35.5	34.8	38.6	39.4	35.3	34.4	38.8	39.7
36.3	36.8	32.5	36.4	40.5	36.6	36.1	38.2	38.4	39.3
41.0	31.8	37.3	33.1	37.0	37.6	37.0	38.7	39.0	35.8
37.0	37.2	40.7	37.4	37.1	37.8	35.9	35.6	36.7	34.5
37.1	40.3	36.7	37.0	33.9	40.1	38.0	35.2	34.8	39.5
39.9	36.9	32.9	33.8	39.8	34.0	36.8	35.0	38.1	36.9

A visual inspection of the data indicates some obvious facts. For example, most of the mileages are in the 30s, with a smaller fraction in the 40s. But it is difficult to provide much additional information on the 100 mileage ratings without resorting to some method of summarizing the data. One such method is a dot plot.

Dot Plots

A MINITAB **dot plot** for the 100 EPA mileage ratings is shown in Figure 2.8. The horizontal axis of the figure is a scale for the quantitative variable in miles per gallon (mpg). The numerical value of each measurement in the data set is located on the horizontal scale by a dot. When data values repeat, the dots are placed above one another, forming a pile at that particular numerical location. As you can see, this dot plot verifies that almost all of the mileage ratings are in the 30s, with most falling between 35 and 40 miles per gallon.

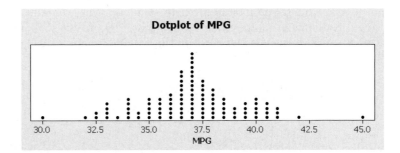

FIGURE 2.8

MINITAB dot plot for
100 EPA mileage ratings

Stem-and-Leaf Display

Another graphical representation of these same data, a MINITAB **stem-and-leaf display**, is shown in Figure 2.9. In this display, the *stem* is the portion of the measurement (mpg) to the left of the decimal point, while the remaining portion, to the right of the decimal point, is the *leaf*.

In Figure 2.9, the stems for the data set are listed in the second column, from the smallest (30) to the largest (44). Then the leaf for each observation is listed to the right, in the row of the display corresponding to the observation's stem. For example, the leaf 3 of the first observation (36.3) in Table 2.2 appears in the row corresponding to the stem 36. Similarly, the leaf 7 for the second observation (32.7) in Table 2.2 appears in the row corresponding to the stem 32, while the leaf 5 for the third observation (40.5) appears in the row corresponding to the stem 40. (The stems and leaves for these first three observations are highlighted in Figure 2.9.) Typically, the leaves in each row are ordered as shown in the MINITAB stem-and-leaf display.

The stem-and-leaf display presents another compact picture of the data set. You can see at a glance that the 100 mileage readings were distributed between 30.0 and 44.9, with most of them falling in stem rows 35 to 39. The 6 leaves in stem row 34 indicate that 6 of the 100 readings were at least 34.0, but less than 35.0. Similarly, the 11 leaves in stem row 35 indicate that 11 of the 100 readings were at least 35.0, but less than 36.0. Only five cars had readings equal to 41 or larger, and only one was as low as 30.

The definitions of the stem and leaf for a data set can be modified to alter the graphical description. For example, suppose we had defined the stem as the tens digit for the gas mileage data, rather than the ones and tens digits. With this definition, the stems and leaves corresponding to the measurements 36.3 and 32.7 would be as follows:

Stem	Leaf	Stem	Leaf
3	6	3	2

Note that the decimal portion of the numbers has been dropped. Generally, only one digit is displayed in the leaf.

```
Stem-and-leaf of MPG   N  = 100
Leaf Unit = 0.10


 1      30  0
 2      31  8
 6      32  5799
12      33  126899
18      34  024588
29      35  01235667899
49      36  01233445566777888999
(21)    37  000011122334456677899
30      38  0122345678
20      39  00345789
12      40  0123557
 5      41  002
 2      42  1
 1      43
 1      44  9
```

FIGURE 2.9

MINITAB stem-and-leaf
display for 100 mileage ratings

If you look at the data, you'll see why we didn't define the stem this way. All the mileage measurements fall into the 30s and 40s, so all the leaves would fall into just two stem rows in this display. The resulting picture would not be nearly as informative as Figure 2.9.

Now Work Exercise 2.25

Biography

JOHN TUKEY (1915–2000)—
The Picasso of Statistics

Like the legendary artist Pablo Picasso, who mastered and revolutionized a variety of art forms during his lifetime, John Tukey is recognized for his contributions to many subfields of statistics. Born in Massachusetts, Tukey was home schooled, graduated with his bachelor's and master's degrees in chemistry from Brown University, and received his Ph.D. in mathematics from Princeton University. While at Bell Telephone Laboratories in the 1960s and early 1970s, Tukey developed exploratory data analysis, a set of graphical descriptive methods for summarizing and presenting huge amounts of data. Many of these tools, including the stem-and-leaf display and the box plot, are now standard features of modern statistical software packages. (In fact, it was Tukey himself who coined the term *software* for computer programs.)

Histograms

An SPSS **histogram** for the 100 EPA mileage readings given in Table 2.2 is shown in Figure 2.10. The horizontal axis of the figure, which gives the miles per gallon for a given automobile, is divided into **class intervals**, commencing with the interval from 30.0–31.5 and proceeding in intervals of equal size to 43.5–45.0 mpg. The vertical axis gives the number (or *frequency*) of the 100 readings that fall into each interval. It appears that about 33 of the 100 cars, or 33%, attained a mileage between 36.0 and 37.5 mpg. This class interval contains the highest frequency, and the intervals tend to contain a smaller number of the measurements as the mileages get smaller or larger.

Histograms can be used to display either the frequency or relative frequency of the measurements falling into the class intervals. The class intervals, frequencies, and relative frequencies for the EPA car mileage data are shown in the summary table, Table 2.3.*

By summing the relative frequencies in the intervals 34.5–36.0, 36.0–37.5, and 37.5–39.0, you can see that 65% of the mileages are between 34.5 and 39.0. Similarly, only 2% of the cars obtained a mileage rating over 42.0. Many other summary statements can be made by further examining the histogram and accompanying summary table. Note that the sum of all class frequencies will always equal the sample size *n*.

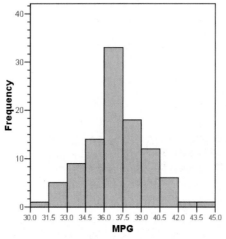

FIGURE 2.10

SPSS histogram for 100 EPA mileage ratings

*SPSS, like many software packages, will classify an observation that falls on the borderline of a class interval into the next-highest interval. For example, the gas mileage of 37.5, which falls on the border between the class intervals 36.0–37.5 and 37.5–39.0, is classified into the 37.5–39.0 class. The frequencies in Table 2.3 reflect this convention.

TABLE 2.3   Class Intervals, Frequencies, and Relative Frequencies for the Car Mileage Data

Class Interval	Frequency	Relative Frequency
30.0–31.5	1	.01
31.5–33.0	5	.05
33.0–34.5	9	.09
34.5–36.0	14	.14
36.0–37.5	33	.33
37.5–39.0	18	.18
39.0–40.5	12	.12
40.5–42.0	6	.06
42.0–43.5	1	.01
43.5–45.0	1	.01
Totals	100	1.00

In interpreting a histogram, consider two important facts. First, the proportion of the total area under the histogram that falls above a particular interval on the *x*-axis is equal to the relative frequency of measurements falling into that interval. For example, the relative frequency for the class interval 36.0–37.5 in Figure 2.10 is .33. Consequently, the rectangle above the interval contains .33 of the total area under the histogram.

Second, imagine the appearance of the relative frequency histogram for a very large set of data (representing, say, a population). As the number of measurements in a data set is increased, you can obtain a better description of the data by decreasing the width of the class intervals. When the class intervals become small enough, a relative frequency histogram will (for all practical purposes) appear as a smooth curve. (See Figure 2.11.)

FIGURE 2.11

The effect of the size of a data set on the outline of a histogram

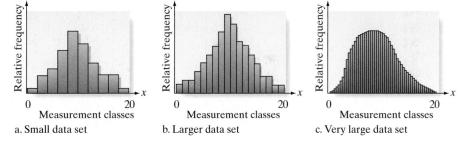

Some recommendations for selecting the number of intervals in a histogram for smaller data sets are given in the following box:

Determining the Number of Classes in a Histogram

Number of Observations in Data Set	Number of Classes
Fewer than 25	5–6
25–50	7–14
More than 50	15–20

Although histograms provide good visual descriptions of data sets—particularly very large ones—they do not let us identify individual measurements. In contrast, each of the original measurements is visible to some extent in a dot plot and is clearly visible in a stem-and-leaf display. The stem-and-leaf display arranges the data in ascending order, so it's easy to locate the individual measurements. For example, in Figure 2.9 we can easily see that two of the gas mileage measurements are equal to 36.3, but we can't see that fact by inspecting the histogram in Figure 2.10. However, stem-and-leaf displays can become unwieldy for very large data sets. A very large number of stems and leaves causes the vertical and horizontal dimensions of the display to become cumbersome, diminishing the usefulness of the visual display.

EXAMPLE 2.2

GRAPHING A
QUANTITATIVE
VARIABLE—
Percentage of Student
Loan Defaults

Problem The data in Table 2.4 give the percentages of each state's total number of college or university student loans that are in default.

a. Create a histogram of these data. Where do most of the default rates lie?

b. Create a stem-and-leaf display for these data. Locate Wyoming's default rate of 2.7 on the graph.

Solution

a. We used SAS to generate a relative frequency histogram in Figure 2.12. Note that seven classes were formed. The classes are identified by their *midpoints* rather than their endpoints. Thus, the first interval has a midpoint of 1.5, the second of 4.5, and so on. The corresponding class intervals based on these midpoints are therefore 0–3.0, 3.0–6.0, etc. Note that the classes with midpoints of 7.5 and 10.5 (ranging from 6.0 to 12.0) contain approximately 60% of the 51 default measurements.

b. We used MINITAB to generate the stem-and-leaf display shown in Figure 2.13. Note that the stem (the *second* column in the printout) has been defined as the number *two* places to the left of the decimal. The leaf (the third column in the printout) is the number *one* place to the left of the decimal.* (The digit to the right of the decimal is not shown.) Thus, the leaf 2 in the stem 0 row (the first row of the printout) represents the default rate of 2.7 for Wyoming. This leaf is shaded on the printout.

LOANDEFAULT

TABLE 2.4 Percentage of Student Loans in Default, by State

State	%	State	%	State	%	State	%
AL	12.0	IL	9.3	MT	6.4	RI	8.8
AK	19.7	IN	6.7	NE	4.9	SC	14.1
AZ	12.1	IA	6.2	NV	10.1	SD	5.5
AR	12.9	KS	5.7	NH	7.9	TN	12.3
CA	11.4	KY	10.3	NJ	12.0	TX	15.2
CO	9.5	LA	13.5	NM	7.5	UT	6.0
CT	8.8	ME	9.7	NY	11.3	VT	8.3
DE	10.9	MD	16.6	NC	15.5	VA	14.4
DC	14.7	MA	8.3	ND	4.8	WA	8.4
FL	11.8	MI	11.4	OH	10.4	WV	9.5
GA	14.8	MN	6.6	OK	11.2	WI	9.0
HI	12.8	MS	15.6	OR	7.9	WY	2.7
ID	7.1	MO	8.8	PA	8.7		

Source: National Direct Student Loan Coalition.

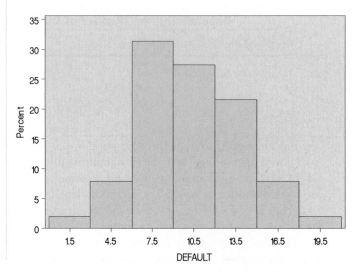

FIGURE 2.12

SAS relative frequency histogram for student loan default rate data

*The first column of the MINITAB stem-and-leaf display represents the cumulative number of measurements from the class interval to the nearest extreme class interval.

```
Stem-and-leaf of DEFAULT   N = 51
Leaf Unit = 1.0


    1      0   2
    5      0   4455
   14      0   666667777
  (12)     0   888888899999
   25      1   000011111
   16      1   2222223
    9      1   4444555
    2      1   6
    1      1   9
```

FIGURE 2.13
MINITAB stem-and-leaf display for student loan default rate data

Look Back As is usually the case with data sets that are not too large (say, fewer than 100 measurements), the stem-and-leaf display provides more detail than the histogram without being unwieldy. For instance, the stem-and-leaf display in Figure 2.13 clearly indicates the values of the individual measurements in the data set. For example, the highest default rate (representing the measurement 19.7 for Alaska) is shown in the last stem row. By contrast, histograms are most useful for displaying very large data sets when the overall shape of the distribution of measurements is more important than the identification of individual measurements.

Now Work Exercise 2.29

Most statistical software packages can be used to generate histograms, stem-and-leaf displays, and dot plots. All three are useful tools for graphically describing data sets. We recommend that you generate and compare the displays whenever you can.

Summary of Graphical Descriptive Methods for Quantitative Data

Dot Plot: The numerical value of each quantitative measurement in the data set is represented by a dot on a horizontal scale. When data values repeat, the dots are placed above one another vertically.

Stem-and-Leaf Display: The numerical value of the quantitative variable is partitioned into a "stem" and a "leaf." The possible stems are listed in order in a column. The leaf for each quantitative measurement in the data set is placed in the corresponding stem row. Leaves for observations with the same stem value are listed in increasing order horizontally.

Histogram: The possible numerical values of the quantitative variable are partitioned into class intervals, each of which has the same width. These intervals form the scale of the horizontal axis. The frequency or relative frequency of observations in each class interval is determined. A vertical bar is placed over each class interval, with the height of the bar equal to either the class frequency or class relative frequency.

Histograms

Using The TI-84/TI-83 Graphing Calculator

I. Making a Histogram from Raw Data

Step 1   *Enter the data.*
Press **STAT** and select **1:Edit**.
Note: If the list already contains data, clear the old data. Use the up arrow to highlight 'L1'. Press **CLEAR ENTER**.
Use the **ARROW** and **ENTER** keys to enter the data set into **L1**.

(continued)

Step 2 *Set up the histogram plot.*

Press **Y** = and **CLEAR** all functions from the Y registers.

Press **2nd** and press **Y** = for **STAT PLOT**.

Press **1** for **Plot 1**.

Set the cursor so that **ON** is flashing, and press **ENTER**.

For **Type**, use the **ARROW** and **ENTER** keys to highlight and select the histogram.

For **Xlist**, choose the column containing the data (in most cases, L1).

Note: Press **2nd 1** for **L1**.

Freq should be set to 1.

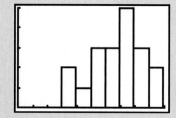

Step 3 *Select your window settings.*

Press **WINDOW** and adjust the settings as follows:

Xmin = lowest class boundary
Xmax = highest class boundary
Xscl = class width
Ymin = 0
Ymax ≥ greatest class frequency
Yscl = 1
Xres = 1

Step 4 *View the graph.*

Press **GRAPH**.

Optional Step *Read class frequencies and class boundaries.*

You can press **TRACE** to read the class frequencies and class boundaries. Use the arrow keys to move between bars.

Example The following figures show TI-84/TI-83 window settings and a histogram of the following sample data:

86, 70, 62, 98, 73, 56, 53, 92, 86, 37, 62, 83, 78, 49, 78, 37, 67, 79, 57

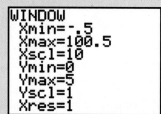

II. Making a Histogram from a Frequency Table

Step 1 *Enter the data.*

Press **STAT** and select **1:Edit**.

Note: If a list already contains data, clear the old data. Use the up arrow to highlight the list name, 'L1' or 'L2'.

Press **CLEAR ENTER**.

Enter the midpoint of each class into **L1**.

Enter the class frequencies or relative frequencies into **L2**.

Step 2 *Set up the histogram plot.*

Press **Y** = and **CLEAR** all functions from the Y registers.

Press **2nd** and **Y** = for **STAT PLOT**.

Press **1** for **Plot 1**.

Set the cursor so that **ON** is flashing, and press **ENTER**.

For **Type**, use the **ARROW** and **ENTER** keys to highlight and select the histogram.

For **Xlist**, choose the column containing the midpoints.

For **Freq**, choose the column containing the frequencies or relative frequencies.

Steps 3–4 *Follow steps 3–4 of Part I.*

Note: To set up the Window for relative frequencies, be sure to set **Ymax** to a value that is greater than or equal to the largest relative frequency.

STATISTICS IN ACTION REVISITED

Interpreting Histograms for the "Eye Cue" Test Data

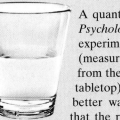

A quantitative variable measured in the *Psychological Science* "water-level task" experiment was the *deviation angle* (measured in degrees) of the judged line from the true surface line (parallel to the tabletop). The smaller the deviation, the better was the task performance. Recall that the researchers want to test the prevailing theory that males will do better than females on judging the correct water level. To check this theory, we accessed the **EYECUE** data file in MINITAB and created two frequency histograms for deviation angle—one for male subjects and one for female subjects. These side-by-side histograms are displayed in Figure SIA2.5.

From the histograms, there is some support for the theory. The histogram for males (the histogram on the right in Figure SIA2.5) shows the center of the distribution at about 0°, with most (about 70%) of the deviation values falling between −10° and 10°. In general, the male subjects tended to perform fairly well on the water-level task. In contrast, the histogram for females (the histogram on the left in Figure SIA2.5) is centered at about 10°, with most (about 70%) of the deviation values falling above 0°. Thus, the histogram for females shows a tendency for female subjects to have greater deviations (and thus poorer performances) in the water-level task than males. In later chapters, we'll learn how to attach a measure of reliability to such an inference.

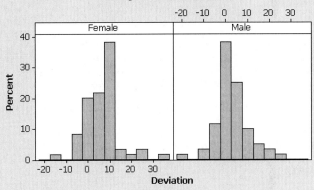

FIGURE SIA2.5

MINITAB histograms for deviation angle in the water-level task: males versus females

Exercises 2.21–2.41

Understanding the Principles

2.21 Explain the difference between a dot plot and a stem-and-leaf display.

2.22 Explain the difference between the stem and the leaf in a stem-and-leaf display.

2.23 In a histogram, what are the class intervals?

2.24 How many classes are recommended in a histogram of a data set with more than 50 observations?

Learning the Mechanics

2.25 Consider the stem-and-leaf display shown here:

NW

Stem	Leaf
5	1
4	457
3	00036
2	1134599
1	2248
0	012

a. How many observations were in the original data set?

b. In the bottom row of the stem-and-leaf display, identify the stem, the leaves, and the numbers in the original data set represented by this stem and its leaves.

c. Re-create all the numbers in the data set, and construct a dot plot.

2.26 Graph the relative frequency histogram for the 500 measurements summarized in the accompanying relative frequency table.

Class Interval	Relative Frequency
.5–2.5	.10
2.5–4.5	.15
4.5–6.5	.25
6.5–8.5	.20
8.5–10.5	.05
10.5–12.5	.10
12.5–14.5	.10
14.5–16.5	.05

2.27 Refer to Exercise 2.26. Calculate the number of the 500 measurements falling into each of the measurement classes. Then graph a frequency histogram of these data.

2.28 Consider the MINITAB histogram shown below.
   a. Is this a frequency histogram or a relative frequency histogram? Explain.
   b. How many class intervals were used in the construction of this histogram?
   c. How many measurements are there in the data set described by this histogram?

MINITAB Output for Exercise 2.28

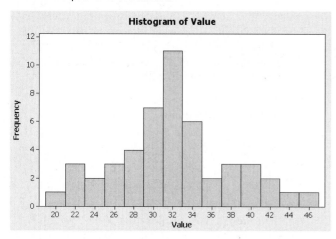

Applying the Concepts—Basic

EARTHQUAKE

2.29 **Earthquake aftershock magnitudes.** Seismologists use the term "aftershock" to describe the smaller earthquakes that follow a main earthquake. Following the Northridge earthquake of January 17, 1994, the Los Angeles area experienced 2,929 aftershocks in a three-week period. The magnitudes (measured on the Richter scale) of these aftershocks were recorded by the U.S. Geological Survey and are saved in the **EARTHQUAKE** file. A MINITAB relative frequency histogram for these magnitudes follows.
   a. Estimate the percentage of the 2,929 aftershocks measuring between 1.5 and 2.5 on the Richter scale.

MINITAB Output for Exercise 2.29

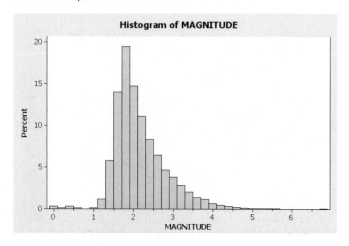

b. Estimate the percentage of the 2,929 aftershocks measuring greater than 3.0 on the Richter scale.

2.30 **Al Qaeda attacks on the United States.** An article in *Studies in Conflict & Terrorism* (Vol. 29, 2006) conducted an empirical analysis of recent incidents involving suicide terrorist attacks. The data in the table that follows are the number of individual suicide bombings or attacks for each of 21 incidents involving an attack against the United States by the Al Qaeda terrorist group. (For example, the infamous September 11, 2001, Al Qaeda attack involved four separate attacks with hijacked airplanes: Two planes crashed into the World Trade Center towers, one battered into the Pentagon, and a fourth plane shattered in a field in Pennsylvania.) Summarize the data with a dot plot. Interpret the results.

ALQAEDA

1	1	2	1	2	4	1	1	1	1	2
3	4	5	1	1	1	2	2	2	1	

Source: Moghadam, A. "Suicide terrorism, occupation, and the globalization of martyrdom: A critique of *Dying to Win*," *Studies in Conflict & Terrorism,* Vol. 29, No. 8, 2006 (Table 3).

2.31 **Reading Japanese books.** Refer to the *Reading in a Foreign Language* (Apr. 2004) experiment to improve the Japanese reading comprehension levels of University of Hawaii students, Exercise 2.7 (p. 34). Fourteen students participated in a 10-week extensive reading program in a second-semester Japanese language course. The number of books read by each student and the student's grade in the course are listed in the following table:

JAPANESE

Number of Books	Course Grade	Number of Books	Course Grade
53	A	30	A
42	A	28	B
40	A	24	A
40	B	22	C
39	A	21	B
34	A	20	B
34	A	16	B

Source: Hitosugi, C. I., and Day, R. R. "Extensive reading in Japanese," *Reading in a Foreign Language,* Vol. 16, No. 1, Apr. 2004 (Table 4).

   a. Construct a stem-and-leaf display for the number of books read by the students.
   b. Highlight (or circle) the leaves in the display that correspond to students who earned an A grade in the course. What inference can you make about these students?

2.32 **Sanitation inspection of cruise ships.** To minimize the potential for gastrointestinal disease outbreaks, all passenger cruise ships arriving at U.S. ports are subject to unannounced sanitation inspections. Ships are rated on a 100-point scale by the Centers for Disease Control and Prevention. A score of 86 or higher indicates that the ship is providing an accepted standard of sanitation. The latest (as of May 2006) sanitation scores for 169 cruise ships are saved in the **SHIPSANIT** file. The first five and last five observations in the data set are listed in the following table:

SHIPSANIT (Selected observations)

Ship Name	Sanitation Score
Adventure of the Seas	95
Albatross	96
Amsterdam	98
Arabella	94
Arcadia	98
.	.
.	.
.	.
Wind Surf	95
Yorktown Clipper	91
Zaandam	98
Zenith	94
Zuiderdam	94

Source: National Center for Environmental Health, Centers for Disease Control and Prevention, May 24, 2006.

a. Generate a stem-and-leaf display of the data. Identify the stems and leaves of the graph.

b. Use the stem-and-leaf display to estimate the proportion of ships that have an accepted sanitation standard.

c. Locate the inspection score of 84 (*Sea Bird*) on the stem-and-leaf display.

Applying the Concepts—Intermediate

2.33 West Nile virus cases. West Nile virus, transmitted to humans through mosquito bites, was first isolated in 1937 in Africa, West Asia, and the Middle East. The first outbreak of the virus in the Western Hemisphere did not occur until 1999 in New York City. The U.S. Geological Survey tracks outbreaks of West Nile virus each year. The number of human cases in 2006 is recorded for each of California's 58 counties in the **WNV** file. (Data for the first and last five counties are shown in the accompanying table.) Use a graphical method to describe the distribution of West Nile virus cases in California counties.

WNV (Selected observations shown)

County	Number
Alameda	1
Alpine	0
Amador	0
Butte	31
Calaveras	0
⋮	⋮
Tulare	6
Tuolomine	0
Ventura	3
Yolo	27
Yuba	5

Source: U.S. Department of the Interior; U.S. Geological Survey, 2006.

2.34 Radioactive lichen. Lichen has a high absorbance capacity for radiation fallout from nuclear accidents. Since lichen is a major food source for Alaskan caribou, and caribou are, in turn, a major food source for many Alaskan villagers, it is important to monitor the level of radioactivity in lichen. Researchers at the University of Alaska, Fairbanks, collect-

ed data on nine lichen specimens at various locations for this purpose. The amount of the radioactive element cesium-137 was measured (in microcuries per milliliter) for each specimen. The data values, converted to logarithms, are given in the following table (note that the closer the value is to zero, the greater is the amount of cesium in the specimen).

LICHEN

Location			
Bethel	−5.50	−5.00	
Eagle Summit	−4.15	−4.85	
Moose Pass	−6.05		
Turnagain Pass	−5.00		
Wickersham Dome	−4.10	−4.50	−4.60

Source: Lichen Radionuclide Baseline Research Project, 2003.

a. Construct a dot plot for the nine measurements.

b. Construct a stem-and-leaf display for the nine measurements.

c. Construct a histogram plot of the nine measurements.

d. Which of the three graphs in parts **a–c**, respectively, is most informative?

e. What proportion of the measurements has a radioactivity level of −5.00 or lower?

2.35 College protests of labor exploitation. The United Students Against Sweatshops (USAS) was formed by students on U.S. and Canadian college campuses in 1999 to protest labor exploitation in the apparel industry. Clark University sociologist Robert Ross analyzed the USAS movement in the *Journal of World-Systems Research* (Winter 2004). Between 1999 and 2000, there were 18 student "sit-ins" for a "sweat-free campus" organized at several universities. The following table gives the duration (in days) of each sit-in, as well as the number of student arrests:

SITIN

Sit-in	Year	University	Duration (days)	Number of Arrests	Tier Ranking
1	1999	Duke	1	0	1st
2	1999	Georgetown	4	0	1st
3	1999	Wisconsin	1	0	1st
4	1999	Michigan	1	0	1st
5	1999	Fairfield	1	0	1st
6	1999	North Carolina	1	0	1st
7	1999	Arizona	10	0	1st
8	2000	Toronto	11	0	1st
9	2000	Pennsylvania	9	0	1st
10	2000	Macalester	2	0	1st
11	2000	Michigan	3	0	1st
12	2000	Wisconsin	4	54	1st
13	2000	Tulane	12	0	1st
14	2000	SUNY Albany	1	11	2nd
15	2000	Oregon	3	14	2nd
16	2000	Purdue	12	0	2nd
17	2000	Iowa	4	16	2nd
18	2000	Kentucky	1	12	2nd

Source: Ross, R. J. S. "From antisweatshop to global justice to antiwar: How the new new left is the same and different from the old new left." *Journal of World-Systems Research*, Vol. X, No. 1, Winter 2004 (Tables 1 and 3).

a. Use a stem-and-leaf display to summarize the data on sit-in duration.

b. Highlight (or circle) the leaves in the display that correspond to sit-ins during which at least one arrest was made. Does the pattern revealed support the theory that sit-ins of longer duration are more likely to lead to arrests?

2.36 Crab spiders hiding on flowers. Crab spiders use camouflage to hide on flowers while lying in wait to prey on other insects. Ecologists theorize that this natural camouflage also enables the spiders to hide from their own predators, such as birds and lizards. Researchers at the French Museum of Natural History conducted a field test of this theory and published the results in *Behavioral Ecology* (Jan. 2005). They collected a sample of 10 adult female crab spiders, each sitting on the yellow central part of a daisy. The chromatic contrast between each spider and the flower it was sitting on was measured numerically with a spectroradiometer, on which higher values indicate a greater contrast (and, presumably, easier detection by predators). The data for the 10 crab spiders are shown in the following table:

SPIDER

57	75	116	37	96	61	56	2	43	32

Data adapted from Thery, M., et al. "Specific color sensitivities of prey and predator explain camouflage in different visual systems," *Behavioral Ecology*, Vol. 16, No. 1, Jan. 2005 (Table 1).

a. Summarize the chromatic contrast measurements for the 10 spiders with a stem-and-leaf display.

b. For birds, the detection threshold is 70. (A contrast of 70 or greater allows the bird to see the spider.) Locate the spiders that can be seen by bird predators by circling their respective contrast values on the stem-and-leaf display.

c. Use the result of part b to make an inference about the likelihood of a bird detecting a crab spider sitting on the yellow central part of a daisy.

2.37 Research on brain specimens. The *postmortem interval* (PMI) is defined as the time elapsed between death and an autopsy. Knowledge of the PMI is considered essential to conducting medical research on human cadavers. The data in the accompanying table are the PMIs of 22 human brain specimens obtained at autopsy in a recent study. (*Brain and Language*, June 1995.) Describe the PMI data graphically with a dot plot. On the basis of the plot, make a summary statement about the PMIs of the 22 human brain specimens.

BRAINPMI

Postmortem Intervals for 22 Human Brain Specimens

5.5	14.5	6.0	5.5	5.3	5.8	11.0	6.1
7.0	14.5	10.4	4.6	4.3	7.2	10.5	6.5
3.3	7.0	4.1	6.2	10.4	4.9		

Source: Hayes, T. L., and Lewis, D. A. "Anatomical specialization of the anterior motor speech area: Hemispheric differences in magnopyramidal neurons." *Brain and Language*, Vol. 49, No. 3, June 1995, p. 292 (Table 1).

2.38 Research on eating disorders. Data from a psychology experiment were reported and analyzed in *The American Statistician* (May 2001). Two samples of female students participated in the experiment. One sample consisted of 11 students known to suffer from the eating disorder bulimia; the other

sample consisted of 14 students with normal eating habits. Each student completed a questionnaire from which a "fear of negative evaluation" (FNE) score was produced. (The higher the score, the greater was the fear of negative evaluation.) The data are displayed in the following table:

BULIMIA

Bulimic students:	21	13	10	20	25	19	16	21	24	13	14			
Normal students:	13	6	16	13	8	19	23	18	11	19	7	10	15	20

Source: Randles, R. H. "On neutral responses (zeros) in the sign test and ties in the Wilcoxon–Mann–Whitney test," *The American Statistician*, Vol. 55, No. 2, May 2001 (Figure 3).

a. Construct a dot plot or stem-and-leaf display for the FNE scores of all 25 female students.

b. Highlight the bulimic students on the graph you made in part **a**. Does it appear that bulimics tend to have a greater fear of negative evaluation? Explain.

c. Why is it important to attach a measure of reliability to the inference made in part **b**?

2.39 Eclipses and occults. Saturn has five satellites that revolve around the planet. *Astronomy* (Aug. 1995) lists 19 different events involving eclipses or occults of Saturnian satellites during the month of August. For each event, the percent of light lost by the eclipsed or occulted satellite at midevent is recorded in the following table:

SATURN

Date	Event	Light Loss (%)
Aug. 2	Eclipse	65
4	Eclipse	61
5	Occult	1
6	Eclipse	56
8	Eclipse	46
8	Occult	2
9	Occult	9
11	Occult	5
12	Occult	39
14	Occult	1
14	Eclipse	100
15	Occult	5
15	Occult	4
16	Occult	13
20	Occult	11
23	Occult	3
23	Occult	20
25	Occult	20
28	Occult	12

Source: ASTRONOMY magazine, Aug. 1995, p. 60.

a. Construct a stem-and-leaf display for light loss percentage of the 19 events.

b. On the stem-and-leaf plot of part **a**, locate the light losses associated with eclipses of Saturnian satellites. (Circle the light losses on the plot.)

c. On the basis of the marked stem-and-leaf display in part **b**, make an inference about which type of event (eclipse or occult) is more likely to lead to a greater light loss.

Applying the Concepts—Advanced

2.40 Comparing SAT scores. Educators are constantly evaluating the efficacy of public schools in the education and training of U.S. students. One quantitative assessment of change over time is the difference in scores on the SAT, which has been used for decades by colleges and universities as one criterion for admission. The **SATSCORES** file contains the average SAT scores for each of the 50 states and the District of Columbia for 1990 and 2005. Selected observations are shown in the following table:

SATSCORES (First five and last two observations)

State	1990	2005
Alabama	1,079	1,126
Alaska	1,015	1,042
Arizona	1,041	1,056
Arkansas	1,077	1,115
California	1,002	1,026
.	.	.
.	.	.
.	.	.
Wisconsin	1,111	1,191
Wyoming	1,072	1,087

Source: College Entrance Examination Board, 2006.

a. Use graphs to display the two SAT score distributions. How have the distributions of state scores changed from 1990 to 2005?

b. As another method of comparing the 1990 and 2005 average SAT scores, compute the **paired difference** by subtracting the 1990 score from the 2005 score for each state. Summarize these differences with a graph.

c. Interpret the graph you made in part **b**. How do your conclusions compare with those of part **a**?

d. Identify the state with the largest improvement in the SAT score between 1990 and 2005.

2.41 Speech listening study. The role that listener knowledge plays in the perception of imperfectly articulated speech was investigated in the *American Journal of Speech–Language Pathology* (Feb. 1995). Thirty female college students, randomly divided into three groups of 10, participated as listeners in the study. All subjects were required to listen to a 48-sentence audiotape of a Korean woman with cerebral palsy, and all were asked to transcribe her entire speech. For the first group of students (the *control group*), the speaker used her normal manner of speaking. For the second group (the *treatment group*), the speaker employed a learned breathing pattern (called breath-group strategy) to improve speech efficiency. The subjects in the third group (the *familiarity group*) also listened to the tape with the breath-group strategy, but only after they had practiced listening twice to another tape in which they were told exactly what the speaker was saying. At the end of the listening–transcribing session, two quantitative variables were measured for each listener: (1) the total number of words transcribed (called the rate of response) and (2) the percentage of words correctly transcribed (called the accuracy score). The data for all 30 subjects are provided in the accompanying table.

a. Use a graphical method to describe the differences in the distributions of response rates among the three groups.

b. Use a graphical method to describe the distribution of accuracy scores for the three listener groups.

LISTEN

Control Group		Treatment Group		Familiarization Group	
Rate of Response	Percent Correct	Rate of Response	Percent Correct	Rate of Response	Percent Correct
250	23.6	254	26.0	193	36.0
230	26.0	178	32.6	223	41.0
197	26.0	139	32.6	232	43.0
238	26.7	249	33.0	214	44.0
174	29.2	236	34.4	269	44.0
263	29.5	231	36.5	256	46.0
275	31.3	161	38.5	224	46.0
193	32.9	255	40.0	225	48.0
204	35.4	275	41.7	288	49.0
168	29.2	181	44.8	244	52.0

Source: Tjaden, K., and Liss, J. M. "The influence of familiarity on judgments of treated speech." *American Journal of Speech–Language Pathology,* Vol. 4, No. 1, Feb. 1995, p. 43 (Table 1).

2.3 Summation Notation

Now that we've examined some graphical techniques for summarizing and describing quantitative data sets, we turn to numerical methods for accomplishing that objective. Before giving the formulas for calculating numerical descriptive measures, let's look at some shorthand notation that will simplify our calculation instructions. Remember that such notation is used for one reason only: to avoid repeating the same verbal descriptions over and over. If you mentally substitute the verbal definition of a symbol each time you read it, you'll soon get used to the symbol.

We denote the measurements of a quantitative data set as

$$x_1, x_2, x_3, \ldots, x_n$$

where x_1 is the first measurement in the data set, x_2 is the second measurement in the data set, x_3 is the third measurement in the data set, ..., and x_n is the nth (and last) measurement in the data set. Thus, if we have five measurements in a set of data, we will write x_1, x_2, x_3, x_4, x_5 to represent the measurements. If the actual numbers are 5, 3, 8, 5, and 4, we have $x_1 = 5$, $x_2 = 3$, $x_3 = 8$, $x_4 = 5$, and $x_5 = 4$.

Most of the formulas we use require a summation of numbers. For example, one sum we'll need to obtain is the sum of all the measurements in the data set, or $x_1 + x_2 + x_3 + \cdots + x_n$. To shorten the notation, we use the symbol $\sum$ for the summation. That is, $x_1 + x_2 + x_3 + \cdots + x_n = \sum_{i=1}^{n} x_i$. Verbally translate $\sum_{i=1}^{n} x_i$ as follows: "The sum of the measurements whose typical member is x_i, beginning with the member x_1 and ending with the member x_n."

Suppose, as in our earlier example, $x_1 = 5$, $x_2 = 3$, $x_3 = 8$, $x_4 = 5$, *and* $x_5 = 4$. Then the sum of the five measurements, denoted $\sum_{i=1}^{5} x_i$, is obtained as follows:

$$\sum_{i=1}^{5} x_i = x_1 + x_2 + x_3 + x_4 + x_5$$
$$= 5 + 3 + 8 + 5 + 4 = 25$$

Another important calculation requires that we square each measurement and then sum the squares. The notation for this sum is $\sum_{i=1}^{n} x_i^2$. For the five measurements, we have

$$\sum_{i=1}^{5} x_i^2 = x_1^2 + x_2^2 + x_3^2 + x_4^2 + x_5^2$$
$$= 5^2 + 3^2 + 8^2 + 5^2 + 4^2$$
$$= 25 + 9 + 64 + 25 + 16 = 139$$

In general, the symbol following the summation sign $\sum$ represents the variable (or function of the variable) that is to be summed.

The Meaning of Summation Notation $\sum_{i=1}^{n} x_i$

Sum the measurements of the variable that appears to the right of the summation symbol, beginning with the first measurement and ending with the nth measurement.

Exercises 2.42–2.45

Learning the Mechanics

Note: *In all exercises,* $\sum$ *represents* $\sum_{i=1}^{n}$.

2.42 A data set contains the observations 5, 1, 3, 2, 1. Find

a. $\sum x$ **b.** $\sum x^2$ **c.** $\sum(x-1)$

d. $\sum(x-1)^2$ **e.** $\left(\sum x\right)^2$

2.43 Suppose a data set contains the observations 3, 8, 4, 5, 3, 4, 6. Find

a. $\sum x$ **b.** $\sum x^2$ **c.** $\sum(x-5)^2$

d. $\sum(x-2)^2$ **e.** $\left(\sum x\right)^2$

2.44 Refer to Exercise 2.42. Find

a. $\sum x^2 - \dfrac{\left(\sum x\right)^2}{5}$ **b.** $\sum(x-2)^2$ **c.** $\sum x^2 - 10$

2.45 A data set contains the observations 6, 0, −2, −1, 3. Find

a. $\sum x$ **b.** $\sum x^2$ **c.** $\sum x^2 - \dfrac{\left(\sum x\right)^2}{5}$

2.4 Numerical Measures of Central Tendency

When we speak of a data set, we refer to either a sample or a population. If statistical inference is our goal, we'll ultimately wish to use sample **numerical descriptive measures** to make inferences about the corresponding measures for a population.

As you'll see, a large number of numerical methods are available to describe quantitative data sets. Most of these methods measure one of two data characteristics:

1. The **central tendency** of the set of measurements—that is, the tendency of the data to cluster, or center, about certain numerical values. (See Figure 2.14a.)
2. The **variability** of the set of measurements—that is, the spread of the data. (See Figure 2.14b.)

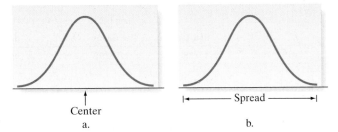

FIGURE 2.14

Numerical descriptive
measures

In this section, we concentrate on **measures of central tendency**. In the next section, we discuss measures of variability.

The most popular and best understood measure of central tendency for a quantitative data set is the *arithmetic mean* (or simply the mean) of the data set.

> **Definition 2.5**
>
> The **mean** of a set of quantitative data is the sum of the measurements, divided by the number of measurements contained in the data set.

In everyday terms, the mean is the average value of the data set and is often used to represent a "typical" value. We denote the **mean of a sample** of measurements by $\bar{x}$ (read "*x*-bar"), and represent the formula for its calculation as shown in the following box:

Formula for a Sample Mean

$$\bar{x} = \frac{\sum_{i=1}^{n} x_i}{n}$$

EXAMPLE 2.3

COMPUTING THE
SAMPLE MEAN

Problem Calculate the mean of the following five sample measurements: 5, 3, 8, 5, 6.

Solution Using the definition of sample mean and the summation notation, we find that

$$\bar{x} = \frac{\sum_{i=1}^{5} x_i}{5} = \frac{5 + 3 + 8 + 5 + 6}{5} = \frac{27}{5} = 5.4$$

Thus, the mean of this sample is 5.4.

Look Back There is no specific rule for rounding when calculating $\bar{x}$, because $\bar{x}$ is specifically defined to be the sum of all measurements, divided by n; that is, it is a specific fraction. When $\bar{x}$ is used for descriptive purposes, it is often convenient to round the calculated value of $\bar{x}$ to the number of significant figures used for the original measurements. When $\bar{x}$ is to be used in other calculations, however, it may be necessary to retain more significant figures.

Now Work Exercise 2.55

EXAMPLE 2.4

FINDING THE
MEAN ON A
PRINTOUT—
Mean Gas Mileage

Problem Calculate the sample mean for the 100 EPA mileages given in Table 2.2.

Solution The mean gas mileage for the 100 cars is denoted

$$\bar{x} = \frac{\sum_{i=1}^{100} x_i}{100}$$

Rather than compute $\bar{x}$ by hand (or even with a calculator), we employed SAS to compute the mean. The SAS printout is shown in Figure 2.15. The sample mean, highlighted on the printout, is $\bar{x} = 36.9940$.

FIGURE 2.15
SAS numerical descriptive
measures for 100 EPA gas
mileages

The MEANS Procedure

Analysis Variable : MPG

Mean	Std Dev	Variance	N	Minimum	Maximum	Median
36.9940000	2.4178971	5.8462263	100	30.0000000	44.9000000	37.0000000

Look Back Given this information, you can visualize a distribution of gas mileage readings centered in the vicinity of $\bar{x} \approx 37$. An examination of the relative frequency histogram (Figure 2.10) confirms that $\bar{x}$ does in fact fall near the center of the distribution.

The sample mean $\bar{x}$ will play an important role in accomplishing our objective of making inferences about populations on the basis of information about the sample. For this reason, we need to use a different symbol for the *mean of a population*—the mean of the set of measurements on every unit in the population. We use the Greek letter μ (mu) for the population mean.

Symbols for the Sample Mean and the Population Mean

In this text, we adopt a general policy of using Greek letters to represent numerical descriptive measures of the population and Roman letters to represent corresponding descriptive measures of the sample. The symbols for the mean are

$$\bar{x} = \text{Sample mean} \qquad \mu = \text{Population mean}$$

We'll often use the sample mean $\bar{x}$ to estimate (make an inference about) the population mean μ. For example, the EPA mileages for the population consisting of *all* cars has a mean equal to some value μ. Our sample of 100 cars yielded mileages with a mean of $\bar{x} = 36.9940$. If, as is usually the case, we don't have access to the measurements for the entire population, we could use $\bar{x}$ as an estimator or approximator for μ. Then we'd need to know something about the reliability of our inference. That is, we'd need to know how accurately we might expect $\bar{x}$ to estimate μ. In Chapter 7, we'll find that this accuracy depends on two factors:

1. The *size of the sample*. The larger the sample, the more accurate the estimate will tend to be.
2. The *variability, or spread, of the data*. All other factors remaining constant, the more variable the data, the less accurate is the estimate.

Another important measure of central tendency is the **median**.

Definition 2.6
The **median** of a quantitative data set is the middle number when the measurements are arranged in ascending (or descending) order.

The median is of most value in describing large data sets. If a data set is characterized by a relative frequency histogram (Figure 2.16), the median is the point on the x-axis such that half the area under the histogram lies above the median and half lies below. [*Note*: In Section 2.2, we observed that the relative frequency associated with a particular interval on the x-axis is proportional to the amount of area under the histogram that lies above the interval.] We denote the *median* of a *sample* by M.

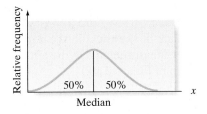

FIGURE 2.16
Location of the median

Calculating a Sample Median M

Arrange the n measurements from the smallest to the largest.

1. If n is odd, M is the middle number.

2. If n is even, M is the mean of the middle two numbers.

EXAMPLE 2.5
COMPUTING
THE MEDIAN

Problem Consider the following sample of n = 7 measurements: 5, 7, 4, 5, 20, 6, 2.
a. Calculate the median M of this sample.
b. Eliminate the last measurement (the 2), and calculate the median of the remaining n = 6 measurements.

Solution
a. The seven measurements in the sample are ranked in ascending order: 2, 4, 5, 5, 6, 7, 20. Because the number of measurements is odd, the median is the middle measurement. Thus, the median of this sample is M = 5.
b. After removing the 2 from the set of measurements, we rank the sample measurements in ascending order as follows: 4, 5, 5, 6, 7, 20. Now the number of measurements is even, so we average the middle two measurements. The median is M = (5 + 6)/2 = 5.5.

Look Back When the sample size n is even (as in part **b**), exactly half of the measurements will fall below the calculated median M. However, when n is odd (as in part **a**), the percentage of measurements that fall below M is approximately 50%. The approximation improves as n increases.

Now Work Exercise 2.52

In certain situations, the median may be a better measure of central tendency than the mean. In particular, the median is less sensitive than the mean to extremely large or small measurements. Note, for instance, that all but one of the measurements in part **a** of Example 2.5 center about x = 5. The single relatively large measurement, x = 20, does not affect the value of the median, 5, but it causes the mean, $\bar{x}$ = 7, to lie to the right of most of the measurements.

As another example of data for which the central tendency is better described by the median than the mean, consider the household incomes of a community being studied by a sociologist. The presence of just a few households with very high incomes will affect the mean more than the median. Thus, the median will provide a more accurate picture of the typical income for the community. The mean could exceed the vast majority of the sample measurements (household incomes), making it a misleading measure of central tendency.

EXAMPLE 2.6
FINDING THE
MEDIAN ON A
PRINTOUT—
Median Gas Mileage

Problem Calculate the median for the 100 EPA mileages given in Table 2.2. Compare the median with the mean computed in Example 2.4.

Solution For this large data set, we again resort to a computer analysis. The median is highlighted on the SAS printout displayed in Figure 2.15 (p. 52). You can see that the median is 37.0. Thus, half of the 100 mileages in the data set fall below 37.0 and half lie above 37.0. Note that the median, 37.0, and the mean, 36.9940, are almost equal, a relationship that indicates a lack of **skewness** in the data. In other words, the data exhibit a tendency to have as many measurements in the left tail of the distribution as in the right tail. (Recall the histogram of Figure 2.10.)

Look Back In general, extreme values (large or small) affect the mean more than the median, since these values are used explicitly in the calculation of the mean. The median is not affected directly by extreme measurements, since only the middle measurement (or two middle measurements) is explicitly used to calculate the median. Consequently, if measurements are pulled toward one end of the distribution, the mean will shift toward that tail more than the median will.

Definition 2.7

A data set is said to be **skewed** if one tail of the distribution has more extreme observations than the other tail.

A comparison of the mean and median gives us a general method for detecting skewness in data sets, as shown in the next box.

Detecting Skewness by Comparing the Mean and the Median

If the data set is skewed to the right, then the median is less than the mean.

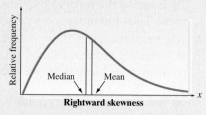

If the data set is symmetric, then the mean equals the median.

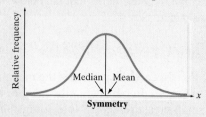

If the data set is skewed to the left, then the mean is less than the median.

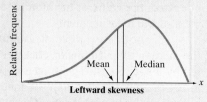

Now Work Exercise 2.51

A third measure of central tendency is the **mode** of a set of measurements.

> **Definition 2.8**
> The **mode** is the measurement that occurs most frequently in the data set.

Therefore, the mode shows where the data tend to concentrate.

EXAMPLE 2.7
FINDING THE
MODE

Problem Each of 10 taste testers rated a new brand of barbecue sauce on a 10-point scale, where 1 = awful and 10 = excellent. Find the mode for the following 10 ratings below:

$$8 \quad 7 \quad 9 \quad 6 \quad 8 \quad 10 \quad 9 \quad 9 \quad 5 \quad 7$$

Solution Since 9 occurs most often (three times), the mode of the ten taste ratings is 9.

Look Back Note that the data are actually qualitative in nature (e.g., "awful," "excellent"). The mode is particularly useful for describing qualitative data. The modal category is simply the category (or class) that occurs most often.

Now Work Exercise 2.53

Because it emphasizes data concentration, the mode is also used with quantitative data sets to locate the region in which much of the data is concentrated. A retailer of men's clothing would be interested in the modal neck size and sleeve length of potential customers. The modal income class of the laborers in the United States is of interest to the U.S. Department of Labor.

For some quantitative data sets, the mode may not be very meaningful. For example, consider the EPA mileage ratings in Table 2.2. A reexamination of the data reveals that the gas mileage of 37.0 occurs most often (four times). However, the mode of 37.0 is not particularly useful as a measure of central tendency.

A more meaningful measure can be obtained from a relative frequency histogram for quantitative data. The measurement class containing the largest relative frequency is called the **modal class**. Several definitions exist for locating the position of the mode within a modal class, but the simplest is to define the mode as the midpoint of the modal class. For example, examine the frequency histogram for the EPA mileage ratings in Figure 2.10 (p. 40). You can see that the modal class is the interval 36.0–37.5. The mode (the midpoint) is 36.75. This modal class (and the mode itself) identifies the area in which the data are most concentrated and, in that sense, is a measure of central tendency. However, for most applications involving quantitative data, the mean and median provide more descriptive information than the mode.

EXAMPLE 2.8
COMPARING THE
MEAN, MEDIAN,
AND MODE—
Executive Salaries

Problem Each year, *Forbes* magazine's "Executive Compensation Scoreboard" lists the total annual pay for CEOs at the 500 largest U.S. firms. The data for the 2005 scoreboard, saved in the **CEOPAY05** file, includes the quantitative variables total annual pay (in millions of dollars) and age. Find the mean, median, and mode for both of these variables. Which measure of central tendency is better for describing the distribution of total annual pay? for describing that of age?

Solution Measures of central tendency for the two variables were obtained with SPSS. The means, medians, and modes are displayed at the top of the SPSS printout shown in Figure 2.17.

 **CEOPAY05**

Statistics

		PAY2005	AGE
N	Valid	497	500
	Missing	3	0
Mean		10.93027	55.63
Median		5.54000	56.00
Mode		1.000ᵃ	57ᵃ

a. Multiple modes exist. The smallest value is shown

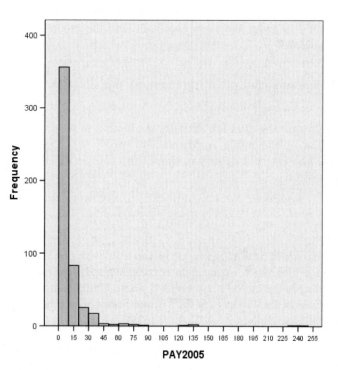

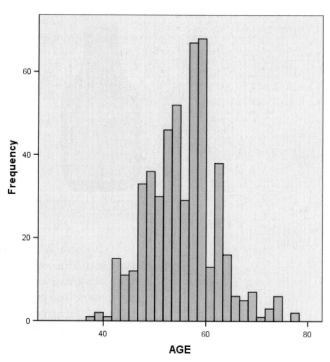

FIGURE 2.17

SPSS analysis of total 2005 pay and age for CEOs in the Executive Compensation Scoreboard

For total annual pay, the mean, median, and mode are $10.93 million, $5.54 million, and $1 million, respectively. Note that the mean is much greater than the median, indicating that the data are highly skewed to the right. This rightward skewness (graphically shown on the histogram for total pay in Figure 2.17) is due to several exceptionally high CEO salaries in 2005. Consequently, we would probably want to use the median, $5.54 million, as the "typical" value for annual pay for CEOs at the 500 largest firms. The mode of $1 million is the total pay value that occurs most often in the data set, but it is not very descriptive of the "center" of the total annual pay distribution.

For age, the mean, median, and mode are 55.63, 56, and 57 years, respectively. All three values are nearly the same, which is typical of symmetric distributions. From the age histogram in Figure 2.17, you can see that the age distribution is nearly symmetric. Consequently, any of the three measures of central tendency could be used to describe the "middle" of the age distribution.

Look Back The choice of which measure of central tendency to use will depend on the properties of the data set analyzed and the application of interest. Consequently, it is vital that you understand how the mean, median, and mode are computed.

Now Work Exercise 2.66

ACTIVITY 2.1: *Real Estate Sales:* **Measures of Central Tendency**

In recent years, the price of real estate in America's major metropolitan areas has skyrocketed. Newspapers usually report recent real-estate sales data in their Saturday editions, both in hard copy and online. These data usually include the actual prices paid for homes, by geographical location, during a certain period, usually a one-week period six to eight weeks earlier, and some summary statistics, which might include comparisons with real-estate sales data in other geographical locations or during other periods.

1. Locate the real-estate sales data in a newspaper for a major metropolitan area. From the information given, identify the period during which the homes listed were sold. Then describe the way the sales prices are organized. Are they categorized by type of home (single family or condominium), by neighborhood or address, by sales price, etc.?

2. What summary statistics and comparisons are provided with the sales data? Describe several types of people who might be interested in the data and how each of the summary statistics and comparisons would be helpful to them. Why are the measures of central tendency that are in fact listed more useful in the real-estate market than are other measures of central tendency?

3. On the basis of the lowest and highest sales prices represented in the data, create 10 intervals of equal size and use these intervals to create a relative frequency histogram for the sales data. Describe the shape of the histogram, and explain how the summary statistics provided with the data are illustrated in the histogram. On the basis of the histogram, describe the "typical" home price.

One-Variable Descriptive Statistics

Using The TI-84/TI-83 Graphing Calculator

Finding Descriptive Statistics

Step 1 *Enter the data.*

Press **STAT** and select **1:Edit**.
Note: If the list already contains data, clear the old data. Use the up arrow to highlight 'L1.' Press **CLEAR ENTER**.
Use the **ARROW** and **ENTER** keys to enter the data set into **L1**.

Step 2 *Calculate descriptive statistics.*

Press **STAT**.
Press the right **ARROW** key to highlight **CALC**.
Press **ENTER** for **1-Var Stats**.
Enter the name of the list containing your data.

Press **2nd 1** for **L1** (or **2nd 2** for **L2** etc.).
Press **ENTER**.
You should see the statistics on your screen. Some of the statistics are off the bottom of the screen. Use the down **ARROW** to scroll through to see the remaining statistics. Use the up **ARROW** to scroll back up.

Example Calculate the descriptive statistics for the sample data set

86, 70, 62, 98, 73, 56, 53, 92, 86, 37, 62, 83, 78, 49, 78, 37, 67, 79, 57

The output screens for this example are as follows:

```
1-Var Stats
 x̄=68.57894737
 Σx=1303
 Σx²=94897
 Sx=17.54142966
 σx=17.07357389
↓n=19
```

```
1-Var Stats
↑n=19
 minX=37
 Q₁=56
 Med=70
 Q₃=83
 maxX=98
■
```

Sorting Data The descriptive statistics do not include the mode. To find the mode, sort your data as follows:
Press **STAT**.
Press **2** for **SORTA(**.

(continued)

Enter the name of the list containing your data. If your data are in **L1**, press **2nd 1**.
Press **ENTER**.
The screen will say **DONE**.
To see the sorted data, press **STAT** and select **1:Edit**.
Scroll down through the list and locate the data value that occurs most frequently.

Exercises 2.46–2.68

Understanding the Principles

2.46 Explain the difference between a measure of central tendency and a measure of variability.

2.47 Give three different measures of central tendency.

2.48 What is the symbol used to represent the sample mean? The population mean?

2.49 What two factors affect the accuracy of the sample mean as an estimate of the population mean?

2.50 Explain the concept of a skewed distribution.

2.51 Describe how the mean compares with the median for a
NW distribution as follows:
   a. Skewed to the left
   b. Skewed to the right
   c. Symmetric

Learning the Mechanics

2.52 Calculate the mean and median of the following grade
NW point averages:

    3.2   2.5   2.1   3.7   2.8   2.0

2.53 Calculate the mode, mean, and median of the following data:
NW   18 10 15 13 17 15 12   15 18 16 11

2.54 Construct one data set consisting of five measurements, and another consisting of six measurements, for which the medians are equal.

2.55 Calculate the mean for samples for which

NW   **a.** $n = 10$, $\sum x = 85$
   b. $n = 16$, $\sum x = 400$
   c. $n = 45$, $\sum x = 35$
   d. $n = 18$, $\sum x = 242$

2.56 Calculate the mean, median, and mode for each of the following samples:
   a. $7, -2, 3, 3, 0, 4$
   b. $2, 3, 5, 3, 2, 3, 4, 3, 5, 1, 2, 3, 4$
   c. $51, 50, 47, 50, 48, 41, 59, 68, 45, 37$

APPLET Applet Exercise 2.1

Use the applet entitled *Mean versus Median* to find the mean and median of each of the three data sets presented in Exercise 2.56. For each data set, set the lower limit to a number less than all of the data, set the upper limit to a number greater than all of the data, and then click on *Update*. Click on the approximate location of each data item on the number line. You can get rid of a point by dragging it to the trash can. To clear the graph between data sets, simply click on the trash can.
   a. Compare the means and medians generated by the applet with those you calculated by hand in Exercise 2.56. If there are differences, explain why the applet might give values slightly different from the hand-calculated values.

   b. Despite providing only approximate values of the mean and median of a data set, describe some advantages of using the applet to find those values.

APPLET Applet Exercise 2.2

Use the applet *Mean versus Median* to illustrate your descriptions in Exercise 2.51. For each part **a**, **b**, and **c**, create a data set with 10 items that has the given property. Using the applet, verify that the mean and median have the relationship you described in Exercise 2.51.

APPLET Applet Exercise 2.3

Use the applet *Mean versus Median* to study the effect that an extreme value has on the difference between the mean and median. Begin by setting appropriate limits and plotting the following data on the number line provided in the applet:

    0 6 7 7 8 8 8 9 9 10

   a. Describe the shape of the distribution and record the value of the mean and median. On the basis of the shape of the distribution, do the mean and median have the relationship that you would expect?
   b. Replace the extreme value of 0 with 2, then 4, and then 6. Record the mean and median each time. Describe what is happening to the mean as 0 is replaced, in turn, by the higher numbers stated. What is happening to the median? How is the difference between the mean and the median changing?
   c. Now replace 0 with 8. What values does the applet give you for the mean and the median? Explain why the mean and the median should now be the same.

Applying the Concepts—Basic

2.57 **Reading Japanese books.** Refer to the *Reading in a Foreign Language* (Apr. 2004) experiment to improve the Japanese reading comprehension levels of 14 University of Hawaii students, Exercise 2.31 (p. 46). The number of books read by each student and the student's course grade are repeated in the table.

JAPANESE

Number of Books	Course Grade	Number of Books	Course Grade
53	A	30	A
42	A	28	B
40	A	24	A
40	B	22	C
39	A	21	B
34	A	20	B
34	A	16	B

Source: Hitosugi, C. I., and Day, R. R. "Extensive reading in Japanese," *Reading in a Foreign Language,* Vol. 16, No. 1, Apr. 2004 (Table 4).

a. Find the mean, median, and mode of the number of books read. Interpret these values.

b. What do the mean and median indicate about the skewness of the distribution of the data?

2.58 Most powerful women in America. *Fortune* (Nov. 14, 2005) published a list of the 50 most powerful women in America. The data on age (in years) and title of each of these 50 women are stored in the **WPOWER50** file. The first five and last two observations of the data are listed in the accompanying table.

a. Find the mean, median, and modal age of these 50 women.

b. What do the mean and median indicate about the skewness of the age distribution?

c. Construct a relative frequency histogram for the age data. What is the modal age class?

WPOWER50 (Selected observations)

Rank	Name	Age	Company	Title
1	Meg Whitman	49	eBay	CEO/Chair
2	Anne Mulcahy	52	Xerox	CEO/Chair
3	Brenda Barnes	51	Sara Lee	CEO/President
4	Oprah Winfrey	51	Harpo	Chair
5	Andrea Jung	47	Avon	CEO/Chair
.	.	.	.	.
.	.	.	.	.
.	.	.	.	.
49	Safra Catz	43	Oracle	President
50	Kathy Cassidy	51	General Electric	Treasurer

Source: Fortune, Nov. 14, 2005.

2.59 Ammonia in car exhaust. Three-way catalytic converters have been installed in new vehicles in order to reduce pollutants from motor vehicle exhaust emissions. However, these converters unintentionally increase the level of ammonia in the air. *Environmental Science & Technology* (Sept. 1, 2000) published a study on the ammonia levels near the exit ramp of a San Francisco highway tunnel. The data in the table represent daily ammonia concentrations (parts per million) on eight randomly selected days during the afternoon drive time in the summer of a recent year.

AMMONIA

1.53	1.50	1.37	1.51	1.55	1.42	1.41	1.48

a. Find the mean daily ammonia level in air in the tunnel.

b. Find the median ammonia level.

c. Interpret the values obtained in parts **a** and **b**.

2.60 Ranking driving performance of professional golfers. A group of Northeastern University researchers developed a new method for ranking the total driving performance of golfers on the Professional Golf Association (PGA) tour. (*The Sport Journal*, Winter 2007.) The method requires knowing a golfer's average driving distance (yards) and driving accuracy (percent of drives that land in the fairway). The values of these two variables are used to compute a driving performance index. Data for the top 40 PGA golfers (ranked by the new method) are

saved in the **PGADRIVER** file. The first five and last five observations are listed in the accompanying table.

a. Find the mean, median, and mode for the 40 driving performance index values.

b. Interpret each of the measures of central tendency calculated in part a.

c. Use the results from part a to make a statement about the type of skewness in the distribution of driving performance indexes. Support your statement with a graph.

PGADRIVER (Selected observations shown)

Rank	Player	Driving Distance (yards)	Driving Accuracy (%)	Driving Performance Index
1	Woods	316.1	54.6	3.58
2	Perry	304.7	63.4	3.48
3	Gutschewski	310.5	57.9	3.27
4	Wetterich	311.7	56.6	3.18
5	Hearn	295.2	68.5	2.82
⋮				⋮
36	Senden	291	66	1.31
37	Mickelson	300	58.7	1.30
38	Watney	298.9	59.4	1.26
39	Trahan	295.8	61.8	1.23
40	Pappas	309.4	50.6	1.17

Source: Wiseman, F. et. al. "A New Method for Ranking Total Driving Performance on the PGA Tour," The Sport Journal, Vol. 10, No. 1, Winter 2007 (Table 2).

2.61 Radioactive lichen. Refer to the University of Alaska study to monitor the level of radioactivity in lichen, presented in Exercise 2.34 (p. 47). The amount of the radioactive element cesium-137 (measured in microcuries per milliliter) for each of nine lichen specimens is repeated in the following table:

LICHEN

Location			
Bethel	− 5.50	− 5.00	
Eagle Summit	− 4.15	− 4.85	
Moose Pass	− 6.05		
Turnagain Pass	− 5.00		
Wickersham Dome	− 4.10	− 4.50	− 4.60

Source: Lichen Radionuclide Baseline Research Project, 2003.

a. Find the mean, median, and mode of the radioactivity levels.

b. Interpret the value of each measure of central tendency calculated in part **a**.

Applying the Concepts—Intermediate

2.62 Recommendation letters for professors. Applicants for an academic position (e.g., assistant professor) at a college or university are usually required to submit at least three letters of recommendation. A study of 148 applicants for an entry-level position in experimental psychology at the University of Alaska at Anchorage revealed that many did not meet the three-letter requirement. (*American Psychologist*, July 1995.) Summary statistics for the number of recommendation letters in each application are as follows:

Mean = 2.28   Median = 3   Mode = 3

Interpret these summary measures.

2.63 Training zoo animals. "The Training Game" is an activity used in psychology in which one person shapes an arbitrary behavior by selectively reinforcing the movements of another person. A group of 15 psychology students at the Georgia Institute of Technology played the game at Zoo Atlanta while participating in an experimental psychology laboratory in which they assisted in the training of animals. (*Teaching of Psychology*, May 1998.) At the end of the session, each student was asked to rate the statement: "'The Training Game is a great way for students to understand the animal's perspective during training." Responses were recorded on a 7-point scale ranging from 1 (strongly disagree) to 7 (strongly agree). The 15 responses were summarized as follows:

$$mean = 5.87, mode = 6.$$

a. Interpret the measures of central tendency in the words of the problem.

b. What type of skewness (if any) is likely to be present in the distribution of student responses? Explain.

2.64 Symmetric or skewed? Would you expect the data sets that follow to possess relative frequency distributions that are symmetric, skewed to the right, or skewed to the left? Explain.

a. The salaries of all persons employed by a large university

b. The grades on an easy test

c. The grades on a difficult test

d. The amounts of time students in your class studied last week

e. The ages of automobiles on a used-car lot

f. The amounts of time spent by students on a difficult examination (maximum time is 50 minutes)

2.65 Children's use of pronouns. Clinical observations suggest that specifically language-impaired (SLI) children have great difficulty with the proper use of pronouns. This phenomenon was investigated and reported in the *Journal of Communication Disorders* (Mar. 1995). Thirty children, all from low-income families, participated in the study. Ten were five-year-old SLI children, 10 were younger (three-year-old) normally developing (YND) children, and 10 were older (five-year-old) normally developing (OND) children. The table contains the gender, deviation intelligence quotient (DIQ), and percentage of pronoun errors observed for each of the 30 subjects.

a. Identify the variables in the data set as quantitative or qualitative.

b. Why is it nonsensical to compute numerical descriptive measures for qualitative variables?

c. Compute measures of central tendency for DIQ for the 10 SLI children.

d. Compute measures of central tendency for DIQ for the 10 YND children.

e. Compute measures of central tendency for DIQ for the 10 OND children.

f. Use the results from parts **c–e** to compare the DIQ central tendencies of the three groups of children. Is it reasonable to use a single number (e.g., mean or median) to describe the center of the DIQ distribution? Or should three "centers" be calculated, one for each of the three groups of children? Explain.

g. Repeat parts **c–f** for the percentage of pronoun errors.

⊙ **SLI**

Subject	Gender	Group	DIQ	Pronoun Errors (%)
1	F	YND	110	94.40
2	F	YND	92	19.05
3	F	YND	92	62.50
4	M	YND	100	18.75
5	F	YND	86	0
6	F	YND	105	55.00
7	F	YND	90	100.00
8	M	YND	96	86.67
9	M	YND	90	32.43
10	F	YND	92	0
11	F	SLI	86	60.00
12	M	SLI	86	40.00
13	M	SLI	94	31.58
14	M	SLI	98	66.67
15	F	SLI	89	42.86
16	F	SLI	84	27.27
17	M	SLI	110	33.33
18	F	SLI	107	0
19	F	SLI	87	0
20	M	SLI	95	0
21	M	OND	110	0
22	M	OND	113	0
23	M	OND	113	0
24	F	OND	109	0
25	M	OND	92	0
26	F	OND	108	0
27	M	OND	95	0
28	F	OND	87	0
29	F	OND	94	0
30	F	OND	98	0

Source: Moore, M. E. "Error analysis of pronouns by normal and language-impaired children." *Journal of Communication Disorders,* Vol. 28, No. 1, Mar. 1995, p. 62 (Table 2), p. 67 (Table 5).

2.66 Mongolian desert ants. The *Journal of Biogeography* (Dec. 2003) published an article on the first comprehensive study of ants in Mongolia (Central Asia). Botanists placed seed baits at 11 study sites and observed the ant species attracted to each site. Some of the data recorded at each study site are provided in the table at the bottom of p. 61.

a. Find the mean, median, and mode for the number of ant species discovered at the 11 sites. Interpret each of these values.

b. Which measure of central tendency would you recommend to describe the center of the number-of-ant-species distribution? Explain.

c. Find the mean, median, and mode for the percentage of total plant cover at the five Dry Steppe sites only.

d. Find the mean, median, and mode for the percentage of total plant cover at the six Gobi Desert sites only.

e. On the basis of the results of parts **c** and **d**, does the center of the distribution for total plant cover percentage appear to be different at the two regions?

Applying the Concepts—Advanced

2.67 **Eye refractive study.** The conventional method of measuring the refractive status of an eye involves three quantities: (1) sphere power, (2) cylinder power, and (3) axis. Optometric researchers studied the variation in these three measures of refraction. (*Optometry and Vision Science*, June 1995.) Twenty-five successive refractive measurements were obtained on the eyes of over 100 university students. The cylinder power measurements for the left eye of one particular student (ID #11) are listed in the table. [*Note*: All measurements are negative values.]

LEFTEYE

.08	.08	1.07	.09	.16	.04	.07	.17	.11
.06	.12	.17	.20	.12	.17	.09	.07	.16
.15	.16	.09	.06	.10	.21	.06		

Source: Rubin, A., and Harris, W. F. "Refractive variation during autorefraction: Multivariate distribution of refractive status." *Optometry and Vision Science*, Vol. 72, No. 6, June 1995, p. 409 (Table 4).

a. Find measures of central tendency for the data and interpret their values.

b. Note that the data contain one unusually large (negative) cylinder power measurement relative to the other measurements in the data set. Find this measurement. (In Section 2.8, we call this value an **outlier**).

c. Delete the outlier of part **b** from the data set and recalculate the measures of central tendency. Which measure is most affected by the deletion of the outlier?

2.68 **Active nuclear power plants.** The U.S. Energy Information Administration monitors all nuclear power plants operating in the United States. The table at right lists the number of active nuclear power plants operating in each of a sample of 20 states.

a. Find the mean, median, and mode of this data set.

b. Eliminate the largest value from the data set and repeat part **a**. What effect does dropping this measurement have on the measures of central tendency found in part **a**?

c. Arrange the 20 values in the table from lowest to highest. Next, eliminate the lowest two values and the highest two values from the data set, and find the mean of the remaining data values. The result is called a *10% trimmed mean*, since it is calculated after removing the highest 10% and the lowest 10% of the data values. What advantages does a trimmed mean have over the regular arithmetic mean?

NUCLEAR

State	Number of Power Plants
Alabama	5
Arizona	3
California	5
Florida	5
Georgia	4
Illinois	13
Kansas	1
Louisiana	2
Massachusetts	2
Mississippi	1
New Hampshire	1
New York	6
North Carolina	5
Ohio	2
Pennsylvania	9
South Carolina	7
Tennessee	2
Texas	3
Vermont	1
Wisconsin	3

Source: Statistical Abstract of the United States, 2007 (Table 918). U.S. Energy Information Administration, *Electric Power Annual.*

GOBIANTS

Site	Region	Annual Rainfall (mm)	Max. Daily Temp. (°C)	Total Plant Cover (%)	Number of Ant Species	Species Diversity Index
1	Dry Steppe	196	5.7	40	3	.89
2	Dry Steppe	196	5.7	52	3	.83
3	Dry Steppe	179	7.0	40	52	1.31
4	Dry Steppe	197	8.0	43	7	1.48
5	Dry Steppe	149	8.5	27	5	.97
6	Gobi Desert	112	10.7	30	49	.46
7	Gobi Desert	125	11.4	16	5	1.23
8	Gobi Desert	99	10.9	30	4	
9	Gobi Desert	125	11.4	56	4	.76
10	Gobi Desert	84	11.4	22	5	1.26
11	Gobi Desert	115	11.4	14	4	.69

Source: Pfeiffer, M., et al. "Community organization and species richness of ants in Mongolia along an ecological gradient from steppe to Gobi desert," *Journal of Biogeography,* Vol. 30, No. 12, Dec. 2003 (Tables 1 and 2).

2.5 Numerical Measures of Variability

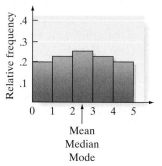

a. Data set 1

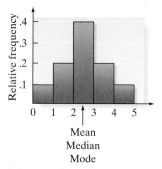

b. Data set 2

FIGURE 2.18

Hypothetical data sets

Measures of central tendency provide only a partial description of a quantitative data set. The description is incomplete without a **measure of the variability**, or **spread**, of the data set. Knowledge of the data set's variability, along with knowledge of its center, can help us visualize the shape of the data set as well as its extreme values.

If you examine the two histograms in Figure 2.18, you'll notice that both hypothetical data sets are symmetric, with equal modes, medians, and means. However, data set 1 (Figure 2.18a) has measurements spread with almost equal relative frequency over the measurement classes, while data set 2 (Figure 2.18b) has most of its measurements clustered about its center. Thus, data set 2 is *less variable* than data set 1. Consequently, you can see that we need a measure of variability as well as a measure of central tendency to describe a data set.

Perhaps the simplest measure of the variability of a quantitative data set is its *range*.

Definition 2.9

The **range** of a quantitative data set is equal to the largest measurement minus the smallest measurement.

The range is easy to compute and easy to understand, but it is a rather insensitive measure of data variation when the data sets are large. This is because two data sets can have the same range and be vastly different with respect to data variation. The phenomenon is demonstrated in Figure 2.18: Both distributions of data shown in the figure have the same range, but most of the measurements in data set 2 tend to concentrate near the center of the distribution. Consequently, the data are much less variable than the data in set 1. Thus, you can see that the range does not always detect differences in data variation for large data sets.

Let's see if we can find a measure of data variation that is more sensitive than the range. Consider the two samples in Table 2.5: Each has five measurements. (We have ordered the numbers for convenience.) Note that both samples have a mean of 3 and also that we have calculated the distance between each measurement and the mean. What information do these distances contain? If they tend to be large in magnitude, as in sample 1, the data are spread out, or highly variable. If the distances are mostly small, as in sample 2, the data are clustered around the mean, $\bar{x}$, and therefore do not exhibit much variability. You can see that these distances, displayed graphically in Figure 2.19, provide information about the variability of the sample measurements.

The next step is to condense the information in these distances into a single numerical measure of variability. Averaging the distances from $\bar{x}$ won't help, because the negative and positive distances cancel; that is, the sum of the deviations (and thus the average deviation) is always equal to zero.

Two methods come to mind for dealing with the fact that positive and negative distances from the mean cancel. The first is to treat all the distances as though they were positive, ignoring the sign of the negative distances. We won't pursue this line of thought because the resulting measure of variability (the mean of the absolute values of the distances) presents analytical difficulties beyond the scope of this text. A second method of eliminating the minus signs associated with the distances is to square them. The quantity we can calculate from the squared distances provides a meaningful description of the variability of a data set and presents fewer analytical difficulties in making inferences.

TABLE 2.5 Two Hypothetical Data Sets

	Sample 1	Sample 2
Measurements	$1, 2, 3, 4, 5$	$2, 3, 3, 3, 4$
Mean	$\bar{x} = \dfrac{1 + 2 + 3 + 4 + 5}{5} = \dfrac{15}{5} = 3$	$\bar{x} = \dfrac{2 + 3 + 3 + 3 + 4}{5} = \dfrac{15}{5} = 3$
Distances of measurement values from $\bar{x}$	$(1 - 3), (2 - 3), (3 - 3), (4 - 3),$ $(5 - 3)$ or $-2, -1, 0, 1, 2$	$(2 - 3), (3 - 3), (3 - 3), (3 - 3),$ $(4 - 3)$ or $-1, 0, 0, 0, 1$

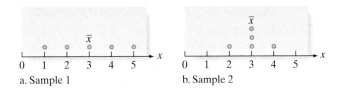

FIGURE 2.19

Dot plots for two data sets

a. Sample 1   b. Sample 2

To use the squared distances calculated from a data set, we first calculate the *sample variance*.

Definition 2.10

The **sample variance** for a sample of n measurements is equal to the sum of the squared distances from the mean, divided by $(n - 1)$. The symbol s^2 is used to represent the sample variance.

Formula for the Sample Variance:

$$s^2 = \frac{\sum_{i=1}^{n}(x_i - \bar{x})^2}{n - 1}$$

Note: A shortcut formula for calculating s^2 is

$$s^2 = \frac{\sum_{i=1}^{n}x_i^2 - \dfrac{\left(\sum_{i=1}^{n}x_i\right)^2}{n}}{n - 1}$$

Referring to the two samples in Table 2.5, you can calculate the variance for sample 1 as follows:

$$s^2 = \frac{(1 - 3)^2 + (2 - 3)^2 + (3 - 3)^2 + (4 - 3)^2 + (5 - 3)^2}{5 - 1}$$

$$= \frac{4 + 1 + 0 + 1 + 4}{4} = 2.5$$

The second step in finding a meaningful measure of data variability is to calculate the *standard deviation* of the data set.

Definition 2.11

The **sample standard deviation**, s, is defined as the positive square root of the sample variance, s^2, or, mathematically,

$$s = \sqrt{s^2}$$

The population variance, denoted by the symbol σ^2 (sigma squared), is the average of the squared distances from the mean, μ, of the measurements on *all* units in the population, and σ (sigma) is the square root of this quantity.

Symbols for Variance and Standard Deviation

s^2 = Sample variance

s = Sample standard deviation

σ^2 = Population variance

σ = Population standard deviation

Notice that, unlike the variance, the standard deviation is expressed in the original units of measurement. For example, if the original measurements are in dollars, the variance is expressed in the peculiar units "dollars squared," but the standard deviation is expressed in dollars.

You may wonder why we use the divisor $(n - 1)$ instead of n when calculating the sample variance. Wouldn't using n seem more logical, so that the sample variance would be the average squared distance from the mean? The trouble is, using n tends to produce an underestimate of the population variance σ^2. So we use $(n - 1)$ in the denominator to provide the appropriate correction for this tendency.* Since sample statistics such as s^2 are used primarily to estimate population parameters such as σ^2, $(n - 1)$ is preferred to n in defining the sample variance.

EXAMPLE 2.9

COMPUTING
MEASURES OF
VARIATION

Problem Calculate the variance and standard deviation of the following sample: 2, 3, 3, 3, 4.

Solution As the number of measurements increases, calculating s^2 and s becomes very tedious. Fortunately, as we show in Example 2.10, we can use a statistical software package (or a calculator) to find these values. If you must calculate these quantities by hand, it is advantageous to use the shortcut formula provided in Definition 2.10.

To do this, we need two summations: $\sum x$ and $\sum x^2$. These can easily be obtained from the following type of tabulation:

x	x^2
2	4
3	9
3	9
3	9
4	16
$\sum x = 15$	$\sum x^2 = 47$

Then we use[†]

$$s^2 = \frac{\sum_{i=1}^{n} x_i^2 - \frac{\left(\sum_{i=1}^{n} x_i\right)^2}{n}}{n - 1} = \frac{47 - \frac{(15)^2}{5}}{5 - 1} = \frac{2}{4} = .5$$

$$s = \sqrt{.5} = .71$$

Look Back As the sample size n increases, these calculations can become very tedious. As the next example shows, we can use the computer to find s^2 and s.

Now Work Exercise 2.76a

EXAMPLE 2.10

FINDING
MEASURES
OF VARIATION
ON A PRINTOUT

Problem Use the computer to find the sample variance s^2 and the sample standard deviation s for the 100 gas mileage readings given in Table 2.2.

Solution The SAS printout describing the gas mileage data is reproduced in Figure 2.20. The variance and standard deviation, highlighted on the printout, are $s^2 = 5.85$ and $s = 2.42$ (rounded to two decimal places).

---

*Appropriate here means that s^2, with a divisor of $(n - 1)$, is an *unbiased estimator* of σ^2. We define and discuss unbiasedness of estimators in Chapter 6.

[†]In calculating s^2, how many decimal places should you carry? Although there are no rules for the rounding procedure, it is reasonable to retain twice as many decimal places in s^2 as you ultimately wish to have in s. If, for example, you wish to calculate s to the nearest hundredth (two decimal places), you should calculate s^2 to the nearest ten-thousandth (four decimal places).

FIGURE 2.20

Reproduction of SAS numerical descriptive measures for 100 EPA mileages

The MEANS Procedure

Analysis Variable : MPG

Mean	Std Dev	Variance	N	Minimum	Maximum	Median
36.9940000	2.4178971	5.8462263	100	30.0000000	44.9000000	37.0000000

You now know that the standard deviation measures the variability of a set of data, and you know how to calculate the standard deviation. The larger the standard deviation, the more variable the data are. The smaller the standard deviation, the less variation there is in the data. But how can we practically interpret the standard deviation and use it to make inferences? This is the topic of Section 2.6.

ACTIVITY 2.2: *Keep the Change:* **Measures of Central Tendency and Variability**

In this activity, we continue our study of the Bank of America *Keep the Change* savings program by looking at the measures of central tendency and variability for the three data sets collected in Activity 1.1 (p. 12).

1. Before performing any calculations, explain why you would expect greater variability in the data set *Purchase Totals* than in *Amounts Transferred*. Then calculate the mean and median of each of these two data sets. Are the mean and median essentially the same for either of these sets? If so, which one? Can you offer an explanation for your results?

2. Make a histogram for each of the data sets *Amounts Transferred* and *Bank Matching*. Describe any properties of the data that are evident in the histograms. Explain why it is more likely that *Bank Matching*, rather than *Amounts Transferred*, is skewed to the right. On the basis of your data and histogram, how concerned does Bank of America need to be about matching the maximum amount of $250 for its customers who are college students?

3. Form a fourth data set, *Mean Amounts Transferred*, by collecting the mean of the data set *Amounts Transferred* for each student in your class. Before performing any calculations, inspect the new data and describe any trends that you notice. Then calculate the mean and standard deviation of *Mean Amounts Transferred*. How close is the mean to $0.50? Without performing further calculations, determine whether the standard deviation of *Amounts Transferred* is less than or greater than the standard deviation of *Mean Amounts Transferred*. Explain.

Keep your results from this activity for use in other activities.

Exercises 2.69–2.85

Understanding the Principles

2.69 What is the range of a data set?

2.70 What is the primary disadvantage of using the range to compare the variability of data sets?

2.71 Describe the sample variance in words rather than with a formula. Do the same with the population variance.

2.72 Can the variance of a data set ever be negative? Explain. Can the variance ever be smaller than the standard deviation? Explain.

2.73 If the standard deviation increases, does this imply that the data are more variable or less variable?

Learning the Mechanics

2.74 Calculate the variance and standard deviation for samples for which

    a. $n = 10, \sum x^2 = 84, \sum x = 20$

    b. $n = 40, \sum x^2 = 380, \sum x = 100$

    c. $n = 20, \sum x^2 = 18, \sum x = 17$

2.75 Calculate the range, variance, and standard deviation for the following samples:

    a. 39, 42, 40, 37, 41

    b. 100, 4, 7, 96, 80, 3, 1, 10, 2

    c. 100, 4, 7, 30, 80, 30, 42, 2

2.76 Calculate the range, variance, and standard deviation for
NW the following samples:

    a. 4, 2, 1, 0, 1

    b. 1, 6, 2, 2, 3, 0, 3

    c. 8, −2, 1, 3, 5, 4, 4, 1, 3

    d. 0, 2, 0, 0, −1, 1, −2, 1, 0, −1, 1, −1, 0, −3, −2, −1, 0, 1

2.77 Using only integers between 0 and 10, construct two data sets with at least 10 observations each such that the two sets have the same mean, but different variances. Construct dot plots for each of your data sets, and mark the mean of each data set on its dot plot.

2.78 Using only integers between 0 and 10, construct two data sets with at least 10 observations each such that the two sets have the same range, but different means. Construct a dot plot for each of your data sets, and mark the mean of each data set on its dot plot.

2.79 Consider the following sample of five measurements: 2, 1, 1, 0, 3.

    a. Calculate the range, s^2, and s.

    b. Add 3 to each measurement and repeat part **a**.

    c. Subtract 4 from each measurement and repeat part **a**.

d. Considering your answers to parts **a**, **b**, and **c**, what seems to be the effect on the variability of a data set of adding the same number to or subtracting the same number from each measurement?

2.80 Compute s^2, and s for each of the data sets listed. Where appropriate, specify the units in which your answer is expressed.
   a. 3, 1, 10, 10, 4
   b. 8 feet, 10 feet, 32 feet, 5 feet
   c. −1, −4, −3, 1, −4, −4
   d. 1/5 ounce, 1/5 ounce, 1/5 ounce, 2/5 ounce, 1/5 ounce, 4/5 ounce

APPLET **Applet Exercise 2.4**

Use the applet entitled *Standard Deviation* to find the standard deviation of each of the four data sets listed in Exercise 2.76. For each data set, set the lower limit to a number less than all of the data, set the upper limit to a number greater than all of the data, and then click on *Update*. Click on the approximate location of each data item on the number line. You can get rid of a point by dragging it to the trash can. To clear the graph between data sets, simply click on the trash can.
   a. Compare the standard deviations generated by the applet with those you calculated by hand in Exercise 2.76. If there are differences, explain why the applet might give values slightly different from the hand-calculated values.
   b. Despite the fact that it provides a slightly different value of the standard deviation of a data set, describe some advantages of using the applet.

APPLET **Applet Exercise 2.5**

Use the applet *Standard Deviation* to study the effect that multiplying or dividing each number in a data set by the same number has on the standard deviation. Begin by setting appropriate limits and plotting the given data on the number line provided in the applet.

$$0 \quad 1 \quad 1 \quad 1 \quad 2 \quad 2 \quad 3 \quad 4$$

   a. Record the standard deviation. Then multiply each data item by 2, plot the new data items, and record the standard deviation. Repeat the process, first multiplying each of the original data items by 3 and then by 4. Describe what happens to the standard deviation as the data items are multiplied by ever higher numbers. Divide each standard deviation by the standard deviation of the original data set. Do you see a pattern? Explain.
   b. Divide each of the original data items by 2, plot the new data, and record the standard deviation. Repeat the process, first dividing each of the original data items by 3 and then by 4. Describe what happens to the standard deviation as the data items are divided by ever higher numbers. Divide each standard deviation by the standard deviation of the original data set. Do you see a pattern? Explain.
   c. Using your results from parts **a** and **b**, describe what happens to the standard deviation of a data set when each of the data items in the set is multiplied or divided

by a fixed number n. Experiment by repeating parts **a** and **b** for other data sets if you need to.

APPLET **Applet Exercise 2.6**

Use the applet *Standard Deviation* to study the effect that an extreme value has on the standard deviation. Begin by setting appropriate limits and plotting the following data on the number line provided in the applet:

$$0 \quad 6 \quad 7 \quad 7 \quad 8 \quad 8 \quad 8 \quad 9 \quad 9 \quad 10$$

   a. Record the standard deviation. Replace the extreme value of 0 with 2, then 4, and then 6. Record the standard deviation each time. Describe what happens to the standard deviation as 0 is replaced by ever higher numbers.
   b. How would the standard deviation of the data set compare with the original standard deviation if the 0 were replaced by 16? Explain.

Applying the Concepts—Basic

⊙ JAPANESE

2.81 **Reading Japanese books.** Refer to the *Reading in a Foreign Language* (Apr. 2004) experiment to improve the Japanese reading comprehension levels of 14 University of Hawaii students, presented in Exercises 2.31 and 2.57 (pp. 46 and 58). The data on number of books read and grade for each student are saved in the **JAPANESE** file.
   a. Find the range, variance, and standard deviation of the number of books read by students who earned an A grade.
   b. Find the range, variance, and standard deviation of the number of books read by students who earned either a B or C grade.
   c. Refer to parts **a** and **b**. Which of the two groups of students has a more variable distribution for number of books read?

⊙ EARTHQUAKE

2.82 **Earthquake aftershock magnitudes.** Refer to Exercise 2.29 (p. 46) and U.S. Geological Survey data on aftershocks from a major California earthquake. The **EARTHQUAKE** file contains the magnitudes (measured on the Richter scale) of 2,929 aftershocks. Following is a MINITAB printout with descriptive statistics pertaining to magnitude:
   a. Locate the range of the magnitudes of the 2,929 aftershocks.
   b. Locate the variance of the magnitudes of the 2,929 aftershocks.
   c. Locate the standard deviation of the magnitudes of the 2,929 aftershocks.
   d. If the target of your interest is these specific 2,929 aftershocks, what symbols should you use to describe the variance and standard deviation?

MINITAB Output for Exercise 2.82

Descriptive Statistics: MAGNITUDE

Variable	N	Mean	StDev	Variance	Minimum	Median	Maximum	Range
MAGNITUDE	2929	2.1197	0.6636	0.4403	0.0000	2.0000	6.7000	6.7000

Variable	Mode	N for Mode
MAGNITUDE	1.8	298

2.83 Ammonia in car exhaust. Refer to the *Environmental Science & Technology* (Sept. 1, 2000) study on the ammonia levels near the exit ramp of a San Francisco highway tunnel, presented in Exercise 2.59 (p. 59). The data (in parts per million) for eight days during afternoon drive time are reproduced in the following table:

AMMONIA

1.53	1.50	1.37	1.51	1.55	1.42	1.41	1.48

a. Find the range of the ammonia levels.
b. Find the variance of the ammonia levels.
c. Find the standard deviation of the ammonia levels.
d. Suppose the standard deviation of the daily ammonia levels during morning drive time at the exit ramp is 1.45 ppm. Which time, morning or afternoon drive time, has the more variable ammonia levels?

Applying the Concepts—Intermediate

WPOWER50

2.84 Most powerful women in America. Refer to Exercise 2.58 (p. 59) and *Fortune's* (Nov. 14, 2002) list of the 50 most powerful women in America. The data are stored in the **WPOWER50** file.

a. Find the range of the ages of these 50 women.

b. Find the variance of the ages of these 50 women.
c. Find the standard deviation of the ages for these 50 women.
d. Suppose the standard deviation of the ages of the most powerful women in Europe is 10 years. For which location, the United States or Europe, is the age data more variable?
e. If the largest age in the data set is omitted, would the standard deviation increase or decrease? Verify your answer.

NUCLEAR

2.85 Active nuclear power plants. Refer to Exercise 2.68 (p. 61) and the U.S. Energy Information Administration's data on the number of nuclear power plants operating in each of 20 states. The data are saved in the **NUCLEAR** file.

a. Find the range, variance, and standard deviation of this data set.
b. Eliminate the largest value from the data set and repeat part **a**. What effect does dropping this measurement have on the measures of variation found in part **a**?
c. Eliminate the smallest and largest value from the data set and repeat part **a**. What effect does dropping both of these measurements have on the measures of variation found in part **a**?

2.6 Interpreting the Standard Deviation

We've seen that if we are comparing the variability of two samples selected from a population, the sample with the larger standard deviation is the more variable of the two. Thus, we know how to interpret the standard deviation on a relative or comparative basis, but we haven't explained how it provides a measure of variability for a single sample.

To understand how the standard deviation provides a measure of variability of a data set, consider a specific data set and answer the following questions: How many measurements are within one standard deviation of the mean? How many measurements are within two standard deviations? For example, look at the 100 mileage-per-gallon readings given in Table 2.2. Recall that $\bar{x} = 36.99$ and $s = 2.42$. Then

$$\bar{x} - s = 34.57 \quad \bar{x} + s = 39.41$$
$$\bar{x} - 2s = 32.15 \quad \bar{x} + 2s = 41.83$$

If we examine the data, we find that 68 of the 100 measurements, or 68%, are in the interval

$$\bar{x} - s \text{ to } \bar{x} + s$$

Similarly, we find that 96, or 96%, of the 100 measurements are in the interval

$$\bar{x} - 2s \text{ to } \bar{x} + 2s$$

We usually write these intervals as

$$(\bar{x} - s, \bar{x} + s) \text{ and } (\bar{x} - 2s, \bar{x} + 2s)$$

Such observations identify criteria for interpreting a standard deviation that apply to *any* set of data, whether a population or a sample. The criteria, expressed as a mathematical theorem and as a rule of thumb, are presented in Tables 2.6 and 2.7. In these tables, we give two sets of answers to the questions of how many measurements fall within one, two, and three standard deviations of the mean. The first, which applies to *any* set of data, is derived from a theorem proved by the Russian mathematician P. L. Chebyshev (1821–1894). The second, which applies to **mound-shaped, symmetric distributions** of data (for which the mean, median, and mode are all about the same), is based upon empirical evidence that has accumulated over the years. However, the percentages given for the intervals in Table 2.7 provide remarkably good approximations even when the distribution of the data is slightly skewed or asymmetric. Note that the rules apply to either population or sample data sets.

TABLE 2.6   Interpreting the Standard Deviation: Chebyshev's Rule

Chebyshev's rule applies to any data set, regardless of the shape of the frequency distribution of the data.

a. It is possible that very few of the measurements will fall within one standard deviation of the mean [i.e., within the interval $(\bar{x} - s, \bar{x} + s)$ for samples and $(\mu - \sigma, \mu + \sigma)$ for populations].

b. At least $\frac{3}{4}$ of the measurements will fall within two standard deviations of the mean [i.e., within the interval $(\bar{x} - 2s, \bar{x} + 2s)$ for samples and $(\mu - 2\sigma, \mu + 2\sigma)$ for populations].

c. At least $\frac{8}{9}$ of the measurements will fall within three standard deviations of the mean [i.e., within the interval $(\bar{x} - 3s, \bar{x} + 3s)$ for samples and $(\mu - 3\sigma, \mu + 3\sigma)$ for populations].

d. Generally, for any number k greater than 1, at least $(1 - 1/k^2)$ of the measurements will fall within k standard deviations of the mean [i.e., within the interval $(\bar{x} - ks, \bar{x} + ks)$ for samples and $(\mu - k\sigma, \mu + k\sigma)$ for populations].

TABLE 2.7   Interpreting the Standard Deviation: The Empirical Rule

The **empirical rule** is a rule of thumb that applies to data sets with frequency distributions that are mound shaped and symmetric, as follows:

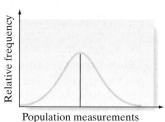

a. Approximately 68% of the measurements will fall within one standard deviation of the mean [i.e., within the interval $(\bar{x} - s, \bar{x} + s)$ for samples and $(\mu - \sigma, \mu + \sigma)$ for populations].

b. Approximately 95% of the measurements will fall within two standard deviations of the mean [i.e., within the interval $(\bar{x} - 2s, \bar{x} + 2s)$ for samples and $(\mu - 2\sigma, \mu + 2\sigma)$ for populations].

c. Approximately 99.7% (essentially all) of the measurements will fall within three standard deviations of the mean [i.e., within the interval $(\bar{x} - 3s, \bar{x} + 3s)$ for samples and $(\mu - 3\sigma, \mu + 3\sigma)$ for populations].

Biography

**PAFNUTY L. CHEBYSHEV (1821–1894)—
The Splendid Russian Mathematician**

P. L. Chebyshev was educated in mathematical science at Moscow University, eventually earning his master's degree. Following his graduation, Chebyshev joined St. Petersburg (Russia) University as a professor, becoming part of the well-known "Petersburg mathematical school." It was here that he proved his famous theorem about the probability of a measurement being within k standard deviations of the mean (Table 2.6). His fluency in French allowed him to gain international recognition in probability theory. In fact, Chebyshev once objected to being described as a "splendid Russian mathematician," saying he surely was a "worldwide mathematician." One student remembered Chebyshev as "a wonderful lecturer" who "was always prompt for class. As soon as the bell sounded, he immediately dropped the chalk and, limping, left the auditorium."

EXAMPLE 2.11
INTERPRETING
THE STANDARD
DEVIATION—
Rat-in-Maze
Experiment

Problem Thirty students in an experimental psychology class use various techniques to train a rat to move through a maze. At the end of the course, each student's rat is timed as it negotiates the maze. The results (in minutes) are listed in Table 2.8. Determine the fraction of the 30 measurements in the intervals $\bar{x} \pm s$, $\bar{x} \pm 2s$, and $\bar{x} \pm 3s$, and compare the results with those predicted in Tables 2.6 and 2.7.

RATMAZE

TABLE 2.8 Times (in Minutes) of 30 Rats Running through a Maze

1.97	.60	4.02	3.20	1.15	6.06	4.44	2.02	3.37	3.65
1.74	2.75	3.81	9.70	8.29	5.63	5.21	4.55	7.60	3.16
3.77	5.36	1.06	1.71	2.47	4.25	1.93	5.15	2.06	1.65

Solution First, we entered the data into the computer and used MINITAB to produce summary statistics. The mean and standard deviation of the sample data, highlighted on the printout shown in Figure 2.21, are

$$\bar{x} = 3.74 \text{ minutes} \qquad s = 2.20 \text{ minutes}$$

rounded to two decimal places.

Descriptive Statistics: RUNTIME

```
Variable   N    Mean   StDev   Minimum   Median   Maximum
RUNTIME    30   3.744  2.198   0.600     3.510    9.700
```

FIGURE 2.21

MINITAB descriptive statistics for rat maze times

Now we form the interval

$$(\bar{x} - s, \bar{x} + s) = (3.74 - 2.20, 3.74 + 2.20) = (1.54, 5.94)$$

A check of the measurements shows that 23 of the times are within this one standard-deviation interval around the mean. This number represents 23/30, or $\approx 77\%$, of the sample measurements.

The next interval of interest is

$$(\bar{x} - 2s, \bar{x} + 2s) = (3.74 - 4.40, 3.74 + 4.40) = (-.66, 8.14)$$

All but two of the times are within this interval, so 28/30, or approximately 93%, are within two standard deviations of $\bar{x}$.

Finally, the three-standard-deviation interval around $\bar{x}$ is

$$(\bar{x} - 3s, \bar{x} + 3s) = (3.74 - 6.60, 3.74 + 6.60) = (-2.86, 10.34)$$

All of the times fall within three standard deviations of the mean.

These one-, two-, and three-standard-deviation percentages (77%, 93%, and 100%) agree fairly well with the approximations of 68%, 95%, and 100% given by the empirical rule (Table 2.7).

Look Back If you look at the MINITAB frequency histogram for this data set in (Figure 2.22), you'll note that the distribution is not really mound shaped, nor is it extremely skewed. Thus, we get reasonably good results from the mound-shaped approximations. Of course, we know from Chebyshev's rule (Table 2.6) that no matter what the shape of the distribution, we would expect at least 75% and at least 89% of the measurements to lie within two and three standard deviations of $\bar{x}$, respectively.

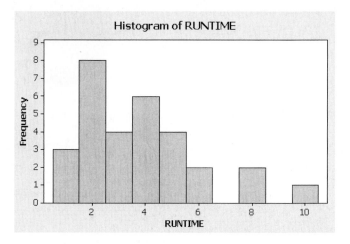

FIGURE 2.22
MINITAB histogram of rat maze times

Now Work Exercise 2.90

EXAMPLE 2.12

CHECKING THE CALCULATION OF THE SAMPLE STANDARD DEVIATION

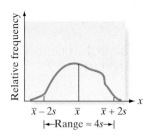

FIGURE 2.23

The relation between the range and the standard deviation

Problem Chebyshev's rule and the empirical rule are useful as a check on the calculation of the standard deviation. For example, suppose we calculated the standard deviation for the gas mileage data (Table 2.2) to be 5.85. Are there any "clues" in the data that enable us to judge whether this number is reasonable?

Solution The range of the mileage data in Table 2.2 is $44.9 - 30.0 = 14.9$. From Chebyshev's rule and the empirical rule, we know that most of the measurements (approximately 95% if the distribution is mound shaped) will be within two standard deviations of the mean. And regardless of the shape of the distribution and the number of measurements, almost all of them will fall within three standard deviations of the mean. Consequently, we would expect the range of the measurements to be between 4 (i.e., $\pm 2s$) and 6 (i.e., $\pm 3s$) standard deviations in length. (See Figure 2.23.) For the car mileage data, this means that s should fall between

$$\frac{\text{Range}}{6} = \frac{14.9}{6} = 2.48 \quad \text{and} \quad \frac{\text{Range}}{4} = \frac{14.9}{4} = 3.73$$

Hence, the standard deviation should not be much larger than 1/4 of the range, particularly for the data set with 100 measurements. Thus, we have reason to believe that the calculation of 5.85 is too large. A check of our work reveals that 5.85 is the variance s^2, not the standard deviation s. (See Example 2.10.) We "forgot" to take the square root (a common error); the correct value is $s = 2.42$. Note that this value is slightly smaller than the range divided by 6 (2.48). The larger the data set, the greater is the tendency for very large or very small measurements (extreme values) to appear, and when they do, the range may exceed six standard deviations.

Look Back In examples and exercises, we'll sometimes use $s \approx \text{range}/4$ to obtain a crude, and usually conservatively large, approximation for s. However, we stress that this is no substitute for calculating the exact value of s when possible.

Now Work Exercise 2.91

In the next example, we use the concepts in Chebyshev's rule and the empirical rule to build the foundation for making statistical inferences.

EXAMPLE 2.13
MAKING A
STATISTICAL
INFERENCE—Car
Battery Guarantee

Problem A manufacturer of automobile batteries claims that the average length of life for its grade A battery is 60 months. However, the guarantee on this brand is for just 36 months. Suppose the standard deviation of the life length is known to be 10 months and the frequency distribution of the life-length data is known to be mound shaped.

a. Approximately what percentage of the manufacturer's grade A batteries will last more than 50 months, assuming that the manufacturer's claim is true?

b. Approximately what percentage of the manufacturer's batteries will last less than 40 months, assuming that the manufacturer's claim is true?

c. Suppose your battery lasts 37 months. What could you infer about the manufacturer's claim?

Solution If the distribution of life length is assumed to be mound shaped with a mean of 60 months and a standard deviation of 10 months, it would appear as shown in Figure 2.24. Note that we can take advantage of the fact that mound-shaped distributions are (approximately) symmetric about the mean, so that the percentages given by the empirical rule can be split equally between the halves of the distribution on each side of the mean.

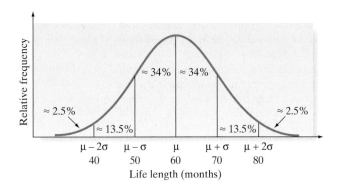

FIGURE 2.24
Battery life-length
distribution: manufacturer's
claim assumed true

For example, since approximately 68% of the measurements will fall within one standard deviation of the mean, the distribution's symmetry implies that approximately (1/2)(68%) = 34% of the measurements will fall between the mean and one standard deviation on each side. This concept is illustrated in Figure 2.24. The figure also shows that 2.5% of the measurements lie beyond two standard deviations in each direction from the mean. This result follows from the fact that if approximately 95% of the measurements fall within two standard deviations of the mean, then about 5% fall outside two standard deviations; if the distribution is approximately symmetric, then about 2.5% of the measurements fall beyond two standard deviations on each side of the mean.

a. It is easy to see in Figure 2.24 that the percentage of batteries lasting more than 50 months is approximately 34% (between 50 and 60 months) plus 50% (greater than 60 months). Thus, approximately 84% of the batteries should have a life exceeding 50 months.

b. The percentage of batteries that last less than 40 months can also be easily determined from Figure 2.24: Approximately 2.5% of the batteries should fail prior to 40 months, assuming that the manufacturer's claim is true.

c. If you are so unfortunate that your grade A battery fails at 37 months, you can make one of two inferences: Either your battery was one of the approximately 2.5% that fail prior to 40 months, or something about the manufacturer's claim is not true. Because the chances are so small that a battery fails before 40 months, you would have good reason to have serious doubts about the manufacturer's claim. A mean smaller than 60 months or a standard deviation longer than 10 months would each increase the likelihood of failure prior to 40 months.*

*The assumption that the distribution is mound shaped and symmetric may also be incorrect. However, if the distribution were skewed to the right, as life-length distributions often tend to be, the percentage of measurements more than two standard deviations below the mean would be even less than 2.5%.

Look Back The approximations given in Figure 2.24 are more dependent on the assumption of a mound-shaped distribution than are the assumptions stated in the empirical rule (Table 2.7), because the approximations in Figure 2.24 depend on the (approximate) symmetry of the mound-shaped distribution. We saw in Example 2.11 that the empirical rule can yield good approximations even for skewed distributions. This will *not* be true of the approximations in Figure 2.24; the distribution *must* be mound shaped and (approximately) symmetric.

Example 2.13 is our initial demonstration of the statistical inference-making process. At this point, you should realize that we'll use sample information (in Example 2.13, your battery's failure at 37 months) to make inferences about the population (in Example 2.13, the manufacturer's claim about the length of life for the population of all batteries). We'll build on this foundation as we proceed.

STATISTICS IN ACTION REVISITED

Interpreting Descriptive Statistics for the "Eye Cue" Test Data

We return to the analysis of data from the *Psychological Science* "water-level task" experiment. The quantitative variable of interest is *deviation angle* (measured in degrees) of the judged line from the true surface line (parallel to the tabletop). Recall that the researchers want to test the theory that males do better than females on judging the correct water level (i.e., that males, in general, have deviation angles closer to 0 than females have). The MINITAB descriptive statistics printout for the EYECUE data is displayed in Figure SIA2.6, with the means and standard deviations highlighted.

The sample mean for females is 6.57 degrees and the mean for males is 2.92 degrees. Our interpretation is that males' judged lines have smaller deviation angles from the true surface line (2.92 degrees, on average) than females' judged lines (6.57 degrees, on average), supporting the theory stated by the researchers.

To interpret the standard deviation, we substitute into the formula Mean ±2(Standard deviation) to obtain the intervals:

Females: $6.57 \pm 2(7.82) = 6.57 \pm 15.64 = (-9.07, 22.21)$
Males: $2.92 \pm 2(7.54) = 2.92 \pm 15.08 = (-12.16, 18.00)$

From Chebyshev's rule (Table 2.6), we know that at least 75% of the females who perform the water task will have deviation angles anywhere between −9.07 and 22.21 degrees. Similarly, we know that at least 75% of the males will have deviation angles anywhere from −12.16 to 18.00 degrees. Note that these ranges indicate that there is very little difference in the spread of the deviation angle distributions of the two groups. However, the ranges also indicate that a male subject is more likely to judge the water line below the actual surface (i.e., with a greater negative deviation angle) than a female is, while a female subject is more likely to judge the water line above the actual surface (i.e., with a greater positive deviation angle) than a male is.

Descriptive Statistics: Deviation

Variable	Gender	N	Mean	StDev	Variance	Minimum	Median	Maximum
Deviation	Female	60	6.57	7.82	61.10	−15.00	7.00	35.00
	Male	60	2.917	7.543	56.891	−18.000	2.000	27.000

FIGURE SIA2.6
MINITAB analysis of deviation angle in the water-level task—males versus females

Exercises 2.86–2.102

Understanding the Principles

2.86 To what kind of data sets can Chebyshev's rule be applied? How about the empirical rule?

2.87 The output from a statistical computer program indicates that the mean and standard deviation of a data set consisting of 200 measurements are $1,500 and $300, respectively.

a. What are the units of measurement of the variable of interest? On the basis of the units, what type of data is this, quantitative or qualitative?

b. What can be said about the number of measurements between $900 and $2,100? between $600 and $2,400? between $1,200 and $1,800? between $1,500 and $2,100?

2.88 For any set of data, what can be said about the percentage of the measurements contained in each of the following intervals?

a. $\bar{x} - s$ to $\bar{x} + s$

b. $\bar{x} - 2s$ to $\bar{x} + 2s$

c. $\bar{x} - 3s$ to $\bar{x} + 3s$

2.89 For a set of data with a mound-shaped relative frequency distribution, what can be said about the percentage of the measurements contained in each of the intervals specified in Exercise 2.88?

Learning the Mechanics

2.90 The following is a sample of 25 measurements:

LM 2_90

7	6	6	11	8	9	11	9	10	8	7	7	5
9	10	7	7	7	7	9	12	10	10	8	6	

NW **a.** Compute $\bar{x}$, s^2, and s for this sample.

b. Count the number of measurements in the intervals $\bar{x} \pm s$, $\bar{x} \pm 2s$, and $\bar{x} \pm 3s$. Express each count as a percentage of the total number of measurements.

c. Compare the percentages found in part **b** with the percentages given by the empirical rule and Chebyshev's rule.

d. Calculate the range and use it to obtain a rough approximation for s. Does the result compare favorably with the actual value for s found in part **a**?

2.91 Given a data set with a largest value of 760 and a smallest value of 135, what would you estimate the standard deviation to be? Explain the logic behind the procedure you used to estimate the standard deviation. Suppose the standard deviation is reported to be 25. Is this number reasonable? Explain.

Applying the Concepts—Basic

WPOWER50

2.92 **Most powerful women in America.** Refer to the *Fortune* (Nov. 14, 2005) list of the 50 most powerful women in America, saved in the WPOWER50 file. In Exercise 2.58 (p. 59) you found the mean age of the 50 women in the data set, and in Exercise 2.84 (p. 67) you found the standard deviation. Use the mean and standard deviation to form an interval that will contain at least 75% of the ages in the data set.

SHIPSANIT

2.93 **Sanitation inspection of cruise ships.** Refer to Exercise 2.32 (p. 46) and the Centers for Disease Control and Prevention listing of the May 2006 sanitation scores for 169 cruise ships. The data are saved in the **SHIPSANIT** file.

a. Find the mean and standard deviation of the sanitation scores.

b. Calculate the intervals $\bar{x} \pm s$, $\bar{x} \pm 2s$, and $\bar{x} \pm 3s$.

c. Find the percentage of measurements in the data set that fall within each of the intervals in part **b**. Do these per-

centages agree with either Chebyshev's rule or the empirical rule?

2.94 **Recommendation letters for professors.** Refer to the *American Psychologist* (July 1995) study of 148 applicants for a position in experimental psychology, presented in Exercise 2.62 (p. 59). Recall that the mean number of recommendation letters included in each application packet was $\bar{x} = 2.28$. The standard deviation was also reported in the article; its value was $s = 1.48$.

a. Sketch the relative frequency distribution for the number of recommendation letters included in each application for the experimental psychology position. (Assume that the distribution is mound shaped and relatively symmetric.)

b. Locate an interval on the distribution found in part **a** that captures approximately 95% of the measurements in the sample.

c. Locate an interval on the distribution found in part **a** that captures almost all the sample measurements.

2.95 **Dentists' use of anesthetics.** A study published in *Current Allergy & Clinical Immunology* (Mar. 2004) investigated allergic reactions of dental patients to local anesthetics. Based on a survey of dental practitioners, the study reported that the mean number of units (ampoules) of local anesthetics used per week by dentists was 79, with a standard deviation of 23. Suppose we want to determine the percentage of dentists who use less than 102 units of local anesthetics per week.

a. Assuming that nothing is known about the shape of the distribution for the data, what percentage of dentists use less than 102 units of local anesthetics per week?

b. Assuming that the data has a mound-shaped distribution, what percentage of dentists use less than 102 units of local anesthetics per week?

Applying the Concepts—Intermediate

NZBIRDS

2.96 **Extinct New Zealand birds.** Refer to the *Evolutionary Ecology Research* (July 2003) study of the patterns of extinction in the New Zealand bird population, presented in Exercise 2.18 (p. 36). Consider the data on the egg length (measured in millimeters) for the 132 bird species saved in the NZBIRDS file.

a. Find the mean and standard deviation of the egg lengths.

b. Form an interval that can be used to predict the egg length of a bird species found in New Zealand.

2.97 **Hand washing versus hand rubbing.** In hospitals, washing the hands with soap is emphasized as the single most important measure in the prevention of infections. As an alternative to hand washing, some hospitals allow health workers to rub their hands with an alcohol-based antiseptic. The *British Medical Journal* (Aug. 17, 2002) reported on a study to compare the effectiveness of washing the hands with soap and rubbing the hands with alcohol. One group of health care workers used hand rubbing, while a second group used hand washing to clean their hands. The bacterial count (number of colony-forming units) on the hand of each worker was recorded. The table gives descriptive statistics on bacteria counts for the two groups of health care workers.

	Mean	Standard Deviation
Hand rubbing	35	59
Hand washing	69	106

a. For hand rubbers, form an interval that contains about 95% of the bacterial counts. (*Note:* The bacterial count cannot be less than 0.)

b. Repeat part **a** for hand washers.

c. On the basis of your results in parts **a** and **b**, make an inference about the effectiveness of the two hand-cleaning methods.

2.98 Sentence complexity study. A study published in *Applied Psycholinguistics* (June 1998) compared the language skills of young children (16–30 months old) from low-income and middle-income families. A total of 260 children—65 in the low-income and 195 in the middle-income group—completed the Communicative Development Inventory (CDI) exam. One of the variables measured on each child was sentence complexity score. Summary statistics for the scores of the two groups are reproduced in the accompanying table. Use this information to sketch a graph of the sentence complexity score distribution for each income group. (Assume that the distributions are mound shaped and symmetric.) Compare the distributions. What can you infer?

	Low Income	Middle Income
Sample Size	65	195
Mean	7.62	15.55
Median	4	14
Standard Deviation	8.91	12.24
Minimum	0	0
Maximum	36	37

Source: Arriaga, R. I. et al. "Scores on the MacArthur Communicative Development Inventory of children from low-income and middle-income families." *Applied Psycholinguistics,* Vol. 19, No. 2, June 1998, p. 217 (Table 7).

2.99 Velocity of Winchester bullets. The *American Rifleman* (June 1993) reported on the velocity of ammunition fired from the FEG P9R pistol, a 9-mm gun manufactured in Hungary. Field tests revealed that Winchester bullets fired from the pistol had a mean velocity (at 15 feet) of 936 feet per second and a standard deviation of 10 feet per second. Tests were also conducted with Uzi and Black Hills ammunition.

a. Describe the velocity distribution of Winchester bullets fired from the FEG P9R pistol.

b. A bullet whose brand is unknown is fired from the FEG P9R pistol. Suppose the velocity (at 15 feet) of the bullet is 1,000 feet per second. Is the bullet likely to be manufactured by Winchester? Explain.

2.100 Speed of light from galaxies. Astronomers theorize that cold dark matter caused the formation of galaxies and clusters of galaxies in the universe. The theoretical model for cold dark matter requires an estimate of the velocities of light emitted from galaxy clusters. *The Astronomical Journal* (July 1995) published a study of observed velocities of galaxies in four different clusters. Galaxy velocity was measured in kilometers per second (km/s), using a spectrograph and high-power telescope.

a. The observed velocities of 103 galaxies located in the cluster named A2142 are summarized in the accompanying histogram. Comment on whether the empirical rule is applicable to describing the velocity distribution for this cluster.

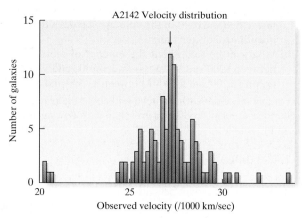

Source: Oegerle, W. R., Hill, J. M., and Fitchett, M. J. "Observations of high dispersion clusters of galaxies: Constraints on cold dark matter." *The Astronomical Journal,* Vol. 110, No. 1, July 1995, p. 37 (Figure 1).

b. The mean and standard deviation of the 103 velocities observed in galaxy cluster A2142 were reported as $\bar{x} = 27{,}117$ km/s and $s = 1{,}280$ km/s, respectively. Use this information to construct an interval that captures approximately 95% of the velocities of the galaxies in the cluster.

c. Recommend a single velocity value to be used in the CDM model for galaxy cluster A2142. Explain your reasoning.

Applying the Concepts—Advanced

2.101 Improving SAT scores. The National Education Longitudinal Survey (NELS) tracks a nationally representative sample of U.S. students from eighth grade through high school and college. Research published in *Chance* (Winter 2001) examined the SAT scores of 265 NELS students who paid a private tutor to help them improve their scores. The table summarizes the changes in both the SAT-Mathematics and SAT-Verbal scores for these students.

	SAT-Math	SAT-Verbal
Mean change in score	19	7
Standard deviation of score changes	65	49

a. Suppose one of the 265 students who paid a private tutor is selected at random. Give an interval that is likely to contain the change in this student's SAT-Math score.

b. Repeat part **a** for the SAT-Verbal score.

c. Suppose the selected student's score increased on one of the SAT tests by 140 points. Which test, the SAT-Math or SAT-Verbal, is the one most likely to have had the 140-point increase? Explain.

2.102 Animal-assisted therapy for heart patients. Recall the *American Heart Association Conference* (Nov. 2005) study to gauge whether animal-assisted therapy can improve the physiological responses of heart failure patients. (See Chapter 1, Study 3, p. 6.) A team of nurses from the UCLA Medical Center randomly divided 76 heart patients into three groups. Each patient in group T was visited by a human volunteer accompanied by a trained dog, each patient in group V was visited by a volunteer only, and the patients in group C were not visited at all. The anxiety level of each patient was measured (in points) both before and after the visits. The accompanying table gives summary statistics for the drop in anxiety level for patients in the three groups. Suppose the anxiety level of a patient selected from the study had a drop of 22.5 points. From which group is the patient more likely to have come? Explain.

	Sample Size	Mean Drop	Std. Dev.
Group T: Volunteer + Trained Dog	26	10.5	7.6
Group V: Volunteer only	25	3.9	7.5
Group C: Control group (no visit)	25	1.4	7.5

Source: Cole, K., et al. "Animal assisted therapy decreases hemodynamics, plasma epinephrine and state anxiety in hospitalized heart failure patients." *American Heart Association Conference*, Dallas, Texas, Nov. 2005.

2.7 Numerical Measures of Relative Standing

We've seen that numerical measures of central tendency and variability describe the general nature of a quantitative data set (either a sample or a population). In addition, we may be interested in describing the *relative* quantitative location of a particular measurement within a data set. Descriptive measures of the relationship of a measurement to the rest of the data are called **measures of relative standing**.

One measure of the relative standing of a measurement is its *percentile ranking*. For example, suppose you scored an 80 on a test and you want to know how you fared in comparison with others in your class. If the instructor tells you that you scored at the 90th percentile, it means that 90% of the grades were lower than yours and 10% were higher. Thus, if the scores were described by the relative frequency histogram in Figure 2.25, the 90th percentile would be located at a point such that 90% of the total area under the relative frequency histogram lies below the 90th percentile and 10% lies above. If the instructor tells you that you scored in the 50th percentile (the median of the data set), 50% of the test grades would be lower than yours and 50% would be higher.

Percentile rankings are of practical value only with large data sets. Finding them involves a process similar to the one used in finding a median. The measurements are ranked in order, and a rule is selected to define the location of each percentile. Since we are interested primarily in interpreting the percentile rankings of measurements (rather than in finding particular percentiles for a data set), we define the *p*th *percentile* of a data set as in Definition 2.12.

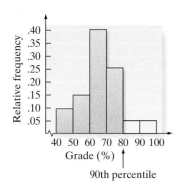

FIGURE 2.25
Location of 90th percentile for test grades

> **Definition 2.12**
>
> For any set of n measurements (arranged in ascending or descending order), the ***p*th percentile** is a number such that $p\%$ of the measurements fall below that number and $(100 - p)\%$ fall above it.

EXAMPLE 2.14
FINDING AND INTERPRETING PERCENTILES

Problem Refer to the student default rates of the 50 states (and the District of Columbia) in Table 2.4. An SPSS printout describing the data is shown in Figure 2.26. Locate the 25th percentile and 95th percentile on the printout, and interpret the associated values.

Solution Both the 25th percentile and 95th percentile are highlighted on the SPSS printout. The values in question are 7.9 and 16.0, respectively. Our interpretations are as follows: 25% of the 51 default rates fall below 7.9 and 95% of the default rates fall below 16.0.

Look Back The method for computing percentiles with small data sets varies according to the software used. As the sample size increases, the percentiles from the different software packages will converge to a single number.

Statistics

DEFAULT

N	Valid	51
	Missing	0
Percentiles	5	4.860
	10	5.760
	25	7.900
	50	9.700
	75	12.300
	90	15.120
	95	16.000

FIGURE 2.26

SPSS percentiles for student default rate data

Now Work Exercise 2.103

Another measure of relative standing in popular use is the *z-score*. As you can see in Definition 2.13, the *z*-score makes use of the mean and standard deviation of a data set in order to specify the relative location of the measurement.

> **Definition 2.13**
>
> The **sample z-score** for a measurement x is
>
> $$z = \frac{x - \bar{x}}{s}$$
>
> The **population z-score** for a measurement x is
>
> $$z = \frac{x - \mu}{\sigma}$$

Note that the *z*-score is calculated by subtracting $\bar{x}$ (or μ) from the measurement x and then dividing the result by s (or σ). The final result, the **z-score**, represents the distance between a given measurement x and the mean, expressed in standard deviations.

EXAMPLE 2.15

FINDING A
z-SCORE

Problem Suppose a sample of 2,000 high school seniors' verbal SAT scores is selected. The mean and standard deviation are

$$\bar{x} = 550 \quad s = 75$$

Suppose Joe Smith's score is 475. What is his sample *z*-score?

Solution Joe Smith's Verbal SAT score lies below the mean score of the 2,000 seniors, as shown in Figure 2.27.

FIGURE 2.27

Verbal SAT scores of high school seniors

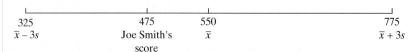

325	475	550	775
$\bar{x} - 3s$	Joe Smith's score	$\bar{x}$	$\bar{x} + 3s$

We compute

$$z = \frac{x - \bar{x}}{s} = \frac{475 - 550}{75} = -1.0$$

which tells us that Joe Smith's score is 1.0 standard deviation *below* the sample mean; in short, his sample *z*-score is −1.0.

Look Back The numerical value of the z-score reflects the relative standing of the measurement. A large positive z-score implies that the measurement is larger than almost all other measurements, whereas a large negative z-score indicates that the measurement is smaller than almost every other measurement. If a z-score is 0 or near 0, the measurement is located at or near the mean of the sample or population.

Now Work Exercise 2.106

We can be more specific if we know that the frequency distribution of the measurements is mound shaped. In this case, the following interpretation of the z-score can be given:

Interpretation of z-Scores for Mound-Shaped Distributions of Data

1. Approximately 68% of the measurements will have a z-score between −1 and 1.

2. Approximately 95% of the measurements will have a z-score between −2 and 2.

3. Approximately 99.7% (almost all) of the measurements will have a z-score between −3 and 3.

Note that this interpretation of z-scores is identical to that given by the empirical rule for mound-shaped distributions (Table 2.7). The statement that a measurement falls into the interval from $(\mu - \sigma)$ to $(\mu + \sigma)$ is equivalent to the statement that a measurement has a population z-score between −1 and 1, since all measurements between $(\mu - \sigma)$ and $(\mu + \sigma)$ are within one standard deviation of μ. These z-scores are displayed in Figure 2.28.

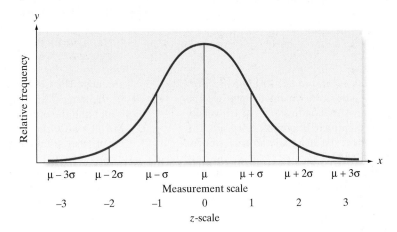

FIGURE 2.28
Population z-scores for a mound-shaped distribution

Exercises 2.103–2.117

Understanding the Principles

2.103 Give the percentage of measurements in a data set that
NW are above and below each of the following percentiles:
    a. 75th percentile
    b. 50th percentile
    c. 20th percentile
    d. 84th percentile

2.104 What is the 50th percentile of a quantitative data set called?

2.105 For mound-shaped data, what percentage of measurements have a z-score between −2 and 2?

Learning the Mechanics

2.106 Compute the z-score corresponding to each of the follow-
NW ing values of x:

    a. $x = 40, s = 5, \bar{x} = 30$
    b. $x = 90, \mu = 89, \sigma = 2$
    c. $\mu = 50, \sigma = 5, x = 50$
    d. $s = 4, x = 20, \bar{x} = 30$
    e. In parts **a–d**, state whether the z-score locates x within a sample or within a population.
    f. In parts **a–d**, state whether each value of x lies above or below the mean and by how many standard deviations.

2.107 Compare the z-scores to decide which of the following x values lie the greatest distance above the mean and the greatest distance below the mean:
    a. $x = 100, \mu = 50, \sigma = 25$
    b. $x = 1, \mu = 4, \sigma = 1$
    c. $x = 0, \mu = 200, \sigma = 100$
    d. $x = 10, \mu = 5, \sigma = 3$

2.108 Suppose that 40 and 90 are two elements of a population data set and that their z-scores are −2 and 3, respectively. Using only this information, is it possible to determine the population's mean and standard deviation? If so, find them. If not, explain why it's not possible.

Applying the Concepts—Basic

2.109 **Drivers stopped by police.** According to the Bureau of Justice Statistics (June 2006), 74% of all licensed drivers stopped by police are 25 years or older. Give a percentile ranking for the age of 25 years in the distribution of all ages of licensed drivers stopped by police.

2.110 **Math scores of eighth graders.** According to the National Center for Education Statistics (2005), scores on a mathematics assessment test for United States eighth graders have a mean of 279, a 25th percentile of 255, a 75th percentile of 304, and a 90th percentile of 324. Interpret each of these numerical descriptive measures.

SHIPSANIT

2.111 **Sanitation inspection of cruise ships.** Refer to the sanitation levels of cruise ships presented in Exercise 2.93 (p. 73) and saved in the SHIPSANIT file.
   a. Give a measure of relative standing for the *Nautilus Explorer's* score of 78. Interpret the result.
   b. Give a measure of relative standing for the *Rotterdam's* score of 98. Interpret the result.

JAPANESE

2.112 **Reading Japanese books.** Refer to the *Reading in a Foreign Language* (Apr. 2004) experiment to improve the Japanese reading comprehension levels of 14 University of Hawaii students, presented in Exercise 2.57 (p. 58). The data on number of books read and grade for each student are saved in the **JAPANESE** file.
   a. Find the mean and standard deviation of the number of books read by students who earned an A grade. Find the z-score for an A student who read 40 books. Interpret the result.
   b. Find the mean and standard deviation of the number of books read by students who earned either a B or a C grade. Find the z-score for a B or C student who read 40 books. Interpret the result.
   c. Refer to parts **a** and **b**. Which of the two groups of students is more likely to have read 40 books? Explain.

Applying the Concepts—Intermediate

NZBIRDS

2.113 **Extinct New Zealand birds.** Refer to the *Evolutionary Ecology Research* (July 2003) study of the patterns of extinction in the New Zealand bird population, presented in Exercise 2.96 (p. 73). Again, consider the data on the egg length (measured in millimeters) for the 132 bird species saved in the NZBIRDS file.
   a. Find the 10th percentile for the egg length distribution and interpret its value.

   b. The moas *(P. australis)* is a bird species with an egg length of 205 millimeters. Find the z-score for this species of bird and interpret its value.

2.114 **Lead in drinking water.** The U.S. Environmental Protection Agency (EPA) sets a limit on the amount of lead permitted in drinking water. The EPA *Action Level* for lead is .015 milligram per liter (mg/L) of water. Under EPA guidelines, if 90% of a water system's study samples have a lead concentration less than .015 mg/L, the water is considered safe for drinking. I (coauthor Sincich) received a report on a study of lead levels in the drinking water of homes in my subdivision. The 90th percentile of the study sample had a lead concentration of .00372 mg/L. Are water customers in my subdivision at risk of drinking water with unhealthy lead levels? Explain.

LEFTEYE

2.115 **Eye refractive study.** Refer to the *Optometry and Vision Science* (June 1995) study of refractive variation in eyes, presented in Exercise 2.67 (p. 61). The 25 cylinder power measurements are saved in the LEFTEYE file.
   a. Find the 10th percentile of cylinder power measurements. Interpret the result.
   b. Find the 95th percentile of cylinder power measurements. Interpret the result.
   c. Calculate the z-score for the cylinder power measurement of −1.07. Interpret the result.

2.116 **Blue versus red exam study.** In a study of how external clues influence performance, psychology professors at the University of Alberta and Pennsylvania State University gave two different forms of a midterm examination to a large group of introductory psychology students. The questions on the exam were identical and in the same order, but one exam was printed on blue paper and the other on red paper. (*Teaching Psychology*, May 1998.) Grading only the difficult questions on the exam, the researchers found that scores on the blue exam had a distribution with a mean of 53% and a standard deviation of 15%, while scores on the red exam had a distribution with a mean of 39% and a standard deviation of 12%. (Assume that both distributions are approximately mound shaped and symmetric.)
   a. Give an interpretation of the standard deviation for the students who took the blue exam.
   b. Give an interpretation of the standard deviation for the students who took the red exam.
   c. Suppose a student is selected at random from the group of students who participated in the study and the student's score on the difficult questions is 20%. Which exam form is the student more likely to have taken, the blue or the red exam? Explain.

Applying the Concepts—Advanced

2.117 **GPAs of students.** At one university, the students are given z-scores at the end of each semester, rather than the traditional GPAs. The mean and standard deviation of all students' cumulative GPAs, on which the z-scores are based, are 2.7 and .5, respectively.
   a. Translate each of the following z-scores to corresponding GPA scores: $z = 2.0$, $z = -1.0$, $z = .5$, $z = -2.5$.

b. Students with z-scores below -1.6 are put on probation. What is the corresponding probationary GPA?

c. The president of the university wishes to graduate the top 16% of the students with *cum laude* honors and the

top 2.5% with *summa cum laude* honors. Where (approximately) should the limits be set in terms of z-scores? In terms of GPAs? What assumption, if any, did you make about the distribution of the GPAs at the university?

2.8 Methods for Detecting Outliers (Optional)

Sometimes it is important to identify inconsistent or unusual measurements in a data set. An observation that is unusually large or small relative to the data values we want to describe is called an **outlier**.

Outliers are often attributable to one of several causes. First, the measurement associated with the outlier may be invalid. For example, the experimental procedure used to generate the measurement may have malfunctioned, the experimenter may have misrecorded the measurement, or the data might have been coded incorrectly in the computer. Second, the outlier may be the result of a misclassified measurement. That is, the measurement belongs to a population different from that from which the rest of the sample was drawn. Finally, the measurement associated with the outlier may be recorded correctly and from the same population as the rest of the sample, but represents a rare (chance) event. Such outliers occur most often when the relative frequency distribution of the sample data is extremely skewed, because a skewed distribution has a tendency to include extremely large or small observations relative to the others in the data set.

Definition 2.14

An observation (or measurement) that is unusually large or small relative to the other values in a data set is called an **outlier**. Outliers typically are attributable to one of the following causes:

1. The measurement is observed, recorded, or entered into the computer incorrectly.

2. The measurement comes from a different population.

3. The measurement is correct, but represents a rare (chance) event.

Two useful methods for detecting outliers, one graphical and one numerical, are **box plots** and z-scores. The box plot is based on the *quartiles* of a data set. **Quartiles** are values that partition the data set into four groups, each containing 25% of the measurements. The *lower quartile* Q_L is the 25th percentile, the *middle quartile* is the median M (the 50th percentile), and the *upper quartile* Q_U is the 75th percentile. (See Figure 2.29.)

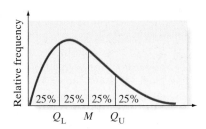

FIGURE 2.29
The quartiles for a data set

Definition 2.15

The **lower quartile** Q_L is the 25th percentile of a data set. The **middle quartile** M is the median. The **upper quartile** Q_U is the 75th percentile.

A box plot is based on the *interquartile range (IQR)* — the distance between the lower and upper quartiles:

$$\text{IQR} = Q_U - Q_L$$

> **Definition 2.16**
>
> The **interquartile range (IQR)** is the distance between the lower and upper quartiles:
>
> $$IQR = Q_U - Q_L$$

An annotated MINITAB box plot for the gas mileage data (see Table 2.2) is shown in Figure 2.30.* Note that a rectangle (the *box*) is drawn, with the bottom and top of the rectangle (the **hinges**) drawn at the quartiles Q_L and Q_U, respectively.

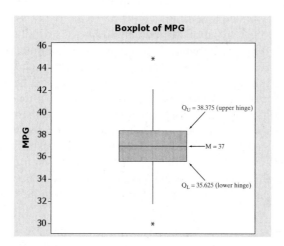

FIGURE 2.30

Annotated MINITAB box plot for EPA gas mileages

By definition, then, the "middle" 50% of the observations—those between Q_L and Q_U—fall inside the box. For the gas mileage data, these quartiles are at 35.625 and 38.375. Thus,

$$IQR = 38.375 - 35.625 = 2.75$$

The median is shown at 37 by a horizontal line within the box.

To guide the construction of the "tails" of the box plot, two sets of limits, called **inner fences** and **outer fences**, are used. Neither set of fences actually appears on the plot. Inner fences are located at a distance of 1.5(IQR) from the hinges. Emanating from the hinges of the box are vertical lines called the **whiskers**. The two whiskers extend to the most extreme observation inside the inner fences. For example, the inner fence on the lower side of the gas mileage box plot is

$$
\begin{aligned}
\text{Lower inner fence} &= \text{Lower hinge} - 1.5(\text{IQR}) \\
&= 35.625 - 1.5(2.75) \\
&= 35.625 - 4.125 = 31.5
\end{aligned}
$$

The smallest measurement *inside* this fence is the second-smallest measurement, 31.8. Thus, the lower whisker extends to 31.8. Similarly, the upper whisker extends to 42.1, the largest measurement inside the upper inner fence:

$$
\begin{aligned}
\text{Upper inner fence} &= \text{Upper hinge} + 1.5(\text{IQR}) \\
&= 38.375 + 1.5(2.75) \\
&= 38.375 + 4.125 = 42.5
\end{aligned}
$$

Values that are beyond the inner fences are deemed *potential outliers* because they are extreme values that represent relatively rare occurrences. In fact, for mound-shaped distributions, less than 1% of the observations are expected to fall outside the inner fences. Two of the 100 gas mileage measurements, 30.0 and 44.9, fall beyond the inner

---

*Although box plots can be generated by hand, the amount of detail required makes them particularly well suited for computer generation. We use computer software to generate the box plots in this section.

fences, one on each end of the distribution. Each of these potential outliers is represented by a common symbol (an asterisk in MINITAB).

The other two imaginary fences, the outer fences, are defined at a distance 3(IQR) from each end of the box. Measurements that fall beyond the outer fences (also represented by an asterisk in MINITAB) are very extreme measurements that require special analysis. Since less than one-hundredth of 1% (.01% or .0001) of the measurements from mound-shaped distributions are expected to fall beyond the outer fences, these measurements are considered to be *outliers*. Because there are no measurements of gas mileage beyond the outer fences, there are no outliers.

Elements of a Box Plot

1. A rectangle (the **box**) is drawn with the ends (the **hinges**) drawn at the lower and upper quartiles (Q_L and Q_U). The median of the data is shown in the box, usually by a line.

2. The points at distances 1.5(IQR) from each hinge mark the **inner fences** of the data set. Lines (the **whiskers**) are drawn from each hinge to the most extreme measurement inside the inner fence. Thus,

$$\text{Lower inner fence} = Q_L - 1.5(\text{IQR})$$

$$\text{Upper inner fence} = Q_U + 1.5(\text{IQR})$$

3. A second pair of fences, the **outer fences**, appears at a distance of 3(IQR) from the hinges. One symbol (e.g., "*") is used to represent measurements falling between the inner and outer fences, and another (e.g., "0") is used to represent measurements that lie beyond the outer fences. Thus, outer fences are not shown unless one or more measurements lie beyond them. We have

$$\text{Lower outer fence} = Q_L - 3(\text{IQR})$$

$$\text{Upper outer fence} = Q_U + 3(\text{IQR})$$

4. The symbols used to represent the median and the extreme data points (those beyond the fences) will vary with the software you use to construct the box plot. (You may use your own symbols if you are constructing a box plot by hand.) You should consult the program's documentation to determine exactly which symbols are used.

Aids to the Interpretation of Box Plots

1. Examine the length of the box. The IQR is a measure of the sample's variability and is especially useful for the comparison of two samples. (See Example 2.16.)

2. Visually compare the lengths of the whiskers. If one is clearly longer, the distribution of the data is probably skewed in the direction of the longer whisker.

3. Analyze any measurements that lie beyond the fences. Less than 5% should fall beyond the inner fences, even for very skewed distributions. Measurements beyond the outer fences are probably outliers, with one of the following explanations:
   a. The measurement is incorrect. It may have been observed, recorded, or entered into the computer incorrectly.
   b. The measurement belongs to a population different from the population that the rest of the sample was drawn from. (See Example 2.17.)
   c. The measurement is correct *and* from the same population as the rest of the sample. Generally, we accept this explanation only after carefully ruling out all others.

Recall that outliers may be incorrectly recorded observations, members of a population different from the rest of the sample, or, at the least, very unusual measurements from the same population. The box plot of Figure 2.30 detected two potential outliers: the two gas mileage measurements beyond the inner fences. When we analyze these measurements, we find that they are correctly recorded. Perhaps they represent mileages that correspond to exceptional models of the car being tested or to unusual gas mixtures. Outlier analysis often reveals useful information of this kind and therefore plays an important role in the statistical inference-making process.

In addition to detecting outliers, box plots provide useful information on the variation in a data set. The elements (and nomenclature) of box plots are summarized in the box on p. 81. Some aids to the interpretation of box plots are also given.

EXAMPLE 2.16
GENERATING BOX PLOTS WITH THE COMPUTER

Problem Use a statistical software package to draw a box plot for the student loan default data shown in Table 2.4. Identify any outliers in the data set.

Solution The MINITAB box plot for the student loan default rates is shown in Figure 2.31. Note that the median appears to be about 9.5, and, with the exception of a single extreme observation, the distribution appears to be symmetric between approximately 3% and 17%. The single outlier is beyond the inner fence, but inside the outer fence. Examination of the data reveals that this observation corresponds to Alaska's default rate of 19.7%.

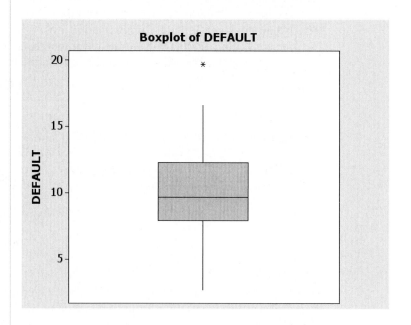

FIGURE 2.31
MINITAB box plot for student loan default rates

Look Back Before removing the outlier from the data set, a good analyst will make a concerted effort to find the cause of the outlier.

Now Work Exercise 2.125

EXAMPLE 2.17
COMPARING BOX PLOTS—Stimulus Reaction Study

Problem A Ph.D. student in psychology conducted a stimulus reaction experiment as a part of her dissertation research. She subjected 50 subjects to a threatening stimulus and 50 to a nonthreatening stimulus. The reaction times of all 100 students, recorded to the nearest tenth of a second, are listed in Table 2.9. Box plots of the two resulting samples of reaction times, generated with SAS, are shown in Figure 2.32. Interpret the box plots.

REACTION

TABLE 2.9 Reaction Times of Students

Nonthreatening Stimulus

2.0	1.8	2.3	2.1	2.0	2.2	2.1	2.2	2.1	2.1
2.0	2.0	1.8	1.9	2.2	2.0	2.2	2.4	2.1	2.0
2.2	2.1	2.2	1.9	1.7	2.0	2.0	2.3	2.1	1.9
2.0	2.2	1.6	2.1	2.3	2.0	2.0	2.0	2.2	2.6
2.0	2.0	1.9	1.9	2.2	2.3	1.8	1.7	1.7	1.8

Threatening Stimulus

1.8	1.7	1.4	2.1	1.3	1.5	1.6	1.8	1.5	1.4
1.4	2.0	1.5	1.8	1.4	1.7	1.7	1.7	1.4	1.9
1.9	1.7	1.6	2.5	1.6	1.6	1.8	1.7	1.9	1.9
1.5	1.8	1.6	1.9	1.3	1.5	1.6	1.5	1.6	1.5
1.3	1.7	1.3	1.7	1.7	1.8	1.6	1.7	1.7	1.7

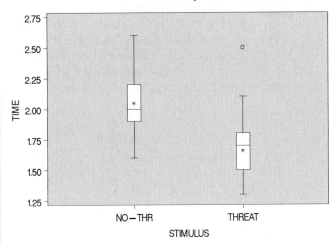

FIGURE 2.32

SAS box plots for reaction-time data

Solution In SAS, the median is represented by the horizontal line through the box, while the asterisk (*) represents the mean. Analysis of the box plots on the same numerical scale reveals that the distribution of times corresponding to the threatening stimulus lies below that of the nonthreatening stimulus. The implication is that the reaction times tend to be faster to the threatening stimulus. Note, too, that the upper whiskers of both samples are longer than the lower whiskers, indicating that the reaction times are positively skewed.

No observations in the two samples fall between the inner and outer fences. However, there is one outlier: the observation of 2.5 seconds corresponding to the threatening stimulus that is beyond the outer fence (denoted by the square symbol in SAS). When the researcher examined her notes from the experiments, she found that the subject whose time was beyond the outer fence had mistakenly been given the nonthreatening stimulus. You can see in Figure 2.32 that his time would have been within the upper whisker if moved to the box plot corresponding to the nonthreatening stimulus. The box plots should be reconstructed, since they will both change slightly when this misclassified reaction time is moved from one sample to the other.

Look Back The researcher concluded that the reactions to the threatening stimulus were faster than those to the nonthreatening stimulus. However, she was asked by her Ph.D. committee whether the results were *statistically significant*. Their question addresses the issue of whether the observed difference between the samples might be attributable to chance or sampling variation rather than to real differences between the populations. To answer this question, the researcher must use inferential statistics rather than graphical descriptions. We discuss how to compare two samples by means of inferential statistics in Chapter 9.

The next example illustrates how z-scores can be used to detect outliers and make inferences.

EXAMPLE 2.18
INFERENCE
USING z-SCORES

Problem Suppose a female bank employee believes that her salary is low as a result of sex discrimination. To substantiate her belief, she collects information on the salaries of her male counterparts in the banking business. She finds that their salaries have a mean of $54,000 and a standard deviation of $2,000. Her salary is $47,000. Does this information support her claim of sex discrimination?

Solution The analysis might proceed as follows: First, we calculate the z-score for the woman's salary with respect to those of her male counterparts. Thus,

$$z = \frac{\$47,000 - \$54,000}{\$2,000} = -3.5$$

The implication is that the woman's salary is 3.5 standard deviations *below* the mean of the male salary distribution. Furthermore, if a check of the male salary data shows that the frequency distribution is mound shaped, we can infer that very few salaries in this distribution should have a z-score less than -3, as shown in Figure 2.33. Clearly, a z-score of -3.5 represents an outlier. Either her salary is from a distribution different from the male salary distribution, or it is a very unusual (highly improbable) measurement from a salary distribution no different from the male distribution.

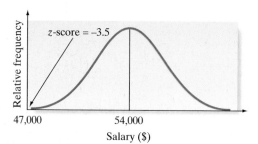

FIGURE 2.33
Male salary distribution

Look Back Which of the two situations do you think prevails? Statistical thinking would lead us to conclude that her salary does not come from the male salary distribution, lending support to the female bank employee's claim of sex discrimination. However, the careful investigator should require more information before inferring that sex discrimination is the cause. We would want to know more about the data collection technique the woman used and more about her competence at her job. Also, perhaps other factors, such as length of employment, should be considered in the analysis.

Now Work Exercise 2.123

Examples 2.17 and 2.18 exemplify an approach to statistical inference that might be called the **rare-event approach**. An experimenter hypothesizes a specific frequency distribution to describe a population of measurements. Then a sample of measurements is drawn from the population. If the experimenter finds it unlikely that the sample came from the hypothesized distribution, the hypothesis is judged to be false. Thus, in Example 2.18, the woman believes that her salary reflects discrimination. She hypothesizes that her salary should be just another measurement in the distribution of her male counterparts' salaries if no discrimination exists. However, it is so unlikely that the sample (in this case, her salary) came from the male frequency distribution that she rejects that hypothesis, concluding that the distribution from which her salary was drawn is different from the distribution for the men.

This rare-event approach to inference making is discussed further in later chapters. Proper application of the approach requires a knowledge of probability, the subject of our next chapter.

Rules of Thumb for Detecting Outliers[*]

Box Plots: Observations falling between the inner and outer fences are deemed *suspect outliers*. Observations falling beyond the outer fence are deemed *highly suspect outliers*.

z-Scores: Observations with z-scores greater than 3 in absolute value are considered *outliers*. For some highly skewed data sets, observations with z-scores greater than 2 in absolute value *may be outliers*.

Box Plots
Using the TI-84/TI-83 Graphing Calculator
Making a Box Plot

Step 1   *Enter the data.*

Press **STAT** and select **1:Edit**.
Note: If the list already contains data, clear the old data. Use the up **ARROW** to highlight 'L1'. Press **CLEAR ENTER**.
Use the **ARROW** and **ENTER** keys to enter the data set into **L1**.

Step 2   *Set up the box plot.*

Press **Y =** and **CLEAR** all functions from the Y registers.
Press **2nd Y =** for **STAT PLOT**.
Press **1** for **Plot 1**.
Set the cursor so that 'ON' is flashing and press **ENTER**.
For **TYPE**, use the right **ARROW** to scroll through the plot icons, and select the boxplot in the middle of the second row.
For **XLIST**, choose **L1**.
Set **FREQ** to **1**.

Step 3   *View the Graph*
Press **ZOOM** and select **9:ZoomStat**.

Optional Step   *Read the five-number summary.*
Press **TRACE**.
Use the left and right **ARROW** keys to move between minX, Q1, Med, Q3, and maxX.

Example   Make a box plot for the data:

86, 70, 62, 98, 73, 56, 53, 92, 86, 37, 62, 83, 78, 49, 78, 37, 67, 79, 57.

The output screen for this example is as follows:

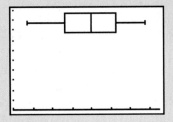

---

[*]The z-score and box plot methods both establish rule-of-thumb limits outside of which a measurement is deemed to be an outlier. Usually, the two methods produce similar results. However, the presence of one or more outliers in a data set can inflate the computed value of s. Consequently, it will be less likely that an errant observation would have a z-score larger than $|3|$. In contrast, the values of the quartiles used to calculate the intervals for a box plot are not affected by the presence of outliers.

STATISTICS IN ACTION REVISITED

Detecting Outliers in the "Eye Cue" Test Data

In the *Psychological Science* "water-level task" experiment, the quantitative variable of interest is *deviation angle* (measured in degrees) of the judged line from the true surface line (parallel to the tabletop). Are there any unusual values of this variable in the **EYECUE** data set? We will employ both the box plot and z-score methods to aid in identifying outliers in the data.

Descriptive statistics for the deviation angles of all 120 experimental subjects, produced with MINITAB, are shown in Figure SIA2.7. To employ the z-score method, we need the mean and standard deviation. These values are highlighted in the figure. Then the three-standard-deviation interval is

$$4.742 \pm 3(7.865) = 4.742 \pm 23.595 = (-18.853, 28.337)$$

If you examine the deviation angles in the **EYECUE** data set, you find that only one of these values, 35,

falls beyond the three-standard-deviation interval. Thus, the data for the subject whose judged line had a deviation angle of 35° (z-score = 3.85) is considered a highly suspect outlier, according to the z-score approach.

A box plot for the data is shown in Figure SIA2.8. Although several suspect outliers (asterisks) are shown on the box plot, there are no highly suspect outliers (zeros) shown. That is, no data points fall beyond the outer fences. Thus, the box plot method does not identify the value of 35 as a highly suspect outlier, only as a suspect outlier.

Before any type of inference is made concerning the population of deviation angles for subjects performing the water-level task, we should consider whether this potential outlier is a legitimate observation (in which case it will remain in the data set for analysis) or is associated with a subject that is not a member of the population of interest (in which case it will be removed from the data set).

Descriptive Statistics: Deviation

Variable	N	Mean	StDev	Minimum	Q1	Median	Q3	Maximum	IQR
Deviation	120	4.742	7.865	-18.000	0.000	4.000	9.000	35.000	9.000

FIGURE SIA2.7

MINITAB descriptive statistics for deviation angle of judged water level

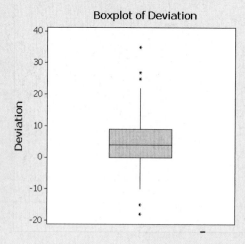

FIGURE SIA2.8

MINITAB box plot for deviation angle of judged water level

Exercises 2.118–2.135

Understanding the Principles

2.118 Define an outlier.

2.119 Define the 25th, 50th, and 75th percentiles of a data set.

2.120 What is the interquartile range?

2.121 What are the hinges of a box plot?

2.122 With mound-shaped data, what proportion of the measurements have z-scores between −3 and 3?

Learning the Mechanics

2.123 A sample data set has a mean of 57 and a standard deviation of 11. Determine whether each of the following sample measurements is an outlier.
NW
a. 65
b. 21
c. 72
d. 98

2.124 Suppose a data set consisting of exam scores has a lower quartile $Q_L = 60$, a median $M = 75$, and an upper quartile $Q_U = 85$. The scores on the exam range from 18 to 100. Without having the actual scores available to you, construct as much of the box plot as possible.

2.125 Consider the following horizontal box plot:

NW

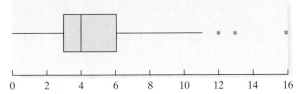

a. What is the median of the data set (approximately)?
b. What are the upper and lower quartiles of the data set (approximately)?
c. What is the interquartile range of the data set (approximately)?
d. Is the data set skewed to the left, skewed to the right, or symmetric?
e. What percentage of the measurements in the data set lie to the right of the median? To the left of the upper quartile?
f. Identify any outliers in the data.

2.126 Consider the following two sample data sets:

LM2_126

Sample A			Sample B		
121	171	158	171	152	170
173	184	163	168	169	171
157	85	145	190	183	185
165	172	196	140	173	206
170	159	172	172	174	169
161	187	100	199	151	180
142	166	171	167	170	188

a. Construct a box plot for each data set.
b. Identify any outliers that may exist in the two data sets.

APPLET Applet Exercise 2.7

Use the applet *Standard Deviation* to determine whether an item in a data set may be an outlier. Begin by setting appropriate limits and plotting the given data on the number line provided in the applet. Here is the data set:

10 80 80 85 85 85 85 90 90 90 90 90 95 95 95 95 100 100

a. The green arrow shows the approximate location of the mean. Multiply the standard deviation given by the applet by 3. Is the data item 10 more than three standard deviations away from the green arrow (the mean)? Can you conclude that the 10 is an outlier?
b. Using the mean and standard deviation from part **a**, move the point at 10 on your plot to a point that appears to be about three standard deviations from the mean. Repeat the process in part **a** for the new plot and the new suspected outlier.
c. When you replaced the extreme value in part **a** with a number that appeared to be within three standard deviations of the mean, the standard deviation got smaller and the mean moved to the right, yielding a new data set whose extreme value was *not* within three standard deviations of the mean. Continue to replace the extreme value

with higher numbers until the new value is within three standard deviations of the mean in the new data set. Use trial and error to estimate the smallest number that can replace the 10 in the original data set so that the replacement is not considered to be an outlier.

Applying the Concepts—Basic

2.127 **Treating psoriasis with the "Doctorfish of Kangal."** Psoriasis is a skin disorder with no known cure and no proven effective pharmacological treatment. An alternative treatment for psoriasis is ichthyotherapy, also known as therapy with the "Doctorfish of Kangal." Fish from the hot pools of Kangal, Turkey, feed on the skin scales of bathers, reportedly reducing the symptoms of psoriasis. In one study, 67 patients diagnosed with psoriasis underwent three weeks of ichthyotherapy. (*Evidence-Based Research in Complementary and Alternative Medicine*, Dec. 2006.) The Psoriasis Area Severity Index (PASI) of each patient was measured both before and after treatment. (The lower the PASI score, the better is the skin condition.) Box plots of the PASI scores, both before (baseline) and after three weeks of ichthyotherapy treatment, are shown in the accompanying diagram.
a. Find the approximate 25th percentile, the median, and the 75th percentile for the PASI scores before treatment.
b. Find the approximate 25th percentile, the median, and the 75th percentile for the PASI scores after treatment.
c. Comment on the effectiveness of ichthyotherapy in treating psoriasis.

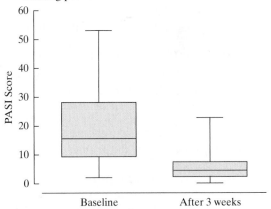

Source: Grassberger, M. and Hoch, W. "Ichthyotherapy as alternative treatment for patients with psoriasis: A pilot study," *Evidence-Based Research in Complementary and Alternative Medicine*, Vol. 3, No. 4, Dec. 2004 (Figure 3).

2.128 **Dentists' use of anesthetics.** Refer to the *Current Allergy & Clinical Immunology* (Mar. 2004) study of the use of local anesthetics in dentistry, presented in Exercise 2.95 (p. 73). Recall that the mean number of units (ampoules) of local anesthetics used per week by dentists was 79, with a standard deviation of 23. Consider a dentist who used 175 units of local anesthetics in a week.
a. Find the z-score for this measurement.
b. Would you consider the measurement to be an outlier? Explain.
c. Give several reasons the outlier may have occurred.

PMI

2.129 **Research on brain specimens.** Refer to the *Brain and Language* data on postmortem intervals (PMIs) of 22 human brain specimens, presented in Exercise 2.37 (p. 48). The mean and standard deviation of the PMI values are 7.3 and 3.18, respectively.

a. Find the z-score for the PMI value of 3.3.

b. Is the PMI value of 3.3 considered an outlier? Explain.

2.130 **Estimating the age of glacial drifts.** Tills are glacial drifts consisting of a mixture of clay, sand, gravel, and boulders. Engineers from the University of Washington's Department of Earth and Space Sciences studied the chemical makeup of buried tills in order to estimate the age of the glacial drifts in Wisconsin. (*American Journal of Science*, Jan. 2005.) The ratio of the elements aluminum (AI) and beryllium (Be) in sediment is related to the duration of burial. The Al–Be ratios for a sample of 26 buried till specimens are given in the accompanying table. An SPSS printout with descriptive statistics for the Al–Be ratio is shown at the bottom of the page.

TILLRATIO

3.75 4.05 3.81 3.23 3.13 3.30 3.21 3.32 4.09 3.90 5.06 3.85 3.88
4.06 4.56 3.60 3.27 4.09 3.38 3.37 2.73 2.95 2.25 2.73 2.55 3.06

Source: Adapted from *American Journal of Science*, Vol. 305, No. 1, Jan. 2005, p. 16 (Table 2).

a. Find and interpret the z-score associated with the largest ratio, the smallest ratio, and the mean ratio.

b. Would you consider the largest ratio to be unusually large? Why or why not?

c. Construct a box plot for the data and identify any outliers.

Applying the Concepts—Intermediate

2.131 **Speech listening study.** Refer to the *American Journal of Speech–Language Pathology* (Feb. 1995) study, presented in Exercise 2.41 (p. 49). Recall that three groups of college students listened to an audiotape of a woman with imperfect speech. The groups were called *control, treatment,* and *familiarity.* At the end of the session, each student transcribed the spoken words.

a. Construct box plots for the percentage of words correctly transcribed (i.e., the accuracy rate) for each group.

b. How do the median accuracy rates compare for the three groups?

c. How do the variabilities of the accuracy rates compare for the three groups?

d. The standard deviations of the accuracy rates are 3.56 for the control group, 5.45 for the treatment group, and 4.46 for the familiarity group. Do the standard deviations agree with the interquartile ranges (part **c**) with regard to the comparison of the variabilities of the accuracy rates?

e. Is there evidence of outliers in any of the three distributions?

WPOWER50

2.132 **Most powerful women in America.** Refer to the *Fortune* (Nov. 14, 2005) ranking of the 50 most powerful women in America, presented in Exercise 2.92 (p. 73). Use side-by-side box plots to compare the ages of the women in three groups, based on their position within the firm: Group 1 (CEO/ Chairman, CEO/president, or CFO/president); Group 2 (CEO, DFO, CFO/EVP, CIO/EVP, chairman, COO, CRO, or president); Group 3 (EVP, executive, founder, SVP, treasurer, or vice chair). Do you detect outliers?

SHIPSANIT

2.133 **Sanitation inspection of cruise ships.** Refer to the data on sanitation levels of cruise ships, presented in Exercise 2.93 (p. 73).

a. Use the box plot method to detect any outliers in the data.

b. Use the z-score method to detect any outliers in the data.

c. Do the two methods agree? If not, explain why.

SATSCORES

2.134 **Comparing SAT Scores.** Refer to Exercise 2.40 (p. 49), in which we compared state average SAT scores in 1990 and 2005. The data are saved in the **SATSCORES** file.

a. Construct side-by-side box plots of the SAT scores for the two years.

b. Compare the variability of the SAT scores for the two years.

c. Are any state SAT scores outliers in either year? If so, identify them.

Applying the Concepts—Advanced

2.135 **Library book checkouts.** A city librarian claims that books have been checked out an average of seven (or more) times in the last year. You suspect he has exaggerated the checkout rate (book usage) and that the mean number of checkouts per book per year is, in fact, less than seven. Using the computerized card catalog, you randomly select one book and find that it has been checked out four times in the last year. Assume that the standard deviation of the number of checkouts per book per year is approximately 1.

a. If the mean number of checkouts per book per year really is 7, what is the z-score corresponding to four?

b. Considering your answer to part **a**, do you have reason to believe that the librarian's claim is incorrect?

c. If you knew that the distribution of the number of checkouts was mound shaped, would your answer to part **b** change? Explain.

d. If the standard deviation of the number of checkouts per book per year were 2 (instead of 1), would your answers to parts **b** and **c** change? Explain.

SPSS Output for Exercise 2.130

Descriptive Statistics

	N	Minimum	Maximum	Mean	Std. Deviation
RATIO	26	2.25	5.06	3.5069	.63439
Valid N (listwise)	26				

2.9 Graphing Bivariate Relationships (Optional)

The claim is often made that the crime rate and the unemployment rate are "highly correlated." Another popular belief is that smoking and lung cancer are "related." Some people even believe that the Dow Jones Industrial Average and the lengths of fashionable skirts are "associated." The words *correlated*, *related*, and *associated* imply a relationship between two variables—in the examples just mentioned, two *quantitative* variables.

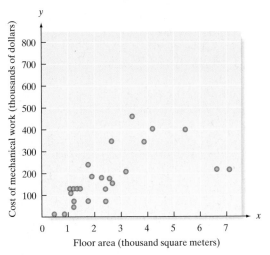

FIGURE 2.34

Scattergram of cost vs. floor area

One way to describe the relationship between two quantitative variables—called a **bivariate relationship**—is to plot the data in a **scattergram** (or **scatterplot**). A scattergram is a two-dimensional plot, with one variable's values plotted along the vertical axis and the others along the horizontal axis. For example, Figure 2.34 is a scattergram relating (1) the cost of mechanical work (heating, ventilating, and plumbing) to (2) the floor area of the building, for a sample of 26 factory and warehouse buildings. Note that the scattergram suggests a general tendency for mechanical cost to increase as building floor area increases.

When an increase in one variable is generally associated with an increase in the second variable, we say that the two variables are "positively related" or "positively correlated."[*] Figure 2.34 implies that mechanical cost and floor area are positively correlated. Alternatively, if one variable has a tendency to decrease as the other increases, we say the variables are "negatively correlated." Figure 2.35 shows several hypothetical scattergrams that portray a positive bivariate relationship (Figure 2.35a), a negative bivariate relationship (Figure 2.35b), and a situation in which the two variables are unrelated (Figure 2.35c).

FIGURE 2.35
Hypothetical bivariate relationships

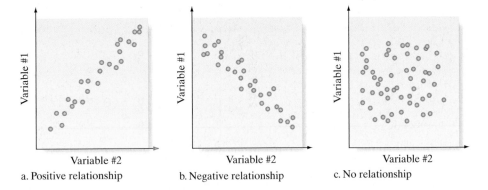

a. Positive relationship   b. Negative relationship   c. No relationship

EXAMPLE 2.19
GRAPHING BIVARIATE DATA

Problem A medical item used to administer to a hospital patient is called a *factor*. For example, factors can be intravenous (IV) tubing, IV fluid, needles, shave kits, bedpans, diapers, dressings, medications, and even code carts. The coronary care unit at Bayonet Point Hospital (St. Petersburg, Florida) recently investigated the relationship between the number of factors administered per patient and the patient's length of stay (in days). Data on these two variables for a sample of 50 coronary care patients are given in Table 2.10. Use a scattergram to describe the relationship between the two variables of interest: number of factors and length of stay.

[*]A formal definition of correlation is given in Chapter 9. There, we will learn that correlation measures the strength of the linear (or straight-line) relationship between two quantitative variables.

MEDFACTORS

TABLE 2.10 Medfactors Data on Patients' Factors and Length of Stay

Number of Factors	Length of Stay (days)	Number of Factors	Length of Stay (days)
231	9	354	11
323	7	142	7
113	8	286	9
208	5	341	10
162	4	201	5
117	4	158	11
159	6	243	6
169	9	156	6
55	6	184	7
77	3	115	4
103	4	202	6
147	6	206	5
230	6	360	6
78	3	84	3
525	9	331	9
121	7	302	7
248	5	60	2
233	8	110	2
260	4	131	5
224	7	364	4
472	12	180	7
220	8	134	6
383	6	401	15
301	9	155	4
262	7	338	8

Source: Bayonet Point Hospital, Coronary Care Unit.

Solution Rather than construct the plot by hand, we resort to a statistical software package. The SPSS plot of the data in Table 2.10, with length of stay (LOS) on the horizontal axis and number of factors (FACTORS) on the vertical axis, is shown in Figure 2.36.

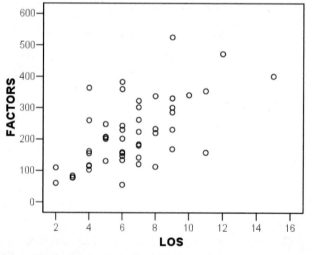

FIGURE 2.36
SPSS scatterplot of data in Table 2.10

Although the plotted points exhibit a fair amount of variation, the scattergram clearly shows an increasing trend. It appears that a patient's length of stay is positively correlated with the number of factors administered to the patient.

Look Back If hospital administrators can be confident that the sample trend shown in Figure 2.36 accurately describes the trend in the population, then they may use this information to improve their forecasts of lengths of stay for future patients.

Now Work Exercise 2.139

The scattergram is a simple, but powerful, tool for describing a bivariate relationship. However, keep in mind that it is only a graph. No measure of reliability can be attached to inferences made about bivariate populations based on scattergrams of sample data. The statistical tools that enable us to make inferences about bivariate relationships are presented in Chapter 11.

Scatterplots
Using the TI-84/TI-83 Graphing Calculator
Making Scatterplots

Step 1 *Enter the data.*
Press **STAT** and select **1:Edit**.
Note: If a list already contains data, clear the old data. Use the up arrow to highlight the list name, 'L1' or 'L2'.
Press **CLEAR ENTER**.
Enter your x-data in L1 and your y-data in L2.

Step 2 *Set up the scatterplot.*
Press **2nd** and **Y =** for **STAT PLOT**.
Press **1** for Plot 1.
Set the cursor so that **ON** is flashing.
For **Type**, use the arrow and Enter keys to highlight and select the scatterplot (first icon in the first row).
For **Xlist**, choose the column containing the x-data.
For **Ylist**, choose the column containing the y-data.

Step 3 *View the scatterplot.*
Press **ZOOM 9** for ZoomStat.

Example The accompanying figures show a table of data entered on the TI-83 and the scatterplot of the data obtained by following the preceding steps.

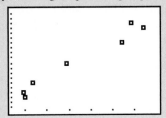

Exercises 2.136–2.150

Understanding the Principles

2.136 For what types of variables, quantitative or qualitative, are scatterplots useful?

2.137 Define a bivariate relationship.

2.138 What is the difference between positive association and negative association as it pertains to the relationship between two variables?

Learning the Mechanics

2.139 Construct a scatterplot for the data in the table that fol-
NW lows. Do you detect a trend?

Variable #1:	5	3	−1	2	7	6	4	0	8
Variable #2:	14	3	10	1	8	5	3	2	12

2.140 Construct a scatterplot for the data in the table that follows. Do you detect a trend?

Variable #1:	.5	1	1.5	2	2.5	3	3.5	4	4.5	5
Variable #2:	2	1	3	4	6	10	9	12	17	17

Applying the Concepts—Basic

2.141 Baseball batting averages versus wins. Baseball wisdom says, "If you can't hit, you can't win." Is the number of games won by a major league baseball team in a season related to the team's batting average? The information in the table, found in the *Baseball Almanac* (2007), shows the number of games won and the batting averages for the 14 teams in the American League for the 2006 Major League Baseball season. Construct a scatterplot of the data. Do you observe a trend?

⊙ ALWINS

Team	Games Won	Batting Ave.
New York	97	.285
Toronto	87	.284
Baltimore	70	.277
Boston	86	.269
Tampa Bay	61	.255
Cleveland	78	.280
Detroit	95	.274
Chicago	90	.280
Kansas City	62	.271
Minnesota	96	.287
Los Angeles	89	.274
Texas	80	.278
Seattle	78	.272
Oakland	93	.260

Source: Baseball Almanac, 2007; www.mlb.com.

2.142 Feeding behavior of fish. Zoologists at the University of Western Australia conducted a study of the feeding behavior of black bream (a type of fish) spawned in aquariums. (*Brain and Behavior Evolution*, April 2000.) In one experiment, the zoologists recorded the number of aggressive strikes of two black bream feeding at the bottom of the aquarium in the 10-minute period following the addition of food. The number of strikes and age of the fish (in days) were recorded approximately each week for nine weeks, as shown in the table.

⊙ BLACKBREAM

Week	Number of Strikes	Age of Fish (days)
1	85	120
2	63	136
3	34	150
4	39	155
5	58	162
6	35	169
7	57	178
8	12	184
9	15	190

Source: Shand, J. et al. "Variability in the location of the retinal ganglion cell area centralis is correlated with ontogenetic changes in feeding behavior in the Blackbream, Acanthopagrus 'butcher'," Brain and Behavior, Vol. 55, No. 4, April 2000 (Figure H).

a. Construct a scattergram of the data, with number of strikes on the *y*-axis and age of the fish on the *x*-axis.

b. Examine the scattergram of part **a.** Do you detect a trend?

2.143 Walking study. A "self-avoiding walk" describes a path in which you never retrace your steps or cross your own path. An "unrooted walk" is a path in which it is impossible to distinguish between the starting point and ending point of the path. The *American Scientist* (July–Aug. 1998) investigated the relationship between self-avoiding and unrooted walks. The table gives the number of unrooted walks and possible number of self-avoiding walks of various lengths, where length is measured as number of steps.

a. Construct a plot to investigate the relationship between total possible number of self-avoiding walks and walk length. What pattern (if any) do you observe?

b. Repeat part **a** for unrooted walks.

⊙ WALK

Walk Length (Number of Steps)	Unrooted Walks	Self-Avoiding Walks
1	1	4
2	2	12
3	4	36
4	9	100
5	22	284
6	56	780
7	147	2,172
8	388	5,916

Source: Hayes, B. "How to avoid yourself." American Scientist, Vol. 86, No. 4, July–Aug. 1998, p. 317 (Figure 5).

2.144 Are geography journals worth their cost? In *Geoforum* (Vol. 37, 2006), Simon Fraser University professor Nicholas Blomley assessed whether the price of a geography journal is correlated with quality. He collected pricing data (cost for a 1-year subscription, in US dollars) for a sample of 28 geography journals. In addition to cost, three other variables were measured: journal impact factor (JIF), defined as the average number of times articles from the journal have been cited; number of citations for a journal over the past five years; and relative price index (RPI), a measure developed by economists. [*Note*: A journal with an RPI less than 1.25 is considered a "good value."] The data for the 28 geography journals are saved in the **GEOJRNL** file. (Selected observations are shown in the accompanying table.)

a. Construct a scatterplot for the variables JIF and cost. Do you detect a trend?

b. Construct a scatterplot for the variables number of cites and cost. Do you detect a trend?

c. Construct a scatterplot for the variables RPI and cost. Do you detect a trend?

⊙ GEOJRNL (Selected observations)

Journal	Cost ($)	JIF	Cites	RPI
J. Econ. Geogr.	468	3.139	207	1.16
Prog. Hum. Geog.	624	2.943	544	0.77
T.I. Brit. Geogr.	499	2.388	249	1.11
Econ. Geogr.	90	2.325	173	0.30
A.A.A. Geogr.	698	2.115	377	0.93
⋮	⋮	⋮	⋮	⋮
Geogr. Anal.	213	0.902	106	0.88
Geogr. J.	223	0.857	81	0.94
Appl. Geogr.	646	0.853	74	3.38

Source: Blomley, N. "Is this journal worth US$1118?" Geoforum, Vol. 37, 2006.

Applying the Concepts—Intermediate

SITIN

2.145 College protests of labor exploitation. Refer to the *Journal of World-Systems Research* (Winter 2004) study of 14 student sit-ins for a "sweat-free campus" at universities in 1999 and 2000, presented in Exercise 2.35 (p. 47). The SITIN file contains data on the duration (in days) of each sit-in, as well as data on the number of student arrests.

    a. Use a scatterplot to graph the relationship between duration and number of arrests. Do you detect a trend?

    b. Repeat part **a**, but graph only the data for sit-ins in which there was at least one arrest. Do you detect a trend?

    c. Comment on the reliability of the trend you detected in part **b**.

2.146 Short-term memory study. Research published in the *American Journal of Psychiatry* (July 1995) attempted to establish a link between hippocampal (brain) volume and short-term verbal memory of patients with combat-related post-traumatic stress disorder (PTSD). A sample of 21 Vietnam veterans with a history of combat-related PTSD participated in the study. Magnetic resonance imaging was used to measure the volume of the right hippocampus (in cubic millimeters) of each subject. Verbal memory retention of each subject was measured by the percent retention subscale of the Wechsler Memory Scale. The data for the 21 patients are plotted in the accompanying scattergram. The researchers "hypothesized that smaller hippocampal volume would be associated with deficits in short-term verbal memory in patients with PTSD." Does the scattergram provide visual evidence to support this theory?

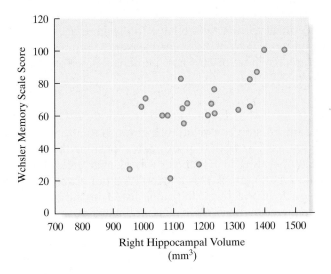

GOBIANTS

2.147 Mongolian desert ants. Refer to the *Journal of Biogeography* (Dec. 2003) study of ants in Mongolia, presented in Exercise 2.66 (p. 60). Data on annual rainfall, maximum daily temperature, percentage of plant cover, number of ant species, and species diversity index recorded at each of 11 study sites are saved in the **GOBIANTS** file.

    a. Construct a scatterplot to investigate the relationship between annual rainfall and maximum daily temperature. What type of trend (if any) do you detect?

    b. Use scatterplots to investigate the relationship that annual rainfall has with each of the other four variables in the data set. Are the other variables positively or negatively related to rainfall?

2.148 Forest fragmentation study. Ecologists classify the cause of forest fragmentation as either anthropogenic (i.e., due to human development activities, such as road construction or logging) or natural in origin (e.g., due to wetlands or wildfire). *Conservation Ecology* (Dec. 2003) published an article on the causes of fragmentation for 54 South American forests. Using advanced high-resolution satellite imagery, the researchers developed two fragmentation indexes for each forest—one for anthropogenic fragmentation and one for fragmentation from natural causes. The values of these two indexes (where higher values indicate more fragmentation) for five of the forests in the sample are shown in the accompanying table. The data for all 54 forests are saved in the **FORFRAG** file.

FORFRAG (Selected observations)

Ecoregion (forest)	Anthropogenic Index	Natural Origin Index
Araucaria moist forests	34.09	30.08
Atlantic Coast *restingas*	40.87	27.60
Bahia coastal forests	44.75	28.16
Bahia interior forests	37.58	27.44
Bolivian *Yungas*	12.40	16.75

Source: Wade, T. G., et al. "Distribution and causes of global forest fragmentation," *Conservation Ecology*, Vol. 72, No. 2, Dec. 2003 (Table 6).

    a. Ecologists theorize that an approximately linear (straight-line) relationship exists between the two fragmentation indexes. Graph the data for all 54 forests. Does the graph support the theory?

    b. Delete the data for the three forests with the largest anthropogenic indexes, and reconstruct the graph of part **a**. Comment on the ecologists' theory.

Applying the Concepts—Advanced

PGADRIVER

2.149 Ranking driving performance of professional golfers. Refers to *The Sport Journal* (Winter, 2007) analysis of a new method for ranking the total driving performance of golfers on the PGA tour, Exercise 2.60 (p. 59). Recall that the method uses both the average driving distance (in yards) and the driving accuracy (percent of drives that land in the fairway). Data on these two variables for the top 40 PGA golfers are saved in the **PGADRIVER** file. A professional golfer is practicing a new swing to increase his average driving distance. However, he is concerned that his driving accuracy will be lower. Is his concern reasonable? Explain.

2.150 **Spreading rate of spilled liquid.** A contract engineer at DuPont Corp. studied the rate at which a spilled volatile liquid will spread across a surface. (*Chemical Engineering Progress*, Jan. 2005.). Suppose that 50 gallons of methanol spills onto a level surface outdoors. The engineer uses derived empirical formulas (assuming a state of turbulence-free convection) to calculate the mass (in pounds) of the spill after a period ranging from 0 to 60 minutes. The calculated mass values are given in the accompanying table. Is there evidence to indicate that the mass of the spill tends to diminish as time increases?

LIQUIDSPILL

Time (minutes)	Mass (pounds)
0	6.64
1	6.34
2	6.04
4	5.47
6	4.94
8	4.44
10	3.98
12	3.55
14	3.15
16	2.79
18	2.45
20	2.14
22	1.86
24	1.60
26	1.37
28	1.17
30	0.98
35	0.60
40	0.34
45	0.17
50	0.06
55	0.02
60	0.00

Source: Barry, J. "Estimating rates of spreading and evaporation of volatile liquids." *Chemical Engineering Progress,* Vol. 101, No. 1. Jan. 2005.

2.10 Distorting the Truth with Descriptive Statistics

A picture may be "worth a thousand words," but pictures can also color messages or distort them. In fact, the pictures displayed in statistics—histograms, bar charts, and other graphical images—are susceptible to distortion, so we have to examine each of them with care. Accordingly, we begin this section by mentioning a few of the pitfalls to watch for in interpreting a chart or a graph. Then we discuss how numerical descriptive statistics can be used to distort the truth.

Graphical Distortions

One common way to change the impression conveyed by a graph is to alter the scale on the vertical axis, the horizontal axis, or both. For example, consider the data on collisions of large marine vessels operating in European waters over a certain five-year period, summarized in Table 2.11. Figure 2.37 is a MINITAB bar graph showing the frequency of collisions for each of the three locations listed in the table. The graph shows that in-port collisions occur more often than collisions at sea or collisions in restricted waters.

 COLLISION

TABLE 2.11   Collisions of Marine Vessels by Location

Location	Number of Ships
At Sea	376
In Restricted Waters	273
In Port	478
TOTAL	1,127

Source: The Dock and Harbour Authority.

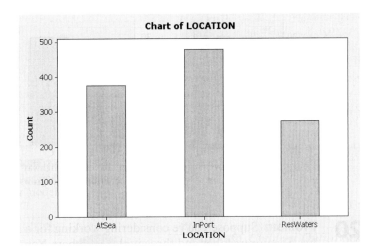

FIGURE 2.37

MINITAB bar graph of vessel collisions by location

Now, suppose you want to use the same data to exaggerate the difference between the number of in-port collisions and the number of collisions in restricted waters. One way to do this is to increase the distance between successive units on the vertical axis—that is, *stretch* the vertical axis by graphing only a few units per inch. A telltale sign of stretching is a long vertical axis, but this indication is often hidden by starting the vertical axis at some point above the origin, 0. Such a graph is shown in the SPSS printout in Figure 2.38. By starting the bar chart at 250 collisions (instead of 0), it appears that the frequency of in-port collisions is many times greater than the frequency of collisions in restricted waters.

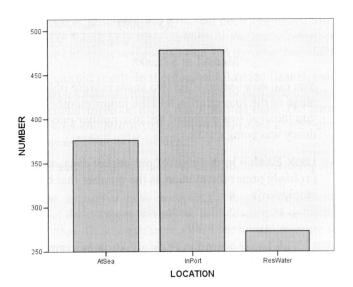

FIGURE 2.38

SPSS bar graph of vessel collisions by location, with adjusted vertical axis

Another method of achieving visual distortion with bar graphs is by making the width of the bars proportional to their height. For example, look at the bar chart in Figure 2.39a, which depicts the percentage of the total number of motor vehicle deaths in a year that occurred on each of four major highways. Now suppose we make both the width and the height grow as the percentage of fatal accidents grows. This change is shown in Figure 2.39b. The distortion is that the reader may tend to equate the *area* of the bars with the percentage of deaths occurring at each highway when, in fact, the true relative frequency of fatal accidents is proportional only to the *height* of the bars.

Although we've discussed only a few of the ways that graphs can be used to convey misleading pictures of phenomena, the lesson is clear: Look at all graphical descriptions of data with a critical eye. In particular, check the axes and the size of the units on each axis. Ignore the visual changes, and concentrate on the actual numerical changes indicated by the graph or chart.

The information in a data set can also be distorted by using numerical descriptive measures, as Example 2.20 shows.

Percentile   75

Pie chart   30

Quartiles*   79

Range   62

Rare-event approach*   84

Scattergram*   89

Scatterplot*   89

Skewness   54

Spread   62

Standard deviation   63

Stem-and-leaf display   39

Symmetric distribution   68

Upper quartile*   79

Variability   51

Variance   63

Whiskers*   80

z-score   76

CHAPTER NOTES

Describing QUALITATIVE Data

1. Identify category classes
2. Determine class frequencies
3. Class relative frequency = (class frequency)/*n*
4. Graph relative frequencies

Pie Chart:

Bar Graph:

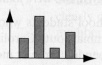

Pareto Diagram:

Graphing QUANTITATIVE Data

1. Identify class intervals
2. Determine class interval frequencies
3. Class interval relative frequency = (class interval frequency)/*n*
4. Graph class interval relative frequencies

Dot Plot:

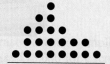

Stem-and-Leaf Display:

1	3
2	2489
3	126678
4	37
5	2

Histogram:

Box plot:

Scatterplot:

Key Symbols

	Sample	Population
Mean:	$\bar{x}$	μ
Variance:	s^2	σ^2
Std. Dev.	s	σ
Median:	M	
Lower Quartile:	Q_L	
Upper Quartile:	Q_U	
Interquartile Range:	IQR	

GUIDE TO SELECTING THE DATA DESCRIPTION METHOD

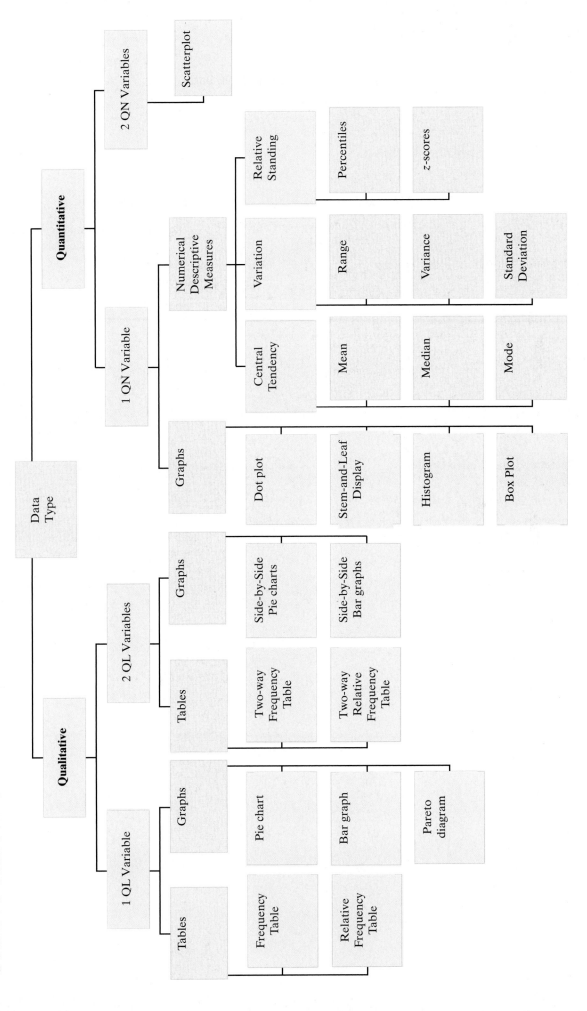

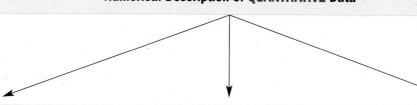

Central Tendency

Mean:

$$\bar{x} = (\Sigma x_i)/n$$

Median: Middle value when data ranked in order

Mode: Value that occurs most often

Variation

Range: Difference between largest and smallest value

Variance:

$$s^2 = \frac{\Sigma(x_i - \bar{x})^2}{n-1} = \frac{\Sigma x_i^2 - \frac{(\Sigma x_i)^2}{n}}{n-1}$$

Std Dev.: $s = \sqrt{s^2}$

Interquartile Range:

$$IQR = Q_U - Q_L$$

Relative Standing

Percentile Score: Percentage of values that fall below *x*-score

z-score: $z = (x - \bar{x})/s$

Rules for Detecting Quantitative Outliers

Interval	Chebyshev's Rule	Empirical Rule
$\bar{x} \pm s$	At least 0%	$\approx 68\%$
$\bar{x} \pm 2s$	At least 75%	$\approx 95\%$
$\bar{x} \pm 3s$	At least 89%	$\approx$ All

Rules for Detecting Quantitative Outliers

Method	Suspect	Highly Suspect
Box plot:	Values between inner & outer fences	Values beyond outer fences
z-score:	$2 < |z| < 3$	$|z| > 3$

SUPPLEMENTARY EXERCISES 2.151–2.192

Note: Starred (*) exercises refer to the optional sections in this chapter.

Understanding the Principles

2.151 Discuss conditions under which the median is preferred to the mean as a measure of central tendency.

2.152 Explain why we generally prefer the standard deviation to the range as a measure of variation.

2.153 Give a situation in which we would prefer using a stem-and-leaf display over a histogram in describing quantitative data graphically.

2.154 Give a situation in which we would prefer using a box plot over z-scores to detect an outlier.

2.155 Give a technique that is used to distort information shown on a graph.

Learning the Mechanics

2.156 Construct a relative frequency histogram for the data summarized in the following table:

Measurement Class	Relative Frequency	Measurement Class	Relative Frequency
.00–.75	.02	5.25–6.00	.15
.75–1.50	.01	6.00–6.75	.12
1.50–2.25	.03	6.75–7.50	.09
2.25–3.00	.05	7.50–8.25	.05
3.00–3.75	.10	8.25–9.00	.04
3.75–4.50	.14	9.00–9.75	.01
4.50–5.25	.19		

2.157 Consider the following three measurements: 50, 70, 80. Find the z-score for each measurement if they are from a population with a mean and standard deviation equal to
 a. $\mu = 60, \sigma = 10$
 b. $\mu = 60, \sigma = 5$
 c. $\mu = 40, \sigma = 10$
 d. $\mu = 40, \sigma = 100$

*2.158 Refer to Exercise 2.157. For parts **a–d**, determine whether the three measurements 50, 70, and 80 are outliers.

2.159 Compute s^2 for data sets with the following characteristics:
 a. $\sum_{i=1}^{n} x_i^2 = 246, \sum_{i=1}^{n} x_i = 63, n = 22$
 b. $\sum_{i=1}^{n} x_i^2 = 666, \sum_{i=1}^{n} x_i = 106, n = 25$
 c. $\sum_{i=1}^{n} x_i^2 = 76, \sum_{i=1}^{n} x_i = 11, n = 7$

2.160 If the range of a set of data is 20, find a rough approximation to the standard deviation of the data set.

2.161 For each of the data sets in parts **a–c**, compute and $\bar{x}, s^2$, and s. If appropriate, specify the units in which your answers are expressed.
 a. 4, 6, 6, 5, 6, 7
 b. −\$1, \$4, −\$3, \$0, −\$3, −\$6
 c. 3/5%, 4/5%, 2/5%, 1/5%, 1/16%
 d. Calculate the range of each data set in parts **a–c**.

2.162 For each of the data sets in parts **a–d**, $\bar{x}$, compute s^2, and s.
 a. 13, 1, 10, 3, 3
 b. 13, 6, 6, 0

 c. 1, 0, 1, 10, 11, 11, 15
 d. 3, 3, 3, 3
 e. For each of the data sets in parts **a–d**, form the interval $\bar{x} \pm 2s$ and calculate the percentage of the measurements that fall into that interval.

*2.163 Construct a scatterplot for the data listed here. Do you detect any trends?

Variable #1:	174	268	345	119	400	520	190	448	307	252
Variable #2:	8	10	15	7	22	31	15	20	11	9

2.164 The data sets in parts **a–c** have been invented to demonstrate that the lower bounds given by Chebyshev's rule are appropriate. Notice that the data are contrived and would not be encountered in a real-life problem.
 a. Consider a data set that contains ten 0s, two 1s, and ten 2s. Calculate $\bar{x}, s^2$, and s. What percentage of the measurements are in the interval $\bar{x} \pm s$? Compare this result with that obtained from Chebyshev's rule.

 b. Consider a data set that contains five 0s, thirty-two 1s, and five 2s. Calculate $\bar{x}, s^2$, and s. What percentage of the measurements are in the interval $\bar{x} \pm 2s$? Compare this result with that obtained from Chebyshev's rule.

 c. Consider a data set that contains three 0s, fifty 1s, and three 2s. Calculate $\bar{x}, s^2$, and s. What percentage of the measurements are in the interval $\bar{x} \pm 3s$? Compare this result with that obtained from Chebyshev's rule.

 d. Draw a histogram for each of the data sets in parts **a**, **b**, and **c**. What do you conclude from these graphs and the answers to parts **a**, **b**, and **c**?

Applying the Concepts—Basic

2.165 **Types of cancer treated.** The Moffitt Cancer Center at the University of South Florida treats over 25,000 patients a year. The accompanying graphic describes the types of cancer treated in Moffitt's patients.
 a. What type of graph is portrayed?
 b. Which type of cancer is treated most often at Moffitt?

 c. What percentage of Moffitt's patients are treated for melanoma, lymphoma, or leukemia?

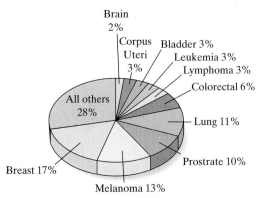

Source: Today's Tomorrows, Annual Report, H. Lee Moffitt Cancer Center & Research Institute, Winter 2002–2003.

MINITAB Output for Exercise 2.167   **Descriptive Statistics: DRIVHEAD**

```
Variable   N    Mean   StDev  Minimum    Q1   Median     Q3  Maximum
DRIVHEAD  98   603.7   185.4    216.0  475.0   605.0  724.3   1240.0
```

CRASH

2.166 Crash tests on new cars. Each year, the National Highway Traffic Safety Administration (NHTSA) crash tests new car models to determine how well they protect the driver and front-seat passenger in a head-on collision. The NHTSA has developed a "star" scoring system for the frontal crash test, with results ranging from one star (*) to five stars (*****). The more stars in the rating, the better the level of crash protection in a head-on collision. The NHTSA crash test results for 98 cars in a recent model year are stored in the data file named **CRASH**. The driver-side star ratings for the 98 cars are summarized in the MINITAB printout below. Use the information in the printout to form a pie chart. Interpret the graph.

Tally for Discrete Variables: DRIVSTAR

DRIVSTAR	Count	Percent
2	4	4.08
3	17	17.35
4	59	60.20
5	18	18.37
N=	98	

2.167 Crash tests on new cars (cont'd). Refer to Exercise 2.166 and the NHTSA crash test data. One quantitative variable recorded by the NHTSA is driver's severity of head injury (measured on a scale from 0 to 1,500). Numerical descriptive statistics for the 98 driver head-injury ratings in the **CRASH** file are displayed in the MINITAB printout at the top of the page.

a. Interpret each of the statistics shown on the printout.
b. Find the z-score for a driver head-injury rating of 408. Interpret the result.

2.168 Competition for bird nest holes. The *Condor* (May 1995) published a study of competition for nest holes among collared flycatchers, a bird species. The authors collected the data for the study by periodically inspecting nest boxes located on the island of Gotland in Sweden. The nest boxes were grouped into 14 discrete locations (called plots). The accompanying table gives the number of flycatchers killed and the number of flycatchers breeding at each plot.

CONDOR

Plot Number	Number Killed	Number of Breeders
1	5	30
2	4	28
3	3	38
4	2	34
5	2	26
6	1	124
7	1	68
8	1	86
9	1	32
10	0	30
11	0	46
12	0	132
13	0	100
14	0	6

Source: Merilä, J., and Wiggins, D. A. "Interspecific competition for nest holes causes adult mortality in the collared flycatcher." *The Condor,* Vol. 97, No. 2, May 1995, p. 447 (Table 4). Cooper Ornithological Society.

a. Calculate the mean, median, and mode for the number of flycatchers killed at the 14 study plots.
b. Interpret the measures of central tendency you found in part **a**.
c. Graphically examine the relationship between number killed and number of breeders. Do you detect a trend?

2.169 Ratings of chess players. The United States Chess Federation (USCF) establishes a numerical rating for each competitive chess player. The USCF rating is a number between 0 and 4,000 that changes over time, depending on the outcome of tournament games. The higher the rating, the better (more successful) the player is. The following table describes the rating distribution of 65,455 players who were active USCF members in 2004:

USCF

Classification	Rating Range	Number of Players
Senior Master	2,800 to 2,899	1
Senior Master	2,700 to 2,799	13
Senior Master	2,600 to 2,699	66
Senior Master	2,500 to 2,599	87
Senior Master	2,400 to 2,499	133
Master	2,300 to 2,399	231
Master	2,200 to 2,299	691
Expert	2,100 to 2,199	783
Expert	2,000 to 2,099	1,516
Class A	1,900 to 1,999	1,907
Class A	1,800 to 1,899	2,682
Class B	1,700 to 1,799	3,026
Class B	1,600 to 1,699	3,437
Class C	1,500 to 1,599	3,582
Class C	1,400 to 1,499	3,386
Class D	1,300 to 1,399	3,139
Class D	1,200 to 1,299	3,153
Class E	1,100 to 1,199	2,973
Class E	1,000 to 1,099	3,021
Class F	900 to 999	3,338
Class F	800 to 899	3,520
Class G	700 to 799	3,829
Class G	600 to 699	3,946
Class H	500 to 599	3,783
Class H	400 to 499	3,544
Class I	300 to 399	3,142
Class I	200 to 299	2,547
Class J	0 to 199	3,979
Total Members		**65,455**
Average Rating		**1,068**

Source: United States Chess Federation; www.uschess.org.

a. Convert the information in the table into a graph that portrays the distribution of USCF ratings.
b. What percentage of players have a USCF rating of 2,000 or higher? (These players are "experts," "masters," or "senior masters.")
c. The mean USCF rating of the 65,455 players is 1,068. Give a practical interpretation of this value.

MTBE

2.170 Groundwater contamination in wells. Refer to the *Environmental Science & Technology* (Jan. 2005) study of the factors related to MTBE contamination in 223 New Hampshire wells, presented in Exercise 2.19 (p. 37). The data are saved in the MTBE file. Two of the many quantitative variables measured for each well are the pH level (standard units) and the MTBE level (micrograms per liter).

   a. Construct a histogram for the pH levels of the sampled wells. From the histogram, estimate the proportion of wells with pH values less than 7.0.

   b. For those wells with detectible levels of MTBE, construct a histogram for the MTBE values. From the histogram, estimate the proportion of contaminated wells with MTBE values that exceed 5 micrograms per liter.

   c. Find the mean and standard deviation for the pH levels of the sampled wells, and construct the interval $\bar{x} \pm 2s$. Estimate the percentage of wells with pH levels that fall within the interval. What rule did you apply to obtain the estimate? Explain.

   d. Find the mean and standard deviation for the MTBE levels of the sampled wells, and construct the interval $\bar{x} \pm 2s$. Estimate the percentage of wells with MTBE levels that fall within the interval. What rule did you apply to obtain the estimate? Explain.

2.171 Public image of mental illness. A survey was conducted to investigate the impact of the mass media on the public's perception of mental illness. (*Health Education Journal*, Sept. 1994.) The media coverage of each of 562 items related to mental health in Scotland was classified into one of five categories: violence to others, sympathetic coverage, harm to self, comic images, and criticism of accepted definitions of mental illness. A summary of the results is provided in the following table:

Media Coverage	Number of Items
Violence to others	373
Sympathetic	102
Harm to self	71
Comic images	12
Criticism of definitions	4
Total	**562**

Source: Philo, G. et al. "The impact of the mass media on public images of mental illness: Media content and audience belief." *Health Education Journal,* Vol. 53, No. 3, Sept. 1994, p. 274 (Table 1).

   a. Construct a relative frequency table for the data.

   b. Display the relative frequencies in a graph.

   c. Discuss the findings.

2.172 Achievement test scores. The distribution of scores on a nationally administered college achievement test has a median of 520 and a mean of 540.

   a. Explain why it is possible for the mean to exceed the median for this distribution of measurements.

   b. Suppose you are told that the 90th percentile is 660. What does this mean?

   c. Suppose you are told that you scored at the 94th percentile. Interpret this statement.

Applying the Concepts—Intermediate

2.173 New-book reviews. *Choice* magazine provides new-book reviews in each issue. A random sample of 375 *Choice* book reviews in American history, geography, and area studies was selected, and the "overall opinion" of the book stated in each review was ascertained. (*Library Acquisitions: Practice and Theory*, Vol. 19, 1995.) Overall opinion was coded as follows: 1 = would not recommend, 2 = cautions or very little recommendation, 3 = little or no preference, 4 = favorable/recommended, 5 = outstanding/significant contribution. The data are summarized in the following table:

Opinion Code	Number of Reviews
1	19
2	37
3	35
4	238
5	46
Total	375

Source: Carlo, P. W., and Natowitz, A. "*Choice* book reviews in American history, geography, and area studies: An analysis for 1988–1993." *Library Acquisitions: Practice & Theory*, Vol. 19, No. 2, 1995, p. 159 (Figure 1).

   a. Use a Pareto diagram to find the opinion that occurred most often. What proportion of the books reviewed received this opinion?

   b. Do you agree with the following statement extracted from the study: "A majority (more than 75%) of books reviewed are evaluated favorably and recommended for purchase"?

2.174 Archaeologists' study of ring diagrams. Archaeologists gain insight into the social life of ancient tribes by measuring the distance (in centimeters) of each artifact found at a site from the "central hearth" (i.e., the middle of an artifact scatter). A graphical summary of these distances, in the form of a histogram, is called a *ring diagram*. (*World Archaeology*, Oct. 1997.) Ring diagrams for two archaeological sites (A and G) in Europe are shown below and on p. 104.

   a. Identify the type of skewness (if any) present in the data collected at the two sites.

   b. Archaeologists have associated unimodal ring diagrams with open-air hearths and multimodal ring diagrams with hearths inside dwellings. Can you identify the type of hearth (*open air or inside dwelling*) that was most likely present at each site?

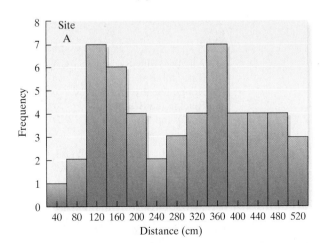

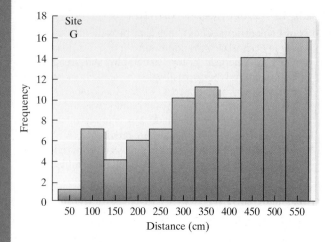

2.175 **Beanie Babies.** Beanie Babies are toy stuffed animals that have become valuable collector's items. *Beanie World Magazine* provided the age, retired status, and value of 50 Beanie Babies. The data are saved in the **BEANIE** file, with several of the observations shown in the table below.

   a. Summarize the retired/current status of the 50 Beanie Babies with an appropriate graph. Interpret the graph.

   b. Summarize the values of the 50 Beanie Babies with an appropriate graph. Interpret the graph.

   *****c.** Use a graph to portray the relationship between a Beanie Baby's value and its age. Do you detect a trend?

   d. According to Chebyshev's rule, what percentage of the age measurements would you expect to find in the intervals $\bar{x} \pm .75s$, $\bar{x} \pm 2.5s$, and $\bar{x} \pm 4s$?

   e. What percentage of the age measurements actually fall into the intervals of part **d**? Compare your results with those of part **d**.

   f. Repeat parts **d** and **e** for value.

2.176 **Control-of-life perception study.** The *locus of control* (LOC) is a measure of one's perception of control over factors affecting one's life. In one study, the LOC was measured for two groups of individuals undergoing weight reduction for treatment of obesity. (*Journal of Psychology*, Mar. 1991.) The LOC mean and standard deviation for a sample of 46 adults were 6.45 and 2.89, respectively, while the LOC mean and standard deviation for a sample of 19

adolescents were 10.89 and 2.48, respectively. A lower score on the LOC scale indicates a perception that internal factors are in control, while a higher score indicates a perception that external factors are in control.

   a. Calculate the one- and two-standard-deviation intervals around the means for each group. Plot these intervals on a line graph, using different colors or symbols to represent each group.

   b. Assuming that the distributions of LOC scores are approximately mound shaped, estimate the numbers of individuals within each interval.

   c. On the basis of your answers to parts **a** and **b**, do you think an inference can be made that *all* adults and adolescents undergoing weight-reduction treatment differ with respect to LOC? What factors did you consider in making this inference? (We'll consider this exercise again in Chapter 9 to show how to measure the reliability of this inference.)

2.177 **Whistling dolphins.** Marine scientists who study dolphin communication have discovered that bottlenose dolphins exhibit an individualized whistle contour known as their

DOLPHIN

Whistle Category	Number of Whistles
Type a	97
Type b	15
Type c	9
Type d	7
Type e	7
Type f	2
Type g	2
Type h	2
Type i	2
Type j	4
Type k	13
Other types	25

Source: McCowan, B., and Reiss, D. "Quantitative comparison of whistle repertoires from captive adult bottlenose dolphins (Delphiniae, *Tursiops truncates*): A re-evaluation of the signature whistle hypothesis," *Ethology,* Vol. 100, No. 3, July 1995, p. 200 (Table 2).

BEANIE (Selected observations)

Name	Age (Months) as of Sept. 1998	Retired (R) Current (C)	Value ($)
1. Ally the Alligator	52	R	55.00
2. Batty the Bat	12	C	12.00
3. Bongo the Brown Monkey	28	R	40.00
4. Blackie the Bear	52	C	10.00
5. Bucky the Beaver	40	R	45.00
⋮	⋮	⋮	⋮
46. Stripes the Tiger (Gold/Black)	40	R	400.00
47. Teddy the 1997 Holiday Bear	12	R	50.00
48. Tuffy the Terrier	17	C	10.00
49. Tracker the Basset Hound	5	C	15.00
50. Zip the Black Cat	28	R	40.00

Source: Beanie World Magazine, Sept. 1998.

signature whistle. A study was conducted to categorize the signature whistles of 10 captive adult bottlenose dolphins in socially interactive contexts. (*Ethology*, July 1995.) A total of 185 whistles was recorded during the study period; each whistle contour was analyzed and assigned to a category by a contour similarity (CS) technique. The results are reported in the table on p. 104. Use a graphical method to summarize the results. Do you detect any patterns in the data that might be helpful to marine scientists?

2.178 Chinese herbal drugs. Platelet-activating factor (PAF) is a potent chemical that occurs in patients suffering from shock, inflammation, hypotension, respiratory, or cardiovascular disorders. A bioassay was undertaken to investigate the potential of 17 traditional Chinese herbal drugs in PAF inhibition. (*Progress in Natural Science*, June 1995.) The prevention of the PAF binding process, measured as a percentage, for each drug is provided in the accompanying table.

PAF

Drug	PAF Inhibition (%)
Hai-feng-teng (Fuji)	77
Hai-feng-teng (Japan)	33
Shan-ju	75
Zhang-yiz-hu-jiao	62
Shi-nan-teng	70
Huang-hua-hu-jiao	12
Hua-nan-hu-jiao	0
Xiao-yie-pa-ai-xiang	0
Mao-ju	0
Jia-ju	15
Xie-yie-ju	25
Da-yie-ju	0
Bian-yie-hu-jiao	9
Bi-bo	24
Duo-mai-hu-jiao	40
Yan-sen	0
Jiao-guo-hu-jiao	31

Source: Guiqiu, H. "PAF receptor antagonistic principles from Chinese traditional drugs." *Progress in Natural Science*, Vol. 5, No. 3, June 1995, p. 301 (Table 1).

a. Construct a stem-and-leaf display of the data.

b. Compute the mean, median, and mode of the inhibition percentages for the 17 herbal drugs. Interpret the results.

c. Locate the median, mean, and mode on the stem-and-leaf display you constructed in part **a**. Do these measures of central tendency appear to locate the center of the data?

d. Compute the range, variance, and standard deviation for the inhibition percentages.

e. Form an interval that is highly likely to capture the inhibition percentage of a Chinese herbal drug.

*2.179 Freckling of superalloy ingots. Freckles are defects that sometimes form during the solidification of alloy ingots. A freckle index has been developed to measure the level of freckling on the ingot. A team of engineers conducted several experiments to measure the freckle index of a certain type of superalloy. (*Journal of Metallurgy*, Sep. 2004.) The data for $n = 18$ alloy tests are shown in the next table.

FRECKLE

30.1	22.0	14.6	16.4	12.0	2.4	22.2	10.0	15.1
12.6	6.8	4.1	2.5	1.4	33.4	16.8	8.1	3.2

Source: Yang, W.H. et al. "A freckle criterion for the solidification of superalloys with a tilted solidification front." *Journal of Metallurgy*, Vol. 56, No. 9, Sep. 2004 (Table IV).

a. Construct a box plot for the data and use it to find any outliers.

b. Find and interpret the z-scores associated with the alloys you identified in part **a**.

SLI

*2.180 Children's use of pronouns. Refer to the *Journal of Communication Disorders* data on the deviation intelligence quotient (DIQ) and percent pronoun errors for 30 children, presented in Exercise 2.65 (p. 60) data are saved in the **SLI** file.

a. Plot all the data to investigate a possible trend between DIQ and proper use of pronouns. What do you observe?

b. Plot the data for the 10 SLI children only. Is there a trend between DIQ and proper use of pronouns?

2.181 Oil spill impact on seabirds. The *Journal of Agricultural, Biological, and Environmental Statistics* (Sept. 2000) published a study on the impact of the *Exxon Valdez* tanker oil spill on the seabird population in Prince William Sound, Alaska. A subset of the data analyzed is stored in the EVOS file. Data were collected on 96 shoreline locations (called transects) of constant width, but variable length. For each transect, the number of seabirds found is recorded, as are the length (in kilometers) of the transect and whether or not the transect was in an oiled area. (The first five and last five observations in the **EVOS** file are listed in the accompanying table.)

a. Identify the variables measured as quantitative or qualitative.

EVOS (Selected observations)

Transect	Seabirds	Length	Oil
1	0	4.06	No
2	0	6.51	No
3	54	6.76	No
4	0	4.26	No
5	14	3.59	No
.	.	.	.
.	.	.	.
.	.	.	.
92	7	3.40	Yes
93	4	6.67	Yes
94	0	3.29	Yes
95	0	6.22	Yes
96	27	8.94	Yes

Source: McDonald, T. L., Erickson, W. P., and McDonald, L. L. "Analysis of count data from before–after control-impact studies," *Journal of Agricultural, Biological, and Environmental Statistics*, Vol. 5, No. 3, Sept. 2000, pp. 277–8 (Table A.1).

b. Identify the experimental unit.

c. Use a pie chart to describe the percentage of transects in oiled and unoiled areas.

*d. Use a graphical method to examine the relationship between observed number of seabirds and transect length.

MINITAB Output for Exercise 2.181 **Descriptive Statistics: Density**

Variable	Oil	N	Mean	StDev	Minimum	Q1	Median	Q3	Maximum
Density	no	36	3.27	6.70	0.000	0.000	0.890	3.87	36.23
	yes	60	3.495	5.968	0.0000	0.000	0.700	5.233	32.836

e. Observed seabird density is defined as observed count divided by length of transect. MINITAB descriptive statistics for seabird densities in unoiled and oiled transects are displayed in the printout above. Assess whether the distribution of seabird densities differs for transects in oiled and unoiled areas.

f. For unoiled transects, give an interval of values that is likely to contain at least 75% of the seabird densities.

g. For oiled transects, give an interval of values that is likely to contain at least 75% of the seabird densities.

h. Which type of transect, an oiled or unoiled one, is more likely to have a seabird density of 16? Explain.

2.182 Hearing loss in senior citizens. Audiologists have developed a rehabilitation program for hearing-impaired patients in a Canadian home for senior citizens. (*Journal of the Academy of Rehabilitative Audiology*, 1994.) Each of the 30 residents of the home were diagnosed for degree and type of sensorineural hearing loss, coded as follows: 1 = hearing within normal limits, 2 = high-frequency hearing loss, 3 = mild loss, 4 = mild-to-moderate loss, 5 = moderate loss, 6 = moderate-to-severe loss, and 7 = severe-to-profound loss. The data are listed in the accompanying table. Use a graph to portray the results. Which type of hearing loss appears to be the most prevalent among nursing home residents?

HEARLOSS

6	7	1	1	2	6	4	6	4	2	5	2	5
1	5	4	6	6	5	5	5	2	5	3	6	4
6	6	4	2								

Source: Jennings, M. B., and Head, B. G. "Development of an ecological audiologic rehabilitation program in a home-for the-aged." *Journal of the Academy of Rehabilitative Audiology*, Vol. 27, 1994, p. 77 (Table 1).

Applying the Concepts—Advanced

2.183 Speed of light from galaxies. Refer to *The Astronomical Journal* study of galaxy velocities, presented in Exercise 2.100 (p. 74). A second cluster of galaxies, named A1775, is thought to be a *double cluster*—that is, two clusters of galaxies in close proximity. Fifty-one velocity observations (in kilometers per second, km/s) from cluster A1775 are listed in the following table:

GALAXY2

22,922	20,210	21,911	19,225	18,792	21,993	23,059
20,785	22,781	23,303	22,192	19,462	19,057	23,017
20,186	23,292	19,408	24,909	19,866	22,891	23,121
19,673	23,261	22,796	22,355	19,807	23,432	22,625
22,744	22,426	19,111	18,933	22,417	19,595	23,408
22,809	19,619	22,738	18,499	19,130	23,220	22,647
22,718	22,779	19,026	22,513	19,740	22,682	19,179
19,404	22,193					

Source: Oegerle, W. R., Hill, J. M., and Fitchett, M. J. "Observations of high dispersion clusters of galaxies: Constraints on cold dark matter." *The Astronomical Journal*, Vol. 110, No. 1, July 1995, p. 34 (Table 1), p. 37 (Figure 1).

a. Use a graphical method to describe the velocity distribution of galaxy cluster A1775.

b. Examine the graph you created in part **a**. Is there evidence to support the double-cluster theory? Explain.

c. Calculate numerical descriptive measures (e.g., mean and standard deviation) for galaxy velocities in cluster A1775. Depending on your answer to part **b**, you may need to calculate two sets of numerical descriptive measures, one for each of the clusters (say, A1775A and A1775B) within the double cluster.

d. Suppose you observe a galaxy velocity of 20,000 km/s. Is this galaxy likely to belong to cluster A1775A or A1775B? Explain.

2.184 Standardized test "average." *US News & World Report* reported on many factors contributing to the breakdown of public education. One study mentioned in the article found that over 90% of the nation's school districts reported that their students were scoring "above the national average" on standardized tests. Using your knowledge of measures of central tendency, explain why the schools' reports are incorrect. Does your analysis change if the term "average" refers to the mean? To the median? Explain what effect this misinformation might have on the perception of the nation's schools.

2.185 Zinc phosphide in pest control. A chemical company produces a substance composed of 98% cracked corn particles and 2% zinc phosphide for use in controlling rat populations in sugarcane fields. Production must be carefully controlled to maintain the 2% zinc phosphide, because too much zinc phosphide will cause damage to the sugarcane and too little will be ineffective in controlling the rat population. Records from past production indicate that the distribution of the actual percentage of zinc phosphide present in the substance is approximately mound shaped, with a mean of 2.0% and a standard deviation of .08%. Suppose one batch chosen randomly actually contains 1.80% zinc phosphide. Does this indicate that there is too little zinc phosphide in today's production? Explain your reasoning.

OILSPILL

2.186 Hull failures of oil tankers. Owing to several major ocean oil spills by tank vessels, Congress passed the 1990 Oil Pollution Act, which requires all tankers to be designed with thicker hulls. Further improvements in the structural design of a tank vessel have been proposed since then, each with the objective of reducing the likelihood of an oil spill and decreasing the amount of outflow in the event of a hull puncture. To aid in this development, *Marine Technology* (Jan. 1995) reported on the spillage amount (in thousands of metric tons) and cause of puncture for 50 recent major oil spills from tankers and carriers. [*Note*: Cause of puncture is classified as either collision (C), fire/explosion (FE), hull failure (HF), or grounding (G).] The data are saved in the **OILSPILL** file.

a. Use a graphical method to describe the cause of oil spillage for the 50 tankers. Does the graph suggest that any one cause is more likely to occur than any other? How is this information of value to the design engineers?

b. Find and interpret descriptive statistics for the 50 spillage amounts. Use this information to form an interval that can be used to predict the spillage amount of the next major oil spill.

2.187 **Risk of jail suicides.** Suicide is the leading cause of death of Americans incarcerated in correctional facilities. To determine what factors increase the risk of suicide in urban jails, a group of researchers collected data on all 37 suicides that occurred over a 15-year period in the Wayne County Jail in Detroit, Michigan. (*American Journal of Psychiatry*, July 1995.) The data on each suicide victim are saved in the **SUICIDE** file. Selected observations are shown in the table at the bottom of the page.

    a. Identify the type (quantitative or qualitative) of each variable measured.

    b. Are suicides at the jail more likely to be committed by inmates charged with murder/manslaughter or with lesser crimes? Illustrate with a graph.

    c. Are suicides at the jail more likely to be committed at night? Illustrate with a graph.

    d. What is the mean length of time an inmate is in jail before committing suicide? What is the median? Interpret these two numbers.

    e. Is it likely that a future suicide at the jail will occur after 200 days? Explain.

    *f. Have suicides at the jail declined over the years? Support your answer with a graph.

2.188 **Land purchase decision.** A buyer for a lumber company must decide whether to buy a piece of land containing 5,000 pine trees. If 1,000 of the trees are at least 40 feet tall, the buyer will purchase the land; otherwise, he won't. The owner of the land reports that the height of the trees has a mean of 30 feet and a standard deviation of 3 feet. On the basis of this information, what is the buyer's decision?

2.189 **Salaries of professional athletes.** The salaries of superstar professional athletes receive much attention in the media. The multimillion-dollar long-term contract is now commonplace among this elite group. Nevertheless, rarely does a season pass without negotiations between one or more of the players' associations and team owners for additional salary and fringe benefits for *all* players in their particular sports.

    a. If a players' association wanted to support its argument for higher "average" salaries, which measure of central tendency do you think it should use? Why?

    b. To refute the argument, which measure of central tendency should the owners apply to the players' salaries? Why?

2.190 **Grades in statistics.** The final grades given by two professors in introductory statistics courses have been carefully examined. The students in the first professor's class had a grade point average of 3.0 and a standard deviation of .2. Those in the second professor's class had grade points with an average of 3.0 and a standard deviation of 1.0. If you had a choice, which professor would you take for this course?

Critical Thinking Challenges

2.191 **The new Hite Report.** In 1968, researcher Shere Hite shocked conservative America with her famous Hite Report on the permissive sexual attitudes of American men and women. Some 20 years later, Hite was surrounded by controversy again with her book *Women and Love: A Cultural Revolution in Progress* (Knopf Press, 1988). In this book, Hite reveals some startling statistics describing how women feel about contemporary relationships:

- Eighty-four percent are not emotionally satisfied with their relationship.
- Ninety-five percent report "emotional and psychological harassment" from their men.
- Seventy percent of those married five years or more are having extramarital affairs.
- Only 13% of those married more than two years are "in love."

Hite conducted the survey by mailing out 100,000 questionnaires to women across the country over a seven-year period. The questionnaires were mailed to a wide variety of organizations, including church groups, women's voting and political groups, women's rights organizations, and counseling and walk-in centers for women. Organizational leaders were asked to circulate the questionnaires to their members. Hite also relied on volunteer respondents who wrote in for copies of the questionnaire. Each questionnaire consisted of 127 open-ended questions, many with numerous subquestions and follow-ups. Hite's instructions read, "It is not necessary to answer every question! Feel free to skip around and answer those questions you choose." Approximately 4,500 completed questionnaires were returned, a response rate of 4.5%. These questionnaires form the data set from which the preceding percentages were determined. Hite claims that the 4,500 women respondents are a representative sample of all women in the United States and, therefore, that the survey

SUICIDE (Selected Observations)

Victim	Days in Jail before Suicide	Marital Status	Race	Murder/Manslaughter Charge	Time of Suicide	Year
1	3	Married	W	Yes	Night	1972
2	4	Single	W	Yes	Night	1987
3	5	Single	NW	Yes	Afternoon	1975
4	7	Widowed	NW	Yes	Night	1981
5	10	Single	NW	Yes	Afternoon	1982
.	.	.	.	.	.	.
.	.	.	.	.	.	.
36	41	Single	NW	No	Night	1985
37	86	Married	W	No	Night	1968

Source: DuRand, C. J., et al. "A quarter century of suicide in a major urban jail: Implications for community psychiatry." *American Journal of Psychiatry,* Vol. 152, No. 7, July 1995, p. 1078 (Table 1).

results imply that vast numbers of women are "suffering a lot of pain in their love relationships with men." Critically assess the survey results. Do you believe they are reliable?

2.192 No Child Left Behind Act. According to the government, federal spending on K–12 education has increased dramatically over the past 20 years, but student performance has stayed essentially the same. Hence, in 2002, President George Bush signed into law the No Child Left Behind Act, a bill that promised improved student achievement for all U.S. children. *Chance* (Fall 2003) reported on a graphic that was designed to support the new legislation. The graphic, obtained from the U.S. Department of Education Web site (www.ed.gov), is reproduced here. The bars in the graph represent annual federal spending on education, in billions of dollars (left-side vertical axis). The horizontal line represents the annual average 4th-grade children's reading ability score (right-side vertical axis). Critically assess the information portrayed in the graph. Does it, in fact, support the government's position that our children are not making classroom improvements despite federal spending on education? Use the following facts (divulged in the *Chance* article) to help you frame your answer: (1) The U.S. student population has also increased dramatically over the past 20 years, (2) 4th-grade reading test scores are designed to have an average of 250 with a standard deviation of 50, and (3) the reading test scores of 7th and 12th graders and the mathematics

scores of 4th graders did improve substantially over the past 20 years.

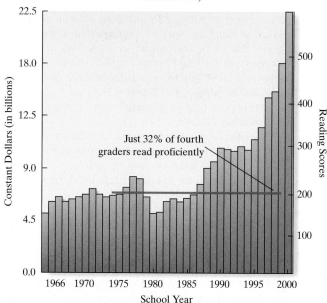

Federal Spending on
K-12 Education
(Elementary and Secondary
Education Act)

SUMMARY ACTIVITY: Describing Data from Popular Sources

We list here several sources of real-life data sets. (Many in the list have been obtained from Wasserman and Bernero's *Statistics Sources,* and many can be accessed via the Internet). This index of data sources is quite complete and is a useful reference for anyone interested in finding almost any type of data. First we list some almanacs:

CBS News Almanac

Information Please Almanac

World Almanac and Book of Facts

United States Government publications are also rich sources of data:

Agricultural Statistics

Digest of Educational Statistics

Handbook of Labor Statistics

Housing and Urban Development Yearbook

Social Indicators

Uniform Crime Reports for the United States

Vital Statistics of the United States

Business Conditions Digest

Economic Indicators

Monthly Labor Review

Survey of Current Business

Bureau of the Census Catalog

Statistical Abstract of the United States

Main data sources are published on an annual basis:

Commodity Yearbook

Facts and Figures on Government Finance

Municipal Yearbook

Standard and Poor's Corporation, Trade and Securities: Statistics

Some sources contain data that are international in scope:

Compendium of Social Statistics

Demographic Yearbook

United Nations Statistical Yearbook

World Handbook of Political and Social Indicators

 Utilizing the data sources listed, sources suggested by your instructor, or your own resourcefulness, find one real-life quantitative data set that stems from an area of particular interest to you.

a. Describe the data set by using a relative frequency histogram, stem-and-leaf display, or dot plot.

b. Find the mean, median, variance, standard deviation, and range of the data set.

c. Use Chebyshev's rule and the empirical rule to describe the distribution of this data set. Count the actual number of observations that fall within one, two, and three standard deviations of the mean of the data set, and compare these counts with the description of the data set you developed in part **b.**

REFERENCES

Huff, D. *How to Lie with Statistics*. New York: Norton, 1954.

Mendenhall, W., Beaver, R. J., and Beaver, B. M. *Introduction to Probability and Statistics*, 12th ed. North Scituate, MA: Duxbury, 2006.

Tufte, E. R. *Envisioning Information*. Cheshire, CT: Graphics Press, 1990.

Tufte, E. R. *Visual Display of Quantitative Information*. Cheshire, CT: Graphics Press, 1983.

Tufte, E.R. *Visual Explanations*. Cheshire, CT: Graphics Press, 1997.

Tukey, J. *Exploratory Data Analysis*. Reading, MA: Addison-Wesley, 1977.

Using Technology

Describing Data with MINITAB

Graphing Data

To obtain graphical descriptions of your data, click on the "Graph" button on the MINITAB menu bar. The resulting menu list appears as shown in Figure 2.M.1. Several of the options covered in this text are "Bar Chart," "Pie Chart," "Scatter Plot," "Histogram," "Dotplot," and "Stem-and-Leaf." Click on the graph of your choice to view the appropriate dialog box. For example, the dialog box for a histogram is shown in Figure 2.M.2. Make the appropriate variable selections and click "OK" to view the graph.

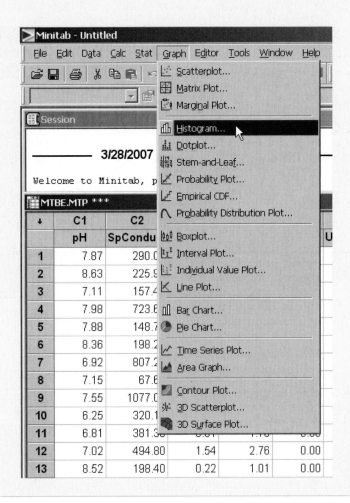

FIGURE 2.M.1

MINITAB menu options for graphing your data

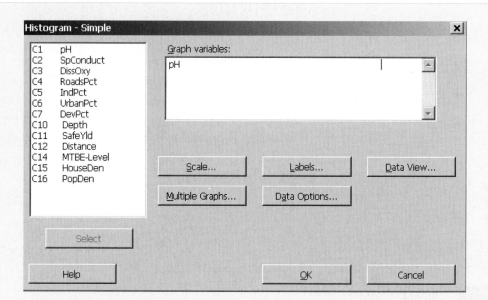

FIGURE 2.M.2
MINITAB histogram dialog box

Numerical Descriptive Statistics

To obtain numerical descriptive measures for a quantitative variable (e.g., mean, standard deviation, etc.), click on the "Stat" button on the main menu bar, then click on "Basic Statistics," and then click on "Display Descriptive Statistics." (See Figure 2.M.3.) The resulting dialog box appears in Figure 2.M.4.

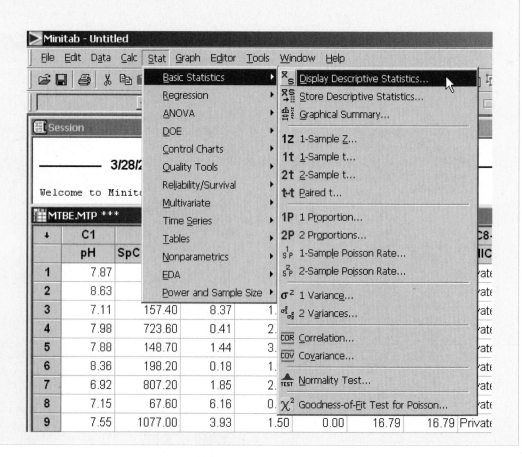

FIGURE 2.M.3
MINITAB options for obtaining descriptive statistics

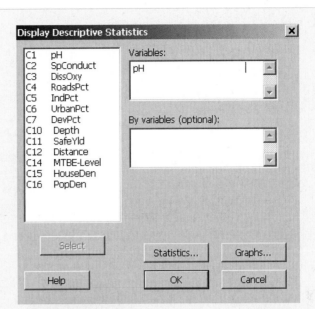

FIGURE 2.M.4
MINITAB descriptive statistics
dialog box

Select the quantitative variables you want to analyze and place them in the "Variables" box. You can control which particular descriptive statistics appear by clicking the "Statistics" button on the dialog box and making your selections. (As an option, you can create histograms and dot plots for the data by clicking the "Graphs" button and making the appropriate selections.) Click on "OK" to view the descriptive statistics printout.

CHAPTER 3

Probability

CONTENTS

3.1 Events, Sample Spaces, and Probability

3.2 Unions and Intersections

3.3 Complementary Events

3.4 The Additive Rule and Mutually Exclusive Events

3.5 Conditional Probability

3.6 The Multiplicative Rule and Independent Events

3.7 Random Sampling

STATISTICS IN ACTION

Lotto Buster! Can You Improve Your Chance of Winning?

USING TECHNOLOGY

Generating a Random Sample with MINITAB

WHERE WE'VE BEEN

- Identified the objective of inferential statistics: to make inferences about a population on the basis of information in a sample
- Introduced graphical and numerical descriptive measures for both quantitative and qualitative data

WHERE WE'RE GOING

- Develop probability as a measure of uncertainty
- Introduce basic rules for finding probabilities
- Use a probability as a measure of reliability for an inference

STATISTICS IN ACTION

Lotto Buster! Can You Improve Your Chance of Winning?

"Welcome to the Wonderful World of Lottery Busters." So began the premier issue of Lottery Buster, *a monthly publication for players of the state lottery games.* Lottery Buster *provides interesting facts and figures on the 42 state lotteries and 2 multistate lotteries currently operating in the United States and, more importantly, tips on how to increase a player's odds of winning the lottery.*

In 1963, New Hampshire became the first state in modern times to authorize a state lottery as an alternative to increasing taxes. (Prior to that time, beginning in 1895, lotteries were banned in America because they were thought to be corrupt.) Since then, lotteries have become immensely popular, for two reasons. First, they lure you with the opportunity to win millions of dollars with a $1 investment, and second, when you lose, at least you believe that your money is going to a good cause. Many state lotteries, like Florida's, earmark a high percentage of lottery revenues to fund state education.

The popularity of the state lottery has brought with it an avalanche of "experts" and "mathematical wizards" (such as the editors of *Lottery Buster*) who provide advice on how to win the lottery—for a fee, of course! Many offer guaranteed "systems" of winning through computer software products with catchy names such as Lotto Wizard, Lottorobics, Win4D, and Lotto-Luck.

For example, most knowledgeable lottery players would agree that the "golden rule" or "first rule" in winning lotteries is *game selection*. State lotteries generally offer three types of games: Instant (scratch-off tickets or online) games, Daily Numbers (Pick-3 or Pick-4), and the weekly Pick-6 Lotto game.

One version of the Instant game involves scratching off the thin opaque covering on a ticket with the edge of a coin to determine whether you have won or lost. The cost of the ticket ranges from 50¢ to $1, and the amount won ranges from $1 to $100,000 in most states to as much as $1 million in others. *Lottery Buster* advises against playing the Instant game because it is "a pure chance play, and you can win only by dumb luck. No skill can be applied to this game."

The Daily Numbers game permits you to choose either a three-digit (Pick-3) or four-digit (Pick-4) number at a cost of $1 per ticket. Each night, the winning number is drawn. If your number matches the winning number, you win a large sum of money, usually $100,000. You do have some control over the Daily Numbers game (since you pick the numbers that you play); consequently, there are strategies available to increase your chances of winning. However, the Daily Numbers game, like the Instant game, is not available for out-of-state play.

To play Pick-6 Lotto, you select six numbers of your choice from a field of numbers ranging from 1 to N, where N depends on which state's game you are playing. For example, Florida's current Lotto game involves picking six numbers ranging from 1 to 53. (See Figure SIA3.1.) The cost of a ticket is $1, and the payoff, if your six numbers match the winning numbers drawn, is $7 million or more, depending on the number

FIGURE SIA3.1
Reproduction of Florida's 6/53 Lotto ticket

of tickets purchased. (To date, Florida has had the largest state weekly payoff, over $200 million.) In addition to capturing the grand prize, you can win second-, third-, and fourth-prize payoffs by matching five, four, and three of the six numbers drawn, respectively. And you don't have to be a resident of the state to play the state's Lotto game.

In this chapter, several Statistics In Action Revisited examples demonstrate how to use the basic concepts of probability to compute the odds of winning a state lottery game and to as-

sess the validity of the strategies suggested by lottery "experts."

Statistics in Action Revisited for Chapter 3

- Computing and Understanding the Probability of Winning Lotto (p. 123)

- The Probability of Winning Lotto with a Wheel System (p. 133)

- The Probability of Winning Cash 3 or Play 4 (p. 147)

Recall that one branch of statistics is concerned with decisions made about a population on the basis of information learned about a sample. You can see how this is accomplished more easily if you understand the relationship between population and sample—a relationship that becomes clearer if we reverse the statistical procedure of making inferences from sample to population. In this chapter, we assume that the population is *known* and calculate the chances of obtaining various samples from the population. Thus, we show that probability is the reverse of statistics: In probability, we use information about the population to infer the probable nature of the sample.

Probability plays an important role in making inferences. Suppose, for example, you have an opportunity to invest in an oil exploration company. Past records show that, out of 10 previous oil drillings (a sample of the company's experiences), all 10 came up dry. What do you conclude? Do you think the chances are better than 50:50 that the company will hit a gusher? Should you invest in this company? Chances are, your answer to these questions will be an emphatic "No!" However, if the company's exploratory prowess is sufficient to hit a producing well 50% of the time, a record of 10 dry wells out of 10 drilled is an event that is just too improbable.

Or suppose you're playing poker with what your opponents assure you is a well-shuffled deck of cards. In three consecutive five-card hands, the person on your right is dealt four aces. On the basis of this sample of three deals, do you think the cards are being adequately shuffled? Again, your answer is likely to be "No," because dealing three hands of four aces is just too improbable if the cards were properly shuffled.

Note that the decisions concerning the potential success of the oil-drilling company and the adequacy of card shuffling both involve knowing the chance—or probability—of a certain sample result. Both situations were contrived so that you could easily conclude that the probabilities of the sample results were small. Unfortunately, the probabilities of many observed sample results aren't so easy to evaluate intuitively. In these cases, we need the assistance of a theory of probability.

3.1 Events, Sample Spaces, and Probability

Let's begin our treatment of probability with straightforward examples that are easily described. With the aid of these simple examples, we can introduce important definitions that will help us develop the notion of probability more easily.

Suppose a coin is tossed once and the up face is recorded. The result we see is called an *observation*, or *measurement*, and the process of making an observation is called an *experiment*. Notice that our definition of experiment is broader than the one used in the physical sciences, which brings to mind test tubes, microscopes, and other laboratory equipment. Statistical experiments may include, in addition to these things, recording an Internet user's preference for a Web browser, recording a voter's opinion on an important political issue, measuring the amount of dissolved oxygen in a polluted river, observing the level of anxiety of a test taker, counting the number of errors in an inventory, and observing the fraction of insects killed by a new insecticide. The point is that a statistical experiment can be almost any act of observation, as long as the outcome is uncertain.

Definition 3.1

An **experiment** is an act or process of observation that leads to a single outcome that cannot be predicted with certainty.

Consider another simple experiment consisting of tossing a die and observing the number on the up face. The six possible outcomes of this experiment are as follows:

1. Observe a 1.
2. Observe a 2.
3. Observe a 3.
4. Observe a 4.
5. Observe a 5.
6. Observe a 6.

Note that if this experiment is conducted once, *you can observe one and only one of these six basic outcomes, and the outcome cannot be predicted with certainty*. Also, these outcomes cannot be decomposed into more basic ones. Because observing the outcome of an experiment is similar to selecting a sample from a population, the basic possible outcomes of an experiment are called **sample points**.*

Definition 3.2

A **sample point** is the most basic outcome of an experiment.

EXAMPLE 3.1

LISTING SAMPLE POINTS—Coin-Tossing Experiment

Problem Two coins are tossed, and their up faces are recorded. List all the sample points for this experiment.

Solution Even for a seemingly trivial experiment, we must be careful when listing the sample points. At first glance, we might expect one of three basic outcomes: Observe two heads; Observe two tails; or Observe one head and one tail. However, further reflection reveals that the last of these, Observe one head and one tail, can be decomposed into two outcomes: Head on coin 1, Tail on coin 2; and Tail on coin 1, Head on coin 2. Thus, we have four sample points:

1. Observe *HH*.
2. Observe *HT*.
3. Observe *TH*.
4. Observe *TT*.

In this list, *H* in the first position means "Head on coin 1," *H* in the second position means "Head on coin 2," and so on.

Look Back Even if the coins are identical in appearance, they are, in fact, two distinct coins. Thus, the sample points must account for this distinction.

Now Work Exercise 3.15a

We often wish to refer to the collection of all the sample points of an experiment. This collection is called the *sample space* of the experiment. For example, there are six sample points in the sample space associated with the die-toss experiment. The sample spaces for the experiments discussed thus far are shown in Table 3.1.

*Alternatively, the term *simple event* can be used.

TABLE 3.1 Experiments and Their Sample Spaces

Experiment: Observe the up face on a coin.
Sample Space: 1. Observe a head.
                  2. Observe a tail.
This sample space can be represented in set notation as a set containing two sample points:
$$S: \quad \{H, T\}$$
Here, H represents the sample point Observe a head and T represents the sample point Observe a tail.

Experiment: Observe the up face on a die.
Sample Space: 1. Observe a 1.
                  2. Observe a 2.
                  3. Observe a 3.
                  4. Observe a 4.
                  5. Observe a 5.
                  6. Observe a 6.
This sample space can be represented in set notation as a set of six sample points:
$$S: \quad \{1, 2, 3, 4, 5, 6\}$$

Experiment: Observe the up faces on two coins.
Sample Space: 1. Observe HH.
                  2. Observe HT.
                  3. Observe TH.
                  4. Observe TT.
This sample space can be represented in set notation as a set of four sample points:
$$S: \quad \{HH, HT, TH, TT\}$$

Definition 3.3

The **sample space** of an experiment is the collection of all its sample points.

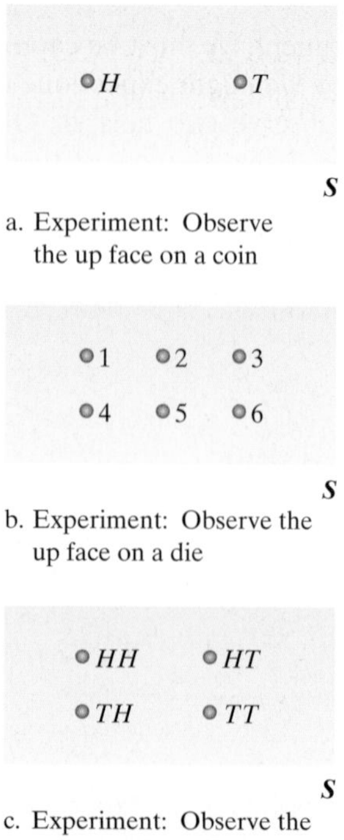

Just as graphs are useful in describing sets of data, a pictorial method for presenting the sample space will often be useful. Figure 3.1 shows such a representation for each of the experiments in Table 3.1. In each case, the sample space is shown as a closed figure, labeled S, containing all possible sample points. Each sample point is represented by a solid dot (i.e., a "point") and labeled accordingly. Such graphical representations are called **Venn diagrams**.

Now that we know that an experiment will result in *only one* basic outcome — called a sample point — and that the sample space is the collection of all possible sample points, we're ready to discuss the probabilities of the sample points. You have undoubtedly used the term *probability* and have some intuitive idea about its meaning. Probability is generally used synonymously with "chance," "odds," and similar concepts. For example, if a fair coin is tossed, we might reason that both the sample points Observe a head and Observe a tail have the same *chance* of occurring. Thus, we might state, "The probability of observing a head is 50%" or "The odds of seeing a head are 50:50." Both of these statements are based on an informal knowledge of probability. We'll begin our treatment of probability by using such informal concepts and then solidify what we mean later.

The probability of a sample point is a number between 0 and 1 which measures the likelihood that the outcome will occur when the experiment is performed. This number is usually taken to be the relative frequency of the occurrence of a sample point in a very long series of repetitions of an experiment.* For example, if we are assigning probabilities to the two sample points in the coin-toss experiment (Observe a head and Observe a tail), we might reason that if we toss a balanced coin a very large number of

a. Experiment: Observe the up face on a coin

b. Experiment: Observe the up face on a die

c. Experiment: Observe the up faces on two coins

FIGURE 3.1

Venn diagrams for the three experiments from Table 3.1

---

*The result derives from an axiom in probability theory called the *law of large numbers*. Phrased informally, the law states that the relative frequency of the number of times that an outcome occurs when an experiment is replicated over and over again (i.e., a large number of times) approaches the true (or theoretical) probability of the outcome.

times, the sample points Observe a head and Observe a tail will occur with the same relative frequency of .5.

Our reasoning is supported by Figure 3.2. The figure plots the relative frequency of the number of times that a head occurs in simulations (by computer) of the toss of a coin N times, where N ranges from as few as 25 tosses to as many as 1,500 tosses. You can see that when N is large (e.g., $N = 1,500$), the relative frequency is converging to .5. Thus, the probability of each sample point in the coin-tossing experiment is .5.

For some experiments, we may have little or no information on the relative frequency of occurrence of the sample points; consequently, we must assign probabilities to the sample points on the basis of general information about the experiment. For example, if the experiment is about investing in a business venture and observing whether it succeeds or fails, the sample space would appear as in Figure 3.3.

We are unlikely to be able to assign probabilities to the sample points of this experiment based on a long series of repetitions, since unique factors govern each performance of this kind of experiment. Instead, we may consider factors such as the personnel managing the venture, the general state of the economy at the time, the rate of success of similar ventures, and any other information deemed pertinent. If we finally decide that the venture has an 80% chance of succeeding, we assign a probability of .8 to the sample point Success. This probability can be interpreted as a measure of our degree of belief in the outcome of the business venture; that is, it is a subjective probability. Notice, however, that such probabilities should be based on expert information that is carefully assessed. If not, we may be misled on any decisions

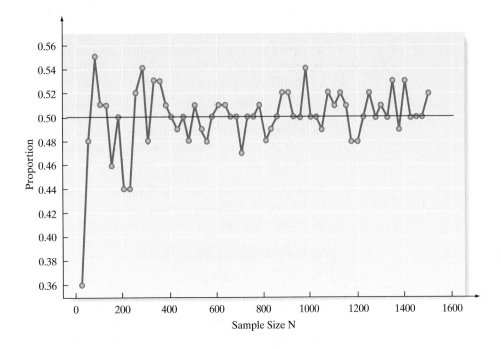

FIGURE 3.2

Proportion of heads in N tosses of a coin

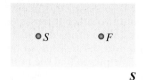

S

FIGURE 3.3

Experiment: invest in a business venture and observe whether it succeeds (S) or fails (F)

based on these probabilities or based on any calculations in which they appear. [*Note:* For a text that deals in detail with the subjective evaluation of probabilities, see Winkler (1972) or Lindley (1985).]

No matter how you assign the probabilities to sample points, the probabilities assigned must obey two rules:

Probability Rules for Sample Points

Let p_i represent the probability of sample point i. Then

1. All sample point probabilities *must* lie between 0 and 1 (i.e., $0 \leq p_i \leq 1$).

2. The probabilities of all the sample points within a sample space *must* sum to 1 (i.e., $\Sigma p_i = 1$).

Assigning probabilities to sample points is easy for some experiments. For example, if the experiment is to toss a fair coin and observe the face, we would probably all agree to assign a probability of $\frac{1}{2}$ to the two sample points, Observe a head and Observe a tail. However, many experiments have sample points whose probabilities are more difficult to assign.

EXAMPLE 3.2

SAMPLE POINT PROBABILITIES— Hotel Guest Room Survey

Problem Many American hotels offer complimentary shampoo in their guest rooms. Suppose you randomly select one hotel from a registry of all hotels in the United States and check whether or not the hotel offers complimentary shampoo. Show how this problem might be formulated in the framework of an experiment with sample points and a sample space. Indicate how probabilities might be assigned to the sample points.

Solution The experiment can be defined as the selection of an American hotel and the observation of whether or not complimentary shampoo is offered in the hotel's guest rooms. There are two sample points in the sample space corresponding to this experiment:

S: {The hotel offers complimentary shampoo.}
N: {The hotel does not offer complimentary shampoo.}

The difference between this and the coin-toss experiment becomes apparent when we attempt to assign probabilities to the two sample points. What probability should we assign to the sample point *S*? If you answer .5, you are assuming that the events *S* and *N* occur with equal likelihood, just like the sample points Heads and Tails in the coin-toss experiment. But assigning sample point probabilities for the hotel shampoo experiment is not so easy. In fact, a recent survey of American hotels found that 80% now offer complimentary shampoo to guests. In that case, it might be reasonable to approximate the probability of the sample point *S* as .8 and that of the sample point *N* as .2.

Look Back Here we see that the sample points are not always equally likely, so assigning probabilities to them can be complicated, particularly for experiments that represent real applications (as opposed to coin- and die-toss experiments).

Now Work Exercise 3.20

Although the probabilities of sample points are often of interest in their own right, it is usually probabilities of *collections* of sample points that are important. Example 3.3 demonstrates this point.

EXAMPLE 3.3

PROBABILITY OF
A COLLECTION OF
SAMPLE POINTS—
Die-Tossing
Experiment

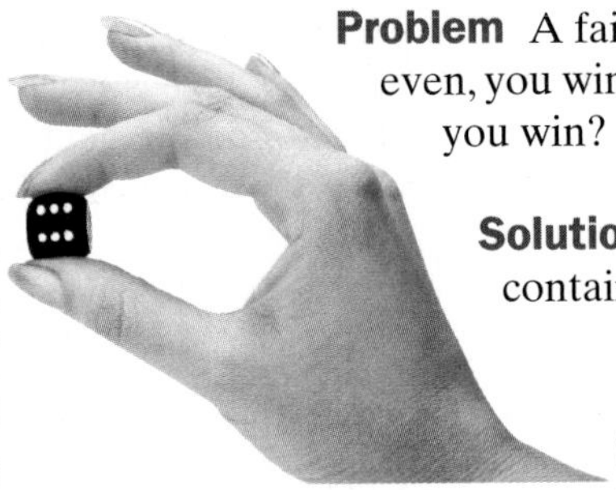

Problem A fair die is tossed, and the up face is observed. If the face is even, you win $1. Otherwise, you lose $1. What is the probability that you win?

Solution Recall that the sample space for this experiment contains six sample points:

$$S: \{1,2,3,4,5,6\}$$

Since the die is balanced, we assign a probability of $\frac{1}{6}$ to each of the sample points in this sample space. An even number will occur if one of the sample points Observe a 2, Observe a 4, or Observe a 6 occurs. A collection of sample points such as this is called an *event*, which we denote by the letter A. Since the event A contains three sample points—all with probability $\frac{1}{6}$—and since no sample points can occur simultaneously, we reason that the probability of A is the sum of the probabilities of the sample points in A. Thus, the probability of A is $\frac{1}{6} + \frac{1}{6} + \frac{1}{6} = \frac{1}{2}$.

Look Back On the basis of this notion of probability, *in the long run*, you will win $1 half the time and lose $1 half the time.

Now Work Exercise 3.14

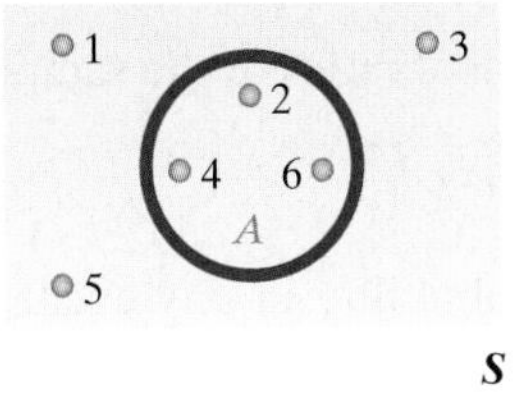

FIGURE 3.4
Die-toss experiment with event
A, observe an even number

Figure 3.4 is a Venn diagram depicting the sample space associated with a die-toss experiment and the event A, Observe an even number. The event A is represented by the closed figure inside the sample space **S**. This closed figure A comprises all of the sample points that make up event A.

To decide which sample points belong to the set associated with an event A, test each sample point in the sample space **S**. If event A occurs, then that sample point is in the event A. For example, the event A, Observe an even number, in the die-toss experiment will occur if the sample point Observe a 2 occurs. By the same reasoning, the sample points Observe a 4 and Observe a 6 are also in event A.

To summarize, we have demonstrated that an event can be defined in words, or it can be defined as a specific set of sample points. This leads us to the following general definition of an *event*:

Definition 3.4
An **event** is a specific collection of sample points.

EXAMPLE 3.4

PROBABILITY OF
AN EVENT—
Coin-Tossing
Experiment

Problem Consider the experiment of tossing two *unbalanced* coins. Because the coins are *not* balanced, their outcomes (H or T) are not equiprobable. Suppose the correct probabilities associated with the sample points are given in the accompanying table. [*Note:* The necessary properties for assigning probabilities to sample points are satisfied.]
Consider the events

A: {Observe exactly one head.}
B: {Observe at least one head.}

Calculate the probability of A and the probability of B.

Sample Point	Probability
HH	$\frac{4}{9}$
HT	$\frac{2}{9}$
TH	$\frac{2}{9}$
TT	$\frac{1}{9}$

Solution Event A contains the sample points HT and TH. Since two or more sample points cannot occur at the same time, we can easily calculate the probability of event A by summing the probabilities of the two sample points. Thus, the probability of observing exactly one head (event A), denoted by the symbol $P(A)$, is

$$P(A) = P(\text{Observe } HT.) + P(\text{Observe } TH.) = \frac{2}{9} + \frac{2}{9} = \frac{4}{9}$$

Similarly, since B contains the sample points HH, HT, and TH, it follows that

$$P(B) = \frac{4}{9} + \frac{2}{9} + \frac{2}{9} = \frac{8}{9}$$

Look Back Again, these probabilities should be interpreted *in the long run*. For example, $P(B) = \frac{8}{9} \approx .89$ implies that if we were to toss two coins an infinite number of times, we would observe at least two heads on about 89% of the tosses.

Now Work Exercise 3.11

The preceding example leads us to a general procedure for finding the probability of an event A:

Probability of an Event

The probability of an event A is calculated by summing the probabilities of the sample points in the sample space for A.

Thus, we can summarize the steps for calculating the probability of any event, as indicated in the next box.

Steps for Calculating Probabilities of Events

1. Define the experiment; that is, describe the process used to make an observation and the type of observation that will be recorded.

2. List the sample points.

3. Assign probabilities to the sample points.

4. Determine the collection of sample points contained in the event of interest.

5. Sum the sample point probabilities to get the probability of the event.

EXAMPLE 3.5

APPLYING THE FIVE STEPS TO FIND A PROBABILITY—
Study of Divorced Couples

Problem The American Association for Marriage and Family Therapy (AAMFT) is a group of professional therapists and family practitioners that treats many of the nation's couples and families. The AAMFT released the findings of a study that tracked the post-divorce history of 100 pairs of former spouses with children. Each divorced couple was classified into one of four groups, nicknamed "perfect pals (PP)," "cooperative colleagues (CC)," "angry associates (AA)," and "fiery foes (FF)." The proportions classified into each group are shown in Table 3.2.

Suppose one of the 100 couples is selected at random.

a. Define the experiment that generated the data in Table 3.2, and list the sample points.

b. Assign probabilities to the sample points.

c. What is the probability that the former spouses are "fiery foes?"

d. What is the probability that the former spouses have at least some conflict in their relationship?

TABLE 3.2 Results of AAMFT Study of Divorced Couples

Group	Proportion
Perfect Pals (PP) (Joint-custody parents who get along well)	.12
Cooperative Colleagues (CC) (Occasional conflict, likely to be remarried)	.38
Angry Associates (AA) (Cooperate on children-related issues only, conflicting otherwise)	.25
Fiery Foes (FF) (Communicate only through children, hostile toward each other)	.25

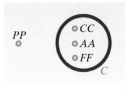

FIGURE 3.5

Venn diagram for AAMFT survey

Solution

a. The experiment is the act of classifying the randomly selected couple. The sample points—the simplest outcomes of the experiment—are the four groups (categories) listed in Table 3.2. They are shown in the Venn diagram in Figure 3.5.

b. If, as in Example 3.1, we were to assign equal probabilities in this case, each of the response categories would have a probability of one-fourth (1/4), or .25. But by examining Table 3.2, you can see that equal probabilities are not reasonable in this case, because the response percentages are not all the same in the four categories. It is more reasonable to assign a probability equal to the response proportion in each class, as shown in Table 3.3.*

TABLE 3.3 Sample Point Probabilities For AAMFT Survey

Sample Point	Probability
PP	.12
CC	.38
AA	.25
FF	.25

c. The event that the former spouses are "fiery foes" corresponds to the sample point FF. Consequently, the probability of the event is the probability of the sample point. From Table 3.3, we find that $P(FF) = .25$. Therefore, there is a .25 probability (or one-fourth chance) that the couple we select are "fiery foes."

d. The event that the former spouses have at least some conflict in their relationship, call it event C, is not a sample point, because it consists of more than one of the response classifications (the sample points). In fact, as shown in Figure 3.5, the event C consists of three sample points: CC, AA, and FF. The probability of C is defined to be the sum of the probabilities of the sample points in C:

$$P(C) = P(CC) + P(AA) + P(FF) = .38 + .25 + .25 = .88$$

Thus, the chance that we observe a divorced couple with some degree of conflict in their relationship is .88—a fairly high probability.

Look Back The key to solving this problem is to follow the steps outlined in the box. We defined the experiment (Step 1) and listed the sample points (Step 2) in part **a.** The assignment of probabilities to the sample points (Step 3) was done in part **b.** For each probability in parts c and d, we identified the collection of sample points in the event (Step 4) and summed their probabilities (Step 5).

Now Work Exercise 3.23

*Since the response percentages were based on a sample of divorced couples, these assigned probabilities are estimates of the true population response percentages. You will learn how to measure the reliability of probability estimates in Chapter 7.

The preceding examples have one thing in common: The number of sample points in each of the sample spaces was small; hence, the sample points were easy to identify and list. How can we manage this when the sample points run into the thousands or millions? For example, suppose you wish to select 5 marines for a dangerous mission from a division of 1,000. Then each different group of 5 marines would represent a sample point. How can you determine the number of sample points associated with this experiment?

One method of determining the number of sample points for a complex experiment is to develop a counting system. Start by examining a simple version of the experiment. For example, see if you can develop a system for counting the number of ways to select 2 marines from a total of 4. If the marines are represented by the symbols M_1, M_2, M_3, and M_4, the sample points could be listed in the following pattern:

$$(M_1, M_2) \qquad (M_2, M_3) \qquad (M_3, M_4)$$
$$(M_1, M_3) \qquad (M_2, M_4)$$
$$(M_1, M_4)$$

Note the pattern and now try a more complex situation—say, sampling 3 marines out of 5. List the sample points and observe the pattern. Finally, see if you can deduce the pattern for the general case. Perhaps you can program a computer to produce the matching and counting for the number of samples of 5 selected from a total of 1,000.

A second method of determining the number of sample points for an experiment is to use **combinatorial mathematics**. This branch of mathematics is concerned with developing counting rules for given situations. For example, there is a simple rule for finding the number of different samples of 5 marines selected from 1,000. This rule, called the **combinations rule**, is given in the box.

Combinations Rule

Suppose a sample of n elements is to be drawn from a set of N elements. Then the number of different samples possible is denoted by $\binom{N}{n}$ and is equal to

$$\binom{N}{n} = \frac{N!}{n!(N-n)!}$$

where

$$n! = n(n-1)(n-2)\cdots(3)(2)(1)$$

and similarly for $N!$ and $(N-n)!$ For example, $5! = 5 \cdot 4 \cdot 3 \cdot 2 \cdot 1$. [*Note:* The quantity $0!$ is defined to be equal to 1.]

EXAMPLE 3.6

USING THE COMBINATIONS RULE—

Selecting 2 Marines from 4

Problem Consider the task of choosing 2 marines from a platoon of 4 to send on a dangerous mission. Use the combinations counting rule to determine how many different selections can be made.

Solution For this example, $N = 4$, $n = 2$, and

$$\binom{4}{2} = \frac{4!}{2!2!} = \frac{4 \cdot 3 \cdot 2 \cdot 1}{(2 \cdot 1)(2 \cdot 1)} = 6$$

Look Back You can see that this answer agrees with the number of sample points listed at the top of the page.

Now Work Exercise 3.13

EXAMPLE 3.7
USING THE
COMBINATIONS
RULE—
Selecting 5
Investments from 20

Problem Suppose you plan to invest equal amounts of money in each of 5 business ventures. If you have 20 ventures from which to make the selection, how many different samples of 5 ventures can be selected from the 20?

Solution For this example, $N = 20$ and $n = 5$. Then the number of different samples of 5 that can be selected from the 20 ventures is

$$\binom{20}{5} = \frac{20!}{5!(20-5)!} = \frac{20!}{5!15!}$$

$$= \frac{20 \cdot 19 \cdot 18 \cdot \cdots \cdot 3 \cdot 2 \cdot 1}{(5 \cdot 4 \cdot 3 \cdot 2 \cdot 1)(15 \cdot 14 \cdot 13 \cdot \cdots \cdot 3 \cdot 2 \cdot 1)} = 15,504$$

Look Back You can see that attempting to list all the sample points for this experiment would be an extremely tedious and time-consuming, if not practically impossible, task.

The combinations rule is just one of a large number of **counting rules** that have been developed by combinatorial mathematicians. This counting rule applies to situations in which the experiment calls for selecting n elements from a total of N elements, without replacing each element before the next is selected. Consult the references to learn about other basic counting rules.

STATISTICS IN ACTION REVISITED

Computing and Understanding the Probability of Winning Lotto

In Florida's state lottery game, called Pick-6 Lotto, you select six numbers of your choice from a set of numbers ranging from 1 to 53. We can apply the combinations rule to determine the total number of combinations of 6 numbers selected from 53 (i.e., the total number of sample points [or possible winning tickets]). Here, $N = 53$ and $n = 6$; therefore, we have

$$\binom{N}{n} = \frac{N!}{n!(N-n)!} = \frac{53!}{6!47!}$$

$$= \frac{(53)(52)(51)(50)(49)(48)(47!)}{(6)(5)(4)(3)(2)(1)(47!)}$$

$$= 22,957,480$$

Now, since the Lotto balls are selected at random, each of these 22,957,480 combinations is equally likely to occur. Therefore, the probability of winning Lotto is

$$P(\text{Win } 6/53 \text{ Lotto}) = 1/(22,957,480) = .00000004356$$

This probability is often stated as follows: The odds of winning the game with a single ticket are 1 in 22,957,480, or 1 in approximately 23 million. For all practical purposes, this probability is 0, implying that you have almost no chance of winning the lottery with a single ticket. Yet each week there is almost always a winner in the Florida Lotto. This apparent contradiction can be explained with the following analogy:

Suppose there is a line of minivans, front to back, from New York City to Los Angeles, California. Based on the distance between the two cities and the length of a standard minivan, there would be approximately 23 million minivans in line. Lottery officials will select, at random, one of the minivans and put a check for $10 million in the glove compartment. For a cost of $1, you may roam the country and select one (and only one) minivan and check the glove compartment. Do you think you will find $10 million in the minivan you choose? You can be almost certain that you won't. But now permit anyone to enter the lottery for $1 and suppose that 50 million people do so. With such a large number of participants, it is very likely that someone will find the minivan with the $10 million—but it almost certainly won't be you! (This example illustrates an axiom in statistics called the law of large numbers. See the footnote at the bottom of p. 116.)

Exercises 3.1–3.32

Understanding the Principles

3.1 What is an experiment?

3.2 What are the most basic outcomes of an experiment called?

3.3 Define the sample space.

3.4 What is a Venn diagram?

3.5 Give two probability rules for sample points.

3.6 What is an event?

3.7 How do you find the probability of an event made up of several sample points?

3.8 Give a scenario where the combinations rule is appropriate for counting the number of sample points.

Learning the Mechanics

3.9 An experiment results in one of the following sample points: $E_1, E_2, E_3, E_4,$ and E_5.
   a. Find $P(E_3)$ if $P(E_1) = .1, P(E_2) = .2, P(E_4) = .1,$ and $P(E_5) = .1$
   b. Find $P(E_3)$ if $P(E_1) = P(E_3), P(E_2) = .1,$ $P(E_4) = .2,$ and $P(E_5) = .1$
   c. Find $P(E_3)$ if $P(E_1) = P(E_2) = P(E_4) = P(E_5) = .1$

3.10 The following Venn diagram describes the sample space of a particular experiment and events A and B:

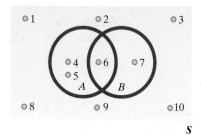

   a. Suppose the sample points are equally likely. Find $P(A)$ and $P(B)$.
   b. Suppose $P(1) = P(2) = P(3) = P(4) = P(5) = \frac{1}{20}$ and $P(6) = P(7) = P(8) = P(9) = P(10) = \frac{3}{20}$. Find $P(A)$ and $P(B)$.

3.11 The sample space for an experiment contains five sample
NW points with probabilities as shown in the table. Find the probability of each of the following events:

   A: {Either 1, 2, or 3 occurs.}
   B: {Either 1, 3, or 5 occurs.}
   C: {4 does not occur.}

Sample Points	Probabilities
1	.05
2	.20
3	.30
4	.30
5	.15

3.12 Compute each of the following:
   a. $\binom{9}{4}$ **b.** $\binom{7}{2}$ **c.** $\binom{4}{4}$
   d. $\binom{5}{0}$ **e.** $\binom{6}{5}$

3.13 Compute the number of ways you can select n elements
NW from N elements for each of the following:
   a. $n = 2, N = 5$
   b. $n = 3, N = 6$
   c. $n = 5, N = 20$

3.14 Two fair dice are tossed, and the up face on each die is
NW recorded.

   a. List the 36 sample points contained in the sample space.
   b. Assign probabilities to the sample points.
   c. Find the probability of observing each of the following events:

   A: {A 3 appears on each of the two dice.}
   B: {The sum of the numbers is even.}
   C: {The sum of the numbers is equal to 7.}
   D: {A 5 appears on at least one of the dice.}
   E: {The sum of the numbers is 10 or more.}

3.15 Two marbles are drawn at random and without replacement from a box containing two blue marbles and three red marbles.
NW **a.** List the sample points.
   b. Assign probabilities to the sample points.
   c. Determine the probability of observing each of the following events:

   A: {Two blue marbles are drawn.}
   B: {A red and a blue marble are drawn.}
   C: {Two red marbles are drawn.}

3.16 Simulate the experiment described in Exercise 3.15, using any five identically shaped objects, two of which are one color and three another. Mix the objects, draw two, record the results, and then replace the objects. Repeat the experiment a large number of times (at least 100). Calculate the proportion of times events A, B, and C occur. How do these proportions compare with the probabilities you calculated in Exercise 3.15? Should these proportions equal the probabilities? Explain.

APPLET **Applet Exercise 3.1**

Use the applet entitled *Simulating the Probability of Rolling a 6* to explore the relationship between the proportion of sixes rolled on several rolls of a die and the theoretical probability of rolling a 6 on a fair die.
   a. To simulate rolling a die one time, click on the *Roll* button on the screen while $n = 1$. The outcome of the roll appears in the list at the right, and the cumulative proportion of sixes for one roll is shown above the graph and as a point on the graph corresponding to 1. Click *Reset* and repeat the process with $n = 1$ several times. What are the possible values of the cumulative proportion of sixes for one roll of a die? Can the cumulative proportion of sixes for one roll of a die equal the theoretical probability of rolling a 6 on a fair die? Explain.
   b. Set $n = 10$ and click the *Roll* button. Repeat this several times, resetting after each time. Record the cumulative proportion of sixes for each roll. Compare the cumulative proportions for $n = 10$ with those for $n = 1$ in part **a**. Which tend to be closer to the theoretical probability of rolling a 6 on a fair die?
   c. Repeat part **b** for $n = 1,000$, comparing the cumulative proportions for $n = 1,000$ with those for $n = 1$ in part **a** and for $n = 10$ in part **b**.
   d. On the basis of your results for parts **a, b,** and **c,** do you believe that we could justifiably conclude that a die is unfair because we rolled it 10 times and didn't roll any sixes? Explain.

APPLET **Applet Exercise 3.2**

Use the applet entitled *Simulating the Probability of a Head with a Fair Coin* to explore the relationship between the proportion of heads on several flips of a coin and the theoretical probability of getting heads on one flip of a fair coin.

a. Repeat parts **a–c** of Applet Exercise 3.1 for the experiment of flipping a coin and the event of getting heads.

b. On the basis of your results for part **a**, do you believe that we could justifiably conclude that a coin is unfair because we flipped it 10 times and didn't roll any heads? Explain.

Applying the Concepts—Basic

3.17 Post office violence. *The Wall Street Journal* (Sept. 1, 2000) reported on an independent study of postal workers and violence at post offices. In a sample of 12,000 postal workers, 600 were physically assaulted on the job in a recent year. Use this information to estimate the probability that a randomly selected postal worker will be physically assaulted on the job during the year.

3.18 Male nannies. In a recent survey conducted by the International Nanny Association (INA) and reported at the INA website (www.nanny.org, 2007), 4,176 nannies were placed in a job last year. Only 24 of the nannies placed were men. Find the probability that a randomly selected nanny who was placed last year is a male nanny (a "mannie").

3.19 USDA chicken inspection. The United States Department of Agriculture (USDA) reports that, under its standard inspection system, one in every 100 slaughtered chickens passes inspection for fecal contamination. (*Tampa Tribune*, Mar. 31, 2000.)

a. If a slaughtered chicken is selected at random, what is the probability that it passes inspection for fecal contamination?

b. The probability of part **a** was based on a USDA study which found that 306 of 32,075 chicken carcasses passed inspection for fecal contamination. Do you agree with the USDA's statement about the likelihood of a slaughtered chicken passing inspection for fecal contamination?

3.20 African rhinos. Two species of rhinoceros native to Africa
NW are black rhinos and white rhinos. In March 2007, the International Rhino Federation estimated that the African rhinoceros population consisted of 3,610 white rhinos and 11,330 black rhinos. Suppose one rhino is selected at random from the African rhino population and its species (black or white) is observed.

a. List the sample points for this experiment.

b. Assign probabilities to the sample points on the basis of the estimates made by the International Rhino Federation.

3.21 Fungi in beech forest trees. Beechwood forests in East Central Europe are being threatened by dynamic changes in land ownership and economic upheaval. The current status of the beech tree species in this area was evaluated by Hungarian university professors in *Applied Ecology and Environmental Research* (Vol. 1, 2003). Of 188 beech trees surveyed, 49 had been damaged by fungi. Depending on the species of fungus, damage will occur on either the trunk, branches, or leaves of the tree. In the damaged trees, the trunk was affected 85% of the time, the leaves 10% of the time, and the branches 5% of the time.

a. Give a reasonable estimate of the probability of a beech tree in East Central Europe being damaged by fungi.

b. A fungus-damaged beech tree is selected at random, and the area (trunk, leaf, or branch) affected is observed. List

the sample points for this experiment, and assign a reasonable probability to each one.

3.22 Chance of rain. Answer the following question posed in the *Atlanta Journal-Constitution* (Feb. 7, 2000): When a meteorologist says, "The probability of rain this afternoon is .4," does it mean that it will be raining 40% of the time during the afternoon?

Applying the Concepts—Intermediate

3.23 Sociology fieldwork methods. Refer to University of New
NW Mexico professor Jane Hood's study of the fieldwork methods used by qualitative sociologists, presented in Exercise 2.14 (p. 36). Recall that she discovered that fieldwork methods could be classified into four distinct categories: Interview, Observation plus Participation, Observation Only, and Grounded Theory. The table that follows, reproduced from *Teaching Sociology* (July 2006) gives the number of sociology field research papers in each category. Suppose we randomly select one of these research papers and determine the method used.

a. List the possible outcomes (sample points) for this experiment.

b. Explain why the sample points are not equally likely to occur.

c. Assign reasonable probabilities to the sample points.

d. Find the probability that the method used is either Interview or Grounded Theory.

FIELDWORK

Fieldwork Method	Number of Papers
Interview	5,079
Observation plus Participation	1,042
Observation Only	848
Grounded Theory	537

Source: Hood, J. C. "Teaching against the text: The case of qualitative methods," *Teaching Sociology*, Vol. 34, July 2006 (Exhibit 2).

3.24 Sex composition patterns of children in families. In having children, is there a genetic factor that causes some families to favor one sex over the other? That is, does having boys or girls "run in the family"? This was the question of interest in *Chance* (Fall 2001). Using data collected on children's sex for over 4,000 American families that had at least two children, the researchers compiled the accompanying table. An American family with at least two children is selected, and the sex composition of the first two children is observed.

a. List the sample points for this experiment.

Sex Composition of First Two Children	Frequency
Boy–Boy	1,085
Boy–Girl	1,086
Girl–Boy	1,111
Girl–Girl	926
TOTAL	4,208

Source: Rodgers, J. L., and Doughty, D. "Does having boys or girls run in the family?" *Chance*, Vol. 14, No. 4, Fall 2001, Table 3.

b. If having a boy is no more likely than having a girl and vice versa, assign a probability to each sample point.

c. Use the information in the table to estimate the sample point probabilities. Do these estimates agree (to a reasonable degree of approximation) with the probabilities you found in part **b**?

d. Make an inference about whether having boys or girls "runs in the family."

3.25 Choosing portable grill displays. University of Maryland marketing professor R. W. Hamilton studied how people attempt to influence the choices of others by offering undesirable alternatives. (*Journal of Consumer Research*, Mar. 2003.) Such a phenomenon typically occurs when family members propose a vacation spot, friends recommend a restaurant for dinner, and realtors show the buyer potential homesites. In one phase of the study, the researcher had each of 124 college students select showroom displays for portable grills. Five different displays (representing five different-sized grills) were available, but only three would be selected. The students were instructed to select the displays to maximize purchases of Grill #2 (a smaller grill).

a. In how many possible ways can the three-grill displays be selected from the 5 displays? List the possibilities.

b. The table shows the grill display combinations and number of each selected by the 124 students. Use this information to assign reasonable probabilities to the different display combinations.

c. Find the probability that a student who participated in the study selected a display combination involving Grill #1.

Grill Display Combination	Number of Students
1–2–3	35
1–2–4	8
1–2–5	42
2–3–4	4
2–3–5	1
2–4–5	34

Source: Hamilton, R. W. "Why do people suggest what they do not want? Using context effects to influence others' choices," *Journal of Consumer Research*, Vol. 29, Mar. 2003, Table 1.

3.26 Perfect SAT scores. The maximum score possible on the SAT is 1600. According to the test developers, the chance that a student scores a perfect 1600 on the SAT is 5 in 10,000.

a. Find the probability that a randomly selected student scores a 1600 on the SAT.

b. In a recent year, 545 students scored a perfect 1600 on the SAT. How is this possible, given the probability calculated in part **a**?

3.27 Jai alai Quinella bet. The Quinella bet at the parimutuel game of jai alai consists of picking the jai alai players that will place first and second in a game, *irrespective of order*. In jai alai, eight players (numbered 1, 2, 3, . . . , 8) compete in every game.

a. How many different Quinella bets are possible?

b. Suppose you bet the Quinella combination 2–7. If the players are of equal ability, what is the probability that you win the bet?

MTBE

3.28 Groundwater contamination in wells. Refer to the *Environmental Science & Technology* (Jan. 2005) study of methyl *tert*-butyl ether (MTBE) contamination in New Hampshire wells, presented in Exercise 2.19 (p. 37). Data collected for a sample of 223 wells are saved in the MTBE file. Recall that each well was classified according to its class (public or private), aquifer (bedrock or unconsolidated), and detectable level of MTBE (below limit or detectable).

a. Consider an experiment in which the class, aquifer, and detectable MTBE level of a well are observed. List the sample points for this experiment. [*Hint:* One sample point is Private/Bedrock/BelowLimit.]

b. Use statistical software to find the number of the 223 wells in each sample point outcome. Then use this information to compute probabilities for the sample points.

c. Find and interpret the probability that a well has a detectable level of MTBE.

Applying the Concepts—Advanced

3.29 Lead bullets as forensic evidence. *Chance* (Summer, 2004) published an article on the use of lead bullets as forensic evidence in a federal criminal case. Typically, the Federal Bureau of Investigation (FBI) will use a laboratory method to match the lead in a bullet found at a crime scene with unexpended lead cartridges found in the possession of a suspect. The value of this evidence depends on the chance of a *false positive*—that is, the probability that the FBI finds a match, given that the lead at the crime scene and the lead in the possession of the suspect are actually from two different "melts," or sources. To estimate the false positive rate, the FBI collected 1,837 bullets that the agency was confident all came from different melts. Then, using its established criteria, the FBI examined every possible pair of bullets and counted the number of matches. According to *Chance*, the FBI found 693 matches. Use this information to compute the chance of a false positive. Is this probability small enough for you to have confidence in the FBI's forensic evidence?

3.30 Matching socks. Consider the following question posed to Marilyn vos Savant in her weekly newspaper column, "Ask Marilyn":

I have two pairs of argyle socks, and they look nearly identical—one navy blue and the other black. [When doing the laundry] my wife matches the socks incorrectly much more often than she does correctly. . . . If all four socks are in front of her, it seems to me that her chances are 50% for a wrong match and 50% for a right match. What do you think?
Source: Parade Magazine, Feb. 27, 1994.

Use your knowledge of probability to answer this question. [*Hint:* List the sample points in the experiment.]

3.31 Post-op nausea study. Nausea and vomiting after surgery are common side effects of anesthesia and painkillers. Six different drugs, varying in cost, were compared for their effectiveness in preventing nausea and vomiting. (*New England Journal of Medicine*, June 10, 2004.) The medical researchers looked at all possible combinations of the drugs as treatments, including a single drug, as well as two-drug, three-drug, four-drug, five-drug, and six-drug combinations.

a. How many two-drug combinations of the six drugs are possible?

b. How many three-drug combinations of the six drugs are possible?

c. How many four-drug combinations of the six drugs are possible?

d. How many five-drug combinations of the six drugs are possible?

e. The researchers stated that a total of 64 drug combinations were tested as treatments for nausea. Verify that there are 64 ways that the six drugs can be combined. (Remember to include the one-drug and six-drug combinations, as well as the control treatment of no drugs.)

3.32 Dominant versus recessive traits. An individual's genetic makeup is determined by the genes obtained from each parent. For every genetic trait, each parent possesses a gene pair, and each parent contributes one-half of this gene pair, with equal probability, to his or her offspring, forming a new gene pair. The offspring's traits (eye color, baldness, etc.) come from this new gene pair, each gene of which possesses some characteristic.

For the gene pair that determines eye color, each gene trait may be one of two types: dominant brown (*B*) or recessive blue (*b*). A person possessing the gene pair *BB* or *Bb* has brown eyes, whereas the gene pair *bb* produces blue eyes.

a. Suppose both parents of an individual are brown eyed, each with a gene pair of type *Bb*. What is the probability that a randomly selected child of this couple will have blue eyes?

b. If one parent has brown eyes, type *Bb*, and the other has blue eyes, what is the probability that a randomly selected child of this couple will have blue eyes?

c. Suppose one parent is brown eyed with a gene pair of type *BB*. What is the probability that a child has blue eyes?

3.2 Unions and Intersections

An event can often be viewed as a composition of two or more other events. Such events, which are called **compound events**, can be formed (composed) in two ways.

Definition 3.5

The **union** of two events *A* and *B* is the event that occurs if either *A* or *B* (or both) occurs on a single performance of the experiment. We denote the union of events *A* and *B* by the symbol $A \cup B$. $A \cup B$ consists of all the sample points that belong to *A or B or both*. (See Figure 3.6a.)

Definition 3.6

The **intersection** of two events *A* and *B* is the event that occurs if both *A* and *B* occur on a single performance of the experiment. We write $A \cap B$ for the intersection of *A* and *B*. $A \cap B$ consists of all the sample points belonging to *both A and B*. (See Figure 3.6b.)

FIGURE 3.6
Venn diagrams for union and intersection

Entire shaded area is $A \cup B$.

a. Union

Shaded area is $A \cap B$.

b. Intersection

EXAMPLE 3.8

PROBABILITIES OF UNIONS AND INTERSECTIONS
—Die-Toss Experiment

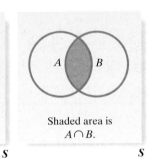

Problem Consider a die-toss experiment in which the following events are defined:

A: {Toss an even number.}
B: {Toss a number less than or equal to 3.}

a. Describe $A \cup B$ for this experiment.

b. Describe $A \cap B$ for this experiment.

c. Calculate $P(A \cup B)$ and $P(A \cap B)$, assuming that the die is fair.

Solution Draw the Venn diagram as shown in Figure 3.7.

a. The union of A and B is the event that occurs if we observe either an even number, a number less than or equal to 3, or both on a single throw of the die. Consequently, the sample points in the event $A \cup B$ are those for which A occurs, B occurs, or both A and B occur. Checking the sample points in the entire sample space, we find that the collection of sample points in the union of A and B is

$$A \cup B = \{1, 2, 3, 4, 6\}$$

b. The intersection of A and B is the event that occurs if we observe *both* an even number and a number less than or equal to 3 on a single throw of the die. Checking the sample points to see which imply the occurrence of *both* events A and B, we see that the intersection contains only one sample point:

$$A \cap B = \{2\}$$

In other words, the intersection of A and B is the sample point Observe a 2.

c. Recalling that the probability of an event is the sum of the probabilities of the sample points of which the event is composed, we have

$$P(A \cup B) = P(1) + P(2) + P(3) + P(4) + P(6)$$

$$= \frac{1}{6} + \frac{1}{6} + \frac{1}{6} + \frac{1}{6} + \frac{1}{6} = \frac{5}{6}$$

and

$$P(A \cap B) = P(2) = \frac{1}{6}$$

Look Back Since the six sample points are equally likely, the probabilities in part **c** are simply the number of sample points in the event of interest, divided by 6.

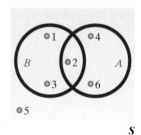

FIGURE 3.7
Venn diagram for die toss

Now Work Exercise 3.43a–d

Unions and intersections can be defined for more than two events. For example, the event $A \cup B \cup C$ represents the union of three events: A, B, and C. This event, which includes the set of sample points in A, B, or C, will occur if any one (or more) of the events A, B, and C occurs. Similarly, the intersection $A \cap B \cap C$ is the event that all three of the events A, B, and C occur. Therefore, $A \cap B \cap C$ is the set of sample points that are in all three of the events A, B, and C.

EXAMPLE 3.9

FINDING PROBABILITIES FROM A TWO-WAY TABLE— Mother's Race Versus Maternal Age

Problem *Family Planning Perspectives* reported on a study of over 200,000 births in New Jersey over a recent two-year period. The study investigated the link between the mother's race and the age at which she gave birth (called maternal age). The percentages of the total number of births in New Jersey, by the maternal age and race classifications, are given in Table 3.4.

This table is called a **two-way table,** since responses are classified according to two variables: maternal age (rows) and race (columns).

TABLE 3.4 Percentage of New Jersey Birth Mothers, by Age and Race

	RACE	
Maternal Age (years)	White	Black
≤17	2%	2%
18–19	3%	2%
20–29	41%	12%
≥30	33%	5%

Source: Reichman, N. E., and Pagnini, D. L. "Maternal age and birth outcomes: Data from New Jersey." *Family Planning Perspectives,* Vol. 29, No. 6, Nov./Dec. 1997, p. 269 (adapted from Table 1).

Define the following event:

A: {A New Jersey birth mother is white.}
B: {A New Jersey mother was a teenager when giving birth.}

a. Find $P(A)$ and $P(B)$.
b. Find $P(A \cup B)$.
c. Find $P(A \cap B)$.

Solution Following the steps for calculating probabilities of events, we first note that the objective is to characterize the race and maternal age distribution of New Jersey birth mothers. To accomplish this objective, we define the experiment to consist of selecting a birth mother from the collection of all New Jersey birth mothers during the two-year period of the study and observing her race and maternal age class. The sample points are the eight different age–race classifications:

E_1: {$\leq$ 17 yrs., white} E_5: {$\leq$ 17 yrs., black}
E_2: {18−19 yrs., white} E_6: {18−19 yrs., black}
E_3: {20−29 yrs., white} E_7: {20−29 yrs., black}
E_4: {$\geq$ 30 yrs., white} E_8: {$\geq$ 30 yrs., black}

Next, we assign probabilities to the sample points. If we blindly select one of the birth mothers, the probability that she will occupy a particular age–race classification is just the proportion, or relative frequency, of birth mothers in that classification. These proportions (as percentages) are given in Table 3.4. Thus,

$P(E_1)$ = Relative frequency of birth mothers in age−race class {$\leq$ 17 yrs., white} = .02
$P(E_2) = .03$
$P(E_3) = .41$
$P(E_4) = .33$
$P(E_5) = .02$
$P(E_6) = .02$
$P(E_7) = .12$
$P(E_8) = .05$

You may verify that the sample point probabilities sum to 1.

a. To find $P(A)$, we first determine the collection of sample points contained in event A. Since A is defined as {white}, we see from Table 3.4 that A contains the four sample points represented by the first column of the table. In other words, the event A consists of the race classification {white} in all four age classifications. The probability of A is the sum of the probabilities of the sample points in A:

$$P(A) = P(E_1) + P(E_2) + P(E_3) + P(E_4) = .02 + .03 + .41 + .33 = .79$$

Similarly, B = {teenage mother, age $\leq$19 years} consists of the four sample points in the first and second rows of Table 3.4:

$$P(B) = P(E_1) + P(E_2) + P(E_5) + P(E_6) = .02 + .03 + .02 + .02 = .09$$

b. The union of events A and B, $A \cup B$, consists of all sample points in *either A or B (or both)*. That is, the union of A and B consists of all birth mothers who are white *or* who gave birth as a teenager. In Table 3.4, this is any sample point found in the first column *or* the first two rows. Thus,

$$P(A \cup B) = .02 + .03 + .41 + .33 + .02 + .02 = .83$$

c. The intersection of events A and B, $A \cap B$, consists of all sample points in *both A and B*. That is, the intersection of A and B consists of all birth mothers who are white *and* who gave birth as a teenager. In Table 3.4, this is any sample point found in the first column *and* the first two rows. Thus,

$$P(A \cap B) = .02 + .03 = .05.$$

Look Back As in previous problems, the key to finding the probabilities of parts **b** and **c** is to identify the sample points that make up the event of interest. In a two-way table such as Table 3.4, the total number of sample points will be equal to the number of rows times the number of columns.

Now Work Exercise 3.45f–g

3.3 Complementary Events

A very useful concept in the calculation of event probabilities is the notion of **complementary events**:

> **Definition 3.7**
>
> The **complement** of an event A is the event that A does *not* occur—that is, the event consisting of all sample points that are not in event A. We denote the complement of A by A^c.

An event A is a collection of sample points, and the sample points included in A^c are those not in A. Figure 3.8 demonstrates this idea. Note from the figure that all sample points in S are included in *either* A or A^c and that *no* sample point is in both A and A^c. This leads us to conclude that the probabilities of an event and its complement *must sum to 1*:

> **Rule of Complements**
>
> The sum of the probabilities of complementary events equals 1; that is,
>
> $$P(A) + P(A^c) = 1.$$

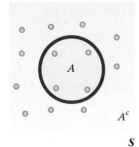

FIGURE 3.8
Venn diagram of complementary events

In many probability problems, calculating the probability of the complement of the event of interest is easier than calculating the event itself. Then, because

$$P(A) + P(A^c) = 1$$

we can calculate $P(A)$ by using the relationship

$$P(A) = 1 - P(A^c).$$

EXAMPLE 3.10

PROBABILITY OF A COMPLEMENTARY EVENT—Coin-Toss Experiment

Problem Consider the experiment of tossing two fair coins. (The sample space for this experiment is shown in Figure 3.9.) Calculate the probability of event A: {Observing at least one head}.

Solution We know that the event A: {Observe at least one head.} consists of the sample points

$$A: \{HH, HT, TH\}$$

The complement of A is defined as the event that occurs when A does not occur. Therefore,

$$A^c: \{\text{Observe no heads.}\} = \{TT\}$$

This complementary relationship is shown in Figure 3.9. Assuming that the coins are balanced, we have

$$P(A^c) = P(TT) = \frac{1}{4}$$

and

$$P(A) = 1 - P(A^c) = 1 - \frac{1}{4} = \frac{3}{4}.$$

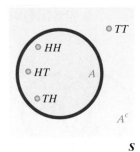

FIGURE 3.9
Complementary events in the toss of two coins

Look Back Note that we can find $P(A)$ by summing the probabilities of the sample points HH, HT, and TH in A. Many times, however, it is easier to find $P(A^c)$ by using the rule of complements.

Now Work Exercise 3.43e–f

EXAMPLE 3.11

APPLYING THE RULE OF COMPLEMENTS— 10 Coin Tosses

Problem A fair coin is tossed 10 times, and the up face is recorded after each toss. What is the probability of event A: {Observe at least one head.}?

Solution We solve this problem by following the five steps for calculating probabilities of events. (See Section 3.1.)

Step 1 Define the experiment. The experiment is to record the results of the 10 tosses of the coin.

Step 2 List the sample points. A sample point consists of a particular sequence of 10 heads and tails. Thus, one sample point is $HHTTTHTHTT$, which denotes head on first toss, head on second toss, tail on third toss, etc. Others are $HTHHHTTTTT$ and $THHTHTHTTH$. Obviously, the number of sample points is very large—too many to list. It can be shown (see Section 3.8) that there are $2^{10} = 1{,}024$ sample points for this experiment.

Step 3 Assign probabilities. Since the coin is fair, each sequence of heads and tails has the same chance of occurring; therefore, all the sample points are equally likely. Then

$$P(\text{Each sample point}) = \frac{1}{1{,}024}$$

Step 4 Determine the sample points in event A. A sample point is in A if at least one H appears in the sequence of 10 tosses. However, if we consider the complement of A, we find that

$$A^c = \{\text{No heads are observed in 10 tosses.}\}$$

Thus, A^c contains only one sample point:

$$A^c: \{TTTTTTTTTT\}$$

and $P(A^c) = \dfrac{1}{1{,}024}$

Step 5 Now we use the relationship of complementary events to find $P(A)$:

$$P(A) = 1 - P(A^c) = 1 - \frac{1}{1{,}024} = \frac{1{,}023}{1{,}024} = .999$$

Look Back With such a high probability, we are virtually certain of observing at least one head in 10 tosses of the coin.

3.4 The Additive Rule and Mutually Exclusive Events

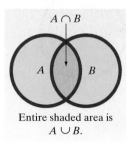

$A \cap B$

A B

Entire shaded area is
$A \cup B$.

S

FIGURE 3.10
Venn diagram of union

In Section 3.2, we saw how to determine which sample points are contained in a union of two sets and how to calculate the probability of the union by summing the separate probabilities of the sample points in the union. It is also possible to obtain the probability of the union of two events by using the *additive rule of probability*.

By studying the Venn diagram in Figure 3.10, you can see that the probability of the union of two events A and B can be obtained by summing $P(A)$ and $P(B)$ and subtracting $P(A \cap B)$. We must subtract $P(A \cap B)$ because the sample point probabilities in $A \cap B$ have been included twice—once in $P(A)$ and once in $P(B)$.

The formula for calculating the probability of the union of two events is given in the next box.

> **Additive Rule of Probability**
>
> The probability of the union of events A and B is the sum of the probability of event A and the probability of event B, minus the probability of the intersection of events A and B; that is
>
> $$P(A \cup B) = P(A) + P(B) - P(A \cap B)$$

EXAMPLE 3.12

APPLYING THE ADDITIVE RULE—Hospital Admissions Study

Problem Hospital records show that 12% of all patients are admitted for surgical treatment, 16% are admitted for obstetrics, and 2% receive both obstetrics and surgical treatment. If a new patient is admitted to the hospital, what is the probability that the patient will be admitted for surgery, for obstetrics, or for both?

Solution Consider the following events:

> A: {A patient admitted to the hospital receives surgical treatment.}
> B: {A patient admitted to the hospital receives obstetrics treatment.}

Then, from the given information,

$$P(A) = .12$$
$$P(B) = .16$$

and the probability of the event that a patient receives both obstetrics and surgical treatment is

$$P(A \cap B) = .02$$

The event that a patient admitted to the hospital receives either surgical treatment, obstetrics treatment, or both is the union $A \cup B$. The probability of which is given by the additive rule of probability:

$$P(A \cup B) = P(A) + P(B) - P(A \cap B)$$
$$= .12 + .16 - .02 = .26$$

Thus, 26% of all patients admitted to the hospital receive either surgical treatment, obstetrics treatment, or both.

Look Back From the information given, it is not possible to list and assign probabilities to all the sample points. Consequently, we cannot proceed through the five-step process (p. 120) for finding $P(A \cup B)$, and we must use the additive rule.

Now Work Exercise 3.41

A very special relationship exists between events A and B when $A \cap B$ contains no sample points. In this case, we call the events A and B *mutually exclusive events*.

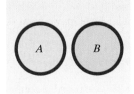

FIGURE 3.11
Venn diagram of mutually exclusive events

Definition 3.8

Events A and B are **mutually exclusive** if $A \cap B$ contains no sample points—that is, if A and B have no sample points in common. For mutually exclusive events,

$$P(A \cap B) = 0$$

Figure 3.11 shows a Venn diagram of two mutually exclusive events. The events A and B have no sample points in common; that is, A and B cannot occur simultaneously, and $P(A \cap B) = 0$. Thus, we have the important relationship given in the next box.

Probability of Union of Two Mutually Exclusive Events

If two events A and B are *mutually exclusive*, the probability of the union of A and B equals the sum of the probability of A and the probability of B; that is, $P(A \cup B) = P(A) + P(B)$.

Caution

The preceding formula is *false* if the events are *not* mutually exclusive. In that case (i.e., two non-mutually exclusive events), you must apply the general additive rule of probability.

EXAMPLE 3.13
UNION OF TWO MUTUALLY EXCLUSIVE EVENTS—Coin-Tossing Experiment

Problem Consider the experiment of tossing two balanced coins. Find the probability of observing *at least* one head.

Solution Define the events

$$A: \{\text{Observe at least one head.}\}$$
$$B: \{\text{Observe exactly one head.}\}$$
$$C: \{\text{Observe exactly two heads.}\}$$

Note that

$$A = B \cup C$$

and that $B \cap C$ contains no sample points. (See Figure 3.12.) Thus, B and C are mutually exclusive, and it follows that

$$P(A) = P(B \cup C) = P(B) + P(C) = \frac{1}{2} + \frac{1}{4} = \frac{3}{4}$$

Look Back Although this example is quite simple, it shows us that writing events with verbal descriptions that include the phrases "at least" or "at most" as unions of mutually exclusive events is very useful. This practice enables us to find the probability of the event by adding the probabilities of the mutually exclusive events.

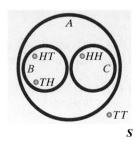

FIGURE 3.12
Venn diagram for coin toss experiment

STATISTICS IN ACTION REVISITED

The Probability of Winning Lotto with a Wheeling System

Refer to Florida's Pick-6 Lotto game, in which you select six numbers of your choice from a field of numbers ranging from 1 to 53. In Section 3.1, we learned that the probability of winning Lotto on a single ticket is only 1 in approximately 23 million. The "experts" at Lotto Buster recommend many strategies for increasing the odds of winning the lottery. One strategy is to employ a *wheeling system*. In a complete wheeling system, you select more than six numbers—say, seven—and play every combination of six of those seven numbers.

For example, suppose you choose to "wheel" the following seven numbers: 2, 7, 18, 23, 30, 32, and 51. Every combination of six of these seven numbers is listed in

(continued)

Table SIA3.1. You can see that there are seven different possibilities. (Use the combinations rule with $N = 7$ and $n = 6$ to verify this.) Thus, we would purchase seven tickets (at a cost of $7) corresponding to these different combinations in a complete wheeling system.

To determine whether this strategy does, in fact, increase our odds of winning, we need to find the probability that one of these seven combinations occurs during the 6/53 Lotto draw. That is, we need to find the probability that either Ticket #1 or Ticket #2 or Ticket #3 or Ticket #4 or Ticket #5 or Ticket #6 or Ticket #7 is the winning combination. Note that this probability is stated with the use of the word *or*, implying a *union* of seven events. Letting T_1 represent the event that Ticket

#1 wins, and defining in a similar fashion, we want to find

$$P(T_1 \text{ or } T_2 \text{ or } T_3 \text{ or } T_4 \text{ or } T_5 \text{ or } T_6 \text{ or } T_7)$$

Recall (Section 3.2) that the 22,957,480 possible combinations in Pick-6 Lotto are mutually exclusive and equally likely to occur. Consequently, the probability of the union of the seven events is simply the sum of the probabilities of the individual events, where each event has probability 1/(22,957,480):

$$P(\text{win Lotto with 7 Wheeled Numbers})$$
$$= P(T_1 \text{ or } T_2 \text{ or } T_3 \text{ or } T_4 \text{ or } T_5 \text{ or } T_6 \text{ or } T_7)$$
$$= 7/(22,957,480) = .0000003$$

In terms of odds, we now have 3 chances in 10 million of winning the Lotto with the complete wheeling system. The "experts" are correct: Our odds of winning Lotto have increased (from 1 in 23 million). However, the probability of winning is so close to 0 that we question whether the $7 spent on lottery tickets is worth the negligible increase in odds. In fact, it can be shown that to increase your chance of winning the 6/53 Lotto to 1 chance in 100 (i.e., .01) by means of a complete wheeling system, you would have to wheel 26 of your favorite numbers—a total of 230,230 combinations at a cost of $230,230!

TABLE SIA3.1 Wheeling the Seven Numbers 2, 7, 18, 23, 30, 32, and 51

Ticket #1	2	7	18	23	30	32
Ticket #2	2	7	18	23	30	51
Ticket #3	2	7	18	23	32	51
Ticket #4	2	7	18	30	32	51
Ticket #5	2	7	23	30	32	51
Ticket #6	2	18	23	30	32	51
Ticket #7	7	18	23	30	32	51

Exercises 3.33–3.57

Understanding the Principles

3.33 Define in words the union of two events.

3.34 Define in words the intersection of two events.

3.35 Define in words the complement of an event.

3.36 State the rule of complements.

3.37 State the additive rule of probability for any two events.

3.38 Define in words mutually exclusive events.

3.39 State the additive rule of probability for mutually exclusive events.

Learning the Mechanics

3.40 Suppose $P(A) = .4$, $P(B) = .7$ and $P(A \cap B) = .3$. Find the following probabilities:
a. $P(B^c)$
b. $P(A^c)$
c. $P(A \cup B)$

3.41 A fair coin is tossed three times, and the events A and B are
NW defined as follows:

A: {At least one head is observed.}
B: {The number of heads observed is odd.}

a. Identify the sample points in the events A, B, $A \cup B$, A^c, and $A \cap B$.
b. Find $P(A)$, $P(B)$, $P(A \cup B)$, $P(A^c)$, and $P(A \cap B)$ by summing the probabilities of the appropriate sample points.
c. Use the additive rule to find $P(A \cup B)$. Compare your answer with the one you obtained in part **b**.
d. Are the events A and B mutually exclusive? Why?

3.42 A pair of fair dice is tossed. Define the following events:

A: {You will roll a 7 (i.e., the sum of the numbers of dots on the upper faces of the two dice is equal to 7).}
B: {At least one of the two dice is showing a 4.}

a. Identify the sample points in the events A, B, $A \cap B$, $A \cup B$, and A^c.
b. Find $P(A)$, $P(B)$, $P(A \cap B)$, $P(A \cup B)$, and $P(A^c)$ and by summing the probabilities of the appropriate sample points.
c. Use the additive rule to find $P(A \cup B)$. Compare your answer with that for the same event in part **b**.
d. Are A and B mutually exclusive? Why?

3.43 Consider the following Venn diagram, where $P(E_1) =$
NW $P(E_2) = P(E_3) = \frac{1}{5}$, $P(E_4) = P(E_5) = \frac{1}{20}$, $P(E_6) = \frac{1}{10}$, $P(E_7) = \frac{1}{5}$:

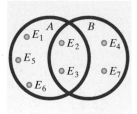

Find each of the following probabilities:
a. $P(A)$ **b.** $P(B)$ **c.** $P(A \cup B)$
d. $P(A \cap B)$ **e.** $P(A^c)$ **f.** $P(B^c)$
g. $P(A \cup A^c)$ **h.** $P(A^c \cap B)$

3.44 Consider the following Venn diagram, where

$P(E_1) = .10$, $P(E_2) = .05$, $P(E_3) = P(E_4) = .2$,
$P(E_5) = .06$, $P(E_6) = .3$, $P(E_7) = .06$, and
$P(E_8) = .03$:

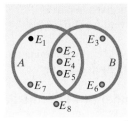

Find the following probabilities:
a. $P(A^c)$ **b.** $P(B^c)$ **c.** $P(A^c \cap B)$
d. $P(A \cup B)$ **e.** $P(A \cap B)$ **f.** $P(A^c \cup B^c)$
g. Are events A and B mutually exclusive? Why?

3.45 The outcomes of two variables are (Low, Medium, High)
NW and (On, Off), respectively. An experiment is conducted in which the outcomes of each of the two variables are observed. The probabilities associated with each of the six possible outcome pairs are given in the following table:

	Low	Medium	High
On	.50	.10	.05
Off	.25	.07	.03

Consider the following events:

A: {On}
B: {Medium or On}
C: {Off and Low}
D: {High}

a. Find $P(A)$.   **b.** Find $P(B)$.
c. Find $P(C)$.   **d.** Find $P(D)$.
e. Find $P(A^c)$.   **f.** Find $P(A \cup B)$.
g. Find $P(A \cap B)$.
h. Consider each possible pair of events taken from the events A, B, C, and D. List the pairs of events that are mutually exclusive. Justify your choices.

3.46 Three fair coins are tossed. We wish to find the probability of the event A: {Observe at least one head.}
a. Express A as the union of three mutually exclusive events. Using the expression you wrote, find the probability of A.
b. Express A as the complement of an event. Using the expression you wrote, find the probability of A.

APPLET Applet Exercise 3.3

Use the applets entitled *Simulating the Probability of Rolling a 6* and *Simulating the Probability of Rolling a 3 or 4* to explore the additive rule of probability.
a. Explain why the applet *Simulating the Probability of Rolling a 6* can also be used to simulate the probability of rolling a 3. Then use the applet with $n = 1000$ to simulate the probability of rolling a 3. Record the cumulative proportion. Repeat the process to simulate the probability of rolling a 4.
b. Use the applet *Simulating the Probability of Rolling a 3 or 4* with $n = 1000$ to simulate the probability of rolling a 3 or 4. Record the cumulative proportion.

c. Add the two cumulative proportions from part **a**. How does this sum compare with the cumulative proportion in part **b**? How does your answer illustrate the additive rule for probability?

APPLET Applet Exercise 3.4

Use the applets entitled *Simulating the Probability of Rolling a 6* and *Simulating the Probability of Rolling a 3 or 4* to simulate the probability of the complement of an event.
a. Explain how the applet *Simulating the Probability of Rolling a 6* can also be used to simulate the probability of the event *rolling a 1, 2, 3, 4, or 5*. Then use the applet with $n = 1000$ to simulate this probability.
b. Explain how the applet *Simulating the Probability of Rolling a 3 or 4* can also be used to simulate the probability of the event *rolling a 1, 2, 5, or 6*. Then use the applet with $n = 1000$ to simulate this probability.
c. Which applet could be used to simulate the probability of the event *rolling a 1, 2, 3, or 4*? Explain.

Applying the Concepts—Basic

3.47 Toxic chemical incidents. *Process Safety Progress* (Sep. 2004) reported on an emergency response system for incidents involving toxic chemicals in Taiwan. The system has logged over 250 incidents since being implemented. The accompanying table gives a breakdown of the locations where these incidents occurred. Consider the location of a toxic chemical incident in Taiwan.
a. List the sample points for this experiment.
b. Assign reasonable probabilities to the sample points.
c. What is the probability that the incident occurs in a school laboratory?
d. What is the probability that the incident occurs in either a chemical or a nonchemical plant?
e. What is the probability that the incident does not occur in transit?

Location	Percent of Incidents
School laboratory	6%
In transit	26%
Chemical plant	21%
Nonchemical plant	35%
Other	12%
TOTAL	100%

Source: Chen, J.R., et al. "Emergency response of toxic chemicals in Taiwan: The system and case studies," *Process Safety Progress*, Vol. 23, No. 3, Sep. 2004 (Figure 5a).

3.48 Scanning errors at Wal-Mart. The National Institute of Standards and Technology (NIST) mandates that for every 100 items scanned through the electronic checkout scanner at a retail store, no more than 2 should have an inaccurate price. A recent study of the accuracy of checkout scanners at Wal-Mart stores in California (*Tampa Tribune*, Nov. 22, 2005) showed that, of the 60 Wal-Mart stores investigated, 52 violated the NIST scanner accuracy standard. If 1 of the 60 Wal-Mart stores is randomly selected, what is the probability that that store does not violate the NIST scanner accuracy standard?

3.49 Gene expression profiling. Gene expression profiling is a state-of-the-art method for determining the biology of cells. In *Briefings in Functional Genomics and Proteomics*

(Dec. 2006), biologists at Pacific Northwest National Laboratory reviewed several gene expression profiling methods. The biologists applied two of the methods (A and B) to data collected on proteins in human mammary cells. The probability that the protein is cross-referenced (i.e., identified) by method A is .41, the probability that the protein is cross-referenced by method B is .42, and the probability that the protein is cross-referenced by both methods is .40.

a. Draw a Venn diagram to illustrate the results of the gene-profiling analysis.

b. Find the probability that the protein is cross-referenced by either method A or method B.

c. On the basis of your answer to part b, find the probability that the protein is not cross-referenced by either method.

3.50 Binge alcohol drinking. A study of binge alcohol drinking by college students was published in the *American Journal of Public Health* (July 1995). Suppose an experiment consists of randomly selecting one of the undergraduate students who participated in the study. Consider the following events:

A: {The student is a binge drinker.}
B: {The student is a male.}
C: {The student lives in a coed dorm.}

Describe each of the following events in terms of unions, intersections, and complements ($A \cup B$, $A \cap B$, A^c, etc.):

a. The student is male and a binge drinker.
b. The student is not a binge drinker.
c. The student is male or lives in a coed dorm.
d. The student is female and not a binge drinker.

3.51 Odds of winning at roulette. *Roulette* is a very popular game in many American casinos. In Roulette, a ball spins on a circular wheel that is divided into 38 arcs of equal length, bearing the numbers $00, 0, 1, 2, \ldots, 35, 36$. The number of the arc on which the ball stops is the outcome of one play of the game. The numbers are also colored in the manner shown in the following table

Red: $1, 3, 5, 7, 9, 12, 14, 16, 18, 19, 21, 23, 25, 27, 30, 32, 34, 36$
Black: $2, 4, 6, 8, 10, 11, 13, 15, 17, 20, 22, 24, 26, 28, 29, 31, 33, 35$
Green: $00, 0$

Players may place bets on the table in a variety of ways, including bets on odd, even, red, black, high, low, etc. Consider the following events:

A: {The outcome is an odd number (00 and 0 are considered neither odd nor even.)}
B: {The outcome is a black number.}
C: {The outcome is a low number (1–18).}

a. Define the event $A \cap B$ as a specific set of sample points.

b. Define the event $A \cup B$ as a specific set of sample points.

c. Find $P(A)$, $P(B)$, $P(A \cap B)$, $P(A \cup B)$, and $P(C)$ by summing the probabilities of the appropriate sample points.

d. Define the event $A \cap B \cap C$ as a specific set of sample points.

e. Use the additive rule to find $P(A \cup B)$. Are events A and B mutually exclusive? Why?

f. Find $P(A \cap B \cap C)$ by summing the probabilities of the sample points given in part d.

g. Define the event $(A \cup B \cup C)$ as a specific set of sample points.

h. Find $P(A \cup B \cup C)$ by summing the probabilities of the sample points given in part g.

Applying the Concepts—Intermediate

3.52 Abortion provider survey. The Alan Guttmacher Institute Abortion Provider Survey is a survey of all 238 known nonhospital abortion providers in the United States (*Perspectives on Sexual and Reproductive Health*, Jan./Feb. 2003.) For one part of the survey, the 358 providers were classified according to case load (number of abortions performed per year) and whether they permit their patients to take the abortion drug misoprostol at home or require the patients to return to the abortion facility to receive the drug. The responses are summarized in the accompanying table. Suppose we select, at random, one of the 358 providers and observe the provider's case load (fewer than 50, or 50 or more) and home use of the drug (yes or no).

Permit Drug at Home	Number of Abortions		
	Fewer than 50	50 or More	Totals
Yes	170	130	300
No	48	10	58
Totals	218	140	358

Source: Henshaw, S. K., and Finer, L. B. "The accessibility of abortion services in the United States, 2001," *Perspectives on Sexual and Reproductive Health*, Vol. 35, No. 1, Jan./Feb. 2003 (Table 4).

a. List all the possible outcomes for this provider.

b. On the basis of the table, assign reasonable probabilities to the outcomes in part a.

c. Find the probability that the provider permits home use of the abortion drug.

d. Find the probability that the provider permits home use of the drug or has a case load of fewer than 50 abortions.

e. Find the probability that the provider permits home use of the drug and has a case load of fewer than 50 abortions.

3.53 Fighting probability of fallow deer bucks. In *Aggressive Behavior* (Jan./Feb. 2007), zoologists investigated the likelihood of fallow deer bucks fighting during the mating season. During a 270-hour observation period, the researchers recorded 205 encounters between two bucks. Of these, 167 involved one buck clearly initiating the encounter with the other. In these 167 initiated encounters, the zoologists kept track of whether or not a physical contact fight occurred and whether the initiator ultimately won or lost the encounter. (The buck that is driven away by the other is considered the loser.) A summary of the 167 initiated encounters is provided in the accompanying table. Suppose we select one of these 167 encounters and note the outcome (fight status and winner).

a. What is the probability that a fight occurs and the initiator wins?

b. What is the probability that no fight occurs?

c. What is the probability that there is no clear winner?

d. What is the probability that a fight occurs or the initiator loses?

e. Are the events "No clear winner" and "initiator loses" mutually exclusive?

	Initiator Wins	No Clear Winner	Initiator Loses	Totals
Fight	26	23	15	64
No Fight	80	12	11	103
Totals	106	35	26	167

Sources: Bartos, L. et al. "Estimation of the probability of fighting in fallow deer (*Dama dama*) during the rut," *Aggressive Behavior*, Vol. 33, Jan./Feb. 2007.

3.54 Stacking in the NBA. In professional sports, *stacking* is a term used to describe the practice of African-American players being excluded from certain positions because of race. To illustrate the stacking phenomenon, the *Sociology of Sport Journal* (Vol. 14, 1997) presented the table shown here. The table summarizes the race and positions of 368 National Basketball Association (NBA) players in 1993. Suppose an NBA player is selected at random from that year's player pool.

	Position			
	Guard	Forward	Center	Totals
White	26	30	28	84
Black	128	122	34	284
Totals	154	152	62	368

a. What is the probability that the player is white?

b. What is the probability that the player is a center?

c. What is the probability that the player is African-American and plays guard?

d. What is the probability that the player is not a guard?

e. What is the probability that the player is white or a center?

3.55 Federal civil trial appeals. The *Journal of the American Law and Economics Association* (Vol. 3, 2001) published the results of a study of appeals of federal civil trials. The accompanying table, extracted from the article, gives a breakdown of 2,143 civil cases that were appealed by either the plaintiff or the defendant. The outcome of the appeal, as well as the type of trial (judge or jury), was determined for each civil case. Suppose one of the 2,143 cases is selected at random and both the outcome of the appeal and the type of trial are observed.

	Jury	Judge	Totals
Plaintiff trial win— reversed	194	71	265
Plaintiff trial win— affirmed/dismissed	429	240	669
Defendant trial win—reversed	111	68	179
Defendant trial win—affirmed/ dismissed	731	299	1,030
Totals	1,465	678	2,143

a. List the sample points for this experiment.

b. Find $P(A)$, where $A = \{$jury trial$\}$.

c. Find $P(B)$, where $B = \{$plaintiff trial win is reversed$\}$.

d. Are A and B mutually exclusive events?

e. Find $P(A^c)$.

f. Find $P(A \cup B)$.

g. Find $P(A \cap B)$.

3.56 Gang research study. The National Gang Crime Research Center (NGCRC) has developed a six-level gang classification system for both adults and juveniles. The NGCRC collected data on approximately 7,500 confined offenders and assigned each a score, using the gang classification system. (*Journal of Gang Research*, Winter 1997.) One of several other variables measured by the NGCRC was whether or not the offender ever carried a homemade weapon (e.g., a knife) while in custody. The table on p. 138 gives the number of confined offenders in each of the gang score and homemade weapon categories. Assume one of the confined offenders is randomly selected.

a. Find the probability that the offender has a gang score of 5.

b. Find the probability that the offender has carried a homemade weapon.

c. Find the probability that the offender has a gang score below 3.

d. Find the probability that the offender has a gang score of 5 and has carried a homemade weapon.

e. Find the probability that the offender has a gang score of 0 or has never carried a homemade weapon.

f. Are the events described in parts **a** and **b** mutually exclusive? Explain.

g. Are the events described in parts **a** and **c** mutually exclusive? Explain.

Applying the Concepts—Advanced

3.57 Galileo's passe-dix game. Passe-dix is a game of chance played with three fair dice. Players bet whether the sum of the faces shown on the dice will be above or below 10. During the late 16th century, the astronomer and mathematician Galileo Galilei was asked by the Grand Duke of Tuscany to explain why "the chance of throwing a total of 9 with three fair dice was less than that of throwing a total of 10." (*Interstat*, Jan. 2004.) The Grand Duke believed that the chance should be the same, since "there are an equal number of partitions of the numbers 9 and 10." Find the flaw in the Grand Duke's reasoning and answer the question posed to Galileo.

Two-way Table for Exercise 3.56

Gang Classification Score	Weapon		
	Yes	No	Totals
0 (Never joined a gang, no close friends in a gang)	255	2,551	2,806
1 (Never joined a gang, 1–4 close friends in a gang)	110	560	670
2 (Never joined a gang, 5 or more friends in a gang)	151	636	787
3 (Inactive gang member)	271	959	1,230
4 (Active gang member, no position of rank)	175	513	688
5 (Active gang member, holds position of rank)	476	831	1,307
Totals	1,438	6,050	7,488

Source: Knox, G. W. et al. "A gang classification system for corrections," *Journal of Gang Research,* Vol. 4, No. 2, Winter 1997, p. 54 (Table 4).

3.5 Conditional Probability

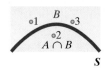

FIGURE 3.13

Reduced sample space for the die-toss experiment: given that event B has occurred

The event probabilities we've been discussing give the relative frequencies of the occurrences of the events when the experiment is repeated a very large number of times. Such probabilities are often called **unconditional probabilities**, because no special conditions are assumed other than those which define the experiment.

Often, however, we have additional knowledge that might affect the outcome of an experiment, so we may need to alter the probability of an event of interest. A probability that reflects such additional knowledge is called the **conditional probability** of the event. For example, we've seen that the probability of observing an even number (event A) on a toss of a fair die is $1/2$. But suppose we're given the information that on a particular throw of the die the result was a number less than or equal to 3 (event B). Would the probability of observing an even number on that throw of the die still be equal to $1/2$? It can't be, because making the assumption that B has occurred reduces the sample space from six sample points to three sample points (namely, those contained in event B). This reduced sample space is as shown in Figure 3.13.

Because the sample points for the die-toss experiment are equally likely, each of the three sample points in the reduced sample space is assigned an equal *conditional probability* of $1/3$. Since the only even number of the three in the reduced sample space B is the number 2 and the die is fair, we conclude that the probability that A occurs *given that B occurs* is $1/3$. We use the symbol $P(A|B)$ to represent the probability of event A given that event B occurs. For the die-toss example,

$$P(A|B) = \frac{1}{3}$$

To get the probability of event A given that event B occurs, we proceed as follows: We divide the probability of the part of A that falls within the reduced sample space B, namely, $P(A \cap B)$, by the total probability of the reduced sample space, namely, $P(B)$. Thus, for the die-toss example with event A: {Observe an even number.} and event B: {Observe a number less than or equal to 3.} we find that

$$P(A|B) = \frac{P(A \cap B)}{P(B)} = \frac{P(2)}{P(1) + P(2) + P(3)} = \frac{1/6}{3/6} = \frac{1}{3}$$

The formula for $P(A|B)$ is true in general:

Conditional Probability Formula

To find the *conditional probability that event A occurs given that event B occurs*, divide the probability that *both A and B* occur by the probability that *B* occurs; that is,

$$P(A|B) = \frac{P(A \cap B)}{P(B)} \qquad [\text{We assume that } P(B) \neq 0.]$$

This formula adjusts the probability of $A \cap B$ from its original value in the complete sample space **S** to a conditional probability in the reduced sample space B. On the one hand, if the sample points in the complete sample space are equally likely, then the formula will assign equal probabilities to the sample points in the reduced sample space, as in the die-toss experiment. On the other hand, if the sample points have unequal probabilities, the formula will assign conditional probabilities proportional to the probabilities in the complete sample space. The latter situation is illustrated in the next three examples.

EXAMPLE 3.14
THE CONDITIONAL PROBABILITY FORMULA— Executives Who Cheat at Golf

Problem To develop programs for business travelers staying at convention hotels, Hyatt Hotels Corp. commissioned a study of executives who play golf. The study revealed that 55% of the executives admitted that they had cheated at golf. Also, 20% of the executives admitted that they had cheated at golf and had lied in business. Given that an executive had cheated at golf, what is the probability that the executive also had lied in business?

Solution Let's define events A and B as follows:

$$A = \{\text{Executive had cheated at golf.}\}$$
$$B = \{\text{Executive had lied in business.}\}$$

From the study, we know that 55% of executives had cheated at golf, so $P(A) = .55$. Now, executives who *both* cheat at golf (event A) *and* lie in business (event B) represent the compound event $A \cap B$. From the study, $P(A \cap B) = .20$. We want to know the probability that an executive lied in business (event B) given that he or she cheated at golf (event A); that is, we want to know the conditional probability $P(B|A)$. Applying the conditional probability formula, we have

$$P(B|A) = \frac{P(A \cap B)}{P(A)} = \frac{.20}{.55} = .364$$

Thus, given that a certain executive had cheated at golf, the probability that the executive also had lied in business is .364.

Look Back One of the keys to applying the formula correctly is to write the information in the study in the form of probability statements involving the events of interest. The word "and" in the clause "cheat at golf *and* lie in business" implies an intersection of the two events A and B. The word "given" in the phrase "*given* that an executive cheats at golf" implies that event A is the given event.

Now Work Exercise 3.73

EXAMPLE 3.15
APPLYING THE CONDITIONAL PROBABILITY FORMULA IN A TWO-WAY TABLE—Smoking and Cancer

Problem Many medical researchers have conducted experiments to examine the relationship between cigarette smoking and cancer. Consider an individual randomly selected from the adult male population. Let A represent the event that the individual smokes, and let A^c denote the complement of A (the event that the individual does not smoke). Similarly, let B represent the event that the individual develops cancer, and let B^c be the complement of that event. Then the four sample points associated with the experiment are shown in Figure 3.14, and their probabilities for a certain section of the United States are given in Table 3.5. Use these sample point probabilities to examine the relationship between smoking and cancer.

Solution One method of determining whether the given probabilities indicate that smoking and cancer are related is to compare the *conditional probability* that an adult male acquires cancer given that he smokes with the conditional probability that an adult male acquires cancer given that he does not smoke [i.e., compare $P(B|A)$ with $P(B|A^c)$].

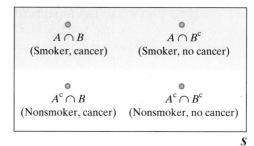

FIGURE 3.14

Sample space for Example 3.15

TABLE 3.5 Probabilities of Smoking and Developing Cancer

	DEVELOPS CANCER	
Smoker	Yes, B	No, B^c
Yes, A	.05	.20
No, A^c	.03	.72

First, we consider the reduced sample space A corresponding to adult male smokers. This reduced sample space is highlighted in Figure 3.14. The two sample points $A \cap B$ and $A \cap B^c$ are contained in this reduced sample space, and the adjusted probabilities of these two sample points are the two conditional probabilities

$$P(B|A) = \frac{P(A \cap B)}{P(A)} \quad \text{and} \quad P(B^c|A) = \frac{P(A \cap B^c)}{P(A)}$$

The probability of event A is the sum of the probabilities of the sample points in A:

$$P(A) = P(A \cap B) + P(A \cap B^c) = .05 + .20 = .25$$

Then the values of the two conditional probabilities in the reduced sample space A are

$$P(B|A) = \frac{.05}{.25} = .20 \quad \text{and} \quad P(B^c|A) = \frac{.20}{.25} = .80$$

These two numbers represent the probabilities that an adult male smoker develops cancer and does not develop cancer, respectively.

In a like manner, the conditional probabilities of an adult male nonsmoker developing cancer and not developing cancer are

$$P(B|A^c) = \frac{P(A^c \cap B)}{P(A^c)} = \frac{.03}{.75} = .04$$

$$P(B^c|A^c) = \frac{P(A^c \cap B^c)}{P(A^c)} = \frac{.72}{.75} = .96$$

Two of the conditional probabilities give some insight into the relationship between cancer and smoking: the probability of developing cancer given that the adult male is a smoker, and the probability of developing cancer given that the adult male is not a smoker. The conditional probability that an adult male smoker develops cancer (.20) is five times the probability that a nonsmoker develops cancer (.04). This relationship does not imply that smoking *causes* cancer, but it does suggest a pronounced link between smoking and cancer.

Look Back Notice that the conditional probabilities $P(B^c|A) = .80$ and $P(B|A) = .20$ are in the same 4-to-1 ratio as the original unconditional probabilities, .20 and .05. The conditional probability formula simply adjusts the unconditional probabilities so that they add to 1 in the reduced sample space A of adult male smokers. Similarly, the conditional probabilities $P(B^c|A^c) = .96$ and $P(B|A^c) = .04$ are in the same 24-to-1 ratio as the unconditional probabilities .72 and .03.

Now Work Exercise 3.63a–b

EXAMPLE 3.16
CONDITIONAL PROBABILITY IN A TWO-WAY TABLE—Consumer Complaints

Problem The Federal Trade Commission's investigation of consumer product complaints has generated much interest on the part of manufacturers in the quality of their products. A manufacturer of an electromechanical kitchen utensil conducted an analysis of a large number of consumer complaints and found that they fell into the six categories shown in Table 3.6. If a consumer complaint is received, what is the probability that the cause of the complaint was the appearance of the product given that the complaint originated during the guarantee period?

TABLE 3.6 Distribution of Product Complaints

	REASON FOR COMPLAINT			
	Electrical	Mechanical	Appearance	Totals
During Guarantee Period	18%	13%	32%	63%
After Guarantee Period	12%	22%	3%	37%
Totals	30%	35%	35%	100%

Solution Let A represent the event that the cause of a particular complaint is the appearance of the product, and let B represent the event that the complaint occurred during the guarantee period. Checking Table 3.6, you can see that $(18 + 13 + 32)\% = 63\%$ of the complaints occur during the guarantee period. Hence, $P(B) = .63$. The percentage of complaints that were caused by appearance and occurred during the guarantee period (the event $A \cap B$) is 32%. Therefore, $P(A \cap B) = .32$.

Using these probability values, we can calculate the conditional probability $P(A|B)$ that the cause of a complaint is appearance given that the complaint occurred during the guarantee time:

$$P(A|B) = \frac{P(A \cap B)}{P(B)} = \frac{.32}{.63} = .51$$

Thus, slightly more than half of the complaints that occurred during the guarantee period were due to scratches, dents, or other imperfections in the surface of the kitchen devices.

Look Back Note that the answer $\frac{.32}{.63}$ is the proportion for the event of interest A (.32), divided by the row total proportion for the given event B (.63). That is, it is the proportion of the time that A occurs together within the given event B.

Now Work Exercise 3.77

3.6 The Multiplicative Rule and Independent Events

The probability of an intersection of two events can be calculated with the *multiplicative rule*, which employs the conditional probabilities we defined in the previous section. Actually, we've already developed the formula in another context. Recall that the conditional probability of B given A is

$$P(B|A) = \frac{P(A \cap B)}{P(A)}$$

Multiplying both sides of this equation by $P(A)$, we obtain a formula for the probability of the intersection of events A and B. This formula is often called the **multiplicative rule of probability**.

Multiplicative Rule of Probability

$P(A \cap B) = P(A)P(B|A)$ or, equivalently, $P(A \cap B) = P(B)P(A|B)$

EXAMPLE 3.17

THE MULTIPLICATIVE RULE—a Famous Psychological Experiment

Problem In a classic psychology study conducted in the early 1960s, Stanley Milgram performed a series of experiments in which a teacher is asked to shock a learner who is attempting to memorize word pairs whenever the learner gives the wrong answer. The shock levels increase with each successive wrong answer. (Unknown to the teacher, the shocks are not real.) Two events of interest are

A: {The teacher "applies" a severe shock (450 volts).}
B: {The learner protests verbally prior to receiving the shock.}

A recent application of Milgram's shock study reveled that $P(B) = .5$ and $P(A|B) = .7$. On the basis of this information, what is the probability that a learner will protest verbally *and* a teacher will apply a severe shock? That is, find $P(A \cap B)$.

Solution We want to calculate $P(A \cap B)$. Using the formula for the multiplicative rule, we obtain

$$P(A \cap B) = P(B) \, P(A|B) = (.5)(.7) = .35$$

Thus, about 35% of the time the learner will give verbal protest and the teacher will apply a severe shock.

Look Back The multiplicative rule can be expressed in two ways: $P(A \cap B) = P(A) \, P(B|A)$, or $P(A \cap B) = P(B) \, P(A|B)$. Select the formula that involves a given event for which you know the probability (e.g., event B in the example).

Now Work Exercise 3.72

Intersections often contain only a few sample points. In this case, the probability of an intersection is easy to calculate by summing the appropriate sample point probabilities. However, the formula for calculating intersection probabilities is invaluable when the intersection contains numerous sample points, as the next example illustrates.

ACTIVITY 3.1: *Exit Polls:* Conditional Probability

Exit polls are conducted in selected locations as voters leave their polling places after voting. In addition to being used to predict the outcome of elections before the votes are counted, these polls seek to gauge tendencies among voters. The results are usually stated in terms of conditional probabilities.

The accompanying table shows the results of exit polling which suggest that men were more likely to vote for George W. Bush, while women were more likely to vote for John F. Kerry, in the 2004 presidential election. In addition, the table suggests that more women than men voted in the election. The six percentages in the last three columns represent conditional probabilities, where the given event is gender.

2004 Presidential Election, Vote by Gender

	Bush	Kerry	Other
Male (46%)	55%	44%	1%
Female (54%)	48%	51%	1%

Source: CNN.com

1. Find similar exit poll results in which voters are categorized by race, income, education, or some other criterion for a recent national, state, or local election. Choose two different examples, and interpret the percentages given as probabilities, or conditional probabilities where appropriate.

2. Use the multiplicative rule of probability to find the probabilities related to the percentages given. [For example, in the accompanying table, find P(Bush and Male).] Then interpret each of these probabilities and use them to determine the total percentage of the electorate that voted for each candidate.

3. Describe a situation in which a political group might use a form of exit polling to gauge voters' opinions on a "hot-button" topic (e.g., global warming). Identify the political group, the "hot-button" topic, the criterion used to categorize voters, and how the voters' opinions will be determined. Then describe how the results will be summarized as conditional probabilities. How might the results of the poll be used to support a particular agenda?

EXAMPLE 3.18

APPLYING THE MULTIPLICATIVE RULE—Study of Welfare Workers

Problem A county welfare agency employs 10 welfare workers who interview prospective food stamp recipients. Periodically, the supervisor selects, at random, the forms completed by two workers and subsequently audits them for illegal deductions. Unknown to the supervisor, three of the workers have regularly been giving illegal deductions to applicants. What is the probability that both of the workers chosen have been giving illegal deductions?

Solution Define the following two events:

A: {First worker selected gives illegal deductions.}
B: {Second worker selected gives illegal deductions.}

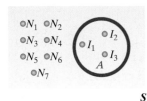

FIGURE 3.15

Venn diagram for finding $P(A)$

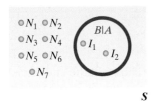

FIGURE 3.16

Venn diagram for finding $P(B|A)$

We want to find the probability that both workers selected have been giving illegal deductions. This event can be restated as {First worker gives illegal deductions *and* second worker gives illegal deductions.}. Thus, we want to find the probability of the intersection $A \cap B$. Applying the multiplicative rule, we have

$$P(A \cap B) = P(A)P(B|A)$$

To find $P(A)$, it is helpful to consider the experiment as selecting 1 worker from the 10. Then the sample space for the experiment contains 10 sample points (representing the 10 welfare workers), in which the 3 workers giving illegal deductions are denoted by the symbol I (I_1, I_2, I_3) and the 7 workers not giving illegal deductions are denoted by the symbol N ($N_1, \ldots, N_7$). The resulting Venn diagram is illustrated in Figure 3.15.

Since the first worker is selected at random from the 10, it is reasonable to assign equal probabilities to the 10 sample points. Thus, each sample point has a probability of $^1/_{10}$. The sample points in event A are $\{I_1, I_2, I_3\}$—the three workers who are giving illegal deductions. Thus,

$$P(A) = P(I_1) + P(I_2) + P(I_3) = \frac{1}{10} + \frac{1}{10} + \frac{1}{10} = \frac{3}{10}$$

To find the conditional probability $P(B|A)$, we need to alter the sample space **S**. Since we know that A has occurred (i.e., the first worker selected is giving illegal deductions), only 2 of the 9 remaining workers in the sample space are giving illegal deductions. The Venn diagram for this new sample space is shown in Figure 3.16.

Each of these nine sample points is equally likely, so each is assigned a probability of $^1/_9$. Since the event $B|A$ contains the sample points $\{I_1, I_2\}$, we have

$$P(B|A) = P(I_1) + P(I_2) = \frac{1}{9} + \frac{1}{9} = \frac{2}{9}$$

Substituting $P(A) = {}^3/_{10}$ and $P(B|A) = {}^2/_9$ into the formula for the multiplicative rule, we find that

$$P(A \cap B) = P(A)P(B|A) = \left(\frac{3}{10}\right)\left(\frac{2}{9}\right) = \frac{6}{90} = \frac{1}{15}$$

Thus, there is a 1-in-15 chance that both workers chosen by the supervisor have been giving illegal deductions to food stamp recipients.

Look Back The key words *both* and *and* in the statement "both A and B occur" imply an intersection of two events. This, in turn, implies that we should *multiply* probabilities to obtain the answer.

Now Work Exercise 3.67

The sample-space approach is only one way to solve the problem posed in Example 3.18. An alternative method employs the concept of a **tree diagram**. Tree diagrams are helpful for calculating the probability of an intersection.

To illustrate, a tree diagram for Example 3.18 is displayed in Figure 3.17. The tree begins at the far left with two branches. These branches represent the two possible outcomes N (no illegal deductions) and I (illegal deductions) for the first worker selected. The unconditional probability of each outcome is given (in parentheses) on the appropriate branch. That is, for the first worker selected, $P(N) = {}^7/_{10}$ and $P(I) = {}^3/_{10}$. (These unconditional probabilities can be obtained by summing sample point probabilities as in Example 3.18.)

The next level of the tree diagram (moving to the right) represents the outcomes for the second worker selected. The probabilities shown here are conditional probabilities, since the outcome for the first worker is assumed to be known. For example, if the first worker is giving illegal deductions (I), the probability that the second worker is also giving illegal deductions (I) is $^2/_9$, because, of the 9 workers left to be selected, only 2 remain who are giving illegal deductions. This conditional probability, $^2/_9$, is shown in parentheses on the bottom branch of Figure 3.17.

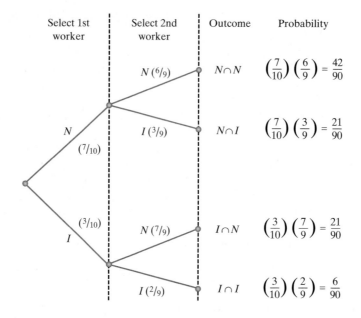

FIGURE 3.17
Tree diagram for
Example 3.18

Finally, the four possible outcomes of the experiment are shown at the end of each of the four tree branches. These events are intersections of two events (outcome of first worker *and* outcome of second worker). Consequently, the multiplicative rule is applied to calculate each probability, as shown in Figure 3.17. You can see that the intersection $\{I \cap I\}$—the event that both workers selected are giving illegal deductions—has probability $^6/_{90} = ^1/_{15}$, which is the same as the value obtained in Example 3.18.

In Section 3.5, we showed that the probability of event A may be substantially altered by the knowledge that an event B has occurred. However, this will not always be the case; in some instances, the assumption that event B has occurred will *not* alter the probability of event A at all. When this occurs, we say that the two events A and B are *independent events*.

Definition 3.9

Events A and B are **independent events** if the occurrence of B does not alter the probability that A has occurred; that is, events A and B are independent if

$$P(A|B) = P(A)$$

When events A and B are independent, it is also true that

$$P(B|A) = P(B)$$

Events that are not independent are said to be **dependent**.

EXAMPLE 3.19

CHECKING FOR INDEPENDENCE —Die-Tossing Experiment

Problem Consider the experiment of tossing a fair die, and let

$$A = \{\text{Observe an even number.}\}$$
$$B = \{\text{Observe a number less than or equal to 4.}\}$$

Are A and B independent events?

Solution The Venn diagram for this experiment is shown in Figure 3.18. We first calculate

$$P(A) = P(2) + P(4) + P(6) = \frac{1}{2}$$

$$P(B) = P(1) + P(2) + P(3) + P(4) = \frac{4}{6} = \frac{2}{3}$$

$$P(A \cap B) = P(2) + P(4) = \frac{2}{6} = \frac{1}{3}$$

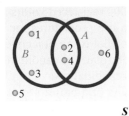

FIGURE 3.18
Venn diagram for die-toss
experiment

Now, assuming that B has occurred, we see that the conditional probability of A given B is

$$P(A|B) = \frac{P(A \cap B)}{P(B)} = \frac{1/3}{2/3} = \frac{1}{2} = P(A)$$

Thus, assuming that the occurence of event B does not alter the probability of observing an even number, that probability remains $\frac{1}{2}$. Therefore, the events A and B are independent.

Look Back   Note that if we calculate the conditional probability of B given A, our conclusion is the same:

$$P(B|A) = \frac{P(A \cap B)}{P(A)} = \frac{1/3}{1/2} = \frac{2}{3} = P(B)$$

Biography

**BLAISE PASCAL (1623–1662)—
Solver of the Chevalier's Dilemma**

As a precocious child growing up in France, Blaise Pascal showed an early inclination toward mathematics. Although his father would not permit Pascal to study mathematics before the age of 15 (removing all math texts from his house), at age 12 Blaise discovered on his own that the sum of the angles of a triangle are two right angles.

Pascal went on to become a distinguished mathematician, as well as a physicist, a theologian, and the inventor of the first digital calculator. Most historians attribute the beginning of the study of probability to the correspondence between Pascal and Pierre de Fermat in 1654. The two solved the Chevalier's Dilemma, a gambling problem related to Pascal by his friend and Paris gambler the Chevalier de Mere. The problem involved determining the expected number of times one could roll two dice without throwing a double 6. (Pascal proved that the "break-even" point was 25 rolls.)

EXAMPLE 3.20
CHECKING FOR INDEPENDENCE
—Consumer Complaint Study

Problem   Refer to the consumer product complaint study in Example 3.16. The percentages of complaints of various types during and after the guarantee period are shown in Table 3.6 (p. 141). Define the following events:

A: {The cause of the complaint is the appearance of the product.}
B: {The complaint occurred during the guarantee term.}

Are A and B independent events?

Solution   Events A and B are independent if $P(A|B) = P(A)$. In Example 3.16, we calculated $P(A|B)$ to be .51, and from Table 3.6, we see that

$$P(A) = .32 + .03 = .35$$

Therefore, $P(A|B)$ is not equal to $P(A)$, and A and B are dependent events.

Now Work Exercise 3.63c

To gain an intuitive understanding of independence, think of situations in which the occurrence of one event does not alter the probability that a second event will occur. For example, new medical procedures are often tested on laboratory animals. The scientists conducting the tests generally try to perform the procedures on the animals so that the results for one animal do not affect the results for the others. That is, the event that the procedure is successful on one animal is *independent* of the result for another. In this way, the scientists can get a more accurate idea of the efficacy of the procedure than if the results were dependent, with the success or failure for one animal affecting the results for other animals.

As a second example, consider an election poll in which 1,000 registered voters are asked their preference between two candidates. Pollsters try to use procedures for selecting a sample of voters so that the responses are independent. That is, the objective of the pollster is to select the sample so that one polled voter's preference for candidate A does not alter the probability that a second polled voter prefers candidate A.

Now consider the world of sports. Do you think the results of a batter's successive trips to the plate in baseball, or of a basketball player's successive shots at the basket, are independent? For example, if a basketball player makes two successive shots, is the probability of making the next shot altered from its value if the result of the first shot is not known? If a player makes two shots in a row, the probability of a third successful shot is likely to be different from what we would assign if we knew nothing about the first two shots. Why should this be so? Research has shown that many such results in sports tend to be *dependent* because players (and even teams) tend to get on "hot" and "cold" streaks, during which their probabilities of success may increase or decrease significantly.

We will make three final points about independence. The first is that the property of independence, unlike the property of mutual exclusivity, cannot be shown on, or gleaned from, a Venn diagram. This means that *you can't trust your intuition*. In general, the only way to check for independence is by performing the calculations of the probabilities in the definition.

The second point concerns the relationship between the properties of mutual exclusivity and independence. Suppose that events A and B are mutually exclusive, as shown in Figure 3.19, and that both events have nonzero probabilities. Are these events independent or dependent? That is, does the assumption that B occurs alter the probability of the occurrence of A? It certainly does, because if we assume that B has occurred, it is impossible for A to have occurred simultaneously. That is, $P(A|B) = 0$. Thus, *mutually exclusive events are dependent events*, since $P(A) \neq P(A|B)$.

The third point is that the probability of the intersection of independent events is very easy to calculate. Referring to the formula for calculating the probability of an intersection, we find that

$$P(A \cap B) = P(A)P(B|A)$$

Thus, since $P(B|A) = P(B)$ when A and B are independent, we have the following useful rule:

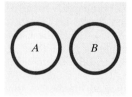

FIGURE 3.19

Mutually exclusive events are dependent events

Probability of Intersection of Two Independent Events

If events A and B are independent, then the probability of the intersection of A and B equals the product of the probabilities of A and B; that is,

$$P(A \cap B) = P(A)P(B)$$

The converse is also true: If $P(A \cap B) = P(A)P(B)$, then events A and B are independent.

In the die-toss experiment, we showed in Example 3.19 that the two events A: {Observe an even number.} and B: {Observe a number less than or equal to 4.} are independent if the die is fair. Thus,

$$P(A \cap B) = P(A)P(B) = \left(\frac{1}{2}\right)\left(\frac{2}{3}\right) = \frac{1}{3}$$

This result agrees with the one we obtained in the example:

$$P(A \cap B) = P(2) + P(4) = \frac{2}{6} = \frac{1}{3}$$

EXAMPLE 3.21
PROBABILITY OF INDEPENDENT EVENTS OCCURRING SIMULTANEOUSLY
—Divorced-Couples Study

Problem Recall from Example 3.5 (p. 120) that the American Association for Marriage and Family Therapy (AAMFT) found that 25% of divorced couples are classified as "fiery foes" (i.e., they communicate through their children and are hostile toward each other).

a. What is the probability that in a sample of 2 divorced couples, both are classified as "fiery foes?"

b. What is the probability that in a sample of 10 divorced couples, all 10 are classified as "fiery foes?"

Solution

a. Let F_1 represent the event that divorced couple 1 is classified as a pair of "fiery foes" and F_2 represent the event that divorced couple 2 is also classified as a pair of "fiery foes." Then the event that *both* couples are "fiery foes" is the intersection of the two events, $F_1 \cap F_2$. On the basis of the AAMFT survey which found that 25% of divorced couples are "fiery foes," we could reasonably conclude that $P(F_1) = .25$ and $P(F_2) = .25$. However, in order to compute the probability of $F_1 \cap F_2$ from the multiplicative rule, we must make the assumption that the two events are independent. Since the classification of any divorced couple is not likely to affect the classification of another divorced couple, this assumption is reasonable. Assuming independence, we have

$$P(F_1 \cap F_2) = P(F_1)P(F_2) = (.25)(.25) = .0625$$

b. To see how to compute the probability that 10 of 10 divorced couples will all be classified as "fiery foes," first consider the event that 3 of 3 couples are "fiery foes." If F_3 represents the event that the third divorced couple are "fiery foes," then we want to compute the probability of the intersection $F_1 \cap F_2 \cap F_3$. Again assuming independence of the classifications, we have

$$P(F_1 \cap F_2 \cap F_3) = P(F_1)P(F_2)P(F_3) = (.25)(.25)(.25) = .015625$$

Similar reasoning leads us to the conclusion that the intersection of 10 such events can be calculated as follows:

$$P(F_1 \cap F_2 \cap F_3 \cap \ldots \cap F_{10}) = P(F_1)P(F_2)\ldots P(F_{10}) = (.25)^{10} = .000001$$

Thus, the probability that 10 of 10 divorced couples sampled are all classified as "fiery foes" is 1 in 1 million, assuming that the probability of each couple being classified as "fiery foes" is .25 and that the classification decisions are independent.

Look Back The very small probability in part **b** makes it extremely unlikely that 10 of 10 divorced couples are "fiery foes." If this event should actually occur, we should question the probability of .25 provided by the AAMFT and used in the calculation—it is likely to be much higher. (This conclusion is another application of the rare-event approach to statistical inference.)

Now Work Exercise 3.88a

STATISTICS IN ACTION REVISITED

The Probability of Winning Cash 3 or Play 4

In addition to the biweekly Lotto 6/53, the Florida Lottery runs several other games. Two popular daily games are Cash 3 and Play 4. In Cash 3, players pay $1 to select three numbers in order, where each number ranges from 0 to 9. If the three numbers selected (e.g., 2–8–4) match exactly the order of the three numbers drawn, the player wins $500. Play 4 is similar to Cash 3, but players must match four numbers (each number ranging from 0 to 9). For a $1 Play 4 ticket (e.g., 3–8–3–0), the player will win $5,000 if the numbers match the order of the four numbers drawn.

During the official drawing for Cash 3, 10 table tennis balls numbered 0, 1, 2, 3, 4, 5, 6, 7, 8, and 9 are placed

(continued)

into each of three chambers. The balls in the first chamber are colored pink, the balls in the second chamber are blue, and the balls in the third chamber are yellow. One ball of each color is randomly drawn, with the official order as pink–blue–yellow. In Play 4, a fourth chamber with orange balls is added, and the official order is pink–blue–yellow–orange. Since the draws of the colored balls are random and independent, we can apply an extension of the probability rule for the intersection of two independent events to find the odds of winning Cash 3 and Play 4. The probability of matching a numbered ball being drawn from a chamber is 1/10; therefore,

$$
\begin{aligned}
P(\text{Win Cash 3}) &= P(\text{match pink } and \text{ match blue}\\
&\quad and \text{ match yellow})\\
&= P(\text{match pink}) \times P(\text{match blue}) \times\\
&\quad P(\text{match yellow})\\
&= (1/10)(1/10)(1/10) = 1/1000 = .001
\end{aligned}
$$

$$
\begin{aligned}
P(\text{Win Play 4}) &= P(\text{match pink } and \text{ match blue}\\
&\quad and \text{ match yellow } and \text{ match orange})\\
&= P(\text{match pink}) \times P(\text{match blue}) \times\\
&\quad P(\text{match yellow}) \times P(\text{match orange})\\
&= (1/10)(1/10)(1/10)\\
&= 1/10,000 = .0001
\end{aligned}
$$

Although the odds of winning one of these daily games is much better than the odds of winning Lotto 6/53, there is still only a 1 in 1,000 chance (for Cash 3) or 1 in 10,000 chance (for Play 4) of winning the daily game. And the payoffs ($500 or $5,000) are much smaller. In fact, it can be shown that you will lose an average of 50¢ every time you play either Cash 3 or Play 4!

ACTIVITY 3.2: *Keep the Change:* Independent Events

Once again we return to the Bank of America *Keep the Change* savings program, presented in Activity 1.1 (p. 12). This time, we look at whether certain events involving purchase totals and amounts transferred to savings are independent. Throughout this activity, the experiment consists of randomly selecting one purchase from a large group of purchases.

1. Define events A and B as follows:

A: {Purchase total ends in $0.25.}
B: {Amount transferred is less than $0.50.}

Explain why events A and B are not independent. Are events A and B mutually exclusive? Use this example to explain the difference between independent events and mutually exclusive events.

2. Now define events A and B in this manner:

A: {Purchase total is greater than $10.}
B: {Amount transferred is less than $0.50.}

Do you believe that these events are independent? Explain your reasoning.

3. To investigate numerically whether the events in Question 2 are independent, we will use the data collected in Activity 1.1. Pool your data with the data from other students or from the entire class so that the combined data set represents at least 100 purchases. Complete the table by counting the number of purchases in each category.

Distribution of Purchases

	Purchase Total		
Transfer Amount	≤ $10	> $10	Totals
$0.00–$0.49			
$0.99–$0.99			
Totals			

Compute appropriate probabilities based on your completed table to test whether the events of Question 2 are independent. If you conclude that the events are not independent, can you explain your conclusion in terms of the original data?

Exercises 3.58–3.93

Understanding the Principles

3.58 Explain the difference between an unconditional probability and a conditional probability.

3.59 Give the multiplicative rule of probability for two independent events.

3.60 Give the formula for finding $P(B|A)$.

3.61 Give the multiplicative rule of probability for any two events.

3.62 Defend or refute each of the following statements:
 a. Dependent events are always mutually exclusive.
 b. Mutually exclusive events are always dependent.
 c. Independent events are always mutually exclusive.

Learning the Mechanics

3.63 For two events A and B, $P(A) = .4, P(B) = .2$, and
NW $P(A \cap B) = .1$.
 a. Find $P(A|B)$.
 b. Find $P(B|A)$.
 c. Are A and B independent events?

3.64 For two events A and B, $P(A) = .4, P(B) = .2$, and $P(A|B) = .6$.
 a. Find $P(A \cap B)$.
 b. Find $P(B|A)$.

3.65 For two independent events A and B, $P(A) = .4$ and $P(B) = .2$.
a. Find $P(A \cap B)$.
b. Find $P(A|B)$.
c. Find $P(A \cup B)$.

3.66 An experiment results in one of three mutually exclusive events A, B, and C. It is known that $P(A) = .30$, $P(B) = .55$, and $P(C) = .15$, Find each of the following probabilities:
a. $P(A \cup B)$ **b.** $P(A \cap B))$
c. $P(A|B)$ **d.** $P(B \cup C))$
e. Are B and C independent events? Explain.

3.67 Consider the experiment defined by the accompanying
NW Venn diagram, with the sample space S containing five sample points. The sample points are assigned the following probabilities: $P(E_1) = .1$, $P(E_2) = .1$, $P(E_3) = .2$, $P(E_4) = .5$, $P(E_5) = .1$.

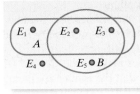

S

a. Calculate $P(A), P(B)$, and $P(A \cap B)$.
b. Suppose we know that event A has occurred, so the reduced sample space consists of the three sample points in A: E_1, E_2, and E_3. Use the formula for conditional probability to determine the probabilities of these three sample points given that A has occurred. Verify that the conditional probabilities are in the same ratio to one another as the original sample point probabilities and that they sum to 1.
c. Calculate the conditional probability $P(B|A)$ in two ways: First, sum $P(E_2|A)$ and $P(E_3|A)$, since these sample points represent the event that B occurs given that A has occurred. Second, use the formula for conditional probability:

$$P(B|A) = \frac{P(A \cap B)}{P(A)}$$

Verify that the two methods yield the same result.

3.68 An experiment results in one of five sample points with the following probabilities: $P(E_1) = .22, P(E_2) = .31$, $P(E_3) = .15, P(E_4) = .22$, and $P(E_5) = .1$. The following events have been defined:

$A: \{E_1, E_3\}$
$B: \{E_2, E_3, E_4\}$
$C: \{E_1, E_5\}$

Find each of the following probabilities:
a. $P(A)$ **b.** $P(B)$ **c.** $P(A \cap B)$
d. $P(A|B)$ **e.** $P(B \cap C)$ **f.** $P(C|B)$
g. Consider each pair of events A and B, A and C, and B and C. Are any of the pairs of events independent? Why?

3.69 Two fair dice are tossed, and the following events are defined:

$A: \{\text{The sum of the numbers showing is odd.}\}$
$B: \{\text{The sum of the numbers showing is 9, 11, or 12.}\}$

Are events A and B independent? Why?

3.70 A sample space contains six sample points and events A, B, and C, as shown in the accompanying Venn diagram. The probabilities of the sample points are $P(1) = .20, P(2) = .05, P(3) = .30, P(4) = .10, P(5) = .10$, and $P(6) = .25$.

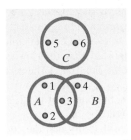

S

a. Which pairs of events, if any, are mutually exclusive? Why?
b. Which pairs of events, if any, are independent? Why?

c. Find $P(A \cup B)$ by adding the probabilities of the sample points and then by using the additive rule. Verify that the answers agree. Repeat for $P(A \cup C)$.

3.71 A box contains two white, two red, and two blue poker chips. Two chips are randomly chosen without replacement, and their colors are noted. Define the following events:

$A: \{\text{Both chips are of the same color.}\}$
$B: \{\text{Both chips are red.}\}$
$C: \{\text{At least one chip is red or white.}\}$

Find $P(B|A)$, $P(B|A^c)$, $P(B|C)$, $P(A|C)$, and $P(C|A^c)$.

APPLET Applet Exercise 3.5

Use the applet entitled *Simulating the Probability of Rolling a 6* to simulate conditional probabilities. Begin by running the applet twice with $n = 10$, without resetting between runs. The data on your screen represent 20 rolls of a die. The diagram above the *Roll* button shows the frequency of each of the six possible outcomes. Use this information to find each of the following probabilities:
a. The probability of 6 given that the outcome is 5 or 6
b. The probability of 6 given that the outcome is even
c. The probability of 4 or 6 given that the outcome is even
d. The probability of 4 or 6 given that the outcome is odd

Applying the Concepts—Basic

3.72 **Treating brain cancer.** According to the Children's Oncolo-
NW gy Group Research Data Center at the University of Florida, 20% of children with neuroblastoma (a form of brain cancer) undergo surgery rather than the traditional treatment of chemotherapy or radiation. The surgery is successful in curing the disease 95% of the time. (*Explore*, Spring 2001.) Consider a child diagnosed with neuroblastoma. What are the chances that the child undergoes surgery and is cured?

3.73 **Speeding linked to fatal car crashes.** According to the Na-
NW tional Highway Traffic and Safety Administration's National Center for Statistics and Analysis (NCSA), "Speeding is one of the most prevalent factors contributing to fatal traffic crashes." (*NHTSA Technical Report*, Aug. 2005.) The probability that speeding is a cause of a fatal

crash is .3. Furthermore, the probability that speeding and missing a curve are causes of a fatal crash is .12. Given that speeding is a cause of a fatal crash, what is the probability that the crash occurred on a curve?

3.74 National Firearms Survey. Refer to the Harvard School of Public Health study of privately held firearm stock in the United States, presented in Exercise 2.6 (p. 34). Recall that in a representative household telephone survey of 2,770 adults, 26% reported that they owned at least one gun. (*Injury Prevention*, Jan. 2007.) The accompanying pie chart summarizes the types of firearms owned by those who own at least one gun. Suppose 1 of the 2,770 adults surveyed is randomly selected.

a. What is the probability that the adult owns at least one gun?

b. Given that the adult does own at least one gun, what is the probability that the adult owns a revolver?

c. What is the probability that the adult owns at least one gun and the gun is a handgun?

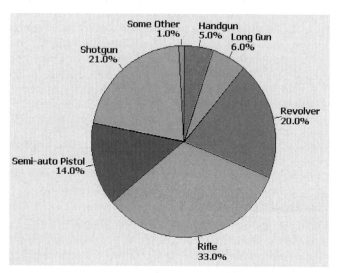

Source: Hepburn, M., Miller, D. A., and Hemenway, D. "The US gun stock: Results from the 2004 national firearms survey," *Injury Prevention*, Vol. 13, No. 1, Jan. 2007 (Figure 1).

3.75 Testing a psychic's ability. Consider an experiment in which 10 identical small boxes are placed side by side on a table. A crystal is placed at random inside one of the boxes. A self-professed "psychic" is asked to pick the box that contains the crystal.

a. If the "psychic" simply guesses, what is the probability that she picks the box with the crystal?

b. If the experiment is repeated seven times, what is the probability that the "psychic" guesses correctly at least once?

c. A group called the Tampa Bay Skeptics recently tested a self-proclaimed "psychic" by administering the preceding experiment seven times. The "psychic" failed to pick the correct box all seven times. (*Tampa Tribune*, Sept. 20, 1998.) What would you infer about this person's psychic ability?

3.76 Herbal medicines survey. Refer to *The American Association of Nurse Anesthetists Journal* (Feb. 2000) study on the use of herbal medicines before surgery, presented in Exercise 1.20 (p. 20). Recall that 51% of surgical patients use herbal medicines against their doctor's advice prior to surgery.

a. What is the probability that a randomly selected surgical patient will use herbal medicines against his or her doctor's advice?

b. What is the probability that in a sample of two independently selected surgical patients, both will use herbal medicines against their doctor's advice?

c. What is the probability that in a sample of five independently selected surgical patients, all five will use herbal medicines against their doctor's advice?

d. Would you expect the event in part **c** to occur? Explain.

3.77 World Series winners. The New York Yankees, a member of the Eastern Division of the American League in Major League Baseball (MLB), recently won three consecutive World Series. The accompanying table summarizes the 16 MLB World Series winners from 1990 to 2006 by division and league. (There was no World Series in 1994, due to a players' strike.) One of these 16 World Series winners is to be chosen at random.

a. Given that the winner is a member of the American League, what is the probability that the winner plays in the Eastern Division?

b. If the winner plays in the Central Division, what is the probability that the winner is a member of the National League?

c. If the winner is a member of the National League, what is the probability that the winner plays in either the Central or Western Division?

		League	
		National	American
Division	Eastern	3	7
	Central	2	2
	Western	1	1

Source: Major League Baseball.

3.78 Study of ancient pottery. Refer to the *Chance* (Fall 2000) study of ancient pottery found at the Greek settlement of Phylakopi, presented in Exercise 2.12 (p. 35). Of the 837 pottery pieces uncovered at the excavation site, 183 were painted. These painted pieces included 14 painted in a curvilinear decoration, 165 painted in a geometric decoration, and 4 painted in a naturalistic decoration. Suppose 1 of the 837 pottery pieces is selected and examined.

a. What is the probability that the pottery piece is painted?

b. Given that the pottery piece is painted, what is the probability that it is painted in a curvilinear decoration?

Applying the Concepts—Intermediate

3.79 Are you really being served red snapper? Red snapper is a rare and expensive reef fish served at upscale restaurants. Federal law prohibits restaurants from serving a cheaper, look-alike variety of fish (e.g., vermillion snapper or lane snapper) to customers who order red snapper. Researchers at the University of North Carolina used DNA analysis to examine fish specimens labeled "red snapper" that were purchased form vendors across the country. (*Nature*, July 15, 2004.) The DNA tests revealed that 77% of the specimens were not red snapper, but the cheaper, look-alike variety of fish.

a. Assuming that the results of the DNA analysis are valid, what is the probability that you are actually served red snapper the next time you order it at a restaurant?

b. If there are five customers at a restaurant, all who have ordered red snapper, what is the probability that at least one customer is actually served red snapper?

3.80 Fighting probability of fallow deer bucks. Refer to the *Aggressive Behavior* (Jan./Feb., 2007) study of fallow deer bucks fighting during the mating season, presented in Exercise 3.53 (p. 136). Recall that researchers recorded 167 encounters between two bucks, one of which clearly initiated the encounter with the other. A summary of the fight status of the initiated encounters is provided in the accompanying table. Suppose we select 1 of these 167 encounters and note the outcome (fight status and winner).

a. Given that a fight occurs, what is the probability that the initiator wins?

b. Given no fight, what is the probability that the initiator wins?

c. Are the events "no fight" and "initiator wins" independent?

	Initiator Wins	No Clear Winner	Initiator Loses	Totals
Fight	26	23	15	64
No Fight	80	12	11	103
Totals	106	35	26	167

Source: Bartos, L. et al. "Estimation of the probability of fighting in fallow deer (*Dama dama*) during the rut," *Aggressive Behavior*, Vol. 33, Jan./Feb., 2007.

NZBIRDS

3.81 Extinct New Zealand birds. Refer to the *Evolutionary Ecology Research* (July 2003) study of the patterns of extinction in the New Zealand bird population, presented in Exercise 2.18 (p. 36). Consider the data on extinction status (extinct, absent from island, present) for the 132 bird species. The data are saved in the **NZBIRDS** file and are summarized in the accompanying MINITAB printout. Suppose you randomly select 10 of the 132 bird species (without replacement) and record the extinction status of each.

a. What is the probability that the first species you select is extinct? (*Note:* Extinct = Yes on the MINITAB printout.)

b. Suppose the first 9 species you select are all extinct. What is the probability that the 10th species you select is extinct?

Tally for Discrete Variables: Extinct

```
Extinct   Count   Percent
Absent      16     12.12
    No      78     59.09
   Yes      38     28.79
    N=     132
```

3.82 Requirements for high school graduation. In Italy, all high school students must take a high school diploma (HSD) exam and write a paper in order to graduate. In *Organizational Behavior and Human Decision Processes* (July 2000), University of Milan researcher L. Macchi provided the following information to a group of college undergraduates. *Fact 1*: In Italy, 360 out of every 1,000 students fail their HSD exam. *Fact 2*: Of those who fail the HSD, 75% also fail the written paper. *Fact 3*: Of those who pass

the HSD, only 20% fail the written paper. Define events *A* and *B* as follows:

$$A = \{\text{The student fails the HSD exam.}\}$$
$$B = \{\text{The student fails the written paper.}\}$$

a. Write Fact 1 as a probability statement involving events *A* and/or *B*.

b. Write Fact 2 as a probability statement involving events *A* and/or *B*.

c. Write Fact 3 as a probability statement involving events *A* and/or *B*.

d. State $P(A \cap B)$ in the words of the problem.

e. Find $P(A \cap B)$.

3.83 Intrusion detection systems. A computer intrusion detection system (IDS) is designed to provide an alarm whenever someone intrudes (e.g., through unauthorized access) into a computer system. A probabilistic evaluation of a system with two independently operating intrusion detection systems (a double IDS) was published in the *Journal of Research of the National Institute of Standards and Technology* (Nov.–Dec. 2003). Consider a double IDS with system A and system B. If there is an intruder, system A sounds an alarm with probability .9 and system B sounds an alarm with probability .95. If there is no intruder, the probability that system A sounds an alarm (i.e., a false alarm) is .2 and the probability that system B sounds an alarm is .1.

a. Use symbols to express the four probabilities just given.

b. If there is an intruder, what is the probability that both systems sound an alarm?

c. If there is no intruder, what is the probability that both systems sound an alarm?

d. Given that there is an intruder, what is the probability that at least one of the systems sounds an alarm?

3.84 Cigar smoking and cancer. The *Journal of the National Cancer Institute* (Feb. 16, 2000) published the results of a study that investigated the association between cigar smoking and death from tobacco-related cancers. Data were obtained for a national sample of 137,243 American men. The results are summarized in the accompanying table. Each male in the study was classified according to his cigar-smoking status and whether or not he died from a tobacco-related cancer.

a. Find the probability that a randomly selected man never smoked cigars and died from cancer.

b. Find the probability that a randomly selected man was a former cigar smoker and died from cancer.

c. Find the probability that a randomly selected man was a current cigar smoker and died from cancer.

d. Given that a male was a current cigar smoker, find the probability that he died from cancer.

e. Given that a male never smoked cigars, find the probability that he died from cancer.

	Died from Cancer		
Cigars	Yes	No	Totals
Never Smoked	782	120,747	121,529
Former Smoker	91	7,757	7,848
Current Smoker	141	7,725	7,866
Totals	1,014	136,229	137,243

Source: Shapiro, J. A., Jacobs, E. J., and Thun, M. J. "Cigar smoking in men and risk of death from tobacco-related cancers." *Journal of the National Cancer Institute*, Vol. 92, No. 4, Feb. 16, 2000 (Table 2).

3.85 Stacking in the NBA. Refer to the *Sociology of Sport Journal* (Vol. 14, 1997) study of "stacking" in the National Basketball Association (NBA), presented in Exercise 3.54 (p. 137). Consider again the table that follows, which summarizes the races and positions of 368 NBA players in 1993. Suppose an NBA player is selected at random from that year's player pool.

	Position			
	Guard	Forward	Center	Totals
White	26	30	28	84
Black	128	122	34	284
Totals	154	152	62	368

a. Given that the player is white, what is the probability that he is a center?

b. Given that the player is African-American, what is the probability that he is a center?

c. Are the events {White player} and {Center} independent?

d. Recall that "stacking" refers to the practice of African-American players being excluded from certain positions because of race. Use your answers to parts **a–c** to make an inference about stacking in the NBA.

3.86 Refer to the *Conservation Ecology* (Dec. 2003) study of the causes of forest fragmentation, presented in Exercise 2.148 (p. 93). Recall that the researchers used advanced high-resolution satellite imagery to develop fragmentation indexes for each forest. A 3×3 grid was superimposed over an aerial photo of the forest, and each square (pixel) of the grid was classified as forest (F), as earmarked for anthropogenic land use (A), or as natural land cover (N). An example of one such grid is shown here. The edges of the grid (where an "edge" is an imaginary line that separates any two adjacent pixels) are classified as F–A, F–N, A–A, A–N, N–N, or F–F edges.

A	A	N
N	F	F
N	F	F

a. Note that there are 12 edges inside the grid. Classify each edge as F–A, F–N, A–A, A–N, N–N or F–F.

b. The researchers calculated the fragmentation index by considering only the F-edges in the grid. Count the number of F-edges. (These edges represent the sample space for the experiment.)

c. If an F-edge is selected at random, find the probability that it is an F–A edge. (This probability is proportional to the anthropogenic fragmentation index calculated by the researchers.)

d. If an F edge is selected at random, find the probability that it is an F–N edge. (This probability is proportional to the natural fragmentation index calculated by the researchers.)

3.87 Detecting traces of TNT. University of Florida researchers in the Department of Materials Science and Engineering have invented a technique that rapidly detects traces of TNT. (*Today*, Spring 2005.) The method, which involves shining a laser on a potentially contaminated object, provides instantaneous results and gives no false positives. In this application, a false positive would occur if the laser detected traces of TNT when, in fact, no TNT were actually present on the object. Let A be the event that the laser light detects traces of TNT. Let B be the event that the object contains no traces of TNT. The probability of a false positive is 0. Write this probability in terms of A and B, using symbols such as "U", " $\cap$ ", and "|".

3.88 Lie detector test. A new type of lie detector called the Computerized Voice Stress Analyzer (CVSA) has been developed. The manufacturer claims that the CVSA is 98% accurate and, unlike a polygraph machine, will not be thrown off by drugs and medical factors. However, laboratory studies by the U.S. Defense Department found that the CVSA had an accuracy rate of 49.8%, slightly less than pure chance. (*Tampa Tribune*, Jan. 10, 1999.) Suppose the CVSA is used to test the veracity of four suspects. Assume that the suspects' responses are independent.

NW **a.** If the manufacturer's claim is true, what is the probability that the CVSA will correctly determine the veracity of all four suspects?

b. If the manufacturer's claim is true, what is the probability that the CVSA will yield an incorrect result for at least one of the four suspects?

c. Suppose that in a laboratory experiment conducted by the U.S. Defense Department on four suspects, the CVSA yielded incorrect results for two of the suspects. Make an inference about the true accuracy rate of the new lie detector.

3.89 Seeded players at a shuffleboard tournament. Shuffleboard is an outdoor game popular with senior citizens. In single-elimination shuffleboard tournaments, players are paired and play a match against each other, with the winners moving on to play another match. The tournament concludes in a final that matches two undefeated players. The winner of the final match is the overall tournament champion. *Chance* (Winter 2006) investigated the probability of winning a shuffleboard tournament with 64 players. As an example, the authors considered a small tournament with only 4 players, named A, B, C, and D. This tournament involves a total of three matches. In the first round, A plays B in one match and C plays D in the other match. Then the winners play in the final round.

a. List the different outcomes of the tournament. (For example, one outcome is "A and C win first-round matches and A wins the final.")

b. If the players are of equal ability, what is the probability that A wins the tournament?

c. Suppose the players are not of equal ability. The accompanying table gives the likelihood of one player defeating the other. (*Example*: The probability that A defeats B is .9.) What is the probability that A wins the tournament?

Outcome	Probability
A defeats B	.9
A defeats C	.7
A defeats D	.6
B defeats C	.1
B defeats D	.2
C defeats D	.4

3.90 Monitoring the quality of power equipment. *Mechanical Engineering* (Feb. 2005) reported on the need for wireless networks to monitor the quality of industrial equipment. For example, consider Eaton Corp., a company that develops distribution products. Eaton estimates that 90% of the electrical switching devices it sells can monitor the quality of the power running through the device. Eaton further estimates that, of the buyers of electrical switching devices capable of monitoring quality, 90% do not wire the equipment up for that purpose. Use this information to estimate the probability that an Eaton electrical switching device is capable of monitoring power quality and is wired up for that purpose.

Applying the Concepts—Advanced

3.91 Strategy in the game Go. Go is one of the oldest and most popular strategic board games in the world, especially in Japan and Korea. The two-player game is played on a flat surface marked with 19 vertical and 19 horizontal lines. The objective is to control territory by placing pieces called stones on vacant points on the board. Players alternate placing their stones. The player using black stones goes first, followed by the player using white stones. [*Note:* The University of Virginia requires MBA students to learn Go to understand how the Japanese conduct business.] *Chance* (Summer 1995) published an article that investigated the advantage of playing first (i.e., using the black stones) in Go. The results of 577 games recently played by professional Go players were analyzed.

a. In the 577 games, the player with the black stones won 319 times and the player with the white stones won 258 times. Use this information to estimate the probability of winning when you play first in Go.

b. Professional Go players are classified by level. Group C includes the top-level players, followed by Group B (middle-level players) and Group A (low-level players). The accompanying table describes the number of games won by the player with the black stones, categorized by level of the black player and level of the opponent. Estimate the probability of winning when you play first in Go for each combination of player and opponent level.

c. If the player with the black stones is ranked higher than the player with the white stones, what is the probability that black wins?

d. Given that the players are of the same level, what is the probability that the player with the black stones wins?

Black Player's Level	Opponent's Level	Number of Wins	Number of Games
C	A	34	34
C	B	69	79
C	C	66	118
B	A	40	54
B	B	52	95
B	C	27	79
A	A	15	28
A	B	11	51
A	C	5	39
Totals		**319**	**577**

Source: Kim, J., and Kim, H. J., "The advantage of playing first in Go," *Chance*, Vol. 8, No. 3, Summer 1995, p. 26 (Table 3).

3.92 Risk of a natural-gas pipeline accident. *Process Safety Progress* (Dec. 2004) published a risk analysis for a natural-gas pipeline between Bolivia and Brazil. The most likely scenario for an accident would be natural-gas leakage from a hole in the pipeline. The probability that the leak ignites immediately (causing a jet fire) is .01. If the leak does not immediately ignite, it may result in the delayed ignition of a gas cloud. Given no immediate ignition, the probability of delayed ignition (causing a flash fire) is .01. If there is no delayed ignition, the gas cloud will disperse harmlessly. Suppose a leak occurs in the natural-gas pipeline. Find the probability that either a jet fire or a flash fire will occur. Illustrate with a tree diagram.

3.93 NBA draft lottery. The National Basketball Association (NBA) utilizes a lottery to determine the order in which teams draft amateur (high school and college) players. Teams with the poorest won–lost records have the best chance of obtaining the first selection (and, presumably, best player) in the draft. Under the current lottery system, 14 table tennis balls numbered 1 through 14 are placed in a drum, and 4 balls are drawn. There are 1,001 possible combinations when 4 balls are drawn out of 14 without regard to their order of selection. Prior to the drawing, 1,000 combinations are assigned to the 13 teams with the worst records—the lottery teams—based on their order of finish during the regular season. (The team with the worst record is assigned 250 of these combinations, the team with the second-worst record is assigned 200 of the combinations, the team with the third-worst record is assigned 157 of the combinations, etc. The accompanying table gives the number of combinations assigned to each lottery team.) Once the 4 balls are drawn to the top of the drum to determine a four-digit combination, the team that has been assigned that combination will receive the first pick. The 4 balls are then placed back in the drum, and the process is repeated to determine the second, third, and subsequent draft picks. (*Note:* If the one unassigned combination is drawn, the balls are drawn to the top again.)

NBA Lottery Team	Number of Combinations
Worst record	250
2nd-worst record	200
3rd-worst record	157
4th-worst record	120
5th-worst record	89
6th-worst record	64
7th-worst record	44
8th-worst record	29
9th-worst record	18
10th-worst record	11
11th-worst record	7
12th-worst record	6
13th-worst record	5

Source: McCann, M. A. "Illegal defense: The irrational economics of banning high school players from the NBA draft," *Virginia Sports and Entertainment Law Journal*, Vol. 3, No. 2, Spring 2004, Table 5.

a. Demonstrate that there are 1,001 four-number combinations of the numbers 1 through 14.

b. For each of the 13 Lottery teams, find the probability that the team obtains the first pick in the draft.

c. Given that the team with the second-worst record obtains the number-1 pick, find the probability that the team with the worst record obtains the second pick of the draft.

d. Given that the team with the third-worst record obtains the number-1 pick, find the probability that the team with the worst record obtains the second pick of the draft.

e. Find the probability that the team with the worst record obtains the second pick of the draft, assuming that it did not obtain the first pick.

3.7 Random Sampling

How a sample is selected from a population is of vital importance in statistical inference because the probability of an observed sample will be used to infer the characteristics of the sampled population. To illustrate, suppose you deal yourself 4 cards from a deck of 52 cards, and all four cards are aces. Do you conclude that your deck is an ordinary bridge deck, containing only four aces, or do you conclude that the deck is stacked with more than four aces? It depends on how the cards were drawn. If the four aces are always placed at the top of a standard bridge deck, drawing four aces is not unusual—it is certain. If, however, the cards are thoroughly mixed, drawing four aces in a sample of 4 cards is highly improbable. The point, of course, is that in order to use the observed sample of 4 cards to draw inferences about the population (the deck of 52 cards), you need to know how the sample was selected from the deck.

One of the simplest and most frequently employed sampling procedures is implied in the previous examples and exercises. It produces what is known as a *random sample*. We learned in Section 1.5 (p. 13) that a random sample is likely to be *representative* of the population that it is selected from.

> **Definition 3.10**
>
> If *n* elements are selected from a population in such a way that every set of *n* elements in the population has an equal probability of being selected, then the *n* elements are said to be a **random sample.***

If a population is not too large and the elements can be numbered on slips of paper, poker chips, etc., you can physically mix the slips of paper or chips and remove *n* elements from the total. The numbers that appear on the chips selected would indicate the population elements to be included in the sample. Since it is often difficult to achieve a thorough mix, such a procedure only provides an approximation to random sampling. Most researchers rely on **random-number generators** to automatically generate a random sample. Random-number generators are available in tabular form, are built into most statistical software packages, and are available on the Internet (e.g., at *www.random.org*).

EXAMPLE 3.22

SELECTING A RANDOM SAMPLE

—5 Households from 100,000

Problem Suppose you wish to randomly sample five households from a population of 100,000 households.

a. How many different samples can be selected?

b. Use a random-number generator to select a random sample.

Solution

a. Since we want to select $n = 5$ objects (households) from $N = 100,000$, we apply the combinations rule of Section 3.1:

$$\binom{N}{n} = \binom{100,000}{5} = \frac{100,000!}{5!\,99,995!}$$

$$= \frac{100,000 \cdot 99,999 \cdot 99,998 \cdot 99,997 \cdot 99,996}{5 \cdot 4 \cdot 3 \cdot 2 \cdot 1}$$

$$= 8.33 \times 10^{22}$$

Thus, there are 83.3 billion trillion different samples of five households that can be selected from 100,000.

b. To ensure that each of the possible samples has an equal chance of being selected, as is required for random sampling, we can employ a **random-number table**, such as

---

*Strictly speaking, this is a *simple random sample*. There are many different types of random samples. The simple random sample is the most common.

Table I of Appendix A. Random-number tables are constructed in such a way that every number occurs with (approximately) equal probability. Furthermore, the occurrence of any one number in a position is independent of any of the other numbers that appear in the table. To use a table of random numbers, number the N elements in the population from 1 to N. Then turn to Table I and select a starting number in the table. Proceeding from this number either across the row or down the column, remove and record n numbers from the table.

To illustrate, first we number the households in the population from 1 to 100,000. Then we turn to a page of Table I—say, the first page. (A partial reproduction of the first page of Table I is shown in Table 3.7.) Now we arbitrarily select a starting number—say, the random number appearing in the third row, second column. This number is 48,360. Then we proceed down the second column to obtain the remaining four random numbers. In this case, we have selected five random numbers, which are highlighted in Table 3.7. Using the first five digits to represent households from 1 to 99,999 and the number 00000 to represent household 100,000, we can see that the households numbered

$$48,360 \quad 93,093 \quad 39,975 \quad 6,907 \quad 72,905$$

should be included in our sample.

Note: Use only the necessary number of digits in each random number to identify the element to be included in the sample. If, in the course of recording the n numbers from the table, you select a number that has already been selected, simply discard the duplicate and select a replacement at the end of the sequence. Thus, you may have to record more than n numbers from the table to obtain a sample of n unique numbers.

TABLE 3.7 Partial Reproduction of Table I in Appendix A

Row/Column	1	2	3	4	5	6
1	10480	15011	01536	02011	81647	91646
2	22368	46573	25595	85393	30995	89198
3	24130	48360	22527	97265	76393	64809
4	42167	93093	06243	61680	07856	16376
5	37670	39975	81837	16656	06121	91782
6	77921	06907	11008	42751	27756	53498
7	99562	72905	56420	69994	98872	31016
8	96301	91977	05463	07972	18876	20922
9	89579	14342	63661	10281	17453	18103
10	85475	36857	53342	53988	53060	59533
11	28918	69578	88231	33276	70997	79936
12	63553	40961	48235	03427	49626	69445
13	09429	93969	52636	92737	88974	33488

Look Back Can we be sure that all 83.3 billion trillion samples have an equal chance of being selected? The fact is, we can't; but to the extent that the random number table contains truly random sequences of digits, the sample should be very close to random.

Now Work Exercise 3.98

Table I in Appendix A is just one example of a random-number generator. For most scientific studies that require a large random sample, computers are used to generate the random sample. The SAS, MINITAB, and SPSS statistical software packages all have easy-to-use random-number generators.

For example, suppose we required a random sample of $n = 50$ households from the population of 100,000 households in Example 3.22. Here, we might employ the MINITAB random-number generator. Figure 3.20 shows a MINITAB printout listing

HOUSE50.MTW ***

	C1	C2	C3	C4	C5
	HouseID1	HouseID2	HouseID3	HouseID4	HouseID5
1	2036	22037	39884	62123	75750
2	3161	22055	44687	62421	80044
3	3930	22226	44689	63749	81857
4	9838	22773	45422	69079	81918
5	9961	22820	48104	69802	91680
6	11623	32989	50226	70895	94987
7	14114	33119	50443	71029	96110
8	17244	35045	50806	72879	97679
9	17598	35922	50813	73962	99232
10	18769	39272	61059	75517	99789
11					

FIGURE 3.20

MINITAB worksheet with random sample of 50 households

50 random numbers (from a population of 100,000). The households with these identification numbers would be included in the random sample.

Recall our discussion of designed experiments in Section 1.5 (p. 13). The twin notions of random selection and randomization are key to conducting good research with a designed experiment. The next example illustrates a basic application.

EXAMPLE 3.23
RANDOMIZATION IN A DESIGNED EXPERIMENT—
Clinical Trial

Problem A designed experiment in the medical field involving human subjects is referred to as a *clinical trial*. One recent clinical trial was designed to determine the potential of using aspirin in preventing heart attacks. Volunteer physicians were randomly divided into two groups: the *treatment* group and the *control* group. Each physician in the treatment group took one aspirin tablet a day for one year, while the physicians in the control group took an aspirin-free placebo made to look identical to an aspirin tablet. Since the physicians did not know to which group—treatment or control—they were assigned, the clinical trial is called a *blind study*. Assume that 20 physicians volunteered for the study. Use a random-number generator to randomly assign half of the physicians to the treatment group and half to the control group.

Solution Essentially, we want to select a random sample of 10 physicians from the 20. The first 10 selected will be assigned to the treatment group; the remaining 10 will be assigned to the control group. (Alternatively, we could randomly assign each physician, one by one, to either the treatment or control group. However, this would not guarantee exactly 10 physicians in each group.)

The MINITAB random-sample procedure was employed, producing the printout shown in Figure 3.21. Numbering the physicians from 1 to 20, we see that physicians 3, 11, 10, 14, 2, 7, 6, 16, 17, and 9 are assigned to receive the aspirin (the treatment). The remaining physicians are assigned the placebo (the control).

ASPIRIN-STUDY.MTW ***

	C1	C2	C3
	Physician	Treatment	
1	1	3	
2	2	11	
3	3	10	
4	4	14	
5	5	2	
6	6	7	
7	7	6	
8	8	16	
9	9	17	
10	10	9	
11	11		
12	12		
13	13		
14	14		
15	15		
16	16		
17	17		
18	18		
19	19		
20	20		
21			

FIGURE 3.21

MINITAB worksheet with random assignment of physicians

Exercises 3.94–3.105

Understanding the Principles

3.94 Define a random sample.

3.95 How are random samples related to representative samples?

3.96 What is a random-number generator?

3.97 Define a blind study.

Learning the Mechanics

3.98 Suppose you wish to sample $n = 2$ elements from a total of **NW** $N = 10$ elements.
   a. Count the number of different samples that can be drawn, first by listing them and then by using combinatorial mathematics.
   b. If random sampling is to be employed, what is the probability that any particular sample will be selected?
   c. Show how to use the random-number table, Table I in Appendix A, to select a random sample of 2 elements from a population of 10 elements. Perform the sampling procedure 20 times. Do any two of the samples contain the same 2 elements? Given your answer to part **b**, did you expect repeated samples?

3.99 Suppose you wish to sample $n = 3$ elements from a total of $N = 600$ elements.
   a. Count the number of different samples by using combinatorial mathematics.
   b. If random sampling is to be employed, what is the probability that any particular sample will be selected?
   c. Show how to use the random-number table, Table I in Appendix A, to select a random sample of 3 elements from a population of 600 elements. Perform the sampling procedure 20 times. Do any two of the samples contain the same 3 elements? Given your answer to part **b**, did you expect repeated samples?
   d. Use the computer to generate a random sample of 3 from the population of 600 elements.

3.100 Suppose that a population contains $N = 200,000$ elements. Use a computer or Table I of Appendix A to select a random sample of $n = 10$ elements from the population. Explain how you selected your sample.

Applying the Concepts—Basic

3.101 **Random-digit dialing.** To ascertain the effectiveness of their advertising campaigns, firms frequently conduct telephone interviews with consumers by using *random-digit dialing*. With this method, a random-number generator mechanically creates the sample of phone numbers to be called.
   a. Explain how the random-number table (Table I of Appendix A) or a computer could be used to generate a sample of seven-digit telephone numbers.
   b. Use the procedure you described in part **a** to generate a sample of 10 seven-digit telephone numbers.
   c. Use the procedure you described in part **a** to generate 5 seven-digit telephone numbers whose first three digits are 373.

3.102 **Census sampling.** In addition to its decennial enumeration of the population, the U.S. Census Bureau regularly samples the population for demographic information such as income, family size, employment, and marital status. Suppose the Bureau plans to sample 1,000 households in a city that has a total of 534,322 households. Show how the Bureau could use the random-number table in Appendix A or a computer to generate the sample. Select the first 10 households to be included in the sample.

Applying the Concepts—Intermediate

3.103 **Study of the effect of songs with violent lyrics.** A series of designed experiments conducted to examine the effects of songs with violent lyrics on aggressive thoughts and hostile feelings. (*Journal of Personality & Social Psychology*, May 2003.) In one of the experiments, college students were randomly assigned to listen to one of four types of songs: (1) a humorous song with violent lyrics, (2) a humorous song with nonviolent lyrics, (3) a nonhumorous song with violent lyrics, and (4) a nonhumorous song with nonviolent lyrics. Assuming that 120 students participated in the study, use a random-number generator to randomly assign 30 students to each of the four experimental conditions.

3.104 **Auditing an accounting system.** In auditing a firm's financial statements, an auditor is required to assess the operational effectiveness of the accounting system. In performing the assessment, the auditor frequently relies on a random sample of actual transactions. (Stickney and Weil, *Financial Accounting: An Introduction to Concepts, Methods, and Uses*, 1994.) A particular firm has 5,382 customer accounts that are numbered from 0001 to 5382.
   a. One account is to be selected at random for audit. What is the probability that account number 3,241 is selected?
   b. Draw a random sample of 10 accounts and explain in detail the procedure you used.
   c. Refer to part **b**. The following are two possible random samples of size 10:

Sample Number 1				
5011	0082	0963	0772	3415
2663	1126	0008	0026	4189

Sample Number 2				
0001	0003	0005	0007	0009
0002	0004	0006	0008	0010

Is one more likely to be selected than the other? Explain.

Applying the Concepts—Advanced

3.105 **Selecting archaeological dig sites.** Archaeologists plan to perform test digs at a location they believe was inhabited several thousand years ago. The site is approximately 10,000 meters long and 5,000 meters wide. They first draw rectangular grids over the area, consisting of lines every 100 meters, creating a total of $100 \cdot 50 = 5,000$ intersections (not counting one of the outer boundaries). The plan is to randomly sample 50 intersection points and dig at the sampled intersections. Explain how you could use a random-number generator to obtain a random sample of 50 intersections. Develop at least two plans: one that numbers the intersections from 1 to 5,000 prior to selection and another that selects the row and column of each sampled intersection (from the total of 100 rows and 50 columns).

TERMS

Additive rule of probability   132
Combinations rule   122
Combinatorial mathematics   122
Complement   130
Complementary events   130
Compound event   127
Conditional probability   138
Counting rule   123
Dependent events   144
Event   119
Experiment   115

Independent events   144
Intersection   127
Law of large numbers   116
Multiplicative rule
  of probability   141
Mutually exclusive events   133
Odds   163
Probability rules for sample
  points   118
Random-number generator   154

Random-number table   154
Random sample   154
Sample point   115
Sample space   116
Tree diagram   143
Two-way table   128
Unconditional probabilities   138
Union   127
Venn diagram   116

GUIDE TO SELECTING PROBABILITY RULES

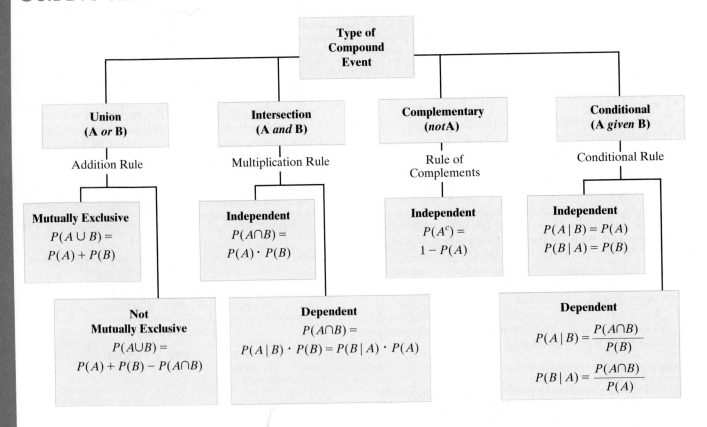

CHAPTER NOTES

Probability Rules for k Sample Points, $S_1, S_2, S_3, \cdots, S_k$
1. $0 \le P(S_i) \le 1$
2. $\Sigma P(S_i) = 1$

Random Sample

All possible such samples have equal probability of being selected

Combinations Rule

Counting number of samples of n elements selected from N elements

$$\binom{N}{n} = \frac{N!}{n!(N-n)!}$$

$$= \frac{N(N-1)(N-2)\dots(N-n+1)}{n(n-1)(n-2)\dots(2)(1)}$$

Key Symbols

S	**Sample Space** (collection of all sample points)
A: $\{1,2\}$	Set of **sample points in event A**
$P(A)$	**Probability** of event A
$A \cup B$	**Union** of events A and B (either A or B occurs)
$A \cap B$	**Intersection** of events A and B (both A and B occur)
A^c	**Complement** of A (event A does not occur)
$A\vert B$	Event A occurs, **given that** event B occurs
$\binom{N}{n}$	Number of **combinations** of N elements taken n at a time
$N!$	**N factorial** $= N(N-1)(N-2)\dots(2)(1)$

SUPPLEMENTARY EXERCISES 3.106–3.142

Understanding the Principles

3.106 Which of the following pairs of events are mutually exclusive?

a. $A = \{$The San Francisco Giants win the World Series next year.$\}$
$B = \{$Barry Bonds, Giants outfielder, hits 75 home runs next year.$\}$

b. $A = \{$Psychiatric patient Tony responds to a stimulus within 5 seconds.$\}$
$B = \{$Psychiatric patient Tony has the fastest stimulus response time of 2.3 seconds.$\}$

c. $A = \{$High school graduate Cindy enrolls at the University of South Florida next year.$\}$
$B = \{$High school graduate Cindy does not enroll in college next year.$\}$

3.107 Use the symbols $\cap$, $\cup$, $\vert$, and ^{c}, to convert the following statements into compound events involving events A and B, where $A = \{$You purchase a notebook computer.$\}$ and $B = \{$You vacation in Europe.$\}$:

a. You purchase a notebook computer or vacation in Europe.
b. You will not vacation in Europe.
c. You purchase a notebook computer and vacation in Europe.
d. Given that you vacation in Europe, you will not purchase a notebook computer.

Learning the Mechanics

3.108 A sample space consists of four sample points $S_1, S_2, S_3,$ and $S_4,$ where $P(S_1) = .2, P(S_2) = .1, P(S_3) = .3,$ and $P(S_4) = .4.$

a. Show that the sample points obey the two probability rules for a sample space.
b. If an event $A = \{S_1, S_4,\}$ find P(A).

3.109 A and B are mutually exclusive events, with $P(A) = .2$ and $P(B) = .3$.

a. Find $P(A\vert B)$.
b. Are A and B independent events?

3.110 For two events A and B, suppose $P(A) = .7,$ $P(B) = .5,$ and $P(A \cap B) = .4$. Find $P(A \cup B)$.

3.111 Given that $P(A \cap B) = .4$ and $P(A\vert B) = .8$, find $P(B)$.

3.112 The accompanying Venn diagram illustrates a sample space containing six sample points and three events, $A, B,$ and $C.$ The probabilities of the sample points are $P(1) = .3, P(2) = .2, P(3) = .1, P(4) = .1, P(5) = .1,$ and $P(6) = .2.$

a. Find $P(A \cap B), P(B \cap C), P(A \cup C), P(A \cup B \cup C),$ $P(B^c), P(A^c \cap B), P(B\vert C),$ and $P(B\vert A).$
b. Are A and B independent? Mutually exclusive? Why?
c. Are B and C independent? Mutually exclusive? Why?

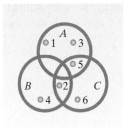

S

3.113 Two events, A and B, are independent, with $P(A) = .3$ and $P(B) = .1.$

a. Are A and B mutually exclusive? Why?
b. Find $P(A\vert B)$ and $P(B\vert A)$.
c. Find $P(A \cup B)$.

3.114 Find the numerical value of

a. 6!   **b.** $\binom{10}{9}$ **c.** $\binom{10}{1}$ **d.** P_2^6

e. $\binom{6}{3}$ **f.** 0!   **g.** P_4^{10} **h.** P_2^{50}

APPLET **Applet Exercise 3.6**

Use the applet entitled *Random Numbers* to generate a list of 50 numbers between 1 and 100, inclusive. Use the list to find each of the following probabilities:

a. The probability that a number chosen from the list is less than or equal to 50
b. The probability that a number chosen from the list is even
c. The probability that a number chosen from the list is both less than or equal to 50 and even
d. The probability that a number chosen from the list is less than or equal to 50, given that the number is even
e. Do your results from parts **a–d** support the conclusion that the events *less than or equal to 50* and *even* are independent? Explain.

Applying the Concepts—Basic

CRASH

3.115 NHTSA new car crash testing. Refer to the National Highway Traffic Safety Administration (NHTSA) crash tests of new car models, presented in Exercise 2.166 (p. 102). Recall that the NHTSA has developed a "star" scoring system, with results ranging from one star (*) to five stars (*****). The more stars in the rating, the better the level of crash protection in a head-on collision. A summary of the driver-side star ratings for 98 cars is reproduced in the accompanying MINITAB printout. Assume that one of the 98 cars is selected at random. State whether each of the following is true or false:

a. The probability that the car has a rating of two stars is 4.
b. The probability that the car has a rating of four or five stars is .7857.
c. The probability that the car has a rating of one star is 0.
d. The car has a better chance of having a two-star rating than of having a five-star rating.

Tally for Discrete Variables: DRIVSTAR

DRIVSTAR	Count	Percent
2	4	4.08
3	17	17.35
4	59	60.20
5	18	18.37
N=	98	

3.116 Sterile couples in Jordan. A sterile family is a couple that has no children by their deliberate choice or because they are biologically infertile. Couples who are childless by chance are not considered to be sterile. Researchers at Yarmouk University (in Jordan) estimated the proportion of sterile couples in that country to be .06. (*Journal of Data Science*, July 2003.) Also, 64% of the sterile couples in Jordan are infertile. Find the probability that a Jordanian couple is both sterile and infertile.

3.117 Selecting a sample. A random sample of five students is to be selected from 50 sociology majors for participation in a special program.

a. In how many different ways can the sample be drawn?

b. Show how the random-number table, Table I of Appendix A, can be used to select the sample of students.

EVOS

3.118 Oil spill impact on seabirds. Refer to the *Journal of Agricultural, Biological, and Environmental Statistics* (Sept. 2000) study of the impact of the *Exxon Valdez* tanker oil spill on the seabird population in Prince William Sound, Alaska, presented in Exercise 2.181 (p. 105). Recall that data were collected on 96 shoreline locations (called transects), and it was determined whether or not the transect was in an oiled area. The data are stored in the file called EVOS. From the data, estimate the probability that a randomly selected transect in Prince William Sound is contaminated with oil.

3.119 Speeding New Jersey drivers. A certain stretch of highway on the New Jersey Turnpike has a posted speed limit of 60 mph. A survey of drivers on this portion of the turnpike revealed the following facts (reported in *The Washington Post*, Aug. 16, 1998):

(1) 14% of the drivers on this stretch of highway were African-Americans
(2) 98% of the drivers were exceeding the speed limit by at least 5 mph (and thus were subject to being stopped by the state police)
(3) Of these violators, 15% of the drivers were African-Americans
(4) Of the drivers stopped for speeding by the New Jersey state police, 35% were African-Americans

a. Convert each of the reported percentages into a probability statement.
b. Find the probability that a driver on this stretch of highway is African-American and exceeding the speed limit.
c. Given that a driver on this stretch of highway is exceeding the speed limit, what is the probability that the driver is not African-American?
d. Given that a driver on this stretch of highway is stopped by the New Jersey state police, what is the probability that the driver is not African-American?
e. For this stretch of highway, are the events {African-American driver and {Driver exceeding the speed limit} independent (to a close approximation)? Explain.
f. Use the probabilities in *The Washington Post* report to make a statement about whether African-Americans are stopped more often for speeding than would be expected on this stretch of turnpike.

3.120 Carbon monoxide poisoning. The *American Journal of Public Health* (July 1995) published a study on unintentional carbon monoxide (CO) poisoning of Colorado residents. A total of 981 cases of CO poisoning was reported during a six-year period. Each case was classified as fatal or nonfatal and by source of exposure. The number of cases occurring in each of the categories is shown in the next table. Assume that 1 of the 981 cases of unintentional CO poisoning is randomly selected.

a. List all sample points for this experiment.
b. What is the set of all sample points called?
c. Let *A* be the event that the CO poisoning is caused by fire. Find $P(A)$.

Source of Exposure	Fatal	Nonfatal	Total
Fire	63	53	116
Auto exhaust	60	178	238
Furnace	18	345	363
Kerosene heater or space heater	9	18	27
Appliance	9	63	72
Other gas-powered motor	3	73	76
Fireplace	0	16	16
Other	3	19	22
Unknown	9	42	51
Total	174	807	981

Source: Cook, M. C., Simon P. A., and Hoffman, R. E. "Unintentional carbon monoxide poisoning in Colorado, 1986 through 1991," *American Journal of Public Health*, Vol. 85, No. 7, July 1995, p. 989 (Table 1).

d. Let B be the event that the CO poisoning is fatal. Find $P(B)$.

e. Let C be the event that the CO poisoning is caused by auto exhaust. Find $P(C)$.

f. Let D be the event that the CO poisoning is caused by auto exhaust and is fatal. Find $P(D)$.

g. Let E be the event that the CO poisoning is caused by fire but is nonfatal. Find $P(E)$.

h. Given that the source of the poisoning is fire, what is the probability that the case is fatal?

i. Given that the case is nonfatal, what is the probability that it is caused by auto exhaust?

j. If the case is fatal, what is the probability that the source is unknown?

k. If the case is nonfatal, what is the probability that the source is not fire or a fireplace?

3.121 **Beach erosional hot spots.** Beaches that exhibit high erosion rates relative to the surrounding beach are defined as *erosional hot spots*. The U.S. Army Corps of Engineers is conducting a study of beach hot spots. Through an online questionnaire, data are collected on six beach hot spots. The data are listed in the next table.

Beach Hot Spot	Beach Condition	Nearshore Bar Condition	Long-Term Erosion Rate (miles/year)
Miami Beach, FL	No dunes/flat	Single, shore parallel	4
Coney Island, NY	No dunes/flat	Other	13
Surfside, CA	Bluff/scarp	Single, shore parallel	35
Monmouth Beach, NJ	Single dune	Planar	Not estimated
Ocean City, NJ	Single dune	Other	Not estimated
Spring Lake, NJ	Not observed	Planar	14

Source: Identification and Characterization of Erosional Hotspots, William & Mary Virginia Institute of Marine Science, U.S. Army Corps of Engineers Project Report, Mar. 18, 2002.

a. Suppose you record the nearshore bar condition of each beach hot spot. Give the sample space for this experiment.

b. Find the probabilities of the sample points in the sample space you defined in part **a**.

c. What is the probability that a beach hot spot has either a planar or single, shore-parallel nearshore bar condition?

d. Now suppose you record the beach condition of each beach hot spot. Give the sample space for this experiment.

e. Find the probabilities of the sample points in the sample space you defined in part **d**.

f. What is the probability that the condition of the beach at a particular beach hot spot is not flat?

3.122 **Chemical insect attractant.** An entomologist is studying the effect of a chemical sex attractant (pheromone) on insects. Several insects are released at a site equidistant from the pheromone under study and a control substance. If the pheromone has an effect, more insects will travel toward it rather than toward the control. Otherwise, the insects are equally likely to travel in either direction. Suppose the pheromone under study has no effect, so that it is equally likely that an insect will move toward the pheromone or toward the control. Suppose five insects are released.

a. List or count the number of different ways the insects can travel.

b. What is the chance that all five travel toward the pheromone?

c. What is the chance that exactly four travel toward the pheromone?

d. What inference would you make if the event in part **c** actually occurs? Explain.

3.123 **Factors identifying urban counties.** *Urban* and *rural* describe geographic areas upon which land-zoning regulations, school district policy, and public-service policy are often set. However, the characteristics of urban and rural areas are not clearly defined. Researchers at the University of Nevada at Reno asked a sample of county commissioners to give their perception of the single most important factor in identifying urban counties. (*Professional Geographer*, Feb. 2000.) In all, five factors were mentioned by the commissioners: total population, agricultural change, presence of industry, growth, and population concentration. The survey results are displayed in the pie chart below. Suppose one of the commissioners is selected at random and the most important factor specified by the commissioner is recorded.

a. List the sample points for this experiment.

b. Assign reasonable probabilities to the sample points.

c. Find the probability that the most important factor specified by the commissioner is population related.

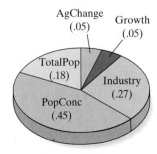

3.124 Study of aggressiveness and birth order. Psychologists tend to believe that there is a relationship between aggressiveness and order of birth. To test this belief, a psychologist chose 500 elementary school students at random and administered each a test designed to measure the student's aggressiveness. Each student was classified according to one of four categories. The percentages of students falling into the four categories are shown here.

	Firstborn	Not Firstborn
Aggressive	15%	15%
Not Aggressive	25%	45%

a. If one student is chosen at random from the 500, what is the probability that the student is firstborn?

b. What is the probability that the student is aggressive?

c. What is the probability that the student is aggressive, given that the student was firstborn?

d. If we have
  A: {The student chosen is aggressive.}
  B: {The student chosen is firstborn.}
  are *A* and *B* independent? Explain.

3.125 Elderly and their roles. Refer to the *Psychology and Aging* (Dec. 2000) study of elderly people and their roles, presented Exercise 2.10 (p. 35). The table summarizing the most salient roles identified by each respondent in a national sample of 1,102 adults 65 years or older is reproduced below.

Most Salient Role	Number
Spouse	424
Parent	269
Grandparent	148
Other relative	59
Friend	73
Homemaker	59
Provider	34
Volunteer, club, church member	36
Total	1,102

Source: Krause, N., and Shaw, B. A. "Role-specific feelings of control and mortality," *Psychology and Aging,* Vol. 15, No. 4, Dec. 2000 (Table 2).

a. What is the probability that a randomly selected elderly person feels that his or her most salient role is that of spouse?

b. What is the probability that a randomly selected elderly person feels that his or her most salient role is that of parent or grandparent?

c. What is the probability that a randomly selected elderly person feels that his or her most salient role does not involve being a spouse or relative of any kind?

3.126 Elderly wheelchair user study. The *American Journal of Public Health* (Jan. 2002) reported on a study of elderly wheelchair users who live at home. A sample of 306 wheelchair users, age 65 or older, were surveyed about whether they had an injurious fall during the year and whether their home featured any one of five structural modifications: bathroom modifications, widened doorways/hallways, kitchen modifications, installed railings, and easy-open doors. The responses are summarized in the accompanying table. Suppose we select, at random, one of the 306 wheelchair users surveyed.

Home Features	Injurious Fall(s)	No Falls	Totals
All 5	2	7	9
At least 1, but not all	26	162	188
None	20	89	109
Totals	48	258	306

Source: Berg, K., Hines, M., and Allen, S. "Wheelchair users at home: Few home modifications and many injurious falls," *American Journal of Public Health,* Vol. 92, No. 1, Jan. 2002 (Table 1).

a. Find the probability that the wheelchair user had an injurious fall.

b. Find the probability that the wheelchair user had all five features installed in the home.

c. Find the probability that the wheelchair user had no falls and none of the features installed in the home.

d. Given no features installed in the home, find the probability of an injurious fall.

3.127 Shooting free throws. In college basketball games, a player may be afforded the opportunity to shoot two consecutive foul shots (free throws).

a. Suppose a player who makes (i.e., scores on) 80% of his foul shots has been awarded two free throws. If the two throws are considered independent, what is the probability that the player makes both shots? exactly one? neither shot?

b. Suppose a player who makes 80% of his or her first attempted foul shots has been awarded two free throws and the outcome on the second shot is dependent on the outcome of the first shot. In fact, if this player makes the first shot, he makes 90% of the second shots; and if he misses the first shot, he makes 70% of the second shots. In this case, what is the probability that the player makes both shots? exactly one? neither shot?

c. In parts **a** and **b**, we considered two ways of *modeling* the probability that a basketball player makes two consecutive foul shots. Which model do you think gives a more realistic explanation of the outcome of shooting foul shots; that is, do you think two consecutive foul shots are independent or dependent? Explain.

3.128 Antibiotic for blood infections. Enterococci are bacteria that cause blood infections in hospitalized patients. One antibiotic used to battle enterococci is vancomycin. A study by the federal Centers for Disease Control and Prevention revealed that 8% of all enterococci isolated in hospitals nationwide were resistant to vancomycin. (*New York Times,* Sept. 12, 1995.) Consider a random sample of three patients with blood infections caused by the enterococci bacterium. Assume that all three patients are treated with the antibiotic vancomycin.

a. What is the probability that all three patients are treated successfully? What assumption did you make concerning the patients?

b. What is the probability that the bacteria resist the antibiotic in at least one patient?

3.129 **Maize seeds.** The genetic origin and properties of maize (modern-day corn) were investigated in *Economic Botany* (Jan.–Mar. 1995). Seeds from maize ears carry either single spikelets or paired spikelets, but not both. Progeny tests on approximately 600 maize ears revealed the following information: 40% of all seeds carry single spikelets, while 60% carry paired spikelets. A seed with single spikelets will produce maize ears with single spikelets 29% of the time and paired spikelets 71% of the time. A seed with paired spikelets will produce maize ears with single spikelets 26% of the time and paired spikelets 74% of the time.

a. Find the probability that a randomly selected maize ear seed carries a single spikelet and produces ears with single spikelets.

b. Find the probability that a randomly selected maize ear seed produces ears with paired spikelets.

3.130 **Alzheimer's gene study.** Scientists have already discovered two genes (on chromosomes 19 and 21) that mutate to cause the early onset of Alzheimer's disease. *Science News* (July 8, 1995) reported on a search for a third gene that causes Alzheimer's. An international team of scientists gathered genetic information from 21 families afflicted by the early-onset form of the disease. In 6 of these families, mutations in the S182 gene on chromosome 14 accounted for the early onset of Alzheimer's.

a. Consider a family afflicted by the early onset of Alzheimer's disease. Find the approximate probability that the researchers will discover mutations in the S182 gene on chromosome 14 for this family.

b. Discuss the reliability of the probability you found in part **a.** How could a more accurate estimate of the probability be obtained?

3.131 **Series and parallel systems.** Consider the two systems shown in the accompanying schematic. System A operates properly only if all three components operate properly. (The three components are said to operate *in series.*) The probability of failure for system A components 1, 2, and 3 are .12, .09, and .11, respectively. Assume that the components operate independently of each other.

System B comprises two subsystems said to operate *in parallel.* Each subsystem has two components that operate in series. System B will operate properly as long as at least one of the subsystems functions properly. The probability of failure for each component in the system is .1. Assume that the components operate independently of each other.

a. Find the probability that System A operates properly.

b. What is the probability that at least one of the components in System A will fail and therefore that the system will fail?

c. Find the probability that System B operates properly.

d. Find the probability that exactly one subsystem in System B fails.

e. Find the probability that System B fails to operate properly.

f. How many parallel subsystems like the two shown here would be required to guarantee that the system would operate properly at least 99% of the time?

Applying the Concepts—Advanced

3.132 **Patient medical instruction sheets.** Physicians and pharmacists sometimes fail to inform patients adequately about the proper application of prescription drugs and the precautions to take in order to avoid potential side effects. One method of increasing patients' awareness of the problem is for physicians to provide patient medication instruction (PMI) sheets. The American Medical Association, however, has found that only 20% of the doctors who frequently prescribe drugs distribute PMI sheets to their patients. Assume that 20% of all patients receive the PMI sheet with their prescriptions and that 12% receive the PMI sheet and are hospitalized because of a drug-related problem. What is the probability that a person will be hospitalized for a drug-related problem, given that the person has received the PMI sheet?

3.133 **Odds of winning a horse race.** Handicappers for horse races express their beliefs about the probability of each horse winning a race in terms of **odds.** If the probability of event E is $P(E)$, then the *odds in favor of E are* $P(E)$ to $1 - P(E)$. Thus, if a handicapper assesses a probability of .25 that Smarty Jones will win the Belmont Stakes, the odds in favor of Smarty Jones are $^{25}/_{100}$ to $^{75}/_{100}$, or 1 to 3. It follows that the *odds against E are* $1 - P(E)$ to $P(E)$, or 3 to 1 against a win by Smarty Jones. In general, if the odds in favor of event E are a to b, then $P(E) = a/(a + b)$.

a. A second handicapper assesses the probability of a win by Smarty Jones to be $^1/_3$. According to the second

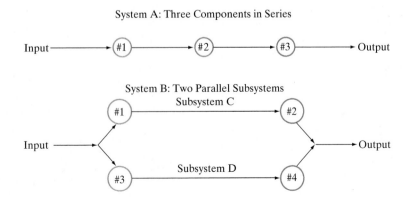

System A: Three Components in Series

Input ——→ #1 ——→ #2 ——→ #3 ——→ Output

System B: Two Parallel Subsystems

Subsystem C

#1 ——→ #2

Input ——→

Subsystem D

#3 ——→ #4

——→ Output

handicapper, what are the odds in favor of a Smarty Jones win?

b. A third handicapper assesses the odds in favor of Smarty Jones to be 1 to 1. According to the third handicapper, what is the probability of a Smarty Jones win?

c. A fourth handicapper assesses the odds against Smarty Jones winning to be 3 to 2. Find this handicapper's assessment of the probability that Smarty Jones will win.

3.134 **Chance of winning at blackjack.** Blackjack, a favorite game of gamblers, is played by a dealer and at least one opponent. At the outset of the game, 2 cards of a 52-card bridge deck are dealt to the player and 2 cards to the dealer. Drawing an ace and a face card is called *blackjack*. If the dealer does not draw a blackjack and the player does, the player wins. If both the dealer and player draw blackjack, a "push" (i.e., a tie) occurs.

a. What is the probability that the dealer will draw a blackjack?

b. What is the probability that the player wins with a blackjack?

3.135 **Finding an organ transplant match.** One of the problems encountered with organ transplants is the body's rejection of the transplanted tissue. If the antigens attached to the tissue cells of the donor and receiver match, the body will accept the transplanted tissue. Although the antigens in identical twins always match, the probability of a match in other siblings is .25, and that of a match in two people from the population at large is .001. Suppose you need a kidney and you have two brothers and a sister.

a. If one of your three siblings offers a kidney, what is the probability that the antigens will match?

b. If all three siblings offer a kidney, what is the probability that all three antigens will match?

c. If all three siblings offer a kidney, what is the probability that none of the antigens will match?

d. Repeat parts **b** and **c**, this time assuming that the three donors were obtained from the population at large.

3.136 **Language impairment in children.** Children who develop unexpected difficulties with the spoken language are often diagnosed as *specifically language impaired* (SLI). A study published in the *Journal of Speech, Language, and Hearing Research* (Dec. 1997) investigated the incidence of SLI in kindergarten children. As an initial screen, each in a national sample of over 7,000 children was given a test for language performance. The percentages of children who passed and failed the screen were 73.8% and 26.2%, respectively. All children who failed the screen were tested clinically for SLI. About one-third of those who passed the screen were randomly selected and also tested for SLI. The percentage of children diagnosed with SLI in the "failed screen" group was 20.5%; the percentage diagnosed with SLI in the "pass screen" group was 2.8%.

a. For this problem, let "pass" represent a child who passed the language performance screen, "fail" represent a child who failed the screen, and "SLI" represent a child diagnosed with SLI. Now find each of the following probabilities: $P(\text{Pass})$, $P(\text{Fail})$, $P(\text{SLI}|\text{Pass})$, and $P(\text{SLI}|\text{Fail})$.

b. Use the probabilities from part **a** to find $P(\text{Pass} \cap \text{SLI})$ and $P(\text{Fail} \cap \text{SLI})$. What probability law did you use to calculate these probabilities?

c. Use the probabilities from part **b** to find $P(\text{SLI})$. What probability law did you use to calculate this probability?

3.137 **Accuracy of pregnancy tests.** Seventy-five percent of all women who submit to pregnancy tests are really pregnant. A certain pregnancy test gives a *false positive* result with probability .02 and a *valid positive result* with probability .99. If a particular woman's test is positive, what is the probability that she really is pregnant? [*Hint:* If A is the event that a woman is pregnant and B is the event that the pregnancy test is positive, then B is the union of the two mutually exclusive events $A \cap B$ and $A^c \cap B$. Also, the probability of a false positive result may be written as $P(B|A^c) = .02$.]

3.138 **Chance of winning at "craps."** A version of the dice game "craps" is played in the following manner. A player starts by rolling two balanced dice. If the roll (the sum of the two numbers showing on the dice) results in a 7 or 11, the player wins. If the roll results in a 2 or a 3 (called *craps*), the player loses. For any other roll outcome, the player continues to throw the dice until the original roll outcome recurs (in which case the player wins) or until a 7 occurs (in which case the player loses).

a. What is the probability that a player wins the game on the first roll of the dice?

b. What is the probability that a player loses the game on the first roll of the dice?

c. If the player throws a total of 4 on the first roll, what is the probability that the game ends (win or lose) on the next roll?

3.139 **Odd Man Out.** Three people play a game called "Odd Man Out." In this game, each player flips a fair coin until the outcome (heads or tails) for one of the players is not the same as that for the other two players. This player is then "the odd man out" and loses the game. Find the probability that the game ends (i.e., either exactly one of the coins will fall heads or exactly one of the coins will fall tails) after only one toss by each player. Suppose one of the players, hoping to reduce his chances of being the odd man out, uses a two-headed coin. Will this ploy be successful? Solve by listing the sample points in the sample space.

Critical Thinking Challenges

3.140 **"Let's Make a Deal."** Marilyn vos Savant, who is listed in the *Guinness Book of World Records Hall of Fame* as having the "Highest IQ," writes a weekly column in the Sunday newspaper supplement *Parade Magazine*. Her column, "Ask Marilyn," is devoted to games of skill, puzzles, and mind-bending riddles. In one issue (*Parade Magazine*, Feb. 24, 1991), vos Savant posed the following question:

Suppose you're on a game show and you're given a choice of three doors. Behind one door is a car; behind the others, goats. You pick a door—say, #1—and the host, who knows what's behind the doors, opens another door—say, #3—which has a goat. He then says to you, "Do you want to pick door #2?" Is it to your advantage to switch your choice?

Marilyn's answer: "Yes, you should switch. The first door has a $1/3$ chance of winning [the car], but the second has a $2/3$ chance [of winning the car]." Predictably, vos Savant's surprising answer elicited thousands of critical letters, many of them from Ph.D. mathematicians, that disagreed with her. Who is correct, the Ph.D.'s or Marilyn?

3.141 Most likely coin-toss sequence. In *Parade Magazine*'s (Nov. 26, 2000) column "Ask Marilyn," the following question was posed: "I have just tossed a [balanced] coin 10 times, and I ask you to guess which of the following three sequences was the result. One (and only one) of the sequences is genuine."

1. H H H H H H H H H H
2. H H T T H T T H H H
3. T T T T T T T T T T

Marilyn's answer to the question posed was "Though the chances of the three specific sequences oc-

curring randomly are equal . . . it's reasonable for us to choose sequence (2) as the most likely genuine result." Do you agree?

3.142 Flawed Pentium computer chip. In October 1994, a flaw was discovered in the Pentium microchip installed in personal computers. The chip produced an incorrect result when dividing two numbers. Intel, the manufacturer of the Pentium chip, initially announced that such an error would occur once in 9 billion division operations, or "once every 27,000 years," for a typical user; consequently, Intel did not immediately offer to replace the chip.

Depending on the procedure, statistical software packages (e.g., SAS) may perform an extremely large number of divisions to produce required output. For heavy users of the software, 1 billion divisions over a short time frame is not unusual. Will the flawed chip be a problem for a heavy SAS user? [*Note:* Two months after the flaw was discovered, Intel agreed to replace all Pentium chips free of charge.]

SUMMARY ACTIVITY: Simulating Probability with a Deck of Cards

Obtain a standard deck of 52 playing cards (the kind commonly used for bridge, poker, or solitaire). An experiment will consist of drawing 1 card at random from the deck of cards and recording which card was observed. This random drawing will be simulated by shuffling the deck thoroughly and observing the top card. Consider the following two events:

A: {The card observed is a heart.}
B: {The card observed is an ace, king, queen, or jack.}

a. Find $P(A)$, $P(B)$, $P(A \cap B)$, and $P(A \cup B)$.
b. Conduct the experiment 10 times, and record the observed card each time. Be sure to return the observed card each time, and shuffle the deck thoroughly before making the draw. After you've observed 10 cards, calculate the proportion of observations that satisfy event A, event B, event

$A \cap B$, event $A \cup B$. Compare the observed proportions with the true probabilities calculated in part **a**.
c. Conduct the experiment 40 more times, to obtain a total of 50 observed cards. Now calculate the proportion of observations that satisfy event A, event B, event $A \cap B$, and event $A \cup B$. Compare these proportions with those found in part **b** and the true probabilities found in part **a**.
d. Conduct the experiment 50 more times, to obtain a total of 100 observations. Compare the observed proportions for the 100 trials with those found previously. What comments do you have concerning the different proportions found in parts **b**, **c**, and **d** compared with the true probabilities found in part **a**? How do you think the observed proportions and true probabilities would compare if the experiment were conducted 1,000 times? 1 million times?

REFERENCES

Bennett, D. J. *Randomness*. Cambridge, MA: Harvard University Press, 1998.

Epstein, R. A. *The Theory of Gambling and Statistical Logic*, rev. ed. New York: Academic Press, 1977.

Feller, W. *An Introduction to Probability Theory and Its Applications*, 3d ed., Vol. 1. New York: Wiley, 1968.

Lindley, D. V. *Making Decisions*, 2d ed. London: Wiley, 1985.

Parzen, E. *Modern Probability Theory and Its Applications*. New York: Wiley, 1960.

Wackerly, D., Mendenhall, W., and Scheaffer, R. L. *Mathematical Statistics with Applications*, 6th ed. Boston: Duxbury, 2002.

Williams, B. *A Sampler on Sampling*. New York: Wiley, 1978.

Winkler, R. L. *An Introduction to Bayesian Inference and Decision*. New York: Holt, Rinehart and Winston, 1972.

Wright, G., and Ayton, P., eds. *Subjective Probability*. New York: Wiley, 1994.

Using Technology

Generating a Random Sample with MINITAB

To obtain a random sample of observations (cases) from a data set stored in the MINITAB worksheet, click on the "Calc" button on the MINITAB menu bar, then click on "Random Data", and finally, click on "Sample from Columns", as shown in Figure 3.M.1. The resulting dialog box appears as shown in Figure 3.M.2. Specify the sample size (i.e., number of rows), the variable(s) to be sampled, and the column(s) in which you want to save the sample. Click "OK", and the MINITAB worksheet will reappear with the values of the variable for the selected (sampled) cases in the column specified.

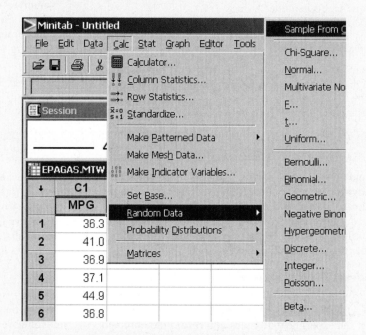

FIGURE 3.M.1
MINITAB menu options for sampling your data

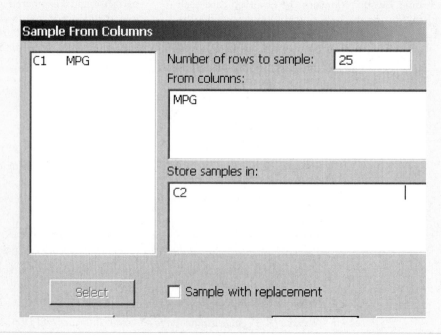

FIGURE 3.M.2
MINITAB options for selecting a random sample for worksheet columns

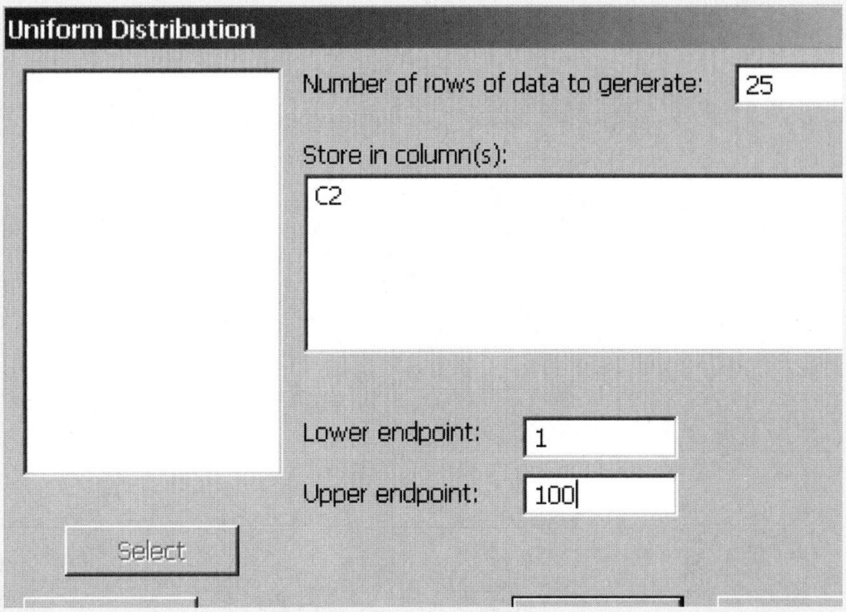

FIGURE 3.M.3
MINITAB options for
selecting a random sample of
cases

In MINITAB, you can also generate a sample of case numbers. From the MINITAB menu, click on the "Calc" button, then click on "Random Data", and finally, click on the "Uniform" option (see Figure 3.M.1). The resulting dialog box appears as shown in Figure 3.M.3. Specify the number of rows (i.e., the sample size) and the column in which the case numbers selected will be stored. Click "OK", and the MINITAB worksheet will reappear with the case numbers for the selected (sampled) cases in the column specified.

[*Note:* If you want the option of generating the same (identical) sample multiple times from the data set, then first click on the "Set Base" option shown in Figure 3.M.1. Specify an integer in the resulting dialog box. If you always select the same integer, MINITAB will select the same sample when you choose the random-sampling options.]

CHAPTER 4

Random Variables and Probability Distributions

CONTENTS

4.1 Two Types of Random Variables

4.2 Probability Distributions for Discrete Random Variables

4.3 The Binomial Distribution

4.4 Probability Distributions for Continuous Random Variables

4.5 The Normal Distribution

4.6 Descriptive Methods for Assessing Normality

4.7 Approximating a Binomial Distribution with a Normal Distribution (Optional)

4.8 Sampling Distributions

4.9 The Central Limit Theorem

STATISTICS IN ACTION
Super Weapons Development—Is the Hit Ratio Optimized?

USING TECHNOLOGY
Binomial Probabilities, Normal Probabilities, and Simulated Sampling Distributions Using MINITAB

WHERE WE'VE BEEN
- Used probability to make an inference about a population from data in an observed sample
- Used probability to measure the reliability of the inference

WHERE WE'RE GOING
- Develop the notion of a random variable
- Learn that many types of numerical data are observed values of discrete random variables
- Study two important types of random variables and their probability models: the binomial and the normal model
- Define a sampling distribution as the probability distribution of a statistic
- Learn that the sampling distribution of $\bar{x}$ follows a normal model

STATISTICS IN ACTION

Super Weapons Development— Is the Hit Ratio Optimized?

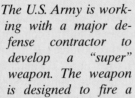

The U.S. Army is working with a major defense contractor to develop a "super" weapon. The weapon is designed to fire a large number of sharp tungsten bullets—called flechettes—with a single shot that will destroy a large number of enemy soldiers. Flechettes are about the size of an average nail, with small fins at one end to stabilize them in flight. Since World War I, when France dropped them in large quantities from aircraft on masses of ground troops, munitions experts have experimented with using flechettes in a variety of guns. The problem with using flechettes as ammunition is accuracy: Current weapons that fire large quantities of flechettes have unsatisfactory hit ratios when fired at long distances.

The defense contractor (not named here for both confidentiality and security reasons) has developed a prototype gun that fires 1,100 flechettes with a single round. In range tests, three 2-feet-wide targets were set up a distance of 500 meters (approximately 1,500 feet) from the weapon. With a number line used as a reference, the centers of the three targets were at 0, 5, and 10 feet, respectively, as shown in Figure SIA4.1. The prototype gun was aimed at the middle target (center at 5 feet) and fired once. The point where each of the 1,100 flechettes landed at the 500-meter distance was measured with the use of a horizontal and vertical grid. For the purposes of this application, only the horizontal measurements are considered. These 1,100 measurements are saved in the **MOAGUN** file. (The data are simulated for confidentiality reasons.) For example, a flechette with a value of $x = 5.5$ hit the middle target, but a flechette with a value of $x = 2.0$ did not hit any of the three targets. (See Figure SIA4.1.)

The defense contractor is interested in the likelihood of any one of the targets being hit by a flechette and, in particular, wants to set the gun specifications to maximize the number of hits. The weapon is designed to have a mean horizontal value equal to the aim point (e.g., $\mu = 5$ feet when aimed at the center target). By changing specifications, the contractor can vary the standard deviation σ. The **MOAGUN** file contains flechette measurements for three different range tests: one with a standard deviation of $\sigma = 1$ foot, one with $\sigma = 2$ feet, and one with $\sigma = 4$ feet.

In this chapter, two Statistics in Action Revisited examples demonstrate how we can use one of the probability models discussed in the chapter—the normal probability distribution—to aid the defense contractor in developing its "super" weapon.

Statistics in Action Revisited

- Using the Normal Model to Maximize the Probability of a Hit with the Super Weapon (p. 209)

- Assessing whether the Normal Distribution Is Appropriate for Modeling the Super Weapon Hit Data (p. 217)

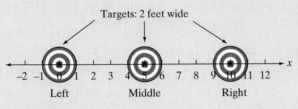

Targets: 2 feet wide

Left   Middle   Right

FIGURE SIA4.1
Target placement on gun range

HH
●
(2)

HT
●
(1)

TT
●
(0)

TH
●
(1)

S

FIGURE 4.1

Venn diagram for coin-tossing experiment

Y̶ou may have noticed that many of the examples of experiments in Chapter 3 generated quantitative (numerical) observations. The unemployment rate, the percentage of voters favoring a particular candidate, the cost of textbooks for a school term, and the amount of pesticide in the discharge waters of a chemical plant are all examples of numerical measurements of some phenomenon. Thus, most experiments have sample points that correspond to values of some numerical variable.

To illustrate, consider the coin-tossing experiment of Chapter 3. Figure 4.1 is a Venn diagram showing the sample points when two coins are tossed and the up faces (heads or tails) of the coins are observed. One possible numerical outcome is the total number of heads observed. This value (0, 1, or 2) is shown in parentheses on the Venn diagram, with one numerical value associated with each sample point. In the jargon of probability, the variable "total number of heads observed in two tosses of a coin" is called a *random variable*.

> **Definition 4.1**
>
> A **random variable** is a variable that assumes numerical values associated with the random outcomes of an experiment, where one (and only one) numerical value is assigned to each sample point.

The term *random variable* is more meaningful than the term *variable* alone, because the adjective *random* indicates that the coin-tossing experiment may result in one of the several possible values of the variable—0, 1, and 2—according to the *random* outcomes of the experiment: *HH, HT, TH,* and *TT.* Similarly, if the experiment is to count the number of customers who use the drive-up window of a bank each day, the random variable (the number of customers) will vary from day to day, partly because of the random phenomena that influence whether customers use the window. Thus, the possible values of this random variable range from 0 to the maximum number of customers the window can serve in a day.

We define two different types of random variables, *discrete* and *continuous*, in Section 4.1. Then we spend the remainder of the chapter discussing one specific type of discrete random variable and one specific type of continuous random variable.

4.1 Two Types of Random Variables

Recall that the sample-point probabilities corresponding to an experiment must sum to 1. Dividing one unit of probability among the sample points in a sample space and, consequently, assigning probabilities to the values of a random variable is not always as easy as the examples in Chapter 3 might lead you to believe. If the number of sample points can be completely listed, the job is straightforward. But if the experiment results in an infinite number of sample points that are impossible to list, the task of assigning probabilities to the sample points is impossible without the aid of a probability model. The next three examples demonstrate the need for different probability models, depending on the number of values that a random variable can assume.

EXAMPLE 4.1

VALUES OF A DISCRETE RANDOM VARIABLE— Wine Ratings

Problem A panel of 10 experts for the *Wine Spectator* (a national publication) is asked to taste a new white wine and assign it a rating of 0, 1, 2, or 3. A score is then obtained by adding together the ratings of the 10 experts. How many values can this random variable assume?

Solution A sample point is a sequence of 10 numbers associated with the rating of each expert. For example, one sample point is

$$\{1, 0, 0, 1, 2, 0, 0, 3, 1, 0\}$$

The random variable assigns a score to each one of these sample points by adding the 10 numbers together. Thus, the smallest score is 0 (if all 10 ratings are 0), and the largest score is 30 (if all 10 ratings are 3). Since every integer between 0 and 30 is a possible score, the random variable denoted by the symbol x can assume 31 values. Note that the value of the random variable for the sample point shown here is $x = 8$.*

Look Back This is an example of a *discrete random variable*, since there is a finite number of distinct possible values. Whenever all the possible values a random variable can assume can be listed (or *counted*), the random variable is *discrete*.

---

*The standard mathematical convention is to use a capital letter (e.g., X) to denote the theoretical random variable. The possible values (or realizations) of the random variable are typically denoted with a lowercase letter (e.g., x). Thus, in Example 4.1, the random variable X can take on the values $x = 0, 1, 2, \ldots, 30$. Since this notation can be confusing for introductory statistics students, we simplify the notation by using the lowercase x to represent the random variable throughout.

EXAMPLE 4.2
VALUES OF A DISCRETE RANDOM VARIABLE—
EPA Application

Problem Suppose the Environmental Protection Agency (EPA) takes readings once a month on the amount of pesticide in the discharge water of a chemical company. If the amount of pesticide exceeds the maximum level set by the EPA, the company is forced to take corrective action and may be subject to penalty. Consider the random variable number x of months before the company's discharge exceeds the EPA's maximum level. What values can x assume?

Solution The company's discharge of pesticide may exceed the maximum allowable level on the first month of testing, the second month of testing, etc. It is possible that the company's discharge will *never* exceed the maximum level. Thus, the set of possible values for the number of months until the level is first exceeded is the set of all positive integers $1, 2, 3, 4, \ldots$.

If we can list the values of a random variable x, even though the list is never ending, we call the list *countable* and the corresponding random variable *discrete*. Thus, the number of months until the company's discharge first exceeds the limit is a *discrete random variable*.

Now Work Exercise 4.7

EXAMPLE 4.3
VALUES OF A CONTINUOUS RANDOM VARIABLE—
Another EPA Application

Problem Refer to Example 4.2. A second random variable of interest is the amount x of pesticide (in milligrams per liter) found in the monthly sample of discharge waters from the same chemical company. What values can this random variable assume?

Solution Unlike the *number* of months before the company's discharge exceeds the EPA's maximum level, the set of all possible values for the *amount* of discharge *cannot* be listed (i.e., is not countable). The possible values for the amount x of pesticide would correspond to the points on the interval between 0 and the largest possible value the amount of the discharge could attain, the maximum number of milligrams that could occupy 1 liter of volume. (Practically, the interval would be much smaller, say, between 0 and 500 milligrams per liter.)

Look Back When the values of a random variable are not countable, but instead correspond to the points on some interval, we call the variable a *continuous random variable*. Thus, the *amount* of pesticide in the chemical plant's discharge waters is a *continuous random variable*.

Now Work Exercise 4.8

> **Definition 4.2**
> Random variables that can assume a *countable* number of values are called **discrete.**

> **Definition 4.3**
> Random variables that can assume values corresponding to any of the points contained in an interval are called **continuous.**

The following are examples of discrete random variables:

1. The number of seizures an epileptic patient has in a given week: $x = 0, 1, 2, \ldots$
2. The number of voters in a sample of 500 who favor impeachment of the president: $x = 0, 1, 2, \ldots, 500$
3. The number of students applying to medical schools this year: $x = 0, 1, 2, \ldots$

4. The change received for paying a bill: $x = 1¢, 2¢, 3¢, \ldots, \$1, \ldots$

5. The number of customers waiting to be served in a restaurant at a particular time: $x = 0, 1, 2, \ldots$

Note that several of the examples of discrete random variables begin with the words *The number of* .... This wording is very common, since the discrete random variables most frequently observed are counts. The following are examples of continuous random variables:

1. The length of time (in seconds) between arrivals at a hospital clinic: $0 \leq x \leq \infty$ (infinity)

2. The length of time (in minutes) it takes a student to complete a one-hour exam: $0 \leq x \leq 60$

3. The amount (in ounces) of carbonated beverage loaded into a 12-ounce can in a can-filling operation: $0 \leq x \leq 12$

4. The depth (in feet) at which a successful oil-drilling venture first strikes oil: $0 \leq x \leq c$, where c is the maximum depth obtainable

5. The weight (in pounds) of a food item bought in a supermarket: $0 \leq x \leq 500$ [*Note:* Theoretically, there is no upper limit on x, but it is unlikely that it would exceed 500 pounds.]

Discrete random variables and their probability distributions are discussed in Section 4.2. Continuous random variables and their probability distributions are the topic of Section 4.4.

Exercises 4.1–4.12

Understanding the Principles

4.1 What is a random variable?

4.2 How do discrete and continuous random variables differ?

Applying the Concepts—Basic

4.3 **Type of Random Variable.** Classify the following random variables according to whether they are discrete or continuous:

a. The number of words spelled correctly by a student on a spelling test

b. The amount of water flowing through the Hoover Dam in a day

c. The length of time an employee is late for work

d. The number of bacteria in a particular cubic centimeter of drinking water

e. The amount of carbon monoxide produced per gallon of unleaded gas

f. Your weight

4.4 **Type of Random Variable.** Identify the following random variables as discrete or continuous:

a. The amount of flu vaccine in a syringe

b. The heart rate (number of beats per minute) of an American male

c. The time it takes a student to complete an examination

d. The barometric pressure at a given location

e. The number of registered voters who vote in a national election

f. Your score on the SAT

4.5 **Type of Random Variable.** Identify the following variables as discrete or continuous:

a. The difference in reaction time to the same stimulus before and after training

b. The number of violent crimes committed per month in your community

c. The number of commercial aircraft near-misses per month

d. The number of winners each week in a state lottery

e. The number of free throws made per game by a basketball team

f. The distance traveled by a school bus each day

4.6 **NHTSA crash tests.** The National Highway Traffic Safety Administration (NHTSA) has developed a driver-side "star" scoring system for crash-testing new cars. Each crash-tested car is given a rating ranging from one star (*) to five stars (*****); the more stars in the rating, the better is the level of crash protection in a head-on collision. Suppose that a car is selected and its driver-side star rating is determined. Let x equal the number of stars in the rating. Is x a discrete or continuous random variable?

4.7 **Customers in line at a Subway shop.** The number of cus-
NW tomers, x, waiting in line to order sandwiches at a Subway shop at noon is of interest to the store manager. What values can x assume? Is x a discrete or continuous random variable?

⊚ GOBIANTS

4.8 **Mongolian desert ants.** Refer to the *Journal of Biogeogra-*
NW *phy* (Dec. 2003) study of ants in Mongolia, presented in Exercise 2.66 (p. 60). Two of the several variables recorded at each of 11 study sites were annual rainfall (in millimeters) and number of ant species. Identify these variables as discrete or continuous.

Applying the Concepts—Intermediate

4.9 **Psychology.** Give an example of a discrete random variable of interest to a psychologist.

4.10 Sociology. Give an example of a discrete random variable of interest to a sociologist.

4.11 Nursing. Give an example of a discrete random variable of interest to a hospital nurse.

4.12 Art history. Give an example of a discrete random variable of interest to an art historian.

4.2 Probability Distributions for Discrete Random Variables

A complete description of a discrete random variable requires that we *specify all the values the random variable can assume* and *the probability associated with each value*. To illustrate, consider Example 4.4.

EXAMPLE 4.4
FINDING A PROBABILITY DISTRIBUTION— Coin-Tossing Experiment

Problem Recall the experiment of tossing two coins (p. 169), and let x be the number of heads observed. Find the probability associated with each value of the random variable x, assuming that the two coins are fair.

Solution The sample space and sample points for this experiment are reproduced in Figure 4.2. Note that the random variable x can assume values 0, 1, 2. Recall (from Chapter 3) that the probability associated with each of the four sample points is $1/4$. Then, identifying the probabilities of the sample points associated with each of these values of x, we have

$$P(x = 0) = P(TT) = \frac{1}{4}$$

$$P(x = 1) = P(TH) + P(HT) = \frac{1}{4} + \frac{1}{4} = \frac{1}{2}$$

$$P(x = 2) = P(HH) = \frac{1}{4}$$

Thus, we now know the values the random variable can assume (0, 1, 2) and how the probability is *distributed over* those values ($1/4, 1/2, 1/4$). This dual specification completely describes the random variable and is referred to as the *probability distribution*, denoted by the symbol $p(x)$.* The probability distribution for the coin-toss example is shown in tabular form in Table 4.1 and in graphic form in Figure 4.3. Since the probability distribution for a discrete random variable is concentrated at specific points (values of x), the graph in Figure 4.3a represents the probabilities as the heights of vertical lines over the corresponding values of x. Although the representation of the probability distribution as a histogram, as in Figure 4.3b, is less precise (since the probability is spread over a unit interval), the histogram representation will prove useful when we approximate probabilities of certain discrete random variables in Section 4.4.

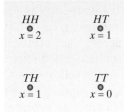

FIGURE 4.2

Venn diagram for the two-coin-toss experiment

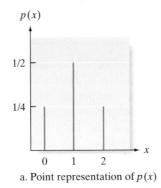

a. Point representation of $p(x)$

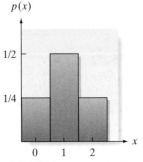

b. Histogram representation of $p(x)$

FIGURE 4.3

Probability distribution for coin-toss experiment: graphical form

| **TABLE 4.1 Probability Distribution for Coin-Toss Experiment: Tabular Form** ||
x	$p(x)$
0	$1/4$
1	$1/2$
2	$1/4$

*In standard mathematical notation, the probability that a random variable X takes on a value x is denoted $P(X = x) = p(x)$. Thus, $P(X = 0) = p(0)$, $P(X = 1) = p(1)$, etc. In this text, we adopt the simpler $p(x)$ notation.

Look Back We could also present the probability distribution for x as a formula, but this would unnecessarily complicate a very simple example. We give the formulas for the probability distributions of some common discrete random variables later in the chapter.

Now Work Exercise 4.20

> **Definition 4.4**
>
> The **probability distribution** of a discrete random variable is a graph, table, or formula that specifies the probability associated with each possible value that the random variable can assume.

Two requirements must be satisfied by all probability distributions for discrete random variables:

> **Requirements for the Probability Distribution of a Discrete Random Variable x**
>
> **1.** $p(x) \geq 0$ for all values of x.
>
> **2.** $\sum p(x) = 1$
>
> where the summation of $p(x)$ is over all possible values of x.*

Now Work Exercise 4.18

Example 4.4 illustrates how the probability distribution for a discrete random variable can be derived, but for many practical situations the task is much more difficult. Fortunately, numerous experiments and associated discrete random variables observed in nature possess identical characteristics. Thus, you might observe a random variable in a psychology experiment that would possess the same probability distribution as a random variable observed in an engineering experiment or a social sample survey. We classify random variables according to type of experiment, derive the probability distribution for each of the different types, and then use the appropriate probability distribution when a particular type of random variable is observed in a practical situation. The probability distributions for most commonly occurring discrete random variables have already been derived. This fact simplifies the problem of finding the probability distributions for random variables, as the next example illustrates.

EXAMPLE 4.5

PROBABILITY DISTRIBUTION USING A FORMULA— Parasitic Fish

Problem The distribution of parasites (tapeworms) found in Mediterranean brill fish was studied in the *Journal of Fish Biology* (Aug. 1990). The researchers showed that the distribution of x, the number of brill that must be sampled until a parasitic infection is found in the digestive tract, can be modeled with the formula $p(x) = (.6)(.4)^{x-1}$, $x = 1, 2, 3$, etc. Find the probability that exactly three brill fish must be sampled before a tapeworm is found in the digestive tract.

Solution We want to find the probability that $x = 3$. Using the formula, we have

$$p(3) = (.6)(.4)^{3-1} = (.6)(.4)^2 = (.6)(.16) = .096$$

Thus, there is about a 10% chance that exactly three fish must be sampled before a tapeworm is found in the digestive tract.

---

*Unless otherwise indicated, summations will always be over all possible values of x.

Look Back The probability of interest can also be derived with the use of the principles of probability developed in Chapter 3. The event of interest is $N_1N_2P_3$, where N_1 represents no parasite found in the first sampled fish, N_2 represents no parasite found in the second sampled fish, and P_3 represents a parasite found in the third sampled fish. The researchers discovered that the probability of finding a parasite in any sampled fish is .6 (and consequently, the probability of not finding a parasite in any sampled fish is .4). Using the multiplicative law of probability for independent events, we find that the probability of interest is $(.4)(.4)(.6) = .096$.

Since probability distributions are analogous to the relative frequency distributions of Chapter 2, it should be no surprise that the mean and standard deviations are useful descriptive measures.

If a discrete random variable x were observed a very large number of times and the data generated were arranged in a relative frequency distribution, the relative frequency distribution would be indistinguishable from the probability distribution for the random variable. Thus, the probability distribution for a random variable is a theoretical model for the relative frequency distribution of a population. To the extent that the two distributions are equivalent (and we will assume that they are), the probability distribution for x possesses a mean μ and a variance σ^2 that are identical to the corresponding descriptive measures for the population. The procedures for finding μ and σ^2 for a random variable follow.

Examine the probability distribution for x (the number of heads observed in the toss of two fair coins) in Figure 4.4. Try to locate the mean of the distribution intuitively. We may reason that the mean μ of this distribution is equal to 1 as follows: In a large number of experiments, $\frac{1}{4}$ should result in $x = 0$ heads, $\frac{1}{2}$ in $x = 1$ head, and $\frac{1}{4}$ in $x = 2$ in heads. Therefore, the average number of heads is

$$\mu = 0(\tfrac{1}{4}) + 1(\tfrac{1}{2}) + 2(\tfrac{1}{4}) = 0 + \tfrac{1}{2} + \tfrac{1}{2} = 1$$

Note that to get the population mean of the random variable x, we multiply each possible value of x by its probability $p(x)$, and then we sum this product over all possible values of x. The *mean of x* is also referred to as the *expected value of x*, denoted $E(x)$.

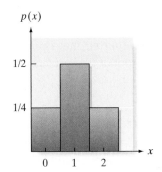

$p(x)$

1/2

1/4

0 1 2 x

FIGURE 4.4
Probability distribution for a two-coin toss

> **Definition 4.5**
> The **mean**, or **expected value**, of a discrete random variable x is
> $$\mu = E(x) = \Sigma x p(x)$$

The term *expected* is a mathematical term and should not be interpreted as it is typically used. Specifically, *a random variable might never be equal to its "expected value."* Rather, the expected value is the mean of the probability distribution, or a measure of its central tendency. You can think of μ as the mean value of x in a *very large* (actually, *infinite*) number of repetitions of the experiment in which the values of x occur in proportions equivalent to the probabilities of x.

EXAMPLE 4.6

FINDING AN EXPECTED VALUE—

An Insurance Application

Problem Suppose you work for an insurance company and you sell a $10,000 one-year term insurance policy at an annual premium of $290. Actuarial tables show that the probability of death during the next year for a person of your customer's age, sex, health, etc., is .001. What is the expected gain (amount of money made by the company) for a policy of this type?

Solution The experiment is to observe whether the customer survives the upcoming year. The probabilities associated with the two sample points, Live and Die, are .999 and .001, respectively. The random variable you are interested in is the gain x, which can assume the values shown in the following table:

Gain x	Sample Point	Probability
$290	Customer lives	.999
−$9,710	Customer dies	.001

If the customer lives, the company gains the $290 premium as profit. If the customer dies, the gain is negative because the company must pay $10,000, for a net "gain" of $(290 − 10,000) = −$9,710$. The expected gain is therefore

$$\mu = E(x) = \Sigma x p(x)$$
$$= (290)(.999) + (-9,710)(.001) = \$280$$

In other words, if the company were to sell a very large number of $10,000 one-year policies to customers possessing the characteristics described, it would (on the average) net $280 per sale in the next year.

Look Back Note that $E(x)$ need not equal a possible value of x. That is, the expected value is $280, but x will equal either $290 or −$9,710 each time the experiment is performed (a policy is sold and a year elapses). The expected value is a measure of central tendency—and in this case represents the average over a very large number of one-year policies—but is not a possible value of x.

Now Work Exercise 4.24d

We learned in Chapter 2 that the mean and other measures of central tendency tell only part of the story about a set of data. The same is true about probability distributions: We need to measure variability as well. Since a probability distribution can be viewed as a representation of a population, we will use the population variance to measure its variability.

The *population variance* σ^2 is defined as the average of the squared distance of x from the population mean μ. Since x is a random variable, the squared distance, $(x - \mu)^2$, is also a random variable. Using the same logic we employed to find the mean value of x, we calculate the mean value of $(x - \mu)^2$ by multiplying all possible values of $(x - \mu)^2$ by $p(x)$ and then summing over all possible x values.* This quantity,

$$E[(x - \mu)^2] = \sum_{\text{all } x} (x - \mu)^2 p(x)$$

is also called the *expected value of the squared distance from the mean;* that is, $\sigma^2 = E[(x - \mu)^2]$. The standard deviation of x is defined as the square root of the variance σ^2.

Definition 4.6

The **variance** of a random variable x is

$$\sigma^2 = E[(x - \mu)^2] = \Sigma(x - \mu)^2 p(x) = \Sigma x^2 p(x) - \mu^2$$

Definition 4.7

The **standard deviation** of a discrete random variable is equal to the square root of the variance, or $\sigma = \sqrt{\sigma^2}$.

Knowing the mean μ and standard deviation σ of the probability distribution of x, in conjunction with Chebyshev's rule (Table 2.6) and the Empirical rule (Table 2.7), we can make statements about the likelihood of values of x falling within the intervals $\mu \pm \sigma, \mu \pm 2\sigma$, and $\mu \pm 3\sigma$. These probabilities are given in the following box:

Chebyshev's Rule and Empirical Rule for a Discrete Random Variable

Let x be a discrete random variable with probability distribution $p(x)$, mean μ, and standard deviation σ. Then, depending on the shape of $p(x)$, the following probability statements can be made:

(continued)

---

*It can be shown that $E[(x - \mu)^2] = E(x^2) - \mu^2$, where $E(x^2) = \Sigma x^2 p(x)$. Note the similarity between this expression and the shortcut formula $\Sigma(x - \bar{x})^2 = \Sigma x^2 - (\Sigma x)^2/n$ given in Chapter 2.

	Chebyshev's Rule	Empirical Rule
	Applies to any probability distribution (see Figure 4.5a)	Applies to probability distributions that are mound shaped and symmetric (see Figure 4.5b)
$P(\mu - \sigma < x < \mu + \sigma)$	≥ 0	$\approx .68$
$P(\mu - 2\sigma < x < \mu + 2\sigma)$	$\geq \frac{3}{4}$	$\approx .95$
$P(\mu - 3\sigma < x < \mu + 3\sigma)$	$\geq \frac{8}{9}$	≈ 1.00

FIGURE 4.5

Shapes of two probability distributions for a discrete random variable x

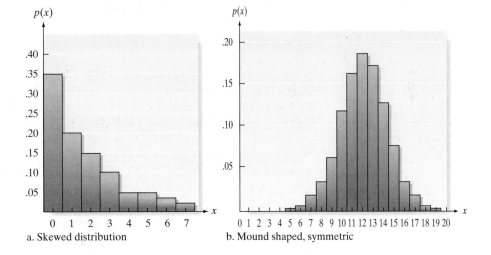

a. Skewed distribution

b. Mound shaped, symmetric

EXAMPLE 4.7

FINDING μ AND σ—Skin Cancer Treatment

Problem Medical research has shown that a certain type of chemotherapy is successful 70% of the time when used to treat skin cancer. Suppose five skin cancer patients are treated with this type of chemotherapy, and let x equal the number of successful cures out of the five. The probability distribution for the number x of successful cures out of five is given in the following table:

x	0	1	2	3	4	5
$p(x)$	.002	.029	.132	.309	.360	.168

a. Find $\mu = E(x)$. Interpret the result.

b. Find $\sigma = \sqrt{E[(x - \mu)^2]}$. Interpret the result.

c. Graph $p(x)$. Locate μ and the interval $\mu \pm 2\sigma$ on the graph. Use either Chebyshev's rule or the empirical rule to approximate the probability that x falls into this interval. Compare your result with the actual probability.

d. Would you expect to observe fewer than two successful cures out of five?

Solution

a. Applying the formula for μ, we obtain

$$\mu = E(x) = \Sigma x p(x)$$
$$= 0(.002) + 1(.029) + 2(.132) + 3(.309) + 4(.360) + 5(.168) = 3.50$$

On average, the number of successful cures out of five skin cancer patients treated with chemotherapy will equal 3.5. Remember that this expected value has meaning only when the experiment—treating five skin cancer patients with chemotherapy—is repeated a large number of times.

b. Now we calculate the variance of x:

$$\sigma^2 = E[(x - \mu)^2] = \Sigma(x - \mu)^2 p(x)$$

$$= (0 - 3.5)^2(.002) + (1 - 3.5)^2(.029) + (2 - 3.5)^2(.132)$$

$$+ (3 - 3.5)^2(.309) + (4 - 3.5)^2(.360) + (5 - 3.5)^2(.168)$$

$$= 1.05$$

Thus, the standard deviation is

$$\sigma = \sqrt{\sigma^2} = \sqrt{1.05} = 1.02$$

This value measures the spread of the probability distribution of x, the number of successful cures out of five. A more useful interpretation is obtained by answering parts **c** and **d**.

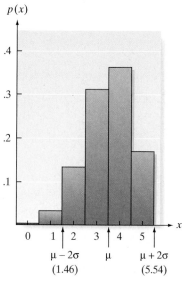

c. The graph of $p(x)$ is shown in Figure 4.6, with the mean μ and the interval $\mu \pm 2\sigma = 3.50 \pm 2(1.02) = 3.50 \pm 2.04 = (1.46, 5.54)$ also indicated. Note particularly that $\mu = 3.5$ locates the center of the probability distribution. Since this distribution is a theoretical relative frequency distribution that is moderately mound shaped (see Figure 4.6), we expect (from Chebyshev's rule) at least 75% and, more likely (from the empirical rule), approximately 95%, of observed x values to fall between 1.46 and 5.54. You can see from the figure that the actual probability that x falls in the interval $\mu \pm 2\sigma$ includes the sum of $p(x)$ for the values $x = 2$, $x = 3$, $x = 4$, and $x = 5$. This probability is $p(2) + p(3) + p(4) + p(5) = .132 + .309 + .360 + .168 = .969$. Therefore, 96.9% of the probability distribution lies within two standard deviations of the mean. This percentage is consistent with both Chebyshev's rule and the Empirical rule.

FIGURE 4.6
Graph of $p(x)$ for Example 4.7

d. Fewer than two successful cures out of five implies that $x = 0$ or $x = 1$. Both of these values of x lie outside the interval $\mu \pm 2\sigma$, and the empirical rule tells us that such a result is unlikely (approximate probability of .05). The exact probability, $P(x \leq 1)$, is $p(0) + p(1) = .002 + .029 = .031$. Consequently, in a single experiment in which five skin cancer patients are treated with chemotherapy, we would not expect to observe fewer than two successful cures.

Now Work Exercise 4.22

Discrete Random Variables

Using The TI-84/TI-83 Graphing Calculator

Calculating the mean and standard deviation of a discrete random variable

Step 1   *Enter the data.*
Press **STAT** and select **1:Edit.**
Note: If the lists already contain data, clear the old data. Use the up **ARROW** to highlight '**L1**'.
Press **CLEAR ENTER**.
Use the up **ARROW** to highlight '**L2**'.
Press **CLEAR ENTER**.
Use the **ARROW** and **ENTER** keys to enter the X-values of the variable into L1.
Use the **ARROW** and **ENTER** keys to enter the probabilities, P(X), into L2.

(continued)

Step 2 *Access the Calc Menu.*
Press **STAT**.
Arrow right to **CALC**.
Select **1-Var Stats**.
Press **ENTER**.

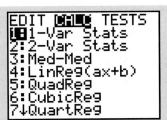

*Press **2nd 1** for **L1**.*
Press **COMMA**.
Press **2nd 2** for **L2**.
Press **ENTER**.

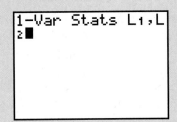

The mean and standard deviation will be displayed on the screen, as will the quartiles, min, and max.

Example Find the mean and standard deviation of the following discrete probability distribution:

X	P(X)
0	.05
1	.10
2	.25
3	.60

First enter the data into **L1** and **L2**.

The output screens for this example are as follows:

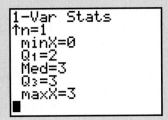

The mean of this discrete random variable is 2.4 and the standard deviation is .86.

Exercises 4.13–4.40

Understanding the Principles

4.13 Give three different ways of representing the probability distribution of a discrete random variable.

4.14 What does the expected value of a random variable represent?

4.15 Will $E(x)$ always be equal to a specific value of the random variable x?

4.16 For a mound-shaped, symmetric distribution, what is the probability that x falls into the interval $\mu \pm 2\sigma$?

Learning the Mechanics

4.17 A discrete random variable x can assume five possible values: 2, 3, 5, 8, and 10. Its probability distribution is shown here:

x	2	3	5	8	10
$p(x)$	.15	.10		.25	.25

   a. What is $p(5)$?
   b. What is the probability that x equals 2 or 10?
   c. What is $P(x \leq 8)$?

4.18 Explain why each of the following is or is not a valid proba-
NW bility distribution for a discrete random variable x:

   a.

x	0	1	2	3
$p(x)$	.2	.3	.3	.2

   b.

x	-2	-1	0
$p(x)$	.25	.50	.20

   c.

x	4	9	20
$p(x)$	-.3	1.0	.3

   d.

x	2	3	5	6
$p(x)$	.15	.20	.40	.35

4.19 The random variable x has the following discrete probability distribution:

x	10	11	12	13	14
$p(x)$	.2	.3	.2	.1	.2

Since the values that x can assume are mutually exclusive events, the event $\{x \leq 12\}$ is the union of three mutually exclusive events:

$$\{x = 10\} \cup \{x = 11\} \cup \{x = 12\}$$

   a. Find $P(x \leq 12)$.   **b.** Find $P(x > 12)$.
   c. Find $P(x \leq 14)$.   **d.** Find $P(x = 14)$.
   e. Find $P(x \leq 11 \text{ or } x > 12)$.

4.20 Toss three fair coins, and let x equal the number of heads
NW observed.
   a. Identify the sample points associated with this experiment, and assign a value of x to each sample point.
   b. Calculate $p(x)$ for each value of x.
   c. Construct a probability histogram for $p(x)$.
   d. What is $P(x = 2 \text{ or } x = 3)$? $\frac{1}{2}$

4.21 Consider the probability distribution shown for the random variable x here:

x	1	2	4	10
$p(x)$	.2	.4	.2	.2

   a. Find $\mu = E(x)$.
   b. Find $\sigma^2 = E[(x - \mu)^2]$.
   c. Find σ.
   d. Interpret the value you obtained for μ.
   e. In this case, can the random variable x ever assume the value μ? Explain.

   f. In general, can a random variable ever assume a value equal to its expected value? Explain.

4.22 Consider the probability distribution shown here:
NW

x	-4	-3	-2	-1	0	1	2	3	4
$p(x)$	.02	.07	.10	.15	.30	.18	.10	.06	.02

   a. Calculate μ, σ^2, and σ.
   b. Graph $p(x)$. Locate μ, $\mu - 2\sigma$, and $\mu + 2\sigma$ on the graph.
   c. What is the probability that x will fall into the interval $\mu \pm 2\sigma$?

4.23 Consider the probability distributions shown here:

x	0	1	2
$p(x)$	.3	.4	.3

y	0	1	2
$p(y)$	.1	.8	.1

   a. Use your intuition to find the mean for each distribution. How did you arrive at your choice?
   b. Which distribution appears to be more variable? Why?
   c. Calculate μ and σ^2 for each distribution. Compare these answers with your answers in parts **a** and **b**.

APPLET Applet Exercise 4.1

Use the applet entitled *Random Numbers* to generate a list of 25 numbers between 1 and 3, inclusive. Let x represent a number chosen from this list.
   a. What are the possible values of x?
   b. Give the probability distribution for x in table form.
   c. Let y be a number randomly chosen from the set $\{1, 2, 3\}$. Give the probability distribution for y in table form.
   d. Compare the probability distributions of x and y in parts **b** and **c**. Why should these distributions be approximately the same?

APPLET Applet Exercise 4.2

Run the applet entitled *Simulating the Probability of a Head with a Fair Coin* 10 times with $n = 2$, resetting between runs, to simulate flipping two coins 10 times. Count and record the number of heads each time. Let x represent the number of heads on a single flip of the coins.
   a. What are the possible values of x?
   b. Use the results of the simulation to find the probability distribution for x in table form.
   c. Explain why the probability of exactly two heads should be close to .25.

Applying the Concepts—Basic

4.24 **NHTSA crash tests.** Refer to the National Highway Traffic Safety Administration (NHTSA) crash tests of new car models, presented in Exercise 4.6 (p. 172). A summary of the driver-side star ratings for the 98 cars in the **CRASH** file is reproduced in the MINITAB printout on p. 181. Assume that 1 of the 98 cars is selected at random, and let x equal the number of stars in the car's driver-side star rating.
   a. Use the information in the printout to find the probability distribution for x.
   b. Find $P(x = 5)$.
   c. Find $P(x \leq 2)$.
NW   **d.** Find $\mu = E(x)$ and interpret the result.

MINITAB output for Exercise 4.24

Tally for Discrete Variables: DRIVSTAR

DRIVSTAR	Count	Percent
2	4	4.08
3	17	17.35
4	59	60.20
5	18	18.37
N=	98	

4.25 Dust mite allergies. A dust mite allergen level that exceeds 2 micrograms per gram ($\mu g/g$) of dust has been associated with the development of allergies. Consider a random sample of four homes, and let x be the number of homes with a dust mite level that exceeds 2 $\mu g/g$. The probability distribution for x, based on a May 2000 study by the National Institute of Environmental Health Sciences, is shown in the following table:

x	0	1	2	3	4
$p(x)$	.09	.30	.37	.20	.04

a. Verify that the probabilities for x in the table sum to 1.
b. Find the probability that three or four of the homes in the sample have a dust mite level that exceeds 2 $\mu g/g$.
c. Find the probability that fewer than two homes in the sample have a dust mite level that exceeds 2 $\mu g/g$.
d. Find $E(x)$. Give a meaningful interpretation of the result.
e. Find σ.
f. Find the exact probability that x is in the interval $\mu \pm 2\sigma$. Compare your answer with Chebyshev's rule and the empirical rule.

4.26 Controlling the water hyacinth. Entomologists are continually searching for new biological agents to control the water hyacinth, one of the world's worst aquatic weeds. An insect that naturally feeds on the water hyacinth is the delphacid. Female delphacids lay anywhere form one to four eggs onto a water hyacinth blade. The *Annals of the Entomological Society of America* (Jan. 2005) published a study of the life cycle of a South American delphacid species. The following table gives the percentages of water hyacinth blades that have one, two, three, and four delphacid eggs:

	One Egg	Two Eggs	Three Eggs	Four Eggs
Percentage of Blades	40	54	2	4

Source: Sosa, A.J., et al. "Life history of *Megamelus scutellaris* with description of immature stages," *Annals of the Entomological Society of America,* Vol. 98, No. 1, Jan. 2005 (adapted from Table 1).

a. One of the water hyacinth blades in the study is randomly selected, and x, the number of delphacid eggs on the blade, is observed. Give the probability distribution of x.
b. What is the probability that the blade has at least three delphacid eggs?
c. Find $E(x)$ and interpret the result.

4.27 Choosing portable grill displays. Refer to the *Journal of Consumer Research* (Mar. 2003) marketing study of influencing consumer choices by offering undesirable alternatives, presented in Exercise 3.25 (p. 126). Recall that each of 124 college students selected showroom displays for portable grills. Five different displays (representing five different-sized grills) were available, but the students were instructed to select only three displays in order to maximize purchases of Grill #2 (a smaller grill). The table that follows shows the grill display combinations and number of each selected by the 124 students. Suppose 1 of the 124 students is selected at random. Let x represent the sum of the grill numbers selected by that student. (This sum is an indicator of the size of the grills selected.)
a. Find the probability distribution for x.
b. What is the probability that x exceeds 10?

Grill Display Combination	Number of Students
1–2–3	35
1–2–4	8
1–2–5	42
2–3–4	4
2–3–5	1
2–4–5	34

Source: Hamilton, R. W. "Why do people suggest what they do not want? Using context effects to influence others' choices," *Journal of Consumer Research*, Vol. 29, Mar. 2003, Table 1.

4.28 Benford's Law of Numbers. Refer to the *American Scientist* (July–Aug. 1998) study of which integer is most likely to occur as the first significant digit in a randomly selected number (Benford's law), presented in Exercise 2.20 (p. 38). The table giving the frequency of each integer selected as the first digit in a six-digit random number is reproduced here:

First Digit	Frequency of Occurrence
1	109
2	75
3	77
4	99
5	72
6	117
7	89
8	62
9	43
Total	743

a. Construct a probability distribution for the first significant digit x.
b. Find $E(x)$.
c. If possible, give a practical interpretation of $E(x)$.

Applying the Concepts—Intermediate

4.29 Perfect SAT score. In Exercise 3.26 (p. 126), you learned that 5 in every 10,000 students who take the SAT score a perfect 1,600. Consider a random sample of three students who take the SAT. Let x equal the number who score a perfect 1,600.
a. Find $p(x)$ for $x = 0, 1, 2, 3$.
b. Graph $p(x)$.   c. Find $P(x \le 1)$.

4.30 Mail rooms contaminated with anthrax. In the fall of 2001, there was a highly publicized outbreak of anthrax cases among U.S. Postal Service workers. In *Chance* (Spring 2002), research statisticians discussed the problem of sampling mail rooms for the presence of anthrax spores. Let x equal the number of mail rooms contaminated with anthrax spores in a random sample of n mail rooms selected from a population of N mail rooms. The researchers showed that the probability distribution for x is given by the formula

$$p(x) = \frac{\binom{k}{x}\binom{N-k}{n-x}}{\binom{N}{n}}$$

where k is the number of contaminated mail rooms in the population. (This probability distribution is known as the *hypergeometric distribution*.) Suppose $N = 100$, $n = 3$, and $k = 20$.

a. Find $p(0)$. **b.** Find $p(1)$.
c. Find $p(2)$. **d.** Find $p(3)$.

4.31 Beach erosional hot spots. Refer to the U.S. Army Corps of Engineers' study of beach erosional hot spots, presented in Exercise 3.121 (p. 161). The data on the nearshore bar condition for six beach hot spots are reproduced in the accompanying table. Suppose you randomly select two of these six beaches and count x, the total number in the sample with a planar nearshore bar condition.

Beach Hotspot	Nearshore Bar Condition
Miami Beach, FL	Single, shore parallel
Coney Island, NY	Other
Surfside, CA	Single, shore parallel
Monmouth Beach, NJ	Planar
Ocean City, NJ	Other
Spring Lake, NJ	Planar

Source: "Identification and Characterization of Erosional Hotspots," William & Mary Virginia Institute of Marine Science, U.S. Army Corps of Engineers Project Report, March 18, 2002.

a. List all possible pairs of beach hot spots that can be selected from the six.
b. Assign probabilities to the outcomes in part **a**.
c. For each outcome in part **a**, determine the value of x.
d. Form a probability distribution table for x.
e. Find the expected value of x. Interpret the result.

4.32 Expected Lotto winnings. The chance of winning Florida's Pick-6 Lotto game is 1 in approximately 23 million. Suppose you buy a $1 Lotto ticket in anticipation of winning the $7 million grand prize. Calculate your expected net winnings for this single ticket. Interpret the result.

4.33 Contaminated gun cartridges. A weapons manufacturer uses a liquid propellant to produce gun cartridges. During the manufacturing process, the propellant can get mixed with another liquid to produce a contaminated cartridge. A University of South Florida statistician hired by the company to investigate the level of contamination in the stored cartridges found that 23% of the cartridges in a particular lot were contaminated. Suppose you randomly sample (without replacement) gun cartridges from this lot until you find a contaminated one. Let x be the number of cartridges sampled until a contaminated one is found. It is known that the probability distribution for x is given by the formula

$$p(x) = (.23)(.77)^{x-1}, \qquad x = 1, 2, 3, \ldots$$

a. Find $p(1)$. Interpret this result.
b. Find $p(5)$. Interpret this result.
c. Find $P(x \geq 2)$. Interpret this result.

4.34 Successful NFL running plays. In his article "American Football" (*Statistics in Sport*, 1998), Iowa State University

statistician Hal Stern evaluates winning strategies of teams in the National Football League (NFL). In a section on estimating the probability of winning a game, Stern used actual NFL play-by-play data to approximate the probabilities associated with certain outcomes (e.g., running plays, short pass plays, and long pass plays). The following table gives the probability distribution for the yardage gained, x, on a running play (a negative gain represents a loss of yards on the play):

x, Yards	Probability	x, Yards	Probability
−4	.020	6	.090
−2	.060	8	.060
−1	.070	10	.050
0	.150	15	.085
1	.130	30	.010
2	.110	50	.004
3	.090	99	.001
4	.070		

a. Find the probability of gaining 10 yards or more on a running play.
b. Find the probability of losing yardage on a running play.

4.35 Federal civil trial appeals. Refer to the *Journal of the American Law and Economics Association* (Vol. 3, 2001) study of appeals of federal civil trials, presented in Exercise 3.55 (p. 137). A breakdown of the 678 civil cases that were originally tried in front of a judge (rather than a jury) and appealed by either the plaintiff or defendant is reproduced in the accompanying table. Suppose each civil case is awarded points (positive or negative) on the basis of the outcome of the appeal for the purpose of evaluating federal judges. If the appeal is affirmed or dismissed, +5 points are awarded. If the appeal of a plaintiff trial win is reversed, −1 point is awarded. If the appeal of a defendant trial win is reversed, −3 points are awarded. Suppose 1 of the 678 cases is selected at random, and the number x of points awarded is determined. Find and graph the probability distribution for x.

Outcome of Appeal	Number of Cases
Plaintiff trial win—reversed	71
Plaintiff trial win—affirmed/ dismissed	240
Defendant trial win—reversed	68
Defendant trial win—affirmed/ dismissed	299
Total	678

4.36 The Showcase Showdown. On the popular television game show *The Price is Right*, contestants can play "The Showcase Showdown." The game involves a large wheel with 20 nickel values, 5, 10, 15, 20, ..., 95, 100, marked on it. Contestants spin the wheel once or twice, with the objective of obtaining the highest total score *without going over a dollar (100)*. [According to the *American Statistician* (Aug. 1995), the optimal strategy for the first spinner in a three-player game is to spin a second time only if the value of the initial spin is 65 or less.] Let x represent the score of a single contestant playing "The Showcase Showdown." Assume a "fair" wheel (i.e., a wheel with equally likely outcomes). If the total of the player's spins exceeds 100, the total score is set to 0.

a. If the player is permitted only one spin of the wheel, find the probability distrubtion for *x*.

b. Refer to part **a**. Find $E(x)$ and interpret this value.

c. Refer to part **a**. Give a range of values within which *x* is likely to fall.

d. Suppose the player will spin the wheel twice, no matter what the outcome of the first spin. Find the probability distribution for *x*.

e. What assumption did you make to obtain the probability distribution in part **d**? Is it a reasonable assumption?

f. Find μ and for σ the probability distribution of part **d**, and interpret the results.

g. Refer to part **d**. What is the probability that in two spins the player's total score exceeds a dollar (i.e., is set to 0)?

h. Suppose the player obtains a 20 on the first spin and decides to spin again. Find the probability distribution for *x*.

i. Refer to part **h**. What is the probability that the player's total score exceeds a dollar?

j. Given that the player obtains a 65 on the first spin and decides to spin again, find the probability that the player's total score exceeds a dollar.

k. Repeat part **j** for different first-spin outcomes. Use this information to suggest a strategy for the one-player game.

4.37 **Expected winnings in roulette.** In the popular casino game of roulette, you can bet on whether the ball will fall in an arc on the wheel colored red, black, or green. You showed (Exercise 3.51, p. 136) that the probability of a red outcome is 18/38, that of a black outcome is 18/38, and that of a green outcome is 2/38. Suppose you make a $5 bet on red. Find your expected net winnings for this single bet. Interpret the result.

Applying the Concepts—Advanced

4.38 **Punnett square for earlobes.** Geneticists use a grid—called a *Punnett square*—to display all possible gene combinations in genetic crosses. (The grid is named for Reginald Punnett, a British geneticist who developed the method in the early 1900s.) The accompanying figure is a Punnett square for a cross involving human earlobes. In humans, free earlobes (E) are dominant over attached earlobes (e). Consequently, the gene pairs EE and Ee will result in free earlobes, while the gene pair ee results in attached earlobes. Consider a couple with genes as shown in the accompanying Punnett square. Suppose the couple has seven children. Let *x* represent the number of children with attached earlobes (i.e., with the gene pair ee). Find the probability distribution of *x*.

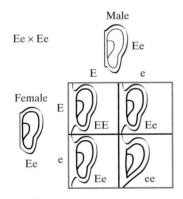

4.39 **Robot-sensor system configuration.** Engineers at Broadcom Corp. and Simon Fraser University collaborated on research involving a robot-sensor system in an unknown environment. (*The International Journal of Robotics Research*, Dec. 2004.) As an example, the engineers presented the three-point, single-link robotic system shown in the accompanying figure. Each point (*A*, *B*, or *C*) in the physical space of the system has either an "obstacle" status or a "free" status. There are two single links in the system: $A \leftrightarrow B$ and $B \leftrightarrow C$. A link has a "free" status if and only if both points in the link are "free"; otherwise the link has an "obstacle" status. Of interest is the random variable *x*: the total number of links in the system that are "free."

a. List the possible values of *x* for the system.

b. The researchers stated that the probability of any point in the system having a "free" status is .5. Assuming that the three points in the system operate independently, find the probability distribution for *x*.

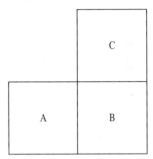

4.40 **Parlay card betting.** Odds makers try to predict which professional and college football teams will win and by how much (the *spread*). If the odds makers do this accurately, adding the spread to the underdog's score should make the final score a tie. Suppose a bookie will give you $6 for every $1 you risk if you pick the winners in three ball games (adjusted by the spread) on a "parlay" card. What is the bookie's expected earnings per dollar wagered? Interpret this value.

4.3 The Binomial Distribution

Many experiments result in *dichotomous* responses (i.e., responses for which there exist two possible alternatives, such as Yes–No, Pass–Fail, Defective–Nondefective, or Male–Female). A simple example of such an experiment is the coin-toss experiment. A coin is tossed a number of times, say, 10. Each toss results in one of two outcomes, Head or Tail, and the probability of observing each of these two outcomes remains the same for each of the 10 tosses. Ultimately, we are interested in the probability distribution of *x*, the number of heads observed. Many other experiments are equivalent to tossing a coin (either balanced or unbalanced) a fixed number *n* of times and observing the number *x* of times that one of the two possible outcomes occurs. Random variables that possess these characteristics are called **binomial random variables.**

Public opinion and consumer preference polls (e.g., the CNN, Gallup, and Harris polls) frequently yield observations on binomial random variables. For example, suppose a sample of 100 students is selected from a large student body and each person is asked whether he or she favors (a Head) or opposes (a Tail) a certain campus issue. Suppose we are interested in x, the number of students in the sample who favor the issue. Sampling 100 students is analogous to tossing the coin 100 times. Thus, you can see that opinion polls which record the number of people who favor a certain issue are real-life equivalents of coin-toss experiments. We have been describing a **binomial experiment,** identified by the following characteristics:

Characteristics of a Binomial Random Variable

1. The experiment consists of n identical trials.
2. There are only two possible outcomes on each trial. We will denote one outcome by S (for Success) and the other by F (for Failure).
3. The probability of S remains the same from trial to trial. This probability is denoted by p, and the probability of F is denoted by $q = 1 - p$.
4. The trials are independent.
5. The binomial random variable x is the number of S's in n trials.

Biography

**JACOB BERNOULLI (1654–1705)—
The Bernoulli Distribution**

Son of a magistrate and spice maker in Basel, Switzerland, Jacob Bernoulli completed a degree in theology at the University of Basel. While at the university, however, he studied mathematics secretly and against the will of his father. Jacob taught mathematics to his younger brother Johan, and they both went on to become distinguished European mathematicians. At first the brothers collaborated on the problems of the time (e.g., calculus); unfortunately, they later became bitter mathematical rivals. Jacob applied his philosophical training and mathematical intuition to probability and the theory of games of chance, where he developed the law of large numbers. In his book *Ars Conjectandi*, published in 1713 (eight years after his death), the binomial distribution was first proposed. Jacob showed that the binomial distribution was a sum of independent 0–1 variables, now known as Bernoulli random variables.

EXAMPLE 4.8
ASSESSING WHETHER X IS A BINOMIAL

Problem For the following examples, decide whether x is a binomial random variable:

a. A university scholarship committee must select two students to receive a scholarship for the next academic year. The committee receives 10 applications for the scholarships—6 from male students and 4 from female students. Suppose the applicants are all equally qualified, so that the selections are randomly made. Let x be the number of female students who receive a scholarship.

b. Before marketing a new product on a large scale, many companies conduct a consumer-preference survey to determine whether the product is likely to be successful. Suppose a company develops a new diet soda and then conducts a taste-preference survey in which 100 randomly chosen consumers state their preferences from among the new soda and the two leading sellers. Let x be the number of the 100 who choose the new brand over the two others.

c. Some surveys are conducted by using a method of sampling other than simple random sampling (defined in Chapter 3). For example, suppose a television cable company plans to conduct a survey to determine the fraction of households in a certain city that would use the cable television service. The sampling method is to choose a city block at random and then survey every household on that block. This sampling technique is called *cluster sampling*. Suppose 10 blocks are so sampled, producing a total of 124 household responses. Let x be the number of the 124 households that would use the television cable service.

Solution

a. In checking the binomial characteristics, a problem arises with independence (characteristic 4 in the preceding box). On the one hand, given that the first student selected is female, the probability that the second chosen is female is $^{3}/_{9}$. On the other hand, given that the first selection is a male student, the probability that the second is female is $^{4}/_{9}$. Thus, the conditional probability of a Success (choosing a female student to receive a scholarship) on the second trial (selection) depends on the outcome of the first trial, and the trials are therefore dependent. Since the trials are *not independent*, this variable is not a binomial random variable.*

b. Surveys that produce dichotomous responses and use random-sampling techniques are classic examples of binomial experiments. In this example, each randomly selected consumer either states a preference for the new diet soda or does not. The sample of 100 consumers is a very small proportion of the totality of potential consumers, so the response of one would be, for all practical purposes, independent of another.** Thus, x is a binomial random variable.

c. This example is a survey with dichotomous responses (Yes or No to the cable service), but the sampling method is not simple random sampling. Again, the binomial characteristic of independent trials would probably not be satisfied. The responses of households within a particular block would be dependent, since households within a block tend to be similar with respect to income, level of education, and general interests. Thus, the binomial model would not be satisfactory for x if the cluster sampling technique were employed.

Look Back Nonbinomial variables with two outcomes on every trial typically occur because they do not satisfy characteristic 3 or characteristic 4 of a binomial distribution listed in the previous box.

Now Work Exercise 4.52a

EXAMPLE 4.9
DERIVING THE BINOMIAL PROBABILITY DISTRIBUTION—
Passing a Physical Fitness Exam

Problem The Heart Association claims that only 10% of U.S. adults over 30 years of age meet the President's Physical Fitness Commission's minimum requirements. Suppose four adults are randomly selected and each is given the fitness test.

a. Use the steps given in Chapter 3 (box on p. 120) to find the probability that none of the four adults passes the test.

b. Find the probability that three of the four adults pass the test.

c. Let x represent the number of the four adults who pass the fitness test. Explain why x is a binomial random variable.

d. Use the answers to parts **a** and **b** to derive a formula for $p(x)$, the probability distribution of the binomial random variable x.

Solution

a. **1.** The first step is to define the experiment. Here we are interested in observing the fitness test results of each of the four adults: pass (S) or fail (F).
   2. Next, we list the sample points associated with the experiment. Each sample point consists of the test results of the four adults. For example, $SSSS$ represents the sample point denoting that all four adults pass, while $FSSS$ represents the sample

---

*This variable follows a *hypergeometric* random variable model. Consult the references to learn about the hypergeometric distribution.
**In most real-life applications of the binomial distribution, the population of interest has a finite number of elements (trials), denoted N. When N is large and the sample size n is small relative to N, say, $n/N \leq .05$, the sampling procedure, for all practical purposes, satisfies the conditions of a binomial experiment.

point denoting that adult 1 fails, while adults 2, 3, and 4 pass the test. The 16 sample points are listed in Table 4.2.

TABLE 4.2 Sample Points for Fitness Test of Example 4.9

SSSS	FSSS	FFSS	SFFF	FFFF
	SFSS	FSFS	FSFF	
	SSFS	FSSF	FFSF	
	SSSF	SFFS	FFFS	
		SFSF		
		SSFF		

---

3. We now assign probabilities to the sample points. Note that each sample point can be viewed as the intersection of four adults' test results, and assuming that the results are independent, the probability of each sample point can be obtained by the multiplicative rule as follows:

$$P(SSSS) = P[(\text{adult 1 passes}) \cap (\text{adult 2 passes})$$
$$\cap (\text{adult 3 passes}) \cap (\text{adult 4 passes})]$$
$$= P(\text{adult 1 passes}) \times P(\text{adult 2 passes})$$
$$\times P(\text{adult 3 passes}) \times P(\text{adult 4 passes})$$
$$= (.1)(.1)(.1)(.1) = (.1)^4 = .0001$$

All other sample-point probabilities are calculated by similar reasoning. For example,

$$P(FSSS) = (.9)(.1)(.1)(.1) = .0009$$

You can check that this reasoning results in sample-point probabilities that add to 1 over the 16 points in the sample space.

4. Finally, we add the appropriate sample-point probabilities to obtain the desired event probability. The event of interest is that all four adults fail the fitness test. In Table 4.2, we find only one sample point, $FFFF$, contained in this event. All other sample points imply that at least one adult passes. Thus,

$$P(\text{All four adults fail}) = P(FFFF) = (.9)^4 = .6561$$

b. The event that three of the four adults pass the fitness test consists of the four sample points in the second column of Table 4.2: $FSSS$, $SFSS$, $SSFS$, and $SSSF$. To obtain the event probability, we add the sample-point probabilities:

$$P(3 \text{ of 4 adults pass}) = P(FSSS) + P(SFSS) + P(SSFS) + P(SSSF)$$
$$= (.1)^3(.9) + (.1)^3(.9) + (.1)^3(.9) + (.1)^3(.9)$$
$$= 4(.1)^3(.9) = .0036$$

Note that each of the four sample-point probabilities is the same, because each sample point consists of three S's and one F; the order does not affect the probability because the adults' test results are (assumed) independent.

c. We can characterize this experiment as consisting of four identical trials: the four test results. There are two possible outcomes to each trial, S or F, and the probability of passing, $p = .1$, is the same for each trial. Finally, we are assuming that each adult's test result is independent of all others, so that the four trials are independent. Then it follows that x, the number of the four adults who pass the fitness test, is a binomial random variable.

d. The event probabilities in parts **a** and **b** provide insight into the formula for the probability distribution $p(x)$. First, consider the event that three adults pass (part **b**). We found that

$P(x = 3) = $ (Number of sample points for which $x = 3$) $\times (.1)^{\text{Number of successes}} \times (.9)^{\text{Number of failures}}$

$= 4(.1)^3(.9)^1$

In general, we can use combinatorial mathematics to count the number of sample points. For example,

Number of sample points for which $x = 3$

$=$ Number of different ways of selecting 3 successes in the 4 trials

$= \binom{4}{3} = \dfrac{4!}{3!(4-3)!} = \dfrac{4 \cdot 3 \cdot 2 \cdot 1}{(3 \cdot 2 \cdot 1) \cdot 1} = 4$

The formula that works for any value of x can be deduced as follows:

$$P(x = 3) = \binom{4}{3}(.1)^3(.9)^1 = \binom{4}{x}(.1)^x(.9)^{4-x}$$

The component $\binom{4}{x}$ counts the number of sample points with x successes, and the component $(.1)^x(.9)^{4-x}$ is the probability associated with each sample point having x successes.

For the general binomial experiment, with n trials and probability p of Success on each trial, the probability of x successes is

$$p(x) = \binom{n}{x}p^x(1 - p)^{n-x}$$

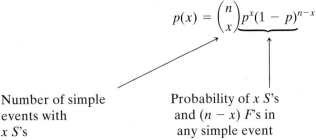

Number of simple
events with
x S's

Probability of x S's
and $(n - x)$ F's in
any simple event

Look Back In theory, you could always resort to the principles developed in this example to calculate binomial probabilities; just list the sample points and sum their probabilities. However, as the number of trials (n) increases, the number of sample points grows very rapidly. (The number of sample points is 2^n.) Thus, we prefer the formula for calculating binomial probabilities, since its use avoids listing sample points.

The **binomial distribution*** is summarized in the following box:

The Binomial Probability Distribution

$$p(x) = \binom{n}{x}p^x q^{n-x} \qquad (x = 0, 1, 2, \ldots, n)$$

where

$p = $ Probability of a success on a single trial

$q = 1 - p$

$n = $ Number of trials

$x = $ Number of successes in n trials

$$\binom{n}{x} = \dfrac{n!}{x!(n - x)!}$$

*The binomial distribution is so named because the probabilities, $p(x)$, $x = 0, 1, \ldots, n$, are terms of the binomial expansion, $(q + p)^n$.

As noted in Chapter 3, the symbol 5! Means $5 \cdot 4 \cdot 3 \cdot 2 \cdot 1 = 120$. Similarly, $n! = n(n-1)(n-2) \cdots 3 \cdot 2 \cdot 1$. (Remember, $0! = 1$.)

EXAMPLE 4.10

APPLYING THE BINOMIAL DISTRIBUTION— Physical Fitness Problem

Problem Refer to Example 4.9. Use the formula for a binomial random variable to find the probability distribution of x, where x is the number of adults who pass the fitness test. Graph the distribution.

Solution For this application, we have $n = 4$ trials. Since a success S is defined as an adult who passes the test, $p = P(S) = .1$ and $q = 1 - p = .9$. Substituting $n = 4$, $p = .1$, and $q = .9$ into the formula for $p(x)$, we obtain

$$p(0) = \frac{4!}{0!(4-0)!}(.1)^0(.9)^{4-0} = \frac{4 \cdot 3 \cdot 2 \cdot 1}{(1)(4 \cdot 3 \cdot 2 \cdot 1)}(.1)^0(.9)^4 = 1(.1)^0(.9)^4 = .6561$$

$$p(1) = \frac{4!}{1!(4-1)!}(.1)^1(.9)^{4-1} = \frac{4 \cdot 3 \cdot 2 \cdot 1}{(1)(3 \cdot 2 \cdot 1)}(.1)^1(.9)^3 = 4(.1)(.9)^3 = .2916$$

$$p(2) = \frac{4!}{2!(4-2)!}(.1)^2(.9)^{4-2} = \frac{4 \cdot 3 \cdot 2 \cdot 1}{(2 \cdot 1)(2 \cdot 1)}(.1)^2(.9)^2 = 6(.1)^2(.9)^2 = .0486$$

$$p(3) = \frac{4!}{3!(4-3)!}(.1)^3(.9)^{4-3} = \frac{4 \cdot 3 \cdot 2 \cdot 1}{(3 \cdot 2 \cdot 1)(1)}(.1)^3(.9)^1 = 4(.1)^3(.9) = .0036$$

$$p(4) = \frac{4!}{4!(4-4)!}(.1)^4(.9)^{4-4} = \frac{4 \cdot 3 \cdot 2 \cdot 1}{(4 \cdot 3 \cdot 2 \cdot 1)(1)}(.1)^4(.9)^0 = 1(.1)^4(.9) = .0001$$

Look Back Note that these probabilities, listed in Table 4.3, sum to 1. A graph of this probability distribution is shown in Figure 4.7 (below).

TABLE 4.3 Probability Distribution for Physical Fitness Example: Tabular Form

x	$p(x)$
0	.6561
1	.2916
2	.0486
3	.0036
4	.0001

FIGURE 4.7

Probability distribution for physical fitness example: graphical form

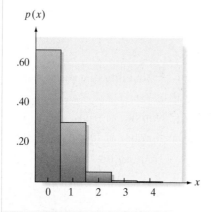

Now Work Exercise 4.45

ACTIVITY 4.1: *Warehouse Club Memberships:* **Exploring a Binomial Random Variable**

Warehouse clubs are retailers that offer lower prices than traditional retailers, but they sell only to customers who have purchased memberships, and often merchandise must be purchased in large quantities. A warehouse club may offer more than one type of membership, such as a regular membership R for a low annual fee and an upgraded membership U for a higher annual fee. The upgraded membership has additional benefits, which might include extended shopping hours, additional discounts on certain products, or cash back on purchases.

A local warehouse club has determined that 20% of its customer base has the upgraded membership.

1. What is the probability $P(U)$ that a randomly chosen customer entering the store has an upgraded membership? What is the probability $P(R)$ that a randomly chosen customer entering the store has a regular (not upgraded) membership?

   In an effort to sell more upgraded memberships, sales associates are placed at the entrance to the store to explain the benefits of the upgraded membership to customers as they enter. Suppose that five customers enter the store in a given period.

2. If 20% of the store's customers have an upgraded membership, how many of the five customers would you expect to have an upgraded membership?

3. Since there are five customers and each customer either has an upgraded membership (U) or does not (R), there are $2^5 = 32$ different possible combinations of membership types among the five customers. List these 32 possibilities.

4. Find the probability of each of the 32 outcomes. Assume that each of the five customers' membership type is independent of each of the other customers' membership type, and use your probabilities $P(U)$ and $P(R)$ with the multiplicative rule [e.g., $P(RRUUR) = P(R)P(R)P(U)P(U)P(R)$].

5. Notice that $P(URRRR) = P(RURRR) = P(RRURR) = P(RRRUR) = P(RRRRU) = P(U)^1P(R)^4$, so that P(exactly one U) = $P(URRRR) + P(RURRR) + P(RRURR) + P(RRRUR) + P(RRRRU) = 5P(U)^1P(R)^4$, where 5 is the number of ways that exactly one U can occur. Find P(exactly one U). Use similar reasoning to establish that P(no U's) = $1P(U)^0P(R)^5$, P(exactly two U's) = $10P(U)^2P(R)^3$, P(exactly three U's) = $10P(U)^3P(R)^2$, P(exactly four U's) = $5P(U)^4P(R)^1$, and P(five U's) = $1P(U)^5P(R)^0$.

6. Let x be the number of upgraded memberships U in a sample of five customers. Use the results of part **5** to write a probability distribution for the random variable x in tabular form. Use the formulas in Section 4.3 to find the mean and standard deviation of the distribution.

7. Calculate np and $\sqrt{npq}$, where $n = 5, p = P(U)$, and $q = P(R)$. How do these numbers compare with the mean and standard deviation of the random variable x in part **6** and with the expected number of customers with upgraded memberships in part **1**?

8. Explain how the characteristics of a binomial random variable are illustrated in this activity.

EXAMPLE 4.11

FINDING μ AND σ—Physical Fitness Problem

Problem Refer to Examples 4.9 and 4.10. Calculate μ and σ, the mean and standard deviation, respectively, of the number of the four adults who pass the test. Interpret the results.

Solution From Section 4.2, we know that the mean of a discrete probability distribution is

$$\mu = \sum x p(x)$$

Referring to Table 4.3, the probability distribution for the number x who pass the fitness test, we find that

$$\mu = 0(.6561) + 1(.2916) + 2(.0486) + 3(.0036) + 4(.0001) = .4$$
$$= 4(.1) = np$$

Thus, in the long run, the average number of adults (out of four) who pass the test is only .4. [*Note: The relationship $\mu = np$ holds in general for a binomial random variable.*]

The variance is

$$\sigma^2 = \Sigma(x - \mu)^2 p(x) = \Sigma(x - .4)^2 p(x)$$

$$= (0 - .4)^2(.6561) + (1 - .4)^2(.2916) + (2 - .4)^2(.0486)$$

$$+ (3 - .4)^2(.0036) + (4 - .4)^2(.0001)$$

$$= .104976 + .104976 + .124416 + .024336 + .001296$$

$$= .36 = 4(.1)(.9) = npq$$

[*Note: The relationship $\sigma^2 = npq$ holds in general for a binomial random variable.*]
Finally, the standard deviation of the number who pass the fitness test is

$$\sigma = \sqrt{\sigma^2} = \sqrt{.36} = .6$$

Applying the empirical rule, we know that approximately 95% of the x values will fall into the interval $\mu \pm 2\sigma = .4 \pm 2(.6) = (-.8, 1.6)$. Since x cannot be negative, we expect (i.e., in the long run) the number of adults out of four who pass the fitness test to be less than 1.6.

Look Back Examining Figure 4.7, you can see that all observations equal to 0 or 1 will fall within the interval $(-.8, 1.6)$. The probabilities corresponding to these values (from Table 4.3) are .6561 and .2916, respectively. Summing them, we obtain $.6561 + .2916 = .9477 \approx .95$. This result, again, supports the empirical rule.

We emphasize that you need not use the expectation summation rules to calculate μ and σ^2 for a binomial random variable. You can find them easily from the formulas $\mu = np$ and $\sigma^2 = npq$.

Mean, Variance, and Standard Deviation for a Binomial Random Variable

Mean: $\mu = np$

Variance: $\sigma^2 = npq$

Standard deviation: $\sigma = \sqrt{npq}$

Using Binomial Tables

Calculating binomial probabilities becomes tedious when n is large. For some values of n and p, the binomial probabilities have been tabulated in Table II of Appendix A. Part of that table is shown in Table 4.4; a graph of the binomial probability distribution for $n = 10$ and $p = .10$ is shown in Figure 4.8.

TABLE 4.4 Reproduction of Part of Table II of Appendix A: Binomial Probabilities for $n = 10$

k \ p	.01	.05	.10	.20	.30	.40	.50	.60	.70	.80	.90	.95	.99
0	.904	.599	.349	.107	.028	.006	.001	.000	.000	.000	.000	.000	.000
1	.996	.914	.736	.376	.149	.046	.011	.002	.000	.000	.000	.000	.000
2	1.000	.988	.930	.678	.383	.167	.055	.012	.002	.000	.000	.000	.000
3	1.000	.999	.987	.879	.650	.382	.172	.055	.011	.001	.000	.000	.000
4	1.000	1.000	.998	.967	.850	.633	.377	.166	.047	.006	.000	.000	.000
5	1.000	1.000	1.000	.994	.953	.834	.623	.367	.150	.033	.002	.000	.000
6	1.000	1.000	1.000	.999	.989	.945	.828	.618	.350	.121	.013	.001	.000
7	1.000	1.000	1.000	1.000	.998	.988	.945	.833	.617	.322	.070	.012	.000
8	1.000	1.000	1.000	1.000	1.000	.988	.989	.954	.851	.624	.264	.086	.004
9	1.000	1.000	1.000	1.000	1.000	1.000	.999	.994	.972	.893	.651	.401	.096

Table II actually contains a total of nine tables, labeled **(a)** through **(i)**, one each corresponding to $n = 5, 6, 7, 8, 9, 10, 15, 20$, and 25, respectively. In each of these tables, the columns correspond to values of p and the rows correspond to values of the random variable x. The entries in the table represent **cumulative binomial probabilities.** For example, the entry in the column corresponding to $p = .10$ and the row corresponding to $x = 2$ is .930 (highlighted), and its interpretation is

$$P(x \le 2) = P(x = 0) + P(x = 1) + P(x = 2) = .930$$

This probability is also highlighted in the graphical representation of the binomial distribution with $n = 10$ and $p = .10$ in Figure 4.8.

You can also use Table II to find the probability that x equals a specific value. For example, suppose you want to find the probability that $x = 2$ in the binomial distribution with $n = 10$ and $p = .10$. This probability is found by subtraction as follows:

$$P(x = 2) = [P(x = 0) + P(x = 1) + P(x = 2)] - [P(x = 0) + P(x = 1)]$$

$$= P(x \leq 2) - P(x \leq 1) = .930 - .736 = .194$$

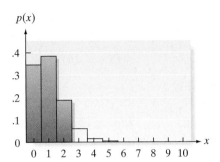

FIGURE 4.8

Binomial probability distribution for $n = 10$ and $p = .10$, with $P(x \leq 2)$ highlighted

The probability that a binomial random variable exceeds a specified value can be found from Table II together with the notion of complementary events. For example, to find the probability that x exceeds 2 when $n = 10$ and $p = .10$, we use

$$P(x > 2) = 1 - P(x \leq 2) = 1 - .930 = .070$$

Note that this probability is represented by the *un*highlighted portion of the graph in Figure 4.8

All probabilities in Table II are rounded to three decimal places. Thus, although none of the binomial probabilities in the table is exactly zero, some are small enough (less than .0005) to round to .000. For example, using the formula to find $P(x = 0)$ when $n = 10$ and $p = .6$, we obtain

$$P(x = 0) = \binom{10}{0}(.6)^0(.4)^{10-0} = .4^{10} = .00010486$$

but this is rounded to .000 in Table II of Appendix A. (See Table 4.4.)

Similarly, none of the table entries is exactly 1.0, but when the cumulative probabilities exceed .9995, they are rounded to 1.000. The row corresponding to the largest possible value for x, $x = n$, is omitted, because all the cumulative probabilities in that row are equal to 1.0 (exactly). For example, in Table 4.4 with $n = 10$, $P(x \leq 10) = 1.0$, no matter what the value of p.

The next example further illustrates the use of Table II.

EXAMPLE 4.12

USING THE BINOMIAL TABLE—

Voting for Mayor

Problem Suppose a poll of 20 voters is taken in a large city. The purpose is to determine x, the number who favor a certain candidate for mayor. Suppose that 60% of all the city's voters favor the candidate.

a. Find the mean and standard deviation of x.

b. Use Table II of Appendix A to find the probability that $x \leq 10$.

c. Use Table II to find the probability that $x > 12$.

d. Use Table II to find the probability that $x = 11$.

e. Graph the probability distribution of x, and locate the interval $\mu \pm 2\sigma$ on the graph.

Solution

a. The number of voters polled is presumably small compared with the total number of eligible voters in the city. Thus, we may treat x, the number of the 20 who favor the mayoral candidate, as a binomial random variable. The value of p is the fraction of the total number of voters who favor the candidate (i.e.,). Therefore, we calculate the mean and variance:

$$\mu = np = 20(.6) = 12$$

$$\sigma^2 = npq = 20(.6)(.4) = 4.8$$

$$\sigma = \sqrt{4.8} = 2.19$$

b. Looking in the row for $k = 10$ and the column for $p = .6$ of Table II (Appendix A) for $n = 20$, we find the value .245. Thus,

$$P(x \le 10) = .245$$

c. To find the probability

$$P(x > 12) = \sum_{x=13}^{20} p(x)$$

we use the fact that for all probability distributions,

$$\sum_{\text{all } x} p(x) = 1.$$

Therefore,

$$P(x > 12) = 1 - P(x \le 12) = 1 - \sum_{x=0}^{12} p(x)$$

Consulting Table II of Appendix A, we find the entry in row $k = 12$, column $p = .6$ to be .584. Thus,

$$P(x > 12) = 1 - .584 = .416$$

d. To find the probability that exactly 11 voters favor the candidate, recall that the entries in Table II are cumulative probabilities and use the relationship

$$P(x = 11) = [p(0) + p(1) + \cdots + p(11)] - [p(0) + p(1) + \cdots + p(10)]$$

$$= P(x \le 11) - P(x \le 10)$$

Then

$$P(x = 11) = .404 - .245 = .159$$

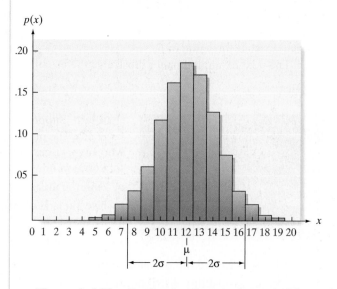

FIGURE 4.9

The binomial probability distribution for x in Example 4.12; $n = 20$ and $p = .6$

e. The probability distribution for x is shown in Figure 4.9. Note that

$$\mu - 2\sigma = 12 - 2(2.2) = 7.6 \qquad \mu + 2\sigma = 12 + 2(2.2) = 16.4$$

The interval $\mu - 2\sigma$ to $\mu + 2\sigma$ also is shown in Figure 4.9. The probability that x falls into the interval $\mu + 2\sigma$ is $P(x = 8, 9, 10, \ldots, 16) = P(x \le 16) - P(x \le 7)$

$.984 - .021 = .963$. Note that this probability is very close to the .95 given by the empirical rule. Thus, we expect the number of voters in the sample of 20 who favor the mayoral candidate to be between 8 and 16.

Now Work Exercise 4.48

Binomial Probabilities
Calculating with the TI-84/TI-83 Graphing Calculator

I. $P(x = k)$

To compute the probability of k successes in n trials where p is the probability of success for each trial, use the **binompdf(** command. *Binompdf* stands for "binomial probability density function." This command is under the **DISTR**ibution menu and has the format **binompdf(n, p, k)**.

Example Compute the probability of 5 successes in 8 trials where the probability of success for a single trial is 40%. In this example, $n = 8$, $p = .4$, and $k = 5$.

Step 1 *Enter the binomial parameter.*
Press **2nd VARS** for **DISTR**.
Press the down **arrow** key until **binompdf** is highlighted.
Press **ENTER**.
After **binompdf(**, type **8, .4, 5**. (*Note*: Be sure to use the comma key between each parameter.)
Press **ENTER**.

You should see

```
binompdf(8,.4,5)
          .12386304
```

Thus, $P(x = 5)$ is about 12.4%.

II. $P(x \le k)$

To compute the probability of k or fewer successes in n trials where the p is probability of success for each trial, use the **binomcdf(** command. *Binomcdf* stands for "binomial *cumulative* probability density function." This command is under the **DISTR**ibution menu and has the format **binomcdf(n, p, k)**.

Example Compute the probability of 5 or fewer successes in 8 trials where the probability of success for a single trial is 40%. In this example, $n = 8$, $p = .4$, and $k = 5$.

Step 1 *Enter the binomial parameters..*
Press **2nd VARS** for **DISTR**.
Press down the **arrow** key until **binomcdf** is highlighted.
Press **ENTER**.
After **binomcdf(**, type **8, .4, 5**.
Press **ENTER**.

You should see

```
binomcdf(8,.4,5)
          .95019264
```

Thus, $P(x \le 5)$ is about 95%.

(continued)

> ### III. $P(x < k), P(x > k), P(x \geq k)$
>
> To find the probability $P(x < k)$ of fewer than k successes, $P(x > k)$ of more than k success-es, or $P(x \geq k)$ of at least k successes, variations of the **binomcdf(** command must be used:
>
> For $P(x < k)$, use **binomcdf($n, p, k-1$)**.
> For $P(x > k)$, use **1 – binomcdf(n, p, k)**.
> For $P(x \geq k)$, use **1 – binomcdf($n, p, k-1$)**.

Exercises 4.41–4.63

Understanding the Principles

4.41 Give the five characteristics of a binomial random variable.

4.42 Give the formula for $p(x)$ for a binomial random variable with $n = 7$ and $p = .2$.

4.43 Consider the following binomial probability distribution:

$$p(x) = \binom{5}{x}(.7)^x(.3)^{5-x} \quad (x = 0, 1, 2, \ldots, 5)$$

a. How many trials (n) are in the experiment?
b. What is the value of p, the probability of success?

Learning the Mechanics

4.44 Compute the following:

a. $\dfrac{6!}{2!(6-2)!}$ b. $\dbinom{5}{2}$ c. $\dbinom{7}{0}$

d. $\dbinom{6}{6}$ e. $\dbinom{4}{3}$

4.45 If x is a binomial random variable, compute $p(x)$ for each of
NW   the following cases:
a. $n = 5, x = 1, p = .2$
b. $n = 4, x = 2, q = .4$
c. $n = 3, x = 0, p = .7$
d. $n = 5, x = 3, p = .1$
e. $n = 4, x = 2, q = .6$
f. $n = 3, x = 1, p = .9$

4.46 Suppose x is a binomial random variable with $n = 5$ and $p = .5$. Compute $p(x)$ for $x = 0, 1, 2, 3, 4$, and 5, using the following two methods:
a. List the sample points (take S for Success and F for Fail-ure on each trial) corresponding to each value of x, as-sign probabilities to each sample point, and obtain $p(x)$ by adding sample–point probabilities.
b. Use the formula for the binomial probability distribu-tion to obtain $p(x)$.

4.47 If x is a binomial random variable, calculate μ, σ^2, and σ for each of the following:
a. $n = 25, p = .5$
b. $n = 80, p = .2$
c. $n = 100, p = .6$
d. $n = 70, p = .9$
e. $n = 60, p = .8$
f. $n = 1,000, p = .04$

4.48 If x is a binomial random variable, use Table II in
NW   Appendix A to find the following probabilities:
a. $P(x = 2)$ for $n = 10, p = .4$
b. $P(x \leq 5)$ for $n = 15, p = .6$
c. $P(x > 1)$ for $n = 5, p = .1$

4.49 If x is a binomial random variable, use Table II in Appendix A to find the following probabilities:
a. $P(x < 10)$ for $n = 25, p = .7$
b. $P(x \geq 10)$ for $n = 15, p = .9$
c. $P(x = 2)$ for $n = 20, p = .2$

4.50 The binomial probability distribution is a family of proba-bility distributions with each single distribution depending on the values of n and p. Assume that x is a binomial ran-dom variable with $n = 4$.
a. Determine a value of p such that the probability distri-bution of x is symmetric.
b. Determine a value of p such that the probability distri-bution of x is skewed to the right.
c. Determine a value of p such that the probability distri-bution of x is skewed to the left.
d. Graph each of the binomial distributions you obtained in parts **a**, **b**, and **c**. Locate the mean for each distribution on its graph.
e. In general, for what values of p will a binomial distribution be symmetric? skewed to the right? skewed to the left?

Applet Exercise 4.3

Use the applets entitled *Simulating the Probability of a Head with an Unfair Coin* $(P(H) = 0.2)$ and *Simulating the Probabili-ty of a Head with an Unfair Coin* $(P(H) = 0.8)$ to study the mean μ of a binomial distribution.
a. Run each applet once with $n = 1,000$ and record the cu-mulative proportions. How does the cumulative propor-tion for each applet compare with the value of $P(H)$ given for the applet?
b. Using the cumulative proportion from each applet as p, compute $\mu = np$ for each applet, where $n = 1,000$. What does the value of μ represent in terms of the re-sults obtained from running each applet in part **a**?
c. In your own words, describe what the mean μ of a bino-mial distribution represents.

Applet Exercise 4.4

Open the applet entitled *Sample from a Population*. On the pull-down menu to the right of the top graph, select *Binary*. Set $n = 10$ as the sample size and repeatedly choose samples from the population. For each sample, record the number of 1's in the sample. Let x be the number of 1's in a sample of size 10. Explain why x is a binomial random variable.

Applet Exercise 4.5

Use the applet entitled *Simulating the Stock Market* to estimate the probability that the stock market will go up each of the next two days. Repeatedly run the applet for $n = 2$, recording the number of ups each time. Use the proportion of 2's among your results as the estimate of the probability. Compare your answer with the binomial probability where $x = 2$, $n = 2$, and $p = 0.5$.

Applying the Concepts—Basic

4.51 Tracking missiles with satellite imagery. The U.S. government has devoted considerable funding to missile defense research over the past 20 years. The latest development is the Space-Based Infrared System (SBIRS), which uses satellite imagery to detect and track missiles. (*Chance*, Summer 2005.) The probability that an intruding object (e.g., a missile) will be detected on a flight track by SBIRS is .8. Consider a sample of 20 simulated tracks, each with an intruding object. Let x equal the number of these tracks on which SBIRS detects the object.
   a. Demonstrate that x is (approximately) a binomial random variable.
   b. Give the values of p and n for the binomial distribution.
   c. Find $P(x = 15)$, the probability that SBIRS will detect the object on exactly 15 tracks.
   d. Find $P(x \geq 15)$, the probability that SBIRS will detect the object on at least 15 tracks.
   e. Find $E(x)$ and interpret the result.

4.52 Analysis of bottled water. Is the bottled water you're drinking really purified water? A four-year study of various brands of bottled water conducted by the Natural Resources Defense Council found that 25% of bottled water is just tap water packaged in a bottle. (*Scientific American*, July 2003.) Consider a sample of five bottled-water brands, and let x equal the number of these brands that use tap water.
NW **a.** Explain why x is (approximately) a binomial random variable.
   b. Give the probability distribution for x as a formula.
   c. Find $P(x = 2)$.
   d. Find $P(x \leq 1)$.

4.53 Belief in an afterlife. A national poll conducted by *The New York Times* (May 7, 2000) revealed that 80% of Americans believe that after you die, some part of you lives on, either in a next life on earth or in heaven. Consider a random sample of 10 Americans and count x, the number who believe in life after death.
   a. Find $P(x = 3)$.   **b.** Find $P(x \leq 7)$.
   c. Find $P(x > 4)$.

4.54 Caesarian births. The American College of Obstetricians and Gynecologists reports that 29% of all births in the United States take place by Caesarian section each year. (*National Vital Statistics Reports*, Sep. 2006.)
   a. In a random sample of 1,000 births, how many, on average, will take place by Caesarian section?
   b. What is the standard deviation of the number of Caesarian section births in a sample of 1,000 births?
   c. Use your answers to parts **a** and **b** to form an interval that is likely to contain the number of Caesarian section births in a sample of 1,000 births.

4.55 Parents who condone spanking. According to a nationwide survey, 60% of parents with young children condone spanking their child as a regular form of punishment. (*Tampa Tribune*, Oct. 5, 2000.) Consider a random sample of three people, each of whom is a parent with young children. Assume that x, the number in the sample who condone spanking, is a binomial random variable.
   a. What is the probability that none of the three parents condones spanking as a regular form of punishment for their children?
   b. What is the probability that at least one condones spanking as a regular form of punishment?
   c. Give the mean and standard deviation of x. Interpret the results.

Applying the Concepts—Intermediate

4.56 Student gambling on sports. A study of gambling activity at the University of West Georgia (UWG) discovered that 60% of the male students wagered on sports the previous year. (*The Sport Journal*, Fall 2006.) Consider a random sample of 50 UWG male students. How many of these students would you expect to have gambled on sports the previous year? Give a range that is likely to include the number of male students who have gambled on sports.

4.57 Fungi in beech forest trees. Refer to the *Applied Ecology and Environmental Research* (Vol. 1, 2003) study of beech trees damaged by fungi, presented in Exercise 3.21 (p. 125). The researchers found that 25% of the beech trees in east central Europe had been damaged by fungi. Consider a sample of 20 beech trees from this area.
   a. What is the probability that fewer than half are damaged by fungi?
   b. What is the probability that more than 15 are damaged by fungi?
   c. How many of the sampled trees would you expect to be damaged by fungi?

4.58 Chickens with fecal contamination. The United States Department of Agriculture (USDA) reports that, under its standard inspection system, one in every 100 slaughtered chickens passes inspection with fecal contamination. (*Tampa Tribune*, Mar. 31, 2000.) In Exercise 3.19 (p. 125), you found the probability that a randomly selected slaughtered chicken passes inspection with fecal contamination. Now find the probability that, in a random sample of 5 slaughtered chickens, at least one passes inspection with fecal contamination.

4.59 Testing a psychic's ESP. Refer to Exercise 3.75 (p. 150) and the experiment conducted by the Tampa Bay Skeptics to see whether an acclaimed psychic has extrasensory perception (ESP). Recall that a crystal is placed, at random, inside 1 of 10 identical boxes lying side by side on a table. The experiment was repeated seven times, and x, the number of correct decisions, was recorded. (Assume that the seven trials are independent.)
   a. If the psychic is guessing (i.e., if the psychic does *not* possess ESP) what is the value of p, the probability of a correct decision on each trial?
   b. If the psychic is guessing, what is the expected number of correct decisions in seven trials?
   c. If the psychic is guessing, what is the probability of no correct decisions in seven trials?
   d. Now suppose the psychic has ESP and $p = .5$. What is the probability that the psychic guesses incorrectly in all seven trials?
   e. Refer to part **d**. Recall that the psychic failed to select the box with the crystal on all seven trials. Is this evidence against the psychic having ESP? Explain.

4.60 Victims of domestic abuse. According to researchers at Johns Hopkins University School of Medicine, 1 in every 3 women has been a victim of domestic abuse. (*Annals of Internal Medicine*, Nov. 1995.) This probability was obtained from a survey of nearly 2,000 adult women residing in Baltimore, Maryland. Suppose we randomly sample 15 women and find that 4 have been abused.

a. What is the probability of observing 4 or more abused women in a sample of 15 if the proportion p of women who are victims of domestic abuse is really $p = \frac{1}{3}$?

b. Many experts on domestic violence believe that the proportion of women who are domestically abused is closer to $p = .10$. Calculate the probability of observing 4 or more abused women in a sample of 15 if $p = .10$.

c. Why might your answers to parts **a** and **b** lead you to believe that $p = \frac{1}{3}$?

Applying the Concepts—Advanced

4.61 Assigning a passing grade. A literature professor decides to give a 20-question true–false quiz to determine who has read an assigned novel. She wants to choose the passing grade such that the probability of passing a student who guesses on every question is less than .05. What score should she set as the lowest passing grade?

4.62 Does having boys run in the family? *Chance* (Fall 2001) reported that the eight men in the Rodgers family produced 24 biological children over four generations. Of these 24 children, 21 were boys and 3 were girls. How likely is it for a family of 24 children to have 21 boys? Use the binomial

distribution and the fact that 50% of the babies born in the United States are male to answer the question. Do you agree with the statement, "Rodgers men produce boys"?

4.63 USGA golf ball specifications. According to the U.S. Golf Association (USGA), "The weight of the [golf] ball shall not be greater than 1.620 ounces avoirdupois (45.93 grams). . . . The diameter of the ball shall not be less than 1.680 inches. . . . The velocity of the ball shall not be greater than 250 feet per second" (USGA, 2006). The USGA periodically checks the specifications of golf balls sold in the United States by randomly sampling balls from pro shops around the country. Two dozen of each kind are sampled, and if more than three do not meet size or velocity requirements, that kind of ball is removed from the USGA's approved-ball list.

a. What assumptions must be made and what information must be known in order to use the binomial probability distribution to calculate the probability that the USGA will remove a particular kind of golf ball from its approved-ball list?

b. Suppose 10% of all balls produced by a particular manufacturer are less than 1.680 inches in diameter, and assume that the number of such balls, x, in a sample of two dozen balls can be adequately characterized by a binomial probability distribution. Find the mean and standard deviation of the binomial distribution.

c. Refer to part **b**. If x has a binomial distribution, then so does the number, y, of balls in the sample that meet the USGA's minimum diameter. [*Note:* $x + y = 24$.] Describe the distribution of y. In particular, what are p, q, and n? Also, find $E(y)$ and the standard deviation of y.

4.4 Continuous Probability Distributions for Continuous Random Variables

The graphical form of the probability distribution for a continuous random variable x is a smooth curve that might appear as shown in Figure 4.10. This curve, a function of x, is denoted by the symbol $f(x)$ and is variously called a **probability density function**, a **frequency function**, or a **probability distribution**.

The areas under a probability distribution correspond to probabilities for x. For example, the area A beneath the curve between the two points a and b, as shown in Figure 4.10, is the probability that x assumes a value between a and $b(a < x < b)$. Because there is no area over a single point, say, $x = a$, it follows that (according to our model) the probability associated with a particular value of x is equal to 0; that is, $P(x = a) = 0$, and hence $P(a < x < b) = P(a \le x \le b)$. In other words, the probability is the same regardless of whether or not you include the endpoints of the interval. Also, because areas over intervals represent probabilities, it follows that the total area under a probability distribution—the total probability assigned to the set of all values of x—should equal 1. Note that probability distributions for continuous random variables possess different shapes, depending on the relative frequency distributions of real data that the probability distributions are supposed to model.

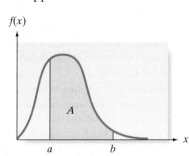

FIGURE 4.10

A probability distribution $f(x)$ for a continuous random variable x

The areas under most probability distributions are obtained by means of calculus or numerical methods.* Because these methods often involve difficult procedures, we give the areas for some of the most common probability distributions in tabular form in Appendix A. Then, to find the area between two values of x, say, $x = a$ and $x = b$, you simply have to consult the appropriate table.

For the continuous random variable presented in the next section, we will give the formula for the probability distribution, along with its mean μ and standard deviation σ. These two numbers will enable you to make some approximate probability statements about a random variable even when you do not have access to a table of areas under the probability distribution.

4.5 The Normal Distribution

One of the most commonly observed continuous random variables has a **bell-shaped probability distribution** (or **bell curve**), as shown in Figure 4.11. It is known as a **normal random variable** and its probability distribution is called a **normal distribution**.

The normal distribution plays a very important role in the science of statistical inference. Moreover, many phenomena generate random variables with probability distributions that are very well approximated by a normal distribution. For example, the error made in measuring a person's blood pressure may be a normal random variable, and the probability distribution for the yearly rainfall in a certain region might be approximated by a normal probability distribution. You can determine the adequacy of the normal approximation to an existing population of data by comparing the relative frequency distribution of a large sample of the data with the normal probability distribution. Methods for detecting disagreement between a set of data and the assumption of normality are presented in Section 4.6.

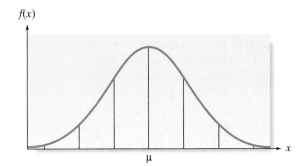

FIGURE 4.11

A normal probability distribution

Biography

CARL F. GAUSS (1777–1855)—
The Gaussian Distribution

The normal distribution began in the 18th century as a theoretical distribution for errors in disciplines in which fluctuations in nature were believed to behave randomly. Although he may not have been the first to discover the formula, the normal distribution was named the Gaussian distribution after Carl Friedrich Gauss. A well-known and re-spected German mathematician, physicist, and astronomer, Gauss applied the normal distribution while studying the motion of planets and stars. Gauss's prowess as a mathematician was exemplified by one of his most important discoveries: At the young age of 22, Gauss constructed a regular 17-gon by ruler and compasses—a feat that was the most major advance in mathematics since the time of the ancient Greeks. In addition to publishing close to 200 scientific papers, Gauss invented the heliograph as well as a primitive telegraph.

---

*Students with a knowledge of calculus should note that the probability that x assumes a value in the interval $a < x < b$ is $P(a < x < b) = \int_a^b f(x)\,dx$, assuming that the integral exists. As with the requirements for a discrete probability distribution, we require that $f(x) \geq 0$ and $\int_{-\infty}^{\infty} f(x)\,dx = 1$.

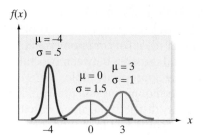

The normal distribution is perfectly symmetric about its mean μ, as can be seen in the examples in Figure 4.12. Its spread is determined by the value of its standard deviation σ.

The formula for the normal probability distribution is shown in the next box. When plotted, this formula yields a curve like that shown in Figure 4.11.

Probability Distribution for a Normal Random Variable x

Probability density function: $f(x) = \dfrac{1}{\sigma\sqrt{2\pi}}e^{-(1/2)[(x-\mu)/\sigma]^2}$

where

$$\mu = \text{Mean of the normal random variable } x$$
$$\sigma = \text{Standard deviation}$$
$$\pi = 3.1416\ldots$$
$$e = 2.71828\ldots$$

$P(x < a)$ is obtained from a table of normal probabilities.

Note that the mean μ and standard deviation σ appear in this formula, so that no separate formulas for μ and σ are necessary. To graph the normal curve, we have to know the numerical values of μ and σ.

Computing the area over intervals under the normal probability distribution is a difficult task.* Consequently, we will use the computed areas listed in Table III of Appendix A. Although there are an infinitely large number of normal curves—one for each pair of values of μ and σ—we have formed a single table that will apply to any normal curve.

Table III is based on a normal distribution with mean $\mu = 0$ and standard deviation $\sigma = 1$, called a *standard normal distribution*. A random variable with a standard normal distribution is typically denoted by the symbol z. The formula for the probability distribution of z is

$$f(z) = \frac{1}{\sqrt{2\pi}}e^{-(1/2)z^2}$$

Figure 4.13 shows the graph of a standard normal distribution.

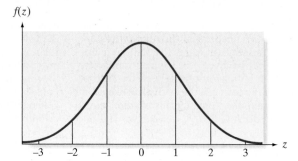

*The student with a knowledge of calculus should note that there is no closed-form expression for $P(a < x < b) = \int_a^b f(x)\,dx$ for the normal probability distribution. The value of this definite integral can be obtained to any desired degree of accuracy by numerical approximation. For this reason, it is tabulated for the user.

> **Definition 4.8**
>
> The **standard normal distribution** is a normal distribution with $\mu = 0$ and $\sigma = 1$. A random variable with a standard normal distribution, denoted by the symbol z, is called a **standard normal random variable**.

Since we will ultimately convert all normal random variables to standard normal variables in order to use Table III to find probabilities, it is important that you learn to use Table III well. A partial reproduction of that table is shown in Table 4.5. Note that the values of the standard normal random variable z are listed in the left-hand column. The entries in the body of the table give the area (probability) between 0 and z. Examples 4.13–4.16 illustrate the use of the table.

TABLE 4.5 Reproduction of Part of Table III in Appendix A

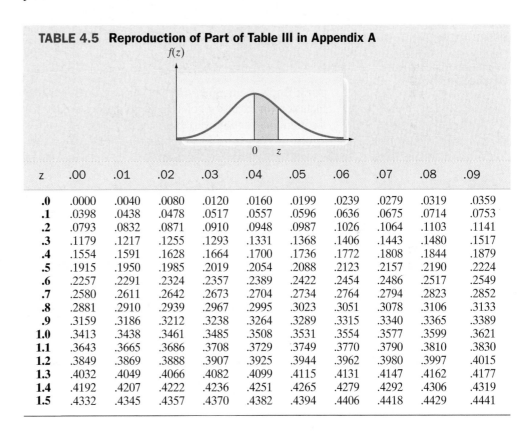

z	.00	.01	.02	.03	.04	.05	.06	.07	.08	.09
.0	.0000	.0040	.0080	.0120	.0160	.0199	.0239	.0279	.0319	.0359
.1	.0398	.0438	.0478	.0517	.0557	.0596	.0636	.0675	.0714	.0753
.2	.0793	.0832	.0871	.0910	.0948	.0987	.1026	.1064	.1103	.1141
.3	.1179	.1217	.1255	.1293	.1331	.1368	.1406	.1443	.1480	.1517
.4	.1554	.1591	.1628	.1664	.1700	.1736	.1772	.1808	.1844	.1879
.5	.1915	.1950	.1985	.2019	.2054	.2088	.2123	.2157	.2190	.2224
.6	.2257	.2291	.2324	.2357	.2389	.2422	.2454	.2486	.2517	.2549
.7	.2580	.2611	.2642	.2673	.2704	.2734	.2764	.2794	.2823	.2852
.8	.2881	.2910	.2939	.2967	.2995	.3023	.3051	.3078	.3106	.3133
.9	.3159	.3186	.3212	.3238	.3264	.3289	.3315	.3340	.3365	.3389
1.0	.3413	.3438	.3461	.3485	.3508	.3531	.3554	.3577	.3599	.3621
1.1	.3643	.3665	.3686	.3708	.3729	.3749	.3770	.3790	.3810	.3830
1.2	.3849	.3869	.3888	.3907	.3925	.3944	.3962	.3980	.3997	.4015
1.3	.4032	.4049	.4066	.4082	.4099	.4115	.4131	.4147	.4162	.4177
1.4	.4192	.4207	.4222	.4236	.4251	.4265	.4279	.4292	.4306	.4319
1.5	.4332	.4345	.4357	.4370	.4382	.4394	.4406	.4418	.4429	.4441

EXAMPLE 4.13

USING THE STANDARD NORMAL TABLE TO FIND $P(-z_0 < z < z_0)$

Problem Find the probability that the standard normal random variable z falls between -1.33 and $+1.33$.

Solution The standard normal distribution is shown again in Figure 4.14. Since all probabilities associated with standard normal random variables can be depicted as areas under the standard normal curve, you should always draw the curve and then equate the desired probability to an area.

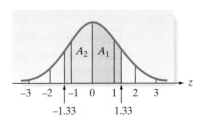

FIGURE 4.14

Areas under the standard normal curve for Example 4.13

FIGURE 4.15

Finding $z = 1.33$ in the standard normal table, Example 4.13

In this example, we want to find the probability that z falls between -1.33 and $+1.33$, which is equivalent to the area between -1.33 and $+1.33$, shown highlighted in Figure 4.14. Table IV gives the area between $z = 0$ and any value of z, so that if we look up $z = 1.33$ (the value in the 1.3 row and .03 column, as shown in Figure 4.15), we find that the area between $z = 0$ and $z = 1.33$ is .4082. This is the area labeled A_1 in Figure 4.14. To find the area A_2 located between $z = 0$ and $z = -1.33$, we note that the symmetry of the normal distribution implies that the area between $z = 0$ and any point to the left is equal to the area between $z = 0$ and the point equidistant to the right. Thus, in this example the area between $z = 0$ and $z = -1.33$ is equal to the area between $z = 0$ and $z = +1.33$. That is,

$$A_1 = A_2 = .4082$$

The probability that z falls between -1.33 and $+1.33$ is the sum of the areas of A_1 and A_2. We summarize in probabilistic notation:

$$P(-1.33 < z < +1.33) = P(-1.33 < z < 0) + P(0 < z \leq 1.33)$$
$$= A_1 + A_2 = .4082 + .4082 = .8164$$

Look Back Remember that "<" and "≤" are equivalent in events involving z, because the inclusion (or exclusion) of a single point does not alter the probability of an event involving a continuous random variable.

Now Work Exercise 4.69 e–f

EXAMPLE 4.14

USING THE STANDARD NORMAL TABLE TO FIND $P(z > z_0)$

Problem Find the probability that a standard normal random variable exceeds 1.64; that is, find $P(z > 1.64)$.

Solution The area under the standard normal distribution to the right of 1.64 is the highlighted area labeled A_1 in Figure 4.16. This area represents the probability that z exceeds 1.64. However, when we look up $z = 1.64$ in Table III, we must remember that the probability given in the table corresponds to the area between $z = 0$ and $z = 1.64$ (the area labeled A_2 in Figure 4.16). From Table III, we find that $A_2 = .4495$. To find the area A_1 to the right of 1.64, we make use of two facts:

1. The standard normal distribution is symmetric about its mean, $z = 0$.
2. The total area under the standard normal probability distribution equals 1.

Taken together, these two facts imply that the areas on either side of the mean, $z = 0$, equal .5; thus, the area to the right of $z = 0$ in Figure 4.16 is $A_1 + A_2 = .5$. Then

$$P(z > 1.64) = A_1 = .5 - A_2 = .5 - .4495 = .0505$$

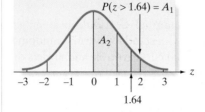

FIGURE 4.16

Areas under the standard normal curve for Example 4.14

Look Back To attach some practical significance to this probability, note that the implication is that the chance of a standard normal random variable exceeding 1.64 is only about .05.

Now Work Exercise 4.70a

EXAMPLE 4.15

USING THE STANDARD NORMAL TABLE TO FIND $P(z < z_0)$

Problem Find the probability that a standard normal random variable lies to the left of .67.

Solution The event sought is shown as the highlighted area in Figure 4.17. We want to find $P(z < .67)$. We divide the highlighted area into two parts: the area A_1 between $z = 0$ and $z = .67$, and the area A_2 to the left of $z = 0$. We must always make such a division when the desired area lies on both sides of the mean ($z = 0$) because Table III contains areas between $z = 0$ and the point you look up. We look up $z = .67$ in Table III to find that $A_1 = .2486$. The symmetry of the standard normal distribution also implies that half the distribution lies on each side of the mean, so the area A_2 to the left of $z = 0$ is .5. Then

$$P(z < .67) = A_1 + A_2 = .2486 + .5 = .7486$$

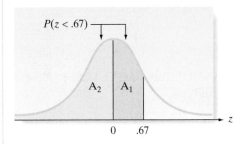

FIGURE 4.17

Areas under the standard normal curve for Example 4.15

Look Back Note that this probability is approximately .75. Thus, about 75% of the time, the standard normal random variable z will fall below .67. This statement implies that $z = .67$ represents the approximate 75th percentile for the distribution.

Now Work Exercise 4.70h

EXAMPLE 4.16

USING THE STANDARD NORMAL TABLE TO FIND $P(|z| > z_0)$

Problem Find the probability that a standard normal random variable exceeds 1.96 in absolute value.

Solution The event sought is shown highlighted in Figure 4.18. We want to find

$$P(|z| > 1.96) = P(z < -1.96 \text{ or } z > 1.96)$$

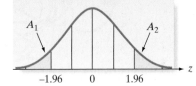

FIGURE 4.18

Areas under the standard normal curve for Example 4.16

Note that the total highlighted area is the sum of the two areas A_1 and A_2—areas that are equal because of the symmetry of the normal distribution.

We look up $z = 1.96$ and find the area between $z = 0$ and $z = 1.96$ to be .4750. Then A_2, the area to the right of 1.96, is $.5 - .4750 = .0250$, so that

$$P(|z| > 1.96) = A_1 + A_2 = .0250 + .0250 = .05$$

Look Back We emphasize, again, the importance of sketching the standard normal curve in finding normal probabilities.

To apply Table III to a normal random variable x with any mean μ and any standard deviation σ, we must first convert the value of x to a z-score. The population z-score for a measurement was defined in Section 2.7 as the *distance* between the measurement and the population mean, divided by the population standard deviation. Thus, the z-score gives the distance between a measurement and the mean in units equal to the standard deviation. In symbolic form, the z-score for the measurement x is

$$z = \frac{x - \mu}{\sigma}$$

Note that when $x = \mu$, we obtain $z = 0$.

An important property of the normal distribution is that if x is normally distributed with any mean and any standard deviation, z is *always* normally distributed with mean 0 and standard deviation 1. That is, z is a standard normal random variable.

Property of Normal Distributions

If x is a normal random variable with mean μ and standard deviation σ, then the random variable z defined by the formula

$$z = \frac{x - \mu}{\sigma}$$

has a standard normal distribution. The value z describes the number of standard deviations between x and μ.

Recall from Example 4.16 that $P(|z| > 1.96) = .05$. This probability, coupled with our interpretation of z, implies that any normal random variable lies more than 1.96 standard deviations from its mean only 5% of the time. Compare this statement with the empirical rule (Chapter 2), which tells us that about 5% of the measurements in mound-shaped distributions will lie beyond two standard deviations from the mean. The normal distribution actually provides the model on which the empirical rule is based, along with much "empirical" experience with real data that often approximately obey the rule, whether drawn from a normal distribution or not.

EXAMPLE 4.17

FINDING A NORMAL PROBABILITY— Cell Phone Application

Problem Assume that the length of time, x, between charges of a cellular phone is normally distributed with a mean of 10 hours and a standard deviation of 1.5 hours. Find the probability that the cell phone will last between 8 and 12 hours between charges.

Solution The normal distribution with mean $\mu = 10$ and $\sigma = 1.5$ is shown in Figure 4.19. The desired probability that the cell phone lasts between 8 and 12 hours is highlighted. In order to find that probability, we must first convert the distribution to a standard normal distribution, which we do by calculating the z-score:

$$z = \frac{x - \mu}{\sigma}$$

The z-scores corresponding to the important values of x are shown beneath the x values on the horizontal axis in Figure 4.19. Note that $z = 0$ corresponds to the mean of $\mu = 10$

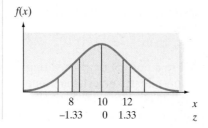

FIGURE 4.19
Areas under the normal curve
for Example 4.17

hours, whereas the *x* values 8 and 12 yield *z*-scores of -1.33 and $+1.33$, respectively. Thus, the event that the cell phone lasts between 8 and 12 hours is equivalent to the event that a standard normal random variable lies between -1.33 and $+1.33$. We found this probability in Example 4.13 (see Figure 4.14) by doubling the area corresponding to $z = 1.33$ in Table III. That is,

$$P(8 \le x \le 12) = P(-1.33 \le z \le 1.33) = 2(.4082) = .8164$$

Now Work Exercise 4.75a

The steps to follow in calculating a probability corresponding to a normal random variable are shown in the following box:

Steps for Finding a Probability Corresponding to a Normal Random Variable

1. Sketch the normal distribution and indicate the mean of the random variable *x*. Then shade the area corresponding to the probability you want to find.

2. Convert the boundaries of the shaded area from *x* values to standard normal random variable *z* values by using the formula

$$z = \frac{x - \mu}{\sigma}$$

Show the *z* values under the corresponding *x* values on your sketch.

3. Use Table III in Appendix A to find the areas corresponding to the z values. If necessary, use the symmetry of the normal distribution to find areas corresponding to negative z values and the fact that the total area on each side of the mean equals .5 to convert the areas from Table III to the probabilities of the event you have shaded.

Standard Normal Probabilities

Using the TI-84/TI-83 Graphing Calculator

Graphing the Area under the Standard Normal Curve

Step 1 *Turn off all plots.*
Press **2nd PRGM** and select **1:ClrDraw**.
Press **ENTER ENTER**, and 'Done' will appear on the screen.
Press **2nd Y=** and select **4:PlotsOff**.
Press **ENTER ENTER**, and 'Done' will appear on the screen.

Step 2 *Set the viewing window. (Recall that almost all of the area under the standard normal curve falls between −5 and 5. A height of 0.5 is a good choice for Ymax.)*
Note: When entering a negative number, be sure to use the negative sign **(−)**, and not the minus sign.
Set **Xmin = −5**
Xmax = 5
Xscl = 1
Ymin = 0

Ymax = .5
Yscl = 0
Xres = 1

Step 3 *View graph.*
Press **2nd VARS**.
Arrow right to **DRAW**.

(continued)

Press **ENTER** to select **1:ShadeNorm(**.
Enter your lower limit.
Press **COMMA**.
Enter your upper limit.
Press **)** Press **ENTER**.
The graph will be displayed along with the area, lower limit, and upper limit.

Example Use the standard normal curve to find the probability that z is less than 1.5.
In this example, set the Window as shown in **Step 2**. For the limits in **Step 3**, use -5
for the lower limit and 1.5 for the upper limit.

The screens for this example are as follows:

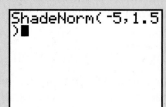
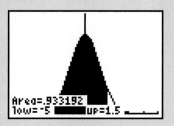

Thus, $P(z < 1.5) = .9332$.

EXAMPLE **4.18**

USING NORMAL
PROBABILITIES
TO MAKE AN
INFERENCE—
Advertised Gas
Mileage

Problem Suppose an automobile manufacturer introduces a new model that has an advertised mean in-city mileage of 27 miles per gallon. Although such advertisements seldom report any measure of variability, suppose you write the manufacturer for the details of the tests and you find that the standard deviation is 3 miles per gallon. This information leads you to formulate a probability model for the random variable x, the in-city mileage for this car model. You believe that the probability distribution of x can be approximated by a normal distribution with a mean of 27 and a standard deviation of 3.

a. If you were to buy this model of automobile, what is the probability that you would purchase one that averages less than 20 miles per gallon for in-city driving? In other words, find $P(x < 20)$.

b. Suppose you purchase one of these new models and it does get less than 20 miles per gallon for in-city driving. Should you conclude that your probability model is incorrect?

Solution

a. The probability model proposed for x, the in-city mileage, is shown in Figure 4.20 . We are interested in finding the area A to the left of 20, since that area corresponds to the probability that a measurement chosen from this distribution falls below 20. In other words, if this model is correct, the area A represents the fraction of cars that can be expected to get less than 20 miles per gallon for in-city driving. To find A, we first calculate the z value corresponding to $x = 20$. That is,

$$z = \frac{x - \mu}{\sigma} = \frac{20 - 27}{3} = -\frac{7}{3} = -2.33$$

Then

$$P(x < 20) = P(z < -2.33)$$

as indicated by the highlighted area in Figure 4.20. Since Table III gives only areas to the right of the mean (and because the normal distribution is symmetric about its mean), we look up 2.33 in Table III and find that the corresponding area is .4901. This is equal to the area between $z = 0$ and $z = -2.33$, so we find that

$$P(x < 20) = A = .5 - .4901 = .0099 \approx .01$$

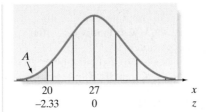

FIGURE 4.20
Area under the normal curve
for Example 4.18

According to this probability model, you should have only about a 1% chance of purchasing a car of this make with an in-city mileage under 20 miles per gallon.

b. Now you are asked to make an inference based on a sample: the car you purchased. You are getting less than 20 miles per gallon for in-city driving. What do you infer? We think you will agree that one of two possibilities exists:

1. The probability model is correct. You simply were unfortunate to have purchased one of the cars in the 1% that get less than 20 miles per gallon in the city.

2. The probability model is incorrect. Perhaps the assumption of a normal distribution is unwarranted, or the mean of 27 is an overestimate, or the standard deviation of 3 is an underestimate, or some combination of these errors occurred. At any rate, the form of the actual probability model certainly merits further investigation.

You have no way of knowing with certainty which possibility is correct, but the evidence points to the second one. We are again relying on the rare-event approach to statistical inference that we introduced earlier. The sample (one measurement in this case) was so unlikely to have been drawn from the proposed probability model that it casts serious doubt on the model. We would be inclined to believe that the model is somehow in error.

Look Back In applying the rare-event approach, the calculated probability must be small (say, less than or equal to .05) in order to infer that the observed event is, indeed, unlikely.

Now Work Exercise 4.84

Nonstandard Normal Probabilities

Using the TI-84/TI-83 Graphing Calculator

I. Finding normal probabilities without a graph

To compute probabilities for a normal distribution, use the **normalcdf(**command. "Normalcdf" stands for "normal cumulative density function." This command is under the **DISTR**ibution menu and has the format **normalcdf(***lower limit, upper limit, mean, standard deviation***)**.

Step 1   *Find the probability.*
Press **2nd VARS** for **DISTR** and select **Normalcdf(**.
After Normalcdf(, type in the lower limit.
Press **COMMA**.
Enter the upper limit.
Press **COMMA**.
Enter the mean.
Press **COMMA**.
Enter the standard deviation.
Press **)**.
Press **ENTER**.
The probability will be displayed on the screen.

(continued)

Example   What is $P(x < 115)$ for a normal distribution with $\mu = 100$ and $\sigma = 10$?

In this example, the lower limit is $-\infty$, the upper limit is 115, the mean is 100, and the standard deviation is 10.

To represent $-\infty$ on the calculator, enter **(−) 1,** press **2nd** and press the **COMMA** key for **EE,** and then press **99.** The screen appears as follows:

```
normalcdf(-1E99,
115,100,10)
        .9331927713
```

II. Finding normal probabilities with a graph

Step 1   *Turn off all plots.*
Press **Y=** and **CLEAR** all functions from the Y registers.
Press **2nd Y=** and select **4:PlotsOff.**
Press **ENTER ENTER**, and 'Done' will appear on the screen.

Step 2   *Set the viewing window. (These values depend on the mean and standard deviation of the data.) Note*: When entering a negative number, be sure to use the negative sign **(−),** not the minus sign.
Press **WINDOW**.
Set **Xmin** $= \mu - 5\sigma$
Xmax $= \mu + 5\sigma$
Xscl $= \sigma$
Ymin $= -.125/\sigma$
Ymax $= .5/\sigma$
Yscl $= 1$
Xres $= 1$

Step 3   *View graph.*
Press **2nd VARS**.
ARROW right to **DRAW**.
Press **ENTER** to select **1:ShadeNorm(.**
Enter the lower limit.
Press **COMMA**.
Enter the upper limit.
Press **COMMA**.
Enter the mean.
Press **COMMA**.
Enter the standard deviation.
Press **)**.
Press **ENTER**.
The graph will be displayed along with the area, lower limit, and upper limit.

Example   What is $P(x < 115)$ for a normal distribution with $\mu = 100$ and $\sigma = 10$?

In this example, the lower limit is $-\infty$, the upper limit is 115, the mean is 100, and the standard deviation is 10.

To represent $-\infty$ on the calculator, enter **(−) 1,** press **2nd** and press the **comma** key for **EE,** and then press **99.** The screens appear as follows:

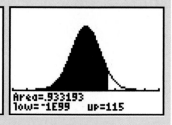

```
WINDOW
 Xmin=50
 Xmax=150
 Xscl=10
 Ymin=-.0125
 Ymax=.05
 Yscl=1■
 Xres=1
```

```
ShadeNorm(-1E99,
115,100,10)
```

Thus $P(x < 115)$ is .9332.

Occasionally you will be given a probability and will want to find the values of the normal random variable that correspond to that probability. For example, suppose the scores on a college entrance examination are known to be normally distributed and a certain prestigious university will consider for admission only those applicants whose scores exceed the 90th percentile of the test score distribution. To determine the minimum score for consideration for admission, you will need to be able to use Table IV in reverse, as demonstrated in the next example.

EXAMPLE 4.19
USING THE NORMAL TABLE IN REVERSE

Problem Find the value of z—call it z_0—in the standard normal distribution that will be exceeded only 10% of the time. That is, find z_0 such that $P(z \geq z_0) = .10$.

Solution In this case, we are given a probability, or an area, and are asked to find the value of the standard normal random variable that corresponds to the area. Specifically, we want to find the value z_0 such that only 10% of the standard normal distribution exceeds z_0. (See Figure 4.21.)

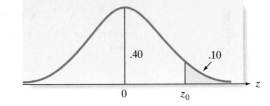

FIGURE 4.21
Areas under the standard normal curve for Example 4.19

We know that the total area to the right of the mean $z = 0$, is .5, which implies that z_0 must lie to the right of 0 ($z_0 > 0$). To pinpoint the value, we use the fact that the area to the right of z_0 is .10, which implies that the area between $z = 0$ and z_0 is $.5 - .1 = .4$. But areas between $z = 0$ and some other z value are exactly the types given in Table III. Therefore, we look up the area .4000 in the body of Table III and find that the corresponding z value is (to the closest approximation) $z_0 = 1.28$. The implication is that the point 1.28 standard deviations above the mean is the 90th percentile of a normal distribution.

Look Back As with earlier problems, it is critical to correctly draw the normal probability sought on the normal curve. The placement of z_0 to the left or right of 0 is the key. Be sure to shade the probability (area) involving z_0. If it does not agree with the probability sought (i.e., if the shaded area is greater than .5 and the probability sought is smaller than .5), then you need to place z_0 on the opposite side of 0.

EXAMPLE 4.20
USING THE NORMAL TABLE IN REVERSE

Problem Find the value of z_0 such that 95% of the standard normal z values lie between $-z_0$ and $+z_0$; that is, find $P(-z_0 \leq z \leq z_0) = .95$.

Solution Here we wish to move an equal distance z_0 in the positive and negative directions from the mean $z = 0$ until 95% of the standard normal distribution is enclosed. This means that the area on each side of the mean will be equal to $\frac{1}{2}(.95) = .475$, as shown in Figure 4.22. Since the area between $z = 0$ and z_0 is .475, we look up .475

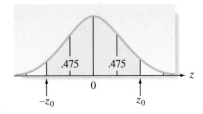

FIGURE 4.22
Areas under the standard normal curve for Example 4.20

in the body of Table III to find the value $z_0 = 1.96$. Thus, as we found in reverse order in Example 4.16, 95% of a normal distribution lies between $+1.96$ and -1.96 standard deviations of the mean.

Now Work Exercise 4.74a

Now that you have learned to use Table III to find a standard normal z value that corresponds to a specified probability, we demonstrate a practical application in Example 4.21.

EXAMPLE 4.21

THE NORMAL TABLE IN REVERSE—College Entrance Exam Application

Problem Suppose the scores x on a college entrance examination are normally distributed with a mean of 550 and a standard deviation of 100. A certain prestigious university will consider for admission only those applicants whose scores exceed the 90th percentile of the distribution. Find the minimum score an applicant must achieve in order to receive consideration for admission to the university.

Solution In this example, we want to find a score x_0 such that 90% of the scores (x values) in the distribution fall below x_0 and only 10% fall above x_0. That is,

$$P(x \leq x_0) = .90$$

Converting x to a standard normal random variable where $\mu = 550$ and $\sigma = 100$, we have

$$P(x \leq x_0) = P\left(z \leq \frac{x_0 - \mu}{\sigma}\right)$$

$$= P\left(z \leq \frac{x_0 - 550}{100}\right) = .90$$

In Example 4.19 (see Figure 4.21), we found the 90th percentile of the standard normal distribution to be $z_0 = 1.28$. That is, we found that $P(z \leq 1.28) = .90$. Consequently, we know that the minimum test score x_0 corresponds to a z-score of 1.28; in other words,

$$\frac{x_0 - 550}{100} = 1.28$$

If we solve this equation for x_0, we find that

$$x_0 = 550 + 1.28(100) = 550 + 128 = 678$$

This x value is shown in Figure 4.23. Thus, the 90th percentile of the test score distribution is 678. That is to say, an applicant must score at least 678 on the entrance exam to receive consideration for admission by the university.

FIGURE 4.22

Area under the normal curve for Example 4.21

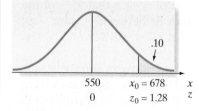

Look Back As the example shows, in practical applications of the normal table in reverse, first find the value of z_0 and then use the z-score formula in reverse to convert the value to the units of x.

Now Work Exercise 4.83e

STATISTICS IN ACTION REVISITED

Using the Normal Model to Maximize the Probability of a Hit with the Super Weapon

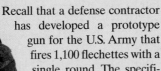

Recall that a defense contractor has developed a prototype gun for the U.S. Army that fires 1,100 flechettes with a single round. The specifications of the weapon are set so that when the gun is aimed at a target 500 meters away, the mean horizontal grid value of the flechettes is equal to the aim point. In the range test, the weapon was aimed at the center target in Figure SIA4.1; thus, $\mu = 5$ feet. For three different tests, the standard deviation was set at $\sigma = 1$ foot, $\sigma = 2$ feet, and $\sigma = 4$ feet. From past experience, the defense contractor has found that the distribution of the horizontal flechette measurements is closely approximated by a normal distribution. Therefore, we can use the normal distribution to find the probability that a single flechette shot from the weapon will hit any one of the three targets. Recall from Figure SIA4.1 that the three targets range from −1 to 1, 4 to 6, and 9 to 11 feet on the horizontal grid.

Consider first the middle target. Letting x represent the horizontal measurement for a flechette shot from the gun, we see that the flechette will hit the target if $4 \leq x \leq 6$. Using the normal probability table (Table 4, Appendix I), we then find that the probability that this flechette will hit the target when $\mu = 5$ and $\sigma = 1$ is

Middle: $P(4 \leq x \leq 6) = P\left(\dfrac{4-5}{1} < z < \dfrac{6-5}{1}\right)$

$\sigma = 1$

$\quad = P(-1 < z < 1)$

$\quad = 2(.3413) = .6826$

Similarly, we find the probabilities that the flechette hits the left and right targets shown in Figure SIA4.1:

Left: $P(-1 \leq x \leq 1) = P\left(\dfrac{-1-5}{1} < z < \dfrac{1-5}{1}\right)$

$\sigma = 1$

$\quad = P(-6 < z < -4) \approx 0$

Right: $P(9 \leq x \leq 11) = P\left(\dfrac{9-5}{1} < z < \dfrac{11-5}{1}\right)$

$\sigma = 1$

$\quad = P(4 < z < 6) \approx 0$

You can see that there is about a 68% chance that a flechette will hit the middle target, but virtually no chance that one will hit the left or right target when the standard deviation is set at 1 foot.

To find these three probabilities for $\sigma = 2$ and $\sigma = 4$, we use the normal probability function in MINITAB. Figure SIA4.2 is a MINITAB worksheet giving the cumulative probabilities of a normal random variable falling below the x values in the first column. The cumulative probabilities for $\sigma = 2$ and $\sigma = 4$ are given in the columns named "sigma2" and "sigma4," respectively.

Using the cumulative probabilities in the figure to find the three probabilities when $\sigma = 2$, we have

Middle: $P(4 \leq x \leq 6) = P(x \leq 6) - P(x \leq 4)$

$\sigma = 2$

$\quad = .6915 - .3085 = .3830$

Left: $P(-1 \leq x \leq 1) = P(x \leq 1) - P(x \leq -1)$

$\sigma = 2$

$\quad = .0227 - .0013 = .0214$

Right: $P(9 \leq x \leq 11) = P(x \leq 11) - P(x \leq 9)$

$\sigma = 2$

$\quad = .9987 - .9773 = .0214$

Thus, when $\sigma = 2$, there is about a 38% chance that a flechette will hit the middle target, a 2% chance that one will hit the left target, and a 2% chance that one will hit the right target. The probability that a flechette will hit either the middle or the left or the right target is simply the sum of these three probabilities (an application of the additive rule of probability). This sum is .3830 + .0214 + .0214 = .4258; consequently, there is about a 42% chance of hitting any one of the three targets when specifications are set so that $\sigma = 2$.

Now we use the cumulative probabilities in Figure SIA4.2 to find the three hit probabilities when $\sigma = 4$:

Middle: $P(4 \leq x \leq 6) = P(x \leq 6) - P(x \leq 4)$

$\sigma = 4$

$\quad = .5987 - .4013 = .1974$

FIGURE SIA4.2

MINITAB worksheet with cumulative normal probabilities

↓	C1	C2	C3	C4	C5
	x	sigma1	sigma2	sigma4	
1	-1	0.00000	0.001350	0.066807	
2	1	0.00003	0.022750	0.158655	
3	4	0.15866	0.308538	0.401294	
4	6	0.84134	0.691462	0.598706	
5	9	0.99997	0.977250	0.841345	
6	11	1.00000	0.998650	0.933193	
7					

(continued)

Left: $P(-1 \le x \le 1) = P(x \le 1) - P(x \le -1)$

$\sigma = 4$

$\qquad\qquad = .1587 - .0668 = .0919$

Right: $P(9 \le x \le 11) = P(x \le 11) - P(x \le 9)$

$\sigma = 4$

$\qquad\qquad = .9332 - .8413 = .0919$

Thus, when $\sigma = 4$, there is about a 20% chance that a flechette will hit the middle target, a 9% chance that one will hit the left target, and a 9% chance that one will hit the right target. The probability that a flechette will hit any one of the three targets is .1974 + .0919 + .0919 = .3812.

These probability calculations reveal a few patterns. First, the probability of hitting the middle target (the target at which the gun is aimed) is reduced as the standard deviation is increased. Obviously, then, if the U.S. Army wants to maximize the chance of hitting the target that the prototype gun is aimed at, it will want specifications set with a small value of σ. But if the Army wants to hit multiple targets with a single shot of the weapon, σ should be increased. With a larger σ, not as many of the flechettes will hit the target aimed at, but more will hit peripheral targets. Whether σ should be set at 4 or 6 (or some other value) depends on how high of a hit rate is required for the peripheral targets.

Exercises 4.64–4.92

Understanding the Principles

4.64 Describe the shape of a normal probability distribution.

4.65 What is the name given to a normal distribution when $\mu = 0$ and $\sigma = 1$?

4.66 If x has a normal distribution with mean μ and standard deviation σ, describe the distribution of $z = (x - \mu)/\sigma$.

Learning the Mechanics

4.67 Find the area under the standard normal distribution between the following pairs of z-scores:
 a. $z = -2.00$ and $z = 0$
 b. $z = -1.00$ and $z = 0$
 c. $z = -1.69$ and $z = 0$
 d. $z = -.58$ and $z = 0$

4.68 Find the area under the standard normal probability distribution between the following pairs of z-scores:
 a. $z = 0$ and $z = 2.00$
 b. $z = 0$ and $z = 1.00$
 c. $z = 0$ and $z = 3$
 d. $z = 0$ and $z = .58$

4.69 Find each of the following probabilities for a standard normal random variable z:
 a. $P(z = 1)$
 b. $P(z \le 1)$
 c. $P(z < 1)$
 d. $P(z > 1)$
 NW **e.** $P(-1 \le z \le 1)$
 NW **f.** $P(-2 \le z \le 2)$
 NW **g.** $P(-2.16 \le z \le .55)$
 NW **h.** $P(-.42 < z < 1.96)$

4.70 Find the following probabilities for the standard normal random variable z:
 NW **a.** $P(z > 1.46)$
 b. $P(z < -1.56)$
 c. $P(.67 \le z \le 2.41)$
 d. $P(-1.96 \le z < -.33)$
 e. $P(z \ge 0)$
 f. $P(-2.33 < z < 1.50)$
 g. $P(z \ge -2.33)$
 NW **h.** $P(z < 2.33)$

4.71 Suppose the random variable x is best described by a normal distribution with $\mu = 25$ and $\sigma = 5$. Find the z-score that corresponds to each of the following x values:
 a. $x = 25$
 b. $x = 30$
 c. $x = 37.5$
 d. $x = 10$
 e. $x = 50$
 f. $x = 32$

4.72 Give the z-score for a measurement from a normal distribution for the following:
 a. 1 standard deviation above the mean
 b. 1 standard deviation below the mean
 c. Equal to the mean
 d. 2.5 standard deviations below the mean
 e. 3 standard deviations above the mean

4.73 Find a value z_0 of the standard normal random variable z such that
 a. $P(z \ge z_0) = .05$
 b. $P(z \ge z_0) = .025$
 c. $P(z \le z_0) = .025$
 d. $P(z \ge z_0) = .10$
 e. $P(z > z_0) = .10$

4.74 Find a value z_0 of the standard normal random variable z such that
 NW **a.** $P(z \le z_0) = .0401$
 b. $P(-z_0 \le z \le z_0) = .95$
 c. $P(-z_0 \le z \le z_0) = .90$
 d. $P(-z_0 \le z \le z_0) = .8740$
 e. $P(-z_0 \le z \le 0) = .2967$
 f. $P(-2 < z < z_0) = .9710$
 g. $P(z \ge z_0) = .5$
 h. $P(z \ge z_0) = .0057$

4.75 Suppose x is a normally distributed random variable with $\mu = 11$ and $\sigma = 2$. Find each of the following:
 NW **a.** $P(10 \le x \le 12)$
 b. $P(6 \le x \le 10)$
 c. $P(13 \le x \le 16)$
 d. $P(7.8 \le x \le 12.6)$
 e. $P(x \ge 13.24)$
 f. $P(x \ge 7.62)$

4.76 Suppose x is a normally distributed random variable with $\mu = 30$ and $\sigma = 8$. Find a value x_0 of the random variable x such that

  a. $P(x \geq x_0) = .5$
  b. $P(x < x_0) = .025$
  c. $P(x > x_0) = .10$
  d. $P(x > x_0) = .95$
  e. 10% of the values of x are less than x_0.
  f. 80% of the values of x are less than x_0.
  g. 1% of the values of x are greater than x_0.

4.77 The random variable x has a normal distribution with standard deviation 25. It is known that the probability that x exceeds 150 is .90. Find the mean μ of the probability distribution.

APPLET Applet Exercise 4.6

Open the applet entitled *Sample from a Population*. On the pull-down menu to the right of the top graph, select *Bell shaped*. The box to the left of the top graph displays the population mean, median, and standard deviation.

  a. Run the applet for each available value of n on the pull-down menu for the sample size. Go from the smallest to the largest value of n. For each value of n, observe the shape of the graph of the sample data and record the mean, median, and standard deviation of the sample.
  b. Describe what happens to the shape of the graph and the mean, median, and standard deviation of the sample as the sample size increases.

Applying the Concepts—Basic

4.78 Dental anxiety study. Psychology students at Wittenberg University completed the Dental Anxiety Scale questionnaire. (*Psychological Reports*, Aug. 1997.) Scores on the scale range from 0 (no anxiety) to 20 (extreme anxiety). The mean score was 11 and the standard deviation was 3.5. Assume that the distribution of all scores on the Dental Anxiety Scale is normal with $\mu = 11$ and $\sigma = 3.5$.

  a. Suppose you score a 16 on the Dental Anxiety Scale. Find the z value for this score.
  b. Find the probability that someone scores between 10 and 15 on the Dental Anxiety Scale.
  c. Find the probability that someone scores above 17 on the Dental Anxiety Scale.

WPOWER50

4.79 Most powerful American women. Refer to the *Fortune* (Nov. 14, 2005) list of the 50 most powerful women in America, presented in Exercise 2.58 (p. 59). Recall that the data on age (in years) of each woman is stored in the **WPOWER50** file. The ages in the data set can be shown to be approximately normally distributed with a mean of 50 years and a standard deviation of 5.3 years. A powerful woman is randomly selected from the data, and her age is observed.

  a. Find the probability that her age will fall between 55 and 60 years.
  b. Find the probability that her age will fall between 48 and 52 years.
  c. Find the probability that her age will be less than 35 years.
  d. Find the probability that her age will exceed 40 years.

4.80 Improving SAT scores. Refer to the *Chance* (Winter 2001) study of students who paid a private tutor to help them improve their SAT scores, presented in Exercise 2.101 (p. 74). The table summarizing the changes in both the SAT-Mathematics and SAT-Verbal scores for these students is reproduced here. Assume that both distributions of SAT score changes are approximately normal.

	SAT-Math	SAT-Verbal
Mean change in score	19	7
Standard deviation of changes in score	65	49

  a. What is the probability that a student increases his or her score on the SAT-Math test by at least 50 points?
  b. What is the probability that a student increases his or her score on the SAT-Verbal test by at least 50 points?

4.81 Transmission delays in wireless technology. Resource Reservation Protocol (RSVP) was originally designed to establish signaling links for stationary networks. In *Mobile Networks and Applications* (Dec. 2003), RSVP was applied to mobile wireless technology (e.g., a PC notebook with wireless LAN card for Internet access). A simulation study revealed that the transmission delay (measured in milliseconds) of an RSVP linked wireless device has an approximate normal distribution with mean $\mu = 48.5$ milliseconds and $\sigma = 8.5$ milliseconds.

  a. What is the probability that the transmission delay is less than 57 milliseconds?
  b. What is the probability that the transmission delay is between 40 and 60 milliseconds?

4.82 Casino gaming. Casino gaming yields over \$35 billion in revenue each year in the United States. In *Chance* (Spring 2005), University of Denver statistician R.C. Hannum discussed the business of casino gaming and its reliance on the laws of probability. Casino games of pure chance (e.g., craps, roulette, baccarat, and keno) always yield a "house advantage." For example, in the game of double-zero roulette, the expected casino win percentage is 5.26% on bets made on whether the outcome will be either black or red. (This percentage implies that for every \$5 bet on black or red, the casino will earn a net of about 25 cents.) It can be shown that in 100 roulette plays on black/red, the average casino win percentage is normally distributed with mean 5.26% and standard deviation 10%. Let x represent the average casino win percentage after 100 bets on black/red in double-zero roulette.

  a. Find $P(x > 0)$. (This is the probability that the casino wins money.)
  b. Find $P(5 < x < 15)$.
  c. Find $P(x < 1)$.
  d. If you observed an average casino win percentage of -25% after 100 roulette bets on black/red, what would you conclude?

CRASH

4.83 NHTSA crash safety tests. Refer to the National Highway Traffic Safety Administration (NHTSA) crash test data for new cars, presented in Exercise 2.166 (p. 102) and saved in the **CRASH** file. One of the variables measured is the severity of a driver's head injury when the car is in a head-on collision with a fixed barrier while traveling at 35 miles per hour. The more points assigned to the head injury rating, the more severe is the injury. The head injury ratings can be shown to be approximately normally distributed with a mean of 605 points and a standard deviation of 185 points. One of the crash-tested cars is randomly selected from the data, and the driver's head injury rating is observed.

a. Find the probability that the rating will fall between 500 and 700 points.

b. Find the probability that the rating will fall between 400 and 500 points.

c. Find the probability that the rating will be less than 850 points.

d. Find the probability that the rating will exceed 1,000 points.

NW **e.** What rating will only 10% of the crash-tested cars exceed?

Applying the Concepts—Intermediate

4.84 **Fitness of cardiac patients.** The physical fitness of a patient
NW is often measured by the patient's maximum oxygen uptake (recorded in milliliters per kilogram, ml/kg). The mean maximum oxygen uptake for cardiac patients who regularly participate in sports or exercise programs was found to be 24.1, with a standard deviation of 6.30. (*Adapted Physical Activity Quarterly*, Oct. 1997.) Assume that this distribution is approximately normal.

a. What is the probability that a cardiac patient who regularly participates in sports has a maximum oxygen uptake of at least 20 ml/kg?

b. What is the probability that a cardiac patient who regularly exercises has a maximum oxygen uptake of 10.5 ml/kg or lower?

c. Consider a cardiac patient with a maximum oxygen uptake of 10.5. Is it likely that this patient participates regularly in sports or exercise programs? Explain.

4.85 **Range of women's heights.** In *Chance* (Winter 2007), Yale Law School professor Ian Ayres published the results of a study he conducted with his son and daughter on whether college students could estimate a range for women's heights. The students were shown a graph of a normal distribution of heights and were asked, "The average height of women over 20 years old in the United States is 64 inches. Using your intuition, please give your best estimate of the range of heights that would include *C*% of women over 20 years old. Please make sure that the center of the range is the average height of 64 inches." The value of *C* was randomly selected as 50%, 75%, 90%, 95%, or 99% for each student surveyed.

a. Give your estimate of the range for $C = 50\%$ of women's heights.

b. Give your estimate of the range for $C = 75\%$ of women's heights.

c. Give your estimate of the range for $C = 90\%$ of women's heights.

d. Give your estimate of the range for $C = 95\%$ of women's heights.

e. Give your estimate of the range for $C = 99\%$ of women's heights.

f. The standard deviation of heights for women over 20 years old is known to be 2.6 inches. Use this information to revise your answers to parts **a–e**.

g. Which value of *C* has the most accurate estimated range? (*Note*: The researchers found that college students were most accurate for $C = 90\%$ and $C = 95\%$.)

4.86 **Alcohol, threats, and electric shocks.** A group of Florida State University psychologists examined the effects of alcohol on the reactions of people to a threat. (*Journal of Abnormal Psychology*, Vol. 107, 1998.) After obtaining a specified blood alcohol level, the psychologists placed experimental subjects in a room and threatened them with electric shocks. Using sophisticated equipment to monitor the subjects' eye movements, the psychologists recorded the startle response (measured in milliseconds) of each subject. The mean and standard deviation of the startle responses were 37.9 and 12.4, respectively. Assume that the startle response *x* for a person with the specified blood alcohol level is approximately normally distributed.

a. Find the probability that *x* is between 40 and 50 milliseconds.

b. Find the probability that *x* is less than 30 milliseconds.

c. Give an interval for *x* centered around 37.9 milliseconds so that the probability that *x* falls in the interval is .95.

d. Ten percent of the experimental subjects have startle responses above what value?

4.87 **Visually impaired students.** The *Journal of Visual Impairment & Blindness* (May–June 1997) published a study of the lifestyles of visually impaired students. Using diaries, the students kept track of several variables, including number of hours of sleep obtained in a typical day. These visually impaired students had a mean of 9.06 hours and a standard deviation of 2.11 hours. Assume that the distribution of the number of hours of sleep for this group of students is approximately normal.

a. Find the probability that a visually impaired student obtains less than 6 hours of sleep on a typical day.

b. Find the probability that a visually impaired student gets between 8 and 10 hours of sleep on a typical day.

c. Twenty percent of all visually impaired students obtain less than how many hours of sleep on a typical day?

4.88 **Length of gestation for pregnant women.** On the basis of data from the National Center for Health Statistics, N. Wetzel used the normal distribution to model the length of gestation for pregnant U.S. women (*Chance*, Spring 2001.) Gestation has a mean length of 280 days with a standard deviation of 20 days.

a. Find the probability that the length of gestation is between 275.5 and 276.5 days. (This estimate is the probability that a woman has her baby 4 days earlier than the "average" due date.)

b. Find the probability that the length of gestation is between 258.5 and 259.5 days. (This estimate is the probability that a woman has her baby 21 days earlier than the "average" due date.)

c. Find the probability that the length of gestation is between 254.5 and 255.5 days. (This estimate is the probability that a woman has her baby 25 days earlier than the "average" due date.)

d. The *Chance* article referenced a newspaper story about three sisters who all gave birth on the same day (March 11, 1998). Karralee had her baby 4 days early, Marrianne had her baby 21 days early, and Jennifer had her baby 25 days early. Use the results from parts **a–c**, to estimate the probability that three women have their babies 4, 21, and 25 days early, respectively. Assume that the births are independent events.

4.89 Forest development following wildfires. *Ecological Applications* (May 1995) published a study on the development of forests following wildfires in the Pacific Northwest. One variable of interest to the researcher was tree diameter at breast height 110 years after the fire. The population of Douglas fir trees was shown to have an approximately normal diameter distribution with $\mu = 50$ centimeters (cm) and $\sigma = 12$ cm. Find the diameter d such that 30% of the Douglas fir trees in the population have diameters that exceed d.

Applying the Concepts—Advanced

4.90 Industrial filling process. The characteristics of an industrial filling process in which an expensive liquid is injected into a container were investigated in the *Journal of Quality Technology* (July 1999). The quantity injected per container is approximately normally distributed with mean 10 units and standard deviation .2 unit. Each unit of fill costs $20. If a container contains less than 10 units (i.e., is underfilled), it must be reprocessed at a cost of $10. A properly filled container sells for $230.

a. Find the probability that a container is underfilled.

b. A container is initially underfilled and must be reprocessed. Upon refilling, it contains 10.6 units. How much profit will the company make on this container?

c. The operations manager adjusts the mean of the filling process upward to 10.5 units in order to make the probability of underfilling approximately zero. Under these conditions, what is the expected profit per container?

4.91 Box plots and the standard normal distribution. What relationship exists between the standard normal distribution and the box-plot methodology (optional Section 2.8) for describing distributions of data by means of quartiles? The answer depends on the true underlying probability distribution of the data. Assume for the remainder of this exercise that the distribution is normal.

a. Calculate the values z_L and z_U of the standard normal random variable z that correspond, respectively, to the hinges of the box plot (i.e., the lower and upper quartiles Q_L and Q_U) of the probability distribution.

b. Calculate the z values that correspond to the inner fences of the box plot for a normal probability distribution.

c. Calculate the z values that correspond to the outer fences of the box plot for a normal probability distribution.

d. What is the probability that an observation lies beyond the inner fences of a normal probability distribution? The outer fences?

e. Can you now better understand why the inner and outer fences of a box plot are used to detect outliers in a distribution? Explain.

4.92 Load on frame structures. In the *Journal of the International Association for Shell and Spatial Structures* (April 2004), Japanese environmental researchers studied the performance of truss-and-frame structures subjected to uncertain loads. The load was assumed to have a normal distribution with a mean of 20 thousand pounds. Also, the probability that the load is between 10 and 30 thousand pounds is .95. On the basis of this information, find the standard deviation of the load distribution.

4.6 Descriptive Methods for Assessing Normality

In the chapters that follow, we learn how to make inferences about the population on the basis of information contained in the sample. Several of these techniques are based on the assumption that the population is approximately normally distributed. Consequently, it will be important to determine whether the sample data come from a normal population before we can apply these techniques properly.

A number of descriptive methods can be used to check for normality. In this section, we consider the four methods summarized in the following box:

Determining whether the Data Are from an Approximately Normal Distribution

1. Construct either a histogram or stem-and-leaf display for the data, and note the shape of the graph. If the data are approximately normal, the shape of the histogram or stem-and-leaf display will be similar to the normal curve shown in Figure 4.11 (i.e., the display will be mound shaped and symmetric about the mean).

2. Compute the intervals $\bar{x} \pm s$, $\bar{x} \pm 2s$, and $\bar{x} \pm 3s$, and determine the percentage of measurements falling into each. If the data are approximately normal, the percentages will be approximately equal to 68%, 95%, and 100%, respectively.

3. Find the interquartile range IQR and standard deviation s for the sample, and then calculate the ratio IQR/s. If the data are approximately normal, then IQR/$s \approx 1.3$.

4. Construct a *normal probability plot* for the data. If the data are approximately normal, the points will fall (approximately) on a straight line.

The first two methods come directly from the properties of a normal distribution established in Section 4.5. Method 3 is based on the fact that for normal distributions, the z values corresponding to the 25th and 75th percentiles are $-.67$ and $.67$, respectively. (See Example 4.15.) Since $\sigma = 1$ for a standard normal distribution,

$$\frac{\text{IQR}}{\sigma} = \frac{Q_U - Q_L}{\sigma} = \frac{.67 - (-.67)}{1} = 1.34$$

The final descriptive method for checking normality is based on a *normal probability plot*. In such a plot, the observations in a data set are ordered from smallest to largest and are then plotted against the expected z-scores of observations calculated under the assumption that the data come from a normal distribution. When the data are, in fact, normally distributed, a linear (straight-line) trend will result. A nonlinear trend in the plot suggests that the data are nonnormal.

Definition 4.9

A **normal probability plot** for a data set is a scatterplot with the ranked data values on one axis and their corresponding expected z-scores from a standard normal distribution on the other axis. [*Note:* Computation of the expected standard normal z-scores are beyond the scope of this text. Therefore, we will rely on available statistical software packages to generate a normal probability plot.]

EXAMPLE 4.22

CHECKING FOR NORMAL DATA— EPA Estimated Gas Mileages

Problem The EPA mileage ratings on 100 cars, first presented in Chapter 2 (p. 38), are reproduced in Table 4.6. Numerical and graphical descriptive measures for the data are shown on the MINITAB and SPSS printouts presented in Figure 4.24a–c. Determine whether the EPA mileage ratings are from an approximate normal distribution.

Solution As a first check, we examine the MINITAB histogram of the data shown in Figure 4.24a. Clearly, the mileages fall into an approximately mound shaped, symmetric distribution centered around the mean of about 37 mpg. Note that a normal curve is superimposed on the figure. Therefore, check #1 in the box indicates that the data are approximately normal.

EPAGAS

TABLE 4.6 EPA Gas Mileage Ratings for 100 Cars (miles per gallon)

36.3	41.0	36.9	37.1	44.9	36.8	30.0	37.2	42.1	36.7
32.7	37.3	41.2	36.6	32.9	36.5	33.2	37.4	37.5	33.6
40.5	36.5	37.6	33.9	40.2	36.4	37.7	37.7	40.0	34.2
36.2	37.9	36.0	37.9	35.9	38.2	38.3	35.7	35.6	35.1
38.5	39.0	35.5	34.8	38.6	39.4	35.3	34.4	38.8	39.7
36.3	36.8	32.5	36.4	40.5	36.6	36.1	38.2	38.4	39.3
41.0	31.8	37.3	33.1	37.0	37.6	37.0	38.7	39.0	35.8
37.0	37.2	40.7	37.4	37.1	37.8	35.9	35.6	36.7	34.5
37.1	40.3	36.7	37.0	33.9	40.1	38.0	35.2	34.8	39.5
39.9	36.9	32.9	33.8	39.8	34.0	36.8	35.0	38.1	36.9

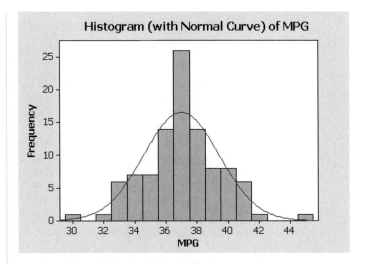

FIGURE 4.24a
MINITAB histogram for gas mileage data

FIGURE 4.24b
MINITAB descriptive statistics for gas mileage data

Descriptive Statistics: MPG

Variable	N	Mean	StDev	Minimum	Q1	Median	Q3	Maximum
MPG	100	36.994	2.418	30.000	35.625	37.000	38.375	44.900

To apply check #2, we obtain $\bar{x} = 37$ and $s = 2.4$ from the MINITAB printout of Figure 4.24b. The intervals $\bar{x} \pm s$, $\bar{x} \pm 2s$, and $\bar{x} \pm 3s$ are shown in Table 4.7, as is the percentage of mileage ratings that fall into each interval. (We obtained these results in Section 2.6, p. 67.) These percentages agree almost exactly with those from a normal distribution.

Check #3 in the box requires that we find the ratio IQR/s. From Figure 4.24b, the 25th percentile (labeled Q_1 by MINITAB) is $Q_L = 35.625$ and the 75th percentile (labeled Q_3 by MINITAB) is $Q_U = 38.375$. Then IQR $= Q_U - Q_L = 2.75$, and the ratio is

$$\frac{\text{IQR}}{s} = \frac{2.75}{2.4} = 1.15$$

Since this value is approximately equal to 1.3, we have further confirmation that the data are approximately normal.

A fourth descriptive method is to interpret a normal probability plot. An SPSS normal probability plot of the mileage data is shown in Figure 4.24c. Notice that the ordered

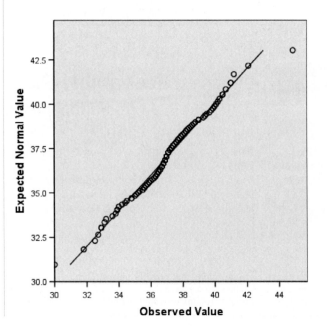

FIGURE 4.24c
SPSS normal probability plot for gas mileage data

TABLE 4.7 Describing the 100 EPA Mileage Ratings

Interval	Percentage in Interval
$\bar{x} \pm s = (34.6, 39.4)$	68
$\bar{x} \pm 2s = (32.2, 41.8)$	96
$\bar{x} \pm 3s = (29.8, 44.2)$	99

mileage values (shown on the horizontal axis) fall reasonably close to a straight line when plotted against the expected values from a normal distribution. Thus, check #4 also suggests that the EPA mileage data are approximately normally distributed.

Look Back The checks for normality given in the box are simple, yet powerful, techniques to apply, but they are only descriptive in nature. It is possible (although unlikely) that the data are nonnormal even when the checks are reasonably satisfied. Thus, we should be careful not to claim that the 100 EPA mileage ratings are, in fact, normally distributed. We can only state that it is reasonable to believe that the data are from a normal distribution.*

Now Work Exercise 4.100

As we will learn in the next chapter, several inferential methods of analysis require the data to be approximately normal. If the data are clearly nonnormal, inferences derived from the method may be invalid. Therefore, it is advisable to check the normality of the data prior to conducting any analysis.

ACTIVITY 4.2: *Keep the Change:* **Assessing Normality**

In this activity, we will once again be working with the data set *Amounts Transferred* from Activity 1.1 on page 12. This activity is designed for small groups or the entire class.

1. Pool your *Amounts Transferred* data with data from other class members or the entire class so that the pooled data set has at least 100 data items. Have someone in the group calculate the mean, standard deviation, and interquartile range of the pooled data set.

2. Divide your small group or class into four subgroups. Have each subgroup apply one of the four methods outlined in this section to assess the normality of the pooled data.

3. Present the findings to the entire class.

Keep the data set from this activity for use in other activities.

Normal Probability Plot

Using the TI-84/T1-83 Graphing Calculator

Graphing a Normal Probability Plot

Step 1 *Enter the data.*
Press **STAT** and select **1:Edit**.
Note: If the list already contains data, clear the old data. Use the up arrow to highlight '**L1**'. Press **CLEAR ENTER**.
Use the **ARROW** and **ENTER** keys to enter the data set into **L1**.

---

*Statistical tests of normality that provide a measure of reliability for the inference are available. However, these tests tend to be very sensitive to slight departures from normality (i.e., they tend to reject the hypothesis of normality for any distribution that is not perfectly symmetrical and mound shaped). Consult the references (see especially Ramsey & Ramsey, 1990) if you want to learn more about these tests.

Step 2   *Set up the Normal Probability Plot.*
Press **Y =** , and **CLEAR** all functions from the Y registers.
Press **2nd** and press **Y =** for **STAT PLOT**.
Press **1** for **Plot 1**.
Set the cursor so that **ON** is flashing.
For **Type**, use the **ARROW** and **ENTER** keys
to highlight and select the last graph in the
bottom row.
For **Data List**, choose the column containing the
data (in most cases, L1). (*Note*: Press **2nd 1**
for **L1**.)
For **Data Axis**, choose **X** and press **ENTER**.

Step 3   *View Plot.*
Press **ZOOM 9**
Your data will be displayed against the expected *z*-scores from a normal
distribution. If you see a "generally" linear relationship, your data set is
approximately normal.

Example   Using a normal probability plot, test whether or not the following data are
normally distributed:

9.7	93.1	33.0	21.2	81.4	51.1
43.5	10.6	12.8	7.8	18.1	12.7

The screen that follows shows the normal probability plot. There is a noticeable
curve to the plot, indicating that the data set is not normally distributed.

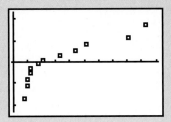

STATISTICS IN ACTION REVISITED

Assessing whether the Normal Distribution Is Appropriate for Modeling the Super Weapon Hit Data

In *Statistics in Action Revisited* in Section 4.5, we used the normal distribution to find the probability that a single flechette from a super weapon that shoots 1,100 flechettes at once hits one of three targets at 500 meters. Recall that for three range tests, the weapon was always aimed at the center target (i.e., the specification mean was set at $\mu = 5$ feet), but the specification standard deviation was varied at $\sigma = 1$ foot, $\sigma = 2$ feet, and $\sigma = 4$ feet. Table SIA4.1 shows the calculated normal probabilities of hitting the three targets for the different values of σ, as well as the actual results of the three range tests. (Recall that the actual data are saved in the **MOAGUN** file.) You can see that the proportion of the 1,100 flechettes that actually hit each target—called the hit ratio—agrees very well with the estimated probability of a hit derived from the normal distribution.

(continued)

TABLE SIA4.1 Summary of Normal Probability Calculations and Actual Range Test Results

Target	Specification	Normal Probability	Actual Number of Hits	Hit Ratio (Hits/1,100)
LEFT (−1 to 1)	$\sigma = 1$	.0000	0	.000
	$\sigma = 2$	.0214	30	.027
	$\sigma = 4$	.0919	73	.066
MIDDLE (4 to 6)	$\sigma = 1$	.6826	764	.695
	$\sigma = 2$	.3820	409	.372
	$\sigma = 4$	.1974	242	.220
RIGHT (9 to 11)	$\sigma = 1$	.0000	0	.000
	$\sigma = 2$	.0214	23	.021
	$\sigma = 4$	.0919	93	.085

Consequently, it appears that our assumption that the horizontal hit measurements are approximately normally distributed is reasonably satisfied. Further evidence that this assumption is satisfied is provided by the MINITAB histograms of the horizontal hit measurements shown in Figures SIA4.3a–c. The normal curves superimposed on the histograms fit the data very well.

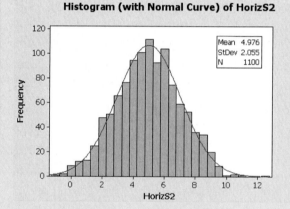

FIGURE SIA4.3a

MINITAB histogram for the horizontal hit measurements when $\sigma = 1$

FIGURE SIA4.3b

MINITAB histogram for the horizontal hit measurements when $\sigma = 2$

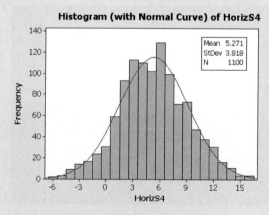

FIGURE SIA4.3c

MINITAB histogram for the horizontal hit measurements when $\sigma = 4$

Exercises 4.93–4.111

Understanding the Principles

4.93 Why is it important to check whether the sample data come from a normal population?

4.94 Give four methods for determining whether the sample data come from a normal population.

4.95 If a population data set is normally distributed, what is the proportion of measurements you would expect to fall within the following intervals?
a. $\mu \pm \sigma$ **b.** $\mu \pm 2\sigma$ **c.** $\mu \pm 3\sigma$

4.96 What is a normal probability plot and how is it used?

Learning the Mechanics

4.97 Normal probability plots for three data sets are shown below. Which plot indicates that the data are approximately normally distributed?

4.98 Consider a sample data set with the following summary statistics: $s = 95, Q_L = 72$, and $Q_U = 195$.
a. Calculate IQR.
b. Calculate IQR/s.
c. Is the value of IQR/s approximately equal to 1.3? What does this imply?

4.99 Examine the sample data in the following table:

LM4_99

32	48	25	135	53	37	5	39	213	165
109	40	1	146	39	25	21	66	64	57
197	60	112	10	155	134	301	304	107	82
35	81	60	95	401	308	180	3	200	59

a. Construct a stem-and-leaf plot to assess whether the data are from an approximately normal distribution.
b. Find the values of Q_L, Q_U, and s for the sample data.

c. Use the results from part **b** to assess the normality of the data.
d. Generate a normal probability plot for the data, and use it to assess whether the data are approximately normal.

4.100 Examine the sample data in the following table:
NW

LM54_100

5.9	5.3	1.6	7.4	8.6	3.2	2.1
4.0	7.3	8.4	5.9	6.7	4.5	6.3
6.0	9.7	3.5	3.1	4.3	3.3	8.4
4.6	8.2	6.5	1.1	5.0	9.4	6.4

a. Construct a stem-and-leaf plot to assess whether the data are from an approximately normal distribution.
b. Compute s for the sample data.
c. Find the values of Q_L and Q_U and the value of s from part **b** to assess whether the data come from an approximately normal distribution.
d. Generate a normal probability plot for the data, and use it to assess whether the data are approximately normal.

Applying the Concepts—Basic

WPOWER50

4.101 Most powerful American women. Refer to the *Fortune* (Nov. 14, 2005) list of the 50 most powerful women in America. In Exercise 4.79 (p. 211), you assumed that the ages (in years) of these women are approximately normally distributed. A MINITAB printout with summary statistics for the age distribution is reproduced at the bottom of the page.
a. Use the relevant statistics on the printout to find the interquartile range IQR.
b. Locate the value of the standard deviation s on the printout.
c. Use the results from parts **a** and **b** to demonstrate that the age distribution is approximately normal.
d. Construct a relative frequency histogram for the age data. Use this graph to support your assumption of normality.

4.102 Galaxy velocity study. Refer to the *Astronomical Journal* (July 1995) study of galaxy velocities, presented in Exercise 2.100 (p. 74). A histogram of the velocities of 103 galaxies located in a particular cluster named A2142 is

Plots for Exercise 4.97

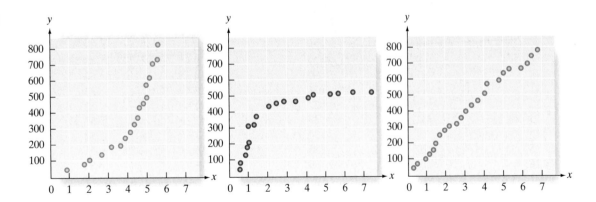

MINITAB Output for Exercise 4.101

Descriptive Statistics: AGE

Variable	N	Mean	StDev	Minimum	Q1	Median	Q3	Maximum
AGE	50	49.880	5.275	39.000	47.000	49.500	53.000	64.000

reproduced below. Comment on whether or not the galaxy velocities are approximately normally distributed.

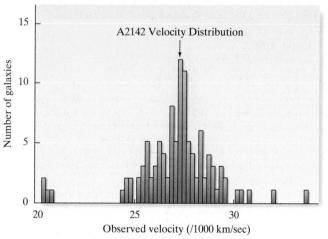

Source: Oegerle, W. R., Hill, J. M., and Fitchett, M. J. "Observations of high dispersion clusters of galaxies: Constraints on cold dark matter," *The Astronomical Journal,* Vol. 110, No. 1, July 1995, p. 37 (Figure 1).

4.103 **Software file updates.** Software configuration management was used to monitor a software engineering team's performance at Motorola, Inc. (*Software Quality Professional,* Nov. 2004.) One of the variables of interest was the number of updates to a file changed because of a problem report. Summary statistics for $n = 421$ files yielded the following results: $\bar{x} = 4.71$, $s = 6.09$, $Q_L = 1$, and $Q_U = 6$. Are these data approximately normally distributed? Explain.

4.104 **Dart-throwing errors.** How accurate are you at the game of darts? Researchers at Iowa State University attempted to develop a probability model for dart throws (*Chance,* Summer 1997). For each of 590 throws made at certain targets on a dart board, the distance from the dart to the target point was measured (to the nearest millimeter). The error distribution for the dart throws is described by the frequency table shown at the bottom of the page.

a. Construct a histogram of the data. Is the error distribution for the dart throws approximately normal?

b. Descriptive statistics for the distances from the target for the 590 throws are as follows:

$$\bar{x} = 24.4 \text{ mm}$$
$$s = 12.8 \text{ mm}$$
$$Q_L = 14 \text{ mm}$$
$$Q_U = 34 \text{ mm}$$

Use this information to decide whether the error distribution is approximately normal.

c. Construct a normal probability plot of the data, and use it to assess whether the error distribution is approximately normal.

4.105 **Estimating glacier elevations.** Digital elevation models (DEMs) are now used to estimate elevations and slopes of remote regions. In *Arctic, Antarctic, and Alpine Research* (May 2004), geographers analyzed reading errors from maps produced by DEMs. Two readers of a DEM map of White Glacier (in Canada) estimated elevations at 400 points in the area. The difference between the elevation estimates of the two readers had a mean of $\mu = .28$ meter and a standard deviation of $\sigma = 1.6$ meters. A histogram of the difference (with a normal histogram superimposed on the graph) is shown on p. 257.

a. On the basis of the histogram, the researchers concluded that the difference between elevation estimates is not normally distributed. Why?

b. Will the interval $\mu \pm 2\sigma$ contain more than 95%, exactly 95%, or less than 95% of the 400 elevation differences? Explain.

DARTS

Distance from Target (mm)	Frequency	Distance from Target (mm)	Frequency	Distance from Target (mm)	Frequency
0	3	21	15	41	7
2	2	22	19	42	7
3	5	23	13	43	11
4	10	24	11	44	7
5	7	25	9	45	5
6	3	26	21	46	3
7	12	27	16	47	1
8	14	28	18	48	3
9	9	29	9	49	1
10	11	30	14	50	4
11	14	31	13	51	4
12	13	32	11	52	3
13	32	33	11	53	4
14	19	34	11	54	1
15	20	35	16	55	1
16	9	36	13	56	4
17	18	37	5	57	3
18	23	38	9	58	2
19	25	39	6	61	1
20	21	40	5	62	2
				66	1

Source: Stern, H. S., and Wilcox, W. "Shooting Darts." *Chance,* Vol. 10, No. 3, Summer 1997, p. 17 (adapted from Figure 2).

Histogram for Exercise 4.105

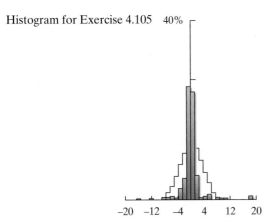

Source: Cogley. J. G., and Jung-Rothenhausler, F. "Uncertainty in digital elevation models of Axel Heiberg Island. Arctic Canada," *Arctic, Antarctic, and Alpine Research,* Vol. 36, No. 2, May 2004 (Figure 3).

Applying the Concepts—Intermediate

HABITAT

4.106 Habitats of endangered species. An evaluation of the habitats of endangered salmon species was performed in *Conservation Ecology* (December 2003).The researchers identified 734 sites (habitats) for Chinook, coho, or steelhead salmon species in Oregon and assigned a habitat quality score to each site. (Scores range from 0 to 36 points, with lower scores indicating poorly maintained or degraded habitats.) The data are saved in the **HABITAT** file. Give your opinion on whether the habitat quality score is normally distributed.

MLB2006-AL & MLB2006-NL

4.107 Baseball batting averages. Major League Baseball (MLB) has two leagues: the American League (AL), which utilizes the designated hitter (DH) to bat for the pitcher, and the National League (NL), which does not allow the DH. A player's batting average is computed by dividing the player's total number of hits by his official number of at bats. The batting averages for all AL and NL players with at least 100 official at bats during the 2006 season are stored in the **MLB2006-AL** and **MLB2006-NL** files, respectively. De-termine whether each batting average distribution is approximately normal.

PGADRIVER

4.108 Ranking the driving performance of professional golfers. Refer to *The Sport Journal* (Winter 2007) article on a new method for ranking the driving performance of PGA golfers, presented in Exercise 2.60 (p. 59). Recall that the method incorporates a golfer's average driving distance (yards) and driving accuracy (percentage of drives that land in the fairway) into a driving performance index. Data on these three variables for the top 40 PGA golfers are saved in the **PGADRIVER** file. Determine which of the variables—driving distance, driving accuracy, and driving performance index—are approximately normally distributed.

CRASH

4.109 NHTSA crash tests. Refer to the National Highway Traffic Safety Administration (NHTSA) crash test data for new cars. In Exercise 4.83 (p. 211), you assumed that the driver's head injury rating is approximately normally distributed. Apply the methods of this chapter to the data saved in the **CRASH** file to support that assumption.

SHIPSANIT

4.110 Cruise ship sanitation scores. Refer to the data on the May 2006 sanitation scores for 151 cruise ships, presented in Exercise 2.32 (p. 46). The data are saved in the **SHIPSANIT** file. Assess whether the sanitation scores are approximately normally distributed.

Applying the Concepts—Advanced

4.111 Language skills study. Refer to the *Applied Psycholinguistics* (June 1998) study of language skills in young children, presented in Exercise 2.98 (p. 74). Recall that the mean sentence complexity score (measured on a 0- to 48-point scale) of low-income children was 7.62, with a standard deviation of 8.91. Demonstrate why the distribution of sentence complexity scores for low-income children is unlikely to be normally distributed.

4.7 Approximating a Binomial Distribution with a Normal Distribution (Optional)

When the discrete binomial random variable (Section 4.3) can assume a large number of values, the calculation of its probabilities may become tedious. To contend with this problem, we provide tables in Appendix A that give the probabilities for some values of n and p, but these tables are by necessity incomplete. Recall that the binomial probability table (Table II) can be used only for $n = 5, 6, 7, 8, 9, 10, 15, 20,$ or 25. To deal with this limitation, we seek approximation procedures for calculating the probabilities associated with a binomial probability distribution.

When n is large, a normal probability distribution may be used to provide a good approximation to the probability distribution of a binomial random variable. To show how this approximation works, we refer to Example 4.12, in which we used the binomial

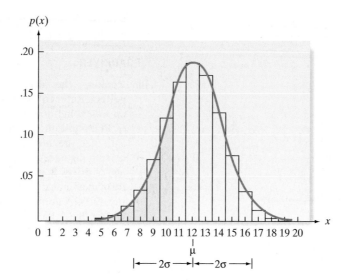

FIGURE 4.25

Binomial distribution for $n = 20$, $p = .6$ and normal distribution with $\mu = 12$, $\sigma = 2.2$

distribution to model the number x of 20 eligible voters who favor a candidate. We assumed that 60% of all the eligible voters favored the candidate. The mean and standard deviation of x were found to be $\mu = 12$ and $\sigma = 2.2$, respectively. The binomial distribution for $n = 20$ and $p = .6$ is shown in Figure 4.25, and the approximating normal distribution with mean $\mu = 12$ and standard deviation $\sigma = 2.2$ is superimposed.

As part of Example 4.12, we used Table II to find the probability that $x \le 10$. This probability, which is equal to the sum of the areas contained in the rectangles (shown in Figure 4.25) that correspond to $p(0), p(1), p(2), \ldots, p(10)$, was found to equal .245. The portion of the normal curve that would be used to approximate the area $p(0) + p(1) + p(2) + \cdots + p(10)$ is highlighted in Figure 4.25. Note that this highlighted area lies to the left of 10.5 (not 10), so we may include all of the probability in the rectangle corresponding to $p(10)$. Because we are approximating a discrete distribution (the binomial) with a continuous distribution (the normal), we call the use of 10.5 (instead of 10 or 11) a **correction for continuity.** That is, we are correcting the discrete distribution so that it can be approximated by the continuous one. The use of the correction for continuity leads to the calculation of the following standard normal z-value:

$$z = \frac{x - \mu}{\sigma} = \frac{10.5 - 12}{2.2} = -.68$$

Using Table III, we find the area between $z = 0$ and $z = .68$ to be .2517. Then the probability that x is less than or equal to 10 is approximated by the area under the normal distribution to the left of 10.5, shown highlighted in Figure 4.25. That is,

$$P(x \le 10) \approx P(z \le -.68) = .5 - P(-.68 < z \le 0) = .5 - .2517 = .2483$$

The approximation differs only slightly from the exact binomial probability, .245. Of course, when tables of exact binomial probabilities are available, we will use the exact value rather than a normal approximation.

The normal distribution will not always provide a good approximation to binomial probabilities. The following is a useful rule of thumb to determine when n is large enough for the approximation to be effective: *The interval $\mu \pm 3\sigma$ should lie within the range of the binomial random variable x (i.e., from 0 to n) in order for the normal approximation to be adequate.* The rule works well because almost all of the normal distribution falls within three standard deviations of the mean, so if this interval is contained within the range of x values, there is "room" for the normal approximation to work.

As shown in Figure 4.26a for the preceding example with $n = 20$ and $p = .6$, the interval $\mu \pm 3\sigma = 12 + 3(2.2) = (5.4, 18.6)$ lies within the range from 0 to 20. However, if we were to try to use the normal approximation with $n = 10$ and $p = .1$, the interval $\mu \pm 3\sigma$ becomes $1 \pm 3(.95)$, or $(-1.85, 3.85)$. As shown in Figure 4.26b, this interval is not contained within the range of x, since $x = 0$ is the lower bound for a binomial random variable. Note in Figure 4.26b that the normal distribution will not "fit" in the range of x; therefore, it will not provide a good approximation to the binomial probabilities.

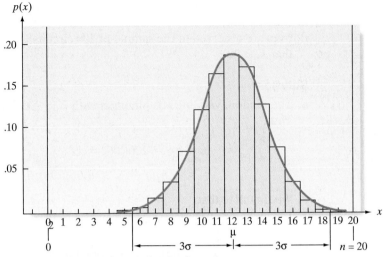

a. $n = 20$, $p = .6$: Normal approximation is good

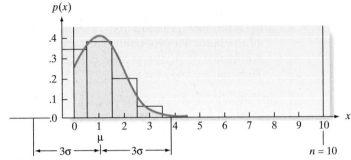

FIGURE 4.26

Rule of thumb for normal approximation to binomial probabilities

b. $n = 10$, $p = .1$: Normal approximation is poor

Biography

**ABRAHAM DE MOIVRE (1667–1754)—
Advisor to Gamblers**

French-born mathematician Abraham de Moivre moved to London when he was 21 years old to escape religious persecution. In England, he earned a living first as a traveling teacher of mathematics and then as an advisor to gamblers, underwriters, and annuity brokers. De Moivre's major contributions to probability theory are contained in two of his books: *The Doctrine of Chances* (1718) and *Miscellanea Analytica* (1730). In these works, he defines statistical independence, develops the formula for the normal probability distribution, and derives the normal curve as an approximation to the binomial distribution. Despite his eminence as a mathematician, de Moivre died in poverty. He is famous for using an arithmetic progression to predict the day of his death.

EXAMPLE 4.23

APPROXIMATING A BINOMIAL PROBABILITY WITH THE NORMAL DISTRIBUTION—

Lot Acceptance Sampling

Problem One problem with any product that is mass produced (e.g., a graphing calculator) is quality control. The process must be monitored or audited to be sure that the output of the process conforms to requirements. One monitoring method is *lot acceptance sampling*, in which items being produced are sampled at various stages of the production process and are carefully inspected. The lot of items from which the sample is drawn is then accepted or rejected on the basis of the number of defectives in the sample. Lots that are accepted may be sent forward for further processing or may be shipped to customers; lots that are rejected may be reworked or scrapped. For example, suppose a manufacturer of calculators chooses 200 stamped circuits from the day's production and determines x, the number of defective circuits in the sample. Suppose that up to a 6% rate of defectives is considered acceptable for the process.

a. Find the mean and standard deviation of x, assuming that the rate of defectives is 6%.
b. Use the normal approximation to determine the probability that 20 or more defectives are observed in the sample of 200 circuits (i.e., find the approximate probability that $x \geq 20$).

Solution

a. The random variable x is binomial with $n = 200$ and the fraction defective $p = .06$. Thus,

$$\mu = np = 200(.06) = 12$$
$$\sigma = \sqrt{npq} = \sqrt{200(.06)(.94)} = \sqrt{11.28} = 3.36$$

We first note that

$$\mu \pm 3\sigma = 12 \pm 3(3.36) = 12 \pm 10.08 = (1.92, 22.08)$$

lies completely within the range from 0 to 200. Therefore, a normal probability distribution should provide an adequate approximation to this binomial distribution.

b. By the rule of complements, $P(x \geq 20) = 1 - P(x \leq 19)$. To find the approximating area corresponding to $x \leq 19$, refer to Figure 4.27. Note that we want to include all the binomial probability histograms from 0 to 19, inclusive. Since the event is of the form $x \leq a$, the proper correction for continuity is $a + .5 = 19 + .5 = 19.5$. Thus, the z value of interest is

$$z = \frac{(a + .5) - \mu}{\sigma} = \frac{19.5 - 12}{3.36} = 2.23$$

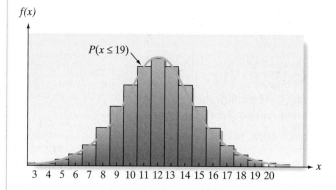

FIGURE 4.27

Normal approximation to the binomial distribution with $n = 200$, $p = .06$

Referring to Table III in Appendix A, we find that the area to the right of the mean, 0, corresponding to $z = 2.23$ (see Figure 4.28) is .4871. So the area $A = P(z \leq 2.23)$ is

$$A = .5 + .4871 = .9871$$

Thus, the normal approximation to the binomial probability we seek is

$$P(x \geq 20) = 1 - P(x \leq 19) \approx 1 - .9871 = .0129$$

In other words, *if, in fact, the true fraction of defectives is .06*, then the probability that 20 or more defectives will be observed in a sample of 200 circuits is extremely small.

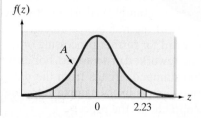

FIGURE 4.28

Standard normal distribution

Look Back If the manufacturer observes $x \geq 20$, the likely reason is that the process is producing more than the acceptable 6% defectives. The lot acceptance sampling procedure is another example of using the rare-event approach to make inferences.

Now Work Exercise 4.115

The steps for approximating a binomial probability by a normal probability are given in the following box:

Using a Normal Distribution to Approximate Binomial Probabilities

1. After you have determined n and p for the binomial distribution, calculate the interval

$$\mu \pm 3\sigma = np \pm 3\sqrt{npq}$$

If the interval lies in the range from 0 to n, the normal distribution will provide a reasonable approximation to the probabilities of most binomial events.

2. Express the binomial probability to be approximated in the form $P(x \leq a)$ or $P(x \leq b) - P(x \leq a)$. For example,

$$P(x < 3) = P(x \leq 2)$$
$$P(x \geq 5) = 1 - P(x \leq 4)$$
$$P(7 \leq x \leq 10) = P(x \leq 10) - P(x \leq 6)$$

3. For each value of interest, a, the correction for continuity is $(a + .5)$ and the corresponding standard normal z value is

$$z = \frac{(a + .5) - \mu}{\sigma} \quad \text{(see Figure 4.29)}$$

4. Sketch the approximating normal distribution and shade the area corresponding to the probability of the event of interest, as in Figure 4.29. Verify that the rectangles you have included in the shaded area correspond to the probability you wish to approximate. Use Table IV and the z value(s) you calculated in step 3 to find the shaded area. This is the approximate probability of the binomial event.

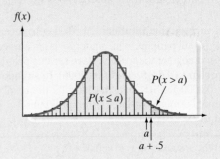

FIGURE 4.29

Approximating binomial probabilities by normal probabilities

Exercises 4.112–4.130

Understanding the Principles

4.112 For large n (say, $n = 100$), why is it advantageous to use the normal distribution to approximate a binomial probability?

4.113 Why do we need a correction for continuity when approximating a binomial probability with the normal distribution?

Learning the Mechanics

4.114 Assume that x is a binomial random variable with n and p as specified in parts **a–f** that follow. For which cases would

it be appropriate to use a normal distribution to approximate the binomial distribution?

    a. $n = 100, p = .01$ **b.** $n = 20, p = .6$
    c. $n = 10, p = .4$ **d.** $n = 1,000, p = .05$
    e. $n = 100, p = .8$ **f.** $n = 35, p = .7$

4.115 Suppose x is a binomial random variable with $p = .4$ and **NW** $n = 25$.

    a. Would it be appropriate to approximate the probability distribution of x with a normal distribution? Explain.

b. Assuming that a normal distribution provides an adequate approximation to the distribution of x, what are the mean and variance of the approximating normal distribution?

c. Use Table II of Appendix A to find the exact value of $P(x \geq 9)$.

d. Use the normal approximation to find $P(x \geq 9)$.

4.116 Assume that x is a binomial random variable with $n = 25$ and $p = .5$. Use Table II of Appendix A and the normal approximation to find the exact and approximate values, respectively, of the following probabilities:

a. $P(x \leq 11)$

b. $P(x \geq 16)$

c. $P(8 \leq x \leq 16)$

4.117 Assume that x is a binomial random variable with $n = 100$ and $p = .40$. Use a normal approximation to find the following:

a. $P(x \leq 35)$ **b.** $P(40 \leq x \leq 50)$

c. $P(x \geq 38)$

4.118 Assume that x is a binomial random variable with $n = 1,000$ and $p = .50$. Find each of the following probabilities:

a. $P(x > 500)$ **b.** $P(490 \leq x < 500)$

c. $P(x > 550)$

Applying the Concepts—Basic

4.119 **Analysis of bottled water.** Refer to the *Scientific American* (July 2003) report on whether bottled water is really purified water, presented in Exercise 4.52 (p. 195). Recall that the Natural Resources Defense Council found that 25% of bottled-water brands fill their bottles with just tap water. In a random sample of 65 bottled-water brands, let x be the number that contain tap water.

a. Find the mean of x.

b. Find the standard deviation of x.

c. Find the z-score for the value $x = 20$.

d. Find the approximate probability that 20 or more of the 65 sampled bottled-water brands will contain tap water.

4.120 **LASIK surgery complications.** According to recent studies, 1% of all patients who undergo laser surgery (i.e., LASIK) to correct their vision have serious post-laser vision problems (*All About Vision*, 2006). In a random sample of 100,000 LASIK patients, let x be the number who experience serious post-laser vision problems.

a. Find $E(x)$.

b. Find Var(x).

c. Find the z-score for $x = 950$.

d. Find the approximate probability that fewer than 950 patients in a sample of 100,000 will experience serious post-laser vision problems.

4.121 **Melanoma deaths.** According to the *American Cancer Society*, melanoma, a form of skin cancer, kills 15% of Americans who suffer from the disease each year. Consider a sample of 10,000 melanoma patients.

a. What are the expected value and variance of x, the number of the 10,000 melanoma patients who die of the affliction this year?

b. Find the probability that x will exceed 1,600 patients per year.

c. Would you expect x, the number of patients dying of melanoma, to exceed 6,500 in any single year? Explain.

4.122 **Caesarian birth study.** In Exercise 4.54 (p. 195), you learned that 29% of all births in the United States occur by Caesarian section each year. (*National Vital Statistics Reports*, Sep. 2006.) In a random sample of 1,000 births this year, let x be the number that occur by Caesarian section.

a. Find the mean of x. (This value should agree with your answer to Exercise 4.54**a**.)

b. Find the standard deviation of x. (This value should agree with your answer to Exercise 4.54**b**.)

c. Find the z-score for the value $x = 200.5$.

d. Find the approximate probability that the number of Caesarian sections in a sample of 1,000 births is less than or equal to 200.

Applying the Concepts—Intermediate

4.123 **Ecotoxicological survival study.** The *Journal of Agricultural, Biological and Environmental Statistics* (Sep. 2000) gave an evaluation of the risk posed by hazardous pollutants. In the experiment, guppies (all the same age and size) were released into a tank of natural seawater polluted with the pesticide dieldrin and the number of guppies surviving after five days was determined. The researchers estimated that the probability of any single guppy surviving is .60. If 300 guppies are released into the polluted tank, estimate the probability that fewer than 100 guppies survive after five days.

4.124 **Parents who condone spanking.** In Exercise 4.55 (p. 195), you learned that 60% of parents with young children condone spanking their child as a regular form of punishment. (*Tampa Tribune*, Oct. 5, 2000.) A child psychologist with 150 parent clients claims that no more than 20 of the parents condone spanking. Do you believe this claim? Explain.

4.125 **Defects in semiconductor wafers.** The computer chips in notebook and laptop computers are produced from semiconductor wafers. Certain semiconductor wafers are exposed to an environment that generates up to 100 possible defects per wafer. The number of defects per wafer, x, was found to follow a binomial distribution if the manufacturing process is stable and generates defects that are randomly distributed on the wafers. (*IEEE Transactions on Semiconductor Manufacturing*, May 1995.) Let p represent the probability that a defect occurs at any one of the 100 points of the wafer. For each of the following cases, determine whether the normal approximation can be used to characterize x:

a. $p = .01$ **b.** $p = .50$ **c.** $p = .90$

4.126 **Victims of domestic abuse.** In Exercise 4.60 (p. 196), you learned that some researchers believe that one in every three women has been a victim of domestic abuse. (*Annals of Internal Medicine*, Nov. 1995.)

a. For a random sample of 150 women, what is the approximate probability that more than half are victims of domestic abuse?

b. For a random sample of 150 women, what is the approximate probability that fewer than 50 are victims of domestic abuse?

c. Would you expect to observe fewer than 30 domestically abused women in a sample of 150? Explain.

4.127 **Fungi in beech forest trees.** Refer to the *Applied Ecology and Environmental Research* (Vol. 1, 2003) study of beech trees damaged by fungi, presented in Exercise 4.57 (p. 195).

Recall that the researchers found that 25% of the beech trees in east central Europe have been damaged by fungi. In an east central European forest with 200 beech trees, how likely is it that more than half of the trees have been damaged by fungi?

Applying the Concepts—Advanced

4.128 Body fat in men. The percentage of fat in the bodies of American men is an approximately normal random variable with mean equal to 15% and standard deviation equal to 2%.
 a. If these values were used to describe the body fat of men in the U.S. Army, and if a measure of 20% or more body fat characterizes the person as obese, what is the approximate probability that a random sample of 10,000 soldiers will contain fewer than 50 who would actually be characterized as obese?
 b. If the army actually were to check the percentage of body fat for a random sample of 10,000 men, and if only 30 contained 20% (or higher) body fat, would you conclude that the army was successful in reducing the percentage of obese men below the percentage in the general population? Explain your reasoning.

4.129 Luggage inspection at Newark airport. According to *New Jersey Business* (Feb. 1996), Newark International Airport's new terminal handles an average of 3,000 international passengers an hour, but is capable of handling twice that number. Also, 80% of arriving international passengers pass through without their luggage being inspected, and the remainder are detained for inspection. The inspection facility can handle 600 passengers an hour without unreasonable delays for the travelers.
 a. When international passengers arrive at the rate of 1,500 per hour, what is the expected number of passengers who will be detained for luggage inspection?
 b. In the future, it is expected that as many as 4,000 international passengers will arrive per hour. When that occurs, what is the expected number of passengers who will be detained for luggage inspection?
 c. Refer to part **b**. Find the approximate probability that more than 600 international passengers will be detained for luggage inspection. (This is also the probability that travelers will experience unreasonable luggage inspection delays.)

4.130 Waiting time at a doctor's office. The median time a patient waits to see a doctor in a large clinic is 20 minutes. On a day when 150 patients visit the clinic, what is the approximate probability that
 a. More than half will wait more than 20 minutes?
 b. More than 85 will wait more than 20 minutes?
 c. More than 60, but fewer than 90, will wait more than 20 minutes?

4.8 Sampling Distributions

In previous sections, we assumed that we knew the probability distribution of a random variable, and using this knowledge, we were able to compute the mean, variance, and probabilities associated with the random variable. However, in most practical applications, the true mean and standard deviation are unknown quantities that have to be estimated. Numerical quantities that describe probability distributions are called *parameters*. Thus, p, the probability of a success in a binomial experiment, and μ and σ, the mean and standard deviation, respectively, of a normal distribution, are examples of parameters.

> **Definition 4.10**
>
> A **parameter** is a numerical descriptive measure of a population. Because it is based on the observations in the population, its value is almost always unknown.

We have also discussed the sample mean $\bar{x}$, sample variance s^2, sample standard deviation s, and the like, which are numerical descriptive measures calculated from the sample. (See Table 4.8 for a list of the statistics covered so far in this text.) We will often use the information contained in these *sample statistics* to make inferences about the parameters of a population.

TABLE 4.8 List of Population Parameters and Corresponding Sample Statistics

	Population Parameter	Sample Statistic
Mean:	μ	$\bar{x}$
Variance:	σ^2	s^2
Standard deviation:	σ	s
Binomial proportion:	p	$\hat{p}$

> ### Definition 4.11
>
> A **sample statistic** is a numerical descriptive measure of a sample. It is calculated from the observations in the sample.

Note that the term *statistic* refers to a *sample* quantity and the term *parameter* refers to a *population* quantity.

Before we can show you how to use sample statistics to make inferences about population parameters, we need to be able to evaluate their properties. Does one sample statistic contain more information than another about a population parameter? On what basis should we choose the "best" statistic for making inferences about a parameter? For example, if we want to estimate a parameter of a population—say, the population mean μ—we can use a number of sample statistics for our estimate. Two possibilities are the sample mean $\bar{x}$ and the sample median M. Which of these do you think will provide a better estimate of μ?

Before answering this question, consider the following example: Toss a fair die, and let x equal the number of dots showing on the up face. Suppose the die is tossed three times, producing the sample measurements 2, 2, 6. The sample mean is then $\bar{x} = 3.33$, and the sample median is $M = 2$. Since the population mean of x is $\mu = 3.5$, you can see that, for this sample of three measurements, the sample mean $\bar{x}$ provides an estimate that falls closer to μ than does the sample median (see Figure 4.29a). Now suppose we toss the die three more times and obtain the sample measurements 3, 4, 6. Then the mean and median of this sample are $\bar{x} = 4.33$ and $M = 4$, respectively. This time M is closer to μ. (See Figure 4.29b.)

FIGURE 4.29

Comparing the sample mean ($\bar{x}$) and sample median (M) as estimators of the population mean (μ)

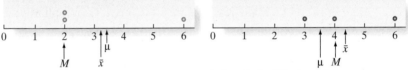

a. Sample 1: $\bar{x}$ is closer than M to μ b. Sample 2: M is closer than $\bar{x}$ to μ

This simple example illustrates an important point: Neither the sample mean nor the sample median will *always* fall closer to the population mean. Consequently, we cannot compare these two sample statistics or, in general, any two sample statistics on the basis of their performance with a single sample. Instead, we need to recognize that sample statistics are themselves random variables, because different samples can lead to different values for the sample statistics. As random variables, sample statistics must be judged and compared on the basis of their probability distributions (i.e., the *collection* of values and associated probabilities of each statistic that would be obtained if the sampling experiment were repeated a *very large number of times*). We will illustrate this concept with another example.

Suppose it is known that in a certain part of Canada the daily high temperature recorded for all past months of January has a mean $\mu = 10°F$ and a standard deviation $\sigma = 5°F$. Consider an experiment consisting of randomly selecting 25 daily high temperatures from the records of past months of January and calculating the sample mean $\bar{x}$. If this experiment were repeated a very large number of times, the value of $\bar{x}$ would vary from sample to sample. For example, the first sample of 25 temperature measurements might have a mean $\bar{x} = 9.8$, the second sample a mean $\bar{x} = 11.4$, the third sample a mean $\bar{x} = 10.5$, etc. If the sampling experiment were repeated a very large number of times, the resulting histogram of sample means would be approximately the probability distribution of $\bar{x}$. If $\bar{x}$ is a good estimator of μ, we would expect the values of $\bar{x}$ to cluster around μ as shown in Figure 4.30. This probability distribution is called a *sampling distribution*, because it is generated by repeating a sampling experiment a very large number of times.

> ### Definition 4.12
>
> The **sampling distribution** of a sample statistic calculated from a sample of n measurements is the probability distribution of the statistic.

In actual practice, the sampling distribution of a statistic is obtained mathematically or (at least approximately) by simulating the sample on a computer, using a procedure similar to that just described.

If $\bar{x}$ has been calculated from a sample of $n = 25$ measurements selected from a population with mean $\mu = 10$ and standard deviation $\sigma = 5$, the sampling distribution (Figure 4.30) provides information about the behavior of $\bar{x}$ in repeated sampling.

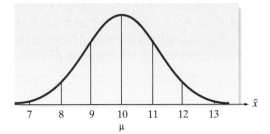

FIGURE 4.30

Sampling distribution for $\bar{x}$ based on a sample of $n = 25$ measurements

For example, the probability that you will draw a sample of 25 measurements and obtain a value of $\bar{x}$ in the interval $9 \leq \bar{x} \leq 10$ will be the area under the sampling distribution over that interval.

Since the properties of a statistic are typified by its sampling distribution, it follows that, to compare two sample statistics, you compare their sampling distributions. For example, if you have two statistics, A and B, for estimating the same parameter (for purposes of illustration, suppose the parameter is the population variance σ^2), and if their sampling distributions are as shown in Figure 4.31, you would prefer statistic A over statistic B. You would do so because the sampling distribution for statistic A centers over σ^2 and has less spread (variation) than the sampling distribution for statistic B. Then, when you draw a single sample in a practical sampling situation, the probability is higher that statistic A will fall nearer σ^2.

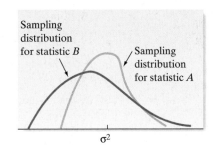

FIGURE 4.31

Two sampling distributions for estimating the population variance σ^2

Remember that, in practice, we will not know the numerical value of the unknown parameter σ^2, so we will not know whether statistic A or statistic B is closer to σ^2 for a particular sample. We have to rely on our knowledge of the theoretical sampling distributions to choose the best sample statistic and then use it sample after sample. The procedure for finding the sampling distribution for a statistic is demonstrated in Example 4.24.

EXAMPLE 4.24

FINDING A
SAMPLING
DISTRIBUTION

Problem Consider a game played with a standard 52-card bridge deck in which you can score 0, 3, or 12 points on any one hand. Suppose the population of points scored per hand is described by the probability distribution shown here. A random sample of $n = 3$ hands is selected from the population.

Points, x	0	3	12
$p(x)$	$\frac{1}{2}$	$\frac{1}{4}$	$\frac{1}{4}$

a. Find the sampling distribution of the sample mean $\bar{x}$.
b. Find the sampling distribution of the sample median M.

Solution Points for every possible sample of $n = 3$ hands are listed in Table 4.9, along with the sample mean and median. The probability of each sample is obtained using the Multiplicative Rule. For example, the probability of the sample $(0, 0, 3)$ is $p(0){\cdot}p(0){\cdot}p(3)$ $= (\frac{1}{2})(\frac{1}{2})(\frac{1}{4}) = \frac{1}{16}$. The probability for each sample is also listed in Table 4.9. Note that the sum of these probabilities is equal to 1.

a. From Table 4.9, you can see that $\bar{x}$ can assume the values 0, 1, 2, 3, 4, 5, 6, 8, 9, and 12. Because $\bar{x} = 0$ occurs in only one sample, $P(\bar{x} = 0) = \frac{8}{64}$. Similarly, $\bar{x} = 1$ occurs in three samples: $(0, 0, 3)$, $(0, 3, 0)$, and $(3, 0, 0)$. Therefore, $P(\bar{x} = 1) = \frac{4}{64} + \frac{4}{64} + \frac{4}{64} = \frac{12}{64}$. Calculating the probabilities of the remaining values of $\bar{x}$ and arranging them in a table, we obtain the following probability distribution:

$\bar{x}$	0	1	2	3	4	5	6	8	9	12
$p(\bar{x})$	$\frac{8}{64}$	$\frac{12}{64}$	$\frac{6}{64}$	$\frac{1}{64}$	$\frac{12}{64}$	$\frac{12}{64}$	$\frac{3}{64}$	$\frac{6}{64}$	$\frac{3}{64}$	$\frac{1}{64}$

TABLE 4.9 All Possible Samples of $n = 3$ Hands of a Card Game, Example 6.1

Possible Samples	$\bar{x}$	M	Probability
0, 0, 0	0	0	$(\frac{1}{2})(\frac{1}{2})(\frac{1}{2}) = \frac{1}{8} = \frac{8}{64}$
0, 0, 3	1	0	$(\frac{1}{2})(\frac{1}{2})(\frac{1}{4}) = \frac{1}{16} = \frac{4}{64}$
0, 0, 12	4	0	$(\frac{1}{2})(\frac{1}{2})(\frac{1}{4}) = \frac{1}{16} = \frac{4}{64}$
0, 3, 0	1	0	$(\frac{1}{2})(\frac{1}{4})(\frac{1}{2}) = \frac{1}{16} = \frac{4}{64}$
0, 3, 3	2	3	$(\frac{1}{2})(\frac{1}{4})(\frac{1}{4}) = \frac{1}{32} = \frac{2}{64}$
0, 3, 12	5	3	$(\frac{1}{2})(\frac{1}{4})(\frac{1}{4}) = \frac{1}{32} = \frac{2}{64}$
0, 12, 0	4	0	$(\frac{1}{2})(\frac{1}{4})(\frac{1}{2}) = \frac{1}{16} = \frac{4}{64}$
0, 12, 3	5	3	$(\frac{1}{2})(\frac{1}{4})(\frac{1}{4}) = \frac{1}{32} = \frac{2}{64}$
0, 12, 12	8	12	$(\frac{1}{2})(\frac{1}{4})(\frac{1}{4}) = \frac{1}{32} = \frac{2}{64}$
3, 0, 0	1	0	$(\frac{1}{4})(\frac{1}{2})(\frac{1}{2}) = \frac{1}{16} = \frac{4}{64}$
3, 0, 3	2	3	$(\frac{1}{4})(\frac{1}{2})(\frac{1}{4}) = \frac{1}{32} = \frac{2}{64}$
3, 0, 12	5	3	$(\frac{1}{4})(\frac{1}{2})(\frac{1}{4}) = \frac{1}{32} = \frac{2}{64}$
3, 3, 0	2	3	$(\frac{1}{4})(\frac{1}{4})(\frac{1}{2}) = \frac{1}{32} = \frac{2}{64}$
3, 3, 3	3	3	$(\frac{1}{4})(\frac{1}{4})(\frac{1}{4}) = \frac{1}{64}$
3, 3, 12	6	3	$(\frac{1}{4})(\frac{1}{4})(\frac{1}{4}) = \frac{1}{64}$
3, 12, 0	5	3	$(\frac{1}{4})(\frac{1}{4})(\frac{1}{2}) = \frac{1}{32} = \frac{2}{64}$
3, 12, 3	6	3	$(\frac{1}{4})(\frac{1}{4})(\frac{1}{4}) = \frac{1}{64}$
3, 12, 12	9	12	$(\frac{1}{4})(\frac{1}{4})(\frac{1}{4}) = \frac{1}{64}$
12, 0, 0	4	0	$(\frac{1}{4})(\frac{1}{2})(\frac{1}{2}) = \frac{1}{16} = \frac{4}{64}$
12, 0, 3	5	3	$(\frac{1}{4})(\frac{1}{2})(\frac{1}{4}) = \frac{1}{32} = \frac{2}{64}$
12, 0, 12	8	12	$(\frac{1}{4})(\frac{1}{2})(\frac{1}{4}) = \frac{1}{32} = \frac{2}{64}$
12, 3, 0	5	3	$(\frac{1}{4})(\frac{1}{4})(\frac{1}{2}) = \frac{1}{32} = \frac{2}{64}$
12, 3, 3	6	3	$(\frac{1}{4})(\frac{1}{4})(\frac{1}{4}) = \frac{1}{64}$
12, 3, 12	9	12	$(\frac{1}{4})(\frac{1}{4})(\frac{1}{4}) = \frac{1}{64}$
12, 12, 0	8	12	$(\frac{1}{4})(\frac{1}{4})(\frac{1}{2}) = \frac{1}{32} = \frac{2}{64}$
12, 12, 3	9	12	$(\frac{1}{4})(\frac{1}{4})(\frac{1}{4}) = \frac{1}{64}$
12, 12, 12	12	12	$(\frac{1}{4})(\frac{1}{4})(\frac{1}{2}) = \frac{1}{64}$

$$\text{Sum} = \frac{64}{64} = 1$$

This is the sampling distribution for $\bar{x}$ because it specifies the probability associated with each possible value of $\bar{x}$.

b. In Table 4.9, you can see that the median M can assume one of the three values: 0, 3, and 12. The value $M = 0$ occurs in 7 different samples. Therefore, $P(M = 0)$ is the sum of the probabilities of these 7 samples; that is, $P(M = 0) = {}^{8}\!/_{64} + {}^{4}\!/_{64} + {}^{4}\!/_{64} + {}^{4}\!/_{64} + {}^{4}\!/_{64} + {}^{4}\!/_{64} = {}^{32}\!/_{64}$. Similarly, $M = 3$ occurs in 13 samples and $M = 12$ occurs in 7 samples, and these probabilities are obtained by summing the probabilities of their respective sample points. Therefore, the probability distribution (i.e., the sampling distribution) for the median M is as follows:

M	0	3	12
$p(M)$	${}^{32}\!/_{64}$	${}^{22}\!/_{64}$	${}^{10}\!/_{64}$

Look Back The sampling distributions of parts **a** and **b** are found by first listing all possible distinct values of the statistic and then calculating the probability of each value. Note that if the values of x were equally likely, then the 27 sample points in Table 6.2 would all have the same probability of occurring, namely, ${}^{1}\!/_{27}$.

Now Work Exercise 4.133

Example 4.24 demonstrates the procedure for finding the exact sampling distribution of a statistic when the number of different samples that could be selected from the population is relatively small. In the real world, populations often consist of a large number of different values, making samples difficult (or impossible) to enumerate. When this situation occurs, we may choose to obtain the approximate sampling distribution for a statistic by simulating the sampling over and over again and recording the proportion of times different values of the statistic occur. Example 4.25 illustrates this procedure.

EXAMPLE 4.25
SIMULATING A SAMPLING DISTRIBUTION— Thickness of Steel Sheets

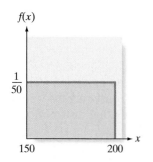

FIGURE 4.32
Uniform distribution for thickness of steel sheets

Problem The rolling machine of a steel manufacturer produces sheets of steel of varying thickness. The thickness of a steel sheet ranges between 150 and 200 millimeters, with the distribution shown in Figure 4.32. (This distribution is known as the **uniform distribution**). Suppose we perform the following experiment over and over again: Randomly sample 11 steel sheets from the production line and record the thickness x of each. Calculate the two sample statistics

$$\bar{x} = \text{Sample mean} = \frac{\Sigma x}{11}$$

$$M = \text{Median} = \text{Sixth sample measurement when the 11 thicknesses are arranged in ascending order}$$

Obtain approximations to the sampling distributions of $\bar{x}$ and M.

Solution We used MINITAB to generate 1,000 samples from the population, each with $n = 11$ observations. Then we computed $\bar{x}$ and M for each sample. Our goal is to obtain approximations to the sampling distributions of $\bar{x}$ and M in order to find out which sample statistic ($\bar{x}$ or M) contains more information about μ. [*Note:* For this particular population, it is known that the population mean is $\mu = 175$ mm. The first 10 of the 1,000 samples generated are presented in Table 4.10. For instance, the first computer-generated sample from the uniform distribution contained the following measurements (arranged in ascending order): 151, 157, 162, 169, 171, 173, 181, 182, 187, 188, and 193 millimeters. The sample mean $\bar{x}$ and median M computed for this sample are

$$\bar{x} = \frac{151 + 157 + \cdots + 193}{11} = 174.0$$

$$M = \text{Sixth ordered measurement} = 173$$

SIMUNI

TABLE 4.10	First 10 Samples of $n = 11$ Thickness Measurements from Uniform Distribution												
Sample					Thickness Measurements							Mean	Median
1	173	171	187	151	188	181	182	157	162	169	193	174.00	173
2	181	190	182	171	187	177	162	172	188	200	193	182.09	182
3	192	195	187	187	172	164	164	189	179	182	173	180.36	182
4	173	157	150	154	168	174	171	182	200	181	187	172.45	173
5	169	160	167	170	197	159	174	174	161	173	160	169.46	169
6	179	170	167	174	173	178	173	170	173	198	187	176.55	173
7	166	177	162	171	154	177	154	179	175	185	193	172.09	175
8	164	199	152	153	163	156	184	151	198	167	180	169.73	164
9	181	193	151	166	180	199	180	184	182	181	175	179.27	181
10	155	199	199	171	172	157	173	187	190	185	150	176.18	173

The MINITAB relative frequency histograms for $\bar{x}$ and M for the 1,000 samples of size $n = 11$ are shown in Figure 4.33. These histograms represent approximations to the true sampling distributions of $\bar{x}$ and M.

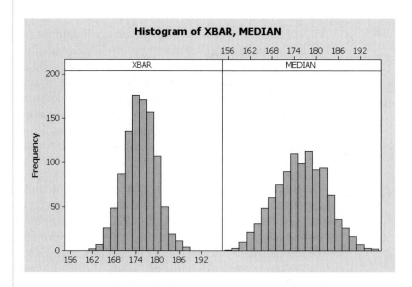

FIGURE 4.33
MINITAB histograms for sample mean and sample median, Example 4.25

Look Back You can see that the values of $\bar{x}$ tend to cluster around μ to a greater extent than do the values of M. Thus, on the basis of the observed sampling distributions, we conclude that $\bar{x}$ contains more information about μ than M does—at least for samples of $n = 11$ measurements from the uniform distribution.

Now Work Exercise 4.138

As noted earlier, many sampling distributions can be derived mathematically, but the theory necessary to do so is beyond the scope of this text. Consequently, when we need to know the properties of a statistic, we will present its sampling distribution and simply describe its properties. Several of the important properties we look for in sampling distributions are discussed in the next section.

Exercises 4.131–4.139

Understanding the Principles

4.131 What is the difference between a population parameter and a sample statistic?

4.132 What is a sampling distribution of a sample statistic?

Learning the Mechanics

4.133
NW The probability distribution shown here describes a population of measurements that can assume values of 0, 2, 4, and 6, each of which occurs with the same relative frequency:

x	0	2	4	6
$p(x)$	$\frac{1}{4}$	$\frac{1}{4}$	$\frac{1}{4}$	

a. List all the different samples of $n = 2$ measurements that can be selected from this population.
b. Calculate the mean of each different sample listed in part **a**.
c. If a sample of $n = 2$ measurements is randomly selected from the population, what is the probability that a specific sample will be selected?
d. Assume that a random sample of $n = 2$ measurements is selected from the population. List the different values of $\bar{x}$ found in part **b**, and find the probability of each. Then give the sampling distribution of the sample mean $\bar{x}$ in tabular form.
e. Construct a probability histogram for the sampling distribution of $\bar{x}$.

4.134 Simulate sampling from the population described in Exercise 4.133 by marking the values of x, one on each of four identical coins (or poker chips, etc.). Place the coins (marked 0, 2, 4, and 6) into a bag, randomly select one, and observe its value. Replace this coin, draw a second coin, and observe its value. Finally, calculate the mean $\bar{x}$ for this sample of $n = 2$ observations randomly selected from the population (Exercise 4.133, part **b**). Replace the coins, mix them, and, using the same procedure, select a sample of $n = 2$ observations from the population. Record the numbers and calculate $\bar{x}$ for this sample. Repeat this sampling process until you acquire 100 values of $\bar{x}$. Construct a relative frequency distribution for these 100 sample means. Compare this distribution with the exact sampling distribution of $\bar{x}$ found in part **e** of Exercise 4.133. [*Note:* The distribution obtained in this exercise is an approximation to the exact sampling distribution. However, if you were to repeat the sampling procedure, drawing two coins not 100 times, but 10,000 times, then the relative frequency distribution for the 10,000 sample means would be almost identical to the sampling distribution of $\bar{x}$ found in Exercise 4.133, part **e**.]

4.135 Consider the population described by the probability distribution shown here:

x	1	2	3	4	5
$p(x)$	.2	.3	.2	.2	.1

The random variable x is observed twice. If these observations are independent, verify that the different samples of size 2 and their probabilities are as follows:

Sample	Probability	Sample	Probability
1, 1	.04	3, 4	.04
1, 2	.06	3, 5	.02
1, 3	.04	4, 1	.04
1, 4	.04	4, 2	.06
1, 5	.02	4, 3	.04
2, 1	.06	4, 4	.04
2, 2	.09	4, 5	.02
2, 3	.06	5, 1	.02
2, 4	.06	5, 2	.03
2, 5	.03	5, 3	.02
3, 1	.04	5, 4	.02
3, 2	.06	5, 5	.01
3, 3	.04		

a. Find the sampling distribution of the sample mean $\bar{x}$.
b. Construct a probability histogram for the sampling distribution of $\bar{x}$.
c. What is the probability that $\bar{x}$ is 4.5 or larger?
d. Would you expect to observe a value of $\bar{x}$ equal to 4.5 or larger? Explain.

4.136 Refer to Exercise 4.135 and find $E(x) = \mu$. Then use the sampling distribution of $\bar{x}$ found in Exercise 4.135 to find the expected value of $\bar{x}$. Note that $E(\bar{x}) = \mu$.

4.137 Refer to Exercise 4.135. Assume that a random sample of $n = 2$ measurements is randomly selected from the population.

a. List the different values that the sample median M may assume, and find the probability of each. Then give the sampling distribution of the sample median.
b. Construct a probability histogram for the sampling distribution of the sample median, and compare it with the probability histogram for the sample mean (Exercise 4.135, part **b**).

4.138
NW In Example 4.25, we use the computer to generate 1,000 samples, each containing $n = 11$ observations, from a uniform distribution over the interval from 0 to 1. Now use the computer to generate 500 samples, each containing $n = 15$ observations, from that same population.

a. Calculate the sample mean for each sample. To approximate the sampling distribution of $\bar{x}$, construct a relative frequency histogram for the 500 values of $\bar{x}$.
b. Repeat part **a** for the sample median. Compare this approximate sampling distribution with the approximate sampling distribution of $\bar{x}$ found in part **a**.

4.139 Consider a population that contains values of x equal to 00, 01, 02, 03, . . . , 96, 97, 98, 99. Assume that these values occur with equal probability. Use the computer to generate 500 samples, each containing $n = 25$ measurements, from this population. Calculate the sample mean $\bar{x}$ and sample variance s^2 for each of the 500 samples.

a. To approximate the sampling distribution of $\bar{x}$, construct a relative frequency histogram for the 500 values of $\bar{x}$.
b. Repeat part **a** for the 500 values of s^2.

4.9 The Central Limit Theorem

Estimating the mean useful life of automobiles, the mean number of crimes per month in a large city, and the mean yield per acre of a new soybean hybrid are practical problems with something in common. In each case, we are interested in making an inference about the mean μ of some population. As we mentioned in Chapter 2, the sample mean $\bar{x}$ is, in general, a good estimator of μ. We now develop pertinent information about the sampling distribution for this useful statistic. We will show that $\bar{x}$ is the minimum-variance unbiased estimator (MVUE) of μ.

EXAMPLE 4.26

DESCRIBING THE SAMPLING DISTRIBUTION OF $\bar{x}$

Problem Suppose a population has the uniform probability distribution given in Figure 4.34. It can be shown that the mean and standard deviation of this probability distribution are, respectively, $\mu = 175$ and $\sigma = 14.43$. Now suppose a sample of 11 measurements is selected from this population. Describe the sampling distribution of the sample mean $\bar{x}$ based on the 1,000 sampling experiments discussed in Example 4.25.

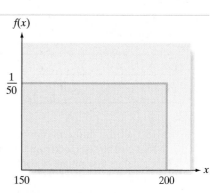

FIGURE 4.34
Sampled uniform population

Solution Recall that in Example 4.25 we generated 1,000 samples of $n = 11$ measurements each. The MINITAB histogram for the 1,000 sample means is shown in Figure 4.35, with a normal probability distribution superimposed. You can see that this normal probability distribution approximates the computer-generated sampling distribution very well.

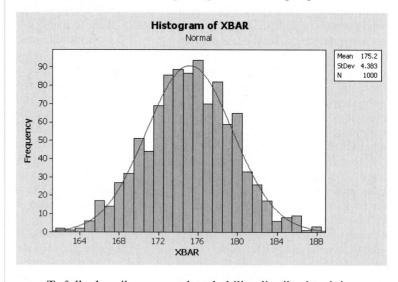

FIGURE 4.35
MINITAB histogram for sample mean in 1,000 samples

To fully describe a normal probability distribution, it is necessary to know its mean and standard deviation. MINITAB gives these statistics for the 1,000 $\bar{x}$'s in the upper right corner of the histogram of Figure 4.35. You can see that the mean is 175.2 and the standard deviation is 4.383.

To summarize our findings based on 1,000 samples, each consisting of 11 measurements from a uniform population, the sampling distribution of $\bar{x}$ appears to be approximately normal with a mean of about 175 and a standard deviation of about 4.38.

Look Back Note that the simulated value $\mu_{\bar{x}} = 175.2$ is very close to $\mu = 175$ for the uniform distribution; that is, the simulated sampling distribution of $\bar{x}$ appears to provide an accurate estimate of μ.

The true sampling distribution of $\bar{x}$ has the properties given in the next box, assuming only that a random sample of n observations has been selected from *any* population.

Properties of the Sampling Distribution of $\bar{x}$

1. The mean of the sampling distribution equals the mean of the sampled population. That is, $\mu_{\bar{x}} = E(\bar{x}) = \mu$.*

2. The standard deviation of the sampling distribution equals

$$\frac{\text{Standard deviation of sampled population}}{\text{Square root of sample size}}$$

That is, $\sigma_{\bar{x}} = \sigma/\sqrt{n}$.**

The standard deviation $\sigma_{\bar{x}}$ is often referred to as the **standard error of the mean**.

You can see that our approximation to $\mu_{\bar{x}}$ in Example 4.26 was precise, since property 1 assures us that the mean is the same as that of the sampled population: 175. Property 2 tells us how to calculate the standard deviation of the sampling distribution of $\bar{x}$. Substituting $\sigma = 14.43$ (the standard deviation of the sampled uniform distribution) and the sample size $n = 11$ into the formula for $\sigma_{\bar{x}}$, we find that

$$\sigma_{\bar{x}} = \frac{\sigma}{\sqrt{n}} = \frac{14.43}{\sqrt{11}} = 4.35$$

Thus, the approximation we obtained in Example 4.26, $\sigma_{\bar{x}} \approx 4.38$, is very close to the exact value, $\sigma_{\bar{x}} = 4.35$.***

What about the shape of the sampling distribution? Two theorems provide this information. One is applicable whenever the original population data are normally distributed. The other, applicable when the sample size n is large, represents one of the most important theoretical results in statistics: the *Central Limit Theorem*.

Theorem 4.1

If a random sample of n observations is selected from a population with a normal distribution, the sampling distribution of $\bar{x}$ will be a normal distribution.

Theorem 4.2: Central Limit Theorem

Consider a random sample of n observations selected from a population (*any* population) with mean μ and standard deviation σ. Then, when n is sufficiently large, the sampling distribution of $\bar{x}$ will be approximately a normal distribution with mean $\mu_{\bar{x}} = \mu$ and standard deviation $\sigma_{\bar{x}} = \sigma/\sqrt{n}$. The larger the sample size, the better will be the normal approximation to the sampling distribution of $\bar{x}$.****

Thus, for sufficiently large samples, the sampling distribution of $\bar{x}$ is approximately normal. How large must the sample size n be so that the normal distribution provides a good approximation to the sampling distribution of $\bar{x}$? The answer depends on the shape of the distribution of the sampled population, as shown by Figure 4.36. Generally speaking, the greater the skewness of the sampled population distribution, the larger the sample size

---

*When this property holds, we say that $\bar{x}$ is an *unbiased* estimator of μ.
**If the sample size n is large relative to the number N of elements in the population (e.g., 5% or more), $\sigma/\sqrt{n}$ must be multiplied by the finite population correction factor $\sqrt{(N - n)/(N - 1)}$. In most sampling situations, this correction factor will be close to 1 and can be ignored.
***It can be shown (proof/omitted) that the value of $\sigma_{\bar{x}}^2$ is the smallest variance of all unbiased estimators of μ; thus, $\bar{x}$ is the MVUE (minimum-variance unbiased estimator) of μ.
****Moreover, because of the Central Limit Theorem, the sum of a random sample of n observations, Σx, will possess a sampling distribution that is approximately normal for large samples. This distribution will have a mean equal to $n\mu$ and a variance equal to $n\sigma^2$. Proof of the Central Limit Theorem is beyond the scope of this book, but it can be found in many mathematical statistics texts.

Biography

PIERRE-SIMON LAPLACE (1749–1827)— The Originator of the Central Limit Theorem

As a boy growing up in Normandy, France, Pierre-Simon Laplace attended a Benedictine priory school. Upon graduation, he entered Caen University to study theology. During his two years there, he discovered his mathematical talents and began his career as an eminent mathematician. In fact, he considered himself the best mathematician in France. Laplace's contributions to mathematics ranged from in- troducing new methods of solving differential equations to presenting complex analyses of motions of astronomical bodies. While study- ing the angles of inclination of comet orbits in 1778, Laplace showed that the sum of the an- gles was normally distributed. Consequently, he is considered to be the originator of the Central Limit Theorem. (A rigorous proof of the theorem, however, was not provided until the early 1930s by another French mathemati- cian, Paul Levy.) Laplace also discovered Bayes's theorem and established Bayesian statistical analysis as a valid approach to many practical problems of his time.

must be before the normal distribution is an adequate approximation to the sampling dis- tribution of $\bar{x}$. For most sampled populations, sample sizes of $n \geq 30$ will suffice for the normal approximation to be reasonable. We will use the normal approximation for the sam- pling distribution of $\bar{x}$ when the sample size is at least 30.

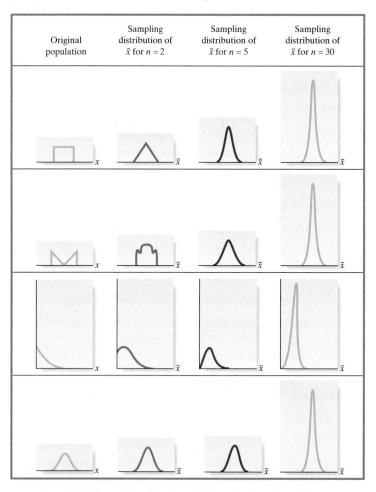

FIGURE 4.36

Sampling distributions of $\bar{x}$ for different populations and different sample sizes

EXAMPLE 4.27
USING THE CENTRAL LIMIT THEOREM TO FIND A PROBABILITY

Problem Suppose we have selected a random sample of $n = 36$ observations from a population with mean equal to 80 and standard deviation equal to 6. It is known that the population is not extremely skewed.

a. Sketch the relative frequency distributions for the population and for the sampling distribution of the sample mean $\bar{x}$.

b. Find the probability that $\bar{x}$ will be larger than 82.

Solution

a. We do not know the exact shape of the population relative frequency distribution, but we do know that it should be centered about $\mu = 80$, its spread should be measured by $\sigma = 6$, and it is not highly skewed. One possibility is shown in Figure 4.37a. From the Central Limit Theorem, we know that the sampling distribution of $\bar{x}$ will be approximately normal, since the sampled population distribution is not extremely skewed. We also know that the sampling distribution will have mean

and standard deviation $\quad \mu_{\bar{x}} = \mu = 80$

$$\sigma_{\bar{x}} = \frac{\sigma}{\sqrt{n}} = \frac{6}{\sqrt{36}} = 1$$

The sampling distribution of $\bar{x}$ is shown in Figure 4.37b.

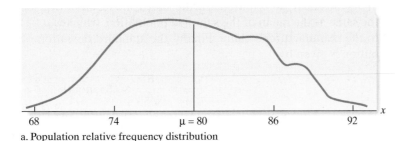

a. Population relative frequency distribution

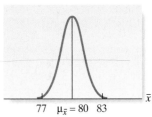

b. Sampling distribution of $\bar{x}$

FIGURE 4.37

A population relative frequency distribution and the sampling distribution for $\bar{x}$

b. The probability that $\bar{x}$ will exceed 82 is equal to the highlighted area in Figure 4.38. To find this area, we need to find the z value corresponding to $\bar{x} = 82$. Recall that the standard normal random variable z is the difference of any normally distributed random variable and its mean, expressed in units of its standard deviation. Since $\bar{x}$ is a normally distributed random variable with mean $\mu_{\bar{x}} = \mu$ and standard deviation $\sigma_{\bar{x}} = \sigma/\sqrt{n}$, it follows that the standard normal z value corresponding to the sample mean $\bar{x}$ is

$$z = \frac{(\text{Normal random variable}) - (\text{Mean})}{\text{Standard Deviation}} = \frac{\bar{x} - \mu_{\bar{x}}}{\sigma_{\bar{x}}}$$

Therefore, for $\bar{x} = 82$, we have

$$z = \frac{\bar{x} - \mu_{\bar{x}}}{\sigma_{\bar{x}}} = \frac{82 - 80}{1} = 2$$

The area A in Figure 4.38 corresponding to $z = 2$ is given in the table of areas under the normal curve (see Table IV of Appendix A) as .4772. Therefore, the tail area corresponding to the probability that $\bar{x}$ exceeds 82 is

$$P(\bar{x} > 82) = P(z > 2) = .5 - .4772 = .0228$$

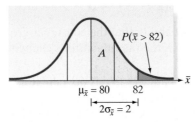

FIGURE 4.38

The sampling distribution of $\bar{x}$

Look Back The key to finding the probability in part **b** is to recognize that the distribution of $\bar{x}$ is normal with $\mu_{\bar{x}} = \mu$ and $\sigma_{\bar{x}} = \sigma/\sqrt{n}$.

Now Work Exercise 4.148

EXAMPLE 4.28

APPLICATION
OF THE
CENTRAL LIMIT
THEOREM—

Testing a
Manufacturer's Claim

Problem A manufacturer of automobile batteries claims that the distribution of the lengths of life of its best battery has a mean of 54 months and a standard deviation of 6 months. Suppose a consumer group decides to check the claim by purchasing a sample of 50 of the batteries and subjecting them to tests that estimate the battery's life.

a. Assuming that the manufacturer's claim is true, describe the sampling distribution of the mean lifetime of a sample of 50 batteries.

b. Assuming that the manufacturer's claim is true, what is the probability that the consumer group's sample has a mean life of 52 or fewer months?

Solution

a. Even though we have no information about the shape of the probability distribution of the lives of the batteries, we can use the Central Limit Theorem to deduce that the sampling distribution for a sample mean lifetime of 50 batteries is approximately normally distributed. Furthermore, the mean of this sampling distribution is the same as the mean of the sampled population, which is $\mu = 54$ months according to the manufacturer's claim. Finally, the standard deviation of the sampling distribution is given by

$$\sigma_{\bar{x}} = \frac{\sigma}{\sqrt{n}} = \frac{6}{\sqrt{50}} = .85 \text{ month}$$

Note that we used the claimed standard deviation of the sampled population, $\sigma = 6$ months. Thus, if we assume that the claim is true, then the sampling distribution for the mean life of the 50 batteries sampled must be as shown in Figure 4.39.

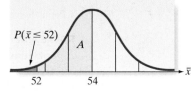

b. If the manufacturer's claim is true, the probability that the consumer group observes a mean battery life of 52 or fewer months for its sample of 50 batteries, $P(\bar{x} \leq 52)$, is equivalent to the highlighted area in Figure 4.39. Since the sampling distribution is approximately normal, we can find this area by computing the standard normal z value:

FIGURE 4.39
Sampling distribution of $\bar{x}$ in Example 4.28 for $n = 50$

$$z = \frac{\bar{x} - \mu_{\bar{x}}}{\sigma_{\bar{x}}} = \frac{\bar{x} - \mu}{\sigma_{\bar{x}}} = \frac{52 - 54}{.85} = -2.35$$

Here, $\mu_{\bar{x}}$, the mean of the sampling distribution of $\bar{x}$, is equal to μ, the mean of the lifetimes of the sampled population, and $\sigma_{\bar{x}}$ is the standard deviation of the sampling distribution of $\bar{x}$. Note that z is the familiar standardized distance (z-score) of Section 2.7, and since $\bar{x}$ is approximately normally distributed, it will possess the standard normal distribution of Section 4.5.

The area A shown in Figure 4.39 between $\bar{x} = 52$ and $\bar{x} = 54$ (corresponding to $z = -2.35$) is found in Table III of Appendix A to be .4906. Therefore, the area to the left of $\bar{x} = 52$ is

$$P(\bar{x} \leq 52) = .5 - A = .5 - .4906 = .0094$$

Thus, the probability that the consumer group will observe a sample mean of 52 or less is only .0094 if the manufacturer's claim is true.

Look Back If the 50 batteries tested do exhibit a mean of 52 or fewer months, the consumer group will have strong evidence that the manufacturer's claim is untrue, because such an event is very unlikely to occur if the claim is true. (This is still another application of the *rare-event approach* to statistical inference.)

Now Work Exercise 4.150

ACTIVITY 4.3: *Keep the Change:* Simulating a Sampling Distribution

In this activity, we will once again be working with the data set *Amounts Transferred*. (See Activity 4.2, p 216.) The activity is designed for small groups or the entire class.

1. Pool your *Amounts Transferred* data with data from other class members or the entire class so that the pooled data set has at least 100 data items. Have someone in the group calculate the mean and the standard deviation of the pooled data set.

2. Devise a convenient way to choose random samples from the pooled data set. You can assign each data item a number beginning with 1 and use a random-number generator to select a sample, or you can write each data item on a small card and then draw cards without looking.

3. Choose a random sample of size $n = 30$ from the pooled data, and find the mean of the sample, called a *sample mean*. Group members should repeat the process of choosing a sample of size $n = 30$ from the pooled data and finding the sample mean until the group has accumulated at least 25 sample means. Call this new data set *Sample Means*.

4. Find the mean and standard deviation of the data set *Sample Means*. Explain how the Central Limit Theorem is illustrated in this activity.

Keep the data set from this activity for use in other activities.

We conclude this section with two final comments on the sampling distribution of $\bar{x}$. First, from the formula $\sigma_{\bar{x}} = \sigma/\sqrt{n}$, we see that the standard deviation of the sampling distribution of $\bar{x}$ gets smaller as the sample size n gets larger. For example, we computed $\sigma_{\bar{x}} = .85$ when $n = 50$ in Example 4.28. However, for $n = 100$, we obtain $\sigma_{\bar{x}} = \sigma/\sqrt{n} = 6/\sqrt{100} = .60$. This relationship will hold true for most of the sample statistics encountered in this text. That is, *the standard deviation of the sampling distribution decreases as the sample size increases*. Consequently, the larger the sample size, the more accurate the sample statistic (e.g., $\bar{x}$) is in estimating a population parameter (e.g., μ). We will use this result in Chapter 5 to help us determine the sample size needed to obtain a specified accuracy of estimation.

Our second comment concerns the Central Limit Theorem. In addition to providing a very useful approximation for the sampling distribution of a sample mean, the Central Limit Theorem offers an explanation for the fact that many relative frequency distributions of data possess mound-shaped distributions. Many of the measurements we take in various areas of research are really means or sums of a large number of small phenomena. For example, a year's growth of a pine seedling is the total of the numerous individual components that affect the plant's growth. Similarly, we can view the length of time a construction company takes to build a house as the total of the times taken to complete a multitude of distinct jobs, and we can regard the monthly demand for blood at a hospital as the total of the many individual patients' needs. Whether or not the observations entering into these sums satisfy the assumptions basic to the Central Limit Theorem is open to question; however, it is a fact that many distributions of data in nature are mound shaped and possess the appearance of normal distributions.

Exercises 4.140–4.161

Understanding the Principles

4.140 What do the symbols $\mu_{\bar{x}}$ and $\sigma_{\bar{x}}$ represent?

4.141 How does the mean of the sampling distribution of $\bar{x}$ relate to the mean of the population from which the sample is selected?

4.142 How does the standard deviation of the sampling distribution of $\bar{x}$ relate to the standard deviation of the population from which the sample is selected?

4.143 State the Central Limit Theorem.

4.144 Will the sampling distribution of $\bar{x}$ always be approximately normally distributed? Explain.

Learning the Mechanics

4.145 Suppose a random sample of n measurements is selected from a population with mean $\mu = 100$ and variance $\sigma^2 = 100$. For each of the following values of n, give the mean and standard deviation of the sampling distribution of the sample mean $\bar{x}$:

a. $n = 4$ **b.** $n = 25$
c. $n = 100$ **d.** $n = 50$
e. $n = 500$ **f.** $n = 1,000$

4.146 Suppose a random sample of $n = 25$ measurements is selected from a population with mean μ and standard deviation σ. For each of the following values of μ and σ, give the values of $\mu_{\bar{x}}$ and $\sigma_{\bar{x}}$:

a. $\mu = 10, \sigma = 3$
b. $\mu = 100, \sigma = 25$
c. $\mu = 20, \sigma = 40$
d. $\mu = 10, \sigma = 100$

4.147 Consider the following probability distribution:

x	1	2	3	8
$p(x)$	.1	.4	.4	.1

a. Find μ, σ^2, and σ.
b. Find the sampling distribution of $\bar{x}$ for random samples of $n = 2$ measurements from this distribution by listing all possible values of $\bar{x}$, and find the probability associated with each.
c. Use the results of part **b** to calculate $\mu_{\bar{x}}$ and $\sigma_{\bar{x}}$. Confirm that $\mu_{\bar{x}} = \mu$ and that $\sigma_{\bar{x}} = \sigma/\sqrt{n} = \sigma/\sqrt{2}$.

4.148 A random sample of $n = 64$ observations is drawn from a
NW population with a mean equal to 20 and standard deviation equal to 16.
a. Give the mean and standard deviation of the (repeated) sampling distribution of $\bar{x}$.
b. Describe the shape of the sampling distribution of $\bar{x}$. Does your answer depend on the sample size?
c. Calculate the standard normal z-score corresponding to a value of $\bar{x} = 16$.
d. Calculate the standard normal z-score corresponding to $\bar{x} = 23$.
e. Find $P(\bar{x} < 16)$.
f. Find $P(\bar{x} > 23)$.
g. Find $P(16 < \bar{x} < 23)$.

4.149 A random sample of $n = 100$ observations is selected from a population with $\mu = 30$ and $\sigma = 16$.
a. Find $\mu_{\bar{x}}$ and $\sigma_{\bar{x}}$.
b. Describe the shape of the sampling distribution of $\bar{x}$.
c. Find $P(\bar{x} \geq 28)$.
d. Find $P(22.1 \leq \bar{x} \leq 26.8)$.
e. Find $P(\bar{x} \leq 28.2)$.
f. Find $P(\bar{x} \geq 27.0)$.

APPLET Applet Exercise 4.7

Open the applet entitled *Sampling Distribution*. On the pull-down menu to the right of the top graph, select *Binary*.
a. Run the applet for the sample size $n = 10$ and the number of samples $N = 1000$. Observe the shape of the graph of the sample proportions, and record the mean, median, and standard deviation of the sample proportions.
b. How does the mean of the sample proportions compare with the mean $\mu = 0.5$ of the original distribution?
c. Use the formula $\sigma = \sqrt{np(1 - p)}$, where $n = 1$ and $p = 0.5$, to compute the standard deviation of the original distribution. Divide the result by $\sqrt{10}$, the square root of the sample size used in the sampling distribution. How does this result compare with the standard deviation of the sample proportions?
d. Explain how the graph of the distribution of sample proportions suggests that the distribution may be approximately normal.
e. Explain how the results of parts **b–d** illustrate the Central Limit Theorem.

APPLET Applet Exercise 4.8

Open the applet entitled *Sampling Distributions*. On the pull-down menu to the right of the top graph, select *Uniform*. The box to the left of the top graph displays the population mean, median, and standard deviation of the original distribution.
a. Run the applet for the sample size $n = 30$ and the number of samples $N = 1000$. Observe the shape of the graph of the sample means, and record the mean, median, and standard deviation of the sample means.
b. How does the mean of the sample means compare with the mean of the original distribution?
c. Divide the standard deviation of the original distribution by $\sqrt{30}$, the square root of the sample size used in the sampling distribution. How does this result compare with the standard deviation of the sample proportions?
d. Explain how the graph of the distribution of sample means suggests that the distribution may be approximately normal.
e. Explain how the results of parts **b–d** illustrate the Central Limit Theorem.

Applying the Concepts—Basic

4.150 **Children's attitude toward reading.** In the journal
NW *Knowledge Quest* (January/February 2002), education professors at the University of Southern California investigated children's attitudes toward reading. One study measured third through sixth graders' attitudes toward recreational reading on a 140-point scale (where higher scores indicate a more positive attitude). The mean score for this population of children was 106 points, with a standard deviation of 16.4 points. Consider a random sample of 36 children from this population, and let $\bar{x}$ represent the mean recreational reading attitude score for the sample.
a. What is $\mu_{\bar{x}}$?
b. What is $\sigma_{\bar{x}}$?
c. Describe the shape of the sampling distribution of $\bar{x}$.
d. Find the z-score for the value $\bar{x} = 100$ points.
e. Find $P(\bar{x} < 100)$.

4.151 **Physical activity of obese young adults.** In a study on the physical activity of young adults, pediatric researchers mea-sured overall physical activity as the total number of registered movements (counts) over a period of time and then computed the number of counts per minute (cpm) for each subject. (*International Journal of Obesity*, January 2007). The study revealed that the overall physical activity of obese young adults has a mean of $\mu = 320$ cpm and a standard deviation of $\sigma = 100$ cpm. (In comparison, the mean for young adults of normal weight is 540 cpm.) In a random sample of $n = 100$ obese young adults, consider the sample mean counts per minute, $\bar{x}$.
a. Describe the sampling distribution of $\bar{x}$.
b. What is the probability that the mean overall physical activity level of the sample is between 300 and 310 cpm?
c. What is the probability that the mean overall physical activity level of the sample is greater than 360 cpm?

4.152 **Research on eating disorders.** Refer to *The American Statistician* (May 2001) study of female students who suffer from bulimia, presented in Exercise 2.38 (p. 48). Recall that each student completed a questionnaire from which a "fear of negative evaluation" (FNE) score was produced. (The higher the score, the greater is the fear of negative evaluation.) Suppose the FNE scores of bulimic students have a distribution with mean $\mu = 18$ and standard deviation $\sigma = 5$. Now, consider a random sample of 45 female students with bulimia.

a. What is the probability that the sample mean FNE score is greater than 17.5?

b. What is the probability that the sample mean FNE score is between 18 and 18.5?

c. What is the probability that the sample mean FNE score is less than 18.5?

4.153 Dentists' use of anesthetics. Refer to the *Current Allergy & Clinical Immunology* (March 2004) study of allergic reactions of dental patients to local anesthetics, presented in Exercise 2.95 (p. 73). The study reported that the mean number of units (ampoules) of local anesthetics used per week by dentists was 79, with a standard deviation of 23. Consider a random sample of 100 dentists, and let $\bar{x}$ represent the mean number of units of local anesthetic used per week for the sample.

a. What is $\mu_{\bar{x}}$?

b. What is $\sigma_{\bar{x}}$?

c. Describe the shape of the sampling distribution of $\bar{x}$.

d. Find the z-score for the value $\bar{x} = 80$ ampoules.

e. Find $P(\bar{x} > 80)$.

4.154 Improving SAT scores. Refer to the *Chance* (Winter 2001) examination of SAT scores of students who pay a private tutor to help them improve their results, presented in Exercise 2.101 (p. 74). On the SAT-Mathematics test, these students had a mean change in score of $+19$ points, with a standard deviation of 65 points. In a random sample of 100 students who pay a private tutor to help them improve their results, what is the likelihood that the change in the sample mean score is less than 10 points?

Applying the Concepts—Intermediate

4.155 Critical part failures in NASCAR vehicles. *The Sport Journal* (Winter, 2007) published a study of critical part failures at NASCAR races. The researchers found that the time x (in hours) until the first critical part failure has a skewed distribution with $\mu = .10$ and $\sigma = .10$. Now, consider a random sample of n = 50 NASCAR races and let $\bar{x}$ represent the sample mean time until the first critical part failure.

a. Find $E(\bar{x})$ and $Var(\bar{x})$.

b. Although x has a skewed distribution, the sampling distribution of $\bar{x}$ is approximately normal. Why?

c. Find the probability that the sample mean time until the first critical part failure exceeds .13 hour.

4.156 Cost of unleaded fuel. According to the American Automobile Association (AAA), the average cost of a gallon of regular unleaded fuel at gas stations in April 2007 was $2.835 (*AAA Fuel Gauge Report*). Assume that the standard deviation of such costs is $.15. Suppose that a random sample of $n = 100$ gas stations is selected from the population and the April 2007 cost per gallon of regular unleaded fuel is determined for each. Consider $\bar{x}$, the sample mean cost per gallon.

a. Calculate $\mu_{\bar{x}}$ and $\sigma_{\bar{x}}$.

b. What is the approximate probability that the sample has a mean fuel cost between $2.84 and $2.86?

c. What is the approximate probability that the sample has a mean fuel cost that exceeds $2.855?

d. How would the sampling distribution of $\bar{x}$ change if the sample size n were doubled from 100 to 200? How do your answers to parts **b** and **c** change when the sample size is doubled?

4.157 Levelness of concrete slabs. Geotechnical engineers use water-level "manometer" surveys to assess the levelness of newly constructed concrete slabs. Elevations are typically measured at eight points on the slab; of interest is the maximum differential between elevations. The *Journal of Performance of Constructed Facilities* (Feb. 2005) published an article on the levelness of slabs in California residential developments. Elevation data collected on over 1,300 concrete slabs *before tensioning* revealed that the maximum differential x has a mean of $\mu = .53$ inch and a standard deviation of $\sigma = .193$ inch. Consider a sample of $n = 50$ slabs selected from those surveyed, and let $\bar{x}$ represent the mean of the sample.

a. Describe fully the sampling distribution of $\bar{x}$.

b. Find $P(\bar{x} > .58)$.

c. The study also revealed that the mean maximum differential of concrete slabs measured *after tensioning and loading* is $\mu = .58$ inch. Suppose the sample data yield $\bar{x} = .59$ inch. Comment on whether the sample mea-surements were obtained before tensioning or after tensioning and loading.

4.158 Susceptibility to hypnosis. The Computer-Assisted Hypnosis Scale (CAHS) is designed to measure a person's susceptibility to hypnosis. In computer-assisted hypnosis, the computer serves as a facilitator of hypnosis by using digitized speech processing coupled with interactive involvement with the hypnotic subject. CAHS scores range from 0 (no susceptibility) to 12 (extremely high susceptibility). A study in *Psychological Assessment* (March 1995) reported a mean CAHS score of 4.59 and a standard deviation of 2.95 for University of Tennessee undergraduates. Assume that $\mu = 4.29$ and $\sigma = 2.95$ for this population. Suppose a psychologist uses the CAHS to test a random sample of 50 subjects.

a. Would you expect to observe a sample mean CAHS score of $\bar{x} = 6$ or higher? Explain.

b. Suppose the psychologist actually observes $\bar{x} = 6.2$. On the basis of your answer to part **a**, make an inference about the population from which the sample was selected.

4.159 Is exposure to a chemical in Teflon-coated cookware hazardous? Perfluorooctanoic acid (PFOA) is a chemical used in Teflon®-coated cookware to prevent food from sticking. The Environmental Protection Agency is investigating the potential risk of PFOA as a cancer-causing agent. (*Science News Online*, August 27, 2005.) It is known that the blood concentration of PFOA in the general population has a mean of $\mu = 6$ parts per billion (ppb) and a standard deviation of $\sigma = 10$ ppb. *Science News Online* reported on tests for PFOA exposure conducted on a sample of 326 people who live near DuPont's Teflon-making Washington (West Virginia) Works facility.

a. What is the probability that the average blood concentration of PFOA in the sample is greater than 7.5 ppb?

b. The actual study resulted in $\bar{x} = 300$ ppb. Use this information to make an inference about the true mean (μ) PFOA concentration for the population that lives near DuPont's Teflon facility.

Applying the Concepts—Advanced

4.160 **Test of Knowledge about Epilepsy.** The Test of Knowledge about Epilepsy (KAE), which is designed to measure attitudes toward persons with epilepsy, uses 20 multiple-choice items, all of which are incorrect. For each person, two scores (ranging from 0 to 20) are obtained: an attitude score (KAE-A) and a general-knowledge score (KAE-GK). On the basis of a large-scale study of college students, the distribution of KAE-A scores has a mean of $\mu = 11.92$ and a standard deviation of $\sigma = 2.95$ while the distribution of KAE-GK scores has a mean of $\mu = 6.35$ and a standard deviation of $\sigma = 2.12$. (*Rehabilitative Psychology*, Spring 1995.) Consider a random sample of 100 college students, and suppose you observe a sample mean KAE score of 6.5. Is this result more

likely to be the mean of the attitude scores (KAE-A) or the general-knowledge scores (KAE-GK)? Explain.

4.161 **Hand washing versus hand rubbing.** Refer to the *British Medical Journal* (August 17, 2002) study comparing the effectiveness of hand washing with soap and hand rubbing with alcohol, presented in Exercise 2.97 (p. 73). Health care workers who used hand rubbing had a mean bacterial count of 35 per hand with a standard deviation of 59. Health care workers who used hand washing had a mean bacterial count of 69 per hand with a standard deviation of 106. In a random sample of 50 health care workers, all using the same method of cleaning their hands, the mean bacterial count per hand ($\bar{x}$) is less than 30. Give your opinion on whether this sample of workers used hand rubbing with alcohol or hand washing with soap.

KEY TERMS

Note: Starred (*) terms are from the optional sections in this chapter.

Bell-shaped probability
  distribution   197
Bell curve   197
Binomial distribution 187
Binomial experiment   184
Binomial random variable   184
Central Limit Theorem   235
Continuous random variable   171
Countable   171
Cumulative binomial probabilities   190
Continuous probability distribution   196

*Correction for continuity   222
Discrete random variable   171
Expected value 175
Mean of a discrete random
  variable   175
Normal distribution   197
Normal probability plot   214
Normal random variable   197
Parameter   227
Probability density function   196
Probability distribution for a discrete
  random variable   174

Random variable   170
Sample statistic   228
Sampling distribution   228
Standard deviation of a discrete
  random variable   176
Standard error of the mean   235
Standard normal distribution   199
Standard normal random variable   199
Uniform distribution   231
Variance of a discrete random
  variable   176

GUIDE TO SELECTING A PROBABILITY DISTRIBUTION

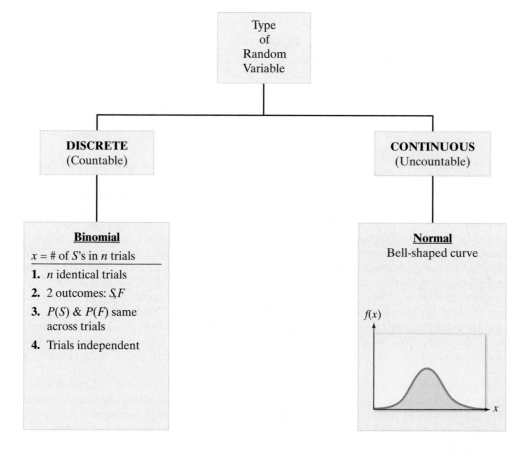

CHAPTER NOTES

Properties of Discrete Probability Distributions

(1) $p(x) \geq 0$

(2) $\sum_{\text{all } x} p(x) = 1$

Properties of Continuous Probability Distributions

(1) $P(x = a) = 0$

(2) $P(a < x < b)$ is area under curve between a and b

Key Symbols

$p(x)$	Probability distribution for discrete random variable x
$f(x)$	Probability distribution for continuous random variable x
S	Outcome of binomial trial denoted "success"
F	Outcome of binomial trial denoted "failure"
P	$P(S)$ in binomial trial
q	$P(F)$ in binomial trial $= 1 - p$
$f(x)$	Probability distribution (density function) for continuous random variable x
e	Constant used in normal probability distributions, $e = 2.71828\ldots$
π	Constant used in normal probability distributions, $\pi = 3.1416\ldots$
$\mu_{\bar{x}}$	True mean of sampling distribution of $\bar{x}$
$\sigma_{\bar{x}}$	True standard deviation of sampling distribution of $\bar{x}$

KEY FORMULAS

Random Variable	Prob. Dist'n	Mean	Variance
General Discrete:	Table, formula, or graph for $p(x)$	$\sum_{\text{all } x} x \cdot p(x)$	$\sum_{\text{all } x} (x - \mu)^2 \cdot p(x)$
Binomial:	$p(x) = \binom{n}{x} p^x q^{n-x}$	np	npq
Normal:	$(c \leq x \leq d)$ $f(x) = \dfrac{1}{\sigma\sqrt{2\pi}} e^{-1/2[(x-\mu)/\sigma]^2}$	σ	σ^2
Standard Normal:	$f(z) = \dfrac{1}{\sqrt{2\pi}} e^{-1/2(z)^2}$ $z = (x - \mu)/\sigma$	$\mu = 0$	$\sigma^2 = 1$

	Mean	Standard Deviation	z-score
Sampling distribution of $\bar{x}$	$\mu_{\bar{x}} = \mu$	$\sigma_{\bar{x}} = \dfrac{\sigma}{\sqrt{n}}$	$z = \dfrac{\bar{x} - \mu_{\bar{x}}}{\sigma_{\bar{x}}} = \dfrac{\bar{x} - \mu}{\sigma/\sqrt{n}}$

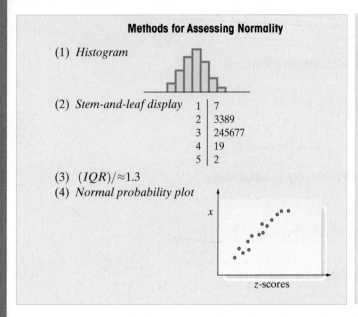

Methods for Assessing Normality

(1) *Histogram*

(2) *Stem-and-leaf display*

1	7
2	3389
3	245677
4	19
5	2

(3) $(IQR)/\approx 1.3$
(4) *Normal probability plot*

x

z-scores

Normal Approximation to Binomial

x is binomial (n, p)

$$P(x \leq a) \approx P\{z < (a + .5) - \mu\}$$

Sampling distribution of a statistic—the theoretical probability distribution of the statistic in repeated sampling

Central limit theorem—the sampling distribution of the sample mean, $\bar{x}$, is approximately normal for large n (e.g., $n \geq 30$).

Generating the Sampling Distribution of $\bar{x}$

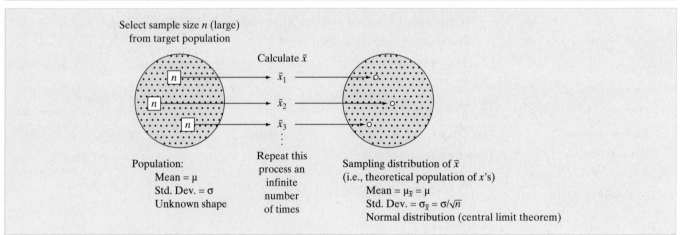

Select sample size n (large)
from target population

Calculate $\bar{x}$

$\bar{x}_1$

$\bar{x}_2$

$\bar{x}_3$

Population:
Mean = μ
Std. Dev. = σ
Unknown shape

Repeat this
process an
infinite
number
of times

Sampling distribution of $\bar{x}$
(i.e., theoretical population of x's)
Mean = $\mu_{\bar{x}} = \mu$
Std. Dev. = $\sigma_{\bar{x}} = \sigma/\sqrt{n}$
Normal distribution (central limit theorem)

SUPPLEMENTARY EXERCISES 4.162–4.195

Note: Starred (*) exercises refer to the optional section in this chapter.

Understanding the Principles

4.162 Which of the following describe discrete random variables, and which describe continuous random variables?

  a. The length of time that an exercise physiologist's program takes to elevate her client's heart rate to 140 beats per minute

  b. The number of crimes committed on a college campus per year

  c. The number of square feet of vacant office space in a large city

  d. The number of voters who favor a new tax proposal

4.163 For each of the following examples, decide whether x is a binomial random variable and explain your decision:

  a. A manufacturer of computer chips randomly selects 100 chips from each hour's production in order to estimate the proportion of defectives. Let x represent the number of defectives in the 100 chips sampled.

  b. Of five applicants for a job, two will be selected. Although all applicants appear to be equally qualified, only three have the ability to fulfill the expectations of the company. Suppose that the two selections are made at random from the five applicants, and let x be the number of qualified applicants selected.

  c. A software developer establishes a support hot line for customers to call in with questions regarding use of the software. Let x represent the number of calls received on the hot line during a specified workday.

   d. Florida is one of a minority of states with no state income tax. A poll of 1,000 registered voters is conducted to determine how many would favor a state income tax in light of the state's current fiscal condition. Let x be the number in the sample who would favor the tax.

4.164 Describe how you could obtain the simulated sampling distribution of a sample statistic.

4.165 **True or False.** The sample mean, $\bar{x}$ will always be equal to $\mu_{\bar{x}}$.

4.166 **True or False.** The sampling distribution of $\bar{x}$ is normally distributed, regardless of the size of the sample n.

Learning the Mechanics

4.167 Suppose x is a binomial random variable with $n = 20$ and $p = .7$.

   a. Find $P(x = 14)$.
   b. Find $P(x \leq 12)$.
   c. Find $P(x > 12)$.
   d. Find $P(9 \leq x \leq 18)$.
   e. Find $P(8 < x < 18)$.
   f. Find μ, σ^2, and σ.
   g. What is the probability that x is in the interval $\mu \pm 2\sigma$?

4.168 Consider the discrete probability distribution shown here:

x	10	12	18	20
$p(x)$	.2	.3	.1	.4

   a. Calculate μ, σ^2, and σ.
   b. What is $P(x < 15)$?
   c. Calculate $\mu \pm 2\sigma$.
   d. What is the probability that x is in the interval $\mu \pm 2\sigma$?

4.169 The random variable x has a normal distribution with $\mu = 70$ and $\sigma = 10$. Find the following probabilities:

   a. $P(x \leq 75)$
   b. $P(x \geq 90)$
   c. $P(60 \leq x \leq 75)$
   d. $P(x > 75)$
   e. $P(x = 75)$
   f. $P(x \leq 95)$

4.170 The random variable x has a normal distribution with $\mu = 40$ and $\sigma^2 = 36$. Find a value of x, say, x_0, such that

   a. $P(x \geq x_0) = .5$
   b. $P(x \leq x_0) = .9911$
   c. $P(x \leq x_0) = .0028$
   d. $P(x \geq x_0) = .0228$
   e. $P(x \leq x_0) = .1003$
   f. $P(x \geq x_0) = .7995$

*4.171 Assume that x is a binomial random variable with $n = 100$ and $p = .5$. Use the normal probability distribution to approximate the following probabilities:

   a. $P(x \leq 48)$
   b. $P(50 \leq x \leq 65)$
   c. $P(x \geq 70)$
   d. $P(55 \leq x \leq 58)$

   e. $P(x = 62)$
   f. $P(x \leq 49 \text{ or } x \geq 72)$

4.172 A random sample of $n = 68$ observations is selected from a population with $\mu = 19.6$ and $\sigma = 3.2$. Approximate each of the following probabilities:

   a. $P(\bar{x} \leq 19.6)$
   b. $P(\bar{x} \leq 19)$
   c. $P(\bar{x} \geq 20.1)$
   d. $P(19.2 \leq \bar{x} \leq 20.6)$

Applying the Concepts—Basic

4.173 **Quit-smoking program.** According to the University of South Florida's Tobacco Research and Intervention Program, only 5% of the nation's cigarette smokers ever enter into a treatment program to help them quit smoking. (*USF Magazine*, Spring 2000.) In a random sample of 200 smokers, let x be the number who enter into a treatment program.

   a. Explain why x is a binomial random variable (to a reasonable degree of approximation).
   b. What is the value of p? Interpret this value.
   c. What is the expected value of x? Interpret this value.

4.174 **Mobile phones with Internet access.** According to a Jupiter/NPD Consumer Survey of young adults (18–24 years of age) who shop online, 20% own a mobile phone with Internet access. (*American Demographics*, May 2002.) In a random sample of 200 young adults who shop online, let x be the number who own a mobile phone with Internet access.

   a. Explain why x is a binomial random variable (to a reasonable degree of approximation).
   b. What is the value of p? Interpret this value.
   c. What is the expected value of x? Interpret this value.

4.175 **Alkalinity of river water.** The alkalinity level of water specimens collected from the Han River in Seoul, Korea, has a mean of 50 milligrams per liter and a standard deviation of 3.2 milligrams per liter. (*Environmental Science & Engineering*, Sept. 1, 2000.) Assume that the distribution of alkalinity levels is approximately normal, and find the probability that a water specimen collected from the river has an alkalinity level

   a. exceeding 45 milligrams per liter.
   b. below 55 milligrams per liter.
   c. between 51 and 52 milligrams per liter.

4.176 **No-hitters in baseball.** In baseball, a "no-hitter" is a regulation nine-inning game in which the pitcher yields no hits to the opposing batters. *Chance* (Summer 1994) reported on a study of no-hitters in Major League Baseball (MLB). The initial analysis focused on the total number of hits yielded per game per team for all nine-inning MLB games. The distribution of hits per 9 innings is approximately normal with mean 8.72 and standard deviation 1.10.

   a. What percentage of nine-inning MLB games result in fewer than six hits?
   b. Demonstrate statistically why a no-hitter is considered an extremely rare occurrence in MLB.

4.177 **Married-women study.** According to *Women's Day* magazine (Jan. 2007), only 50% of married women would marry their current husbands again if given the chance. Consider the number x in a random sample of 10 married women who would marry their husbands again. Identify the discrete probability distribution that best models the distribution of x. Explain.

*4.178 Sickle-cell anemia. Eight percent of the African-American population is known to carry the trait for sickle-cell anemia. (Sickle Cell-Information Center, Atlanta, GA) If 1,000 African-Americans are sampled at random, what is the approximate probability that
  a. More than 175 carry the trait?
  b. Fewer than 140 carry the trait?

4.179 New book reviews. In Exercise 2.173 (p. 103), you read about a study of book reviews in American history, geography, and area studies published in *Choice* magazine. (*Library Acquisitions: Practice and Theory*, Vol. 19, 1995.) The overall rating stated in each review was ascertained and recorded as follows: 1 = would not recommend, 2 = cautious or very little recommendation, 3 = little or no preference, 4 = favorable/recommended, 5 = outstanding/significant contribution. Based on a sample of 375 book reviews, the probability distribution of rating, x, follows.

Booking Rating x	$p(x)$
1	.051
2	.099
3	.093
4	.635
5	.122

  a. Is this a valid probability distribution?
  b. What is the probability that a book reviewed in *Choice* has a rating of 1?
  c. What is the probability that a book reviewed in *Choice* has a rating of at least 4?
  d. What is the probability that a book reviewed in *Choice* has a rating of 2 or 3?
  e. Find $E(x)$.
  f. Interpret the result, part **e**.

4.180 Salaries of travel professionals. According to *Business Travel News* (July 17, 2006), the average salary of a travel management professional is $98,500. Assume that the standard deviation of such salaries is $30,000. Consider a random sample of 50 travel management professionals, and let $\bar{x}$ represent the mean salary for the sample.
  a. What is $\mu_{\bar{x}}$?
  b. What is $\sigma_{\bar{x}}$?
  c. Describe the shape of the sampling distribution of $\bar{x}$.
  d. Find the z-score for the value $\bar{x} = \$89{,}500$.
  e. Find $P(\bar{x} > 89{,}500)$.

4.181 Violence and stress. Interpersonal violence (e.g., rape) generally leads to psychological stress for the victim. *Clinical Psychology Review* (Vol. 15, 1995) reported on the results of all recently published studies of the relationship between interpersonal violence and psychological stress. The distribution of the time elapsed between the violent incident and the initial sign of stress has a mean of 5.1 years and a standard deviation of 6.1 years. Consider a random sample of $n = 150$ victims of interpersonal violence. Let $\bar{x}$ represent the mean time elapsed between the violent act and the first sign of stress for the sampled victims.
  a. Give the mean and standard deviation of the sampling distribution of $\bar{x}$.
  b. Will the sampling distribution of $\bar{x}$ be approximately normal? Explain.

  c. Find $P(\bar{x} > 5.5)$.
  d. Find $P(4 < \bar{x} < 5)$.

Applying the Concepts—Intermediate

4.182 Factors identifying urban counties. Refer to the *Professional Geographer* (Feb. 2000) study of urban and rural counties, presented in Exercise 3.123 (p. 161). Forty-five percent of county commissioners in Nevada feel that "population concentration" is the single most important factor used in identifying urban counties. Suppose 10 county commissioners are selected at random to serve on a Nevada review board that will consider redefining urban and rural areas. What is the probability that more than half of the commissioners will specify "population concentration" as the single most important factor used in identifying urban counties?

4.183 Cholesterol levels in psychiatric patients. A study of serum cholesterol levels in psychiatric patients of a maximum-security forensic hospital revealed that cholesterol level is approximately normally distributed with a mean of 208 milligrams per deciliter (mg/dL) and a standard deviation of 25 mg/dL. (*Journal of Behavioral Medicine*, Feb. 1995.) Prior research has shown that patients who exhibit violent behavior have cholesterol levels below 200 mg/dL.
  a. What is the probability of observing a psychiatric patient with a cholesterol level below 200 mg/dL?
  b. For three randomly selected patients, what is the probability that at least one will have a cholesterol level below 200 mg/dL?

4.184 Comparison of exam scores: red versus blue exam. Refer to the *Teaching Psychology* (May 1998) study of how external clues influence performance, presented in Exercise 2.116 (p. 78). Recall that two different forms of a midterm psychology examination were given, one printed on blue paper and the other on red paper. Grading only the difficult questions, the researchers found that scores on the blue exam had a distribution with a mean of 53% and a standard deviation of 15%, while scores on the red exam had a distribution with a mean of 39% and a standard deviation of 12%. Assuming that both distributions are approximately normal, on which exam is a student more likely to score below 20% on the difficult questions, the blue one or the red one? (Compare your answer with that of Exercise 2.116**c**.)

4.185 Parents' behavior at a gym meet. *Pediatric Exercise Science* (Feb. 1995) published an article on the behavior of parents at competitive youth gymnastic meets. On the basis of a survey of the gymnasts, the researchers estimated the probability of a parent "yelling" at his or her child before, during, or after the meet as .05. In a random sample of 20 parents attending a gymnastic meet, find the probability that at least 1 parent yells at his or her child before, during, or after the meet.

4.186 Study of children with developmental delays. It is well known that children with developmental delays (i.e., mild mental retardation) are slower, cognitively, than normally developing children. Are their social skills also lacking? A study compared the social interactions of the two groups of children in a controlled playground environment. (*American Journal on Mental Retardation*, Jan. 1992.) One variable of interest was the number of intervals of "no play" by each child. Children with developmental delays had a mean of 2.73 intervals of "no play" and a standard deviation of 2.58 intervals. Based on this information, is it possible for the variable of interest to be normally distributed? Explain.

4.187 Galaxy velocity study. Recall *The Astronomical Journal* (July 1995) study of galaxy velocity, Exercise 4.102 (p. 219). The observed velocity of a galaxy located in galaxy cluster A2142 was found to have a normal distribution with a mean of 27,117 kilometers per second (km/s) and a standard deviation of 1,280 km/s. A galaxy with a velocity of 24,350 km/s is observed. Comment on the likelihood of this galaxy being located in cluster A2142.

4.188 Supercooling temperature of frogs. Many species of terrestrial tree frogs that hibernate at or near the ground surface can survive prolonged exposure to low winter temperatures. In freezing conditions, the frog's body temperature, called its *supercooling temperature*, remains relatively higher because of an accumulation of glycerol in its body fluids. A study in *Science* revealed that the supercooling temperature of terrestrial frogs frozen at $-6°C$ has a relative frequency distribution with a mean of $-2°C$ and a standard deviation of $.3°C$. Consider the mean supercooling temperature $\bar{x}$ of a random sample of $n = 42$ terrestrial frogs frozen at $-6°C$.

  a. Find the probability that $\bar{x}$ exceeds $-2.05°C$.

  b. Find the probability that $\bar{x}$ falls between $-2.20°C$ and $-2.10°C$.

4.189 Surface roughness of pipe. The journal *Anti-Corrosion Methods and Materials* (Vol. 50, 2003) published a study of the surface roughness of oil field pipes. A scanning probe instrument was used to measure the surface roughness x (in micrometers) of 20 sampled sections of coated interior pipe. Consider the sample mean $\bar{x}$.

  a. Assume that the surface roughness distribution has a mean of $\mu = 1.8$ micrometers and a standard deviation of $\sigma = .5$ micrometer. Use this information to find the probability that $\bar{x}$ exceeds 1.85 micrometers.

  b. The sample data from the study is listed in the table. Compute $\bar{x}$.

  c. On the basis of the result from part **b**, comment on the validity of the assumptions made in part **a**.

ROUGHPIPE

1.72	2.50	2.16	2.13	1.06	2.24	2.31	2.03	1.09	1.40
2.57	2.64	1.26	2.05	1.19	2.13	1.27	1.51	2.41	1.95

Source: Farshad, F. & Pesacreta, T. "Coated pipe interior surface roughness as measured by three scanning probe instruments," *Anti-Corrosion Methods and Materials*, Vol. 50, No. 1, 2003 (Table III).

4.190 Countries that allow a free press. The degree to which democratic and nondemocratic countries attempt to control the news media was examined in the *Journal of Peace Research* (Nov. 1997). Between 1948 and 1996, 80% of all democratic regimes allowed a free press. In contrast, over the same period, 10% of all nondemocratic regimes allowed a free press.

  a. In a random sample of 50 democratic regimes, how many would you expect to allow a free press? Give a range that is highly likely to include the number of democratic regimes with a free press.

  b. In a random sample of 50 nondemocratic regimes, how many would you expect to allow a free press? Give a range that is highly likely to include the number of nondemocratic regimes with a free press.

4.191 Rubidium in metamorphic rock. A chemical analysis of metamorphic rock in western Turkey was reported in *Geological Magazine* (May 1995). The trace amount (in parts per million) of the element rubidium was measured for 20 rock specimens. The data are listed here. Assess whether the sample data come from a normal population.

RUBIDIUM

164	286	355	308	277	330	323	370	241	402
301	200	202	341	327	285	277	247	213	424

Source: Bozkurt, E. et al. "Geochemistry and the tectonic significance of augen gneisses from the southern Menderes Massif (West Turkey)." *Geological Magazine,* Vol. 132, No. 3, May 1995, p. 291 (Table 1).

Applying the Concepts—Advanced

4.192 Reliability of a "One-Shot" Device. A "one-shot" device can be used only once; after use, the device (e.g., a nuclear weapon, space shuttle, automobile air bag) either is destroyed or must be rebuilt. The destructive nature of a one-shot device makes repeated testing either impractical or too costly. Hence, the reliability of such a device must be determined with minimal testing. Consider a one-shot device that has some probability p of failure. Of course, the true value of p is unknown, so designers will specify a value of p which is the largest defective rate that they are willing to accept. Designers will conduct n tests of the device and determine the success or failure of each test. If the number of observed failures, x, is less than or equal to some specified value k, then the device is considered to have the desired failure rate. Consequently, the designers want to know the minimum sample size n needed so that observing K or fewer defectives in the sample will demonstrate that the true probability of failure for the one-shot device is no greater than p.

  a. Suppose the desired failure rate for a one-shot device is $p = .10$. Suppose also that designers will conduct $n = 20$ tests of the device and conclude that the device is performing to specifications if $K = 1$ (i.e., if 1 or no failure is observed in the sample). Find $P(x \leq 1)$.

  b. In reliability analysis, $1 - P(x \leq K)$ is often called the *level of confidence* for concluding that the true failure rate is less than or equal to p. Find the level of confidence for the one-shot device described in part **a**. In your opinion, is this an acceptable level? Explain.

  c. Demonstrate that the confidence level can be increased by either (1) increasing the sample size n or (2) decreasing the number K of failures allowed in the sample.

  d. Typically, designers want a confidence level of .90, .95, or .99. Find the values of n and K to use so that the designers can conclude with at least 95% confidence that the failure rate for the one-shot device of part **a** is no greater than $p = .10$.

[Note: The U.S. Department of Defense Reliability Analysis Center (DoD RAC) provides designers with free access to tables and toolboxes that give the minimum sample size n *required to obtain a desired confidence level for a specified number of observed failures in the sample.]*

4.193 Dye discharged in paint. A machine used to regulate the amount of dye dispensed for mixing shades of paint can be set so that it discharges an average of μ milliliters (mL) of dye per can of paint. The amount of dye discharged is known to have a normal distribution with a standard deviation of .4 mL. If more than 6 mL of dye are discharged when making a certain shade of blue paint, the shade is unacceptable. Determine the setting for μ so that only 1% of the cans of paint will be unacceptable.

4.194 **Fecal pollution at Huntington Beach.** The state of California mandates fecal indicator bacteria monitoring at all public beaches. When the concentration of fecal bacteria in the water exceeds a certain limit (400 colony-forming units of fecal coliform per 100 milliliters), local health officials must post a sign (called surf zone posting) warning beachgoers of potential health risks upon entering the water. For fecal bacteria, the state uses a single-sample standard; that is, if the fecal limit is exceeded in a single sample of water, surf zone posting is mandatory. This single-sample standard policy has led to a recent rash of beach closures in California.

Joon Ha Kim and Stanley B. Grant, engineers at the University of California at Irvine, conducted a study of the surf water quality at Huntington Beach in California and reported the results in *Environmental Science & Technology* (September 2004). The researchers found that beach closings were occurring despite low pollution levels in some instances, while in others signs were not posted when the fecal limit was exceeded. They attributed these "surf zone posting errors" to the variable nature of water quality in the surf zone (for example, fecal bacteria concentration tends to be higher during ebb tide and at night) and the inherent time delay between when a water sample is collected and when a sign is posted or removed. In order to prevent posting errors, the researchers recommend using an averaging method, rather than a single sample, to determine unsafe water quality. (For example, one simple averaging method is to take a random sample of multiple water specimens and compare the average fecal bacteria level of the sample with the limit of 400 cfu/100 mL in order to determine whether the water is safe.)

Discuss the pros and cons of using the single-sample standard versus the averaging method. Part of your discussion should address the probability of posting a sign when in fact the water is safe and the probability of posting a sign when in fact the water is unsafe. (Assume that the fecal bacteria concentrations of water specimens at Huntington Beach follow an approximately normal distribution.)

4.195 **IQs and *The Bell Curve*.** In their controversial book *The Bell Curve* (Free Press, 1994), Professors Richard J.

Herrnstein (a Harvard psychologist who died while the book was in production) and Charles Murray (a political scientist at MIT) explored, as the subtitle states, "intelligence and class structure in American life." *The Bell Curve* employs statistical analyses heavily in an attempt to support the authors' positions. Since the book's publication, many expert statisticians have raised doubts about the authors' statistical methods and the inferences drawn from them. (See, for example, "Wringing *The Bell Curve*: A cautionary tale about the relationships among race, genes, and IQ," *Chance*, Summer 1995.) One of the many controversies sparked by the book is the authors' tenet that level of intelligence (or lack thereof) is a cause of a wide range of intractable social problems, including constrained economic mobility. The measure of intelligence chosen by the authors is the well-known intelligent quotient (IQ). Numerous tests have been developed to measure IQ; Herrnstein and Murray use the Armed Forces Qualification Test (AFQT), originally designed to measure the cognitive ability of military recruits. Psychologists traditionally treat IQ as a random variable having a normal distribution with mean $\mu = 100$ and standard deviation $\sigma = 15$.

In their book, Herrnstein and Murray refer to five cognitive classes of people defined by percentiles of the normal distribution. Class I ("very bright") consists of those with IQs above the 95th percentile; Class II ("bright") are those with IQs between the 75th and 95th percentiles; Class III ("normal") includes IQs between the 25th and 75th percentiles; Class IV ("dull") are those with IQs between the 5th and 25th percentiles; and Class V ("very dull") are IQs below the 5th percentile.

a. Assuming that the distribution of IQ is accurately represented by the normal curve, determine the proportion of people with IQs in each of the five cognitive classes defined by Herrnstein and Murray.

b. Although Herrnstein and Murray define the cognitive classes in terms of percentiles, they stress that IQ scores should be compared with z-scores, not percentiles. In other words, it is more informative to give the difference in z-scores for two IQ scores than it is to give the difference in percentiles. Do you agree?

c. Researchers have found that scores on many intelligence tests are decidedly nonnormal. Some distributions are skewed toward higher scores, others toward lower scores. How would the proportions in the five cognitive classes defined in part **a** differ for an IQ distribution that is skewed right? Skewed left?

SUMMARY ACTIVITY: Demonstration of Central Limit Theorem

To understand the Central Limit Theorem and sampling distribution, consider the following experiment: Toss four identical coins, and record the number of heads observed. Then repeat this experiment four more times, so that you end up with a total of five observations for the random variable x, the number of heads when four coins are tossed.

Now derive and graph the probability distribution for x, assuming that the coins are balanced. Note that the mean of this distribution is $\mu = 2$ and the standard deviation is $\sigma = 1$. This probability distribution represents the one from which you are drawing a random sample of five measurements.

Next, calculate the mean $\bar{x}$ of the five measurements; that is, calculate the mean number of heads you observed in five repetitions of the experiment. Although you have repeated the basic experiment five times, you have only one observed value

of $\bar{x}$. To derive the probability distribution, or sampling distribution, of $\bar{x}$ empirically, you have to repeat the entire process (of tossing four coins five times) many times. Do it 100 times.

The approximate sampling distribution of $\bar{x}$ can be derived theoretically by making use of the central limit theorem. We expect at least an approximate normal probability distribution with a mean $\mu = 2$ and a standard deviation.

$$\sigma_{\bar{x}} = \frac{\sigma}{\sqrt{n}} = \frac{1}{\sqrt{5}} = .45$$

Count the number of your 100 $\bar{x}$'s that fall in to each of the intervals in the accompanying figure. Use the normal probability distribution with $\mu = 2$ and $\sigma_{\bar{x}} = .45$ to calculate the expected number of the 100 $\bar{x}$'s in each of the intervals. How closely does the theory describe your experimental results?

Interval $\leftarrow 1 \rightarrow$	Interval $\leftarrow 2 \rightarrow$	Interval $\leftarrow 3 \rightarrow$	Interval $\leftarrow 4 \rightarrow$	Interval $\leftarrow 5 \rightarrow$	Interval $\leftarrow 6 \rightarrow$
	1.10	1.55	2	2.45	2.90
	$\mu - 2\sigma_{\bar{x}}$	$\mu - \sigma_{\bar{x}}$	μ	$\mu + \sigma_{\bar{x}}$	$\mu + 2\sigma_{\bar{x}}$

REFERENCES

Hogg, R. V., and Craig, A. T. *Introduction to Mathematical Statistics*, 5th ed. Englewood Cliffs, NJ: Prentice Hall, 1995.

Larsen, R. J., and Marx, M. L. *An Introduction to Mathematical Statistics and Its Applications*, 3d ed. Upper Saddle River, NJ: Prentice Hall, 2001.

Lindgren, B. W. *Statistical Theory*, 3d ed. New York: Macmillan, 1976.

Mood, A. M., Graybill, F. A., and Boes, D. C. *Introduction to the Theory of Statistics*, 3d ed. New York: McGraw-Hill, 1974.

Parzen, E. *Modern Probability Theory and Its Applications*. New York: Wiley, 1960.

Ramsey, P. P., and Ramsey, P. H. "Simple tests of normality in small samples." *Journal of Quality Technology*, Vol. 22, 1990.

Ross, S. M. *Stochastic Processes*, 2d ed. New York: Wiley, 1996.

Wackerly, D., Mendenhall, W., and Scheaffer, R.L. *Mathematical Statistics with Applications*, 6th ed. North Scituate, MA: Duxbury, 2002.

Winkler, R. L., and Hays, W. *Statistics: Probability, Inference, and Decision*, 2d ed. New York: Holt, Rinehart and Winston, 1975.

Using Technology: Binomial Probabilities, Normal Probabilities, and Simulated Sampling Distributions with MINITAB

Binomial Probabilities:

To obtain binomial probabilities with the use of MINITAB, first enter the values of x that you desire probabilities for in a column (e.g., C1) on the MINITAB worksheet. Now click on the "Calc" button on the MINITAB menu bar, next click on "Probability Distributions," and then click on "Binomial", as shown in Figure 4.M.1. The resulting dialog box appears as shown in Figure 4.M.2. Select either "Probabilities" or "Cumulative probabilities," specify the parameters of the distribution (e.g., sample size n

(continued)

and probability of success p), and enter C1 in the "Input column." When you click "OK", the binomial probabilities for the values of x (saved in C1) will appear on the MINITAB worksheet.

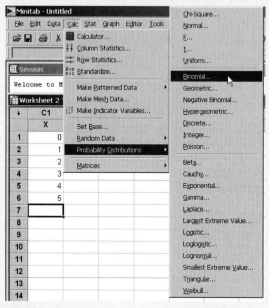

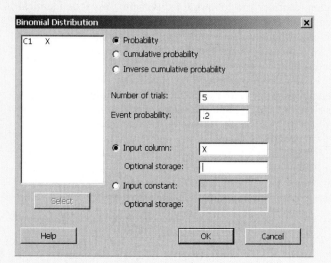

FIGURE 4.M.2
MINITAB binomial distribution dialog box

FIGURE 4.M.1
MINITAB menu options for obtaining probabilities

Normal Probabilities:

To obtain cumulative probabilities for the normal random variable with MINITAB, first click on the "Calc" button on the MINITAB menu bar, then click on "Probability Distributions," and finally click on "Normal", as shown in Figure 4.M.3. The resulting dialog box appears as shown in Figure 4.M.4. Select "Cumulative probability," specify the parameters of the distribution (e.g., the mean μ and standard deviation σ), and enter the value of x in the "Input constant" box. When you click "OK," the cumulative normal probability for the value of x will appear in the MINITAB session window.

To obtain a normal probability plot by using MINITAB, click on the "Graph" button on the MINITAB menu bar. Then click on "Probability Plot", as shown in Figure 4.M.5. Select "Single" (for one variable) on the next box, and the dialog box

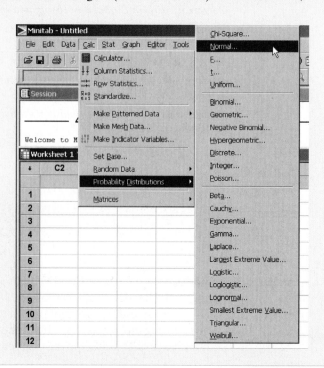

FIGURE 4.M.3
MINITAB menu options for obtaining normal probabilities

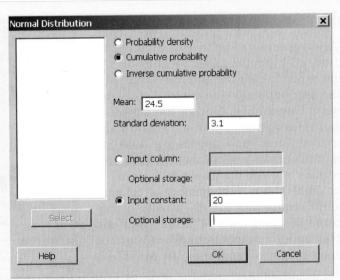

FIGURE 4.M.4
MINITAB normal distribution
dialog box

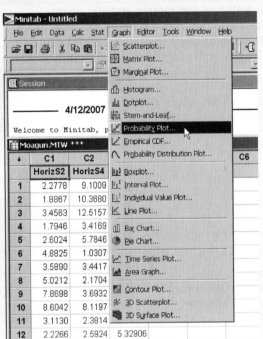

FIGURE 4.M.5
MINITAB options for a
normal probability plot

will appear as shown in Figure 4.M.6. Specify the variable of interest in the "Graph variables" box. Then click the "Distribution" button and select the "Normal" option. Click "OK" to return to the Probability Plot dialog box, and then click "OK" to generate the normal probability plot.

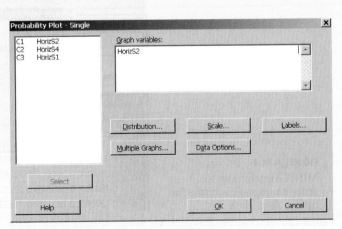

FIGURE 4.M.6
MINITAB probability plot
dialog box

(continued)

CHAPTER 5

Inferences Based on a Single Sample

Estimation with Confidence Intervals

CONTENTS

5.1 Identifying the Target Parameter

5.2 Large-Sample Confidence Interval for a Population Mean

5.3 Small-Sample Confidence Interval for a Population Mean

5.4 Large-Sample Confidence Interval for a Population Proportion

5.5 Determining the Sample Size

STATISTICS IN ACTION

Speed—Can a High School Football Player Improve His Sprint Time?

USING TECHNOLOGY

Confidence Intervals with MINITAB

WHERE WE'VE BEEN

■ Learned that populations are characterized by numerical descriptive measures—called *parameters*

■ Found that decisions about population parameters are based on *statistics* computed from the sample

■ Discovered that *inferences* about parameters are subject to uncertainty and that this uncertainty is reflected in the *sampling distribution* of a statistic

WHERE WE'RE GOING

■ Estimate a population parameter (means or proportion) on the basis of a large sample selected from the population

■ Use the sampling distribution of a statistic to form a confidence interval for the population parameter

■ Show how to select the proper sample size for estimating a population parameter

STATISTICS IN ACTION

Speed—Can a High School Football Player Improve His Sprint Time?

The game of football requires both skill and speed. Consider that one of the key "statistics" for National Football League (NFL) coaches in evaluating college players for the NFL draft is a player's 40-yard sprint time. There are a plethora of drills available to football coaches that have been successfully used to aid a player's skill development. However, many coaches are under the impression that "you can't teach speed." Michael Gray and Jessica Sauerbeck, researchers at Northern Kentucky University, designed and tested a speed training program for junior varsity and varsity high school football players (The Sport Journal, *Winter 2004*).

The training program was carried out over a five-week period and incorporated a number of speed improvement drills, including 50-yard sprints run at varying speeds, high-knee running sprints, butt-kick sprints, "crazy legs" straddle runs, quick-feet drills, jumping, power skipping, and all-out sprinting. A sample of 38 high school athletes participated in the study. Each participant was timed in a 40-yard sprint prior to the start of the program and timed again after completing the program. The decrease in times (measured in seconds) was recorded for each athlete. These data, saved in the **SPRINT** file, are shown in Table SIA5.1. [*Note*: A negative decrease implies that the athlete's time after completion of the program was higher than his time prior to training.] The goal of the research is to demonstrate that the training program is effective in improving 40-yard sprint times.

In this chapter, several Statistics in Action Revisited examples demonstrate how confidence intervals can be used to evaluate the effectiveness of the speed training program.

Statistics in Action Revisited

- Estimating the Mean Decrease in Sprint Time (p. 261)
- Estimating the Proportion of Sprinters Who Improve after Speed Training (p. 280)
- Determining the Number of Athletes to Participate in the Training Program (p. 287)

SPRINT

TABLE SIA5.1 Decrease in 40-Yard Sprint Times for 38 Football Players

−.01	.10	.10	.24	.25	.05	.28	.25	.20	.14
.32	.34	.30	.09	.05	0.00	.04	.17	0.00	.21
.15	.30	.02	.12	.14	.10	.08	.50	.36	.10
.01	.90	.34	.38	.44	.08	0.00	0.00		

Source: Gray, M., and Sauerbeck, J. A. "Speed training program for high school football players," *The Sport Journal*, Vol. 7, No. 1, Winter 2004 (Table 2).

5.1 Identifying the Target Parameter

In this chapter, our goal is to estimate the value of an unknown population parameter, such as a population mean or a proportion from a binomial population. For example, we might want to know the mean gas mileage for a new car model, the average expected life of a flat-screen computer monitor, or the proportion of Iraq War veterans with post-traumatic stress syndrome.

You'll see that different techniques are used for estimating a mean or proportion, depending on whether a sample contains a large or small number of measurements. Nevertheless, our objectives remain the same: We want to use the sample information to estimate the population parameter of interest (called the *target parameter*) and to assess the reliability of the estimate.

> **Definition 5.1**
>
> The unknown population parameter (e.g., mean or proportion) that we are interested in estimating is called the **target parameter.**

Often, there are one or more key words in the statement of the problem that indicate the appropriate target parameter. Some key words associated with the two parameters covered in this section are listed in the following box:

Determining the Target Parameter

Parameter	Key Words or Phrases	Type of Data
μ	Mean; average	Quantitative
p	Proportion; percentage; fraction; rate	Qualitative

For the examples given in the first paragraph of this section, the words *mean* in "mean gas mileage" and *average* in "average life expectancy" imply that the target parameter is the population mean μ. The word *proportion* in "proportion of Iraq War veterans with post-traumatic stress syndrome" indicates that the target parameter is the binomial proportion p.

In addition to key words and phrases, the type of data (quantitative or qualitative) collected is indicative of the target parameter. With quantitative data, you are likely to be estimating the mean of the data. With qualitative data with two outcomes (success or failure), the binomial proportion of successes is likely to be the parameter of interest.

We consider a method of estimating a population mean on the basis of a large sample in Section 5.2 and a small sample in Section 5.3. Estimating a population proportion is presented in Section 5.4. Finally, we show how to determine the sample sizes necessary for reliable estimates of the target parameters in Section 5.5.

5.2 Large-Sample Confidence Interval for a Population Mean

Suppose a large hospital wants to estimate the average length of time patients remain in the hospital. Hence, the hospital's target parameter is the population mean μ. To accomplish this objective, the hospital administrators plan to randomly sample 100 of all previous patients' records and to use the sample mean $\bar{x}$ of the lengths of stay to estimate μ, the mean of *all* patients' visits. The sample mean $\bar{x}$ represents a *point estimator* of the population mean μ. How can we assess the accuracy of this large-sample point estimator?

According to the central limit theorem, the sampling distribution of the sample mean is approximately normal for large samples, as shown in Figure 5.1.

Let us calculate the interval

$$\bar{x} \pm 2\sigma_{\bar{x}} = \bar{x} \pm \frac{2\sigma}{\sqrt{n}}$$

That is, we form an interval four standard deviations wide—from two standard deviations below the sample mean to two standard deviations above the mean. Prior to drawing the sample, what are the chances that this interval will enclose μ, the population mean?

To answer this question, refer to Figure 5.1. If the 100 measurements yield a value of $\bar{x}$ that falls between the two lines on either side of μ

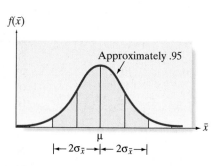

$f(\bar{x})$

Approximately .95

μ

$|\!\leftarrow 2\sigma_{\bar{x}} \rightarrow\!|\!\leftarrow 2\sigma_{\bar{x}} \rightarrow\!|$

FIGURE 5.1

Sampling distribution of $\bar{x}$

(i.e., within two standard deviations of μ), then the interval $\bar{x} \pm 2\sigma_{\bar{x}}$ will contain μ; if $\bar{x}$ falls outside these boundaries, the interval $\bar{x} \pm 2\sigma_{\bar{x}}$ will not contain μ. Since the area under the normal curve (the sampling distribution of $\bar{x}$) between these boundaries is about .95 (more precisely, from Table III in Appendix A, the area is .9544), the interval $\bar{x} \pm 2\sigma_{\bar{x}}$ will contain μ with a probability approximately equal to .95.

For instance, consider the lengths of time spent in the hospital for 100 patients, shown in Table 5.1. A SAS printout of summary statistics for the sample of 100 lengths of stay is shown in Figure 5.2.

From the top of the printout, we find that $\bar{x} = 4.53$ days and $s = 3.68$ days. To achieve our objective, we must construct the interval

$$\bar{x} \pm 2\sigma_{\bar{x}} = 4.53 \pm 2\frac{\sigma}{\sqrt{100}}$$

But now we face a problem. You can see that without knowing the standard deviation σ of the original population—that is, the standard deviation of the lengths of stay of *all* patients—we cannot calculate this interval. However, since we have a large sample ($n = 100$ measurements), we can approximate the interval by using the sample standard deviation s to approximate σ. Thus,

$$\bar{x} \pm 2\frac{\sigma}{\sqrt{100}} \approx \bar{x} \pm 2\frac{s}{\sqrt{100}} = 4.53 \pm 2\left(\frac{3.68}{10}\right) = 4.53 \pm .74$$

That is, we estimate that the mean length of stay in the hospital for all patients falls into the interval from 3.79 to 5.27 days. (This interval is highlighted at the bottom of the SAS printout in Figure 5.2. Differences are due to rounding.)

HOSPLOS

TABLE 5.1   Lengths of Stay (in Days) for 100 Patients

2	3	8	6	4	4	6	4	2	5
8	10	4	4	4	2	1	3	2	10
1	3	2	3	4	3	5	2	4	1
2	9	1	7	17	9	9	9	4	4
1	1	1	3	1	6	3	3	2	5
1	3	3	14	2	3	9	6	6	3
5	1	4	6	11	22	1	9	6	5
2	2	5	4	3	6	1	5	1	6
17	1	2	4	5	4	4	3	2	3
3	5	2	3	3	2	10	2	4	2

```
Sample Statistics for LOS

      N           Mean          Std. Dev.        Std. Error
   ---------------------------------------------------------
     100          4.53             3.68              0.37

Hypothesis Test

     Null hypothesis:     Mean of LOS  =   0
     Alternative:         Mean of LOS  ^= 0

              t Statistic        Df        Prob > t
           ---------------------------------------
                 12.318          99         <.0001


95 % Confidence Interval for the Mean

        Lower Limit:                   3.80
        Upper Limit:                   5.26
```

FIGURE 5.2

SAS printout with summary statistics and 95% confidence interval for data on 100 hospital stays

Can we be sure that μ, the true mean, is in the interval from 3.79 to 5.27? We cannot be certain, but we can be reasonably confident that it is. This confidence is derived from the knowledge that if we were to draw repeated random samples of 100 measurements from this population and form the interval $\bar{x} \pm 2\sigma_{\bar{x}}$ each time, approximately 95% of the intervals would contain μ. We have no way of knowing (without looking at all the patients' records) whether our sample interval is one of the 95% that contain μ or one of the 5% that do not, but the odds certainly favor its containing μ. Consequently, the interval from 3.79 to 5.27 provides a reliable estimate of the mean length of patient stay in the hospital.

The formula that tells us how to calculate an interval estimate on the basis of sample data is called an *interval estimator* or *confidence interval*. The probability, .95, that measures the confidence we can place in the interval estimate is called a *confidence coefficient*. The percentage, 95%, is called the *confidence level* for the interval estimate. It is not usually possible to assess precisely the reliability of point estimators, because they are single points rather than intervals. So, since we prefer to use estimators for which a measure of reliability can be calculated, we will generally use interval estimators.

> ### Definition 5.2
>
> An **interval estimator** (or **confidence interval**) is a formula that tells us how to use sample data to calculate an interval that estimates a population parameter.

> ### Definition 5.3
>
> The **confidence coefficient** is the probability that an interval estimator encloses the population parameter—that is, the relative frequency with which the interval estimator encloses the population parameter when the estimator is used repeatedly a very large number of times. The **confidence level** is the confidence coefficient expressed as a percentage.

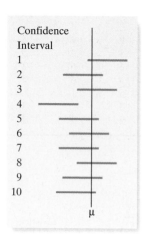

FIGURE 5.3

Confidence intervals for μ: 10 samples

Now we have seen how an interval can be used to estimate a population mean. When we use an interval estimator, we can usually calculate the probability that the estimation *process* will result in an interval that contains the true value of the population mean. That is, the probability that the interval contains the parameter in repeated usage is usually known. Figure 5.3 shows what happens when 10 different samples are drawn from a population and a confidence interval for μ is calculated from each. The location of μ is indicated by the vertical line in the figure. Ten confidence intervals, each based on one of 10 samples, are shown as horizontal line segments. Note that the confidence intervals move from sample to sample, sometimes containing μ and other times missing μ. *If our confidence level is 95%, then in the long run, 95% of our sample confidence intervals will contain μ.*

Suppose you wish to choose a confidence coefficient other than .95. Notice in Figure 5.1 that the confidence coefficient .95 is equal to the total area under the sampling distribution, less .05 of the area, which is divided equally between the two tails. Using this idea, we can construct a confidence interval with any desired confidence coefficient by increasing or decreasing the area (call it α) assigned to the tails of the sampling distribution. (See Figure 5.4.) For example, if we place the area $\alpha/2$ in each tail and if $z_{\alpha/2}$ is the z value such that $\alpha/2$ will lie to its right, then the confidence interval with confidence coefficient is $(1 - \alpha)$ is

$$\bar{x} \pm z_{\alpha/2}\sigma_{\bar{x}}$$

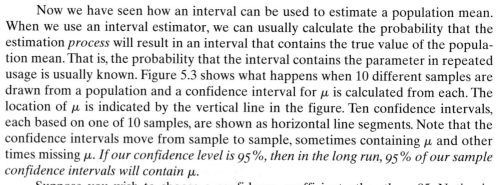

FIGURE 5.4

Locating $z_{\alpha/2}$ on the standard normal curve

Biography

**JERZY NEYMAN (1894–1981)—
Speaking Statistics with a Polish Accent**

Polish-born Jerzy Neyman was educated at the University of Kharkov (Russia) in elementary mathematics, but taught himself graduate mathematics by studying journal articles on the subject. After receiving his doctorate in 1924 from the University of Warsaw (Poland), Neyman accepted a position at University College (London). There, he developed a friendship with Egon Pearson; Neyman and Pearson together developed the theory of hypothesis testing (Chapter 8). In a 1934 talk to the Royal Statistical Society, Neyman first pro-posed the idea of interval estimation, which he called "confidence intervals." (It is interesting that Neyman rarely receives credit in text-books as the originator of the confidence interval procedure.) In 1938, he emigrated to the United States and went to the University of California at Berkeley, where he built one of the strongest statistics departments in the country. Jerzy Neyman is considered one of the great founders of modern statistics. He was a superb teacher and innovative researcher who loved his students, always sharing his ideas with them. Neyman's influence on those he met is best expressed by a quote from prominent statistician David Salsburg: "We have all learned to speak statistics with a Polish accent."

Definition 5.4

The value z_α is defined as the value of the standard normal random variable z such that the area α will lie to its right. In other words, $P(z > z_\alpha) = \alpha$.

To illustrate, for a confidence coefficient of .90, we have $(1 - \alpha) = .90$, $\alpha = .10$, and $\alpha/2 = .05$; $z_{.05}$ is the z value that locates area .05 in the upper tail of the sampling distribution. Recall that Table III in Appendix A gives the areas between the mean and a specified z-value. Since the total area to the right of the mean is .5, we find that $z_{.05}$ will be the z value corresponding to an area of $.5 - .05 = .45$ to the right of the mean. (See Figure 5.5.) This z value is $z_{.05} = 1.645$.

Confidence coefficients used in practice usually range from .90 to .99. The most commonly used confidence coefficients with corresponding values of α and $z_{\alpha/2}$ are shown in Table 5.2.

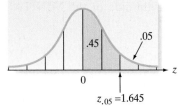

FIGURE 5.5

The z value ($z_{.05}$) corresponding to an area equal to .05 in the upper tail of the z-distribution

TABLE 5.2 Commonly Used Values of $z_{\alpha/2}$

Confidence Level $100(1 - \alpha)$	α	$\alpha/2$	$z_{\alpha/2}$
90%	.10	.05	1.645
95%	.05	.025	1.96
99%	.01	.005	2.575

Now Work Exercise 5.7

Large-Sample $100(1 - \alpha)\%$ Confidence Interval for μ

The large-sample $100(1 - \alpha)\%$ confidence interval for μ is

$$\bar{x} \pm z_{\alpha/2}\sigma_{\bar{x}} = \bar{x} \pm z_{\alpha/2}\frac{\sigma}{\sqrt{n}}$$

where $z_{\alpha/2}$ is the z value with an area $\alpha/2$ to its right (see Figure 7.4) and $\sigma_{\bar{x}} = \sigma/\sqrt{n}$. The parameter σ is the standard deviation of the sampled population and n is the sample size.

Note: When σ is unknown (as is almost always the case) and n is large (say, $n \geq 30$), the confidence interval is approximately equal to

$$\bar{x} \pm z_{\alpha/2}\left(\frac{s}{\sqrt{n}}\right)$$

where s is the sample standard deviation.

Conditions Required for a Valid Large-Sample Confidence Interval for μ

1. A random sample is selected from the target population.
2. The sample size n is large (i.e., $n \geq 30$). (Due to the central limit theorem, this condition guarantees that the sampling distribution of $\bar{x}$ is approximately normal.)

EXAMPLE 5.1 A LARGE-SAMPLE CONFIDENCE INTERVAL FOR μ—Mean Number of Unoccupied Seats per Flight

AIRNOSHOWS

Problem Unoccupied seats on flights cause airlines to lose revenue. Suppose a large airline wants to estimate its average number of unoccupied seats per flight over the past year. To accomplish this, the records of 225 flights are randomly selected, and the number of unoccupied seats is noted for each of the sampled flights. (The data are saved in the **AIRNOSHOWS** file.) Descriptive statistics for the data are displayed in the MINITAB printout of Figure 5.6.

Estimate μ, the mean number of unoccupied seats per flight during the past year, using a 90% confidence interval.

Solution The general form of the 90% confidence interval for a population mean is

$$\bar{x} \pm z_{\alpha/2}\sigma_{\bar{x}} = \bar{x} \pm z_{.05}\sigma_{\bar{x}} = \bar{x} \pm 1.645\left(\frac{\sigma}{\sqrt{n}}\right)$$

From Figure 7.6, we find (after rounding) that $\bar{x} = 11.6$. Since we do not know the value of σ (the standard deviation of the number of unoccupied seats per flight for all flights of the year), we use our best approximation—the sample standard deviation, $s = 4.1$ shown on the MINITAB printout.Then the 90% confidence interval is approximately

$$11.6 \pm 1.645\left(\frac{4.1}{\sqrt{225}}\right) = 11.6 \pm .45$$

or from 11.15 to 12.05. That is, at the 90% confidence level, we estimate the mean number of unoccupied seats per flight to be between 11.15 and 12.05 during the sampled year. This result is verified (except for rounding) on the right side of the MINITAB printout in Figure 5.6.

FIGURE 5.6
MINITAB printout with descriptive statistics and 90% confidence interval for Example 5.1

Variable	N	Mean	StDev	SE Mean	90% CI
NOSHOWS	225	11.5956	4.1026	0.2735	(11.1438, 12.0473)

Look Back We stress that the confidence level for this example, 90%, refers to the procedure used. If we were to apply that procedure repeatedly to different samples, approximately 90% of the intervals would contain μ. Although we do not know for sure whether this particular interval (11.15, 12.05) is one of the 90% that contain μ or one of the 10% that do not, our knowledge of probability gives us "confidence" that the interval contains μ.

Now Work Exercise 5.11

The interpretation of confidence intervals for a population mean is summarized in the next box.

Interpretation of a Confidence Interval for a Population Mean

When we form a $100(1 - \alpha)\%$ confidence interval for μ, we usually express our confidence in the interval with a statement such as "We can be $100(1 - \alpha)\%$ confident that μ lies between the lower and upper bounds of the confidence interval," where, for a particular application, we substitute the appropriate numerical values for the level of

(continued)

confidence and for the lower and upper bounds. *The statement reflects our confidence in the estimation process, rather than in the particular interval that is calculated from the sample data.* We know that repeated application of the same procedure will result in different lower and upper bounds on the interval. Furthermore, we know that $100(1 - \alpha)\%$ of the resulting intervals will contain μ. There is (usually) no way to determine whether any particular interval is one of those which contain μ or one of those which do not. However, unlike point estimators, confidence intervals have some measure of reliability—the confidence coefficient—associated with them. For that reason, they are generally preferred to point estimators.

Sometimes, the estimation procedure yields a confidence interval that is too wide for our purposes. In this case, we will want to reduce the width of the interval to obtain a more precise estimate of μ. One way to accomplish that is to decrease the confidence coefficient, $1 - \alpha$. For example, consider the problem of estimating the mean length of stay, μ for hospital patients. Recall that for a sample of 100 patients, $\bar{x} = 4.53$ days and $s = 3.68$ days. A 90% confidence interval for μ is

$$\bar{x} \pm 1.645(\sigma/\sqrt{n}) \approx 4.53 \pm (1.645)(3.68)/\sqrt{100} = 4.53 \pm .61$$

or (3.92, 5.14). You can see that this interval is narrower than the previously calculated 95% confidence interval, (3.79, 5.27). Unfortunately, we also have "less confidence" in the 90% confidence interval. An alternative method used to decrease the width of an interval without sacrificing "confidence" is to increase the sample size n. We demonstrate this method in Section 5.5.

STATISTICS IN ACTION REVISITED

Estimating the Mean Decrease in Sprint Time

Refer to the speed training program for junior varsity and varsity high school football players described on p. 255. Recall that the five-week training program incorporated a number of speed improvement drills for a sample of 38 high school athletes. The time in a 40-yard sprint run both before and after the training program was recorded for each athlete, and the decreases in time (measured in seconds) were saved in the **SPRINT** file (Table SIA5.1). Is the training program really effective in improving 40-yard sprint times?

One way to answer this question is to form a confidence interval for the true mean decrease in sprint time for all athletes who participate in the speed training program. If the true decrease is positive (i.e., if the "before" mean time is greater than the "after" mean time), then we conclude that the training program is effective. Since the sample size is large ($n = 38$), we can apply the large-sample methodology of this section.

The data are stored as a MINITAB worksheet and the software used to find a large-sample 95% confidence interval for the population mean decrease in sprint times. The MINITAB printout is displayed in Figure SIA5.1.

The interval, highlighted on the printout, is (.128, .247). Note that the entire interval falls above 0. Consequently, we are 95% confident that the true mean decrease in sprint times is positive. It appears that the speed training program is, in fact, effective in improving the average 40-yard sprint times of high school athletes. Note, however, that the mean decrease in time ranges from a minimum of .128 second to a maximum of .247 second. Although this decrease might be deemed critically important to world-class sprinters, it would probably be unnoticeable to high school football players.

Variable	N	Mean	StDev	SE Mean	95% CI
DecrTime	38	0.187895	0.181423	0.029431	(0.128262, 0.247527)

FIGURE SIA5.1

MINITAB confidence interval for speed training study

Confidence Interval for a Population Mean (known σ or $n \geq 30$)

Using the TI-84/TI-83 Graphing Calculator

Creating a Confidence Interval for a Population Mean

Step 1   *Enter the data. (Skip to Step 2 if you have summary statistics, not raw data.)*

Press **STAT** and select **1:Edit**.
Note: If the list already contains data, clear the old data. Use the up **ARROW** to highlight 'L1'.
Press **CLEAR ENTER**.
Use the **ARROW** and **ENTER** keys to enter the data set into **L1**.

Step 2   *Access the Statistical Tests Menu*

Press **STAT**.
Arrow right to **TESTS**.
Arrow down to **Zinterval**.
Press **ENTER**.

Step 3   *Choose "**Data**" or "**Stats**". ("Data" is selected when you have entered the raw data into a List. "Stats" is selected when you are given only the mean, standard deviation, and sample size.)*

Press **ENTER**.
If you selected "Data", enter a value for σ.
(The best approximation is s, the sample standard deviation.)
Set **List** to **L1**.
Set **Freq** to **1**.
Set **C-Level** to the confidence level.
Arrow down to "**Calculate**".
Press **ENTER**.

If you selected "Stats", enter a value for σ.
(The best approximation is s, the sample standard deviation.)
Enter the sample mean and sample size.
Set **C-Level** to the confidence level.
Arrow down to "**Calculate**".
Press **ENTER**.
(The screen at the right is set up for an example with a standard deviation of 20, a mean of 200, and a sample size of 40.)
The confidence interval will be displayed along with the sample mean and the sample size.

Example   Use the the following information to compute a 90% confidence interval for the mean number of unoccupied seats per flight:

$$\overline{x} = 11.6 \text{ seats}, \qquad s = 4.1 \text{ seats}, \qquad n = 225 \text{ flights}$$

As you can see from the screen, our 90% confidence interval is (11.5, 12.05): You will also notice that the output displays the sample mean and sample size.

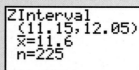

Exercises 5.1–5.24

Understanding the Principles

5.1 Define the target parameter.

5.2 What is the confidence coefficient in a 90% confidence interval for μ?

5.3 Explain the difference between an interval estimator and a point estimator for μ.

5.4 Explain what is meant by the statement "We are 95% confident that an interval estimate contains μ."

5.5 Will a large-sample confidence interval be valid if the population from which the sample is taken is not normally distributed? Explain.

5.6 What conditions are required to form a valid large-sample confidence interval for μ?

Learning the Mechanics

5.7 Find $z_{\alpha/2}$ for each of the following:
NW **a.** $\alpha = .10$ **b.** $\alpha = .01$
    c. $\alpha = .05$ **d.** $\alpha = .20$

5.8 What is the confidence level of each of the following confidence intervals for μ?

a. $\bar{x} \pm 1.96 \left(\dfrac{\sigma}{\sqrt{n}} \right)$

b. $\bar{x} \pm 1.645 \left(\dfrac{\sigma}{\sqrt{n}} \right)$

c. $\bar{x} \pm 2.575 \left(\dfrac{\sigma}{\sqrt{n}} \right)$

d. $\bar{x} \pm 1.28 \left(\dfrac{\sigma}{\sqrt{n}} \right)$

e. $\bar{x} \pm .99 \left(\dfrac{\sigma}{\sqrt{n}} \right)$

5.9 A random sample of n measurements was selected from a population with unknown mean μ and standard deviation σ. Calculate a 95% confidence interval for μ for each of the following situations:
a. $n = 75$, $\bar{x} = 28$, $s^2 = 12$
b. $n = 200$, $\bar{x} = 102$, $s^2 = 22$
c. $n = 100$, $\bar{x} = 15$, $s = .3$
d. $n = 100$, $\bar{x} = 4.05$, $s = .83$
e. Is the assumption that the underlying population of measurements is normally distributed necessary to ensure the validity of the confidence intervals in parts **a–d**? Explain.

5.10 A random sample of 90 observations produced a mean $\bar{x} = 25.9$ and a standard deviation $s = 2.7$.
a. Find a 95% confidence interval for the population mean μ.
b. Find a 90% confidence interval for μ.
c. Find a 99% confidence interval for μ.

5.11 A random sample of 100 observations from a normally dis-
NW tributed population possesses a mean equal to 83.2 and a standard deviation equal to 6.4.
a. Find a 95% confidence interval for μ.
b. What do you mean when you say that a confidence coefficient is .95?
c. Find a 99% confidence interval for μ.
d. What happens to the width of a confidence interval as the value of the confidence coefficient is increased while the sample size is held fixed?

e. Would your confidence intervals of parts **a** and **c** be valid if the distribution of the original population were not normal? Explain.

5.12 The mean and standard deviation of a random sample of n measurements are equal to 33.9 and 3.3, respectively.
a. Find a 95% confidence interval for μ if $n = 100$.
b. Find a 95% confidence interval for μ if $n = 400$.
c. Find the widths of the confidence intervals you calculated in parts **a** and **b**. What is the effect on the width of a confidence interval of quadrupling the sample size while holding the confidence coefficient fixed?

APPLET **Applet Exercise 5.1**

Use the applet entitled *Confidence Intervals for a Mean (the impact of confidence level)* to investigate the situation in Exercise 5.11 further. For this exercise, assume that $\mu = 83.2$ is the population mean and $\sigma = 6.4$ is the population standard deviation.
a. Using $n = 100$ and the normal distribution with mean and standard deviation as just given, run the applet one time. How many of the 95% confidence intervals contain the mean? How many would you expect to contain the mean? How many of the 99% confidence intervals contain the mean? How many would you expect to contain the mean?
b. Which confidence level has a greater frequency of intervals that contain the mean? Is this result what you would expect? Explain.
c. Without clearing, run the applet several more times. What happens to the proportion of 95% confidence intervals that contain the mean as you run the applet more and more? What happens to the proportion of 99% confidence intervals that contain the mean as you run the applet more and more? Interpret these results in terms of the meanings of the 95% confidence interval and the 99% confidence interval.
d. Change the distribution to *right skewed*, clear, and run the applet several more times. Do you get the same results as in part **c?** Would you change your answer to part **e** of Exercise 5.11? Explain.

APPLET **Applet Exercise 5.2**

Use the applet entitled *Confidence Intervals for a Mean (the impact of confidence level)* to investigate the effect of the sample size on the proportion of confidence intervals that contain the mean when the underlying distribution is skewed. Set the distribution to *right skewed*, the mean to 10, and the standard deviation to 1.
a. Using $n = 30$, run the applet several times without clearing. What happens to the proportion of 95% confidence intervals that contain the mean as you run the applet more and more? What happens to the proportion of 99% confidence intervals that contain the mean as you run the applet more and more? Do the proportions seem to be approaching the values that you would expect?
b. Clear and run the applet several times, using $n = 100$. What happens to the proportions of 95% confidence intervals and 99% confidence intervals that contain the mean this time? How do these results compare with your results in part **a?**
c. Clear and run the applet several times, using $n = 1000$. How do the results compare with your results in parts **a** and **b?**

d. Describe the effect of sample size on the likelihood that a confidence interval contains the mean for a skewed distribution.

Applying the Concepts—Basic

5.13 Latex allergy in health-care workers. Health-care workers who use latex gloves with glove powder on a daily basis are particularly susceptible to developing a latex allergy. Symptoms of a latex allergy include conjunctivitis, hand eczema, nasal congestion, a skin rash, and shortness of breath. Each in a sample of 46 hospital employees who were diagnosed with latex allergy based on a skin-prick test reported on their exposure to latex gloves. (*Current Allergy & Clinical Immunology,* March 2004.) Summary statistics for the number of latex gloves used per week are $\bar{x} = 19.3$ and $s = 11.9$.

a. Give a point estimate for the average number of latex gloves used per week by all health-care workers with a latex allergy.

b. Form a 95% confidence interval for the average number of latex gloves used per week by all health-care workers with a latex allergy.

c. Give a practical interpretation of the interval you found in part **b**.

d. Give the conditions required for the interval in part **b** to be valid.

5.14 Personal networks of older adults. In sociology, a personal network is defined as the people with whom you make frequent contact. The Living Arrangements and Social Networks of Older Adults (LSN) research program used a stratified random sample of men and women born between 1908 and 1937 to gauge the size of the personal network of older adults. Each adult in the sample was asked to "please name the people (e.g., in your neighborhood) you have frequent contact with and who are also important to you." Based on the number of people named, the personal network size for each adult was determined. The responses of 2,819 adults in the LSN sample yielded the following statistics on network size: $\bar{x} = 14.6$; $s = 9.8$. (*Sociological Methods & Research,* August 2001.)

a. Give a point estimate for the mean personal network size of all older adults.

b. Form a 95% confidence interval for the mean personal network size of all older adults.

c. Give a practical interpretation of the interval you found in part **b**.

d. Give the conditions required for the interval in part **b** to be valid.

PONDICE

5.15 Albedo of ice melt ponds. Refer to the National Snow and Ice Data Center (NSIDC) collection of data on the albedo, depth, and physical characteristics of ice-melt ponds in the Canadian Arctic, presented in Exercise 2.11 (p. 35). Albedo is the ratio of the light reflected by the ice to that received by it. (High albedo values give a white appearance to the ice.) Visible albedo values were recorded for a sample of 504 ice-melt ponds located in the Barrow Strait in the Canadian Arctic; these data are saved in the **PONDICE** file.

a. Find a 90% confidence interval for the true mean visible albedo value of all Canadian Arctic ice ponds.

b. Give both a practical and a theoretical interpretation of the interval.

c. Recall from Exercise 2.11 that the type of ice for each pond was classified as first-year ice, multiyear ice, or landfast ice. Find 90% confidence intervals for the mean visible albedo for each of the three types of ice. Interpret the intervals.

NZBIRDS

5.16 Extinct New Zealand birds. Refer to the *Evolutionary Ecology Research* (July 2003) study of the patterns of extinction in the New Zealand bird population, presented in Exercise 2.96 (p. 73). Suppose you are interested in estimating the mean egg length (in millimeters) for the New Zealand bird population.

a. What is the target parameter?

b. Recall that the egg lengths for 132 bird species are saved in the **NZBIRDS** file. Obtain a random sample of 50 egg lengths from the data set.

c. Find the mean and standard deviation of the 50 egg lengths you obtained in part **b**.

d. Use the information from part **c** to form a 99% confidence interval for the true mean egg length of a bird species found in New Zealand.

e. Give a practical interpretation of the interval you found in part **d**.

Applying the Concepts—Intermediate

PERAGGR

5.17 Personality and aggressive behavior. How does personality impact aggressive behavior? A team of university psychologists conducted a review of studies that examined the relationship between personality and aggressive behavior (*Psychological Bulletin*, Vol. 132, 2006). One variable of interest to the researchers was the difference between the aggressive behavior level of individuals in the study who scored high on a personality test and those who scored low on the test. This variable, standardized to be between -7 and 7, was called "effect size". (A large positive effect size indicates that those who score high on the personality test are more aggressive than those who score low.) The researchers collected the effect sizes for a sample of $n = 109$ studies published in psychology journals. This data is saved in the **PERAGGR** file. A dot plot and summary statistics for effect size are shown in the MINITAB printouts on p. 265. Of interest to the researchers is the true mean effect size μ for all psychological studies of personality and aggressive behavior.

a. Identify the parameter of interest to the researchers.

b. Examine the dot plot. Does effect size have a normal distribution? Explain why your answer is irrelevant to the subsequent analysis.

c. Locate a 95% confidence interval for μ on the accompanying printout. Interpret the result.

d. If the true mean effect size exceeds 0, then the researchers will conclude that in the population, those who score high on a personality test are more aggressive than those who score low. Can the researchers draw this conclusion? Explain.

5.18 Sentence complexity study. Refer to the *Applied Psycholinguistics* (June 1998) study of language skills of low-income children, presented in Exercise 2.98 (p. 74). Each in a sample of 65 low-income children was administered the Communicative Development Inventory (CDI) exam. The

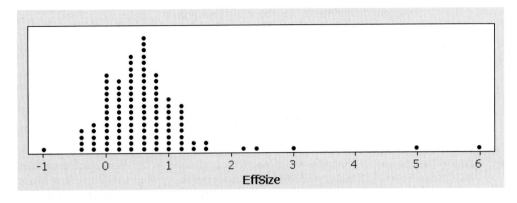

MINITAB Output for
Exercise 5.17

Variable	N	Mean	StDev	SE Mean	95% CI
EffSize	109	0.6477	0.8906	0.0853	(0.4786, 0.8167)

sentence complexity scores had a mean of 7.62 and a standard deviation of 8.91.

a. Construct a 90% confidence interval for the mean sentence complexity score of all low-income children.

b. Interpret the interval you found in part **a** in the words of the problem.

c. Suppose we know that the true mean sentence complexity score of middle-income children is 15.55. Is there evidence that the true mean for low-income children differs from 15.55? Explain.

5.19 Colored string preferred by chickens. Animal behaviorists have discovered that the more domestic chickens peck at objects placed in their environment, the healthier the chickens seem to be. White string has been found to be a particularly attractive pecking stimulus. In one experiment, 72 chickens were exposed to a string stimulus. Instead of white string, blue-colored string was used. The number of pecks each chicken took at the blue string over a specified interval of time was recorded. Summary statistics for the 72 chickens were $\bar{x} = 1.13$ pecks and $s = 2.21$ pecks. (*Applied Animal Behaviour Science*, October 2000.)

a. Use a 99% confidence interval to estimate the population mean number of pecks made by chickens pecking at blue string. Interpret the result.

b. Previous research has shown that $\mu = 7.5$ pecks if chickens are exposed to white string. Based on the results you found in part **a**, is there evidence that chickens are more apt to peck at white string than blue string? Explain.

5.20 Velocity of light from galaxies. Refer to *The Astronomical Journal* (July 1995) study of the velocity of light emitted from a galaxy in the universe, presented in Exercise 2.100 (p. 74). A sample of 103 galaxies located in the galaxy cluster A2142 had a mean velocity of $\bar{x} = 27,117$ kilometers per second (km/s) and a standard deviation of $s = 1,280$ km/s. Suppose your goal is to make an inference about the population mean light velocity of galaxies in cluster A2142.

a. In part **b** of Exercise 2.100, you constructed an interval that captured approximately 95% of the galaxy velocities in the cluster. Explain why that interval is inappropriate for the inference you are asked to make here.

b. Construct a 95% confidence interval for the mean light velocity emitted from all galaxies in cluster A2142. Interpret the result.

5.21 Improving SAT scores. Refer to the *Chance* (Winter 2001) and National Education Longitudinal Survey (NELS) study

of 265 students who paid a private tutor to help them improve their SAT scores, presented in Exercise 2.101 (p. 74). The changes in both the SAT-Mathematics and SAT-Verbal scores for these students are reproduced in the following table:

	SAT-Math	SAT-Verbal
Mean change in score	19	7
Standard deviation of score changes	65	49

a. Construct and interpret a 95% confidence interval for the population mean change in SAT-Mathematics score for students who pay a private tutor.

b. Repeat part **a** for the population mean change in SAT-Verbal score.

c. Suppose the true population mean change in score on one of the SAT tests for all students who paid a private tutor is 15. Which of the two tests, SAT-Mathematics or SAT-Verbal, is most likely to have this mean change? Explain.

5.22 Attention time given to twins. Psychologists have found that twins, in their early years, tend to have lower IQs and pick up language more slowly than nontwins. (*Wisconsin Twin Research Newsletter*, Winter 2004.) The slower intellectual growth of most twins may be caused by benign parental neglect. Suppose it is desired to estimate the mean attention time given to twins per week by their parents. A sample of 50 sets of $2\frac{1}{2}$-year-old twin boys is taken, and at the end of 1 week, the attention time given to each pair is recorded. The data (in hours) are listed in the following table:

ATTIMES

20.7	16.7	22.5	12.1	2.9
23.5	6.4	1.3	39.6	35.6
10.9	7.1	46.0	23.4	29.4
44.1	13.8	24.3	9.3	3.4
15.7	46.6	10.6	6.7	5.4
14.0	20.7	48.2	7.7	22.2
20.3	34.0	44.5	23.8	20.0
43.1	14.3	21.9	17.5	9.6
36.4	0.8	1.1	19.3	14.6
32.5	19.1	36.9	27.9	14.0

Find a 90% confidence interval for the mean attention time given to all twin boys by their parents. Interpret the confidence interval.

Applying the Concepts—Advanced

5.23 Study of undergraduate problem drinking. In *Alcohol & Alcoholism* (Jan./Feb. 2007), psychologists at the University of Pennsylvania compared the levels of alcohol consumption of male and female freshman students. Each student was asked to estimate the amount of alcohol (beer, wine, or liquor) they consume in a typical week. Summary statistics for 128 males and 184 females are provided in the accompanying table.

a. For each gender, find a 95% confidence interval for mean weekly alcohol consumption.

b. Prior to sampling, what is the probability that at least one of the two confidence intervals will not contain the population mean it estimates. Assume that the two intervals are independent.

c. Based on the two confidence intervals, what inference can you make about which gender consumes the most alcohol, on average, per week? (*Caution:* In Chapter 7, we will learn about a more valid method of comparing population means.)

	Males	Females
Sample size, n	128	184
Mean (ounces), $\bar{x}$	16.79	10.79
Standard deviation, s	13.57	11.53

Source: Leeman, R. F., Fenton, M., & Volpicelli, J.R. "Impaired control and undergraduate problem drinking," *Alcohol & Alcoholism*, Vol. 42, No. 1, Jan/Feb. 2007 (Table 1).

5.24 Study of cockroach growth. According to scientists, the cockroach has had 300 million years to develop a resistance to destruction. In a study conducted by researchers for S. C. Johnson & Son, Inc. (manufacturers of Raid® and Off®), 5,000 roaches (the expected number in a roach-infested house) were released in the Raid test kitchen. One week later, the kitchen was fumigated and 16,298 dead roaches were counted, a gain of 11,298 roaches for the 1-week period. Assume that none of the original roaches died during the 1-week period and that the standard deviation of x, the number of roaches produced per roach in a 1-week period, is 1.5. Use the number of roaches produced by the sample of 5,000 roaches to find a 95% confidence interval for the mean number of roaches produced per week for each roach in a typical roach-infested house.

5.3 Small-Sample Confidence Interval for a Population Mean

Federal legislation requires pharmaceutical companies to perform extensive tests on new drugs before they can be marketed. Initially, a new drug is tested on animals. If the drug is deemed safe after this first phase of testing, the pharmaceutical company is then permitted to begin human testing on a limited basis. During this second phase, inferences must be made about the safety of the drug on the basis of information obtained from very small samples.

Suppose a pharmaceutical company must estimate the average increase in blood pressure of patients who take a certain new drug. Assume that only six patients (randomly selected from the population of all patients) can be used in the initial phase of human testing. The use of a *small sample* in making an inference about μ presents two immediate problems when we attempt to use the standard normal z as a test statistic.

Problem 1   The shape of the sampling distribution of the sample mean $\bar{x}$ (and the z-statistic) now depends on the shape of the population that is sampled. We can no longer assume that the sampling distribution of $\bar{x}$ is approximately normal, because the central limit theorem ensures normality only for samples that are sufficiently large.

Solution to Problem 1   The sampling distribution of $\bar{x}$ (and z) is exactly normal even for relatively small samples if the sampled population is normal. It is approximately normal if the sampled population is approximately normal.

Problem 2   The population standard deviation σ is almost always unknown. Although it is still true that $\sigma_{\bar{x}} = \sigma/\sqrt{n}$, the sample standard deviation s may provide a poor approximation for σ when the sample size is small.

Solution to Problem 2   Instead of using the standard normal statistic

$$z = \frac{\bar{x} - \mu}{\sigma_{\bar{x}}} = \frac{\bar{x} - \mu}{\sigma/\sqrt{n}}$$

which requires knowledge of, or a good approximation to, σ, we define and use the statistic

$$t = \frac{\bar{x} - \mu}{s/\sqrt{n}}$$

in which the sample standard deviation s replaces the population standard deviation σ.

Biography

**WILLIAM S. GOSSET (1876–1937)—
Student's *t*-Distribution**

At the age of 23, William Gosset earned a degree in chemistry and mathematics at prestigious Oxford University. He was immediately hired by the Guinness Brewing Company in Dublin, Ireland, for his expertise in chemistry. However, Gosset's mathematical skills allowed him to solve numerous practical problems associated with brewing beer. For example, Gosset applied the Poisson distribution to model the number of yeast cells per unit volume in the fermentation process. His most important discovery was the *t*-distribution in 1908. Since most applied researchers worked with small samples, Gosset was interested in the behavior of the mean in the case of small samples. He tediously took numerous small sets of numbers, calculated the mean and standard deviation of each, obtained their *t*-ratio, and plotted the results on graph paper. The shape of the distribution was always the same: the *t*-distribution. Under company policy, employees were forbidden to publish their research results, so Gosset used the pen name *Student* to publish a paper on the subject. Hence, the distribution has been called Student's *t*-distribution.

If we are sampling from a normal distribution, the ***t*-statistic** has a sampling distribution very much like that of the *z*-statistic: mound shaped, symmetric, and with mean 0. The primary difference between the sampling distributions of *t* and *z* is that the *t*-statistic is more variable than the *z*, a property that follows intuitively when you realize that *t* contains two random quantities ($\overline{x}$ and *s*), whereas *z* contains only one ($\overline{x}$).

The actual amount of variability in the sampling distribution of *t* depends on the sample size *n*. A convenient way of expressing this dependence is to say that the *t* statistic has $(n - 1)$ **degrees of freedom (df).*** Recall that the quantity $(n - 1)$ is the divisor that appears in the formula for s^2. This number plays a key role in the sampling distribution of s^2 and appears in discussions of other statistics in later chapters. In particular, the smaller the number of degrees of freedom associated with the *t*-statistic, the more variable will be its sampling distribution.

In Figure 5.7, we show both the sampling distribution of *z* and the sampling distribution of a *t*-statistic with 4 df. You can see that the increased variability of the *t*-statistic means that the *t*-value, t_α, that locates an area α in the upper tail of the *t*-distribution is larger than the corresponding value, z_α. For any given value of α the *t*-value, t_α, increases as the number of degrees of freedom (df) decreases. Values of *t* that will be used in forming small-sample confidence intervals of μ are given in Table IV of Appendix A. A partial reproduction of this table is shown in Table 5.3.

Note that t_α values are listed for various degrees of freedom, where α refers to the tail area under the *t*-distribution to the right of t_α. For example, if we want the *t*-value with an area of .025 to its right and 4 df, we look in the table under the column $t_{.025}$ for the entry in the row corresponding to 4 df. This entry is $t_{.025} = 2.776$, as shown in Figure 5.8. The corresponding standard normal *z*-score is $z_{.025} = 1.96$.

Note that the last row of Table IV, where df = ∞ (infinity), contains the standard normal *z*-values. This follows from the fact that as the sample size *n* grows very large, *s* becomes closer to σ and thus *t* becomes closer in distribution to *z*. In fact, when df = 29, there is little difference between corresponding tabulated values of *z* and *t*. Thus, we choose

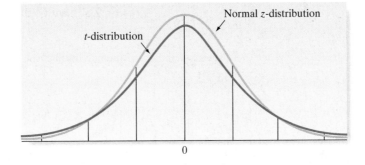

FIGURE 5.7

Standard normal (*z*) distribution and *t*-distribution with 4 df

*Since degrees of freedom are related to the sample size *n,* it is helpful to think of the number of degrees of freedom as the amount of information in the sample available for estimating the target parameter.

TABLE 5.3   Reproduction of Part of Table IV in Appendix A

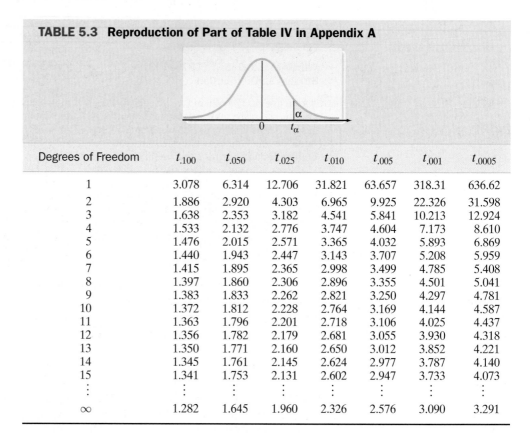

Degrees of Freedom	$t_{.100}$	$t_{.050}$	$t_{.025}$	$t_{.010}$	$t_{.005}$	$t_{.001}$	$t_{.0005}$
1	3.078	6.314	12.706	31.821	63.657	318.31	636.62
2	1.886	2.920	4.303	6.965	9.925	22.326	31.598
3	1.638	2.353	3.182	4.541	5.841	10.213	12.924
4	1.533	2.132	2.776	3.747	4.604	7.173	8.610
5	1.476	2.015	2.571	3.365	4.032	5.893	6.869
6	1.440	1.943	2.447	3.143	3.707	5.208	5.959
7	1.415	1.895	2.365	2.998	3.499	4.785	5.408
8	1.397	1.860	2.306	2.896	3.355	4.501	5.041
9	1.383	1.833	2.262	2.821	3.250	4.297	4.781
10	1.372	1.812	2.228	2.764	3.169	4.144	4.587
11	1.363	1.796	2.201	2.718	3.106	4.025	4.437
12	1.356	1.782	2.179	2.681	3.055	3.930	4.318
13	1.350	1.771	2.160	2.650	3.012	3.852	4.221
14	1.345	1.761	2.145	2.624	2.977	3.787	4.140
15	1.341	1.753	2.131	2.602	2.947	3.733	4.073
$\vdots$	$\vdots$	$\vdots$	$\vdots$	$\vdots$	$\vdots$	$\vdots$	$\vdots$
∞	1.282	1.645	1.960	2.326	2.576	3.090	3.291

the arbitrary cutoff of $n = 30$ (df = 29) to distinguish between large-sample and small-sample inferential techniques.

Returning to the example of testing a new drug, suppose that the six test patients have the blood pressure increases (measured in points) shown in Table 5.4. How can we use this information to construct a 95% confidence interval for μ, the mean increase in blood pressure associated with the new drug for all patients in the population?

BPINCR

**TABLE 5.4 Blood
Pressure Increases
(Points) for Six Patients**

1.7	3.0	.8	3.4	2.7	2.1

First, we know that we are dealing with a sample too small to assume, by the central limit theorem, that the sample mean $\bar{x}$ is approximately normally distributed. That is, we do not get the normal distribution of $\bar{x}$ "automatically" from the central limit theorem when the sample size is small. Instead, we must assume that the measured variable, in this case the increase in blood pressure, is normally distributed in order for the distribution of $\bar{x}$ to be normal.

Second, unless we are fortunate enough to know the population standard deviation σ, which in this case represents the standard deviation of *all* the patients' increases in blood pressure when they take the new drug, we cannot use the standard normal z-statistic to form our confidence interval for μ. Instead, we must use the t-distribution, with $(n - 1)$ degrees of freedom.

In this case, $n - 1 = 5$ df, and the t-value is found in Table 5.3 to be

$$t_{.025} = 2.571 \text{ with 5 df}$$

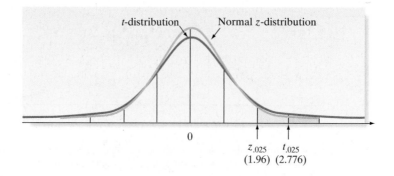

FIGURE 5.8

The $t_{.025}$ value in a t-distribution with 4 df, and the corresponding $z_{.025}$ value

Descriptives

			Statistic	Std. Error
BPINCR	Mean		2.283	.3877
	95% Confidence Interval for Mean	Lower Bound	1.287	
		Upper Bound	3.280	
	5% Trimmed Mean		2.304	
	Median		2.400	
	Variance		.902	
	Std. Deviation		.9496	
	Minimum		.8	
	Maximum		3.4	
	Range		2.6	
	Interquartile Range		1.6	
	Skewness		-.573	.845
	Kurtosis		-.389	1.741

FIGURE 5.9

SPSS confidence interval for mean blood pressure increase

Recall that the large-sample confidence interval would have been of the form

$$\bar{x} \pm z_{\alpha/2}\sigma_{\bar{x}} = \bar{x} \pm z_{\alpha/2}\frac{\sigma}{\sqrt{n}} = \bar{x} \pm z_{.025}\frac{\sigma}{\sqrt{n}}$$

where 95% is the desired confidence level. To form the interval for a small sample from *a normal distribution, we simply substitute t for z and s for σ in the preceding formula, yielding*

$$\bar{x} \pm t_{\alpha/2}\frac{s}{\sqrt{n}}$$

An SPSS printout showing descriptive statistics for the six blood pressure increases is displayed in Figure 5.9. Note that $\bar{x} = 2.283$ and $s = .950$. Substituting these numerical values into the confidence interval formula, we get

$$2.283 \pm (2.571)\left(\frac{.950}{\sqrt{6}}\right) = 2.283 \pm .997$$

or 1.286 to 3.280 points. Note that this interval agrees (except for rounding) with the confidence interval highlighted on the SPSS printout in Figure 5.9.

We interpret the interval as follows: We can be 95% confident that the mean increase in blood pressure associated with taking this new drug is between 1.286 and 3.28 points. As with our large-sample interval estimates, our confidence is in the *process*, not in this particular interval. We know that if we were to repeatedly use this estimation procedure, 95% of the confidence intervals produced would contain the true mean μ, *assuming that the probability distribution of changes in blood pressure from which our sample was selected is normal.* The latter assumption is necessary for the small-sample interval to be valid.

What price did we pay for having to utilize a small sample to make the inference? First, we had to assume that the underlying population is normally distributed, and if the assumption is invalid, our interval might also be invalid.* Second, we had to form the interval by using a *t* value of 2.571 rather than a *z* value of 1.96, resulting in a wider interval to achieve the same 95% level of confidence. If the interval from 1.286 to 3.28 is too wide to be of much use, we know how to remedy the situation: Increase the number of patients sampled in order to decrease the width of the interval (on average).

Now Work Exercise 5.27

---

*By *invalid*, we mean that the probability that the procedure will yield an interval that contains μ is not equal to $(1 - \alpha)$. Generally, if the underlying population is approximately normal, the confidence coefficient will approximate the probability that the interval contains μ.

The procedure for forming a small-sample confidence interval is summarized in the accompanying boxes.

Small-Sample Confidence Interval* for μ

The small-sample confidence interval for μ is

$$\bar{x} \pm t_{\alpha/2}\left(\frac{s}{\sqrt{n}}\right)$$

where $t_{\alpha/2}$ is based on $(n - 1)$ degrees of freedom.

Conditions Required for a Valid Small-Sample Confidence Interval for μ

1. A random sample is selected from the target population.
2. The population has a relative frequency distribution that is approximately normal.

EXAMPLE 5.2

A SMALL-SAMPLE CONFIDENCE INTERVAL FOR μ—Destructive Sampling

Problem Some quality control experiments require *destructive sampling* (i.e., the test to determine whether the item is defective destroys the item) in order to measure a particular characteristic of the product. The cost of destructive sampling often dictates small samples. Suppose a manufacturer of printers for personal computers wishes to estimate the mean number of characters printed before the printhead fails. The printer manufacturer tests $n = 15$ printheads and records the number of characters printed until failure for each. These 15 measurements (in millions of characters) are listed in Table 5.5, followed by a MINITAB summary statistics printout in Figure 5.10.

a. Form a 99% confidence interval for the mean number of characters printed before the printhead fails. Interpret the result.

b. What assumption is required for the interval you found in part **a** to be valid? Is that assumption reasonably satisfied?

 PRINTHEAD

TABLE 5.5   Number of Characters (in Millions) for $n = 15$ Printhead Tests

1.13	1.55	1.43	.92	1.25	1.36	1.32	.85	1.07	1.48	1.20	1.33	1.18	1.22	1.29

FIGURE 5.10

MINITAB printout with descriptive statistics and 99% confidence interval for Example 5.2

```
Variable    N     Mean     StDev    SE Mean        99% CI
NUMCHAR    15   1.23867   0.19316  0.04987   (1.09020, 1.38714)
```

Solution

a. For this small sample ($n = 15$), we use the *t*-statistic to form the confidence interval. We use a confidence coefficient of .99 and $n - 1 = 14$ degrees of freedom to find $t_{\alpha/2}$ in Table VI:

$$t_{\alpha/2} = t_{.005} = 2.977$$

[*Note:* The small sample forces us to extend the interval almost three standard deviations (of $\bar{x}$) on each side of the sample mean in order to form the 99% confidence interval.] From the MINITAB printout shown in Figure 5.10, we find that $\bar{x} = 1.24$

---

*The procedure given in the box assumes that the population standard deviation σ is unknown, which is almost always the case. If σ is known, we can form the small-sample confidence interval just as we would a large-sample confidence interval, using a standard normal z-value instead of t. However, we must still assume that the underlying population is approximately normal.

and $s = .19$. Substituting these (rounded) values into the confidence interval formula, we obtain

$$\bar{x} \pm t_{.005}\left(\frac{s}{\sqrt{n}}\right) = 1.24 \pm 2.977\left(\frac{.19}{\sqrt{15}}\right)$$

$$= 1.24 \pm .15 \quad \text{or} \quad (1.09, 1.39)$$

This interval is highlighted in Figure 5.10.

Our interpretation is as follows: The manufacturer can be 99% confident that the printhead has a mean life of between 1.09 and 1.39 million characters. If the manufacturer were to advertise that the mean life of its printheads is (at least) 1 million characters, the interval would support such a claim. Our confidence is derived from the fact that 99% of the intervals formed in repeated applications of this procedure will contain μ.

b. Since n is small, we must assume that the number of characters printed before the printhead fails is a random variable from a normal distribution. That is, we assume that the population from which the sample of 15 measurements is selected is distributed normally. One way to check this assumption is to graph the distribution of data in Table 5.5. If the sample data are approximately normally distributed, then the population from which the sample is selected is very likely to be normal. A MINITAB stem-and-leaf plot for the sample data is displayed in Figure 5.11. The distribution is mound shaped and nearly symmetric. Therefore, the assumption of normality appears to be reasonably satisfied.

Stem-and-Leaf Display: NUMBER

```
Stem-and-leaf of NUMBER   N  = 15
Leaf Unit = 0.010


    1    8    5
    2    9    2
    3   10    7
    5   11    38
  (4)   12    0259
    6   13    236
    3   14    38
    1   15    5
```

FIGURE 5.11

MINITAB stem-and-leaf display of data in Table 5.4

Look Back Other checks for normality, such as a normal probability plot and the ratio IQR/s, may also be used to verify the normality condition.

Now Work Exercise 5.35

We have emphasized throughout this section that an assumption that the population is approximately normally distributed is necessary for making small-sample inferences about μ when the t-statistic is used. Although many phenomena do have approximately normal distributions, it is also true that many random phenomena have distributions that are not normal or even mound shaped. Empirical evidence acquired over the years has shown that the t-distribution is rather insensitive to moderate departures from normality. That is, the use of the t-statistic when sampling from slightly skewed mound-shaped populations generally produces credible results; however, for cases in which the distribution is distinctly nonnormal, we must either take a large sample or use a *nonparametric method*.

What Do You Do When the Population Relative Frequency Distribution Departs Greatly from Normality?

Answer: Use the nonparametric statistical method of optional Section 6.6.

Confidence Interval for a Population Mean ($n < 30$)

Using the TI-84/TI-83 Graphing Calculator

Creating a Confidence Interval for a Population Mean

Step 1   *Enter the data. (Skip to Step 2 if you have summary statistics, not raw data.)*

Press **STAT** and select **1:Edit**.
Note: If the list already contains data, clear the old data. Use the up ARROW to highlight "L1".
Press **CLEAR ENTER**.
Use the **ARROW** and **ENTER** keys to enter the data set into **L1**.

```
EDIT CALC TESTS
2↑T-Test…
3:2-SampZTest…
4:2-SampTTest…
5:1-PropZTest…
6:2-PropZTest…
7:ZInterval…
8↓TInterval…
```

Step 2   *Access the Statistical Tests Menu*

Press **STAT**.
Arrow right to **TESTS**.
Arrow down to **Tinterval**.
Press **ENTER**.

Step 3   *Choose "**Data**" or "**Stats**" ("Data" is selected when you have entered the raw data into a List. "Stats" is selected when you are given only the mean, standard deviation, and sample size.)*

Press **ENTER**.
If you selected "Data", set **List** to **L1**.
Set **Freq** to **1**.
Set **C-Level** to the confidence level.
Arrow down to "**Calculate**".
Press **ENTER**.

If you selected "Stats", enter the mean, standard deviation, and sample size.
Set **C-Level** to the confidence level.
Arrow down to "**Calculate**".
Press **ENTER**.
(The screen below is set up for an example with a mean of 100 and a standard deviation of 10.)
The confidence interval will be displayed with the mean, standard deviation, and sample size.

```
TInterval
 Inpt:DATA Stats
 List:L1
 Freq:1
 C-Level:.95
 Calculate
```

```
TInterval
 Inpt:Data Stats
 x̄:100
 Sx:10
 n:19
 C-Level:.99
 Calculate
```

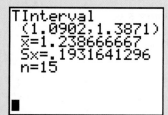

```
TInterval
 (1.0902,1.3871)
 x̄=1.238666667
 Sx=.1931641296
 n=15

█
```

Example   Compute a 99% confidence interval for the mean, using the 15 pieces of data given in Example 7.2:

1.13	1.55	1.43	0.92	1.25
1.36	1.32	0.85	1.07	1.48
1.20	1.33	1.18	1.22	1.29

As you can see from the screen, our 99% confidence interval is **(1.0902, 1.3871)**.
    You will also notice that the output displays the mean, standard deviation, and sample size.

Exercises 5.25–5.43

Understanding the Principles

5.25 State the two problems (and corresponding solutions) that arise with using a small sample to estimate μ.

5.26 Compare the shapes of the z- and t-distributions.

5.27 Explain the differences in the sampling distributions of $\bar{x}$ for
NW large and small samples under the following assumptions:
    a. The variable of interest, x, is normally distributed.
    b. Nothing is known about the distribution of the variable x.

Applet Exercise 5.3

Use the applet entitled *Confidence Intervals for a Mean (the impact of not knowing the standard deviation)* to compare proportions of z-intervals and t-intervals that contain the mean for a population that is normally distributed.

a. Using $n = 5$ and the normal distribution with mean 50 and standard deviation 10, run the applet several times. How do the proportions of z-intervals and t-intervals that contain the mean compare?

b. Repeart part **a** first for $n = 10$ and then for $n = 20$. Compare your results with those you obtained in part **a**.

c. Describe any patterns you observe between the proportion of z-intervals that contain the mean and the proportion of t-intervals that contain the mean as the sample size increases.

Applet Exercise 5.4

Use the applet entitled *Confidence Intervals for a Mean (the impact of not knowing the standard deviation)* to compare proportions of z-intervals and t-intervals that contain the mean for a population with a skewed distribution.

a. Using $n = 5$ and the right-skewed distribution with mean 50 and standard deviation 10, run the applet several times. How do the proportions of z-intervals and t-intervals that contain the mean compare?

b. Repeat part **a** first for $n = 10$ and then for $n = 20$. Compare your results with those you obtained in part **a**.

c. Describe any patterns you observe between the proportion of z-intervals that contain the mean and the proportion of t-intervals that contain the mean as the sample size increases.

d. How does the skewness of the underlying distribution affect the proportions of z-intervals and t-intervals that contain the mean?

Learning the Mechanics

5.28 Suppose you have selected a random sample of $n = 7$ measurements from a normal distribution. Compare the standard normal z-values with the corresponding t-values if you were forming the following confidence intervals:

a. 80% confidence interval
b. 90% confidence interval
c. 95% confidence interval
d. 98% confidence interval
e. 99% confidence interval
f. Use the table values you obtained in parts **a–e** to sketch the z- and t-distributions. What are the similarities and differences?

5.29 Let t_0 be a specific value of t. Use Table IV in Appendix A to find t_0 values such that the following statements are true:

a. $P(t \geq t_0) = .025$, where df $= 10$
b. $P(t \geq t_0) = .01$, where df $= 17$
c. $P(t \leq t_0) = .005$, where df $= 6$
d. $P(t \leq t_0) = .05$, where df $= 13$

5.30 Let t_0 be a particular value of t. Use Table IV of Appendix A to find t_0 values such that the following statements are true:

a. $P(-t_0 < t < t_0) = .95$, where df $= 16$
b. $P(t \leq -t_0 \text{ or } t \geq t_0) = .05$, where df $= 16$
c. $P(t \leq t_0) = .05$, where df $= 16$
d. $P(t \leq -t_0 \text{ or } t \geq t_0) = .10$, where df $= 12$
e. $P(t \leq -t_0 \text{ or } t \geq t_0) = .01$, where df $= 8$

5.31 The following random sample was selected from a normal distribution: 4, 6, 3, 5, 9, 3.

a. Construct a 90% confidence interval for the population mean μ.

b. Construct a 95% confidence interval for the population mean μ.

c. Construct a 99% confidence interval for the population mean μ.

d. Assume that the sample mean $\bar{x}$ and sample standard deviation s remain exactly the same as those you just calculated, but that they are based on a sample of $n = 25$ observations rather than $n = 6$ observations. Repeat parts **a–c**. What is the effect of increasing the sample size on the width of the confidence intervals?

5.32 The following sample of 16 measurements was selected from a population that is approximately normally distributed:

⊙ **LM5_32**

91	80	99	110	95	106	78	121	106	100
97	82	100	83	115	104				

a. Construct an 80% confidence interval for the population mean.

b. Construct a 95% confidence interval for the population mean, and compare the width of this interval with that of part **a**.

c. Carefully interpret each of the confidence intervals, and explain why the 80% confidence interval is narrower.

Applying the Concepts—Basic

5.33 **Al Qaeda attacks on the United States.** Refer to the *Studies in Conflict & Terrorism* (Vol. 29, 2006) analysis of recent incidents involving suicide terrorist attacks, presented in Exercise 2.30 (p. 46). Data on the number of individual suicide bombings or attacks for each in a sample of 21 incidents involving an attack against the United States by the Al Qaeda terrorist group are reproduced in the accompanying table.

a. Find the mean and standard deviation of the sample data.

b. Describe the population from which the sample is selected.

c. Use the information from part **a** to find a 90% confidence interval for the mean μ of the population.

d. Give a practical interpretation of the result you obtained in part **c**.

e. In repeatedly sampling, where intervals similar to the one computed in part **c** are generated, what proportion of the intervals will enclose the true value of μ?

⊙ **ALQAEDA**

1	1	2	1	2	4	1	1	1	1	2	3	4	5	1	1	1	2	2	2	1

Source: Moghadam, A. "Suicide terrorism, occupation, and the globalization of martyrdom: A critique of *Dying to Win*," *Studies in Conflict & Terrorism*, Vol. 29, No. 8, 2006 (Table 3).

5.34 **Assessing the bending strength of a wooden roof.** The white wood material used for the roof of an ancient Japanese temple is imported from Northern Europe. The wooden roof must withstand as much as 100 centimeters of snow in the winter. Architects at Tohoku University (in Japan) conducted a study to estimate the mean bending strength of the white wood roof (*Journal of the International Association for*

MINITAB Output for Exercise 5.35

```
Variable   N     Mean      StDev    SE Mean       95% CI
CESIUM     9   0.009027  0.004854  0.001618  (0.005296, 0.012759)
```

Shell and Spatial Structures, Aug. 2004). A sample of 25 pieces of the imported wood was tested and yielded the following statistics on breaking strength (in MPa): $\bar{x} = 75.4$, $s = 10.9$. Estimate the true mean breaking strength of the white wood with a 90% confidence interval. Interpret the result.

LICHEN

5.35 Radioactive lichen. Refer to the Lichen Radionuclide Baseline Research project at the University of Alaska, presented in Exercise 2.34 (p. 47). Recall that the researchers collected 9 lichen specimens and measured the amount (in microcuries per milliliter) of the radioactive element cesium-137 for each. (The natural logarithms of the data values are saved in the **LICHEN** file.) A MINITAB printout with summary statistics for the actual data is shown above.

a. Give a point estimate for the mean amount of cesium in lichen specimens collected in Alaska.

b. Give the *t*-value used in a small-sample 95% confidence interval for the true mean amount of cesium in Alaskan lichen specimens.

c. Use the result you obtained in part **b** and the values of $\bar{x}$ and *s* shown on the MINITAB printout to form a 95% confidence interval for the true mean amount of cesium in Alaskan lichen specimens.

d. Check the interval you found in part **c** with the 95% confidence interval shown on the MINITAB printout.

e. Give a practical interpretation for the interval you obtained in part **c**.

GOBIANTS

5.36 Rainfall and desert ants. Refer to the *Journal of Biogeography* (December 2003) study of ants and their habitat in the desert of Central Asia, presented in Exercise 2.66 (p. 60). Recall that botanists randomly selected five sites in the Dry Steppe region and six sites in the Gobi Desert where ants were observed. One of the variables of interest is the annual rainfall (in millimeters) at each site. (The data are saved in the **GOBIANTS** file.) Summary statistics for the annual rainfall at each site are provided in the SAS printout below.

a. Give a point estimate for the average annual rainfall amount at ant sites in the Dry Steppe region of Central Asia.

b. Give the *t*-value used in a small-sample 90% confidence interval for the true average annual rainfall amount at ant sites in the Dry Steppe region.

c. Use the result you obtained in part **b** and the values of $\bar{x}$ and *s* shown on the SAS printout to form a 90% confidence interval for the target parameter.

d. Give a practical interpretation for the interval you found in part **c**.

e. Use the data in the **GOBIANTS** file to check the validity of the confidence interval you found in part **c**.

f. Repeat parts **a–e** for the Gobi Desert region of Central Asia.

5.37 Oven cooking study. A group of Harvard University School of Public Health researchers studied the impact of cooking on the size of indoor air particles. (*Environmental Science & Technology*, September 1, 2000.) The decay rate (measured in $\mu m/hour$) for fine particles produced from oven cooking or toasting was recorded on six randomly selected days. The six measurements obtained are as follows:

DECAY

| .95 | .83 | 1.20 | .89 | 1.45 | 1.12 |

Source: Abt, E., et al. "Relative contribution of outdoor and indoor particle sources to indoor concentrations." *Environmental Science & Technology,* Vol. 34, No. 17, Sept. 1, 2000 (Table 3).

a. Find and interpret a 95% confidence interval for the true average decay rate of fine particles produced from oven cooking or toasting.

b. Explain what the phrase "95% confident" implies in the interpretation of part **a**.

c. What must be true about the distribution of the population of decay rates for the inference you made in part **a** to be valid?

Applying the Concepts—Intermediate

5.38 Studies on treating Alzheimer's disease. Alzheimer's disease is a progressive disease of the brain. Much research has been conducted on how to treat Alzheimer's. The journal *eCAM* (November 2006) published an article that critiqued the quality of the methodology used in studies on Alzheimer treatment. For each in a sample of 13 studies, the quality of the methodology was measured on the Wong scale, with scores ranging from 9 (low quality) to 27 (high quality). The data are shown in the table on p. 275. Estimate, with a 99% confidence interval, the mean quality μ of all studies on the treatment of Alzheimer's disease. Interpret the result.

SAS Output for Exercise 5.36

The MEANS Procedure

Analysis Variable : RAIN

REGION	N Obs	Mean	Std Dev
DryStepp	5	183.4000000	20.6470337
Gobi	6	110.0000000	15.9749804

TREATAD

22	21	18	19	20	15	19	20	15	20	17	20	21

Source: Chiappelli, F., et al. "Evidence-based research in complementary and alternative medicine III: Treatment of patients with Alzheimer's disease," *eCAM*, Vol. 3, No. 4, Nov. 2006 (Table 1).

5.39 Reproduction of bacteria-infected spider mites. Zoologists in Japan investigated the reproductive traits of spider mites with a bacteria infection. (*Heredity*, Jan. 2007.) Male and female pairs of infected spider mites were mated in a laboratory and the number of eggs produced by each female recorded. Summary statistics for several samples are provided in the accompanying table. Note that, in some samples, one or both infected spider mites were treated with antibiotic prior to mating.

a. For each type of female–male pair, construct and interpret a 90% confidence interval for the population mean number of eggs produced by the female spider mite.

b. Identify the type of female–male pair that appears to produce the highest mean number of eggs.

Female– Male Pairs	Sample Size	Mean # of Eggs	Standard Deviation
Both untreated	29	20.9	3.34
Male treated	23	20.3	3.50
Female treated	18	22.9	4.37
Both treated	21	18.6	2.11

Source: Gotoh, T., Noda, H., & Ito, S. "*Cardinium* symbionts cause cytoplasmic incompatibility in spider mites," *Heredity*, Vol. 98, No. 1, Jan. 2007 (Table 2).

5.40 Minimizing tractor skidding distance. In planning for a new forest road to be used for tree harvesting, planners must select the location that will minimize tractor skidding distance. In the *Journal of Forest Engineering* (July 1999), researchers wanted to estimate the true mean skidding distance along a new road in a European forest. The skidding distances (in meters) were measured at 20 randomly selected road sites. These values are given in the accompanying table.

a. Estimate, with a 95% confidence interval, the true mean skidding distance of the road.

b. Give a practical interpretation of the interval you found in part **a**.

c. What conditions are required for the inference you made in part **b** to be valid? Are these conditions reasonably satisfied?

d. A logger working on the road claims that the mean skidding distance is at least 425 meters. Do you agree?

SKIDDING

488	350	457	199	285	409	435	574	439	546
385	295	184	261	273	400	311	312	141	425

Source: Tujek, J., & Pacola, E. "Algorithms for skidding distance modeling on a raster Digital Terrain Model," *Journal of Forest Engineering*, Vol. 10, No. 1, July 1999 (Table 1).

5.41 Research on brain specimens. In Exercise 2.37 (p. 48), you learned that the postmortem interval (PMI) is the elapsed time between death and the performance of an autopsy on the cadaver. *Brain and Language* (June 1995) reported on the PMIs of 22 randomly selected human brain specimens obtained at autopsy. The data are reproduced in the following table:

BRAINPMI

5.5	14.5	6.0	5.5	5.3	5.8	11.0	6.4
7.0	14.5	10.4	4.6	4.3	7.2	10.5	6.5
3.3	7.0	4.1	6.2	10.4	4.9		

Source: Hayes, T. L., and Lewis, D. A. "Anatomical specialization of the anterior motor speech area: Hemispheric differences in magnopyramidal neurons," *Brain and Language*, Vol. 49, No. 3, June 1995, p. 292 (Table 1).

a. Construct a 95% confidence interval for the true mean PMI of human brain specimens obtained at autopsy.

b. Interpret the interval you found in part **a**.

c. What assumption is required for the interval from part **a** to be valid? Is this assumption satisfied? Explain.

d. What is meant by the phrase "95% confidence"?

5.42 Eating disorders in females. The "fear of negative evaluation" (FNE) scores for 11 bulimic female students and 14 normal female students, first presented in Exercise 2.38 (p. 48) are reproduced at the bottom of the page. (Recall that the higher the score, the greater is the FNE.)

a. Construct a 95% confidence interval for the mean FNE score of the population of bulimic female students. Interpret the result.

b. Construct a 95% confidence interval for the mean FNE score of the population of normal female students. Interpret the result.

c. What assumptions are required for the intervals of parts **a** and **b** to be statistically valid? Are these assumptions reasonably satisfied? Explain.

Applying the Concepts—Advanced

5.43 Study on waking sleepers early. Scientists have discovered increased levels of the hormone adrenocorticotropin in people just before they awake from sleeping (*Nature*, January 7, 1999). In the study described, 15 subjects were monitored during their sleep after being told that they would be woken at a particular time. One hour prior to the designated wake-up time, the adrenocorticotropin level (pg/mL) was measured in each, with the following results:

$$\bar{x} = 37.3 \qquad s = 13.9$$

a. Use a 95% confidence interval to estimate the true mean adrenocorticotropin level of sleepers one hour prior to waking.

b. Interpret the interval you found in part **a** in the words of the problem.

c. The researchers also found that if the subjects were woken three hours earlier than they anticipated, the average

BULIMIA

Bulimic students	21	13	10	20	25	19	16	21	24	13	14	
Normal students	13	6	16	13	8	19	23	18	11	19	7	10	15	20

Source: Randles, R. H. "On neutral responses (zeros) in the sign test and ties in the Wilcoxon–Mann–Whitney Test." *The American Statistician*, Vol. 55, No. 2, May 2001 (Figure 3).

adrenocorticotropin level was 25.5 pg/mL. Assume that $\mu = 25.5$ for all sleepers who are woken three hours earlier than expected. Use the interval from part **a** to make an inference about the mean adrenocorticotropin level of

sleepers under two conditions: one hour before the anticipated wake-up time and three hours before the anticipated wake-up time.

5.4 Large-Sample Confidence Interval for a Population Proportion

The number of public-opinion polls has grown at an astounding rate in recent years. Almost daily, the news media report the results of some poll. Pollsters regularly determine the percentage of people in favor of the president's welfare-reform program, the fraction of voters in favor of a certain candidate, the fraction of customers who favor a particular brand of wine, and the proportion of people who smoke cigarettes. In each case, we are interested in estimating the percentage (or proportion) of some group with a certain characteristic. In this section, we consider methods for making inferences about population proportions when the sample is large.

EXAMPLE 5.3
ESTIMATING A POPULATION PROPORTION—
Fraction Who Trust the President

Problem Public-opinion polls are conducted regularly to estimate the fraction of U.S. citizens who trust the president. Suppose 1,000 people are randomly chosen and 637 answer that they trust the president. How would you estimate the true fraction of *all* U.S. citizens who trust the president?

Solution What we have really asked is how you would estimate the probability p of success in a binomial experiment in which p is the probability that a person chosen trusts the president. One logical method of estimating p for the population is to use the proportion of successes in the sample. That is, we can estimate p by calculating

$$\hat{p} = \frac{\text{Number of people sampled who trust the president}}{\text{Number of people sampled}}$$

where $\hat{p}$ is read "p hat." Thus, in this case,

$$\hat{p} = \frac{637}{1,000} = .637$$

Look Back To determine the reliability of the estimator $\hat{p}$, we need to know its sampling distribution. That is, if we were to draw samples of 1,000 people over and over again, each time calculating a new estimate $\hat{p}$, what would be the frequency distribution of all the $\hat{p}$-values? The answer lies in viewing $\hat{p}$ as the average, or mean, number of successes per trial over the n trials. If each success is assigned a value equal to 1 and each failure is assigned a value of 0, then the sum of all n sample observations is x, the total number of successes, and $\hat{p} = x/n$ is the average, or mean, number of successes per trial in the n trials. The central limit theorem tells us that the relative frequency distribution of the sample mean for any population is approximately normal for sufficiently large samples.

Now Work Exercise 5.51a

The repeated sampling distribution of $\hat{p}$ has the characteristics listed in the next box and shown in Figure 5.12.

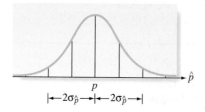

FIGURE 5.12
Sampling distribution of $\hat{p}$

Sampling Distribution of $\hat{p}$

1. The mean of the sampling distribution of $\hat{p}$ is p; that is, $\hat{p}$ is an unbiased estimator of p.

2. The standard deviation of the sampling distribution of $\hat{p}$ is $\sqrt{pq/n}$; that is, $\sigma_p = \sqrt{pq/n}$, where $q = 1 - p$.

3. For large samples, the sampling distribution of $\hat{p}$ is approximately normal. A sample size is considered large if the interval $\hat{p} \pm 3\sigma_{\hat{p}}$ does not include 0 or 1. [*Note:* This requirement is almost equivalent to that given in Section 5.5 for approximating a binomial distribution with a normal one. The difference is that we assumed p to be known in Section 5.5; now we are trying to make inferences about an unknown p, so we use $\hat{p}$ to estimate p in checking the adequacy of the normal approximation.]

The fact that $\hat{p}$ is a "sample mean fraction of successes" allows us to form confidence intervals about p in a manner that is completely analogous to that used for large-sample estimation of μ.

Large-Sample Confidence Interval for p

The large-sample confidence interval for p is

$$\hat{p} \pm z_{\alpha/2}\sigma_{\hat{p}} = \hat{p} \pm z_{\alpha/2}\sqrt{\frac{pq}{n}} \approx \hat{p} \pm z_{\alpha/2}\sqrt{\frac{\hat{p}\hat{q}}{n}}$$

where $\hat{p} = \dfrac{x}{n}$ and $\hat{q} = 1 - \hat{p}$

Note: When n is large, $\hat{p}$ can approximate the value of p in the formula for $\sigma_{\hat{p}}$.

Conditions Required for a Valid Large-Sample Confidence Interval for p

1. A random sample is selected from the target population.

2. The sample size n is large. (This condition will be satisfied if both $n\hat{p} \geq 15$ and $n\hat{q} \geq 15$. Note that $n\hat{p}$ and $n\hat{q}$ are simply the number of successes and number of failures, respectively, in the sample.

Thus, if 637 of 1,000 U.S. citizens say they trust the president, a 95% confidence interval for the proportion of *all* U.S. citizens who trust the president is

$$\hat{p} \pm z_{\alpha/2}\sigma_{\hat{p}} = .637 \pm 1.96\sqrt{\frac{pq}{1,000}}$$

where $q = 1 - p$. Just as we needed an approximation for σ in calculating a large-sample confidence interval for μ, we now need an approximation for p. As Table 5.6 shows, the approximation for p does not have to be especially accurate, because the value of $\sqrt{pq}$ needed for the confidence interval is relatively insensitive to changes in p. Therefore, we can use $\hat{p}$ to approximate p. Keeping in mind that $\hat{q} = 1 - \hat{p}$, we substitute these values into the formula for the confidence interval:

$$\hat{p} \pm 1.96\sqrt{pq/1,000} \approx \hat{p} \pm 1.96\sqrt{\hat{p}\hat{q}/1,000}$$

$$= .637 \pm 1.96\sqrt{(.637)(.363)/1,000} = .637 \pm .030$$

$$= (.607, .667)$$

TABLE 5.6   Values of pq for Several Different Values of p

P	pq	$\sqrt{pq}$
.5	.25	.50
.6 or .4	.24	.49
.7 or .3	.21	.46
.8 or .2	.16	.40
.9 or .1	.09	.30

Then we can be 95% confident that the interval from 60.7% to 66.7% contains the true percentage of *all* U.S. citizens who trust the president. That is, in repeated constructions of confidence intervals, approximately 95% of all samples would produce confidence intervals that enclose p. Note that the guidelines for interpreting a confidence interval about μ also apply to interpreting a confidence interval for p, because p is the "population mean fraction of successes" in a binomial experiment.

EXAMPLE 5.4
A LARGE-SAMPLE CONFIDENCE INTERVAL FOR
p—Proportion Optimistic about the Economy

Problem Many public polling agencies conduct surveys to determine the current consumer sentiment concerning the state of the economy. For example, the Bureau of Economic and Business Research (BEBR) at the University of Florida conducts quarterly surveys to gauge consumer sentiment in the Sunshine State. Suppose that BEBR randomly samples 484 consumers and finds that 257 are optimistic about the state of the economy. Use a 90% confidence interval to estimate the proportion of all consumers in Florida who are optimistic about the state of the economy. Based on the confidence interval, can BEBR infer that the majority of Florida consumers are optimistic about the economy?

Solution The number x of the 484 sampled consumers who are optimistic about the Florida economy is a binomial random variable if we can assume that the sample was randomly selected from the population of Florida consumers and that the poll was conducted identically for each consumer sampled.

The point estimate of the proportion of Florida consumers who are optimistic about the economy is

$$\hat{p} = \frac{x}{n} = \frac{257}{484} = .531$$

We first check to be sure that the sample size is sufficiently large that the normal distribution provides a reasonable approximation to the sampling distribution of $\hat{p}$. We require the number of successes in the sample, $n\hat{p}$, and the number of failures, $n\hat{q}$, both to be at least 15. Since the number of successes is $n\hat{p} = 257$ and the number of failures is $n\hat{q} = 227$, we may conclude that the normal approximation is reasonable.

We now proceed to form the 90% confidence interval for p, the true proportion of Florida consumers who are optimistic about the state of the economy:

$$\hat{p} \pm z_{\alpha/2}\sigma_{\hat{p}} = \hat{p} \pm z_{\alpha/2}\sqrt{\frac{pq}{n}} \approx \hat{p} \pm z_{\alpha/2}\sqrt{\frac{\hat{p}\hat{q}}{n}}$$

$$= .531 \pm 1.645\sqrt{\frac{(.531)(.469)}{484}} = .531 \pm .037 = (.494, .568)$$

(This interval is also shown on the MINITAB printout of Figure 5.13.) Thus, we can be 90% confident that the proportion of all Florida consumers who are confident about the economy is between .494 and .568. As always, our confidence stems from the fact that 90% of all similarly formed intervals will contain the true proportion p and not from any knowledge about whether this particular interval does.

FIGURE 5.13
Portion of MINITAB printout
with 90% confidence interval
for *p*

```
Sample     X    N   Sample p            90% CI
1        257  484   0.530992    (0.493681, 0.568303)
```

Can we conclude on the basis of this interval that the majority of Florida consumers are optimistic about the economy? If we wished to use this interval to infer that a majority is optimistic, the interval would have to support the inference that *p* exceeds .5—that is, that more than 50% of Florida consumers are optimistic about the economy. Note that the interval contains some values below .5 (as low as .494) as well as some above .5 (as high as .568). Therefore, we cannot conclude on the basis of this 90% confidence interval that the true value of *p* exceeds .5.

Look Back If the entire confidence interval fell above .5 (e.g., an interval from .52 to .54), then we could conclude (with 90% confidence) that the true proportion of consumers who are optimistic exceeds .5.

Now Work Exercise 5.51b–c

Caution

Unless n *is extremely large, the large-sample procedure presented in this section performs poorly when* p *is near 0 or 1.* For example, suppose you want to estimate the proportion of people who die from a bee sting. This proportion is likely to be near 0 (say, $p \approx .001$). Confidence intervals for *p* based on a sample of size $n = 50$ will probably be misleading.

To overcome this potential problem, an *extremely* large sample size is required. Since the value of *n* required to satisfy "extremely large" is difficult to determine, statisticians (see Agresti & Coull, 1998) have proposed an alternative method, based on the Wilson (1927) point estimator of *p*. **Wilson's adjustment for estimating *p*** is outlined in the next box. Researchers have shown that this confidence interval works well for any *p*, even when the sample size *n* is very small.

Adjusted (1 − α) 100% Confidence Interval for a Population Proportion *p*

An adjusted confidence interval for *p* is

$$\tilde{p} \pm z_{\alpha/2}\sqrt{\frac{\tilde{p}(1 - \tilde{p})}{n + 4}}$$

where $\tilde{p} = \dfrac{x + 2}{n + 4}$ is the adjusted sample proportion of observations with the characteristic of interest, *x* is the number of successes in the sample, and *n* is the sample size.

EXAMPLE 5.5

USING THE
ADJUSTED
CONFIDENCE
INTERVAL
PROCEDURE—
Proportion Who
Are Victims of a
Violent Crime

Problem According to *True Odds: How Risk Affects Your Everyday Life* (Walsh, 1997), the probability of being the victim of a violent crime is less than .01. Suppose that, in a random sample of 200 Americans, 3 were victims of a violent crime. Use a 95% confidence interval to estimate the true proportion of Americans who were victims of a violent crime.

Solution Let *p* represent the true proportion of Americans who were victims of a violent crime. Since *p* is near 0, an "extremely large" sample is required to estimate its value by the usual large-sample method. Since we are unsure whether the sample size of 200 is large enough, we will apply Wilson's adjustment outlined in the box.

The number of "successes" (i.e., number of victims of a violent crime) in the sample is $x = 3$. Therefore, the adjusted sample proportion is

$$\tilde{p} = \frac{x + 2}{n + 4} = \frac{3 + 2}{200 + 4} = \frac{5}{204} = .025$$

c. Repeat part **b**, keeping $p = .1$ and increasing the sample size by 50 until you find a sample size that yields a similar proportion of the 99% confidence intervals containing p as that in part **a**.

d. Based on your results, describe how the value of p affects the sample size needed to guarantee a certain level of confidence.

Learning the Mechanics

5.47 A random sample of size $n = 196$ yielded $\hat{p} = .64$.

   a. Is the sample size large enough to use the methods of this section to construct a confidence interval for p? Explain.

   b. Construct a 95% confidence interval for p.

   c. Interpret the 95% confidence interval.

   d. Explain what is meant by the phrase "95% confidence interval."

5.48 A random sample of size $n = 144$ yielded $\hat{p} = .76$.

   a. Is the sample size large enough to use the methods of this section to construct a confidence interval for p? Explain.

   b. Construct a 90% confidence interval for p.

   c. What assumption is necessary to ensure the validity of this confidence interval?

5.49 For the binomial sample information summarized in each part, indicate whether the sample size is large enough to use the methods of this chapter to construct a confidence interval for p.

   a. $n = 500, \hat{p} = .05$

   b. $n = 100, \hat{p} = .05$

   c. $n = 10, \hat{p} = .5$

   d. $n = 10, \hat{p} = .3$

5.50 A random sample of 50 consumers taste-tested a new snack food. Their responses were coded (0: do not like; 1: like; 2: indifferent) and recorded as follows:

SNACK

1	0	0	1	2	0	1	1	0	0
0	1	0	2	0	2	2	0	0	1
1	0	0	0	0	1	0	2	0	0
0	1	0	0	1	0	0	1	0	1
0	2	0	0	1	1	0	0	0	1

   a. Use an 80% confidence interval to estimate the proportion of consumers who like the snack food.

   b. Provide a statistical interpretation for the confidence interval you constructed in part **a**.

Applying the Concepts—Basic

5.51 **National Firearms Survey.** Refer to the Harvard School of Public Health survey to determine the size and composition of privately held firearm stock in the United States, presented in Exercise 2.6 (p. 34). Recall that, in a representative household telephone survey of 2,770 adults, 26% reported that they own at least one gun. (*Injury Prevention*, Jan. 2007.) The researchers want to estimate the true percentage of adults in the United States that own at least one gun.

   a. Identify the population of interest to the researchers.

   b. Identify the parameter of interest to the researchers.

   c. Compute an estimate of the population parameter.

   d. Form a 99% confidence interval around the estimate.

   e. Interpret the confidence interval practically.

   f. Explain the meaning of the phrase "99% confident."

5.52 **Are you really being served red snapper?** Refer to the *Nature* (July 15, 2004) study of fish specimens labeled "red snapper," presented in Exercise 3.79 (p. 150). Recall that federal law prohibits restaurants from serving a cheaper look-alike variety of fish (e.g., vermillion snapper or lane snapper) to customers who order red snapper. In an effort to estimate the true proportion of fillets that are really red snapper, a team of University of North Carolina (UNC) researchers analyzed the meat from each in a sample of 22 "red snapper" fish fillets purchased from vendors across the United States. DNA tests revealed that 17 of the 22 fillets (or 77%) were not red snapper, but the cheaper look-alike variety of fish.

   a. Identify the parameter of interest to the UNC researchers.

   b. Explain why a large-sample confidence interval is inappropriate to apply in this study.

   c. Use Wilson's adjustment to construct a 95% confidence interval for the parameter of interest.

   d. Give a practical interpretation of the confidence interval.

5.53 **What we do when we are sick at home.** *USA Today* (Feb. 15, 2007) reported on the results of an opinion poll in which adults were asked what one thing they are most likely to do when they are home sick with a cold or the flu. In the survey, 63% said that they are most likely to sleep and 18% said that they would watch television. Although the sample size was not reported, typically opinion polls include approximately 1,000 randomly selected respondents.

   a. Assuming a sample size of 1,000 for this poll, construct a 95% confidence interval for the true percentage of all adults who would choose to sleep when they are at home sick.

   b. If the true percentage of adults who would choose to sleep when they are at home sick is 70%, would you be surprised? Explain.

5.54 **Scary-movie study.** According to a University of Michigan study, many adults have experienced lingering "fright" effects from a scary movie or TV show they saw as a teenager. (*Tampa Tribune*, March 10, 1999.) In a survey of 150 college students, 39 said they still experience "residual anxiety" from a scary TV show or movie.

   a. Give a point estimate $\hat{p}$ for the true proportion of college students who experience "residual anxiety" from a scary TV show or movie.

   b. Find a 95% confidence interval for p.

   c. Interpret the interval you found in part **b**.

5.55 **Ancient Greek pottery.** Refer to the *Chance* (Fall 2000) study of 837 pieces of pottery found at the ancient Greek settlement at Phylakopi, presented in Exercise 2.12 (p. 35). Of the 837 pieces, 183 were painted with either a curvilinear, geometric, or naturalistic decoration. Find a 90% confidence interval for the population proportion of all pottery artifacts at Phylakopi that are painted. Interpret the resulting interval.

Applying the Concepts—Intermediate

PONDICE

5.56 **Characteristics of ice-melt ponds.** Refer to the University of Colorado study of ice-melt ponds in the Canadian Arctic, presented in Exercise 2.11 (p. 35). Environmental engineers are using data collected by the National Snow and Ice Data Center to learn how climate affects the sea ice. Data on 504 ice melt ponds are saved in the **PONDICE** file. Of these 504 melt ponds, 88 were classified as having "first-year ice." Recall that the researchers estimated that about 17% of melt ponds in the Canadian Arctic have first-year ice. Use the methodology of this chapter to estimate, with 90% confidence, the percentage of all ice-melt ponds in the Canadian Arctic that have first-year ice. Give a practical interpretation of the results.

BIBLE

5.57 Do you believe in the Bible? Refer to the National Opinion Research Center's General Social Survey (GSS), presented in Exercise 2.17 (p. 36). Data on the approximately 2,800 Americans who participated in the 2004 GSS are saved in the **BIBLE** file. Recall that one question in the survey asked about a person's belief in the Bible. Suppose you want to estimate the proportion of all Americans who believe that the Bible is the actual word of God and is to be taken literally. (*Note:* The variable "Bible1" contains the responses to this question.)

a. Use a 95% confidence interval to estimate the proportion of interest.

b. Give a practical interpretation of the interval you used in part **a**.

c. Discuss how the survey methodology can affect the validity of the results.

5.58 Exercise workout dropouts. Researchers at the University of Florida's Department of Exercise and Sport Sciences conducted a study of variety in exercise workouts. (*Journal of Sport Behavior*, 2001.) A sample of 120 men and women were randomly divided into three groups, with 40 people per group. Group 1 members varied their exercise routine in workouts, group 2 members performed the same exercise at each workout, and group 3 members had no set schedule or regulations for their workouts.

a. By the end of the study, 15 people had dropped out of the first exercise group. Estimate the dropout rate for exercisers who vary their routine in workouts. Use a 90% confidence interval and interpret the result.

b. By the end of the study, 23 people had dropped out of the third exercise group. Estimate the dropout rate for exercisers who have no set schedule for their workouts. Use a 90% confidence interval and interpret the result.

5.59 Whistling dolphins. In Exercise 2.177 (p. 104) you learned about the signature whistles of bottlenose dolphins. Marine scientists categorize signature whistles by type—type a, type b, type c, etc. In one study of a sample of 185 whistles emitted from bottlenose dolphins in captivity, 97 were categorized as type a whistles (*Ethology*, July 1995).

a. Estimate the true proportion of bottlenose dolphin signature whistles that are type a whistles. Use a 99% confidence interval.

b. Interpret the interval you used in part **a**.

5.60 Studies on treating Alzheimer's disease. Refer to the journal *eCAM* (November 2006) assessment of the quality of the methodology used in studies of the treatment for Alzheimer's disease, presented in Exercise 5.38 (p. 274). Data on the quality of the methodology (using the Wong scale) for each in a sample of 13 studies are reproduced in the accompanying table. According to the researchers, a study with a Wong score below 18 used a methodology that "fails to support the author's conclusions" about the treatment of Alzheimer's. Use Wilson's adjustment to estimate the proportion of all studies on the treatment of Alzheimer's disease with a Wong score below 18. Construct a 99% confidence interval around the estimate and interpret the result.

TREATAD

22	21	18	19	20	15	19	20	15	20	17	20	21

Source: Chiappelli, F., et al. "Evidence-based research in complementary and alternative medicine III: Treatment of patients with Alzheimer's disease", *eCAM*, Vol. 3, No. 4, Nov. 2006 (Table 1).

Applying the Concepts—Advanced

5.61 Latex allergy in health-care workers. Refer to the *Current Allergy & Clinical Immunology* (March 2004) study of health-care workers who use latex gloves, Presented in Exercise 5.13 (p. 264). In addition to the 46 hospital employees who were diagnosed with a latex allergy on the basis of a skin-prick test, another 37 health-care workers were diagnosed with the allergy by means of a latex-specific serum test. Of these 83 workers with a confirmed latex allergy, only 36 suspected that they had the allergy when they were asked about it on a questionnaire. Make a statement about the likelihood that a health-care worker with a latex allergy suspects that he or she actually has the allergy. Attach a measure of reliability to your inference.

5.5 Determining the Sample Size

Recall from Section 1.5 that one way to collect the relevant data for a study used to make inferences about a population is to implement a designed (planned) experiment. Perhaps the most important design decision faced by the analyst is to determine the size of the sample. We show in this section that the appropriate sample size for making an inference about a population mean or proportion depends on the desired reliability.

Estimating a Population Mean

Consider the example from Section 5.2 in which we estimated the mean length of stay for patients in a large hospital. A sample of 100 patients' records produced the 95% confidence interval $\bar{x} \pm 2\sigma_{\bar{x}} \approx 4.53 \pm .74$. Consequently, our estimate $\bar{x}$ was within .74 day of the true mean length of stay, μ, for all the hospital's patients at the 95% confidence level. That is, the 95% confidence interval for μ was 2(.74) = 1.48 days wide when 100 accounts were sampled. This is illustrated in Figure 5.14a.

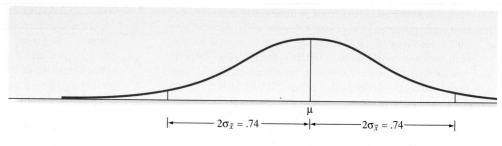

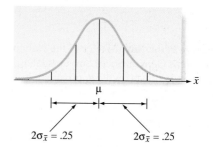

FIGURE 5.14

Relationship between sample size and width of confidence interval: hospital-stay example

a. $n = 100$

b. $n = 867$

Now suppose we want to estimate μ to within .25 day with 95% confidence. That is, we want to narrow the width of the confidence interval from 1.48 days to .50 day, as shown in Figure 5.14b. How much will the sample size have to be increased to accomplish this? If we want the estimator $\bar{x}$ to be within .25 day of μ, then we must have

$$2\sigma_{\bar{x}} = .25 \quad \text{or, equivalently,} \quad 2\left(\frac{\sigma}{\sqrt{n}}\right) = .25$$

Note that we are using $\sigma_{\bar{x}} = \dfrac{\sigma}{\sqrt{n}}$ in the formula, since we are dealing with the sampling distribution of $\bar{x}$ (the estimator of μ).

The necessary sample size is obtained by solving this equation for n. To do that, we need an approximation of σ. We have an approximation from the initial sample of 100 patients' records—namely, the sample standard deviation $s = 3.68$. Thus,

$$2\left(\frac{\sigma}{\sqrt{n}}\right) \approx 2\left(\frac{s}{\sqrt{n}}\right) = 2\left(\frac{3.68}{\sqrt{n}}\right) = .25$$

$$\sqrt{n} = \frac{2(3.68)}{.25} = 29.44$$

$$n = (29.44)^2 = 866.71 \approx 867$$

Approximately 867 patients' records will have to be sampled to estimate the mean length of stay, μ, to within .25 day with (approximately) 95% confidence. The confidence interval resulting from a sample of this size will be approximately .50 day wide. (See Figure 5.14b.)

In general, we express the reliability associated with a confidence interval for the population mean μ by specifying the **sampling error** within which we want to estimate μ with $100(1 - \alpha)\%$ confidence. The sampling error (denoted SE) is then equal to the half-width of the confidence interval, as shown in Figure 5.15.

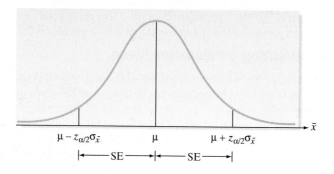

FIGURE 5.15

Specifying the sampling error SE as the half-width of a confidence interval

The procedure for finding the sample size necessary to estimate μ with a specific sampling error is given in the following box:

Determination of Sample Size for $100(1 - \alpha)$% Confidence Intervals for μ

In order to estimate μ with a sampling error SE and with $100(1 - \alpha)$% confidence, the required sample size is found as follows:

$$z_{\alpha/2}\left(\frac{\sigma}{\sqrt{n}}\right) = SE$$

The solution for n is given by the equation

$$n = \frac{(z_{\alpha/2})^2\sigma^2}{(SE)^2}$$

The value of σ is usually unknown. It can be estimated by the standard deviation s from a previous sample. Alternatively, we may approximate the range R of observations in the population and (conservatively) estimate $\sigma \approx R/4$. In any case, you should round the value of n obtained *upward* to ensure that the sample size will be sufficient to achieve the specified reliability.

EXAMPLE 5.6
SAMPLE SIZE FOR ESTIMATING μ—
Mean Inflation Pressure of Footballs

Problem Suppose the manufacturer of official NFL footballs uses a machine to inflate the new balls to a pressure of 13.5 pounds. When the machine is properly calibrated, the mean inflation pressure is 13.5 pounds, but uncontrollable factors cause the pressures of individual footballs to vary randomly from about 13.3 to 13.7 pounds. For quality control purposes, the manufacturer wishes to estimate the mean inflation pressure to within .025 pound of its true value with a 99% confidence interval. What sample size should be specified for the experiment?

Solution We desire a 99% confidence interval that estimates μ with a sampling error of SE = .025 pound. For a 99% confidence interval, we have $z_{\alpha/2} = z_{.005} = 2.575$. To estimate σ, we note that the range of observations is $R = 13.7 - 13.3 = .4$ and we use $\sigma \approx R/4 = .1$. Next, we employ the formula derived in the box to find the sample size n:

$$n = \frac{(z_{\alpha/2})^2\sigma^2}{(SE)^2} \approx \frac{(2.575)^2(.1)^2}{(.025)^2} = 106.09$$

We round this up to $n = 107$. Realizing that σ was approximated by $R/4$, we might even advise that the sample size be specified as $n = 110$ to be more certain of attaining the objective of a 99% confidence interval with a sampling error of .025 pound or less.

Look Back To determine the value of the sampling error SE, look for the value that follows the key words "estimate μ to within...."

Now Work Exercise 5.75

Sometimes the formula will lead to a solution that indicates a small sample size is sufficient to achieve the confidence interval goal. Unfortunately, the procedures and assumptions for small samples differ from those for large samples, as we discovered in Section 5.3. Therefore, if the formulas yield a small sample size, one simple strategy is to select a sample size $n = 30$.

Estimating a Population Proportion

The method just outlined is easily applied to a population proportion p. For example, in Section 5.4 a pollster used a sample of 1,000 U.S. citizens to calculate a 95% confidence interval for the proportion who trust the president, obtaining the interval

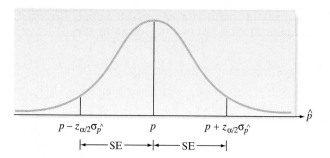

FIGURE 5.16

Specifying the sampling error SE of a confidence interval for a population proportion p

.637 $\pm$ .03. Suppose the pollster wishes to estimate more precisely the proportion who trust the president, say, to within .015 with a 95% confidence interval.

The pollster wants a confidence interval for p with a sampling error SE = .015. The sample size required to generate such an interval is found by solving the following equation for n:

$$z_{\alpha/2}\sigma_{\hat{p}} = \text{SE} \quad \text{or} \quad z_{\alpha/2}\sqrt{\frac{pq}{n}} = .015 \quad \text{(see Figure 5.16)}$$

Since a 95% confidence interval is desired, the appropriate z value is $z_{\alpha/2} = z_{.025} = 1.96$. We must approximate the value of the product pq before we can solve the equation for n. As shown in Table 5.6 (p. 328), the closer the values of p and q to .5, the larger is the product pq. Thus, to find a conservatively large sample size that will generate a confidence interval with the specified reliability, we generally choose an approximation of p close to .5. In the case of the proportion of U.S. citizens who trust the president, however, we have an initial sample estimate of $\hat{p} = .637$. A conservatively large estimate of pq can therefore be obtained by using, say, $p = .60$. We now substitute into the equation and solve for n:

$$1.96\sqrt{\frac{(.60)(.40)}{n}} = .015$$

$$n = \frac{(1.96)^2(.60)(.40)}{(.015)^2} = 4,097.7 \approx 4,098$$

The pollster must sample about 4,098 U.S. citizens to estimate the percentage who trust the president with a confidence interval of width .03.

The procedure for finding the sample size necessary to estimate a population proportion p with a specified sampling error SE is given in the following box:

Determination of Sample Size for $100(1 - \alpha)\%$ Confidence Interval for p

In order to estimate a binomial probability p with sampling error SE and with $100(1 - \alpha)\%$ confidence, the required sample size is found by solving the following equation for n:

$$z_{\alpha/2}\sqrt{\frac{pq}{n}} = \text{SE}$$

The solution for n can be written as follows:

$$n = \frac{(z_{\alpha/2})^2(pq)}{(\text{SE})^2}$$

Since the value of the product pq is unknown, it can be estimated by the sample fraction of successes, $\hat{p}$, from a previous sample. Remember (Table 5.6) that the value of pq is at its maximum when p equals .5, so you can obtain conservatively large values of n by approximating p by .5 or values close to .5. In any case, you should round the value of n obtained *upward* to ensure that the sample size will be sufficient to achieve the specified reliability.

EXAMPLE 5.7

SAMPLE SIZE FOR ESTIMATING p — Fraction of Defective Cell Phones

Problem Suppose a large telephone manufacturer that entered the postregulation market quickly has an initial problem with excessive customer complaints and consequent returns of the phones for repair or replacement. The manufacturer wants to estimate the magnitude of the problem in order to design a quality control program. How many telephones should be sampled and checked in order to estimate the fraction defective, p, to within .01 with 90% confidence?

Solution In order to estimate p to within .01 of its true value, we set the half-width of the confidence interval equal to SE = .01, as shown in Figure 5.17.

FIGURE 5.17
Specified reliability for estimate of fraction defective in Example 5.7

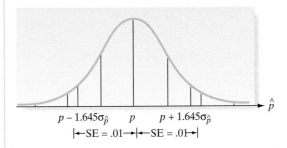

The equation for the sample size n requires an estimate of the product pq. We could most conservatively estimate $pq = .25$ (i.e., use $p = .5$), but this estimate may be too conservative. By contrasts, a value of .1, corresponding to 10% defective, will probably be conservatively large for this application. The solution is therefore

$$n = \frac{(z_{\alpha/2})^2(pq)}{(SE)^2} = \frac{(1.645)^2(.1)(.9)}{(.01)^2} = 2{,}435.4 \approx 2{,}436$$

Thus, the manufacturer should sample 2,436 telephones in order to estimate the fraction defective, p, to within .01 with 90% confidence.

Look Back Remember that this answer depends on our approximation of pq, for which we used .09. If the fraction defective is closer to .05 than .10, we can use a sample of 1,286 telephones (check this) to estimate p to within .01 with 90% confidence.

Now Work Exercise 5.74

The cost of sampling will also play an important role in the final determination of the sample size to be selected to estimate either μ or p. Although more complex formulas can be derived to balance the reliability and cost considerations, we will solve for the necessary sample size and note that the sampling budget may be a limiting factor. (Consult the references on page 346 for a more complete treatment of this problem.)

STATISTICS IN ACTION REVISITED

Determining the Number of Athletes Required to Participate in the Training Program for a Valid Estimate of p

In the previous Statistics in Action applications in this chapter, we used confidence intervals (1) to estimate μ, the population mean decrease in sprint time for athletes who participate in the speed training program (p. 311), and (2) to esti-

mate p, the population proportion of athletes who improve their sprint times after participating in the speed training program (p. 330). The confidence interval for p was fairly wide due to the relatively small number of athletes ($n = 38$) who participated in the program. In order to find a valid, narrower confidence interval for

(continued)

the true proportion, the researchers would need to include more athletes in the study.

How many high school athletes should be sampled to estimate p to within, say, .03 with 95% confidence? Here, we have the sampling error SE = .03 and $z_{.025} = 1.96$ (since, for a 95% confidence interval, $\alpha = .05$ and $\alpha/2 = .025$). From our analysis on page 330, the estimated proportion is $\hat{p} = .87$. Substituting these values into the formula given in the box on page 336, we obtain

$$n = \frac{(z_{.025})^2(\hat{p})(1 - \hat{p})}{(SE)^2} = \frac{(1.96)^2(.87)(.13)}{(.03)^2} = 482.76$$

Consequently, the researchers would have to include 483 high school athletes in the training program in order to find a valid 95% confidence interval for p to within 3% of its true value.

Running the training program with this large number of athletes, however, may be impractical. To reduce the size of the sample, the researchers could choose to increase the sampling error (i.e., the width) of the confidence interval for p or decrease the confidence level $(1 - \alpha)$. However, since both large- and small-sample confidence intervals are available for estimating μ, the researchers can rely on the inference made from the confidence interval for the mean decrease in sprint time (p. 311).

Exercises 5.62–5.82

Understanding the Principles

5.62 How does the sampling error SE compare with the width of a confidence interval?

5.63 **True or false.** For a specified sampling error SE, increasing the confidence level $(1 - \alpha)$ will lead to a larger n in determining the sample size.

5.64 **True or false.** For a fixed confidence level $(1 - \alpha)$, increasing the sampling error SE will lead to a smaller n in determining the sample size.

Learning the Mechanics

5.65 If you wish to estimate a population mean to within .2 with a 95% confidence interval and you know from previous sampling that σ^2 is approximately equal to 5.4, how many observations would you have to include in your sample?

5.66 If nothing is known about p, .5 can be substituted for p in the sample-size formula for a population proportion. But when this is done, the resulting sample size may be larger than needed. Under what circumstances will using $p = .5$ in the sample-size formula yield a sample size larger than is needed to construct a confidence interval for p with a specified bound and a specified confidence level?

5.67 Suppose you wish to estimate a population mean correct to within .15 with a confidence level of .90. You do not know σ^2, but you know that the observations will range in value between 31 and 39.
  a. Find the approximate sample size that will produce the desired accuracy of the estimate. You wish to be conservative to ensure that the sample size will be ample for achieving the desired accuracy of the estimate. [*Hint:* Using your knowledge of data variation from Section 2.5, assume that the range of the observations will equal 4σ.]
  b. Calculate the approximate sample size, making the less conservative assumption that the range of the observations is equal to 6σ.

5.68 In each case, find the approximate sample size required to construct a 95% confidence interval for p that has sampling error SE = .06.
  a. Assume that p is near .3.

  b. Assume that you have no prior knowledge about p, but you wish to be certain that your sample is large enough to achieve the specified accuracy for the estimate.

5.69 The following is a 90% confidence interval for p: (.26, .54). How large was the sample used to construct this interval?

5.70 It costs you $10 to draw a sample of size $n = 1$ and measure the attribute of interest. You have a budget of $1,200.
  a. Do you have sufficient funds to estimate the population mean for the attribute of interest with a 95% confidence interval 4 units in width? Assume that $\sigma = 12$.
  b. If a 90% confidence level were used, would your answer to part **a** change? Explain.

5.71 Suppose you wish to estimate the mean of a normal population with a 95% confidence interval and you know from prior information that $\sigma^2 \approx 1$.
  a. To see the effect of the sample size on the width of the confidence interval, calculate the width of the confidence interval for $n = 16, 25, 49, 100,$ and 400.
  b. Plot the width as a function of sample size n on graph paper. Connect the points by a smooth curve, and note how the width decreases as n increases.

Applying the Concepts—Basic

5.72 **Radioactive lichen.** Refer to the Alaskan Lichen Radionuclide Baseline Research study, presented in Exercise 5.35 (p. 274). In a sample of $n = 9$ lichen specimens, the researchers found the mean and standard deviation of the amount of the radioactive element, cesium-137, that was present to be .009 and .005 microcurie per milliliter, respectively. Suppose the researchers want to increase the sample size in order to estimate the mean μ to within .001 microcurie per milliliter of its true value, using a 95% confidence interval.
  a. What is the confidence level desired by the researchers?
  b. What is the sampling error desired by the researchers?

  c. Compute the sample size necessary to obtain the desired estimate.

5.73 **Scanning errors at Wal-Mart.** The National Institute for Standards and Technology (NIST) studied the accuracy of checkout scanners at Wal-Mart stores in California. NIST

Key Symbols

θ	General population parameter (theta)
μ	Population mean
σ	Population standard deviation
p	Population proportion; P (Success) in binomial trial
q	$1 - p$
$\bar{x}$	Sample mean (estimator of μ)
$\hat{p}$	Sample proportion (estimator of p)
$\mu_{\bar{x}}$	Mean of the population sampling distribution of $\bar{x}$
$\sigma_{\bar{x}}$	Standard deviation of the sampling distribution of $\bar{x}$
$\sigma_{\hat{p}}$	Standard deviation of the sampling distribution of $\hat{p}$
SE	Sampling error in estimation
α	$(1 - \alpha)$ represents the confidence coefficient
$z_{\alpha/2}$	z-value used in a $100(1 - \alpha)\%$ large-sample confidence interval
$t_{\alpha/2}$	Student's t-value used in a $100(1 - \alpha)\%$ small-sample confidence interval

Confidence Interval: An interval that encloses an unknown population parameter with a certain level of confidence, $(1 - \alpha)$

Confidence Coefficient: The probability $(1 - \alpha)$ that a randomly selected confidence interval encloses the true value of the population parameter.

Key Words for Identifying the Target Parameter:

μ—Mean, Average

p—Proportion, Fraction, Percentage, Rate, Probability

Commonly Used z-values for a Large-Sample Confidence Interval:

90% CI:	$(1 - \alpha) = .10$	$z_{.05} = 1.645$
95% CI:	$(1 - \alpha) = .05$	$z_{.025} = 1.96$
99% CI:	$(1 - \alpha) = .01$	$z_{.005} = 2.575$

Determining the Sample Size n:

Estimating μ: $\qquad n = (z_{\alpha/2})^2(\sigma^2)/(\text{SE})^2$

Estimating p: $\qquad n = (z_{\alpha/2})^2(pq)/(\text{SE})^2$

Illustrating the Notion of "95% Confidence"

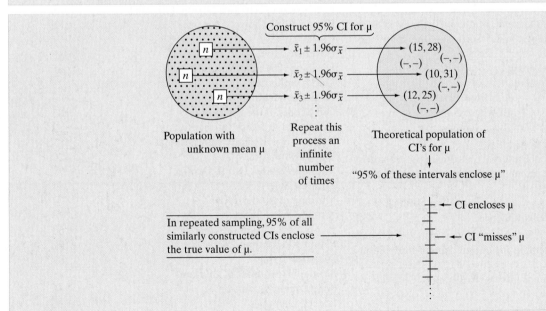

SUPPLEMENTARY EXERCISES 5.83–5.113

Note: List the assumptions necessary for the valid implementation of the statistical procedures you use in solving all these exercises.

Understanding the Principles

5.83 For each of the following, identify the target parameter as μ or p.

  a. Average score on the SAT

  b. Mean time waiting at a supermarket checkout lane

  c. Proportion of voters in favor of legalizing marijuana

  d. Percentage of NFL players who have ever made the Pro Bowl

  e. Dropout rate of American college students

5.84 Interpret the phrase "95% confident" in the following statement: "We are 95% confident that the proportion of all PCs with a computer virus falls between .12 and .18."

5.85 In each of the following instances, determine whether you would use a z- or t-statistic (or neither) to form a 95% confidence interval; then look up the appropriate z or t value.

  a. Random sample of size $n = 21$ from a normal distribution with unknown mean μ and standard deviation σ

  b. Random sample of size $n = 175$ from a normal distribution with unknown mean μ and standard deviation σ

  c. Random sample of size $n = 12$ from a normal distribution with unknown mean and standard deviation $\sigma = 5$

  d. Random sample of size $n = 65$ from a distribution about which nothing is known

  e. Random sample of size $n = 8$ from a distribution about which nothing is known

Learning the Mechanics

5.86 Let represent a particular value of t from Table IV of Appendix A. Find the table values such that the following statements are true:

  a. $P(t \leq t_0) = .05$, where df $= 17$

  b. $P(t \geq t_0) = .005$, where df $= 14$

  c. $P(t \leq -t_0 \text{ or } t \geq t_0) = .10$, where df $= 6$

  d. $P(t \leq -t_0 \text{ or } t \geq t_0) = .01$, where df $= 22$

5.87 In a random sample of 400 measurements, 227 possess the characteristic of interest, A.

  a. Use a 95% confidence interval to estimate the true proportion p of measurements in the population with characteristic A.

  b. How large a sample would be needed to estimate p to within .02 with 95% confidence?

5.88 A random sample of 225 measurements is selected from a population, and the sample mean and standard deviation are $\bar{x} = 32.5$ and $s = 30.0$, respectively.

  a. Use a 99% confidence interval to estimate the mean of the population, μ.

  b. How large a sample would be needed to estimate μ to within .5 with 99% confidence?

  c. What is meant by the phrase "99% confidence" as it is used in this exercise?

Applying the Concepts—Basic

5.89 **CDC health survey.** The Centers for Disease Control and Prevention (CDCP) in Atlanta, Georgia, conduct an annual survey of the general health of the U.S. population as part of their Behavioral Risk Factor Surveillance System. Using random-digit dialing, the CDCP telephones U.S. citizens over 18 years of age and asks them the following four questions:

  (1) Is your health generally excellent, very good, good, fair, or poor?

  (2) How many days during the previous 30 days was your physical health not good because of injury or illness?

  (3) How many days during the previous 30 days was your mental health not good because of stress, depression, or emotional problems?

  (4) How many days during the previous 30 days did your physical or mental health prevent you from performing your usual activities?

Identify the parameter of interest.

5.90 **Tax-exempt charities.** Donations to tax-exempt organizations such as the Red Cross, the Salvation Army, the YMCA, and the American Cancer Society not only go to the stated charitable purpose, but are used to cover fundraising expenses and overhead. The accompanying table lists the charitable commitment (i.e., the percentage of expenses that goes toward the stated charitable purpose) for a sample of 30 charities.

CHARITY

Organization	Charitable Commitment
American Cancer Society	62%
American National Red Cross	91
Big Brothers Big Sisters of America	77
Boy Scouts of America National Council	81
Boys & Girls Clubs of America	81
CARE	91
Covenant House	15
Disabled American Veterans	65
Ducks Unlimited	78
Feed The Children	90
Girl Scouts of the USA	83
Goodwill Industries International	89
Habitat for Humanity International	81
Mayo Foundation	26
Mothers Against Drunk Drivers	71
Multiple Sclerosis Association of America	56
Museum of Modern Art	79
Nature Conservancy	77
Paralyzed Veterans of America	50
Planned Parenthood Federation	81
Salvation Army	84
Shriners Hospital for Children	95
Smithsonian Institution	87
Special Olympics	72
Trust for Public Land	88
United Jewish Appeal/Federation—NY	75
United States Olympic Committee	78
United Way of New York City	85
WGBH Educational Foundation	81
YMCA of the USA	80

Source: "Look Before You Give," Forbes, Dec. 27, 1999, pp. 206–216.

a. Give a point estimate for the mean charitable commitment of tax-exempt organizations.

b. Construct a 98% confidence interval for the mean charitable commitment.

c. What assumption(s) must hold for the method of estimation used in part **b** to be appropriate?

d. Why is the confidence interval of part **b** a better estimator of the mean charitable commitment than the point estimator of part **a**?

5.91 Cell phone use by drivers. In a July 2001 research note, the U.S. Department of Transportation reported the results of the *National Occupant Protection Use Survey*. One focus of the survey was to determine the level of cell phone use by drivers while they are in the act of driving a motor passenger vehicle. Data collected by observers at randomly selected intersections across the country revealed that in a sample of 1,165 drivers, 35 were using their cell phone.

a. Give a point estimate of p, the true driver cell phone use rate (i.e., the true proportion of drivers who are using a cell phone while driving).

b. Compute a 95% confidence interval for p.

c. Give a practical interpretation of the interval you found in part **b**.

5.92 "Made in the USA" survey. Refer to Exercise 2.13 (p. 36) and the *Journal of Global Business* (Spring 2002) survey to determine what "Made in the USA" means to consumers. Recall that 106 shoppers at a shopping mall in Muncie, Indiana, responded to the question " 'Made in the USA' means what percentage of U.S. labor and materials?" Sixty-four shoppers answered "100%."

a. Define the population of interest in the survey.

b. What is the characteristic of interest in the population?

c. Use a 90% confidence interval to estimate the true proportion of consumers who believe that "Made in the USA" means 100% U.S. labor and materials.

d. Give a practical interpretation of the interval you used in part **c**.

e. Explain what the phrase "90% confidence" means for this interval.

5.93 "Made in the USA" survey (cont'd). Refer to Exercise 5.92. Suppose the researchers want to increase the sample size in order to estimate the true proportion p to within .05 of its true value with a 90% confidence interval.

a. What is the confidence level desired by the researchers?

b. What is the sampling error desired by the researchers?

c. Compute the sample size necessary to obtain the desired estimate.

5.94 Water pollution testing. The EPA wants to test a randomly selected sample of n water specimens and estimate μ, the mean daily rate of pollution produced by a mining operation. If the EPA wants a 95% confidence interval estimate with a sampling error of 1 milligram per liter (mg/L), how many water specimens are required in the sample? Assume that prior knowledge indicates that pollution readings in water samples taken during a day are approximately normally distributed with a standard deviation equal to 5 (mg/L).

5.95 Crop weights of pigeons. The *Australian Journal of Zoology* (Vol. 43, 1995) reported on a study of the diets and water requirements of spinifex pigeons. Sixteen pigeons were cap-

tured in the desert and the crop (i.e., stomach) contents of each examined. The accompanying table reports the weight (in grams) of dry seed in the crop of each pigeon. Find a 99% confidence interval for the average weight of dry seeds in the crops of spinifex pigeons inhabiting the Western Australian desert. Interpret the result.

PIGEONS

.457	3.751	.238	2.967	2.509	1.384	1.454	.818
.335	1.436	1.603	1.309	.201	.530	2.144	.834

Source: Excerpted from Williams, J. B., Bradshaw, D., and Schmidt, L. "Field metabolism and water requirements of spinifex pigeons *(Geophaps plumifera)* in Western Australia," *Australian Journal of Zoology,* Vol. 43, No. 1, 1995, p. 7 (Table 2).

5.96 Ammonia in car exhaust. Refer to the *Environmental Science & Technology* (September 1, 2000) study on the ammonia levels near the exit ramp of a San Francisco highway tunnel, presented in Exercise 2.59 (p. 59). The ammonia concentration (parts per million) data for eight randomly selected days during the afternoon drive time are reproduced in the accompanying table. Find a 99% confidence interval for the population mean daily ammonia level in air in the tunnel. Interpret your result.

AMMONIA

1.53	1.50	1.37	1.51	1.55	1.42	1.41	1.48

5.97 Homeless in the United States. The *American Journal of Orthopsychiatry* (July 1995) published an article on the prevalence of homelessness in the United States. A sample of 487 U.S. adults were asked to respond to the question "Was there ever a time in your life when you did not have a place to live?" A 95% confidence interval for the true proportion who were or are homeless, based on the survey data, was determined to be (.045, .091). Interpret the result fully.

Applying the Concepts—Intermediate

5.98 Hearing loss study. Patients with normal hearing in one ear and unaided sensorineural hearing loss in the other are characterized as suffering from unilateral hearing loss. In a study reported in the *American Journal of Audiology* (March 1995), eight patients with unilateral hearing loss were fitted with a special wireless hearing aid in the "bad" ear. The absolute sound pressure level (SPL) was then measured near the eardrum of the ear when noise was produced at a frequency of 500 hertz. The SPLs of the eight patients, recorded in decibels, are as follows:

SOUND

73.0	80.1	82.8	76.8	73.5	74.3	76.0	68.1

Construct and interpret a 90% confidence interval for the true mean SPL of patients with unilateral hearing loss when noise is produced at a frequency of 500 hertz.

5.99 Brown-bag lunches at work. In a study reported in *The Wall Street Journal* (April 4, 1999), the Tupperware Corporation surveyed 1,007 U.S. workers. Of the people surveyed, 665 indicated that they take their lunch to work with them. Of these 665 taking their lunch, 200 reported that they take it in brown bags.

a. Find a 95% confidence interval estimate of the population proportion of U.S. workers who take their lunch to work with them. Interpret the interval.

b. Consider the population of U.S. workers who take their lunch to work with them. Find a 95% confidence interval estimate of the population proportion who take brown-bag lunches. Interpret the interval.

5.100 Role importance for the elderly. Refer to the *Psychology and Aging* (December 2000) study of the roles that elderly people feel are the most important to them, presented in Exercise 2.10 (p. 35). Recall that in a national sample of 1,102 adults 65 years or older, 424 identified "spouse" as their most salient role. Use a 95% confidence interval to estimate the true percentage of adults 65 years or older who feel that being a spouse is their most salient role. Give a practical interpretation of this interval.

5.101 Inbreeding of tropical wasps. Tropical swarm-founding wasps rely on female workers to raise their offspring. One possible explanation for this strange behavior is inbreeding, which increases relatedness among the wasps, presumably making it easier for the workers to pick out their closest relatives as propagators of their own genetic material. To test this theory, 197 swarm-founding wasps were captured in Venezuela, frozen at $-70°C$, and then subjected to a series of genetic tests (*Science*, November 1988). The data were used to generate an inbreeding coefficient x for each wasp specimen, with the following results: $\bar{x} = .044$ and $s = .884$.

a. Construct a 99% confidence interval for the mean inbreeding coefficient of this species of wasp.

b. A coefficient of 0 implies that the wasp has no tendency to inbreed. Use the confidence interval you constructed in part **a** to make an inference about the tendency for this species of wasp to inbreed.

OILSPILL

5.102 Tanker oil spills. Refer to the *Marine Technology* (January 1995) study of the causes of 50 recent major oil spills from tankers and carriers, presented in Exercise 2.186 (p. 106). The data are stored in the **OILSPILL** file.

a. Give a point estimate for the proportion of major oil spills that are caused by hull failure.

b. Form a 95% confidence interval for the estimate you found in part **a**. Interpret the result.

5.103 Time to solve a math programming problem. *IEEE Transactions* (June 1990) presented a hybrid algorithm for solving a polynomial zero–one mathematical programming problem. The algorithm incorporates a mixture of pseudo-

MATHCPU

.045	1.055	.136	1.894	.379	.136	.336	.258	1.070
.506	.088	.242	1.639	.912	.412	.361	8.788	.579
1.267	.567	.182	.036	.394	.209	.445	.179	.118
.333	.554	.258	.182	.070	3.985	.670	3.888	.136
.091	.600	.291	.327	.130	.145	4.170	.227	.064
.194	.209	.258	3.046	.045	.049	.079		

Source: Snyder, W. S., and Chrissis, J. W. "A hybrid algorithm for solving zero-one mathematical programming problems," *IEEE Transactions,* Vol. 22, No. 2, June 1990, p. 166 (Table 1).

Boolean concepts and time-proven implicit enumeration procedures. Fifty-two random problems were solved by the hybrid algorithm; the times to solution (CPU time in seconds) are listed in the accompanying table.

a. Estimate, with 95% confidence, the mean solution time for the hybrid algorithm. Interpret the result.

b. How many problems must be solved to estimate the mean μ to within .25 second with 95% confidence?

5.104 Psychological study of participation and satisfaction. The relationship between an employee's participation in the performance appraisal process and his or her subsequent reactions toward the appraisal was investigated in the *Journal of Applied Psychology* (August 1998). In Chapter 9, we will discuss a quantitative measure of the relationship between two variables, called the coefficient of correlation, r. The researchers obtained r for a sample of 34 studies that examined the relationship between participation in the appraisal and subsequent satisfaction with the appraisal. These correlations are listed in the accompanying table. (Values of r near $+1$ reflect a strong positive relationship between the variables.) Find a 95% confidence interval for the mean of the data, and interpret it in the words of the problem.

CORR34

.50	.58	.71	.46	.63	.66	.31	.35	.51	.06	.35	.19
.40	.63	.43	.16	−.08	.51	.59	.43	.30	.69	.25	.20
.39	.20	.51	.68	.74	.65	.34	.45	.31	.27		

Source: Cawley, B. D., Keeping, L. M., and Levy, P. E. "Participation in the performance appraisal process and employee reactions: A meta-analytic review of field investigations." *Journal of Applied Psychology,* Vol. 83, No. 4, Aug. 1998, pp. 632–633 (Appendix).

SUICIDE

5.105 Jail suicide study. In Exercise 2.187 (p. 107) you considered data on suicides in correctional facilities. The data were published in the *American Journal of Psychiatry* (July 1995). The researchers wanted to know what factors increase the risk of suicide by those incarcerated in urban jails. The data on 37 suicides that occurred in Wayne County Jail, Detroit, Michigan, are saved in the **SUICIDE** file. Answer each of the questions that follow, using a 95% confidence interval for the target parameter implied in the question. Interpret each confidence interval fully.

a. What proportion of suicides at the jail are committed by inmates charged with murder or manslaughter?

b. What proportion of suicides at the jail are committed at night?

c. What is the average length of time an inmate is in jail before committing suicide?

d. What percentage of suicides are committed by white inmates?

5.106 Recommendation letters for professors. Refer to the *American Psychologist* (July 1995) study of applications for university positions in experimental psychology, presented in Exercise 2.62 (p. 59). One of the objectives of the analysis was to identify problems and errors (e.g., not submitting at least three letters of recommendation) in the application packages. In a sample of 148 applications for the positions, 30 failed to provide *any* letters of recommendation. Estimate, with 90% confidence, the true fraction of applicants for a university position in experimental psychology that fail to provide letters of recommendation. Interpret the result.

5.107 **Salmonella in ice cream bars.** Recently, a case of salmonella (bacterial) poisoning was traced to a particular brand of ice cream bar, and the manufacturer removed the bars from the market. Despite this response, many consumers refused to purchase *any* brand of ice cream bars for some time after the event (McClave, personal correspondence). One manufacturer conducted a survey of consumers 6 months after the poisoning. A sample of 244 ice cream bar consumers was contacted, and 23 indicated that they would not purchase ice cream bars because of the potential for food poisoning.

   a. What is the point estimate of the true fraction of the entire market who refuse to purchase bars 6 months after the poisoning?

   b. Is the sample size large enough to use the normal approximation for the sampling distribution of the estimator of the binomial probability? Justify your response.

   c. Construct a 95% confidence interval for the true proportion of the market who still refuse to purchase ice cream bars 6 months after the event.

   d. Interpret both the point estimate and confidence interval in terms of this application.

5.108 **Salmonella in ice cream bars (cont'd).** Refer to Exercise 5.107. Suppose it is now 1 year after the poisoning was traced to ice cream bars. The manufacturer wishes to estimate the proportion who still will not purchase bars to within .02, using a 95% confidence interval. How many consumers should be sampled?

5.109 **Removing a soil contaminant.** A common hazardous compound found in contaminated soil is benzo(a)pyrene [B(a)p]. An experiment was conducted to determine the effectiveness of a method designed to remove B(a)p from soil. (*Journal of Hazardous Materials*, June 1995.) Three soil specimens contaminated with a known amount of B(a)p were treated with a toxin that inhibits microbial growth. After 95 days of incubation, the percentage of B(a)p removed from each soil specimen was measured. The experiment produced the following summary statistics: $\bar{x} = 49.3$ and $s = 1.5$.

   a. Use a 99% confidence interval to estimate the mean percentage of B(a)p removed from a soil specimen in which the toxin was used.

   b. Interpret the interval in terms of this application.

   c. What assumption is necessary to ensure the validity of this confidence interval?

   d. How many soil specimens must be sampled to estimate the mean percentage removed to within .5, using a 99% confidence interval?

5.110 **Material safety data sheets.** For over 20 years, the Occupational Safety & Health Administration has required companies that handle hazardous chemicals to complete material safety data sheets (MSDSs). These sheets have been criticized for being too hard to understand and complete by workers. Although improvements were implemented in 1990, a recent study of 150 MSDSs revealed that only 11% were completed satisfactorily. (*Chemical & Engineering News*, Feb. 7, 2005.) Find an interval that

contains the true proportion of all MSDSs that are completed satisfactorily, with 95% confidence. Would it surprise you if the true proportion was as high as 20%? Explain.

Applying the Concepts—Advanced

5.111 **IMA salary survey.** Each year, *Management Accounting* reports the results of a salary survey of the members of the Institute of Management Accountants (IMA). One year, the 2,112 members responding had a salary distribution with a 20th percentile of $35,100, a median of $50,000, and an 80th percentile of $73,000.

   a. Use this information to determine the minimum sample size that could be used in next year's survey to estimate the mean salary of IMA members to within $2,000 with 98% confidence. [*Hint*: To estimate s, first apply Chebyshev's theorem to find k such that at least 60% of the data fall within k standard deviations of μ. Then find $s \approx$ (80th percentile –20th percentile)/k.]

   b. Explain how you estimated the standard deviation required for the calculation of the sample size.

   c. List any assumptions you make.

5.112 **Air bags pose danger for children.** By law, all new cars must be equipped with both driver-side and passenger-side safety air bags. There is concern, however, over whether air bags pose a danger for children sitting on the passenger side. In a National Highway Traffic Safety Administration (NHTSA) study of 55 people killed by the explosive force of air bags, 35 were children seated on the front-passenger side. (*Wall Street Journal*, January 22, 1997.) This study led some car owners with the information from children to disconnect the passenger-side air bag.

   a. Use the study to estimate the risk of an air bag fatality on a child seated on the front passenger seat.

   b. NHTSA investigators determined that 24 of 35 children killed by the air bags were not wearing seat belts or were improperly restrained. How does this information affect your assessment of the risk of an air bag fatality?

Critical Thinking Challenge

5.113 **Scallops, sampling, and the law.** In *Interfaces* (March–April 1995), the case involved a ship that fishes for scallops off the coast of New England. In order to protect baby scallops from being harvested, the U.S. Fisheries and Wildlife Service requires that "the average meat per scallop weigh at least $\frac{1}{36}$ of a pound." The ship was accused of violating this weight standard. Bennett lays out the scenario:

   The vessel arrived at a Massachusetts port with 11,000 bags of scallops, from which the harbormaster randomly selected 18 bags for weighing. From each such bag, his agents took a large scoopful of scallops; then, to estimate the bag's average meat per scallop, they divided the total weight of meat in the scoopful by the number of scallops it contained. Based on the 18 [numbers] thus generated, the harbormaster estimated that each of the ship's scallops possessed an average $\frac{1}{39}$ of a pound of meat (that is, they were about seven percent lighter than the minimum requirement). Viewing this outcome as conclusive evidence that the weight standard had been violated, federal authorities at once confiscated 95 percent of the catch (which they then sold at *auction). The fishing voyage was thus transformed into a financial catastrophe for its participants.*

The actual scallop weight measurements for each of the 18 sampled bags are listed in the accompanying table. For ease of exposition, Bennett expressed each number as a multiple of $\frac{1}{36}$ of a pound, the minimum permissible average weight per scallop. Consequently, numbers below 1 indicate individual bags that do not meet the standard.

The ship's owner filed a lawsuit against the federal government, declaring that his vessel had fully complied with the weight standard. A Boston law firm was hired to represent the owner in legal proceedings and Bennett was retained by the firm to provide statistical litigation support and, if necessary, expert witness testimony.

a. Recall that the harbormaster sampled only 18 of the ship's 11,000 bags of scallops. One of the questions the lawyers asked Bennett was "Can a reliable estimate of the mean weight of all the scallops be ob-

tained from a sample of size 18?" Give your opinion on this issue.

b. As stated in the article, the government's decision rule is to confiscate a catch if the sample mean weight of the scallops is less than $\frac{1}{36}$ of a pound. Do you see any flaws in this rule?

c. Develop your own procedure for determining whether a ship is in violation of the minimum-weight restriction. Apply your rule to the data. Draw a conclusion about the ship in question.

SCALLOPS

.93	.88	.85	.91	.91	.84	.90	.98	.88
.89	.98	.87	.91	.92	.99	1.14	1.06	.93

Source: Bennett, A. "Misapplications review: Jail terms." *Interfaces,* Vol. 25, No. 2, Mar.–Apr. 1995, p. 20.

SUMMARY ACTIVITY: Conducting a Pilot Study

Choose a population pertinent to your major area of interest—a population that has an unknown mean or, if the population is binomial, that has an unknown probability of success. For example, a marketing major may be interested in the proportion of consumers who prefer a certain product. A sociology major may be interested in estimating the proportion of people in a certain socioeconomic group or the mean income of people living in a particular part of a city. A political science major may wish to estimate the proportion of an electorate in favor of a certain candidate, a certain amendment, or a certain presidential policy. A pre-med student might want to find the average length of time patients stay in the hospital or the average number of people treated daily in the emergency

room. We could continue with examples, but the point should be clear: Choose something of interest to you.

Define the parameter you want to estimate and conduct a *pilot study* to obtain an initial estimate of the parameter of interest and, more importantly, an estimate of the variability associated with the estimator. A pilot study is a small experiment (perhaps 20 to 30 observations) used to gain some information about the population of interest. The purpose of the study is to help plan more elaborate future experiments. Using the results of your pilot study, determine the sample size necessary to estimate the parameter to within a reasonable bound (of your choice) with a 95% confidence interval.

REFERENCES

Agresti, A., and Coull, B. A. "Approximate is better than 'exact' for interval estimation of binomial proportions." *The American Statistician*, Vol. 52, No. 2, May 1998, pp. 119–126.

Cochran, W. G. *Sampling Techniques*, 3d ed. New York: Wiley, 1977.

Freedman, D., Pisani, R., and Purves, R. *Statistics*. New York: Norton, 1978.

Kish, L. *Survey Sampling*. New York: Wiley, 1965.

Mendenhall, W., Beaver, R. J., and Beaver, B. *Introduction to Probability and Statistics*, 11th ed. North Scituate, MA: Duxbury, 2002.

Wilson, E. G. "Probable inference, the law of succession, and statistical inference." *Journal of the American Statistical Association*, Vol. 22, 1927, pp. 209–212.

Using Technology

Confidence Intervals with MINITAB

MINITAB can be used to obtain one-sample confidence intervals for both a population mean and a population proportion. To use a previously created sample data set to generate a confidence interval for the mean, first access the MINITAB data worksheet. Next, click on the "Stat" button on the MINITAB menu bar, and then click on "Basic Statistics" and "1-sample t," as shown in Figure 5.M.1. The resulting dialog box appears as shown in Figure 5.M.2. Click on "Samples in Columns," and then specify the quantitative variable of interest in the open box. Now click on the "Options" button at the bottom of the dialog box, and specify the confidence level in the resulting dialog box as shown in Figure 5.M.3. Click "OK" to return to the "1-Sample t" dialog box, and then click "OK" again to produce the confidence interval.

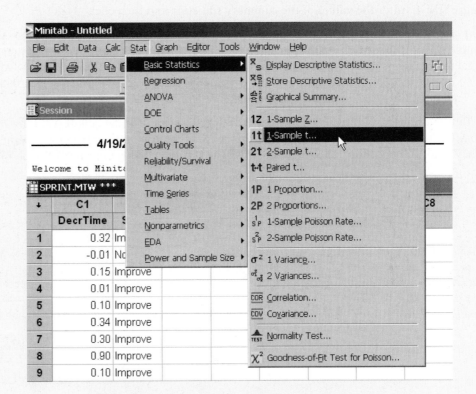

FIGURE 5.M.1
MINITAB menu options for a confidence interval for μ

FIGURE 5.M.2
MINITAB one-sample t dialog box

(continued)

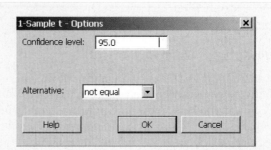

FIGURE 5.M.3
MINITAB one-sample
t options

If you want to produce a confidence interval for the mean from summary information (e.g., the sample mean, sample standard deviation, and sample size), then click on "Summarized data" in the "1-Sample t" dialog box as shown in Figure 5.M.4. Enter the values of the summary statistics, and then click "OK."

[*Important Note*: The MINITAB 1-Sample t procedure uses the *t*-statistic to generate the confidence interval. When the sample size *n* is small, this is the appropriate method. When the sample size *n* is large, the *t*-value will be approximately equal to the large-sample *z*-value and the resulting interval will still be valid. If you have a large sample and you know the value of the population standard deviation σ (which is rarely the case), then select "1-Sample Z" from the "Basic Statistics" menu options (see Figure 5.M.1) and make the appropriate selections.]

To use a previously created sample data set to generate a confidence interval for a population proportion, first access the MINITAB data worksheet. Next, click on the "Stat" button on the MINITAB menu bar, and then click on "Basic Statistics" and "1 Proportion". (See Figure 5.M.1.) The resulting dialog box appears as shown in Figure 5.M.5. Click on "Sam-

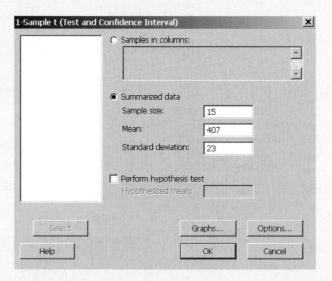

FIGURE 5.M.4
MINITAB one-sample
t dialog box with summary
statistics

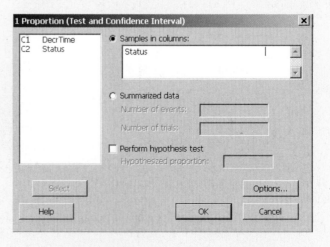

FIGURE 5.M.5
MINITAB one-sample
proportion dialog box

ples in columns," and then specify the qualitative variable of interest in the open box. Now click on the "Options" button at the bottom of the dialog box, and specify the confidence level in the resulting dialog box as shown in Figure 5.M.6. Also, check the "Use test and interval based on normal distribution" box at the bottom. Click "OK" to return to the "1 Proportion" dialog box, and then click "OK" again to produce the confidence interval.

If you want to produce a confidence interval for a proportion from summary information (e.g., the number of success-es and the sample size), then click on "Summarized data" in the "1 Proportion" dialog box. (See Figure 5.M.5.) Enter the value for the number of trials (i.e., the sample size) and the number of events (i.e., the number of successes), and then click "OK."

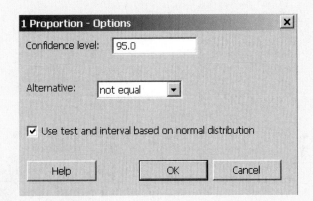

FIGURE 5.M.6

MINITAB one-sample proportion options

CHAPTER 6

Inferences Based on a Single Sample

Tests of Hypothesis

CONTENTS

6.1 The Elements of a Test of Hypothesis

6.2 Large-Sample Test of Hypothesis about a Population Mean

6.3 Observed Significance Levels: *p*-Values

6.4 Small-Sample Test of Hypothesis about a Population Mean

6.5 Large-Sample Test of Hypothesis about a Population Proportion

6.6 A Nonparametric Test about a Population Median

STATISTICS IN ACTION

Diary of a KLEENEX® User—How Many Tissues in a Box?

USING TECHNOLOGY

One-Sample Tests of Hypothesis with MINITAB

WHERE WE'VE BEEN

■ Used sample information to provide a *point estimate* of a population parameter

■ Used the sampling distribution of a statistic to assess the reliability of an estimate through a *confidence interval*

WHERE WE'RE GOING

■ Test a specific value of a population parameter (mean or proportion)—called a *test of hypothesis*

■ Provide a measure of reliability for the hypothesis test—called the *significance level* of the test

STATISTICS IN ACTION

Diary of a KLEENEX® User—How Many Tissues in a Box?

In 1924, Kimberly-Clark Corporation invented a facial tissue for removing cold cream and began marketing it as KLEENEX® brand tissues. Today, KLEENEX® is recognized as the top-selling brand of tissue in the world. A wide variety of KLEENEX® products is available, ranging from extra-large tissues to tissues with lotion. Over the past 80 years, Kimberly-Clark Corporation has packaged the tissues in boxes of different sizes and shapes and varied the number of tissues packaged in each box. For example, currently a family-size box contains 144 two-ply tissues, a cold-care box contains 70 tissues (coated with lotion), and a convenience pocket pack contains 15 miniature tissues.

How does Kimberly-Clark Corp. decide how many tissues to put in each box? According to the *Wall Street Journal,* marketing experts at the company use the results of a survey of KLEENEX® customers to help determine how many tissues are packed in a box. In the mid-1980s, when Kimberly-Clark Corp. developed the cold-care box, designed especially for people who have a cold, the company conducted their initial survey of customers for this purpose. Hun-

dreds of customers were asked to keep count of their KLEENEX® use in diaries. According to the *Wall Street Journal* report, the survey results left "little doubt that the company should put 60 tissues in each box." The number 60 was "the average number of times people blow their nose during a cold." (*Note:* In 2000, the company increased the number of tissues packaged in a cold-care box to 70.)

From summary information provided in the *Wall Street Journal* (September 21, 1984) article, we constructed a data set that represents the results of a survey similar to the one just described. In the data file named **TISSUES,** we recorded the number of tissues used by each of 250 consumers during a period when they had a cold. We apply the hypothesis-testing methodology presented in this chapter to that data set in several Statistics in Action Revisited examples.

Statistics in Action Revisited

- Identifying the Key Elements of a Hypothesis Test Relevant to the KLEENEX® Survey (p. 307)

- Testing a Population Mean in the KLEENEX® Survey (p. 320)

- Testing a Population Proportion in the KLEENEX® Survey (p. 333)

Suppose you wanted to determine whether the mean level of a driver's blood alcohol exceeds the legal limit after two drinks, or whether the majority of registered voters approve of the president's performance. In both cases, you are interested in making an inference about how the value of a parameter relates to a specific numerical value. Is it less than, equal to, or greater than the specified number? This type of inference, called a **test of hypothesis,** is the subject of this chapter.

We introduce the elements of a test of hypothesis in Section 6.1. We then show how to conduct a large-sample test of hypothesis about a population mean in Sections 6.2 and 6.3. In Section 6.4, we utilize small samples to conduct tests about means and, in optional Section 6.6 we consider an alternative nonparametric test. Large-sample tests about binomial probabilities are the subject of Section 6.5.

6.1 The Elements of a Test of Hypothesis

Suppose building specifications in a certain city require that the average breaking strength of residential sewer pipe be more than 2,400 pounds per foot of length (i.e., per linear foot). Each manufacturer that wants to sell pipe in that city must demonstrate that its product meets the specification. Note that we are interested in making an inference about the mean μ of a population. However, in this example we are less interested in estimating the value of μ than we are in testing a *hypothesis* about its value. That is, *we want to decide whether the mean breaking strength of the pipe exceeds 2,400 pounds per linear foot.*

The method used to reach a decision is based on the rare-event concept explained in earlier chapters. We define two hypotheses: (1) The **null hypothesis** represents the status quo to the party performing the sampling experiment; it is the hypothesis that will be supported unless the data provide convincing evidence that it is false. (2) The **alternative,** or **research, hypothesis** will be accepted only if the data provide convincing evidence of its truth. From the point of view of the city conducting the tests, the null hypothesis is that the manufacturer's pipe does *not* meet specifications unless the tests provide convincing evidence otherwise. The null and alternative hypotheses are therefore

Null hypothesis (H_0): $\mu \leq 2{,}400$
> (i.e., the manufacturer's pipe does not meet specifications)

Alternative (research) hypothesis (H_a): $\mu > 2{,}400$
> (i.e., the manufacturer's pipe meets specifications)

> **Now Work Exercise 6.8**

How can the city decide when enough evidence exists to conclude that the manufacturer's pipe meets specifications? Since the hypotheses concern the value of the population mean μ, it is reasonable to use the sample mean $\bar{x}$ to make the inference, just as we did when we formed confidence intervals for μ in Sections 5.2 and 5.3. The city will conclude that the pipe meets specifications only when the sample mean $\bar{x}$ convincingly indicates that the population mean exceeds 2,400 pounds per linear foot.

"Convincing" evidence in favor of the alternative hypothesis will exist when the value of $\bar{x}$ exceeds 2,400 by an amount that cannot be readily attributed to sampling variability. To decide, we compute a **test statistic,** which is the z-value that measures the distance between the value of $\bar{x}$ and the value of μ specified in the null hypothesis. When the null hypothesis contains more than one value of μ, as in this case (H_0: $\mu \leq 2{,}400$), we use the value of μ closest to the values specified in the alternative hypothesis. The idea is that if the hypothesis that μ *equals* 2,400 can be rejected in favor of $\mu > 2{,}400$, then μ *less than or equal to* 2,400 can certainly be rejected. Thus, the test statistic is

$$z = \frac{\bar{x} - 2{,}400}{\sigma_{\bar{x}}} = \frac{\bar{x} - 2{,}400}{\sigma/\sqrt{n}}$$

Note that a value of $z = 1$ means that $\bar{x}$ is 1 standard deviation above $\mu = 2{,}400$, a value of $z = 1.5$ means that $\bar{x}$ is 1.5 standard deviations above $\mu = 2{,}400$, etc. How large must z be before the city can be convinced that the null hypothesis can be rejected in favor of the alternative hypothesis and conclude that the pipe meets specifications?

If you examine Figure 6.1, you will note that the chance of observing $\bar{x}$ more than 1.645 standard deviations above 2,400 is only .05—*if in fact the true mean μ is 2,400.* Thus, if the sample mean is more than 1.645 standard deviations above 2,400, either H_0 is true and a relatively rare event has occurred (one with .05 probability) or H_a is true and the population mean exceeds 2,400. Since we would most likely reject the notion that a rare event has occurred, we would reject the null hypothesis ($\mu \leq 2{,}400$) and conclude that the alternative hypothesis ($\mu > 2{,}400$) is true.

Now, what is the probability that the procedure just set forth will lead us to an incorrect decision? Such an incorrect decision—deciding that the null hypothesis is false when in fact it is true—is called a **Type I decision error.** As indicated in Figure 6.1, the risk of making a Type I error is denoted by the symbol α. That is,

$\alpha = P$ (Type I error)

> $= P$ (Rejecting the null hypothesis when in fact the null hypothesis is true)

In this example,

$$\alpha = P\,(z > 1.645 \text{ when in fact } \mu = 2{,}400) = .05$$

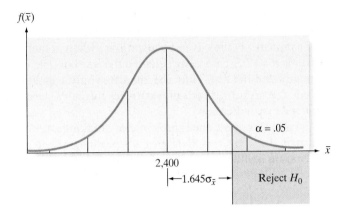

FIGURE 6.1

The sampling distribution of $\bar{x}$, assuming that $\mu = 2,400$

We now summarize the elements of the test:

H_0: $\mu \leq 2,400$ (Mean breaking strength is less than or equal to 2,400 pounds.)

H_a: $\mu > 2,400$ (Mean breaking strength exceeds 2,400 pounds.)

Test statistic: $z = \dfrac{\bar{x} - 2,400}{\sigma_{\bar{x}}}$

Rejection region: $z > 1.645$, which corresponds to $\alpha = .05$

Note that the **rejection region** refers to the values of the test statistic for which we will *reject the null hypothesis.*

To illustrate the use of the test, suppose we test 50 sections of sewer pipe and find the mean and standard deviation for these 50 measurements to be, respectively,

$\bar{x} = 2,460$ pounds per linear foot   and $s = 200$ pounds per linear foot

As in the case of estimation, we can use s to approximate σ when s is calculated from a large set of sample measurements.

The test statistic is

$$z = \frac{\bar{x} - 2,400}{\sigma_{\bar{x}}} = \frac{\bar{x} - 2,400}{\sigma/\sqrt{n}} \approx \frac{\bar{x} - 2,400}{s/\sqrt{n}}$$

Substituting $\bar{x} = 2,460$, $n = 50$, and $s = 200$, we have

$$z \approx \frac{2,460 - 2,400}{200/\sqrt{50}} = \frac{60}{28.28} = 2.12$$

Therefore, the sample mean lies $2.12\sigma_{\bar{x}}$ above the hypothesized value of $\mu = 2,400$, as shown in Figure 6.2. Since this value of z exceeds 1.645, it falls into the rejection region. That is, we reject the null hypothesis that $\mu = 2,400$ and conclude that $\mu > 2,400$. Thus, it appears that the company's pipe has a mean strength that exceeds 2,400 pounds per linear foot.

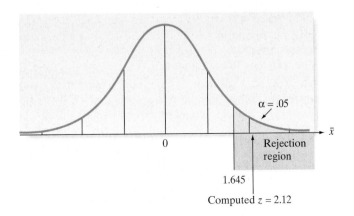

FIGURE 6.2

Location of the test statistic for a test of the hypothesis H_0: $\mu \leq 2,400$

How much faith can be placed in this conclusion? What is the probability that our statistical test could lead us to reject the null hypothesis (and conclude that the company's pipe meets the city's specifications) when in fact the null hypothesis is true? The answer is $\alpha = .05$. That is, we selected the level of risk, α, of making a Type I error when we constructed the test. Thus, the chance is only 1 in 20 that our test would lead us to conclude the manufacturer's pipe satisfies the city's specifications when in fact the pipe does *not* meet specifications.

Now, suppose the sample mean breaking strength for the 50 sections of sewer pipe turned out to be $\bar{x} = 2{,}430$ pounds per linear foot. Assuming that the sample standard deviation is still $s = 200$, we now find that the test statistic is

$$z = \frac{2{,}430 - 2{,}400}{2/\sqrt{50}} = \frac{30}{28.28} = 1.06$$

Therefore, the sample mean $\bar{x} = 2{,}430$ is only 1.06 standard deviations above the null hypothesized value of $\mu = 2{,}400$. As shown in Figure 6.3, this value does not fall into the rejection region ($z > 1.645$). Therefore, we know that we cannot reject H_0 if we use $\alpha = .05$. Even though the sample mean exceeds the city's specification of 2,400 by 30 pounds per linear foot, it does not exceed the specification by enough to provide *convincing* evidence that the *population mean* exceeds 2,400.

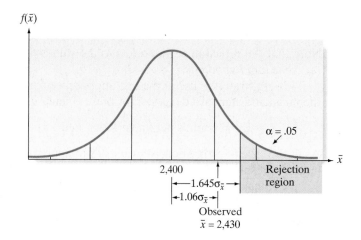

FIGURE 6.3

Location of the test statistic when $\bar{x} = 2{,}430$

Should we accept the null hypothesis $H_0: \mu \leq 2{,}400$ and conclude that the manufacturer's pipe does not meet specifications? To do so would be to risk a **Type II error:** concluding that the null hypothesis is true (the pipe does not meet specifications) when in fact it is false (the pipe does meet specifications). We denote the probability of committing a Type II error by β. It is well known that β is often difficult to determine precisely. Consequently, rather than make a decision (accept H_0) for which the probability of error (β) is unknown, we avoid the potential Type II error by avoiding the conclusion

Biography

EGON S. PEARSON (1895–1980)—
The Neyman–Pearson Lemma

Egon Pearson was the only son of noteworthy British statistician Karl Pearson. (See Biography, p. 446). As you might expect, Egon developed an interest in the statistical methods developed by his father and, upon completing graduate school, accepted a position to work for Karl in the Department of Applied Statistics at University College, London. Egon is best known for his collaboration with Jerzy Neyman (see Biography, p. 446) on the development of the theory of hypothesis testing. One of the basic concepts in the Neyman–Pearson approach was that of the "null" and "alternative" hypotheses. Their famous "Neyman–Pearson" lemma was published in *Biometrika* in 1928. Egon Pearson made numerous other contributions to statistics and was known as an excellent teacher and lecturer. In his last major work, Egon fulfilled a promise made to his father by publishing an annotated version of Karl Pearson's lectures on the early history of statistics.

TABLE 6.1 Conclusions and Consequences for a Test of Hypothesis

	True State of Nature	
Conclusion	H_0 True	H_a True
Accept H_0 (Assume H_0 True)	Correct decision	Type II error (probability β)
Reject H_0 (Assume H_a True)	Type I error (probability α)	Correct decision

that the null hypothesis is true. Instead, we will simply state that *the sample evidence is insufficient to reject H_0 at $\alpha = .05$*. Since the null hypothesis is the "status quo" hypothesis, the effect of not rejecting H_0 is to maintain the status quo. In our pipe-testing example, the effect of having insufficient evidence to reject the null hypothesis that the pipe does not meet specifications is probably to prohibit the utilization of the manufacturer's pipe unless and until there is sufficient evidence that the pipe does meet specifications. That is, until the data indicate convincingly that the null hypothesis is false, we usually maintain the status quo implied by its truth.

Table 6.1 summarizes the four possible outcomes of a test of hypothesis. The "true state of nature" columns in the table refer to the fact that either the null hypothesis H_0 is true or the alternative hypothesis H_a is true. Note that the true state of nature is unknown to the researcher conducting the test. The "conclusion" rows in the table refer to the action of the researcher, assuming that he or she will conclude either that H_0 is true or that H_a is true, on the basis of the results of the sampling experiment. Note that a Type I error can be made *only* when the alternative hypothesis is accepted (equivalently, when the null hypothesis is rejected) and a Type II error can be made *only* when the null hypothesis is accepted. Our policy will be to make a decision only when we know the probability of making the error which corresponds to that decision. Since α is usually specified by the analyst (typically, $\alpha = .05$ in research), we will generally be able to reject H_0 (accept H_a) when the sample evidence supports that decision. However, since β is usually *not* specified, *we will generally avoid the decision to accept H_0, preferring instead to state that the sample evidence is insufficient to reject H_0 when the test statistic is not in the rejection region.*

Caution

Be careful not to "accept H_0" when conducting a test of hypothesis, since the measure of reliability, $\beta = P(\text{Type II error})$, is almost always unknown. If the test statistic does not fall into the rejection region, it is better to state the conclusion as "insufficient evidence to reject H_0."*

The elements of a test of hypothesis are summarized in the next box. Note that the first four elements are all specified *before* the sampling experiment is performed. In no case will the results of the sample be used to determine the hypotheses: The data are collected to *test* the predetermined hypotheses, not to formulate them.

Elements of a Test of Hypothesis

1. *Null hypothesis* (H_0): A theory about the values of one or more population parameters. The theory generally represents the status quo, which we adopt until it is proven false. By convention, the theory is stated as H_0: parameter = value.

2. *Alternative (research) hypothesis* (H_a): A theory that contradicts the null hypothesis. The theory generally represents that which we will accept only when sufficient evidence exists to establish its truth.

(continued)

---

*In many practical applications of hypothesis testing, nonrejection leads the researcher to behave as if the null hypothesis were accepted. Accordingly, the distinction between "accept H_0" and "fail to reject H_0" is frequently blurred in practice.

3. *Test statistic:* A sample statistic used to decide whether to reject the null hypothesis.

4. *Rejection region:* The numerical values of the test statistic for which the null hypothesis will be rejected. The rejection region is chosen so that the probability is α that it will contain the test statistic when the null hypothesis is true, thereby leading to a Type I error. The value of α is usually chosen to be small (e.g., .01, .05, or .10) and is referred to as the **level of significance** of the test.

5. *Assumptions:* Clear statements of any assumptions made about the population(s) being sampled.

6. *Experiment and calculation of test statistic:* Performance of the sampling experiment and determination of the numerical value of the test statistic.

7. *Conclusion:*
   a. If the numerical value of the test statistic falls into the rejection region, we reject the null hypothesis and conclude that the alternative hypothesis is true. We know that the hypothesis-testing process will lead to this conclusion incorrectly (a Type I error) only $100\alpha\%$ of the time when H_0 is true.
   b. If the test statistic does not fall into the rejection region, we do not reject H_0. Thus, we reserve judgment about which hypothesis is true. We do not conclude that the null hypothesis is true because we do not (in general) know the probability β that our test procedure will lead to an incorrect acceptance of H_0 (a Type II error).

As with confidence intervals, the methodology for testing hypotheses varies with the target population parameter. In this chapter, we develop methods for testing a population mean and a population proportion. Some key words and the type of data associated with these target parameters are listed in the following box:

Determining the Target Parameter

Parameter	Key Words or Phrases	Type of Data
μ	Mean; average	Quantitative
p	Proportion; percentage; fraction; rate	Qualitative

ACTIVITY 6.1: *Challenging a Claim:* **Tests of Hypotheses**

Use the Internet or a newspaper or magazine to find an example of a claim made by a political or special-interest group about some characteristic (e.g., favor gun control) of the U.S. population. In this activity, you represent a rival group that believes the claim may be false.

1. In your example, what kinds of evidence might exist which would cause one to suspect that the claim might be false and therefore worthy of a statistical study? Be specific. If the claim were false, how would consumers be hurt?

2. Explain the steps necessary to reject the group's claim at level α. State the null and alternative hypotheses. If you reject the claim, does it mean that the claim is false?

3. If you reject the claim when the claim is actually true, what type of error has occurred? What is the probability of this error occurring?

4. If you were to file a lawsuit against the group based on your rejection of its claim, how might the group use your results to defend itself?

STATISTICS IN ACTION REVISITED

Identifying the Key Elements of a Hypothesis Test Relevant to the KLEENEX® Survey

In Kimberly-Clark Corporation's survey of people with colds, each of 250 customers was asked to keep count of his or her use of KLEENEX® tissues in diaries. One goal of the company was to determine how many tissues to package in a cold-care box of KLEENEX®; consequently, the total number of tissues used was recorded for each person surveyed. Since number of tissues is a quantitative variable, the parameter of interest is either μ, the mean number of tissues used by all customers with colds, or σ^2, the variance of the number of tissues used.

Now, according to a *Wall Street Journal* report, there was "little doubt that the company should put 60 tissues" in a cold-care box of KLEENEX® tissues. This statement was based on a claim made by marketing experts that 60 is the average number of times a person will blow his or her nose during a cold. The key word *average* implies that the target parameter is μ, and the marketers are claiming that $\mu = 60$. Suppose we disbelieve the claim that $\mu = 60$, believing instead that the population mean is smaller than 60 tissues. In order to test the claim against our belief, we set up the following null and alternative hypotheses:

$$H_0: \mu = 60 \qquad H_a: \mu < 60$$

We'll conduct this test in the next Statistics in Action Revisited, on p. 320.

Exercises 6.1–6.17

Understanding the Principles

6.1 Which hypothesis, the null or the alternative, is the status quo hypothesis? Which is the research hypothesis?

6.2 Which element of a test of hypothesis is used to decide whether to reject the null hypothesis in favor of the alternative hypothesis?

6.3 What is the level of significance of a test of hypothesis?

6.4 What is the difference between Type I and Type II errors in hypothesis testing? How do α and β relate to Type I and Type II errors?

6.5 List the four possible results of the combinations of decisions and true states of nature for a test of hypothesis.

6.6 We (generally) reject the null hypothesis when the test statistic falls into the rejection region, but we do not accept the null hypothesis when the test statistic does not fall into the rejection region. Why?

6.7 If you test a hypothesis and reject the null hypothesis in favor of the alternative hypothesis, does your test prove that the alternative hypothesis is correct? Explain.

Applet Exercise 6.1

Use the applet entitled *Hypotheses Test for a Mean* to investigate the frequency of Type I and Type II errors. For this exercise, use $n = 100$ and the normal distribution with mean 50 and standard deviation 10.

a. Set the null mean equal to 50 and the alternative to *not equal*. Run the applet one time. How many times was the null hypothesis rejected at level .05? In this case, the null hypothesis is true. Which type of error occurred each time the true null hypothesis was rejected? What is the probability of rejecting a true null hypothesis at level .05? How does the proportion of times the null hypothesis was rejected compare with this probability?

b. Clear the applet, then set the null mean equal to 47, and keep the alternative at *not equal*. Run the applet one

time. How many times was the null hypothesis *not* rejected at level .05? In this case, the null hypothesis is false. Which type of error occurred each time the null hypothesis was *not* rejected? Run the applet several more times without clearing. Based on your results, what can you conclude about the probability of failing to reject the null hypothesis for the given conditions?

Applying the Concepts—Basic

6.8 Calories in school lunches. A University of Florida economist conducted a study of Virginia elementary school lunch menus. During the state-mandated testing period, school lunches averaged 863 calories. (*National Bureau of Economic Research*, November 2002.) The economist claimed that after the testing period end, the average caloric content of Virginia school lunches dropped significantly. Set up the null and alternative hypothesis to test the economist's claim.

6.9 A camera that detects liars. According to *New Scientist* (January 2, 2002), a new thermal imaging camera that detects small temperature changes is now being used as a polygraph device. The United States Department of Defense Polygraph Institute (DDPI) claims that the camera can detect liars correctly 75% of the time by monitoring the temperatures of their faces. Give the null hypothesis for testing the claim made by the DDPI.

6.10 Effectiveness of online courses. The Sloan Survey of Online Learning, "Making the Grade: Online Education in the United States, 2006," reported that 60% of college presidents believe that their online education courses are as good as or superior to courses that utilize traditional face-to-face instruction. (*Inside Higher Ed*, Nov. 2006.) Give the null hypothesis for testing the claim made by the Sloan Survey.

6.11 Use of herbal therapy. According to the *Journal of Advanced Nursing* (January 2001), 45% of senior women (i.e., women over the age of 65) use herbal therapies to prevent or treat health problems. Also senior women who use herbal therapies use an average of 2.5 herbal products in a year.

a. Give the null hypothesis for testing the first claim by the journal.

b. Give the null hypothesis for testing the second claim by the journal.

6.12 Infant's listening time. *Science* (January 1, 1999) reported that the mean listening time of 7-month-old infants exposed to a three-syllable sentence (e.g., "ga ti ti") is 9 seconds. Set up the null and alternative hypotheses for testing the claim.

6.13 DNA-reading tool for quick identification of species. A biologist and a zoologist at the University of Florida were the first scientists to test the effectiveness of a high-tech handheld device designed to instantly identify the DNA of an animal species. (*PLOS Biology*, Dec. 2005.) They used the DNA-reading device on tissue samples collected from mollusks with brightly colored shells. The scientists discovered that the error rate of the device is less than 5 percent. Set up the null and alternative hypotheses if you want to support the findings.

Applying the Concepts—Intermediate

6.14 Susceptibility to hypnosis. The Computer-Assisted Hypnosis Scale (CAHS) is designed to measure a person's susceptibility to hypnosis. CAHS scores range from 0 (no susceptibility to hypnosis) to 12 (extremely high susceptibility to hypnosis). *Psychological Assessment* (March 1995) reported that University of Tennessee undergraduates had a mean CAHS score of $\mu = 4.6$. Suppose you want to test whether undergraduates at your college or university are more susceptible to hypnosis than University of Tennessee undergraduates.

a. Set up H_0 and H_a for the test.

b. Describe a Type I error for this test.

c. Describe a Type II error for this test.

6.15 Mercury levels in wading birds. According to a University of Florida wildlife ecology and conservation researcher, the average level of mercury uptake in wading birds in the Everglades has declined over the past several years. (*UF News*, December 15, 2000.) Five years ago, the average level was 15 parts per million.

a. Give the null and alternative hypotheses for testing whether the average level today is less than 15 ppm.

b. Describe a Type I error for this test.

c. Describe a Type II error for this test.

Applying the Concepts—Advanced

6.16 Jury trial outcomes. Sometimes, the outcome of a jury trial defies the "commonsense" expectations of the general public (e.g., the O. J. Simpson verdict in the "Trial of the Century"). Such a verdict is more acceptable if we understand that the jury trial of an accused murderer is analogous to the statistical hypothesis-testing process. The null hypothesis in a jury trial is that the accused is innocent. (The status quo hypothesis in the U.S. system of justice is innocence, which is assumed to be true until proven *beyond a reasonable doubt*.) The alternative hypothesis is guilt, which is accepted only when sufficient evidence exists to establish its truth. If the vote of the jury is unanimous in favor of guilt, the null hypothesis of innocence is rejected and the court concludes that the accused murderer is guilty. Any vote other than a unanimous one for guilt results in a "not guilty" verdict. The court never accepts the null hypothesis; that is, the court never declares the accused "innocent." A "not guilty" verdict (as in the O. J. Simpson case) implies that the court could not find the defendant guilty *beyond a reasonable doubt.*

a. Define Type I and Type II errors in a murder trial.

b. Which of the two errors is the more serious? Explain.

c. The court does not, in general, know the values of α and β, but ideally, both should be small. One of these probabilities is assumed to be smaller than the other in a jury trial. Which one, and why?

d. The court system relies on the belief that the value of α is made very small by requiring a unanimous vote before guilt is concluded. Explain why this is so.

e. For a jury prejudiced against a guilty verdict as the trial begins, will the value of α increase or decrease? Explain.

f. For a jury prejudiced against a guilty verdict as the trial begins, will the value of β increase or decrease? Explain.

6.17 Intrusion detection systems. Refer to the *Journal of Research of the National Institute of Standards and Technology* (November–December 2003) study of a computer intrusion detection system (IDS), presented in Exercise 3.83 (p. 151). Recall that an IDS is designed to provide an alarm whenever unauthorized access (e.g., an intrusion) to a computer system occurs. The probability of the system giving a false alarm (i.e., providing a warning when, in fact, no intrusion occurs) is defined by the symbol α, while the probability of a missed detection (i.e., no warning given, when, in fact, an intrusion occurs) is defined by the symbol β. These symbols are used to represent Type I and Type II error rates, respectively, in a hypothesis-testing scenario.

a. What is the null hypothesis H_0?

b. What is the alternative hypothesis H_a?

c. According to actual data on the EMERALD system collected by the Massachusetts Institute of Technology Lincoln Laboratory, only 1 in 1,000 computer sessions with no intrusions resulted in a false alarm. For the same system, the laboratory found that only 500 of 1,000 intrusions were actually detected. Use this information to estimate the values of α and β.

6.2 Large-Sample Test of Hypothesis about a Population Mean

In Section 6.1, we learned that the null and alternative hypotheses form the basis for a test-of-hypothesis inference. The null and alternative hypotheses may take one of several forms. In the sewer pipe example, we tested the null hypothesis that the population mean strength of the pipe is less than or equal to 2,400 pounds per linear foot against the alternative hypothesis that the mean strength exceeds 2,400. That is, we tested

$$H_0: \mu \leq 2,400$$
$$H_a: \mu > 2,400$$

This is a **one-tailed** (or **one-sided**) **statistical test,** because the alternative hypothesis specifies that the population parameter (the population mean μ, in this example) is strictly greater than a specified value (2,400, in this example). If the null hypothesis had been H_0: $\mu \geq 2,400$ and the alternative hypothesis had been H_a: $\mu < 2,400$, the test would still be one sided, because the parameter is still specified to be on "one side" of the null-hypothesis value. Some statistical investigations seek to show that the population parameter is *either larger or smaller* than some specified value. Such an alternative hypothesis is called a **two-tailed** (or **two-sided**) **hypothesis.**

While alternative hypotheses are always specified as strict inequalities, such as $\mu < 2,400$, $\mu > 2,400$, or $\mu \neq 2,400$, null hypotheses are usually specified as equalities, such as $\mu = 2,400$. *Even when the null hypothesis is an inequality, such as $\mu \leq 2,400$, we specify H_0: $\mu = 2,400$, reasoning that if sufficient evidence exists to show that H_a: $\mu > 2,400$ is true when tested against H_0: $\mu = 2,400$, then surely sufficient evidence exists to reject $\mu < 2,400$ as well.* Therefore, the null hypothesis is specified as the value of μ closest to a one-sided alternative hypothesis and as the only value *not* specified in a two-tailed alternative hypothesis. The steps for selecting the null and alternative hypotheses are summarized in the following box:

Steps for Selecting the Null and Alternative Hypotheses

1. Select the *alternative hypothesis* as that which the sampling experiment is intended to establish. The alternative hypothesis will assume one of three forms:

   a. **One tailed, upper tailed**   *Example: H_a: $\mu > 2,400$*
   b. **One tailed, lower tailed**   *Example: H_a: $\mu < 2,400$*
   c. **Two tailed**   *Example: H_a: $\mu \neq 2,400$*

2. Select the *null hypothesis* as the status quo—that which will be presumed true unless the sampling experiment conclusively establishes the alternative hypothesis. The null hypothesis will be specified as that parameter value closest to the alternative in one-tailed tests and as the complementary (or only unspecified) value in two-tailed tests.

   Example: H_0: $\mu = 2,400$

The rejection region for a **two-tailed test** differs from that for a one-tailed test. When we are trying to detect departure from the null hypothesis in *either* direction, we must establish a rejection region in both tails of the sampling distribution of the test statistic. Figures 6.4a and 6.4b show the one-tailed rejection regions for lower- and upper-tailed tests, respectively. The two-tailed rejection region is illustrated in Figure 6.4c. Note that a rejection region is established in each tail of the sampling distribution for a two-tailed test.

The rejection regions corresponding to typical values selected for α are shown in Table 6.2 for one- and two-tailed tests. Note that the smaller α you select, the more evidence (the larger z) you will need before you can reject H_0.

FIGURE 6.4
Rejection regions corresponding to one- and two-tailed tests

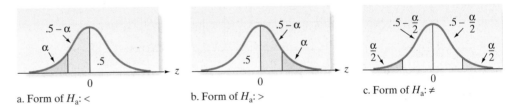

a. Form of H_a: $<$ b. Form of H_a: $>$ c. Form of H_a: $\neq$

TABLE 6.2 Rejection Regions for Common Values of α

	Alternative Hypotheses		
	Lower Tailed	Upper Tailed	Two Tailed
$\alpha = .10$	$z < -1.28$	$z > 1.28$	$z < -1.645$ or $z > 1.645$
$\alpha = .05$	$z < -1.645$	$z > 1.645$	$z < -1.96$ or $z > 1.96$
$\alpha = .01$	$z < -2.33$	$z > 2.33$	$z < -2.575$ or $z > 2.575$

EXAMPLE 6.1

SETTING UP A HYPOTHESIS TEST FOR μ—Mean Drug Response Time

Problem The effect of drugs and alcohol on the nervous system has been the subject of considerable research. Suppose a research neurologist is testing the effect of a drug on response time by injecting 100 rats with a unit dose of the drug, subjecting each rat to a neurological stimulus, and recording its response time. The neurologist knows that the mean response time for rats not injected with the drug (the "control" mean) is 1.2 seconds. She wishes to test whether the mean response time for drug-injected rats differs from 1.2 seconds. Set up the test of hypothesis for this experiment, using $\alpha = .01$.

Solution Since the neurologist wishes to detect whether the mean response time μ for drug-injected rats differs from the control mean of 1.2 seconds in *either* direction—that is, $\mu < 1.2$ or $\mu > 1.2$—we conduct a two-tailed statistical test. Following the procedure for selecting the null and alternative hypotheses, we specify as the alternative hypothesis that the mean differs from 1.2 seconds, since determining whether the drug-injected mean differs from the control mean is the purpose of the experiment. The null hypothesis is the presumption that drug-injected rats have the same mean response time as control rats unless the research indicates otherwise. Thus,

H_0: $\mu = 1.2$ (Mean response time is 1.2 seconds)

H_a: $\mu \neq 1.2$ (Mean response time is less than 1.2 or greater than 1.2 seconds)

The test statistic measures the number of standard deviations between the observed value of $\bar{x}$ and the null-hypothesized value $\mu = 1.2$:

$$\text{Test statistic:} \quad z = \frac{\bar{x} - 1.2}{\sigma_{\bar{x}}}$$

The rejection region must be designated to detect a departure from $\mu = 1.2$ in *either* direction, so we will reject H_0 for values of z that are either too small (negative) or too large (positive). To determine the precise values of z that constitute the rejection region, we first select α, the probability that the test will lead to incorrect rejection of the null hypothesis. Then we divide α equally between the lower and upper tail of the distribution of z, as shown in Figure 6.5. In this example, $\alpha = .01$, so $\alpha/2 = .005$ is placed in each tail. The areas in the tails correspond to $z = -2.575$ and $z = 2.575$, respectively (from Table 6.2), so

$$\text{Rejection region:} \; z < -2.575 \; \text{or} \; z > 2.575 \qquad \text{(see Figure 6.5)}$$

Assumptions: Since the sample size of the experiment is large enough ($n > 30$), the central limit theorem will apply, and no assumptions need be made about the population of response time measurements. The sampling distribution of the sample mean response of 100 rats will be approximately normal, regardless of the distribution of the individual rats' response times.

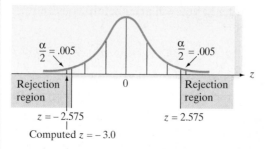

FIGURE 6.5

Two-tailed rejection region: $\alpha = .01$

Look Back Note that the test is set up *before* the sampling experiment is conducted. The data are not used to develop the test. Evidently, the neurologist wants to conclude that the mean response time for the drug-injected rats differs from the control mean only when the evidence is very convincing, because the value of α has been set quite low at .01. If the experiment results in the rejection of H_0, she can be 99% confident that the mean response time of the drug-injected rats differs from the control mean.

Now Work Exercise 6.21

Once the test is set up, she is ready to perform the sampling experiment and conduct the test. The test is performed in Example 6.2.

EXAMPLE 6.2
CARRYING OUT A HYPOTHESIS TEST FOR μ— Mean Drug Response Time

Problem Refer to the neurological response-time test set up in Example 6.1. The sample of 100 drug-injected rats yielded the results (in seconds) shown in Table 6.3. Use these data to conduct the test of hypothesis.

DRUGRAT

TABLE 6.3 Drug Response Times for 100 Rats, Example 8.2

1.90	2.17	0.61	1.17	0.66	1.86	1.41	1.30	0.70	0.56
2.00	1.27	0.98	1.55	0.64	0.60	1.55	0.93	0.48	0.39
0.86	1.19	0.79	1.37	1.31	0.85	0.71	1.21	1.23	0.89
1.84	0.80	0.64	1.08	0.74	0.93	1.71	1.05	1.44	0.42
0.70	0.54	1.40	1.06	0.54	0.17	0.98	0.89	1.28	0.68
0.98	1.14	1.16	1.64	1.16	1.01	1.09	0.77	1.58	0.99
0.57	0.27	0.51	1.27	1.81	0.88	0.31	0.92	0.93	1.66
0.21	0.79	0.94	0.45	1.19	1.60	0.14	0.99	1.08	1.57
0.55	1.65	0.81	1.00	2.55	1.96	1.31	1.88	1.51	1.48
0.61	0.05	1.21	0.48	1.63	1.45	0.22	0.49	1.29	1.40

Solution To carry out the test, we need to find the values of $\bar{x}$ and s. These values, $\bar{x} = 1.05$ and $s = .5$, are shown (highlighted) on the MINITAB printout of Figure 6.6. Now we substitute these sample statistics into the test statistic and obtain

$$z = \frac{\bar{x} - 1.2}{\sigma_{\bar{x}}} = \frac{\bar{x} - 1.2}{\sigma/\sqrt{n}} \approx \frac{1.05 - 1.2}{.5/\sqrt{100}} = -3.0$$

The implication is that the sample mean, 1.05, is (approximately) three standard deviations below the null-hypothesized value of 1.2 in the sampling distribution of $\bar{x}$. You can see in Figure 6.5 that this value of z is in the lower-tail rejection region, which consists of all values of $z < -2.575$. This sampling experiment provides sufficient evidence to reject H_0 and conclude, at the $\alpha = .01$ level of significance, that the mean response time for drug-injected rats differs from the control mean of 1.2 seconds. It appears that the rats receiving an injection of the drug have a mean response time that is less than 1.2 seconds.

FIGURE 6.6

MINITAB descriptive statistics for fill amounts, Example 6.2

Descriptive Statistics: TIME

Variable	N	Mean	StDev	Minimum	Q1	Median	Q3	Maximum
TIME	100	1.0517	0.4982	0.0500	0.6650	0.9950	1.4000	2.5500

Look Back Three points about the test of hypothesis in this example apply to all statistical tests:

1. Since z is less than -2.575, it is tempting to state our conclusion at a significance level lower than $\alpha = .01$. We resist this temptation because the level of α is determined *before* the sampling experiment is performed. If we decide that we are willing to tolerate a 1% Type I error rate, the result of the sampling experiment should have no effect on that decision. *In general, the same data should not be used both to set up and to conduct the test.*

2. When we state our conclusion at the .01 level of significance, we are referring to the failure rate of the *procedure*, not the result of this particular test. We know that the test procedure will lead to the rejection of the null hypothesis only 1% of the time when in fact $\mu = 1.2$. *Therefore, when the test statistic falls into the rejection region, we infer that the alternative $\mu \neq 1.2$ is true and express our confidence in the procedure by quoting either the α level of significance or the $100(1 - \alpha)\%$ confidence level.*

3. Although a test may lead to a "statistically significant" result (i.e., rejecting H_0 at significance level α, as in the preceding test), it may not be "practically significant." For example, suppose the neurologist tested $n = 100{,}000$ drug-injected rats, resulting in $\bar{x} = 1.1995$ and $s = .05$. Now a two-tailed hypothesis test of H_0: $\mu = 1.2$ results in a test statistic of

$$z = \frac{(1.1995 - 1.2)}{.05/\sqrt{100{,}000}} = -3.16$$

This result at $\alpha = .05$ leads us to "reject H_0" and conclude that the mean μ is "statistically different" from 1.2. However, for all practical purposes, the sample mean $\bar{x} = 1.1995$ and the hypothesized mean $\mu = 1.2$ are the same. Because the result is not "practically significant," the neurologist is not likely to consider a unit dose of the drug as an inhibitor to response time in rats. Consequently, *not all "statistically significant" results are "practically significant."*

Now Work Exercise 6.26

The setup of a large-sample test of hypothesis about a population mean is summarized in the next box. Both the one- and two-tailed tests are shown.

Large-Sample Test of Hypothesis about μ

One-Tailed Test	**Two-Tailed Test**
H_0: $\mu = \mu_0$	H_0: $\mu = \mu_0$
H_a: $\mu < \mu_0$ (or H_a: $\mu > \mu_0$)	H_a: $\mu \neq \mu_0$
Test statistic: $z = \dfrac{\bar{x} - \mu_0}{\sigma_{\bar{x}}}$	Test statistic: $z = \dfrac{\bar{x} - \mu_0}{\sigma_{\bar{x}}}$
Rejection region: $z < -z_\alpha$ (or $z > z_\alpha$ when H_a: $\mu > \mu_0$)	Rejection region: $z < -z_{\alpha/2}$ or $z > z_{\alpha/2}$
where z_α is chosen so that $P(z < -z_\alpha) = \alpha$	where $z_{\alpha/2}$ is chosen so that $P(z > z_{\alpha/2}) = \alpha/2$

Note: μ_0 is the symbol for the numerical value assigned to μ under the null hypothesis.

Conditions Required for a Valid Large-Sample Hypothesis Test for μ

1. A random sample is selected from the target population.
2. The sample size n is large (i.e., $n \geq 30$). (Due to the central limit theorem, this condition guarantees that the test statistic will be approximately normal regardless of the shape of the underlying probability distribution of the population.)

Once the test has been set up, the sampling experiment is performed and the test statistic is calculated. The following box contains possible **conclusions** for a test of hypothesis, depending on the result of the sampling experiment.

Possible Conclusions for a Test of Hypothesis

1. If the calculated test statistic falls into the rejection region, reject H_0 and conclude that the alternative hypothesis H_a is true. State that you are rejecting H_0 at the α level of significance. Remember that the confidence is in the testing *process*, not the particular result of a single test.

(continued)

> 2. If the test statistic does not fall into the rejection region, conclude that the sampling experiment does not provide sufficient evidence to reject H_0 at the α level of significance. [Generally, we will not "accept" the null hypothesis unless the probability β of a Type II error has been calculated. (See optional Section 6.6.)]

Exercises 6.18–6.35

Understanding the Principles

6.18 Explain the difference between a one-tailed and a two-tailed test.

6.19 What conditions are required for a valid large-sample test for μ?

6.20 For what values of the test statistic do you reject H_0? fail to reject H_0?

Learning the Mechanics

6.21 For each of the following rejection regions, sketch the sampling distribution for z and indicate the location of the rejection region.
   a. $z > 1.96$
   b. $z > 1.645$
   c. $z > 2.575$
   d. $z < -1.28$
   e. $z < -1.645$ or $z > 1.645$
   f. $z < -2.575$ or $z > 2.575$
   g. For each of the rejection regions specified in parts **a–f**, what is the probability that a Type I error will be made?

6.22 Suppose you are interested in conducting the statistical test of H_0: $\mu = 200$ against H_a: $\mu > 200$, and you have decided to use the following decision rule: Reject H_0 if the sample mean of a random sample of 100 items is more than 215. Assume that the standard deviation of the population is 80.
   a. Express the decision rule in terms of z.
   b. Find α, the probability of making a Type I error, by using this decision rule.

6.23 A random sample of 100 observations from a population with standard deviation 60 yielded a sample mean of 110.
   a. Test the null hypothesis that $\mu = 100$ against the alternative hypothesis that $\mu > 100$, using $\alpha = .05$. Interpret the results of the test.
   b. Test the null hypothesis that $\mu = 100$ against the alternative hypothesis that $\mu \neq 100$, using $\alpha = .05$. Interpret the results of the test.
   c. Compare the results of the two tests you conducted. Explain why the results differ.

6.24 A random sample of 64 observations produced the following summary statistics: $\bar{x} = .323$ and $s^2 = .034$.
   a. Test the null hypothesis that $\mu = .36$ against the alternative hypothesis that $\mu < .36$, using $\alpha = .10$.
   b. Test the null hypothesis that $\mu = .36$ against the alternative hypothesis that $\mu \neq .36$, using $\alpha = .10$. Interpret the result.

APPLET Applet Exercise 6.2

Use the applet entitled *Hypotheses Test for a Mean* to investigate the effect of the underlying distribution on the proportion of Type I errors. For this exercise, take $n = 100$, mean $= 50$, standard deviation $= 10$, null mean $= 50$, and alternative $<$.

a. Select the normal distribution and run the applet several times without clearing. What happens to the proportion of times the null hypothesis is rejected at the .05 level as the applet is run more and more times?
b. Clear the applet and then repeat part **a**, using the right-skewed distribution. Do you get similar results? Explain.
c. Describe the effect that the underlying distribution has on the probability of making a Type I error.

APPLET Applet Exercise 6.3

Use the applet entitled *Hypotheses Test for a Mean* to investigate the effect of the underlying distribution on the proportion of Type II errors. For this exercise, take $n = 100$, mean $= 50$, standard deviation $= 10$, null mean $= 52$, and alternative $<$.

a. Select the normal distribution and run the applet several times without clearing. What happens to the proportion of times the null hypothesis is rejected at the .01 level as the applet is run more and more times? Is this what you would expect? Explain.
b. Clear the applet and then repeat part **a**, using the right-skewed distribution. Do you get similar results? Explain.
c. Describe the effect that the underlying distribution has on the probability of making a Type II error.

APPLET Applet Exercise 6.4

Use the applet entitled *Hypotheses Test for a Mean* to investigate the effect of the null mean on the probability of making a Type II error. For this exercise, take $n = 100$, mean $= 50$, standard deviation $= 10$, and alternative $<$ with the normal distribution. Set the null mean to 55 and run the applet several times without clearing. Record the proportion of Type II errors that occurred at the .01 level. Clear the applet and repeat for null means of 54, 53, 52, and 51. What can you conclude about the probability of a Type II error as the null mean gets closer to the actual mean? Can you offer a reasonable explanation for this behavior?

Applying the Concepts—Basic

6.25 **Teacher perceptions of child behavior.** *Developmental Psychology* (Mar. 2003) published a study on teacher perceptions of the behavior of elementary school children. Teachers rated the aggressive behavior of a sample of 11,160 New York City public school children by responding to the statement "This child threatens or bullies others in order to get his/her own way." Responses were measured on a scale ranging from 1 (*never*) to 5 (*always*). Summary statistics for the sample of 11,160 children were reported as $\bar{x} = 2.15$ and $s = 1.05$. Let μ represent the mean response for the population of all New York City public school children. Suppose you want to test H_0: $\mu = 3$ against H_a: $\mu \neq 3$.
   a. In the words of the problem, define a Type I error and a Type II error.

b. Use the sample information to conduct the test at a significance level of $\alpha = .05$.

c. Conduct the test from part **b** at a significance level of $\alpha = .10$.

6.26 **Latex allergy in health-care workers.** Refer to the *Current Allergy & Clinical Immunology* (March 2004) study of $n = 46$ hospital employees who were diagnosed with a latex allergy from exposure to the powder on latex gloves, presented in Exercise 5.13 (p. 264). The number of latex gloves used per week by the sampled workers is summarized as follows: $\bar{x} = 19.3$ and $s = 11.9$. Let μ represent the mean number of latex gloves used per week by all hospital employees. Consider testing $H_0: \mu = 20$ against $H_a: \mu < 20$.

a. Give the rejection region for the test at a significance level of $\alpha = .01$.

b. Calculate the value of the test statistic.

c. Use the results from parts **a** and **b** to draw the appropriate conclusion.

6.27 **Heart rate during laughter.** Laughter is often called "the best medicine," since studies have shown that laughter can reduce muscle tension and increase oxygenation of the blood. In the *International Journal of Obesity* (Jan., 2007), researchers at Vanderbilt University investigated the physiological changes that accompany laughter. Ninety subjects (18–34 years old) watched film clips designed to evoke laughter. During the laughing period, the researchers measured the heart rate (beats per minute) of each subject, with the following summary results: $\bar{x} = 73.5$, $s = 6$. It is well known that the mean resting heart rate of adults is 71 beats per minute. At $\alpha = .05$, is there sufficient evidence to indicate that the true mean heart rate during laughter exceeds 71 beats per minute?

6.28 **Analyzing remote-sensing data to identify type of land cover.** Geographers use remote-sensing data from satellite pictures to identify urban land cover as either grassland, commercial, or residential. In *Geographical Analysis* (Oct. 2006), researchers from Arizona State, Florida State, and Louisiana State Universities collaborated on a new method for analyzing remote-sensing data. A satellite photograph of an urban area was divided into 4 × 4-meter areas (called pixels). Of interest is a numerical measure of the distribution of the sizes of gaps, or holes, in the pixel, a property called *lacunarity*. The mean and standard deviation of the lacunarity measurements for a sample of 100 pixels randomly selected from a specific urban area are 225 and 20, respectively. It is known that the mean lacunarity measurement for all grassland pixels is 220. Do the data suggest that the area sampled is grassland? Test at $\alpha = .01$.

Applying the Concepts—Intermediate

6.29 **Post-traumatic stress of POWs.** *Psychological Assessment* (March 1995) published the results of a study of World War II aviators captured by German forces after having been shot down. Having located a total of 239 World War II aviator POW survivors, the researchers asked each veteran to participate in the study; 33 responded to the letter of invitation. Each of the 33 POW survivors was administered the Minnesota Multiphasic Personality Inventory, one component of which measures level of post-traumatic stress disorder (PTSD). [*Note:* The higher the score, the higher is the level of PTSD.] The aviators produced a mean PTSD score of $\bar{x} = 9.00$ and a standard deviation of $s = 9.32$.

a. Set up the null and alternative hypotheses for determining whether the true mean PTSD score of all World War II aviator POWs is less than 16. [*Note:* The value 16 represents the mean PTSD score established for Vietnam POWs.]

b. Conduct the test from part **a**, using $\alpha = .10$. What are the practical implications of the test?

c. Discuss the representativeness of the sample used in the study and its ramifications.

6.30 **Point spreads of NFL games.** During the National Football League (NFL) season, Las Vegas oddsmakers establish a point spread on each game for betting purposes. For example, the Indianapolis Colts were established as 7-point favorites over the Chicago Bears in the 2007 Super Bowl. The final scores of NFL games were compared against the final point spreads established by the oddsmakers in *Chance* (Fall 1998). The difference between the outcome of the game and the point spread (called a point-spread error) was calculated for 240 NFL games. The mean and standard deviation of the point-spread errors are $\bar{x} = -1.6$ and $s = 13.3$. Use this information to test the hypothesis that the true mean point-spread error for all NFL games is 0. Conduct the test at $\alpha = .01$ and interpret the result.

6.31 **Bone fossil study.** Humerus bones from the same species of animal tend to have approximately the same length-to-width ratios. When fossils of humerus bones are discovered, archeologists can often determine the species of animal by examining the length-to-width ratios of the bones. It is known that species A exhibits a mean ratio of 8.5. Suppose 41 fossils of humerus bones were unearthed at an archeological site in East Africa, where species A is believed to have lived. (Assume that the unearthed bones were all from the same unknown species.) The length-to-width ratios of the bones were calculated and listed, as shown in the following table:

⊙ **BONES**

10.73	8.89	9.07	9.20	10.33	9.98	9.84	9.59
8.48	8.71	9.57	9.29	9.94	8.07	8.37	6.85
8.52	8.87	6.23	9.41	6.66	9.35	8.86	9.93
8.91	11.77	10.48	10.39	9.39	9.17	9.89	8.17
8.93	8.80	10.02	8.38	11.67	8.30	9.17	12.00
9.38							

a. Test whether the population mean ratio of all bones of this particular species differs from 8.5. Use $\alpha = .01$.

b. What are the practical implications of the test you conducted in part **a**?

6.32 **Cooling method for gas turbines.** During periods of high demand for electricity—especially in the hot summer months—the power output from a gas turbine engine can drop dramatically. One way to counter this drop in power is by cooling the inlet air to the turbine. An increasingly popular cooling method uses high-pressure inlet fogging. The performance of a sample of 67 gas turbines augmented with high-pressure inlet fogging was investigated in the *Journal of Engineering for Gas Turbines and Power* (Jan. 2005). One measure of performance is heat rate (kilojoules per kilowatt per hour). Heat rates for the 67 gas turbines, saved in the **GASTURBINE** file, are listed in the next table. Suppose that a standard gas turbine has, on average, a heat rate of 10,000 kJ/kWh.

a. Conduct a test to determine whether the mean heat rate of gas turbines augmented with high-pressure inlet fogging exceeds 10,000 kJ/kWh. Use $\alpha = .05$.

b. Identify a Type I error for this study. Identify a Type II error.

GASTURBINE

14622	13196	11948	11289	11964	10526	10387	10592	10460	10086
14628	13396	11726	11252	12449	11030	10787	10603	10144	11674
11510	10946	10508	10604	10270	10529	10360	14796	12913	12270
11842	10656	11360	11136	10814	13523	11289	11183	10951	9722
10481	9812	9669	9643	9115	9115	11588	10888	9738	9295
9421	9105	10233	10186	9918	9209	9532	9933	9152	9295
16243	14628	12766	8714	9469	11948	12414			

6.33 Salaries of postgraduates. The *Economics of Education Review* (Vol. 21, 2002) published a paper on the relationship between education level and earnings. The data for the research were obtained from the National Adult Literacy Survey of over 25,000 respondents. The survey revealed that males with a postgraduate degree had a mean salary of $61,340 (with standard error $s_{\bar{x}} = \$2,185$) while females with a postgraduate degree had a mean of $32,227 (with standard error $s_{\bar{x}} = \$932$).

a. The article reports that a 95% confidence interval for μ_M, the population mean salary of all males with postgraduate degrees, is ($57,050, $65,631). Based on this interval, is there evidence that μ_M differs from $60,000? Explain.

b. Use the summary information to test the hypothesis that the true mean salary of males with postgraduate degrees differs from $60,000. Take $\alpha = .05$. (*Note*: $s_{\bar{x}} = s/\sqrt{n}$)

c. Explain why the inferences in parts **a** and **b** agree.

d. The article reports that a 95% confidence interval for μ_F, the population mean salary of all females with postgraduate degrees, is ($30,396, $34,058). Based on this interval, is there evidence that μ_F differs from $33,000? Explain.

e. Use the summary information to test the hypothesis that the true mean salary of females with postgraduate degrees differs from $33,000. Take (*Note*:

f. Explain why the inferences in parts **d** and **e** agree.

6.34 Cyanide in soil. *Environmental Science & Technology* (October 1993) reported on a study of contaminated soil in the Netherlands. Seventy-two 400-gram soil specimens were sampled, dried, and analyzed for the con taminant cyanide. The cyanide concentration [in milligrams per kilogram (mg/kg) of soil] of each soil specimen was determined by an infrared microscopic method. The sample resulted in a mean cyanide level of $\bar{x} = 84$ mg/kg and a standard deviation of $s = 80$ mg/kg.

a. Test the hypothesis that the true mean cyanide level in soil in the Netherlands exceeds 100 mg/kg. Use $\alpha = .10$.

b. Would you reach the same conclusion as in part **a** if you took $\alpha = .05$? $\alpha = .01$? Why can the conclusion of a test change when the value of α is changed?

Applying the Concepts—Advanced

6.35 Social interaction of mental patients. The *Community Mental Health Journal* (Aug. 2000) presented the results of a survey of over 6,000 clients of the Department of Mental Health and Addiction Services (DMHAS) in Connecticut. One of the many variables measured for each mental health patient was frequency of social interaction (on a 5-point scale, where 1 = very infrequently, 3 = occasionally, and 5 = very frequently). The 6,681 clients who were evaluated had a mean social interaction score of 2.95 with a standard deviation of 1.10.

a. Conduct a hypothesis test (at $\alpha = .01$) to determine whether the true mean social interaction score of all Connecticut mental health patients differs from 3.

b. Examine the results of the study from a practical view, and then discuss why "statistical significance" does not always imply "practical significance."

c. Because the variable of interest is measured on a 5-point scale, it is unlikely that the population of ratings will be normally distributed. Consequently, some analysts may perceive the test from part **a** to be invalid and search for alternative methods of analysis. Defend or refute this position.

6.3 Observed Significance Levels: *p*-Values

According to the statistical test procedure described in Section 6.2, the rejection region and, correspondingly, the value of α are selected prior to conducting the test and the conclusions are stated in terms of rejecting or not rejecting the null hypothesis. A second method of presenting the results of a statistical test reports the extent to which the test statistic disagrees with the null hypothesis and leaves to the reader the task of deciding whether to reject the null hypothesis. This measure of disagreement is called the *observed significance level* (or *p-value*) for the test.

> ### Definition 6.1
> The **observed significance level,** or **p-value,** for a specific statistical test is the probability (assuming that H_0 is true) of observing a value of the test statistic that is at least as contradictory to the null hypothesis, and supportive of the alternative hypothesis, as the actual one computed from the sample data.

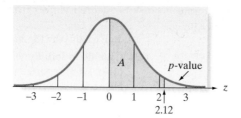

FIGURE 6.7
Finding the p-value for an
upper-tailed test when $z = 2.12$

For example, the value of the test statistic computed for the sample of $n = 50$ sections of sewer pipe was $z = 2.12$. Since the test is one tailed [i.e., the alternative (research) hypothesis of interest is H_a: $\mu > 2{,}400$], values of the test statistic even more contradictory to H_0 than the one observed would be values larger than $z = 2.12$. Therefore, the observed significance level (p-value) for this test is

$$p\text{-value} = P(z \geq 2.12)$$

or, equivalently, the area under the standard normal curve to the right of $z = 2.12$. (See Figure 6.7)

The area A in Figure 6.7 is given in Table III in Appendix A as .4830. Therefore, the upper-tail area corresponding to $z = 2.12$ is

$$p\text{-value} = .5 - .4830 = .0170$$

Consequently, we say that these test results are "statistically significant"; that is, they disagree (rather strongly) with the null hypothesis H_0: $\mu = 2{,}400$ and favor H_a: $\mu > 2{,}400$. Hence, the probability of observing a z value as large as 2.12 is only .0170 if in fact the true value of μ is 2,400.

If you are inclined to select $\alpha = .05$ for this test, then you would reject the null hypothesis because the p-value for the test, .0170, is less than .05. In contrast, if you choose $\alpha = .01$, you would not reject the null hypothesis, because the p-value for the test is larger than .01. Thus, the use of the observed significance level is identical to the test procedure described in the preceding sections, except that the choice of α is left to you.

The steps for calculating the p-value corresponding to a test statistic for a population mean are given in the following box:

> ### Steps for Calculating the p-Value for a Test of Hypothesis
>
> **1.** Determine the value of the test statistic z corresponding to the result of the sampling experiment.
>
>   **a.** If the test is one-tailed, the p-value is equal to the tail area beyond z in the same direction as the alternative hypothesis. Thus, if the alternative hypothesis is of the form $>$, the p-value is the area to the right of, or above, the observed z-value. Conversely, if the alternative is of the form $<$, the p-value is the area to the left of, or below, the observed z-value. (See Figure 6.8.)
>
>   **b.** If the test is two tailed, the p-value is equal to twice the tail area beyond the observed z-value in the direction of the sign of z. That is, if z is positive, the p-value is twice the area to the right of, or above, the observed z value. Conversely, if z is negative, the p-value is twice the area to the left of, or below, the observed z-value. (See Figure 6.9.)

FIGURE 6.8
Finding the p-value for a one-tailed test

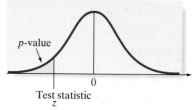

a. Lower–tailed test, H_a: $\mu < \mu_0$

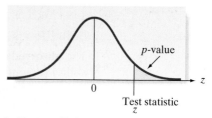

b. Upper–tailed test, H_a: $\mu > \mu_0$

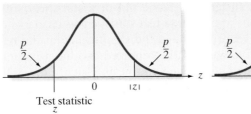

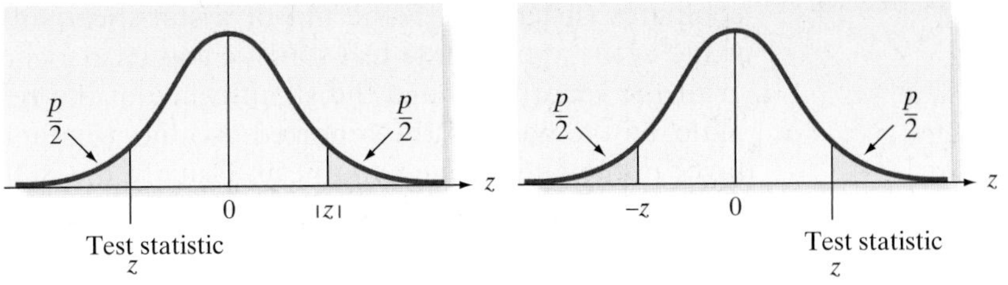

FIGURE 6.9

Finding the *p*-value for a two-tailed test: *p*-value = 2(*p*/2)

a. Test statistic *z* negative

b. Test statistic *z* positive

EXAMPLE 6.3
COMPUTING A
p-VALUE—Test
on Mean Drug
Response Time

Problem Find the observed significance level for the test of the mean response time for drug-injected rats in Examples 6.1 and 6.2.

Solution Example 8.1 presented a two-tailed test of the hypothesis

$$H_0: \mu = 1.2 \text{ seconds}$$

against the alternative hypothesis

$$H_a: \mu \neq 1.2 \text{ seconds}$$

The observed value of the test statistic in Example 6.2 was $z = -3.0$, and any value of z less than -3.0 or greater than $+3.0$ (because this is a two-tailed test) would be even more contradictory to H_0. Therefore, the observed significance level for the test is

$$p\text{-value} = P(z < -3.0 \text{ or } z > +3.0)$$

Thus, we calculate the area below the observed *z*-value, $z = -3.0$, and double it. Consulting Table III in Appendix A, we find that $P(z < -3.0) = .5 - .4987 = .0013$. Therefore, the *p*-value for this two-tailed test is

$$2P(z < -3.0) = 2(.0013) = .0026$$

This *p*-value can also be obtained with statistical software. The *p*-value is shown (highlighted) on the SAS printout of Figure 6.10.

```
Sample Statistics for TIME

    N          Mean       Std. Dev.      Std. Error
-----------------------------------------------------------
   100         1.05         0.50            0.05

Hypothesis Test

    Null hypothesis:    Mean of TIME =  1.2
    Alternative:        Mean of TIME ^= 1.2

    With a specified known standard deviation of 0.5

            Z Statistic       Prob > Z
            -----------       --------
             -3.000            0.0027
```

FIGURE 6.10

SAS test of mean response time, Example 6.3

Look Back We can interpret this *p*-value as a strong indication that the mean reaction time of drug-injected rats differs from the control mean ($\mu \neq 1.2$), since we would observe a test statistic this extreme or more extreme only 26 in 10,000 times if the drug-injected mean were equal to the control mean ($\mu = 1.2$). The extent to which the mean differs from 1.2 could be better determined by calculating a confidence interval for μ.

Now Work Exercise 6.42

When publishing the results of a statistical test of hypothesis in journals, case studies, reports, etc., many researchers make use of *p*-values. Instead of selecting α beforehand and then conducting a test, as outlined in this chapter, the researcher computes (usually with the aid of a statistical software package) and reports the value of the appropriate test statistic and its associated *p*-value. It is left to the reader of the report to judge the significance of the result (i.e., the reader must determine, on the basis of the reported *p*-value, whether to reject the null hypothesis in favor of the alternative hypothesis. Usually, the null hypothesis is rejected if the observed significance level is *less than* the fixed significance level α chosen by the reader. The inherent advantage of reporting test results in this manner is twofold: (1) Readers are permitted to select the maximum value of α that they would be willing to tolerate if they actually carried out a standard test of hypothesis in the manner outlined in this chapter, and (2) a measure of the degree of significance of the result (i.e., the *p*-value) is provided.

Reporting Test Results as *p*-Values: How to Decide Whether to Reject H_0

1. Choose the maximum value of α that you are willing to tolerate.
2. If the observed significance level (*p*-value) of the test is less than the chosen value of α, reject the null hypothesis. Otherwise, do not reject the null hypothesis.

EXAMPLE 6.4
USING *p*-VALUES— Test of Mean Hospital Length of Stay

Problem The lengths of stay (in days) for 100 randomly selected hospital patients, first presented in Table 5.1, are reproduced in Table 6.4. Suppose we want to test the hypothesis that the true mean length of stay (LOS) at the hospital is less than 5 days; that is,

H_0: $\mu = 5$ (Mean LOS is 5 days.)
H_a: $\mu < 5$ (Mean LOS is less than 5 days.)

Assuming that $\sigma = 3.68$, use the data in the table to conduct the test at $\alpha = .05$.

 HOSPLOS

TABLE 6.4 Lengths of Stay for 100 Hospital Patients

2	3	8	6	4	4	6	4	2	5
8	10	4	4	4	2	1	3	2	10
1	3	2	3	4	3	5	2	4	1
2	9	1	7	17	9	9	9	4	4
1	1	1	3	1	6	3	3	2	5
1	3	3	14	2	3	9	6	6	3
5	1	4	6	11	22	1	9	6	5
2	2	5	4	3	6	1	5	1	6
17	1	2	4	5	4	4	3	2	3
3	5	2	3	3	2	10	2	4	2

Solution The data were entered into a computer and MINITAB was used to conduct the analysis. The MINITAB printout for the lower-tailed test is displayed in Figure 6.11. Both the test statistic, $z = -1.28$, and the *p*-value of the test, $p = .101$, are highlighted on the MINITAB printout. Since the *p*-value exceeds our selected α value, $\alpha = .05$, we

---

One-Sample Z: LOS

```
Test of mu = 5 vs < 5
The assumed standard deviation = 3.68
```

FIGURE 6.11
MINITAB lower-tailed test of mean LOS, Example 6.4

					95% Upper		
Variable	N	Mean	StDev	SE Mean	Bound	Z	P
LOS	100	4.530	3.678	0.368	5.135	-1.28	0.101

cannot reject the null hypothesis. Hence, there is insufficient evidence (at $\alpha = .05$) to conclude that the true mean LOS at the hospital is less than 5 days.

Now Work Exercise 6.49

Note: Some statistical software packages (e.g., SPSS) will conduct only two-tailed tests of hypothesis. For these packages, you obtain the *p*-value for a one-tailed test as shown in the next box.

Converting a Two-Tailed *p*-Value from a Printout to a One-Tailed *p*-Value:

$$p = \frac{\text{Reported } p\text{-value}}{2} \quad \text{if} \begin{cases} H_a \text{ is of form} & > & \text{and } z \text{ is positive} \\ H_a \text{ is of form} & < & \text{and } z \text{ is negative} \end{cases}$$

$$p = 1 - \left(\frac{\text{Reported } p\text{-value}}{2} \right) \quad \text{if} \begin{cases} H_a \text{ is of form} & > & \text{and } z \text{ is negative} \\ H_a \text{ is of form} & < & \text{and } z \text{ is positive} \end{cases}$$

Hypothesis Test for a Population Mean (Large-Sample Case)

Using The TI-84/TI-83 Graphing Calculator

Step 1 *Enter the Data. (Skip to Step 2 if you have summary statistics, not raw data.)*
Press **STAT** and select **1: Edit**.
Note: If the list already contains data, clear the old data. Use the up **ARROW** to highlight "**L1**".
Press **CLEAR ENTER**.
Use the **ARROW** and **ENTER** keys to enter the data set into **L1**.

Step 2 *Access the Statistical Tests Menu.*

Press **STAT**.
Arrow right to **TESTS**.
Press **ENTER** to select **Z-Test**.

Step 3 *Choose "**Data**" or "**Stats**". ("Data" is selected when you have entered the raw data into a List. "Stats" is selected when you are given only the mean, standard deviation, and sample size.)*

Press **ENTER**.
If you selected "Data", enter the values for the hypothesis test, where $\mu_0 = $ the value for μ in the null hypothesis and $\sigma = $ assumed value of the population standard deviation.
Set **List** to **L1**.
Set **Freq** to **1**.
Use the **ARROW** to highlight the appropriate alternative hypothesis.
Press **ENTER**.
Arrow down to "**Calculate**".
Press **ENTER**.

If you selected "Stats", enter the values for the hypothesis test, where $\mu_0 = $ the value for μ in the null hypothesis and $\sigma = $ assumed value of the population standard deviation.
Enter the sample mean and sample size.
Use the **ARROW** to highlight the appropriate alternative hypothesis.
Press **ENTER**.
Arrow down to "**Calculate**".
Press **ENTER**.

The chosen test will be displayed, as will the z-test statistic, the *p*-value, the sample mean, and the sample size.

Example A manufacturer claims that the average life expectancy of a particular model of light bulb is at least 10,000 hours, with $\sigma = 1,000$ hours. A simple random sample

(continued)

of 40 bulbs shows a sample mean of 9,755 hours. Using $\sigma = .05$, test the manufacturer's claim.

For this problem, the hypotheses will be

$$H_0: \mu \geq 10{,}000$$
$$H_a: \mu < 10{,}000$$

The screens are shown at the right:

As you can see, the p-value is 0.061. Since $p > .05$, **do not** reject H_0.

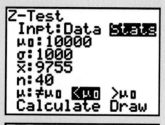

STATISTICS IN ACTION REVISITED

Testing a Population Mean in the KLEENEX® Survey

Refer to Kimberly-Clark Corporation's survey of 250 people who kept a count of their use of KLEENEX® tissues in diaries (p. 301). We want to test the claim made by marketing experts that $\mu = 60$ is the average number of tissues used by people with colds against our belief that the population mean is smaller than 60 tissues. That is, we want to test

$$H_0: \mu = 60 \qquad H_a: \mu < 60$$

We will select $\alpha = .05$ as the level of significance for the test.

The survey results for the 250 sampled KLEENEX® users are stored in the **TISSUES** data file. A MINITAB analysis of the data yielded the printout displayed in Figure SIA6.1

The observed significance level of the test, highlighted on the printout, is p-value = 018. Since this p-value is less than $\alpha = .05$, we have sufficient evidence to reject H_0; therefore, we conclude that the mean number of tissues used by a person with a cold is less than 60 tissues. [*Note:* If we conduct the same test, but with $\alpha = .01$ as the level of significance, we would have insufficient evidence to reject H_0, since p-value = .018 is greater than $\alpha = .01$. Thus, at $\alpha = .01$, there is no evidence to support our alternative hypothesis that the population mean is less than 60.]

```
Test of mu = 60 vs < 60


                                            95%
                                          Upper
Variable    N    Mean    StDev  SE Mean   Bound      T       P
NUMUSED   250  56.6760  25.0343  1.5833  59.2900  -2.10  0.018
```

FIGURE SIA6.1
MINITAB test of $\mu = 60$ for KLEENEX® survey

Exercises 6.36–6.53

Understanding the Principles

6.36 How does the observed significance level (p-value) of a test differ from the value of α?

6.37 In general, do large p-values or small p-values support the alternative hypothesis H_a?

6.38 If a hypothesis test using $\alpha = .05$ were conducted, for which of the following p-values would the null hypothesis be rejected?

a. .06   **b.** .10
c. .01   **d.** .001
e. .251   **f.** .042

6.39 For each pair consisting of α and an observed significance level (p-value), indicate whether the null hypothesis would be rejected.

a. $\alpha = .05$, p-value $= .10$
b. $\alpha = .10$, p-value $= .05$

c. $\alpha = .01$, *p*-value $= .001$

d. $\alpha = .025$, *p*-value $= .05$

e. $\alpha = .10$, *p*-value $= .45$

Learning the Mechanics

6.40 An analyst tested the null hypothesis $\mu \geq 20$ against the alternative hypothesis $\mu < 20$. The analyst reported a *p*-value of .06. What is the smallest value of α for which the null hypothesis would be rejected?

6.41 In a test of H_0: $\mu = 100$ against H_a: $\mu > 100$, the sample data yielded the test statistic $z = 2.17$. Find the *p*-value for the test.

6.42 In a test of H_0: $\mu = 100$ against H_a: $\mu \neq 100$, the sample
NW data yielded the test statistic $z = 2.17$. Find the *p*-value for the test.

6.43 In a test of the hypothesis H_0: $\mu = 50$ versus H_a: $\mu > 50$, a sample of $n = 100$ observations possessed mean $\bar{x} = 49.4$ and standard deviation $s = 4.1$. Find and interpret the *p*-value for this test.

6.44 In a test of the hypothesis H_0: $\mu = 10$ versus H_a: $\mu \neq 10$, a sample of $n = 50$ observations possessed mean $\bar{x} = 10.7$ and standard deviation $s = 3.1$. Find and interpret the *p*-value for this test.

6.45 Consider a test of H_0: $\mu = 75$ performed with the computer. SPSS reports a two-tailed *p*-value of .1032. Make the appropriate conclusion for each of the following situations:

a. H_a: $\mu < 75$, $z = -1.63$, $\alpha = .05$

b. H_a: $\mu < 75$, $z = 1.63$, $\alpha = .10$

c. H_a: $\mu > 75$, $z = 1.63$, $\alpha = .10$

d. H_a: $\mu \neq 75$, $z = -1.63$, $\alpha = .01$

Applying the Concepts—Basic

6.46 **Teacher perceptions of child behavior.** Refer to the *Developmental Psychology* (Mar. 2003) study on the aggressive behavior of elementary school children, presented in Exercise 6.25 (p. 313). Recall that you tested H_0: $\mu = 3$ against H_a: $\mu \neq 3$, where μ is the mean level of aggressiveness for the population of all New York City public school children, as perceived by their teachers and based on the summary statistics $n = 11,160$, $\bar{x} = 2.15$, and $s = 1.05$.

a. Compute the *p*-value of the test.

b. Compare the *p*-value with $\alpha = .10$ and make the appropriate conclusion.

6.47 **Latex allergy in health-care workers.** Refer to the *Current Allergy & Clinical Immunology* (March 2004) study of latex allergy in health-care workers, presented on Exercise 6.26 (p. 314). Recall that you tested H_0: $\mu = 20$ against H_a: $\mu < 20$, where μ is the mean number of latex gloves used per week by all hospital employees, based on the summary statistics $n = 46$, $\bar{x} = 19.3$, and $s = 11.9$.

a. Compute the *p*-value of the test.

b. Compare the *p*-value with $\alpha = .01$ and make the appropriate conclusion.

HYBRIDCARS

6.48 **Prices of hybrid cars.** *BusinessWeek.com* provides consumers with retail prices of new cars at dealers from across the country. The July 2006 prices for the hybrid Toyota Prius were obtained from a sample of 160 dealers. These 160 prices are saved in the **HYBRIDCARS** file.

a. Give the null and alternative hypotheses for testing whether the mean July 2006 dealer price of the Toyota Prius, μ, differs from \$25,000.

b. Data from the **HYBRIDCARS** file were analyzed with MINITAB. The resulting printout is shown below. Find and interpret the *p*-value of the test.

c. State the appropriate conclusion for the test from part **a** if $\alpha = .05$.

d. State the appropriate conclusion for the test from part **a** if $\alpha = .01$.

6.49 **Bone fossil study.** In Exercise 6.31 (p. 314), you tested
NW H_0: $\mu = 8.5$ versus H_a: $\mu \neq 8.5$, where μ is the population mean length-to-width ratio of humerus bones of a particular species of animal. A SAS printout for the hypothesis test is shown at the bottom of the page. Locate the *p*-value on the printout and interpret it.

Applying the Concepts—Intermediate

6.50 **Colored string preferred by chickens.** Refer to the *Applied Animal Behaviour Science* (October 2000) study of domestic chickens exposed to a pecking stimulus, presented in Exercise 5.19 (p. 265). Recall that the average number of

MINITAB output for
Exercise 6.48

```
Test of mu = 25000 vs not = 25000

Variable    N     Mean    StDev   SE Mean       95% CI          T      P
PRICE     160   25476.7  2429.8    192.1   (25097.3, 25856.1)  2.48  0.014
```

SAS printout for Exercise 6.49

```
Sample Statistics for LWRATIO

    N        Mean       Std. Dev.     Std. Error
---------------------------------------------------
   41        9.26         1.20          0.19

Hypothesis Test

    Null hypothesis:     Mean of LWRATIO  =  8.5
    Alternative:         Mean of LWRATIO ^=  8.5

    With a specified known standard deviation of 1.2

         Z Statistic      Prob > Z
         -----------      --------
            4.042          <.0001
```

pecks a chicken takes at a white string over a specified time interval is known to be $\mu = 7.5$ pecks. In an experiment in which 72 chickens were exposed to blue string, the average number of pecks was $\bar{x} = 1.13$ pecks, with a standard deviation of $s = 2.21$ pecks.

a. On average, are chickens more apt to peck at white string than at blue string? Conduct the appropriate test of hypothesis, using $\alpha = .05$.

b. Compare your answer to part **a** with your answer to Exercise 5.19**b**.

c. Find the p-value for the test and interpret it.

6.51 Feminizing human faces. Research published in *Nature* (August 27, 1998) revealed that people are more attracted to "feminized" faces, regardless of gender. In one experiment, 50 human subjects viewed both a Japanese female and a Caucasian male face on a computer. Using special computer graphics, each subject could morph the faces (by making them more feminine or more masculine) until they attained the "most attractive" face. The level of feminization x (measured as a percentage) was measured.

a. For the Japanese female face, $\bar{x} = 10.2\%$ and $s = 31.3\%$. The researchers used this sample information to test the null hypothesis of a mean level of feminization equal to 0%. Verify that the test statistic is equal to 2.3.

b. Refer to part **a**. The researchers reported the p-value of the test as $p \approx .02$. Verify and interpret this result.

c. For the Caucasian male face, $\bar{x} = 15.0\%$ and $s = 25.1\%$. The researchers reported the test statistic (for the test of the null hypothesis stated in part **a**) as 4.23, with an associated p-value of approximately 0. Verify and interpret these results.

6.52 Post-traumatic stress of POWs. Refer to the *Psychological Assessment* study of World War II aviator POWs, presented in Exercise 6.29 (p. 314). You tested whether the true mean post-traumatic stress disorder score of World War II aviator POWs is less than 16. Recall that $\bar{x} = 9.00$ and $s = 9.32$ for a sample of $n = 33$ POWs.

a. Compute the p-value of the test.

b. Refer to part **a**. Would the p-value have been larger or smaller if $\bar{x}$ had been larger?

Applying the Concepts—Advanced

6.53 Ages of cable TV shoppers. In a paper presented at the 2000 Conference of the International Association for Time Use Research, professor Margaret Sanik of Ohio State University reported the results of her study on American cable TV viewers who purchase items from one of the home-shopping channels. She found that the average age of these cable TV shoppers was 51 years. Suppose you want to test the null hypothesis $H_0: \mu = 51$, using a sample of $n = 50$ cable TV shoppers.

a. Find the p-value of a two-tailed test if $\bar{x} = 52.3$ and $s = 7.1$.

b. Find the p-value of an upper-tailed test if $\bar{x} = 52.3$ and $s = 7.1$.

c. Find the p-value of a two-tailed test if $\bar{x} = 52.3$ and $s = 10.4$.

d. For each of the tests in parts **a–c**, give a value of α that will lead to a rejection of the null hypothesis.

e. If $\bar{x} = 52.3$, give a value of s that will yield a p-value of .01 or less for a one-tailed test.

6.4 Small-Sample Test of Hypothesis about a Population Mean

Most water-treatment facilities monitor the quality of their drinking water on an hourly basis. One variable monitored is pH, which measures the degree of alkalinity or acidity in the water. A pH below 7.0 is acidic, one above 7.0 is alkaline, and a pH of 7.0 is neutral. One water-treatment plant has a target pH of 8.5. (Most try to maintain a slightly alkaline level.) The mean and standard deviation of 1 hour's test results, based on 17 water samples at this plant, are

$$\bar{x} = 8.42 \qquad s = .16$$

Does this sample provide sufficient evidence that the mean pH level in the water differs from 8.5?

This inference can be placed in a test-of-hypothesis framework. We establish the target pH as the null hypothesized value and then utilize a two-tailed alternative that the true mean pH differs from the target:

$$H_0: \mu = 8.5 \text{ (Mean pH level is 8.5.)}$$

$$H_a: \mu \neq 8.5 \text{ (Mean pH level differs from 8.5.)}$$

Recall from Section 5.3 that when we are faced with making inferences about a population mean from the information in a small sample, two problems emerge:

1. The normality of the sampling distribution for $\bar{x}$ does not follow from the central limit theorem when the sample size is small. We must assume that the distribution of measurements from which the sample was selected is approximately normally distributed in order to ensure the approximate normality of the sampling distribution of $\bar{x}$.

2. If the population standard deviation σ is unknown, as is usually the case, then we cannot assume that s will provide a good approximation for σ when the sample size is small. Instead, we must use the *t*-distribution rather than the standard normal *z*-distribution to make inferences about the population mean μ.

Therefore, as the test statistic of a small-sample test of a population mean, we use the *t*-statistic:

$$Test\ statistic: t = \frac{\bar{x} - \mu_0}{s/\sqrt{n}} = \frac{\bar{x} - 8.5}{s/\sqrt{n}}$$

In this equation, μ_0 is the null-hypothesized value of the population mean μ. In the example here, $\mu_0 = 8.5$.

To find the rejection region, we must specify the value of α, the probability that the test will lead to rejection of the null hypothesis when it is true, and then consult the *t* table (Table VI of Appendix A). With $\alpha = .05$, the two-tailed rejection region is

$$Rejection\ region: t_{\alpha/2} = t_{.025} = 2.120 \text{ with } n - 1 = 16 \text{ degrees of freedom}$$
$$Reject\ H_0 \text{ if } t < -2.120 \text{ or } t > 2.120$$

The rejection region is shown in Figure 6.12.

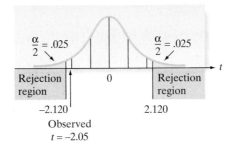

FIGURE 6.12

Two-tailed rejection region for small-sample *t*-test

We are now prepared to calculate the test statistic and reach a conclusion:

$$t = \frac{\bar{x} - \mu_0}{s/\sqrt{n}} = \frac{8.42 - 8.50}{.16/\sqrt{17}} = \frac{-.08}{.039} = -2.05$$

Since the calculated value of *t* does not fall into the rejection region (Figure 6.12), we cannot reject H_0 at the $\alpha = .05$ level of significance. Thus, the water-treatment plant should not conclude that the mean pH differs from the 8.5 target on the basis of the sample evidence.

It is interesting to note that the calculated *t* value, -2.05, is *less than* the .05-level *z*-value, -1.96. The implication is that if we had *incorrectly* used a *z* statistic for this test, we would have rejected the null hypothesis at the .05 level and concluded that the mean pH level differs from 8.5. The important point is that the statistical procedure to be used must always be closely scrutinized and all the assumptions understood. Many statistical lies are the result of misapplications of otherwise valid procedures.

The technique for conducting a small-sample test of hypothesis about a population mean is summarized in the following boxes:

Small-Sample Test of Hypothesis About μ

One-Tailed Test	**Two-Tailed Test**
$H_0: \mu = \mu_0$	$H_0: \mu = \mu_0$
$H_a: \mu < \mu_0$ (or $H_a: \mu > \mu_0$)	$H_a: \mu \neq \mu_0$
Test statistic: $t = \dfrac{\bar{x} - \mu_0}{s/\sqrt{n}}$	Test statistic: $t = \dfrac{\bar{x} - \mu_0}{s/\sqrt{n}}$
Rejection region: $t < -t_\alpha$	Rejection region: $t < -t_{\alpha/2}$ or $t > t_{\alpha/2}$

(or $t > t_\alpha$ when $H_a: \mu > \mu_0$)
where t_α and $t_{\alpha/2}$ are based on $(n - 1)$ degrees of freedom

> **Conditions Required for a Valid Small-Sample Hypothesis Test for** μ
>
> **1.** A random sample is selected from the target population.
> **2.** The population from which the sample is selected has a distribution that is approximately normal.

EXAMPLE 6.5

A SMALL-SAMPLE
TEST FOR μ—
Does a New Engine
Meet Air-Pollution
Standards?

Problem A major car manufacturer wants to test a new engine to determine whether it meets new air-pollution standards. The mean emission μ of all engines of this type must be less than 20 parts per million of carbon. Ten engines are manufactured for testing purposes, and the emission level of each is determined. The data (in parts per million) are listed in Table 6.5.

 EMISSIONS

TABLE 6.5 Emission Levels for Ten Engines

15.6	16.2	22.5	20.5	16.4	19.4	19.6	17.9	12.7	14.9

Do the data supply sufficient evidence to allow the manufacturer to conclude that this type of engine meets the pollution standard? Assume that the manufacturer is willing to risk a Type I error with probability $\alpha = .01$.

Solution The manufacturer wants to support the research hypothesis that the mean emission level μ for all engines of this type is less than 20 parts per million. The elements of this small-sample one-tailed test are as follows:

$$H_0: \mu = 20 \text{ (Mean emission level is 20 ppm.)}$$
$$H_a: \mu < 20 \text{ (Mean emission level is less than 20 ppm.)}$$
$$Test\ statistic: t = \frac{\bar{x} - 20}{s/\sqrt{n}}$$

Rejection region: For $\alpha = .01$ and df $= n - 1 = 9$, the one-tailed rejection region (see Figure 6.13) is $t < -t_{.01} = -2.821$.

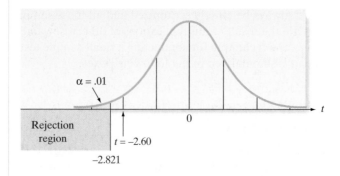

FIGURE 6.13
A t-distribution with 9 df and the rejection region for Example 6.5

Assumption: The relative frequency distribution of the population of emission levels for all engines of this type is approximately normal. Based on the shape of the MINITAB stem-and-leaf display of the data shown in Figure 6.14, this assumption appears to be reasonably satisfied.

To calculate the test statistic, we analyzed the **EMISSIONS** data with MINITAB. The MINITAB printout is shown at the bottom of Figure 6.14. From the printout, we ob-

Stem-and-Leaf Display: E-LEVEL

```
Stem-and-leaf of E-LEVEL    N = 10
Leaf Unit = 1.0


    1    1   2
    3    1   45
   (3)   1   667
    4    1   99
    2    2   0
    1    2   2
```

One-Sample T: E-LEVEL

```
Test of mu = 20 vs < 20
```

| | | | | | 95% Upper | | |
Variable	N	Mean	StDev	SE Mean	Bound	T	P
E-LEVEL	10	17.5700	2.9522	0.9336	19.2814	-2.60	0.014

FIGURE 6.14

MINITAB analysis of 10 emission levels, Example 6.5

tain $\bar{x} = 17.57$ and $s = 2.95$. Substituting these values into the test statistic formula and rounding, we get

$$t = \frac{\bar{x} - 20}{s/\sqrt{n}} = \frac{17.57 - 20}{2.95/\sqrt{10}} = -2.60$$

Since the calculated t falls outside the rejection region (see Figure 6.13), the manufacturer cannot reject H_0. There is insufficient evidence to conclude that $\mu < 20$ parts per million and that the new type of engine meets the pollution standard.

Look Back Are you satisfied with the reliability associated with this inference? The probability is only $\alpha = .01$ that the test would support the research hypothesis if in fact it were false.

Now Work Exercise 6.59a–b

EXAMPLE 6.6
THE p-VALUE FOR A SMALL-SAMPLE TEST OF μ

Problem Find the observed significance level for the test described in Example 6.5. Interpret the result.

Solution The test of Example 6.5 was a lower-tailed test: $H_0: \mu = 20$ versus $H_a: \mu < 20$. Since the value of t computed from the sample data was $t = -2.60$, the observed significance level (or p-value) for the test is equal to the probability that t would assume a value less than or equal to -2.60 if in fact H_0 were true. This is equal to the area in the lower tail of the t-distribution (highlighted in Figure 6.15).

One way to find this area (i.e., the p-value for the test) is to consult the t-table (Table IV in Appendix A). Unlike the table of areas under the normal curve, Table IV gives only the t values corresponding to the areas .100, .050, .025, .010, .005, .001, and .0005. Therefore, we can only approximate the p-value for the test. Since the

FIGURE 6.15

The observed significance level for the test of Example 6.5

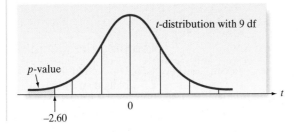

observed t value was based on nine degrees of freedom, we use the df $= 9$ row in Table IV and move across the row until we reach the t values that are closest to the observed $t = -2.60$. [*Note:* We ignore the minus sign.] The t values corresponding to p-values of .010 and .025 are 2.821 and 2.262, respectively. Since the observed t value falls between $t_{.010}$ and $t_{.025}$, the p-value for the test lies between .010 and .025. In other words, $.010 < p$-value $< .025$. Thus, we would reject the null hypothesis $H_0: \mu = 20$ parts per million for any value of α larger than .025 (the upper bound of the p-value).

A second, more accurate, way to obtain the p-value is to use a statistical software package to conduct the test of hypothesis. Both the test statistic (-2.60) and the p-value $(.014)$ are highlighted on the MINITAB printout of Figure 6.14.

You can see that the actual p-value of the test falls within the bounds obtained from Table IV. Thus, the two methods agree and we will reject $H_0: \mu = 20$ in favor of $H_a: \mu < 20$ for any α level larger than .025.

> **Now Work Exercise 6.59c**

Small-sample inferences typically require more assumptions and provide less information about the population parameter than do large-sample inferences. Nevertheless, the t-test is a method of testing a hypothesis about a population mean of a normal distribution when only a small number of observations is available.

> **What Can Be Done if the Population Relative Frequency Distribution Departs Greatly from Normal?**
>
> *Answer:* Use the nonparametric statistical method described in optional Section 6.6.

ACTIVITY 6.2: *Keep the Change:* Tests of Hypotheses

In this activity, we will test claims that the mean amount transferred for any single purchase is $0.50 and that the mean amount that Bank of America matches for a customer during the first 90 days of enrollment is at least $25. We will be working with data sets from Activity 1.1 (p. 12) and Activity 4.3 (p. 239).

1. On the basis of the assumption that all transfer amounts between $0.00 and $0.99 seem to be equally likely to occur, one may conclude that the mean of the amounts transferred is about $0.50. Explain how someone who doesn't believe this conclusion would use a test of hypothesis to argue that the conclusion is false.

2. Suppose that your original data set *Amounts Transferred* from Activity 1.1 represents a random sample of amounts transferred for all Bank of America customers' purchases. Does your sample meet the requirements for performing either a large-sample or a small-sample test of hypothesis about the population mean? Explain. If your data meet the criteria for one of the tests, perform that test at $\alpha = .05$.

3. Use the pooled data set of *Amounts Transferred* from Activity 4.3 to represent a random sample of amounts transferred for all Bank of America customers' purchases. Explain how the conditions for the large-sample test of hypothesis about a population mean are met. Then perform the test at $\alpha = .05$. Do your results suggest that the mean may be something other than $0.50? Explain.

4. A friend suggests to you that the mean amount the bank matches for a customer during the first 90 days is at least $25. Explain how you could use a test of hypothesis to argue that your friend is wrong.

5. Suppose that your data set *Bank Matching* from Activity 1.1 represents a random sample of all Bank of America customers' bank matching. Perform an appropriate test of hypothesis at $\alpha = .05$. against your friend's claim. If necessary, assume that the underlying distribution is normal. Does the test provide evidence that your friend's claim is false?

Keep the results from this activity for use in other activities.

Exercises 6.54–6.72

Understanding the Principles

6.54 In what ways are the distributions of the z-statistic and t-statistic alike? How do they differ?

6.55 Under what circumstances should you use the t-distribution in testing a hypothesis about a population mean?

Learning the Mechanics

6.56 For each of the following rejection regions, sketch the sampling distribution of t and indicate the location of the rejection region on your sketch:
  a. $t > 1.440$, where df $= 6$

b. $t < -1.782$, where df $= 12$

c. $t < -2.060$ or $t > 2.060$, where df $= 25$

6.57 For each of the rejection regions defined in Exercise 6.56, what is the probability that a Type I error will be made?

6.58 A random sample of n observations is selected from a normal population to test the null hypothesis that $\mu = 10$. Specify the rejection region for each of the following combinations of H_a, α, and n:

   a. H_a: $\mu \neq 10$; $\alpha = .05$; $n = 14$

   b. H_a: $\mu > 10$; $\alpha = .01$; $n = 24$

   c. H_a: $\mu > 10$; $\alpha = .10$; $n = 9$

   d. H_a: $\mu < 10$; $\alpha = .01$; $n = 12$

   e. H_a: $\mu \neq 10$; $\alpha = .10$; $n = 20$

   f. H_a: $\mu < 10$; $\alpha = .05$; $n = 4$

6.59 A sample of five measurements, randomly select ed from a
NW normally distributed population, resulted in the following summary statistics: $\bar{x} = 4.8$, $s = 1.3$.

   a. Test the null hypothesis that the mean of the population is 6 against the alternative hypothesis, $\mu < 6$. Use $\alpha = .05$.

   b. Test the null hypothesis that the mean of the population is 6 against the alternative hypothesis, $\mu \neq 6$. Use $\alpha = .05$.

   c. Find the observed significance level for each test.

6.60 The following sample of six measurements was randomly selected from a normally distributed population: $1, 3, -1, 5, 1, 2$.

   a. Test the null hypothesis that the mean of the population is 3 against the alternative hypothesis, $\mu < 3$. Use $\alpha = .05$.

   b. Test the null hypothesis that the mean of the population is 3 against the alternative hypothesis, $\mu \neq 3$. Use $\alpha = .05$.

   c. Find the observed significance level for each test.

6.61 Suppose you conduct a t-test for the null hypothesis H_0: $\mu = 1,000$ versus the alternative hypothesis H_a: $\mu > 1,000$, based on a sample of 17 observations. The test results are $t = 1.89$, p-value $= .038$.

   a. What assumptions are necessary for the validity of this procedure?

   b. Interpret the results of the test.

   c. Suppose the alternative hypothesis had been the two-tailed H_a: $\mu \neq 1,000$. If the t-statistic were unchanged, what would the p-value be for this test? Interpret the p-value for the two-tailed test.

Applying the Concepts—Basic

JAPANESE

6.62 Reading Japanese books. Refer to the *Reading in a Foreign Language* (April 2004) experiment to improve the Japanese reading comprehension levels of University of Hawaii students presented in Exercise 2.31 (p. 46). Recall that 14 students participated in a 10-week extensive reading program in a second-semester Japanese course. The data on number of books read by each student are saved in the **JAPANESE** file. An SPSS printout giving descriptive statistics for the data is shown below.

   a. State the null and alternative hypotheses for determining whether the average number of books read by all students who participated in the extensive reading program exceeds 25.

   b. Find the rejection region for the test, using $\alpha = .05$.

   c. Compute the test statistic.

   d. State the appropriate conclusion for the test.

   e. What conditions are required for the test results to be valid?

6.63 A new dental bonding agent. When bonding teeth, orthodontists must maintain a dry field. A new bonding adhesive (called "Smartbond") has been developed to eliminate the necessity of a dry field. However, there is concern that the new bonding adhesive is not as strong as the current standard, a composite adhesive. (*Trends in Biomaterials & Artificial Organs*, January 2003.) Tests on a sample of 10 extracted teeth bonded with the new adhesive resulted in a mean breaking strength (after 24 hours) of $\bar{x} = 5.07$ Mpa and a standard deviation of $s = .46$ Mpa, where Mpa = megapascal, a measure of force per unit area. Orthodontists want to know if the true mean breaking strength of the new bonding adhesive is less than 5.70 Mpa, the mean breaking strength of the composite adhesive.

   a. Set up the null and alternative hypotheses for the test.

   b. Find the rejection region for the test, using $\alpha = .01$.

   c. Compute the test statistic.

   d. State the appropriate conclusion for the test.

   e. What conditions are required for the test results to be valid?

6.64 Al Qaeda attacks on the United States. Refer to the *Studies in Conflict & Terrorism* (Vol. 29, 2006) analysis of recent incidents involving suicide terrorist attacks, presented in Exercise 5.33 (p. 273). Data on the number of individual suicide bombings that occurred in each of 21 sampled Al Qaeda attacks against the United States are reproduced in the table on p. 328.

   a. Do the data indicate that the true mean number of suicide bombings for all Al Qaeda attacks against the United States differs from 2.5?. Use $\alpha = .10$ and the MINITAB printout (p. 328) to answer the question.

   b. In Exercise 5.33, you found a 90% confidence interval for the mean μ of the population. This interval is also shown on the MINITAB printout. Answer the question in part **a** on the basis of the 90% confidence interval.

   c. Do the inferences derived from the test (part **a**) and confidence interval (part **b**) agree? Explain why or why not.

SPSS output for Exercise 6.62

Descriptive Statistics

	N	Minimum	Maximum	Mean	Std. Deviation
BOOKS	14	16	53	31.64	10.485
Valid N (listwise)	14				

MINITAB output for
Exercise 6.64

One-Sample T: ATTACKS

Test of mu = 2.5 vs not = 2.5

Variable	N	Mean	StDev	SE Mean	90% CI	T	P
ATTACKS	21	1.857	1.195	0.261	(1.407, 2.307)	-2.46	0.023

d. What assumption about the data must be true for the inferences to be valid?

e. Use a graph to check whether the assumption you made in part **d** is reasonably satisfied. Comment on the validity of the inference.

ALQAEDA

1 1 2 1 2 4 1 1 1 1 2 3 4 5 1 1 1 2 2 2 1

Source: Moghadam, A. "Suicide terrorism, occupation, and the globalization of martyrdom: A critique of *Dying to Win*," *Studies in Conflict & Terrorism*, Vol. 29, No. 8, 2006 (Table 3).

6.65 Dental anxiety study. Refer to the *Psychological Reports* (August 1997) study of college students who completed the Dental Anxiety Scale, presented in Exercise 4.78 (p. 211). Recall that scores range from 0 (no anxiety) to 20 (extreme anxiety). Summary statistics for the scores of the 27 students who completed the questionnaire are $\bar{x} = 10.7$ and $s = 3.6$. Conduct a test of hypothesis to determine whether the mean Dental Anxiety Scale score for the population of college students differs from $\mu = 11$. Use $\alpha = .05$.

6.66 Crab spiders hiding on flowers. Refer to the *Behavioral Ecology* (Jan. 2005) experiment on crab spiders' use of camouflage to hide from predators (e.g., birds) on flowers, presented in Exercise 2.36 (p. 48). Researchers at the French Museum of Natural History collected a sample of 10 adult female crab spiders, each sitting on the yellow central part of a daisy, and measured the chromatic contrast between each spider and the flower. The data (for which higher values indicate a greater contrast, and, presumably, an easier detection by predators) are shown in the accompanying table. The researchers discovered that a contrast of 70 or greater allows birds to see the spider. Of interest is whether or not the true mean chromatic contrast of crab spiders on daisies is less than 70.

SPIDER

57	75	116	37	96	61	56	2	43	32

Data adapted from Thery, M., et al. "Specific color sensitivities of prey and predator explain camouflage in different visual systems," *Behavioral Ecology*, Vol. 16, No. 1, Jan. 2005 (Table 1).

a. Define the parameter of interest, μ

b. Set up the null and alternative hypotheses of interest.

c. Find $\bar{x}$ and s for the sample data, and then use these values to compute the test statistic.

d. Give the rejection region for $\alpha = .10$.

e. State the appropriate conclusion in the words of the problem.

Applying the Concepts—Intermediate

LICHEN

6.67 Radioactive lichen. Refer to the 2003 Lichen Radionuclide Baseline Research project to monitor the level of radioactivity in lichen, presented in Exercise 5.35 (p. 274). Recall that University of Alaska researchers collected nine lichen specimens and measured the amount of the radioactive element cesium-137 (in microcuries per milliliter) in each specimen. (The natural logarithms of the data values are saved in the **LICHEN** file.) Assume that in previous years the mean cesium amount in lichen was $\mu = .003$ microcurie per milliliter. Is there sufficient evidence to indicate that the mean amount of cesium in lichen specimens differs from this value? Use the SAS printout below to conduct a complete test of hypothesis at $\alpha = .10$.

6.68 Testing a mosquito repellent. A study was conducted to evaluate the effectiveness of a new mosquito repellent designed by the U.S. Army to be applied as camouflage face paint. (*Journal of the Mosquito Control Association*, June 1995.) The repellent was applied to the forearms of five volunteers who then were exposed to 15 active mosquitos for a 10-hour period. The percentage of the forearm surface area protected from bites (called percent repellency) was calculated for each of the five volunteers. For one color of paint (loam), the following summary statistics were obtained:

$$\bar{x} = 83\% \quad s = 15\%$$

a. The new repellent is considered effective if it provides a repellency of at least 95 percent. Conduct a test to determine whether the mean repellency of the new mosquito repellent is less than 95 percent. Use $\alpha = .10$.

b. What assumptions are required for the hypothesis test in part **a** to be valid?

SAS Output for Exercise 6.67

Sample Statistics for CESIUM

N	Mean	Std. Dev.	Std. Error
9	0.0090	0.0049	0.0016

Hypothesis Test

Null hypothesis: Mean of CESIUM = 0.003
Alternative: Mean of CESIUM ^= 0.003

t Statistic	Df	Prob > t
3.725	8	0.0058

6.69 Minimizing tractor skidding distance. Refer to the *Journal of Forest Engineering* (July 1999) study of minimizing tractor skidding distances along a new road in a European forest, presented in Exercise 5.40 (p. 275). The skidding distances (in meters) were measured at 20 randomly selected road sites. The data are repeated in the accompanying table. Recall that a logger working on the road claims that the mean skidding distance is at least 425 meters. Is there sufficient evidence to refute this claim? Use $\alpha = .10$.

SKIDDING

488	350	457	199	285	409	435	574	439	546
385	295	184	261	273	400	311	312	141	425

Source: Tujek, J. & Pacola, E. "Algorithms for skidding distance modeling on a raster Digital Terrain Model," *Journal of Forest Engineering,* Vol. 10, No. 1, July 1999 (Table 1).

6.70 Mongolian desert ants. Refer to the *Journal of Biogeography* (December 2003) study of ants in Mongolia (central

GOBIANTS

Site	Region	Number of Ant Species
1	Dry Steppe	3
2	Dry Steppe	3
3	Dry Steppe	52
4	Dry Steppe	7
5	Dry Steppe	5
6	Gobi Desert	49
7	Gobi Desert	5
8	Gobi Desert	4
9	Gobi Desert	4
10	Gobi Desert	5
11	Gobi Desert	4

Source: Pfeiffer, M., et al. "Community organization and species richness of ants in Mongolia along an ecological gradient from steppe to Gobi desert," *Journal of Biogeography,* Vol. 30, No. 12, Dec. 2003.

Asia), presented in Exercise 2.66 (p. 60). Recall that botanists placed seed baits at 11 study sites and observed the number of ant species attracted to each site. A portion of the data is provided in the accompanying table. Do these data indicate that the average number of ant species at Mongolian desert sites differs from 5? Conduct the appropriate test at $\alpha = .05$. Are the conditions required for a valid test satisfied?

6.71 Head trauma study. The SCL-90-R is a 90-item symptom inventory checklist designed to reflect the psychological status of an individual. Each symptom (e.g., obsessive–compulsive behavior) is scored on a scale of 0 (none) to 4 (extreme). The total of these scores yields an individual's Positive Symptom Total (PST). "Normal" individuals are known to have a mean PST of about 40. The *Journal of Head Trauma Rehabilitation* (April 1995) reported that a sample of 23 patients diagnosed with mild to moderate traumatic brain injury had a mean PST score of $\bar{x} = 48.43$ and a standard deviation of $s = 20.76$. Is there sufficient evidence to claim that the true mean PST score of all patients with mild-to-moderate traumatic brain injury exceeds the "normal" value of 40? Test, using $\alpha = .05$.

Applying the Concepts—Advanced

6.72 Lengths of great white sharks. One of the most feared predators in the ocean is the great white shark. It is known that the white shark grows to a mean length of 21 feet; however, one marine biologist believes that great white sharks off the Bermuda coast grow much longer owing to unusual feeding habits. To test this claim, some full-grown great white sharks were captured off the Bermuda coast, measured, and then set free. However, because the capture of sharks is difficult, costly, and very dangerous, only three specimens were sampled. Their lengths were 24, 20, and 22 feet. Do these data support the marine biologist's claim?

6.5 Large-Sample Test of Hypothesis about a Population Proportion

Inferences about population proportions (or percentages) are often made in the context of the probability p of "success" for a binomial distribution. We saw how to use large samples from binomial distributions to form confidence intervals for p in Section 5.4. We now consider tests of hypotheses about p.

Consider, for example, a method currently used by doctors to screen women for breast cancer. The method fails to detect cancer in 20% of the women who actually have the disease. Suppose a new method has been developed that researchers hope will detect cancer more accurately. This new method was used to screen a random sample of 140 women known to have breast cancer. Of these, the new method failed to detect cancer in 12 women. Does this sample provide evidence that the failure rate of the new method differs from the one currently in use?

We first view this experiment as a binomial one with 140 screened women as the trials and failure to detect breast cancer as "Success" (in binomial terminology). Let p represent the probability that the new method fails to detect the breast cancer on the one hand, if the new method is no better than the current one, then the failure rate is $p = .2$. On the other hand, if the new method is either better or worse than the current method, then the failure rate is either smaller or larger than 20%; that is, $p \neq .2$.

We can now place the problem in the context of a test of hypothesis:

$$H_0: p = .2$$

$$H_a: p \neq .2$$

Recall that the sample proportion $\hat{p}$ is really just the sample mean of the outcomes of the individual binomial trials and, as such, is approximately normally distributed (for large samples) according to the central limit theorem. Thus, for large samples we can use the standard normal z as the test statistic:

Test statistic: $z = \dfrac{\text{Sample proportion} - \text{Null hypothesized proportion}}{\text{Standard deviation of sample proportion}}$

$$= \dfrac{\hat{p} - p_0}{\sigma_{\hat{p}}}$$

Note that we used the symbol p_0 to represent the null hypothesized value of p.

We use the standard normal distribution to find the appropriate rejection region for the specified value of α. Using $\alpha = .05$, the two-tailed rejection region is

$$z < -z_{\alpha/2} = -z_{.025} = -1.96 \quad \text{or} \quad z > z_{\alpha/2} = z_{.025} = 1.96$$

(See Figure 6.16.)

We are now prepared to calculate the value of the test statistic. Before doing so, however, we want to be sure that the sample size is large enough to ensure that the normal approximation for the sampling distribution of $\hat{p}$ is reasonable. To check this, we calculate a three-standard-deviation interval around the null-hypothesized value p_0, which is assumed to be the true value of p until our test procedure proves otherwise. Recall that $\sigma_{\hat{p}} = \sqrt{pq/n}$ and that we need an estimate of the product pq in order to calculate a numerical value of the test statistic z. Since the

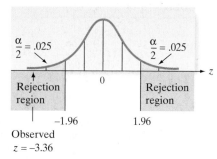

FIGURE 6.16

Rejection region for breast cancer example

null-hypothesized value is generally the value that is accepted until proven otherwise, we use the value of $p_0 q_0$ (where $q_0 = 1 - p_0$) to estimate pq in the calculation of z. Thus,

$$\sigma_{\hat{p}} = \sqrt{\dfrac{pq}{n}} \approx \sqrt{\dfrac{p_0 q_0}{n}} = \sqrt{\dfrac{(.2)(.8)}{140}} = .034$$

and the three-standard-deviation interval around p_0 is

$$p_0 \pm 3\sigma_{\hat{p}} \approx .2 \pm 3(.034) = (.098, .302)$$

As long as this interval does not contain 0 or 1 (i.e., as long as it is completely contained in the interval from 0 to 1), as is the case here, the normal distribution will provide a reasonable approximation for the sampling distribution of $\hat{p}$.

Returning to the hypothesis test at hand, the proportion of the screenings that failed to detect breast cancer is

$$\hat{p} = \dfrac{12}{140} = .086$$

Finally, we calculate the number of standard deviations (the z value) between the sampled and hypothesized values of the binomial proportion:

$$z = \dfrac{\hat{p} - p_0}{\sigma_{\hat{p}}} = \dfrac{\hat{p} - p_0}{\sqrt{p_0 q_0 / n}} = \dfrac{.086 - .2}{.034} = \dfrac{-.114}{.034} = -3.36$$

The implication is that the observed sample proportion is (approximately) 3.36 standard deviations below the null-hypothesized proportion, .2 (Figure 6.16). Therefore, we reject the null hypothesis, concluding at the .05 level of significance that the true failure rate of the new method for detecting breast cancer differs from .20. Since $\hat{p} = .086$, it appears that the new method is better (i.e., has a smaller failure rate) than the method currently in use. (To estimate the magnitude of the failure rate for the new method, a confidence interval can be constructed.)

The test of hypothesis about a population proportion p is summarized in the next box. Note that the procedure is entirely analogous to that used for conducting large-sample tests about a population mean.

Large-Sample Test of Hypothesis about p

One-Tailed Test	**Two-Tailed Test**
$H_0: p = p_0$	$H_0: p = p_0$
$H_a: p < p_0$ (or $H_a: p > p_0$)	$H_a: p \neq p_0$
Test statistic: $z = \dfrac{\hat{p} - p_0}{\sigma_{\hat{p}}}$	Test statistic: $z = \dfrac{\hat{p} - p_0}{\sigma_{\hat{p}}}$

where p_0 = hypothesized value of p, $\sigma_{\hat{p}} = \sqrt{p_0 q_0/n}$, and $q_0 = 1 - p_0$

Rejection region: $z < -z_\alpha$	Rejection region: $z < -z_{\alpha/2}$
(or $z > z_\alpha$ when $H_a: p > p_0$)	or $z > z_{\alpha/2}$

Conditions Required for a Valid Large-Sample Hypothesis Test for p

1. A random sample is selected from a binomial population.
2. The sample size n is large. (This condition will be satisfied if np_0 and nq_0 are both at least 15.)

EXAMPLE 6.7

A HYPOTHESIS TEST FOR p—

Proportion of Defective Batteries

Problem The reputations (and hence sales) of many businesses can be severely damaged by shipments of manufactured items that contain a large percentage of defectives. For example, a manufacturer of alkaline batteries may want to be reasonably certain that less than 5% of its batteries are defective. Suppose 300 batteries are randomly selected from a very large shipment; each is tested and 10 defective batteries are found. Does this outcome provide sufficient evidence for the manufacturer to conclude that the fraction defective in the entire shipment is less than .05? Use $\alpha = .01$.

Solution The objective of the sampling is to determine whether there is sufficient evidence to indicate that the fraction defective, p, is less than .05. Consequently, we will test the null hypothesis that $p = .05$ against the alternative hypothesis that $p < .05$. The elements of the test are

$H_0: p = .05$ (Fraction of defective batteries equals .05.)
$H_a: p < .05$ (Fraction of defective batteries is less than .05.)

Test statistic: $z = \dfrac{\hat{p} - p_0}{\sigma_{\hat{p}}}$

Rejection region: $z < -z_{.01} = -2.33$ (see Figure 6.17)

Before conducting the test, we check to determine whether the sample size is large enough to use the normal approximation to the sampling distribution *of $\hat{p}$*. Since $np_0 = (300)(.05) = 15$ and $nq_0 = (300)(.95) = 285$ are both at least 15, the normal approximation will be adequate.

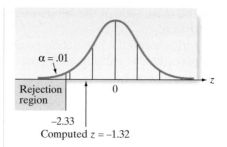

FIGURE 6.17

Rejection region for
Example 6.7

We now calculate the test statistic:

$$z = \frac{\hat{p} - .05}{\sigma_{\hat{p}}} = \frac{(10/300) - .05}{\sqrt{p_0 q_0/n}} = \frac{.03333 - .05}{\sqrt{p_0 q_0/300}}$$

Notice that we use p_0 to calculate $\sigma_{\hat{p}}$ because, in contrast to calculating $\sigma_{\hat{p}}$ for a confidence interval, the test statistic is computed on the assumption that the null hypothesis is true—that is, $p = p_0$. Therefore, substituting the values for $\hat{p}$ and p_0 into the z statistic, we obtain

$$z \approx \frac{-.01667}{\sqrt{(.05)(.95)/300}} = \frac{-.01667}{.0126} = -1.32$$

As shown in Figure 6.17, the calculated z-value does not fall into the rejection region. Therefore, there is insufficient evidence at the .01 level of significance to indicate that the shipment contains less than 5% defective batteries.

Now Work Exercise 6.76a–b

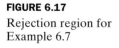

EXAMPLE 6.8

FINDING THE
p-VALUE FOR A
TEST ABOUT A
POPULATION
PROPORTION p

Problem In Example 6.7, we found that we did not have sufficient evidence at the $\alpha = .01$ level of significance to indicate that the fraction defective, p, of alkaline batteries was less than $p = .05$. How strong was the weight of evidence favoring the alternative hypothesis ($H_a: p < .05$)? Find the observed significance level (p-value) for the test.

Solution The computed value of the test statistic z was $z = -1.32$. Therefore, for this lower-tailed test, the observed significance level is

$$p\text{-value} = P(z \le -1.32)$$

This lower-tail area is shown in Figure 6.18. The area between $z = 0$ and $z = 1.32$ is given in Table III in Appendix A as .4066. Therefore, the observed significance level is $.5 - .4066 = .0934$.

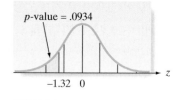

FIGURE 6.18

The observed significance
level for Example 6.8

Note: The p-value can also be obtained with statistical software. The MINITAB printout shown in Figure 6.19 gives the p-value (highlighted).

Test and CI for One Proportion

Test of p = 0.05 vs p < 0.05

				95% Upper		
Sample	X	N	Sample p	Bound	Z-Value	P-Value
1	10	300	0.033333	0.050380	-1.32	0.093

Using the normal approximation.

FIGURE 6.19

MINITAB lower-tailed test
of p, Example 6.8

Look Back Note that this probability is quite small. Although we did not reject H_0: $p = .05$ at $\alpha = .01$, the probability of observing a z-value as small as or smaller than -1.35 is only .0885 if in fact H_0 is true. Therefore, we would reject H_0 if we choose $\alpha = .10$ (since the observed significance level is less than .10), and we would not reject H_0 (the conclusion of Example 6.7) if we choose $\alpha = .05$ or $\alpha = .01$.

> **Now Work Exercise 6.76c**

Small-sample test procedures are also available for p, although most surveys use samples that are large enough to employ the large-sample tests presented in this section. A test of proportions that can be applied to small samples is discussed in Chapter 8.

STATISTICS IN ACTION REVISITED

Testing a Population Proportion in the KLEENEX® Survey

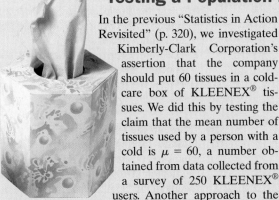

In the previous "Statistics in Action Revisited" (p. 320), we investigated Kimberly-Clark Corporation's assertion that the company should put 60 tissues in a cold-care box of KLEENEX® tissues. We did this by testing the claim that the mean number of tissues used by a person with a cold is $\mu = 60$, a number obtained from data collected from a survey of 250 KLEENEX® users. Another approach to the problem is to consider the proportion of KLEENEX® users who use fewer than 60 tissues when they have a cold. Now the population parameter of interest is p, the proportion of all KLEENEX® users who use fewer than 60 tissues when they have a cold.

Kimberly-Clark Corporation's belief that the company should put 60 tissues in a cold-care box will be supported if half of the KLEENEX® users surveyed use less than 60 tissues and half use more than 60 tissues (i.e., if $p = .5$). Is there evidence to indicate that the population proportion differs from .5? To answer this question, we set up the following null and alternative hypotheses:

$$H_0: p = .5 \quad H_a: p \neq .5$$

Recall that the survey results for the 250 sampled KLEENEX® users are stored in the **TISSUES** data file. In addition to the number of tissues used by each person, the file contains a qualitative variable—called USED60—representing whether the person used fewer or more than 60 tissues. (The values of USED60 in the data set are "BELOW" or "ABOVE.") A MINITAB analysis of this variable yielded the printout displayed in Figure SIA6.2.

On the MINITAB printout, x represents the number of the 250 people with colds that used fewer than 60 tissues. Note that $x = 143$. This value is used to compute the test statistic $z = 2.28$, highlighted on the printout. The p-value of the test, also highlighted on the printout, is p-value = .023. Since this value is less than $\alpha = .05$, there is sufficient evidence (at $\alpha = .05$) to reject H_0; we conclude that the proportion of all KLEENEX® users who use fewer than 60 tissues when they have a cold differs from .5. However, if we test at $\alpha = .01$, there is insufficient evidence to reject H_0. Consequently, our choice of α (as in the previous "Statistics in Action Revisited") is critical to our decision.

```
Test of p = 0.5 vs p not = 0.5

Event = BELOW


Variable     X    N   Sample p        95% CI          Z-Value  P-Value
USED60      143  250  0.572000  (0.510666, 0.633334)    2.28    0.023
```

FIGURE SIA6.2
MINITAB test of $p = .5$ for KLEENEX® survey

Exercises 6.73–6.91

Understanding the Principles

6.73 What type of data, quantitative or qualitative, is typically associated with making inferences about a population proportion p?

6.74 What conditions are required for a valid large-sample test for p?

Learning the Mechanics

6.75 For the binomial sample sizes and null-hypothesized values of p in each part, determine whether the sample size is large enough to use the normal approximation methodology presented in this section to conduct a test of the null hypothesis H_0: $p = p_0$.

a. $n = 500$, $p_0 = .05$
b. $n = 100$, $p_0 = .99$
c. $n = 50$, $p_0 = .2$
d. $n = 20$, $p_0 = .2$
e. $n = 10$, $p_0 = .4$

6.76 Suppose a random sample of 100 observations from a binomial population gives a value of $\hat{p} = .69$ and you wish to test the null hypothesis that the population parameter p is equal to .75 against the alternative hypothesis that p is less than .75.

a. Noting that $\hat{p} = .69$, what does your intuition tell you? Does the value of $\hat{p}$ appear to contradict the null hypothesis?

b. Use the large-sample z-test to test H_0: $p = .75$ against the alternative hypothesis H_a: $p < .75$. Use $\alpha = .05$. How do the test results compare with your intuitive decision from part **a**?

c. Find and interpret the observed significance level of the test you conducted in part **b**.

6.77 Suppose the sample in Exercise 6.76 has produced $\hat{p} = .84$ and we wish to test H_0: $p = .9$ against the alternative H_a: $p < .9$.

a. Calculate the value of the z statistic for this test.

b. Note that the numerator of the z statistic $(\hat{p} - p_0 = .84 - .90 = -.06)$ is the same as for Exercise 8.76. Considering this, why is the absolute value of z for this exercise larger than that calculated in Exercise 8.76?

c. Complete the test, using $\alpha = .05$, and interpret the result.

d. Find the observed significance level for the test and interpret its value.

6.78 A random sample of 100 observations is selected from a binomial population with unknown probability of success, p. The computed value of $\hat{p}$ is equal to .74.

a. Test H_0: $p = .65$ against H_a: $p > .65$. Use $\alpha = .01$.

b. Test H_0: $p = .65$ against H_a: $p > .65$. Use $\alpha = .10$.

c. Test H_0: $p = .90$ against H_a: $p \neq .90$. Use $\alpha = .05$.

d. Form a 95% confidence interval for p.

e. Form a 99% confidence interval for p.

SNACK

6.79 Refer to Exercise 5.50 (p. 282), in which 50 consumers taste-tested a new snack food.

a. Test H_0: $p = .5$ against H_a: $p > .5$, where p is the proportion of customers who do not like the snack food. Use $\alpha = .10$.

b. Report the observed significance level of your test.

APPLET **Applet Exercise 6.5**

Use the applet entitled *Hypotheses Test for a Proportion* to investigate the relationships between the probabilities of Type I and Type II errors occurring at levels .05 and .01. For this exercise, use $n = 100$, true $p = 0.5$, and alternative *not equal*.

a. Set null $p = .5$. What happens to the proportion of times the null hypothesis is rejected at the .05 level and at the .01 level as the applet is run more and more times? What type of error has occurred when the null hypothesis is rejected in this situation? Based on your results, is this type of error more likely to occur at level .05 or at level .01? Explain.

b. Set null $p = .6$. What happens to the proportion of times the null hypothesis is *not* rejected at the .05 level and at the .01 level as the applet is run more and more times?

What type of error has occurred when the null hypothesis is *not* rejected in this situation? Based on your results, is this type of error more likely to occur at level .05 or at level .01? Explain.

c. Use your results from parts **a** and **b** to make a general statement about the probabilities of Type I and Type II errors at levels .05 and .01.

APPLET **Applet Exercise 6.6**

Use the applet entitled *Hypotheses Test for a Proportion* to investigate the effect of the true population proportion p on the probability of a Type I error occurring. For this exercise, use $n = 100$ and alternative *not equal*.

a. Set true $p = .5$ and null $p = .5$. Run the applet several times, and record the proportion of times the null hypothesis is rejected at the .01 level.

b. Clear the applet and repeat part **a** for true $p = .1$ and null $p = .1$. Then repeat one more time for true $p = .01$ and null $p = .01$.

c. Based on your results from parts **a** and **b**, what can you conclude about the probability of a Type I error occurring as the true population proportion gets closer to 0?

Applying the Concepts—Basic

6.80 **"Made in the USA" survey.** Refer to the *Journal of Global Business* (Spring 2002) study of what "Made in the USA" means to consumers, presented in Exercise 2.13 (p. 36). Recall that 64 of 106 randomly selected shoppers believed that "Made in the USA" means that 100% of labor and materials are from the United States. Let p represent the true proportion of consumers who believe "Made in the USA" means that 100% of labor and materials are from the United States.

a. Calculate a point estimate for p.

b. A claim is made that $p = .70$. Set up the null and alternative hypothesis to test this claim.

c. Calculate the test statistic for the test you carried out in part **b**.

d. Find the rejection region for the test if $\alpha = .01$.

e. Use the results from parts **c** and **d** to draw the appropriate conclusion.

6.81 **Accuracy of price scanners at Wal-Mart.** Refer to Exercise 5.73 (p. 288) and the study of the accuracy of checkout scanners at Wal-Mart Stores in California. Recall that the National Institute for Standards and Technology (NIST) mandates that, for every 100 items scanned through the electronic checkout scanner at a retail store, no more than 2 should have an inaccurate price. A study of random items purchased at California Wal-Mart stores found that 8.3% had the wrong price (*Tampa Tribune*, Nov. 22, 2005). Assume that the study included 1,000 randomly selected items.

a. Identify the population parameter of interest in the study.

b. Set up H_0 and H_a for a test to determine whether the true proportion of items scanned at California Wal-Mart stores exceeds the 2% NIST standard.

c. Find the test statistic and rejection region (at $\alpha = .05$) for the test.

d. Give a practical interpretation of the test.

e. What conditions are required for the inference made in part **d** to be valid? Are these conditions met?

6.82 Killing insects with low oxygen. A group of Australian entomological toxicologists investigated the impact of exposure to low oxygen on the mortality of insects. (*Journal of Agricultural, Biological, and Environmental Statistics*, Sep. 2000.) Thousands of adult rice weevils were placed in a chamber filled with wheat grain, and the chamber was exposed to nitrogen gas for 4 days. Insects were assessed as dead or alive 24 hours after exposure. At the conclusion of the experiment, 31,386 weevils, were dead and 35 weevils were found alive. Previous studies had shown a 99% mortality rate in adult rice weevils exposed to carbon dioxide for 4 days. For this study, the parameter of interest is the mortality rate for all adult rice weevils exposed to nitrogen.

a. Give a point estimate of the population parameter.

b. Set up H_0 and H_a for testing whether the true mortality rate for weevils exposed to nitrogen is higher than 99%.

c. Conduct the test from part **b** using $\alpha = .10$. Draw the appropriate conclusion in the words of the problem.

d. What conditions are required for the inference made in part **c** to be valid? Do they appear to be satisfied?

6.83 Single-parent families. Examining data collected on 835 males from the National Youth Survey (a longitudinal survey of a random sample of U.S. households), researchers at Carnegie Mellon University found that 401 of the male youths were raised in a single-parent family. (*Sociological Methods & Research*, February 2001.) Does this information allow you to conclude that more than 45% of male youths are raised in a single-parent family? Test at $\alpha = .05$.

6.84 Graduation rates of student–athletes. Are student–athletes at Division I universities poorer students than nonathletes? The National Collegiate Athletic Association (NCAA) measures the academic outcomes of student–athletes with the Graduation Success Rate (GSR)—the percentage of eligible athletes who graduate within six years of entering college. According to the NCAA, the GSR for all scholarship athletes at Division I institutions is 63% (*Inside Higher Ed*, Nov. 10, 2006.) It is well known that the GSR for all students at Division I colleges is 60%.

a. Suppose the NCAA report was based on a sample of 500 student–athletes, of which 315 graduated within six years. Is this sufficient information to conclude that the GSR for all scholarship athletes at Division I institutions differs from 60%? Test, using $\alpha = .01$.

b. The GSR statistics were also broken down by gender and sport. For example, men's Division I college basketball players had a GSR of 42% (compared with a known GSR of 58% for all male college students). Suppose this statistic was based on a sample of 200 male basketball players, of which 84 graduated within six years. Is this sufficient information to conclude that the GSR for all male basketball players at Division I institutions differs from 58%? Test, using $\alpha = .01$.

Applying the Concepts—Intermediate

6.85 Federal civil trial appeals. Refer to the *Journal of the American Law and Economics Association* (Vol. 3, 2001) study of appeals of federal civil trials, presented in Exercise 3.55 (p. 137). A breakdown of 678 civil cases that were originally tried in front of a judge and appealed by either the plaintiff or the defendant is reproduced in the accompanying table. Do the data provide sufficient evidence to indicate that the percentage of civil cases appealed that are actually reversed is less than 25%? Test, using $\alpha = .01$.

Outcome of Appeal	Number of Cases
Plaintiff trial win—reversed	71
Plaintiff trial win—affirmed/dismissed	240
Defendant trial win—reversed	68
Defendant trial win—affirmed/dismissed	299
TOTAL	678

6.86 Verbs and double-object datives. Any sentence that contains an animate noun as a direct object and another noun as a second object (e.g., "Sue offered Ann a cookie") is termed a double-object dative (DOD). The connection between certain verbs and a DOD was investigated in *Applied Psycholinguistics* (June 1998). The subjects were 35 native English speakers who were enrolled in an introductory English composition course at a Hawaiian community college. After viewing pictures of a family, each subject was asked to write a sentence about the family, using a specified verb. Of the 35 sentences using the verb "buy," 10 had a DOD structure. Conduct a test to determine whether the true fraction of sentences with the verb "buy" that are DODs is less than $1/3$. Use $\alpha = .05$.

6.87 Astronomy students and the Big Bang Theory. Indiana University professors investigated first-year college students' knowledge of astronomy. (*Astronomy Education Review*, Vol. 2, 2003.) One concept of interest was the Big Bang theory of the creation of the universe. In a sample of 148 freshmen students, 37 believed that the Big Bang theory accurately described the creation of planetary systems. Based on this information, would you be willing to state that more than 20% of all freshmen college students believe in the Big Bang theory? How confident are you of your decision?

6.88 Study of lunar soil. *Meteoritics* (March 1995) reported the results of a study of lunar soil evolution. Data were obtained from the *Apollo 16* mission to the moon, during which a 62-cm core was extracted from the soil near the landing site. Monomineralic grains of lunar soil were separated out and examined for coating with dust and glass fragments. Each grain was then classified as coated or uncoated. Of interest is the "coat index"—that is, the proportion of grains that are coated. According to soil evolution theory, the coat index will exceed .5 at the top of the core, equal .5 in the middle of the core, and fall below .5 at the bottom of the core. Use the summary data in the accompanying table to test each part of the three-part theory. Use $\alpha = .05$ for each test.

	Location (depth)		
	Top (4.25 cm)	Middle (28.1 cm)	Bottom (54.5 cm)
Number of grains sampled	84	73	81
Number coated	64	35	29

Source: Basu, A., and McKay, D. S. "Lunar soil evolution processes and Apollo 16 core 60013/60014," *Meteoritics*, Vol. 30, No. 2, Mar. 1995, p. 166 (Table 2).

6.89 Effectiveness of skin cream. Pond's Age-Defying Complex, a cream with alpha-hydroxy acid, advertises that it can reduce wrinkles and improve the skin. In a study published in *Archives of Dermatology* (June 1996), 33 middle-aged women used a cream with alpha-hydroxy acid for 22 weeks. At the end of the study period, a dermatologist judged whether each woman exhibited any improvement in the condition of her skin. The results for the 33 women (where I = improved skin and N = no improvement) are listed in the accompanying table.

  a. Do the data provide sufficient evidence to conclude that the cream will improve the skin of more than 60% of middle-aged women? Test, using $\alpha = .05$.

  b. Find and interpret the *p*-value of the test.

SKINCREAM

I	I	N	I	N	N	I	I	I	I	I	I
N	I	I	I	N	I	I	I	N	I	N	I
I	I	I	I	I	N	I	I	N			

Applying the Concepts—Advanced

6.90 Parents who condone spanking. In Exercise 4.55 (p. 195) you read about a nationwide survey which claimed that 60% of parents with young children condone spanking their child as a regular form of punishment. (*Tampa Tribune*, October 5, 2000.) In a random sample of 100 parents with young children, how many parents would need to say that they condone spanking as a form of punishment in order to refute the claim?

6.91 Testing the placebo effect. The *placebo effect* describes the phenomenon of improvement in the condition of a patient taking a placebo—a pill that looks and tastes real, but contains no medically active chemicals. Physicians at a clinic in La Jolla, California, gave what they thought were drugs to 7,000 asthma, ulcer, and herpes patients. Although the doctors later learned that the drugs were really placebos, 70% of the patients reported an improved condition. (*Forbes*, May 22, 1995.) Use this information to test (at $\alpha = .05$) the placebo effect at the clinic. Assume that if the placebo is ineffective, the probability of a patient's condition improving is .5.

6.6 A Nonparametric Test About a Population Median (Optional)

In Sections 6.2–6.4, we utilized the *z*- and *t*-statistics for testing hypotheses about a population mean. The *z*-statistic is appropriate for large random samples selected from "general" populations—that is, samples with few limitations on the probability distribution of the underlying population. The *t*-statistic was developed for small-sample tests in which the sample is selected at random from a *normal* distribution. The question is, How can we conduct a test of hypothesis when we have a small sample from a *nonnormal* distribution?

The answer is: Use a *distribution-free* procedure that requires fewer or less stringent assumptions about the underlying population—called a *nonparametric* method.

> **Definition 6.2**
>
> **Distribution-free tests** are statistical tests that do not rely on any underlying assumptions about the probability distribution of the sampled population.

> **Definition 6.3**
>
> The branch of inferential statistics devoted to distribution-free tests is called **nonparametrics**.

The **sign test** is a relatively simple nonparametric procedure for testing hypotheses about the central tendency of a nonnormal probability distribution. Note that we used the phrase *central tendency* rather than *population mean*. This is because the sign test, like many nonparametric procedures, provides inferences about the population *median* rather than the population mean μ. Denoting the population median by the Greek letter η, we know (Chapter 2) that η is the 50th percentile of the distribution (Figure 6.19) and, as such, is less affected by the skewness of the distribution and the presence of outliers (extreme observations). Since the nonparametric test must be suitable for all distributions, not just the normal, it is reasonable for nonparametric tests to focus on the more robust (less sensitive to extreme values) measure of central tendency; the median.

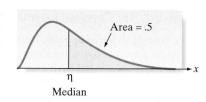

FIGURE 6.19

Location of the population median, η

e. $P(x \geq 15)$ when $n = 25$ and $p = .5$. Also, use the normal approximation to calculate this probability, and then compare the approximation with the exact value.

6.95 Consider the following sample of 10 measurements:

 LM6_95

8.4	16.9	15.8	12.5	10.3	4.9	12.9	9.8	23.7	7.3

Use these data, the binomial tables (Table II, Appendix A), and $\alpha = .05$ to conduct each of the following sign tests:

a. $H_0: \eta = 9$ versus $H_a: \eta > 9$
b. $H_0: \eta = 9$ versus $H_a: \eta \neq 9$
c. $H_0: \eta = 20$ versus $H_a: \eta < 20$
d. $H_0: \eta = 20$ versus $H_a: \eta \neq 20$
e. Repeat each of the preceding tests, using the normal approximation to the binomial probabilities. Compare the results.
f. What assumptions are necessary to ensure the validity of each of the preceding tests?

6.96 Suppose you wish to conduct a test of the research hypothesis
NW that the median of a population is greater than 80. You randomly sample 25 measurements from the population and determine that 16 of them exceed 80. Set up and conduct the appropriate test of hypothesis at the .10 level of significance. Be sure to specify all necessary assumptions.

Applying the Concepts—Basic

 LICHEN

6.97 Radioactive lichen. Refer to the 2003 Lichen Radionuclide Baseline Research project to monitor the level of radioactivity in lichen, Exercise 6.67 (p. 328). Recall that University of Alaska researchers collected 9 lichen specimens and measured the amount of the radioactive element cesium-137 (in microcuries per milliliter) in each specimen. (The natural logarithms of the data values, saved in the **LICHEN** file, are listed in the next table.) In Exercise 6.67, you used the t-statistic to test whether the mean cesium amount in lichen differs from $\mu = .003$ microcurie per milliliter. Use the MINITAB printout below to conduct an alternative nonparametric test at $\alpha = .10$. Does the result agree with that of the t-test from Exercise 6.67?

 LICHEN

Location			
Bethel	−5.50	−5.00	
Eagle Summit	−4.15	−4.85	
Moose Pass	−6.05		
Turnagain Pass	−5.00		
Wickersham Dome	−4.10	−4.50	−4.60

Source: Lichen Radionuclide Baseline Research project, 2003.

MINITAB output for Exercise 6.97

Sign Test for Median: CESIUM

```
Sign test of median =  0.00300 versus not =  0.00300

        N  Below  Equal  Above       P   Median
CESIUM  9      1      0      8  0.0391  0.00783
```

6.98 Caffeine in Starbucks coffee. Scientists at the University of Florida College of Medicine investigated the level of caffeine in 16-ounce cups of Starbucks coffee. (*Journal of Analytical Toxicology*, Oct. 2003.) In one phase of the experiment, cups of Starbucks Breakfast Blend (a mix of Latin American coffees) were purchased on six consecutive days from a single specialty coffee shop. The amount of caffeine in each of the six cups (measured in milligrams) is provided in the following table:

 STARBUCKS

564	498	259	303	300	307

a. Suppose the scientists are interested in determining whether the median amount of caffeine in Breakfast Blend coffee exceeds 300 milligrams. Set up the null and alternative hypotheses of interest.

b. How many of the cups in the sample have a caffeine content that exceeds 300 milligrams?
c. Assuming that $p = .5$, use the binomial table in Appendix A to find the probability that at least 4 of the 6 cups have caffeine amounts that exceed 300 milligrams.
d. On the basis of the probability you found in part **c**, what do you conclude about H_0 and H_a? (Use $\alpha = .05$.)

6.99 Ammonia in car exhaust. Refer to the *Environmental Science & Technology* (Sept. 1, 2000) study of ammonia levels near the exit ramp of a San Francisco highway tunnel, presented in Exercise 2.59 (p. 59). The daily ammonia concentrations (parts per million) on eight randomly selected days during afternoon drive time are reproduced in the accompanying table. Suppose you want to determine whether the median daily ammonia concentration for all afternoon drive-time days exceeds 1.5 ppm.

 AMMONIA

1.53	1.50	1.37	1.51	1.55	1.42	1.41	1.48

a. Set up the null and alternative hypotheses for the test.
b. Find the value of the test statistic.
c. Find the p-value of the test.
d. Give the appropriate conclusion (in the words of the problem) if $\alpha = .05$.

6.100 Quality of white shrimp. In *The American Statistician* (May 2001), the nonparametric sign test was used to analyze data on the quality of white shrimp. One measure of shrimp quality is cohesiveness. Since freshly caught shrimp are usually stored on ice, there is concern that cohesiveness will deteriorate after storage. For a sample of 20 newly caught

---

For example, increasing numbers of both private and public agencies are requiring their employees to submit to tests for substance abuse. One laboratory that conducts such testing has developed a system with a normalized measurement scale in which values less than 1.00 indicate "normal" ranges and values equal to or greater than 1.00 are indicative of potential substance abuse. The lab reports a normal result as long as the median level for an individual is less than 1.00. Eight independent measurements of each individual's sample are made. One individual's results are shown in Table 6.6.

 SUBABUSE

TABLE 6.6	Substance Abuse Test Results						
.78	.51	3.79	.23	.77	.98	.96	.89

If the objective is to determine whether the *population* median (i.e., the true median level if an infinitely large number of measurements were made on the same individual sample) is less than 1.00, we establish that as our alternative hypothesis and test

$$H_0: \eta = 1.00$$
$$H_a: \eta < 1.00$$

The one-tailed sign test is conducted by counting the number of sample measurements that "favor" the alternative hypothesis—in this case, the number that are less than 1.00. If the null hypothesis is true, we expect approximately half of the measurements to fall on each side of the hypothesized median, and if the alternative is true, we expect significantly more than half to favor the alternative—that is, to be less than 1.00. Thus,

Test statistic: S = Number of measurements less than 1.00, the null hypothesized median

If we wish to conduct the test at the $\alpha = .05$ level of significance, the rejection region can be expressed in terms of the observed significance level, or p-value, of the test:

Rejection region: p − value $\leq .05$

In this example, $S = 7$ of the 8 measurements are less than 1.00. To determine the observed significance level associated with that outcome, we note that the number of measurements less than 1.00 is a binomial random variable (check the binomial characteristics presented in Chapter 4), and *if H_0 is true,* the binomial probability p that a measurement lies below (or above) the median 1.00 is equal to .5 (Figure 6.19). What is the probability that a result is *as contrary to or more contrary to H_0* than the one observed? That is, what is the probability that 7 *or more* of 8 binomial measurements will result in Success (be less than 1.00) if the probability of Success is .5? Binomial Table II in Appendix A (with $n = 8$ and $p = .5$) indicates that

$$P(x \geq 7) = 1 - P(x \leq 6) = 1 - .965 = .035$$

Thus, the probability that at least 7 of 8 measurements would be less than 1.00 *if the true median were 1.00* is only .035. The p-value of the test is therefore .035.

This p-value can also be obtained from a statistical software package. The MINITAB printout of the analysis is shown in Figure 6.20, with the p-value highlighted. Since $p = .035$ is less than $\alpha = .05$, we conclude that this sample provides sufficient evidence to reject the null hypothesis. The implication of this rejection is that the laboratory can conclude at the $\alpha = .05$ level of significance that the true median level for the individual tested is less than

FIGURE 6.20
MINITAB printout of sign test

Sign Test for Median: READING

```
Sign test of median =  1.000 versus < 1.000

         N  Below  Equal  Above       P   Median
READING  8      7      0      1  0.0352  0.8350
```

1.00. However, we note that one of the measurements, with a value of 3.79, greatly exceeds the others and deserves special attention. This large measurement is an outlier that would make the use of a t-test and its concomitant assumption of normality dubious. The only assumption necessary to ensure the validity of the sign test is that the probability distribution of measurements is continuous.

The use of the sign test for testing hypotheses about population medians is summarized in the following box:

Sign Test for a Population Median η

One-Tailed Test

$H_0: \eta = \eta_0$
$H_a: \eta > \eta_0$ [or $H_a: \eta < \eta_0$]
Test statistic:
S = Number of sample measurements greater than η_0 [or S = number of measurements less than η_0]

Two-Tailed Test

$H_0: \eta = \eta_0$
$H_a: \eta \neq \eta_0$
Test statistic:
S = Larger of S_1 and S_2, where S_1 is the number of measurements less than η_0 and S_2 is the number of measurements greater than η_0

[*Note:* Eliminate observations from the analysis that are exactly equal to the hypothesized median, η_0.]

Observed significance level:
p-value = $P(x \geq S)$

Observed significance level:
p-value = $2P(x \geq S)$

where x has a binomial distribution with parameters n and $p = .5$. (Use Table II, Appendix A.)

Rejection region: Reject H_0 if p-value $\leq \alpha$

Conditions Required for a Valid Application of the Sign Test

The sample is selected randomly from a continuous probability distribution.
[*Note:* No assumptions need to be made about the shape of the probability distribution.]

Recall that the normal probability distribution provides a good approximation of the binomial distribution when the sample size is large. For tests about the median of a distribution, the null hypothesis implies that $p = .5$, and the normal distribution provides a good approximation if $n \geq 10$. (Samples with $n \geq 10$ satisfy the condition that $np \pm 2\sqrt{npq}$ is contained in the interval from 0 to n.) Thus, we can use the standard normal z-distribution to conduct the sign test for large samples. The large-sample sign test is summarized in the next box.

Large-Sample Sign Test for a Population Median η

One-Tailed Test

$H_0: \eta = \eta_0$
$H_a: \eta > \eta_0$ [or $H_a: \eta < \eta_0$]

Two-Tailed Test

$H_0: \eta = \eta_0$
$H_a: \eta \neq \eta_0$

Test statistic: $z = \dfrac{(S - .5) - .5n}{.5\sqrt{n}}$

[*Note:* S is calculated as shown in the previous box. We subtract .5 from S as the "correction for continuity." The null-hypothesized mean value is $np = .5n$, and the standard deviation is

$$\sqrt{npq} = \sqrt{n(.5)(.5)} = .5\sqrt{n}$$

(See Section 4.7 for details on the normal approximation to the binomial distribution.)]

Rejection region: $z > z_\alpha$

Rejection region: $z > z_{\alpha/2}$

where tabulated z-values can be found in Table III, Appendix A.

EXAMPLE 6.9

SIGN TEST APPLICATION—
Failure Times of CD Players

Problem A manufacturer of compact disc (CD) players has established that the median time to failure for its players is 5,250 hours of utilization. A sample of 20 CD players from a competitor is obtained, and the players are tested continuously until each fails. The 20 failure times range from 5 hours (a "defective" player) to 6,575 hours, and 14 of the 20 exceed 5,250 hours. Is there evidence that the median failure time of the competitor's product differs from 5,250 hours? Use $\alpha = .10$.

Solution The null and alternative hypotheses of interest are

$$H_0: \eta = 5{,}250 \text{ hours}$$
$$H_a: \eta \neq 5{,}250 \text{ hours}$$

Since $n \geq 10$, we use the standard normal z statistic:

$$\text{Test statistic: } z = \frac{(S - .5) - .5n}{.5\sqrt{n}}$$

Here, S is the maximum of S_1 (the number of measurements greater than 5,250) and S_2 (the number of measurements less than 5,250). Also,

$$\text{Rejection region: } z > 1.645, \text{ where } z_{\alpha/2} = z_{.05} = 1.645$$

Assumptions: The probability distribution of the failure times is continuous (time is a continuous variable), but nothing is assumed about its shape.

Since the number of measurements exceeding 5,250 is $S_2 = 14$, it follows that the number of measurements less than 5,250 is $S_1 = 6$. Consequently, $S = 14$, the greater of S_1 and S_2. The calculated z statistic is therefore

$$z = \frac{(S - .5) - .5n}{.5\sqrt{n}} = \frac{13.5 - 10}{.5\sqrt{20}} = \frac{3.5}{2.236} = 1.565$$

The value of z is not in the rejection region, so we cannot reject the null hypothesis at the $\alpha = .10$ level of significance.

Look Back The manufacturer should not conclude, on the basis of this sample, that its competitor's CD players have a median failure time that differs from 5,250 hours. The manufacturer will not "accept H_0," however, since the probability of a Type II error is unknown.

Now Work Exercise 6.96

The one-sample nonparametric sign test for a median provides an alternative to the t-test for small samples from nonnormal distributions. However, if the distribution is approximately normal, the t-test provides a more powerful test about the central tendency of the distribution.

Exercises 6.92–6.106

Understanding the Principles

6.92 Under what circumstances is the sign test preferred to the t-test for making inferences about the central tendency of a population?

6.93 What is the probability that a randomly selected observation exceeds the
   a. Mean of a normal distribution?
   b. Median of a normal distribution?
   c. Mean of a nonnormal distribution?
   d. Median of a nonnormal distribution?

Learning the Mechanics

6.94 Use Table II of Appendix A to calculate the following binomial probabilities:
   a. $P(x \geq 6)$ when $n = 7$ and $p = .5$
   b. $P(x \geq 5)$ when $n = 9$ and $p = .5$
   c. $P(x \geq 8)$ when $n = 8$ and $p = .5$
   d. $P(x \geq 10)$ when $n = 15$ and $p = .5$. Also, use the normal approximation to calculate this probability, and then compare the approximation with the exact value.

white shrimp, cohesiveness was measured both before and after storage on ice for two weeks. The difference in the cohesiveness measurements (before minus after) was obtained for each shrimp. If storage has no effect on cohesiveness, the population median of the differences will be 0. If cohesiveness deteriorates after storage, the population median of the differences will be positive.

a. Set up the null and alternative hypotheses to test whether cohesiveness will deteriorate after storage.

b. In the sample of 20 shrimp, there were 13 positive differences. Use this value to find the *p*-value of the test.

c. Make the appropriate conclusion (in the words of the problem) if $\alpha = .05$.

6.101 **Crab spiders hiding on flowers.** Refer to the *Behavioral Ecology* (Jan. 2005) field study on the natural camouflage of crab spiders, presented in Exercise 2.36 (p. 48). Ecologists collected a sample of 10 adult female crab spiders, each sitting on the yellow central part of a daisy, and measured the chromatic contrast between each spider and the flower. The contrast values for the 10 crab spiders are reproduced in the next table. (*Note:* The lower the contrast, the more difficult it is for predators to see the crab spider on the flower.) Recall that a contrast of 70 or greater allows bird predators to see the spider. Consider a test to determine whether the population median chromatic contrast of spiders on flowers is less than 70.

SPIDER

57	75	116	37	96	61	56	2	43	32

Data adapted from Thery, M., et al. "Specific color sensitivities of prey and predator explain camouflage in different visual systems," *Behavioral Ecology*, Vol. 16, No. 1, Jan. 2005 (Table 1).

a. State the null and alternative hypotheses for the test of interest.

b. Calculate the value of the test statistic.

c. Find the *p*-value for the test.

d. At $\alpha = .10$, what is the appropriate conclusion? State your answer in the words of the problem.

Applying the Concepts—Intermediate

6.102 **Al Qaeda attacks on the United States.** Refer to the *Studies in Conflict & Terrorism* (Vol. 29, 2006) analysis of recent incidents involving suicide terrorist attacks, presented in Exercise 2.30 (p. 46). The data in the next column are the number of individual suicide bombings attacks for each in a sample of 21 recent incidents involving an attack against the United States by the Al Qaeda terrorist group. A counterterrorism expert claims that more than half of all Al Qaeda attacks against the United States involve two or fewer suicide bombings. Is there evidence to support this claim? Test at $\alpha = .05$.

ALQAEDA

1	1	2	1	2	4	1	1	1	1	2
3	4	5	1	1	1	2	2	2	1	

Source: Moghadam, A. "Suicide terrorism, occupation, and the globalization of martyrdom: A critique of *Dying to Win*," *Studies in Conflict & Terrorism*, Vol. 29, No. 8, 2006 (Table 3).

6.103 **Biting rates of flies.** The biting rate of a particular species of fly was investigated in a study reported in the *Journal of the American Mosquito Control Association* (Mar. 1995). Biting rate was defined as the number of flies biting a volunteer during 15 minutes of exposure. The species of fly being investigated was known to have a median biting rate of 5 bites per 15 minutes on Stanbury Island, Utah. However, it is theorized that the median biting rate is higher in bright, sunny weather. To test this theory, 122 volunteers were exposed to the flies during a sunny day on Stanbury Island. Of these volunteers, 95 experienced biting rates greater than 5.

a. Set up the null and alternative hypotheses for the test.

b. Calculate the approximate *p*-value of the test. [*Hint:* Use the normal approximation for a binomial probability.]

c. Make the appropriate conclusion at $\alpha = .01$.

6.104 **Study of guppy migration.** In a study of the excessive transitory migration of guppy populations, 40 adult female guppies were placed in the left compartment of an experimental aquarium tank divided in half by a glass plate. After the plate was removed, the numbers of fish passing through the slit from the left compartment to the right one, and vice versa, were monitored every minute for 30 minutes. (*Zoological Science*, Vol. 6, 1989.) If an equilibrium is reached, the researchers would expect the median number of fish remaining in the left compartment to be 20. The data for the 30 observations (i.e., numbers of fish in the left compartment at the end of each 1-minute interval) are shown in the table at the bottom of the page. Use the large-sample sign test to determine whether the median is less than 20. Test, using $\alpha = .05$.

6.105 **Freckling of superalloy ingots.** Refer to the *Journal of Metallurgy* (Sep. 2004) study of freckling of superalloy ingots, presented in Exercise 2.179 (p. 105). Recall that freckles are defects that sometimes form during the solidification of the ingot. The freckle index for each of $n = 18$ superalloy ingots is shown in the next table. In the population of superalloy ingots, is there evidence to say that 50% of the ingots have a freckle index of 10 or higher? Test, using $\alpha = .01$.

FRECKLE

30.1	22.0	14.6	16.4	12.0	2.4	22.2	10.0	15.1
12.6	6.8	4.1	2.5	1.4	33.4	16.8	8.1	3.2

Source: Yang, W. H., et al. "A freckle criterion for the solidification of superalloys with a tilted solidification front," *Journal of Metallurgy*, Vol. 56, No. 9, Sep. 2004 (Table IV).

GUPPY

16	11	12	15	14	16	18	15	13	15
14	14	16	13	17	17	14	22	18	19
17	17	20	23	18	19	21	17	21	17

Source: Terami, H., and Watanabe, M. "Excessive transitory migration of guppy populations. III. Analysis of perception of swimming space and a mirror effect," *Zoological Science*, Vol. 6, 1989, p. 977 (Figure 2).

6.106 **Minimizing tractor skidding distance.** Refer to the *Journal of Forest Engineering* (July 1999) study of minimizing tractor skidding distances along a new road in a European forest, presented in Exercise 6.69 (p. 329). The skidding distances (in meters) were measured at 20 randomly selected road sites. The data are repeated in the accompanying table. In Exercise 6.69, you conducted a test of hypothesis for the population mean skidding distance. Now conduct a test to determine whether the population median skidding distance is more than 400 meters. Use $\alpha = .10$.

SKIDDING

488	350	457	199	285	409	435	574	439	546
385	295	184	261	273	400	311	312	141	425

Source: Tujek, J. & Pacola, E. "Algorithms for skidding distance modeling on a raster Digital Terrain Model," *Journal of Forest Engineering*, Vol. 10, No. 1, July 1999 (Table 1).

KEY TERMS

Note: Starred (*) terms are from the optional sections in this chapter.

Alternative (research) hypothesis   302
Conclusion   312
*Distribution-free tests   336
Level of significance   306
Lower-tailed test   309

*Nonparametrics   336
Null hypothesis   302
Observed significance level (*p-value*)   315
One-tailed (one-sided) statistical test   309
Rejection region   303
*Sign test   336

Test of hypothesis   301
Test statistic   302
Two-tailed (two-sided) hypothesis   309
Two-tailed test   309
Type I decision error   302
Type II error   304
Upper-tailed test   309

CHAPTER NOTES

Key Words for Identifying the Target Parameter:

μ—Mean, Average
p—Proportion, Fraction, Percentage, Rate, Probability
η—Median

Elements of a Hypothesis Test:

1. *Null hypothesis (H_o)*
2. *Alternative hypothesis (H_a)*
3. *Test statistic ($z, t, or X^2$)*
4. *Significance level (α)*
5. *p-value*
6. *Conclusion*

Forms of Alternative Hypothesis:

Lower tailed: H_a: $\mu_0 < 50$
Upper tailed: H_a: $\mu_0 > 50$
Two tailed: H_a: $\mu_0 \neq 50$

Sign Test

A **nonparametric** (distribution-free) test for a population median.

Key Symbols

μ	Population mean
p	Population proportion, P(Success), in binomial trial
σ^2	Population variance
η	Population median
$\overline{x}$	Sample mean (estimator of μ)
$\hat{p}$	Sample proportion (estimator of p)
s^2	Sample variance (estimator of σ^2)
H_0	Null hypothesis
H_a	Alternative hypothesis
α	Probability of a Type I error
β	Probability of a Type II error

Type I Error = Reject H_0 when H_0 is true (occurs with probability α)

Type II Error = Accept H_0 when H_0 is false (occurs with probability β)

Using *p*-values to Decide:

1. Choose significance level (α)
2. Obtain *p*-value of the test
3. If $\alpha > p$-value, Reject H_0

GUIDE TO SELECTING A ONE-SAMPLE HYPOTHESIS TEST

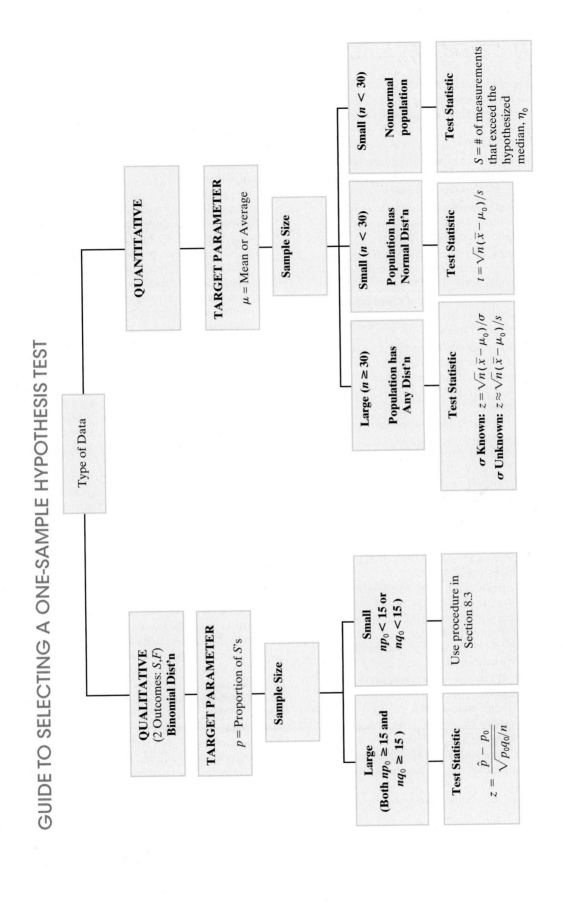

Type of Data

QUALITATIVE
(2 Outcomes: S, F)
Binomial Dist'n

TARGET PARAMETER
p = Proportion of S's

Sample Size

Large
(Both $np_0 \geq 15$ and $nq_0 \geq 15$)

Test Statistic
$z = \dfrac{\hat{p} - p_0}{\sqrt{p_0 q_0 / n}}$

Small
$np_0 < 15$ or $nq_0 < 15$

Use procedure in Section 8.3

QUANTITATIVE

TARGET PARAMETER
μ = Mean or Average

Sample Size

Large ($n \geq 30$)
Population has Any Dist'n

Test Statistic
σ **Known:** $z = \sqrt{n}\,(\bar{x} - \mu_0)/\sigma$
σ **Unknown:** $z \approx \sqrt{n}\,(\bar{x} - \mu_0)/s$

Small ($n < 30$)
Population has Normal Dist'n

Test Statistic
$t = \sqrt{n}\,(\bar{x} - \mu_0)/s$

Small ($n < 30$)
Nonnormal population

Test Statistic
S = # of measurements that exceed the hypothesized median, η_0

SUPPLEMENTARY EXERCISES 6.107–6.135

Note: List the assumptions necessary for the valid implementation of the statistical procedures you use in solving all these exercises. Starred (*) exercises refer to the optional section in this chapter.

Understanding the Principles

6.107 *Complete the following statement:* The smaller the p-value associated with a test of hypothesis, the stronger is the support for the _____ hypothesis. Explain your answer.

6.108 Specify the differences between a large-sample and small-sample test of hypothesis about a population mean μ. Focus on the assumptions and test statistics.

6.109 Which of the elements of a test of hypothesis can and should be specified *prior* to analyzing the data that are to be utilized to conduct the test?

6.110 If the rejection of the null hypothesis of a particular test would cause your firm to go out of business, would you want α to be small or large? Explain.

6.111 *Complete the following statement:* The larger the p-value associated with a test of hypothesis, the stronger is the support for the _____ hypothesis. Explain your answer.

Learning the Mechanics

6.112 A random sample of 20 observations selected from a normal population produced $\bar{x} = 72.6$ and $s^2 = 19.4$.
  a. From a 90% confidence interval for the population mean.
  b. Test H_0: $\mu = 80$ against H_a: $\mu < 80$. Use $\alpha = .05$.
  c. Test H_0: $\mu = 80$ against H_a: $\mu \neq 80$. Use $\alpha = .01$.
  d. Form a 99% confidence interval for μ.
  e. How large a sample would be required to estimate μ to within 1 unit with 95% confidence?

6.113 A random sample of $n = 200$ observations from a binomial population yields $\hat{p} = .29$.
  a. Test H_0: $p = .35$ against H_a: $p < .35$. Use $\alpha = .05$.
  b. Test H_0: $p = .35$ against H_a: $p \neq .35$. Use $\alpha = .05$.
  c. Form a 95% confidence interval for p.
  d. Form a 99% confidence interval for p.
  e. How large a sample would be required to estimate p to within .05 with 99% confidence?

6.114 A random sample of 175 measurements possessed a mean of $\bar{x} = 8.2$ and a standard deviation of $s = .79$.
  a. Form a 95% confidence interval for μ.
  b. Test H_0: $\mu = 8.3$ against H_a: $\mu \neq 8.3$. Use $\alpha = .05$.
  c. Test H_0: $\mu = 8.4$ against H_a: $\mu \neq 8.4$. Use $\alpha = .05$.

*6.115 In a sample of $n = 10$ measurements randomly selected from a nonnormal population, seven of the measurements exceed 150.
  a. Find the p-value for testing H_0: $\eta = 150$ against H_a: $\eta > 150$.
  b. Make the appropriate conclusion at $\alpha = .05$.

6.116 A t-test is conducted for the null hypothesis H_0: $\mu = 10$ versus the alternative hypothesis H_a: $\mu > 10$ for a random sample of $n = 17$ observations. The test results are $t = 1.174$ and p-value $= .1288$.
  a. Interpret the p-value.
  b. What assumptions are necessary for the validity of this test?
  c. Calculate and interpret the p-value, assuming that the alternative hypothesis was instead H_a: $\mu \neq 10$.

Applying the Concepts—Basic

6.117 **FDA mandatory new-drug testing.** When a new drug is formulated, the pharmaceutical company must subject it to lengthy and involved testing before receiving the necessary permission from the Food and Drug Administration (FDA) to market the drug. The FDA requires the pharmaceutical company to provide substantial evidence that the new drug is safe for potential consumers.
  a. If the new-drug testing were to be placed in a test-of-hypothesis framework, would the null hypothesis be that the drug is safe or unsafe? the alternative hypothesis?
  b. Given the choice of null and alternative hypotheses in part **a**, describe Type I and Type II errors in terms of this application. Define α and β in terms of this application.
  c. If the FDA wants to be very confident that the drug is safe before permitting it to be marketed, is it more important that α or β be small? Explain.

6.118 **Cell phone use by drivers.** Refer to the U.S. Department of Transportation (July 2001) study of the level of cell phone use by drivers while they are in the act of driving a motor passenger vehicle, presented in Exercise 5.91 (p. 293). Recall that in a random sample of 1,165 drivers selected across the country, 35 were found using their cell phone.
  a. Conduct a test (at $\alpha = .05$) to determine whether p, the true driver cell phone use rate, differs from .02.
  b. Does the conclusion, you drew in part **a** agree with the inference you derived from the 95% confidence interval for p in Exercise 7.91? Explain why or why not.

6.119 **Sleep deprivation study.** In a British study, 12 healthy college students deprived of one night's sleep received an array of tests intended to measure their thinking time, fluency, flexibility, and originality of thought. The overall test scores of the sleep-deprived students were compared with the average score expected from students who received their accustomed sleep (*Sleep*, January 1989). Suppose the overall scores of the 12 sleep-deprived students had a mean of $\bar{x} = 63$ and a standard deviation of 17. (Lower scores are associated with a decreased ability to think creatively.)
  a. Test the hypothesis that the true mean score of sleep-deprived subjects is less than 80, the mean score of subjects who received sleep prior to taking the test. Use $\alpha = .05$.
  b. What assumption is required for the hypothesis test of part **a** to be valid?
  *c. Suppose that 8 of the 12 students had scores below 80. Use the sign test to determine if the median score is less than 80. (Use $\alpha = .05$.)

6.120 **The "Pepsi challenge."** "Take the Pepsi Challenge" was a marketing campaign used by the Pepsi-Cola Company. Coca-Cola drinkers participated in a blind taste test in which they tasted unmarked cups of Pepsi and Coke and were asked to select their favorite. Pepsi claimed that "in recent blind taste tests, more than half the Diet Coke drinkers surveyed said they preferred the taste of Diet Pepsi." (*Consumer's Research*, May 1993.) Suppose 100 Diet Coke drinkers took the Pepsi Challenge and 56 preferred the taste of Diet Pepsi. Test the hypothesis that more than half of all Diet Coke drinkers will select Diet Pepsi in a blind taste test. Use $\alpha = .05$.

6.121 Masculinizing human faces. Refer to the *Nature* (August 27, 1998) study of facial characteristics that are deemed attractive, presented in Exercise 6.51 (p. 322). In another experiment, 67 human subjects viewed side by side an image of a Caucasian male face and the same image 50% masculinized. Each subject was asked to select the facial image they deemed more attractive. Fifty-eight of the 67 subjects felt that masculinization of face shape decreased attractiveness of the male face. The researchers used this sample information to test whether the subjects showed a preference for either the unaltered or the morphed male face.

a. Set up the null and alternative hypotheses for this test.

b. Compute the test statistic.

c. The researchers reported a *p*-value ≈ 0 for the test. Do you agree?

d. Make the appropriate conclusion in the words of the problem. Use $\alpha = .01$.

*6.122 Lunch at McDonald's. According to the National Restaurant Association, hamburgers are the number-one selling fast-food item in the United States. An economist studying the fast-food buying habits of Americans paid graduate students to stand outside two suburban Boston McDonald's restaurants and ask departing customers whether they spent more than $2.25 on hamburger products for their lunch. Twenty answered yes, 50 said no, and 10 refused to answer the question. (*Newark Star-Ledger*, Mar. 17, 1997.)

a. Apply the sign test to determine (at $\alpha = 05$); whether the median amount spent for hamburgers at lunch at McDonald's is less than $2.25.

b. Does your conclusion apply to all Americans who eat lunch at McDonald's? Justify your answer.

c. What assumptions must hold to ensure the validity of your test in part **a**?

6.123 Alkalinity of river water. In Exercise 4.175 (p. 245), you learned that the mean alkalinity level of water specimens collected from the Han River in Seoul, Korea, is 50 milligrams per liter. (*Environmental Science & Engineering*, September 1, 2000.) Consider a random sample of 100 water specimens collected from a tributary of the Han River. Suppose the mean and standard deviation of the alkalinity levels for the sample are, respectively, $\bar{x} = 67.8$ mpl and $s = 14.4$ mpl. Is there sufficient evidence (at $\alpha = .01$) to indicate that the population mean alkalinity level of water in the tributary exceeds 50 mpl?

Applying the Concepts—Intermediate

6.124 Errors in medical tests. Medical tests have been developed to detect many serious diseases. A medical test is designed to minimize the probability that it will produce a "false positive" or a "false negative." A false positive is a positive test result for an individual who does not have the disease, whereas a false negative is a negative test result for an individual who does have the disease.

a. If we treat a medical test for a disease as a statistical test of hypothesis, what are the null and alternative hypotheses for the medical test?

b. What are the Type I and Type II errors for the test? Relate each to false positives and false negatives.

c. Which of these errors has graver consequences? Considering this error, is it more important to minimize α or β? Explain.

6.125 Crop weights of pigeons. In Exercise 5.95 (p. 293), you analyzed data from a study of the diets of spinifex pigeons. The data, extracted from the *Australian Journal of Zoology*, are reproduced in the accompanying table. Each measurement in the data set represents the weight (in grams) of the contents of the crop of a spinifex pigeon. Conduct a test at $\alpha = .05$ to determine whether the mean weight of the crops of all spinifex pigeons differs from 1 gram. Test the validity of any assumptions you make.

PIGEONS

.457	3.751	.238	2.967	2.509	1.384	1.454	.818
.335	1.436	1.603	1.309	.201	.530	2.144	.834

6.126 Cracks in highway pavement. Using van-mounted state-of-the-art video technology the Mississippi Department of Transportation collected data on the number of cracks (called *crack intensity*) in an undivided two-lane highway. (*Journal of Infrastructure Systems*, March 1995.) The mean number of cracks found in a sample of eight 50-meter sections of the highway was $\bar{x} = .210$, with a variance of $s^2 = .011$. Suppose the American Association of State Highway and Transportation Officials (AASHTO) recommends a maximum mean crack intensity of .100 for safety purposes.

a. Test the hypothesis that the true mean crack intensity of the Mississippi highway exceeds the AASHTO recommended maximum. Use $\alpha = .01$.

b. Define a Type I error and a Type II error for this study.

6.127 Inbreeding of tropical wasps. Refer to the *Science* (November 1988) study of inbreeding in tropical swarm-founding wasps, presented in Exercise 5.101 (p. 294). A sample of 197 wasps, captured, frozen, and subjected to a series of genetic tests, yielded a sample mean inbreeding coefficient of $\bar{x} = .044$ with a standard deviation of $s = .884$. Recall that if the wasp has no tendency to inbreed, the true mean inbreeding coefficient μ for the species will equal 0.

a. Test the hypothesis that the true mean inbreeding coefficient μ for this species of wasp exceeds 0. Use $\alpha = .05$.

b. Compare the inference you made in part **a** with the inference you obtained in Exercise 5.101, using a confidence interval. Do the inferences agree? Explain.

*6.128 Fluoride in drinking water. Many water treatment facilities supplement the natural fluoride concentration with hydrofluosilicic acid in order to reach a target concentration of fluoride in drinking water. Certain levels are thought to enhance dental health, but very high concentrations can be dangerous. Suppose that one such treatment plant targets .75 milligrams per liter (mg/L) for their water. The plant tests 25 samples each day to determine whether the median level differs from the target.

a. Set up the null and alternative hypotheses.

b. Set up the test statistic and rejection region using $\alpha = .10$.

c. Explain the implication of a Type I error in the context of this application. A Type II error.

d. Suppose that one day's samples result in 18 values that exceed .75 mg/L. Conduct the test and state the appropriate conclusion in the context of this application.

e. When it was suggested to the plant's supervisor that a *t*-test should be used to conduct the daily test, she replied that the probability distribution of the fluoride concentrations was "heavily skewed to the right." Show graphically what she meant by this, and explain why this is a reason to prefer the sign test to the *t*-test.

6.129 **PCB in plant discharge.** The EPA sets a limit of 5 parts per million (ppm) on PCB (polychlorinated biphenyl, a dangerous substance) in water. A major manufacturing firm producing PCB for electrical insulation discharges small amounts from the plant. The company management, attempting to control the PCB in its discharge, has given instructions to halt production if the mean amount of PCB in the effluent exceeds 3 ppm. A random sample of 50 water specimens produced the following statistics: $\bar{x} = 3.1$ ppm and $s = .5$ ppm.

a. Do these statistics provide sufficient evidence to halt the production process? Use $\alpha = .01$.

b. If you were the plant manager, would you want to use a large or a small value for α for the test in part **a**?

6.130 **Psychological study of obese adolescents.** Research reported in the *Journal of Psychology* (March 1991) studied the personality characteristics of obese individuals. One variable, the locus of control (LOC), measures the individual's degree of belief that he or she has control over situations. High scores on the LOC scale indicate *less* perceived control. For one sample of 19 obese adolescents, the mean LOC score was 10.89, with a standard deviation of 2.48. Suppose we wish to test whether the mean LOC score for all obese adolescents exceeds 10, the average for "normal" individuals.

a. Specify the null and alternative hypotheses for this test.

b. Conduct the test mentioned in part **a**, and make the proper conclusion.

6.131 **Choosing portable grill displays.** Refer to the *Journal of Consumer Research* (March 2003) experiment on influencing the choices of others by offering undesirable alternatives, presented in Exercise 3.25 (p. 126). Recall that each of 124 college students selected three portable grills out of five to display on the showroom floor. The students were instructed to include Grill #2 (a smaller-sized grill) and select the remaining two grills in the display to maximize purchases of Grill #2. If the six possible grill display combinations (1–2–3, 1–2–4, 1–2–5, 2–3–4, 2–3–5, and 2–4–5) are selected at random, then the proportion of students selecting any display will be $1/6 = .167$. One theory tested by the researcher is that the students will tend to choose the three-grill display so that Grill #2 is a compromise between a more desirable and a less desirable grill. Of the 124 students, 85 students selected a three-grill display that was consistent with that theory. Use this information to test the theory proposed by the researcher at $\alpha = .05$.

Applying the Concepts—Advanced

6.132 **NCAA March Madness.** For three weeks each March, the National Collegiate Athletic Association (NCAA) holds its annual men's basketball championship tournament. The 64 best college basketball teams in the nation play a single-elimination tournament—a total of 63 games—to determine the NCAA champion. Tournament followers, from hard-core gamblers to the casual fan who enters the office betting pool, have a strong interest in handicapping the games. To provide insight into this phenomenon, statisticians Hal Stern and Barbara Mock analyzed data from 13 previous NCAA tournaments and published their results in *Chance* (Winter 1998). The results of first-round games are summarized in the table on p. 347.

a. A common perception among fans, the media, and gamblers is that the higher seeded team has a better than 50–50 chance of winning a first-round game. Is there evidence to support this perception? Conduct the appropriate test for each matchup. What trends do you observe?

b. Is there evidence to support the claim that a 1-, 2-, 3-, or 4-seeded team will win by an average of more than 10 points in first-round games? Conduct the appropriate test for each matchup.

c. Is there evidence to support the claim that a 5-, 6-, 7-, or 8-seeded team will win by an average of less than 5 points in first-round games? Conduct the appropriate test for each matchup.

d. The researchers also calculated the difference between the outcome of the game (victory margin, in points) and the point spread established by Las Vegas oddsmakers for a sample of 360 recent NCAA tournament games. The mean difference is .7 and the standard deviation of the difference is 11.3. If the true mean difference is 0, then the point spread can be considered a good predictor of the outcome of the game. Use this sample information to test the hypothesis that the point spread, on average, is a good predictor of the victory margin in NCAA tournament games.

6.133 **Polygraph test error rates.** A group of physicians subjected the *polygraph* (or *lie detector*) to the same careful testing given to medical diagnostic tests. They found that if 1,000 people were subjected to the polygraph and 500 told the truth and 500 lied, the polygraph would indicate that approximately 185 of the truth tellers were liars and that approximately 120 of the liars were truth tellers (*Discover*, 1986).

a. In the application of a polygraph test, an individual is presumed to be a truth teller (H_0) until "proven" a liar (H_a). In this context, what is a Type I error? A Type II error?

b. According to the study, what is the probability (approximately) that a polygraph test will result in a Type I error? A Type II error?

Critical Thinking Challenges

6.134 **The Hot Tamale caper.** "Hot Tamales" are chewy, cinnamon-flavored candies. A bulk vending machine is

Summary of First-Round NCAA Tournament Games, 1985–1997

Matchup (Seeds)	Number of Games	Number Won by favorite (Higher Seed)	Margin of Victory (Points)	
			Mean	Standard Deviation
1 vs 16	52	52	22.9	12.4
2 vs 15	52	49	17.2	11.4
3 vs 14	52	41	10.6	12.0
4 vs 13	52	42	10.0	12.5
5 vs 12	52	37	5.3	10.4
6 vs 11	52	36	4.3	10.7
7 vs 10	52	35	3.2	10.5
8 vs 9	52	22	−2.1	11.0

Source: Stern, H. S., and Mock, B. "College basketball upsets: Will a 16-seed ever beat a 1-seed?" *Chance,* Vol. 11, No. 1, Winter 1998, p. 29 (Table 3).

known to dispense, on average, 15 Hot Tamales per bag. *Chance* (Fall 2000) published an article on a classroom project in which students were required to purchase bags of Hot Tamales from the machine and count the number of candies per bag. One student group claimed it purchased five bags that had the following candy counts: 25, 23, 21, 21, and 20. There was some question as to whether the students had fabricated the data. Use a hypothesis test to gain insight into whether or not the data collected by the students were fabricated. Use a level of significance that gives the benefit of the doubt to the students.

6.135 Verifying voter petitions. To get their names on the ballot of a local election, political candidates often must obtain petitions bearing the signatures of a minimum number of registered voters. In Pinellas County, Florida, a certain political candidate obtained petitions with 18,200 signatures. (*St. Petersburg Times*, April 7, 1992.) To verify that the names on the petitions were signed by actual registered voters, election officials randomly sampled 100 of the signatures and checked each for authenticity. Only 2 were invalid signatures.

a. Is 98 out of 100 verified signatures sufficient to believe that more than 17,000 of the total 18,200 signatures are valid? Use $\alpha = .01$.

b. Repeat part **a** if only 16,000 valid signatures are required.

SUMMARY ACTIVITY: Testing the "Efficient Market" Theory

The "efficient market" theory postulates that the best predictor of a stock's price at some point in the future is the current price of the stock (with some adjustments for inflation and transaction costs, which we shall assume to be negligible for the purpose of this exercise). To test the theory, select a random sample of 25 stocks on the New York Stock Exchange and record the closing prices on the last days of two recent consecutive months. Calculate the increase or decrease in the stock price over the 1-month period.

a. Define μ as the mean change in price of all stocks over a 1-month period. Set up the appropriate null and alternative hypotheses in terms of μ.

b. Use the sample of 25 stock price differences to conduct the test of hypothesis established in part **a**. Take $\alpha = .05$.

c. Tabulate the number of rejections and nonrejections of the null hypothesis in your class. If the null hypothesis were true, how many rejections of the null hypothesis would you expect among those in your class? How does this expectation compare with the actual number of nonrejections? What does the result of this exercise indicate about the efficient market theory?

REFERENCES

Daniel, W. W. *Applied Nonparametric Statistics,* 2nd ed. Boston: PWS-Kent, 1990.

Hollander, M. and Wolfe, D. A. *Nonparametric Statistical Methods,* 7th ed. Ames: Iowa State University Press, 1980.

Snedecor, G. W., and Cochran, W. G. *Statistical Methods,* 7th ed. Ames: Iowa State University Press, 1980.

Wackerly, D., Mendenhall, W., and Scheaffer, R. *Mathematical Statistics with Applications,* 6th ed. North Scituate, MA: Duxbury, 2002.

Using Technology

One-Sample Tests of Hypothesis with MINITAB

MINITAB can be used to obtain one-sample tests for a population mean, a population proportion, and a population median.

Testing μ: To generate a hypothesis test for the mean with the use of a previously created sample data set, first access the MINITAB data worksheet. Next, click on the "Stat" button on the MINITAB menu bar, and then click on "Basic Statistics" and "1-Sample t," as shown in Figure 6.M.1. The dialog box shown in Figure 6.M.2 appears. Click on "Samples in Columns"; then specify the quantitative variable of interest in the open box. Next, check "Perform hypothesis test" and specify the value of μ_0 for the null hypothesis in the open box. Now click on the "Options" button at the bottom of the dialog box and specify the form of the alternative hypothesis, as shown in Figure 6.M.3. Click "OK" to return to the "1-Sample t" dialog box, and then click "OK" again to produce the hypothesis test.

If you want to produce a test for the mean from summary information (e.g., the sample mean, sample standard deviation, and sample size), then click on "Summarized data" in the "1-Sample t" dialog box, enter the values of the summary statistics and μ_0, and then click "OK."

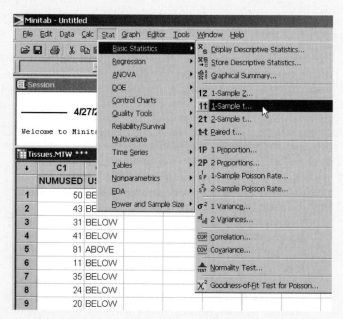

FIGURE 6.M.1
MINITAB menu options for a test about μ

[*Important Note*: The MINITAB one-sample *t*-procedure uses the *t*-statistic to generate the hypothesis test. When the sample size *n* is small, this is the appropriate method. When the sample size *n* is large, the *t*-value will be approximately equal to the large-sample *z*-value and the resulting test will still be valid. If you have a large sample and you know the value of the population standard deviation σ (which is rarely the case), then select "1-sample Z" from the "Basic Statistics" menu options (see Figure 6.M.1) and make the appropriate selections.]

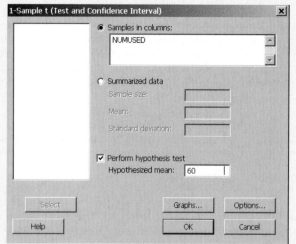

FIGURE 6.M.2
MINITAB one-sample t dialog box

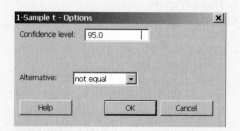

FIGURE 6.M.3
MINITAB one-sample t options

Testing *p*: To generate a test for a population proportion with the use of a previously created sample data set, first access the MINITAB data worksheet. Next, click on the "Stat" button on the MINITAB menu bar, and then click on "Basic Statistics" and "1 Proportion." (See Figure 6.M.1.) The dialog box shown in Figure 6.M.4 appears. Click on "Samples in Columns"; then specify the qualitative variable of interest in the open box. Select "Perform hypothesis test" and enter the null-hypothesis value p_0 in the open box. Click on the "Options" button at the bottom of the dialog box and specify the form of the alternative hypothesis in the resulting dialog box, as shown in Figure 6.M.5. Also, check the "Use test and interval based on normal distribution" box at the bottom. Click "OK" to return to the "1-Proportion" dialog box, and then click "OK" again to produce the test results.

If you want to produce a test for a proportion from summary information (e.g., the number of successes and the sample size), then click on "Summarized data" in the "1-Proportion" dialog box. (See Figure 6.M.4.) Enter the value for the number of trials (i.e., the sample size) and the number of events (i.e., the number of successes), and then click "OK."

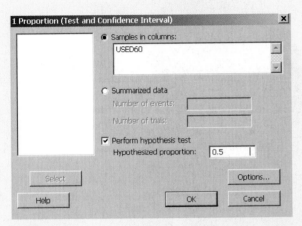

FIGURE 6.M.4
MINITAB one-proportion dialog box

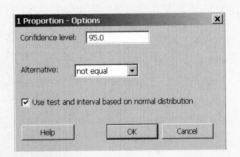

FIGURE 6.M.5
MINITAB one-proportion options

Sign Test

The MINITAB worksheet file with the sample data should contain a single quantitative variable. Click on the "Stat" button on the MINITAB menu bar, and then click on "Nonparametrics" and "1-Sample Sign", as shown in Figure 6.M.6.

The dialog box shown in Figure 6.M.7 then appears. Enter the quantitative variable to be analyzed in the "Variables" box. Select the "Test median" option, and specify the hypothesized value of the median and the form of the alternative hypothesis ("not equal", "less than", or "greater than"). Click "OK" to generate the MINITAB printout.

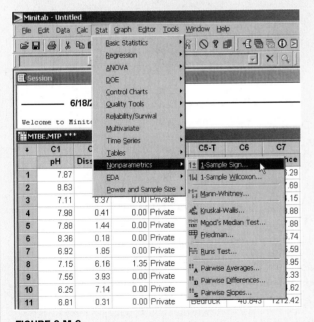

FIGURE 6.M.6
MINITAB nonparametric menu options

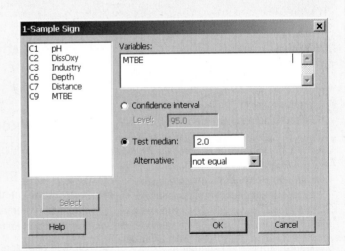

FIGURE 6.M.7
MINITAB 1-sample sign dialog box

Comparing Population Means

CONTENTS

7.1 Comparing Two Population Means: Independent Sampling

7.2 Comparing Two Population Means: Paired Difference Experiments

7.3 Determining the Sample Size

7.4 A Nonparametric Test for Comparing Two Populations: Independent Sampling (Optional)

7.5 A Nonparametric Test for Comparing Two Populations: Paired Difference Experiment (Optional)

7.6 Comparing Three or More Population Means: Analysis of Variance (Optional)

STATISTICS IN ACTION
On the Trail of the Cockroach: Do Roaches Travel at Random?

USING TECHNOLOGY
Comparing Means Using MINITAB

WHERE WE'VE BEEN

■ Explored two methods for making statistical inferences: *confidence intervals* and *tests of hypotheses*

■ Studied confidence intervals and tests for a single population mean μ and a single population proportion p

■ Learned how to select the sample size necessary to stimate a population parameter with a specified margin of error

WHERE WE'RE GOING

■ Learn how to compare two populations using confidence intervals and tests of hypotheses.

■ Apply these inferential methods to problems where we want to compare two (or more) population means.

■ Determine the sizes of the samples necessary to estimate the difference between two population means with a specified margin of error.

STATISTICS IN ACTION

On the Trail of the Cockroach: Do Roaches Travel at Random?

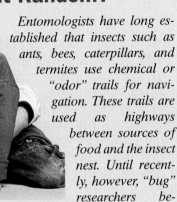

Entomologists have long established that insects such as ants, bees, caterpillars, and termites use chemical or "odor" trails for navigation. These trails are used as highways between sources of food and the insect nest. Until recently, however, "bug" researchers believed that the navigational behavior of cockroaches scavenging for food was random and not linked to a chemical trail.

One of the first researchers to challenge the "random-walk" theory for cockroaches was professor and entomologist Dini Miller of Virginia Tech University. According to Miller, "The idea that roaches forage randomly means that they would have to come out of their hiding places every night and bump into food and water by accident. But roaches never seem to go hungry." Since cockroaches had never before been evaluated for trail following behavior, Miller designed an experiment to test a cockroach's ability to follow a trail of their fecal material (*Explore,* Research at the University of Florida, Fall 1998).

First, Dr. Miller developed a methanol extract from roach feces—called a pheromone. She theorized that "pheromones are communication devices between cockroaches. If you have an infestation and have a lot of fecal material around, it advertises, 'Hey, this is a good cockroach place.'" Then she created a chemical trail with the pheromone on a strip of white chromatography paper and placed the paper at the bottom of a plastic, V-shaped container, 122 square centimeters in size. German cockroaches were released into the container at the beginning of the trail, one at a time, and a video surveillance camera was used to monitor the roach's movements.

In addition to the trail containing the fecal extract (the treatment), a trail using methanol only was created. This second trail served as a control to compare back against the treated trail. Because Dr. Miller also wanted to determine if trail-following ability differed among cockroaches of different age, sex, and reproductive status, four roach groups were utilized in the experiment: adult males, adult females, gravid (pregnant) females, and nymphs (immatures). Twenty roaches of each type were randomly assigned to the treatment trail and ten of each type were randomly assigned to the control trail. Thus, a total of 120 roaches were used in the experiment.

The movement pattern of each cockroach tested was translated into xy coordinates every one-tenth of a second by the Dynamic Animal Movement Analyzer (DAMA) program. Miller measured the perpendicular distance of each xy-coordinate from the trail and then averaged these distances, or deviations, for each cockroach. The average trail deviations (measured in pixels, where 1 pixel equals approximately 2 centimeters) for each of the 120 cockroaches in the study are stored in the data file named **ROACH.**

We apply the statistical methodology presented in this chapter to the cockroach data in the following Statistics in the Action Revisited sections.

Statistics in Action Revisited

- Comparing Roach Trails: Mean Treatment vs. Mean Control (p. 364)

- Comparing Treatment and Control Roach trails using a Nonparametric Test (p. 391)

- Analysis of Variance on the Cockroach Data (p. 413)

Many experiments involve a comparison of population means. For instance, a sociologist may want to estimate the difference in mean life expectancy between innercity and suburban residents. A consumer group may want to test whether two major brands of food freezers differ in the average amount of electricity they use. A professional golfer might be interested in comparing the mean driving distances of several competing brands of golf balls. In this chapter, we consider techniques for using two (or more) samples to compare the populations from which they were selected.

7.1 Comparing Two Population Means: Independent Sampling

In this section we develop both large-sample and small-sample methodologies for comparing two population means. In the large-sample case we use the z-statistic, while in the small-sample case we use the t-statistic.

Large Samples:

EXAMPLE 7.1

A LARGE-SAMPLE CONFIDENCE INTERVAL FOR $(\mu_1 - \mu_2)$: Comparing Mean Weight Loss for Two Diets

Problem A dietitian has developed a diet that is low in fats, carbohydrates, and cholesterol. Although the diet was initially intended to be used by people with heart disease, the dietitian wishes to examine the effect this diet has on the weights of obese people. Two random samples of 100 obese people each are selected, and one group of 100 is placed on the low-fat diet. The other 100 are placed on a diet that contains approximately the same quantity of food, but is not as low in fats, carbohydrates, and cholesterol. For each person, the amount of weight lost (or gained) in a three-week period is recorded. The data, saved in the **DIETSTUDY** file, are listed in Table 7.1. Form a 95% confidence interval for the difference between the population mean weight losses for the two diets. Interpret the result.

 DIETSTUDY

TABLE 7.1 Diet Study Data, Example 7.1

Weight Losses for Low-Fat Diet

8	10	10	12	9	3	11	7	9	2
21	8	9	2	2	20	14	11	15	6
13	8	10	12	1	7	10	13	14	4
8	12	8	10	11	19	0	9	10	4
11	7	14	12	11	12	4	12	9	2
4	3	3	5	9	9	4	3	5	12
3	12	7	13	11	11	13	12	18	9
6	14	14	18	10	11	7	9	7	2
16	16	11	11	3	15	9	5	2	6
5	11	14	11	6	9	4	17	20	10

Weight Losses for Regular Diet

6	6	5	5	2	6	10	3	9	11
14	4	10	13	3	8	8	13	9	3
4	12	6	11	12	9	8	5	8	7
6	2	6	8	5	7	16	18	6	8
13	1	9	8	12	10	6	1	0	13
11	2	8	16	14	4	6	5	12	9
11	6	3	9	9	14	2	10	4	13
8	1	1	4	9	4	1	1	5	6
14	0	7	12	9	5	9	12	7	9
8	9	8	10	5	8	0	3	4	8

Solution Recall that the general form of a large-sample confidence interval for a single mean μ is $\bar{x} \pm z_{\alpha/2}\sigma_{\bar{x}}$. That is, we add and subtract $z_{\alpha/2}$ standard deviations of the sample estimate $\bar{x}$ to the value of the estimate. We employ a similar procedure to form the confidence interval for the difference between two population means.

Let μ_1 represent the mean of the conceptual population of weight losses for all obese people who could be placed on the low-fat diet. Let μ_2 be similarly defined for the other diet. We wish to form a confidence interval for $(\mu_1 - \mu_2)$. An intuitively appealing estimator for $(\mu_1 - \mu_2)$ is the difference between the sample means, $(\bar{x}_1 - \bar{x}_2)$. Thus, we will form the confidence interval of interest with

$$(\bar{x}_1 - \bar{x}_2) \pm z_{\alpha/2}\sigma_{(\bar{x}_1 - \bar{x}_2)}$$

Group Statistics

	DIET	N	Mean	Std. Deviation	Std. Error Mean
WTLOSS	LOWFAT	100	9.31	4.668	.467
	REGULAR	100	7.40	4.035	.404

Independent Samples Test

		Levene's Test for Equality of Variances		t-test for Equality of Means					95% Confidence Interval of the Difference	
		F	Sig.	t	df	Sig. (2-tailed)	Mean Difference	Std. Error Difference	Lower	Upper
WTLOSS	Equal variances assumed	1.367	.244	3.095	198	.002	1.910	.617	.693	3.127
	Equal variances not assumed			3.095	193.940	.002	1.910	.617	.693	3.127

FIGURE 7.1

SPSS analysis of diet study data.

Assuming that the two samples are independent, we write the standard deviation of the difference between the sample means as

$$\sigma_{(\bar{x}_1 - \bar{x}_2)} = \sqrt{\frac{\sigma_1^2}{n_1} + \frac{\sigma_2^2}{n_2}} \approx \sqrt{\frac{s_1^2}{n_1} + \frac{s_2^2}{n_2}}$$

Summary statistics for the diet data are displayed at the top of the SPSS printout shown in Figure 7.1. Note that $\bar{x}_1 = 9.31$, $\bar{x}_2 = 7.40$, $s_1 = 4.67$, and $s_2 = 4.04$. Using these values and observing that $\alpha = .05$ and $z_{.025} = 1.96$, we find that the 95% confidence interval is, approximately,

$$(9.31 - 7.40) \pm 1.96 \sqrt{\frac{(4.67)^2}{100} + \frac{(4.04)^2}{100}} = 1.91 \pm (1.96)(.62) = 1.91 \pm 1.22$$

or $(.69, 3.13)$. This interval (rounded) is highlighted in Figure 7.1.

Using this estimation procedure over and over again for different samples, we know that approximately 95% of the confidence intervals formed in this manner will enclose the difference in population means $(\mu_1 - \mu_2)$. Therefore, we are highly confident that the mean weight loss for the low-fat diet is between .69 and 3.13 pounds more than the mean weight loss for the other diet. With this information, the dietitian better understands the potential of the low-fat diet as a weight-reduction diet.

Look Back If the confidence interval for $(\mu_1 - \mu_2)$ contains 0 [e.g., $(-2.5, 1.3)$], then it is possible for the difference between the population means to be 0 (i.e., $\mu_1 - \mu_2 = 0$). In this case, we could not conclude that a significant difference exists between the mean weight losses for the two diets.

Now Work Exercise 7.6a

The justification for the procedure used in Example 7.1 to estimate $(\mu_1 - \mu_2)$ relies on the properties of the sampling distribution of $(\bar{x}_1 - \bar{x}_2)$. The performance of the estimator in repeated sampling is pictured in Figure 7.2, and its properties are summarized in the following box:

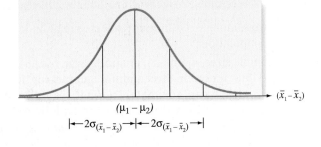

FIGURE 7.2

Sampling distribution of $(\bar{x}_1 - \bar{x}_2)$

Properties of the Sampling Distribution of $(\bar{x}_1 - \bar{x}_2)$

1. The mean of the sampling distribution of $(\bar{x}_1 - \bar{x}_2)$ is $(\mu_1 - \mu_2)$.
2. If the two samples are independent, the standard deviation of the sampling distribution is

$$\sigma_{(\bar{x}_1 - \bar{x}_2)} = \sqrt{\frac{\sigma_1^2}{n_1} + \frac{\sigma_2^2}{n_2}}$$

where σ_1^2 and σ_2^2 are the variances of the two populations being sampled and n_1 and n_2 are the respective sample sizes. We also refer to $\sigma_{(\bar{x}_1 - \bar{x}_2)}$ as the **standard error** of the statistic $(\bar{x}_1 - \bar{x}_2)$.
3. By the central limit theorem, the sampling distribution of $(\bar{x}_1 - \bar{x}_2)$ is approximately normal *for large samples*.

In Example 7.1, we noted the similarity in the procedures for forming a large-sample confidence interval for one population mean and a large-sample confidence interval for the difference between two population means. When we are testing hypotheses, the procedures are again similar. The general large-sample procedures for forming confidence intervals and testing hypotheses about $(\mu_1 - \mu_2)$ are summarized in the following boxes:

Large Sample Confidence Interval for $(\mu_1 - \mu_2)$

$$(\bar{x}_1 - \bar{x}_2) \pm z_{\alpha/2}\sigma_{(\bar{x}_1 - \bar{x}_2)} = (\bar{x}_1 - \bar{x}_2) \pm z_{\alpha/2}\sqrt{\frac{\sigma_1^2}{n_1} + \frac{\sigma_2^2}{n_2}}$$

$$\approx (\bar{x}_1 - \bar{x}_2) \pm z_{\alpha/2}\sqrt{\frac{s_1^2}{n_1} + \frac{s_2^2}{n_2}}$$

Large-Sample Test of Hypothesis for $(\mu_1 - \mu_2)$

One-Tailed Test

$H_0: (\mu_1 - \mu_2) = D_0$
$H_a: (\mu_1 - \mu_2) < D_0$
[or $H_a: (\mu_1 - \mu_2) > D_0$]

Two-Tailed Test

$H_0: (\mu_1 - \mu_2) = D_0$
$H_a: (\mu_1 - \mu_2) \neq D_0$

where D_0 = Hypothesized difference between the means (this difference is often hypothesized to be equal to 0)

Test statistic:

$$z = \frac{(\bar{x}_1 - \bar{x}_2) - D_0}{\sigma_{(\bar{x}_1 - \bar{x}_2)}} \quad \text{where} \quad \sigma_{(\bar{x}_1 - \bar{x}_2)} = \sqrt{\frac{\sigma_1^2}{n_1} + \frac{\sigma_2^2}{n_2}} \approx \sqrt{\frac{s_1^2}{n_1} + \frac{s_2^2}{n_2}}$$

Rejection region: $z < -z_\alpha$
[or $z > z_\alpha$ when
$H_a: (\mu_1 - \mu_2) > D_0$]

Rejection region: $|z| > z_{\alpha/2}$

Conditions Required for Valid Large-Sample Inferences about $(\mu_1 - \mu_2)$

1. The two samples are randomly selected in an independent manner from the two target populations.
2. The sample sizes, n_1 and n_2, are both large (i.e., $n_1 \geq 30$ and $n_2 \geq 30$). (By the central limit theorem, this condition guarantees that the sampling distribution of $(\bar{x}_1 - \bar{x}_2)$ will be approximately normal, regardless of the shapes of the underlying probability distributions of the populations. Also, s_1^2 and s_2^2 will provide good approximations to σ_1^2 and σ_2^2 when both samples are large.)

EXAMPLE 7.2

A LARGE-SAMPLE TEST FOR $(\mu_1 - \mu_2)$: Comparing Mean Weight Loss for Two Diets:

Problem Refer to the study of obese people on a low-fat diet and a regular diet presented in Example 7.1. Another way to compare the mean weight losses for the two different diets is to conduct a test of hypothesis. Use the information on the SPSS printout shown in Figure 7.1 to conduct the test. Take $\alpha = .05$.

Solution Again, we let μ_1 and μ_2 represent the population mean weight losses of obese people on the low-fat diet and regular diet, respectively. If one diet is more effective in reducing the weights of obese people, then either $\mu_1 < \mu_2$ or $\mu_2 < \mu_1$; that is, $\mu_1 \neq \mu_2$.

Thus, the elements of the test are as follows:

$H_0: (\mu_1 - \mu_2) = 0$ (i.e., $\mu_1 = \mu_2$; note that $D_0 = 0$ for this hypothesis test)

$H_a: (\mu_1 - \mu_2) \neq 0$ (i.e., $\mu_1 \neq \mu_2$)

Test statistic: $z = \dfrac{(\bar{x}_1 - \bar{x}_2) - D_0}{\sigma_{(\bar{x}_1 - \bar{x}_2)}} = \dfrac{\bar{x}_1 - \bar{x}_2 - 0}{\sigma_{(\bar{x}_1 - \bar{x}_2)}}$

Rejection region: $z < -z_{\alpha/2} = -1.96$ or $z > z_{\alpha/2} = 1.96$ (see Figure 7.3)

Substituting the summary statistics given in Figure 7.1 into the test statistic, we obtain

$$z = \frac{(\bar{x}_1 - \bar{x}_2) - 0}{\sigma_{(\bar{x}_1 - \bar{x}_2)}} = \frac{9.31 - 7.40}{\sqrt{\dfrac{\sigma_1^2}{n_1} + \dfrac{\sigma_2^2}{n_2}}}$$

$$\approx \frac{1.91}{\sqrt{\dfrac{s_1^2}{n_1} + \dfrac{s_2^2}{n_2}}} = \frac{1.91}{\sqrt{\dfrac{(4.67)^2}{100} + \dfrac{(4.04)^2}{100}}} = \frac{1.91}{.617} = 3.09$$

[*Note:* The value of the test statistic is highlighted in the SPSS printout of Figure 7.1.]

As you can see in Figure 7.3, the calculated z-value clearly falls into the rejection region. Therefore, the samples provide sufficient evidence, at $\alpha = .05$, for the dietitian to conclude that the mean weight losses for the two diets differ.

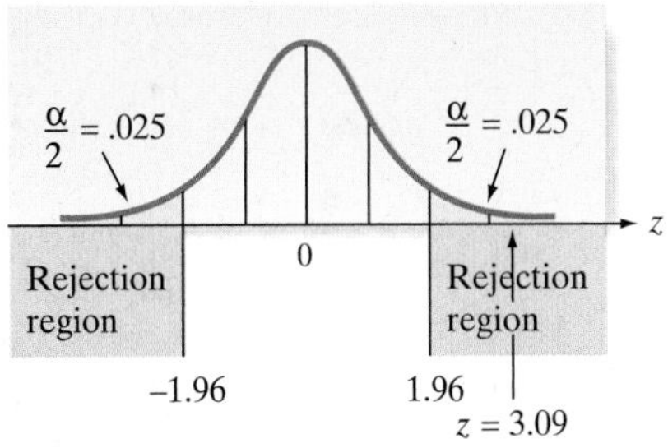

FIGURE 7.3
Rejection region for Example 7.2

Look Back This conclusion agrees with the inference drawn from the 95% confidence interval in Example 7.1. However, the confidence interval provides more information on the mean weight losses. From the hypothesis test, we know only that the two means differ; that is, $\mu_1 \neq \mu_2$. From the confidence interval in Example 7.1, we found that the mean weight loss μ_1 of the low-fat diet was between .69 and 3.13 pounds more than the mean weight loss μ_2 of the regular diet. In other words, the test tells us that the means differ, but the confidence interval tells us how large the difference is. Both inferences are made with the same degree of reliability—namely, 95% confidence (or at $\alpha = .05$).

EXAMPLE 7.3

THE p-VALUE FOR A TEST OF $(\mu_1 - \mu_2)$

Problem Find the observed significance level for the test in Example 7.2. Interpret the result.

Solution The alternative hypothesis in Example 7.2, $H_a: \mu_1 - \mu_2 \neq 0$, required a two-tailed test using

$$z = \frac{\bar{x}_1 - \bar{x}_2}{\sigma_{(\bar{x}_1 - \bar{x}_2)}}$$

as a test statistic. Since the z-value calculated from the sample data was 3.09, the observed significance level (p-value) for the two-tailed test is the probability of observing a value of z at least as contradictory to the null hypothesis as $z = 3.09$; that is,

$$p\text{-value} = 2 \cdot P(z \geq 3.09)$$

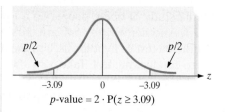

FIGURE 7.4

The observed significance level
for Example 7.2

p-value $= 2 \cdot P(z \geq 3.09)$

This probability is computed under the assumption that H_0 is true and is equal to the highlighted area shown in Figure 7.4.

The tabulated area corresponding to $z = 3.09$ in Table III of Appendix A is .4990. Therefore,

$$P(z \geq 3.09) = .5 - .4990 = .0010$$

and the observed significance level for the test is

$$p\text{-value} = 2(.001) = .002$$

Since our selected α value, .05, exceeds this p-value, we have sufficient evidence to reject $H_0: \mu_1 - \mu_2 = 0$.

Look Back The p-value of the test is more easily obtained from a statistical software package. The p-value is highlighted at the bottom of the SPSS printout shown in Figure 7.1. This value agrees with our calculated p-value.

Now Work Exercise 7.6b

Small Samples

In comparing two population means with small samples (say, $n_1 < 30$ and $n_2 < 30$), the methodology of the previous three examples is invalid. The reason? When the sample sizes are small, estimates of σ_1^2 and σ_2^2 are unreliable and the central limit theorem (which guarantees that the z statistic is normal) can no longer be applied. But as in the case of a single mean (Section 6.4), we use the familiar Student's t-distribution.

To use the t-distribution, both sampled populations must be approximately normally distributed with equal population variances, and the random samples must be selected independently of each other. The assumptions of normality and equal variances imply relative frequency distributions for the populations that would appear as shown in Figure 7.5.

Since we assume that the two populations have equal variances ($\sigma_1^2 = \sigma_2^2 = \sigma^2$), it is reasonable to use the information contained in both samples to construct a **pooled sample estimator** σ^2 for use in confidence intervals and test statistics. Thus, if s_1^2 and s_2^2 are the two sample variances (each estimating the variance σ^2 common to both populations), the pooled estimator of σ^2, denoted as s_p^2, is

$$s_p^2 = \frac{(n_1 - 1)s_1^2 + (n_2 - 1)s_2^2}{(n_1 - 1) + (n_2 - 1)} = \frac{(n_1 - 1)s_1^2 + (n_2 - 1)s_2^2}{n_1 + n_2 - 2}$$

or

$$s_p^2 = \frac{\overbrace{\sum(x_1 - \bar{x}_1)^2}^{\text{From sample 1}} + \overbrace{\sum(x_2 - \bar{x}_2)^2}^{\text{From sample 2}}}{n_1 + n_2 - 2}$$

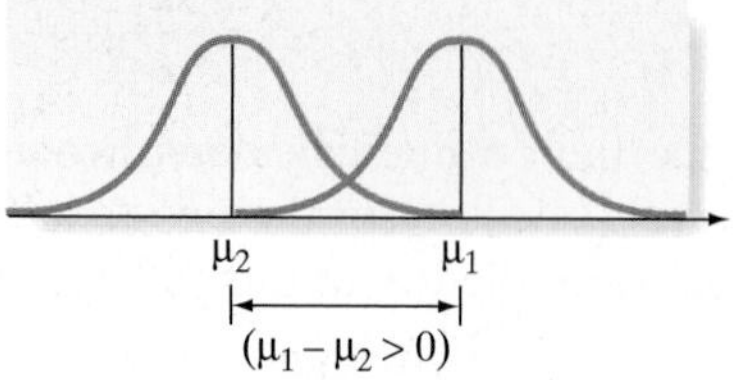

FIGURE 7.5

Assumptions for the two-
sample t: (1) normal
populations; (2) equal
variances

μ_2 μ_1

$(\mu_1 - \mu_2 > 0)$

where x_1 represents a measurement from sample 1 and x_2 represents a measurement from sample 2. Recall that the term *degrees of freedom* was defined in Section 5.3 as 1 less than the sample size. Thus, in this case, we have $(n_1 - 1)$ degrees of freedom for sample 1 and $(n_2 - 1)$ degrees of freedom for sample 2. Since we are pooling the information on σ^2 obtained from both samples, the number of degrees of freedom associated with the pooled variance s_p^2 is equal to the sum of the numbers of degrees of freedom for the two samples, namely, the denominator of s_p^2; that is, $(n_1 - 1) + (n_2 - 1) = n_1 + n_2 - 2$.

Note that the second formula given for s_p^2 shows that the pooled variance is simply a *weighted average* of the two sample variances s_1^2 and s_2^2. The weight given each variance is proportional to its number of degrees of freedom. If the two variances have the same number of degrees of freedom (i.e., *if the sample sizes are equal*), *then the pooled variance is a simple average of the two sample variances*. The result is an average, or "pooled," variance that is a better estimate of σ^2 than either s_1^2 or s_2^2 alone.

Biography

BRADLEY EFRON (1938–PRESENT)— The Bootstrap Method

Bradley Efron was raised in St. Paul, Minnesota, the son of a truck driver who was the amateur statistician for his bowling and baseball leagues. Efron received a B.S. in mathematics from the California Institute of Technology in 1960, but, by his own admission, had no talent for modern abstract math. His interest in the science of statistics developed after he read a book by Harold Cramer from cover to cover. Efron went to the University of Stanford to study statistics, and he earned his Ph.D. there in 1964. He has been a faculty member in Stan-

ford's Department of Statistics since 1966. Over his career, Efron has received numerous awards and prizes for his contributions to modern statistics, including the MacArthur Prize Fellow (1983), the American Statistical Association Wilks Medal (1990), and the Parzen Prize for Statistical Innovation (1998). In 1979, Efron invented a method—called the *bootstrap*—of estimating and testing population parameters in situations in which either the sampling distribution is unknown or the assumptions are violated. The method involves repeatedly taking samples of size n (with replacement) from the original sample and calculating the value of the point estimate. Efron showed that the sampling distribution of the estimator is simply the frequency distribution of the bootstrap estimates.

Both the confidence interval and the test-of-hypothesis procedures for comparing two population means with small samples are summarized in the following boxes:

Small-Sample Confidence Interval for $(\mu_1 - \mu_2)$: Independent Samples

$$(\bar{x}_1 - \bar{x}_2) \pm t_{\alpha/2}\sqrt{s_p^2\left(\frac{1}{n_1} + \frac{1}{n_2}\right)}$$

where $s_p^2 = \dfrac{(n_1 - 1)s_1^2 + (n_2 - 1)s_2^2}{n_1 + n_2 - 2}$

and $t_{\alpha/2}$ is based on $(n_1 + n_2 - 2)$ degrees of freedom.

Small-Sample Test of Hypothesis for $(\mu_1 - \mu_2)$: Independent Samples

One-Tailed Test	**Two-Tailed Test**
$H_0: (\mu_1 - \mu_2) = D_0$	$H_0: (\mu_1 - \mu_2) = D_0$
$H_a: (\mu_1 - \mu_2) < D_0$	$H_a: (\mu_1 - \mu_2) \neq D_0$
[or $H_a: (\mu_1 - \mu_2) > D_0$]	

(continued)

Test statistic:

$$t = \frac{(\bar{x}_1 - \bar{x}_2) - D_0}{\sqrt{s_p^2 \left(\dfrac{1}{n_1} + \dfrac{1}{n_2} \right)}}$$

Rejection region: $t < -t_\alpha$

[or $t > t_\alpha$ when

$H_a: (\mu_1 - \mu_2) > D_0$]

Rejection region: $|t| > t_{\alpha/2}$

where t_α and $t_{\alpha/2}$ are based on $(n_1 + n_2 - 2)$ degrees of freedom.

Conditions Required for Valid Small-Sample Inferences about $(\mu_1 - \mu_2)$

1. The two samples are randomly selected in an independent manner from the two target populations.
2. Both sampled populations have distributions that are approximately normal.
3. The population variances are equal (i.e., $\sigma_1^2 = \sigma_2^2$).

EXAMPLE 7.4

A SMALL-SAMPLE CONFIDENCE INTERVAL FOR $(\mu_1 - \mu_2)$: Comparing Two Methods of Teaching

Problem Suppose you wish to compare a new method of teaching reading to "slow learners" with the current standard method. You decide to base your comparison on the results of a reading test given at the end of a learning period of six months. Of a random sample of 22 "slow learners," 10 are taught by the new method and 12 are taught by the standard method. All 22 children are taught by qualified instructors under similar conditions for the designated six-month period. The results of the reading test at the end of this period are given in Table 7.2.

 READING

TABLE 7.2 Reading Test Scores for Slow Learners

New Method				Standard Method			
80	80	79	81	79	62	70	68
76	66	71	76	73	76	86	73
70	85			72	68	75	66

a. Use the data in the table to estimate the true mean difference between the test scores for the new method and the standard method. Use a 95% confidence interval.

b. Interpret the interval you found in part **a.**

c. What assumptions must be made in order that the estimate be valid? Are they reasonably satisfied?

Solution

a. For this experiment, let μ_1 and μ_2 represent the mean reading test scores of "slow learners" taught with the new and standard methods, respectively. Then the objective is to obtain a 95% confidence interval for $(\mu_1 - \mu_2)$.

The first step in constructing the confidence interval is to obtain summary statistics (e.g., $\bar{x}$ and s) on reading test scores for each method. The data of Table 7.2 were entered into a computer, and SAS was used to obtain these descriptive statistics. The SAS printout appears in Figure 7.6. Note that $\bar{x}_1 = 76.4$, $s_1 = 5.8348$, $\bar{x}_2 = 72.333$, and $s_2 = 6.3437$.

```
               Two Sample t-test for the Means of SCORE within METHOD

Sample Statistics

     Group            N       Mean      Std. Dev.    Std. Error
     ------------------------------------------------------------
     NEW             10       76.4        5.8348        1.8451
     STD             12    72.33333       6.3437        1.8313

Hypothesis Test

     Null hypothesis:       Mean 1 - Mean 2 =  0
     Alternative:           Mean 1 - Mean 2 ^= 0

     If Variances Are      t statistic       Df        Pr > t
     ------------------------------------------------------------
     Equal                    1.552          20         0.1364
     Not Equal                1.564        19.77        0.1336

95% Confidence Interval for the Difference between Two Means

               Lower Limit       Upper Limit
               -----------       -----------
                  -1.40             9.53
```

FIGURE 7.6

SAS printout for Example 7.4

Next, we calculate the pooled estimate of variance to obtain

$$s_p^2 = \frac{(n_1 - 1)s_1^2 + (n_2 - 1)s_2^2}{n_1 + n_2 - 2}$$

$$= \frac{(10 - 1)(5.8348)^2 + (12 - 1)(6.3437)^2}{10 + 12 - 2} = 37.45$$

where s_p^2 is based on $(n_1 + n_2 - 2) = (10 + 12 - 2) = 20$ degrees of freedom. Also, we find $t_{\alpha/2} = t_{.025} = 2.086$ (based on 20 degrees of freedom) from Table VI of Appendix A.

Finally, the 95% confidence interval for $(\mu_1 - \mu_2)$, the difference between mean test scores for the two methods, is

$$(\bar{x}_1 - \bar{x}_2) \pm t_{\alpha/2}\sqrt{s_p^2\left(\frac{1}{n_1} + \frac{1}{n_2}\right)} = (76.4 - 72.33) \pm t_{.025}\sqrt{37.45\left(\frac{1}{10} + \frac{1}{12}\right)}$$

$$= 4.07 \pm (2.086)(2.62)$$

$$= 4.07 \pm 5.47$$

or $(-1.4, 9.54)$. This interval agrees (except for rounding) with the one shown at the bottom of the SAS printout of Figure 7.6.

b. The interval can be interpreted as follows: With a confidence coefficient equal to .95, we estimate that the difference in mean test scores between using the new method of teaching and using the standard method falls into the interval from −1.4 to 9.54. In other words, we estimate (with 95% confidence) the mean test score for the new method to be anywhere from 1.4 points less than, to 9.54 points more than, the mean test score for the standard method. Although the sample means seem to suggest that the new method is associated with a higher mean test score, there is insufficient evidence to indicate that $(\mu_1 - \mu_2)$ differs from 0 because the interval includes 0 as a possible value for $(\mu_1 - \mu_2)$. To demonstrate a difference in mean test scores (if it exists), you could increase the sample size and thereby narrow the width of the confidence interval for $(\mu_1 - \mu_2)$. Alternatively, you can design the experiment differently. This possibility is discussed in the next section.

c. To use the small-sample confidence interval properly, the following assumptions must be satisfied:

1. The samples are randomly and independently selected from the populations of "slow learners" taught by the new method and the standard method.

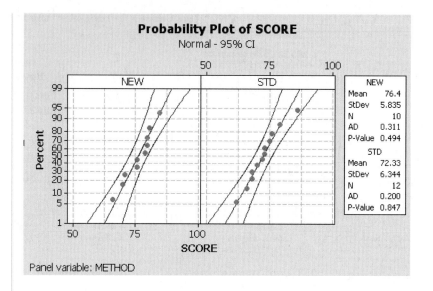

FIGURE 7.7

MINITAB normal probability plots for Example 7.4

2. The test scores are normally distributed for both teaching methods.
3. The variance of the test scores is the same for the two populations; that is, $\sigma_1^2 = \sigma_2^2$.

On the basis of the information provided about the sampling procedure in the description of the problem, the first assumption is satisfied. To check the plausibility of the remaining two assumptions, we resort to graphical methods. Figure 7.7 is a MINITAB printout that gives normal probability plots for the test scores of the two samples of "slow learners." The near straight-line trends on both plots indicate that the distributions of the scores are approximately mound shaped and symmetric. Consequently, each sample data set appears to come from a population that is approximately normal.

One way to check the third assumption is to examine box plots of the sample data.* Figure 7.8 is a MINITAB printout that shows side-by-side vertical box plots of the test scores in the two samples. Recall from Section 2.9 that the box plot represents the "spread" of a data set. The two box plots appear to have about the same spread; thus, the samples appear to come from populations with approximately the same variance.

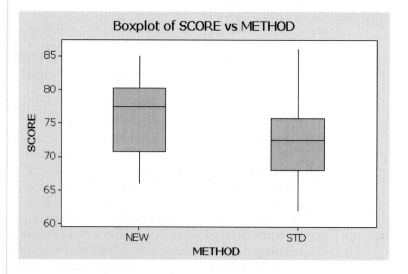

FIGURE 7.8

MINITAB box plots for Example 7.4

Look Back All three assumptions, then, appear to be reasonably satisfied for this application of the small-sample confidence interval.

Now Work Exercise 7.9

*Another, more formal, method is to test the null hypothesis $H_0: \sigma_1^2 = \sigma_2^2$. Consult the chapter references to learn more about this test.

The two-sample t-statistic is a powerful tool for comparing population means when the assumptions are satisfied. It has also been shown to retain its usefulness when the sampled populations are only approximately normally distributed. And when the sample sizes are equal, the assumption of equal population variances can be relaxed. That is, if $n_1 = n_2$, then σ_1^2 and σ_2^2 can be quite different, and the test statistic will still possess, approximately, a Student's t-distribution. In the case where $\sigma_1^2 \neq \sigma_2^2$ and $n_1 \neq n_2$, an approximate small-sample confidence interval or test can be obtained by modifying the number of degrees of freedom associated with the t-distribution.

Confidence Interval for $(\mu_1 - \mu_2)$

Using the TI-84/TI-83 Graphing Calculator

Step 1   *Enter the data. (Skip to Step 2 if you have summary statistics, not raw data.)*

Press **STAT** and select **1:Edit.**
Note: If the lists already contain data, clear the old data. Use the up ARROW to highlight '**L1**'.
Press **CLEAR ENTER.**
Use the up ARROW to highlight '**L2**'
Press **CLEAR ENTER.**
Use the **ARROW** and **ENTER** keys to enter the first data set into **L1.**
Use the **ARROW** and **ENTER** keys to enter the second data set into **L2.**

Step 2   *Access the Statistical Tests Menu.*

Press **STAT.**
Arrow right to **TESTS.**
Arrow down to **2-SampTInt.**
Press **ENTER.**

```
EDIT CALC TESTS
4↑2-SampTTest…
5:1-PropZTest…
6:2-PropZTest…
7:ZInterval…
8:TInterval…
9:2-SampZInt…
0:2-SampTInt…
```

Step 3   *Choose "**Data**" or "**Stats**". ("Data" is selected when you have entered the raw data into the Lists. "Stats" is selected when you are given only the means, standard deviations, and sample sizes.)*

Press **ENTER.**
If you selected "Data", set **List1** to **L1** and **List2** to **L2.**
Set **Freq1** to **1** and set **Freq2** to **1.**
Set **C-Level** to the confidence level.
If you are assuming that the two populations have equal variances, select **Yes** for **Pooled.**
If you are not assuming equal variances, select **No.**
Press **ENTER.**
Arrow down to "**Calculate**".
Press **ENTER.**

If you selected "Stats", enter the means, standard deviations, and sample sizes.
Set **C-Level** to the confidence level.
If you are assuming that the two populations have equal variances, select **Yes** for **Pooled.**
If you are not assuming equal variances, select **No.**
Press **ENTER.**
Arrow down to "**Calculate**".
Press **ENTER.**

(The accompanying screen is set up for an example with a mean of 100, a standard deviation of 10, and a sample size of 15 for the first data set and a mean of 105, a standard deviation of 12, and a sample size of 18 for the second data set.)

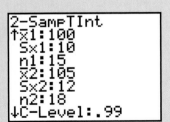

(continued)

The confidence interval will be displayed with the number of degrees of freedom, the sample statistics, and the pooled standard deviation (when appropriate).

Example Compute a 95% confidence interval for $(\mu_1 - \mu_2)$, using the following data (in this example, assume that the population variances are equal):

Group 1: 65 58 78 60 68 69 66 70 53 71 63 63

Group 2: 62 53 36 34 56 50 42 57 46 68 48 42 52 53 43

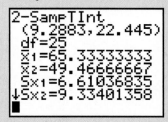 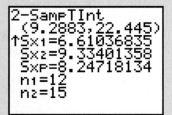

As you can see from the screens, the 95% confidence interval for $(\mu_1 - \mu_2)$ is **(9.2883, 22.445)**.

Also, notice that the output includes the means, standard deviations, sample sizes, and pooled standard deviation.

Hypothesis Test for $(\mu_1 - \mu_2)$

Using the TI-84/TI-83 Graphing Calculator

Step 1 *Enter the data. (Skip to Step 2 if you have summary statistics, not raw data.)*

Press **STAT** and select **1:Edit**.
Note: If the lists already contain data, clear the old data. Use the up **ARROW** to highlight 'L1'.
Press **CLEAR ENTER**.
Use the up **ARROW** to highlight 'L2'.
Press **CLEAR ENTER**.
Use the **ARROW** and **ENTER** keys to enter the first data set into **L1**.
Use the **ARROW** and **ENTER** keys to enter the second data set into **L2**.

Step 2 *Access the Statistical Tests Menu.*

Press **STAT**.
Arrow right to **TESTS**.
Arrow down to **2-SampTTest**.
Press **ENTER**.

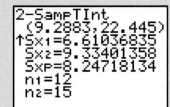

Step 3 *Choose "**Data**" or "**Stats**". ("Data" is selected when you have entered the raw data into the Lists. "Stats" is selected when you are given only the means, standard deviations, and sample sizes.)*

Press **ENTER**.
If you selected "Data", set **List1** to **L1** and **List2** to **L2**.
Set **Freq1** to **1** and set **Freq2** to **1**.
Use the **ARROW** to highlight the appropriate alternative hypothesis.
Press **ENTER**.
If you are assuming that the two populations have equal variances, select **Yes** for **Pooled**.
If you are not assuming equal variances, select **No**.
Press **ENTER**.
Arrow down to "**Calculate**".
Press **ENTER**.

If you selected "Stats", enter the means, standard deviations, and sample sizes.

Use the **ARROW** to highlight the appropriate alternative hypothesis.
Press **ENTER**.
If you are assuming that the two populations have equal variances, select **Yes** for **Pooled**.
If you are not assuming equal variances, select **No**.
Press **ENTER**.
Arrow down to "**Calculate**".
Press **ENTER**.
(The screen that follows is set up for an example with a mean of 100, a standard deviation of 10, and a sample size of 15 for the first data set and a mean of 120, a standard deviation of 12, and a sample size of 18 for the second data set.)

The results of the hypothesis test will be displayed with the *p*-value, the number of degrees of freedom, the sample statistics, and the pooled standard deviation (when appropriate).

Example   Test the hypotheses with the following data (in this example, assume that the population variances are equal):

Group 1: 65 58 78 60 68 69 66 70 53 71 63 63

Group 2: 62 53 36 34 56 50 42 57 46 68 48 42 52 53 43

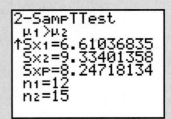

As you can see from the screens, $t = 4.967$ and the *p*-value is 0.00002. Also, notice the output includes the means, standard deviations, sample sizes, and pooled standard deviation.

The next box gives the approximate small-sample procedures to use when the assumption of equal variances is violated. The test for the case of "unequal sample sizes" is based on Satterthwaite's (1946) approximation.

Approximate Small-Sample Procedures when $\sigma_1^2 \neq \sigma_2^2$

Equal Sample Sizes ($n_1 = n_2 = n$)

Confidence interval:
$$(\bar{x}_1 - \bar{x}_2) \pm t_{\alpha/2}\sqrt{(s_1^2 + s_2^2)/n}$$

Test statistic for H_0: $(\mu_1 - \mu_2) = 0$:
$$t = (\bar{x}_1 - \bar{x}_2)/\sqrt{(s_1^2 + s_2^2)/n}$$

where t is based on $\nu = n_1 + n_2 - 2 = 2(n - 1)$ degrees of freedom.

Unequal Sample Sizes ($n_1 \neq n_2$)

Confidence interval:
$$(\bar{x}_1 - \bar{x}_2) \pm t_{\alpha/2}\sqrt{(s_1^2/n_1) + (s_2^2/n_2)}$$

Test statistic for H_0: $(\mu_1 - \mu_2) = 0$:
$$t = (\bar{x}_1 - \bar{x}_2)/\sqrt{(s_1^2/n_1) + (s_2^2/n_2)}$$

(continued)

where t is based on degrees of freedom equal to

$$\nu = \frac{(s_1^2/n_1 + s_2^2/n_2)^2}{\dfrac{(s_1^2/n_1)^2}{n_1 - 1} + \dfrac{(s_2^2/n_2)^2}{n_2 - 1}}$$

Note: The value of ν will generally not be an integer. Round ν down to the nearest integer to use the t-table.

When the assumptions are not clearly satisfied, you can select larger samples from the populations or you can use available nonparametric statistical tests.

What Should You Do if the Assumptions Are Not Satisfied?

Answer: If you are concerned that the assumptions are not satisfied, use the nonparametric Wilcoxon rank sum test for independent samples to test for a shift in population distributions. (See Optional Section 7.4).

STATISTICS IN ACTION REVISITED

Comparing Roach Trails—Mean Treatment versus Mean Control

Consider the experiment designed to investigate the trail-following ability of German cockroaches (p. 351). Recall that an entomologist created a chemical trail with either a methanol extract from roach feces or just methanol (control). Cockroaches were then released into a container at the beginning of the trail, one at a time, and a video surveillance camera was used to monitor the roach's movements. The movement pattern of each cockroach was measured by its average trail deviation (in pixels) and the data stored in the **ROACH** file.

One of the entomologist's theories is that cockroaches, on average, will tend to follow the feces extract trail closer than they would follow the control trail. If we let μ_1 represent the population mean trail deviation of cockroaches assigned to the control trail and μ_2 represent the population mean trail deviation of cockroaches following the feces extract trail, then the speculation is that $\mu_1 > \mu_2$. Of interest, then, is the one-tailed test, $H_0:(\mu_1 - \mu_2) = 0$ versus $H_a: (\mu_1 - \mu_2) > 0$. The experiment was designed so that both samples are large ($n_1 = 40$ roaches and $n_2 = 80$ roaches); thus, the large-sample z-test is appropriate and no assumptions about the trail deviation distributions are required. We conducted this test using MINITAB. The output is shown in Figure SIA7.1.

```
Two-Sample T-Test and CI: DEVIATE, TRAIL

Two-sample T for DEVIATE

TRAIL     N   Mean   StDev   SE Mean
Control   40  64.5   37.7      6.0
Extract   80  22.8   23.9      2.7


Difference = mu (Control) - mu (Extract)
Estimate for difference:  41.7338
95% CI for difference:  (28.6303, 54.8372)
T-Test of difference = 0 (vs not =): T-Value = 6.38  P-Value = 0.000  DF = 55
```

FIGURE SIA7.1
MINITAB Comparison of Means for Two Cockroach Trails

The two-tailed p-value of the test (.000) is highlighted on Figure SIA7.1. Consequently, the one-tailed p-value is .000/2 = 0. Therefore, at α = .05, there is sufficient evidence to conclude that the mean trail deviation of cockroaches following the control is significantly greater than the mean trail deviation of cockroaches following the feces extract trail. Thus, there is strong support for the entomologist's theory. In fact, a 95% confidence interval for $\mu_1 - \mu_2$ (highlighted on the MINITAB printout) is (28.6, 54.8). This implies that the feces extract trail mean is anywhere from 28.6 to 54.8 pixels less than the control mean (with 95% confidence).

Exercises 7.1–7.27

Understanding the Principles

7.1 Describe the sampling distribution of $(\bar{x}_1 - \bar{x}_2)$ when the samples are large.

7.2 To use the t-statistic to test for a difference between the means of two populations, what assumptions must be made about the two populations? About the two samples?

7.3 Two populations are described in each of the cases that follow. In which cases would it be appropriate to apply the small-sample t-test to investigate the difference between the population means?
  a. Population 1: Normal distribution with variance σ_1^2
  Population 2: Skewed to the right with variance $\sigma_2^2 = \sigma_1^2$

  b. Population 1: Normal distribution with variance σ_1^2
  Population 2: Normal distribution with variance $\sigma_2^2 \neq \sigma_1^2$

  c. Population 1: Skewed to the left with variance σ_1^2
  Population 2: Skewed to the left with variance $\sigma_2^2 = \sigma_1^2$

  d. Population 1: Normal distribution with variance σ_1^2
  Population 2: Normal distribution with variance $\sigma_2^2 = \sigma_1^2$

  e. Population 1: Uniform distribution with variance σ_1^2
  Population 2: Uniform distribution with variance $\sigma_2^2 = \sigma_1^2$

7.4 A confidence interval for $(\mu_1 - \mu_2)$ is $(-10, 4)$. Which of the following inferences is correct?
  a. $\mu_1 > \mu_2$ **b.** $\mu_1 < \mu_2$
  c. $\mu_1 = \mu_2$ **d.** no significant difference between means

7.5 A confidence interval for $(\mu_1 - \mu_2)$ is $(-10, -4)$. Which of the following inferences is correct?
  a. $\mu_1 > \mu_2$ **b.** $\mu_1 < \mu_2$
  c. $\mu_1 = \mu_2$ **d.** no significant difference between means

Learning the Mechanics

7.6 In order to compare the means of two populations, independent random samples of 400 observations are selected
NW from each population, with the following results:

Sample 1	Sample 2
$\bar{x}_1 = 5{,}275$	$\bar{x}_2 = 5{,}240$
$s_1 = 150$	$s_2 = 200$

  a. Use a 95% confidence interval to estimate the difference between the population means $(\mu_1 - \mu_2)$. Interpret the confidence interval.
  b. Test the null hypothesis H_0: $(\mu_1 - \mu_2) = 0$ versus the alternative hypothesis H_a: $(\mu_1 - \mu_2) \neq 0$. Give the significance level of the test, and interpret the result.

  c. Suppose the test in part **b** were conducted with the alternative hypothesis H_a: $(\mu_1 - \mu_2) > 0$. How would your answer to part **b** change?
  d. Test the null hypothesis H_0: $(\mu_1 - \mu_2) = 25$ versus the alternative H_a: $(\mu_1 - \mu_2) \neq 25$. Give the significance level, and interpret the result. Compare your answer with that obtained from the test conducted in part **b**.
  e. What assumptions are necessary to ensure the validity of the inferential procedures applied in parts **a–d**?

7.7 Independent random samples of 100 observations each are chosen from two normal populations with the following means and standard deviations:

Population 1	Population 2
$\mu_1 = 14$	$\mu_2 = 10$
$\sigma_1 = 4$	$\sigma_2 = 3$

Let $\bar{x}_1$ and $\bar{x}_2$ denote the two sample means.
  a. Give the mean and standard deviation of the sampling distribution of $\bar{x}_1$.
  b. Give the mean and standard deviation of the sampling distribution of $\bar{x}_2$.
  c. Suppose you were to calculate the difference $(\bar{x}_1 - \bar{x}_2)$ between the sample means. Find the mean and standard deviation of the sampling distribution of $(\bar{x}_1 - \bar{x}_2)$.
  d. Will the statistic $(\bar{x}_1 - \bar{x}_2)$ be normally distributed? Explain.

7.8 Assume that $\sigma_1^2 = \sigma_2^2 = \sigma^2$. Calculate the pooled estimator of σ^2 for each of the following cases:
  a. $s_1^2 = 200$, $s_2^2 = 180$, $n_1 = n_2 = 25$
  b. $s_1^2 = 25$, $s_2^2 = 40$, $n_1 = 20$, $n_2 = 10$
  c. $s_1^2 = .20$, $s_2^2 = .30$, $n_1 = 8$, $n_2 = 12$
  d. $s_1^2 = 2{,}500$, $s_2^2 = 1{,}800$, $n_1 = 16$, $n_2 = 17$
  e. Note that the pooled estimate is a weighted average of the sample variances. To which of the variances does the pooled estimate fall nearer in each of cases **a–d**?

7.9 Independent random samples from normal populations
NW produced the following results:

LM7_9

Sample 1	Sample 2
1.2	4.2
3.1	2.7
1.7	3.6
2.8	3.9
3.0	

a. Calculate the pooled estimate of σ^2.
b. Do the data provide sufficient evidence to indicate that $\mu_2 > \mu_1$? Test, using $\alpha = .10$.
c. Find a 90% confidence interval for $(\mu_1 - \mu_2)$.
d. Which of the two inferential procedures, the test of hypothesis in part **b** or the confidence interval in part **c**, provides more information about $(\mu_1 - \mu_2)$?

7.10 Two independent random samples have been selected, 100 observations from population 1 and 100 from population 2. Sample means $\bar{x}_1 = 70$ and $\bar{x}_2 = 50$ were obtained. From previous experience with these populations, it is known that the variances are $\sigma_1^2 = 100$ and $\sigma_2^2 = 64$.
a. Find $\sigma_{(\bar{x}_1 - \bar{x}_2)}$.
b. Sketch the approximate sampling distribution $(\bar{x}_1 - \bar{x}_2)$, assuming that $(\mu_1 - \mu_2) = 5$.
c. Locate the observed value of $(\bar{x}_1 - \bar{x}_2)$ on the graph you drew in part **b**. Does it appear that this value contradicts the null hypothesis H_0: $(\mu_1 - \mu_2) = 5$?
d. Use the z-table to determine the rejection region for the test of H_0: $(\mu_1 - \mu_2) = 5$ against H_a: $(\mu_1 - \mu_2) \neq 5$. Use $\alpha = .05$. $|z| > 1.96$
e. Conduct the hypothesis test of part **d** and interpret your result.
f. Construct a 95% confidence interval for $(\mu_1 - \mu_2)$. Interpret the interval.
g. Which inference provides more information about the value of $(\mu_1 - \mu_2)$, the test of hypothesis in part **e** or the confidence interval in part **f**?

Applying the Concepts—Basic

7.11 **Children's recall of TV ads.** Marketing professors at Robert Morris and Kent State Universities examined children's recall and recognition of television advertisements. (*Journal of Advertising,* Spring 2006.) Two groups of children were shown a 60-second commercial for Sunkist FunFruit Rock-n-Roll Shapes. One group (the A/V group) was shown the ad with both audio and video; the second group (the video-only group) was shown only the video portion of the commercial. Following the viewing, the children were asked to recall 10 specific items from the ad. The number of items recalled correctly by each child is summarized in the accompanying table. The researchers theorized that "children who receive an audiovisual presentation will have the same level of mean recall of ad information as those who receive only the visual aspects of the ad."

Video-Only Group	A/V Group
$n_1 = 20$	$n_2 = 20$
$\bar{x}_1 = 3.70$	$\bar{x}_2 = 3.30$
$s_1 = 1.98$	$s_2 = 2.13$

Source: Maher, J. K., Hu, M. Y., and Kolbe, R. H. "Children's recall of television ad elements," *Journal of Advertising,* Vol. 35, No. 1, Spring 2006 (Table 1).

a. Set up the appropriate null and alternative hypotheses to test the researchers' theory.
b. Find the value of the test statistic.
c. Give the rejection region for $\alpha = .10$.
d. Make the appropriate inference. What can you say about the researchers' theory?
e. The researchers' reported the p-value of the test as p-value $= .62$. Interpret this result.
f. What conditions are required for the inference to be valid?

7.12 **Index of Biotic Integrity.** The Ohio Environmental Protection Agency used the Index of Biotic Integrity (IBI) to measure the biological condition, or "health," of an aquatic region. The IBI is the sum of metrics that measure the presence, abundance, and health of fish in the region. (Higher values of the IBI correspond to healthier fish populations.) Researchers collected IBI measurements for sites located in different Ohio river basins. (*Journal of Agricultural, Biological, and Environmental Sciences,* June 2005.) Summary data for two river basins, Muskingum and Hocking, are given in the accompanying table.
a. Use a 90% confidence interval to compare the mean IBI values of the two river basins. Interpret the interval.

b. Conduct a test of hypothesis (at $\alpha = .10$) to compare the mean IBI values of the two river basins. Explain why the result will agree with the inference you derived from the 90% confidence interval in part **a**.

River Basin	Sample Size	Mean	Standard Deviation
Muskingum	53	.035	1.046
Hocking	51	.340	.960

Source: Boone, E. L., Keying, Y., and Smith, E. P. "Evaluating the relationship between ecological and habitat conditions using hierarchical models," *Journal of Agricultural, Biological, and Environmental Sciences,* Vol. 10, No. 2, June 2005 (Table 1).

7.13 **Reading Japanese books.** Refer to the *Reading in a Foreign Language* (Apr. 2004) experiment to improve the Japanese reading comprehension levels of University of Hawaii students, presented in Exercise 2.31 (p. 46). Recall that 14 students participated in a 10-week extensive reading program in a second-semester Japanese course. The numbers of books read by each student and the student's course grade are repeated in the following table:

JAPANESE

Number of Books	Course Grade	Number of Books	Course Grade
53	A	30	A
42	A	28	B
40	A	24	A
40	B	22	C
39	A	21	B
34	A	20	B
34	A	16	B

Source: Hitosugi, C. I., and Day, R. R. "Extensive Reading in Japanese," *Reading in a Foreign Language,* Vol. 16, No. 1, Apr. 2004 (Table 4).

a. Consider two populations of students who participate in the reading program prior to taking a second-semester Japanese course: those who earn an A grade and those who earn a B or C grade. Of interest is the difference in the mean number of books read by the two populations of students. Identify the parameter of interest in words and in symbols.
b. Form a 95% confidence interval for the target parameter identified in part **a**.
c. Give a practical interpretation of the confidence interval you formed in part **b**.
d. Compare the inference in part **c** with the inference you derived from stem-and-leaf plots in Exercise 2.31b.

7.14 Rating service at five-star hotels. A study published in the *Journal of American Academy of Business, Cambridge* (March 2002) examined whether the perception of the quality of service at five-star hotels in Jamaica differed by gender. Hotel guests were randomly selected from the lobby and restaurant areas and asked to rate 10 service-related items (e.g., "the personal attention you received from our employees"). Each item was rated on a five-point scale (1 = "much worse than I expected," 5 = "much better than I expected"), and the sum of the items for each guest was determined. A summary of the guest scores are provided in the following table:

Gender	Sample Size	Mean Score	Standard Deviation
Males	127	39.08	6.73
Females	114	38.79	6.94

a. Construct a 90% confidence interval for the difference between the population mean service-rating scores given by male and female guests at Jamaican five-star hotels. .29 ± 1.452

b. Use the interval you constructed in part **a** to make an inference about whether the perception of the quality of service at five-star hotels in Jamaica differs by gender.

7.15 Heights of grade school repeaters. Are children who repeat a grade in elementary school shorter, on average, than their peers? To answer this question, researchers compared the heights of Australian schoolchildren who repeated a grade with the heights of those who did not. (*Archives of Disease in Childhood*, Apr. 2000.) All height measurements were standardized with the use of *z*-scores. A summary of the results, by gender, is shown in the following table:

	Never Repeated	Repeated a Grade
Boys	$n = 1{,}349$	$n = 86$
	$\bar{x} = .30$	$\bar{x} = -.04$
	$s = .97$	$s = 1.17$
Girls	$n = 1{,}366$	$n = 43$
	$\bar{x} = .22$	$\bar{x} = .26$
	$s = 1.04$	$s = .94$

Source: Wake, M., Coghlan, D., and Hesketh, K. "Does height influence progression through primary school grades?" *The Archives of Disease in Childhood*, Vol. 82, Apr. 2000 (Table 3).

a. Set up the null and alternative hypothesis for determining whether the average height of Australian boys who repeated a grade is less than the average height of boys who never repeated.

b. Conduct the test you set up part **a**, using $\alpha = .05$.

c. Repeat parts **a** and **b** for Australian girls.

7.16 Short-term memory study. A group of University of Florida psychologists investigated the effects of age and gender on the short-term memory of adults. (*Cognitive Aging Conference*, Apr. 1996.) Each in a sample of 152 adults was asked to place 20 common household items (e.g., eyeglasses, keys, hat, hammer) into the rooms of a computer-image house. After performing some unrelated activities, each subject was asked to recall the locations of the objects they had placed. The number of correct responses (out of 20) was recorded.

a. The researchers theorized that women would have a higher mean recall score than men. Set up the null and alternative hypotheses to test this theory.

b. Refer to part **a**. The 43 men in the study had a mean recall score of 13.5, while the 109 women had a mean recall score of 14.4. The observed significance level for comparing these two means was found to be $p = .0001$. Interpret this value.

c. The researchers also hypothesized that younger adults would have a higher mean recall score than older adults. Set up H_0 and H_a to test this theory.

d. The observed significance level for the test of part **c** was reported as $p = .0001$. Interpret this result.

7.17 Bulimia study. The "fear of negative evaluation" (FNE) scores for 11 female students known to suffer from the eating disorder bulimia and 14 female students with normal eating habits, first presented in Exercise 2.38 (p. 48), are reproduced on the bottom of the page. (Recall that the higher the score, the greater is the fear of a negative evaluation.)

a. Find a 95% confidence interval for the difference between the population means of the FNE scores for bulimic and normal female students. Interpret the result.

b. What assumptions are required for the interval of part **a** to be statistically valid? Are these assumptions reasonably satisfied? Explain.

Applying the Concepts—Intermediate

7.18 Patent infringement case. *Chance* (Fall 2002) described a lawsuit charging Intel Corp. with infringing on a patent for an invention used in the automatic manufacture of computer chips. In response, Intel accused the inventor of adding material to his patent notebook after the patent was witnessed and granted. The case rested on whether a patent witness' signature was written on top of or under key text in the notebook. Intel hired a physicist who used an X-ray beam to measure the relative concentrations of certain elements (e.g., nickel, zinc, potassium) at several spots on the notebook page. The zinc measurements for three notebook locations—on a text line, on a witness line, and on the intersection of the witness and text line—are provided in the following table:

PATENT

Text line:	.335	.374	.440			
Witness line:	.210	.262	.188	.329	.439	.397
Intersection:	.393	.353	.285	.295	.319	

BULIMIA

Bulimic students	21	13	10	20	25	19	16	21	24	13	14			
Normal students	13	6	16	13	8	19	23	18	11	19	7	10	15	20

Source: Randles, R. H. "On neutral responses (zeros) in the sign test and ties in the Wilcoxon-Mann-Whitney test." *The American Statistician*, Vol. 55, No. 2, May 2001 (Figure 3).

a. Use a test or a confidence interval (at α = .05) to compare the mean zinc measurement for the text line with the mean for the intersection.

b. Use a test or a confidence interval (at α = .05) to compare the mean zinc measurement for the witness line with the mean for the intersection.

c. From the results you obtained in parts **a** and **b**, what can you infer about the mean zinc measurements at the three notebook locations?

d. What assumptions are required for the inferences to be valid? Are they reasonably satisfied?

7.19 How do you choose to argue? Educators frequently lament weaknesses in students' oral and written arguments. In *Thinking and Reasoning* (Oct. 2006), researchers at Columbia University conducted a series of studies to assess the cognitive skills required for successful arguments. One study focused on whether students would choose to argue by weakening the opposing position or by strengthening the favored position. (For example, suppose you are told you would do better at basketball than soccer, but you like soccer. An argument that weakens the opposing position is "You need to be tall to play basketball," An argument that strengthens the favored position is "With practice, I can become really good at soccer.") A sample of 52 graduate students in psychology was equally divided into two groups. Group 1 was presented with 10 items such that the argument always attempts to strengthens the favored position. Group 2 was presented with the same 10 items, but in this case the argument always attempts to weaken the nonfavored position. Each student then rated the 10 arguments on a five-point scale from very weak (1) to very strong (5). The variable of interest was the sum of the 10 item scores, called the *total rating*. Summary statistics for the data are shown in the accompanying table. Use the methodology of this chapter to compare the mean total ratings for the two groups at α = .05. Give a practical interpretation of the results in the words of the problem.

	Group 1 (support favored position)	Group 2 (weaken opposing position)
Sample size	26	26
Mean	28.6	24.9
Standard deviation	12.5	12.2

Source: Kuhn, D., and Udell, W. "Coordinating own and other perspectives in argument," *Thinking and Reasoning,* October 2006.

7.20 Pig castration study. Two methods of castrating male piglets were investigated in *Applied Animal Behaviour Science* (Nov. 1, 2000). Method 1 involved an incision in the spermatic cords, while Method 2 involved pulling and severing the cords. Forty-nine male piglets were randomly allocated to one of the two methods. During castration, the researchers measured the number of high-frequency

	Method 1	Method 2
Sample size	24	25
Mean number of squeals	.74	.70
Standard deviation	.09	.09

Source: Taylor, A. A., and Weary, D. M. "Vocal responses of piglets to castration: Identifying procedural sources of pain," *Applied Animal Behaviour Science,* Vol. 70, No. 1, November 1, 2000.

vocal responses (squeals) per second over a 5-second period. The data are summarized in the accompanying table. Conduct a test of hypothesis to determine whether the population mean number of high-frequency vocal responses differs for piglets castrated by the two methods. Use α = .05.

7.21 Mongolian desert ants. Refer to the *Journal of Biogeography* (Dec. 2003) study of ants in Mongolia (central Asia), presented in Exercise 2.66 (p. 60). Recall that botanists placed seed baits at 5 sites in the Dry Steppe region and 6 sites in the Gobi Desert and observed the number of ant species attracted to each site. These data are listed in the next table. Is there evidence to conclude that a difference exists between the average number of ant species found at sites in the two regions of Mongolia? Draw the appropriate conclusion, using α = .05.

GOBIANTS

Site	Region	Number of Ant Species
1	Dry Steppe	3
2	Dry Steppe	3
3	Dry Steppe	52
4	Dry Steppe	7
5	Dry Steppe	5
6	Gobi Desert	49
7	Gobi Desert	5
8	Gobi Desert	4
9	Gobi Desert	4
10	Gobi Desert	5
11	Gobi Desert	4

Source: Pfeiffer, M., et al. "Community organization and species richness of ants in Mongolia along an ecological gradient from steppe to Gobi desert," *Journal of Biogeography,* Vol. 30, No. 12, Dec. 2003.

7.22 Accuracy of mental maps. To help students organize global information about people, places, and environments, geographers encourage them to develop "mental maps" of the world. A series of lessons was designed to aid students in the development of mental maps. (*Journal of Geography,* May/June 1997.) In one experiment, a class of 24 seventh-grade geography students was given mental map lessons, while a second class of 20 students received traditional instruction. All of the students were asked to sketch a map of the world, and each portion of the map was evaluated for accuracy on a five-point scale (1 = low accuracy, 5 = high accuracy).

a. The mean accuracy scores of the two groups of seventh-graders were compared with the use of a test of hypothesis. State H_0 and H_a for a test to determine whether the mental map lessons improve a student's ability to sketch a world map.

b. What assumptions (if any) are required for the test to be statistically valid?

c. The observed significance level of the test for comparing the mean accuracy scores for continents drawn is .0507. Interpret this result.

d. The observed significance level of the test for comparing the mean accuracy scores for labeling oceans is .7371. Interpret the result.

e. The observed significance level of the test for comparing the mean accuracy scores for the entire map is .0024. Interpret the result.

7.23 Masculinity and crime. The *Journal of Sociology* (July 2003) published a study on the link between the level of masculinity and criminal behavior in men. Using a sample of newly incarcerated men in Nebraska, the researcher identified 1,171 violent events and 532 events in which violence was avoided that the men were involved in. (A violent event involved the use of a weapon, throwing of objects, punching, choking, or kicking. An event in which violence was avoided included pushing, shoving, grabbing, or threats of violence that did not escalate into a violent event.) Each of the sampled men took the Masculinity–Femininity Scale (MFS) test to determine his level of masculinity, based on common male stereotyped traits. MFS scores ranged from 0 to 56 points, with lower scores indicating a more masculine orientation. One goal of the research was to compare the mean MFS scores for two groups of men: those involved in violent events and those who avoided violent events.

a. Identify the target parameter for this study.

b. The sample mean MFS score for the violent-event group was 44.50, while the sample mean MFS score for the avoided-violent-event group was 45.06. Is this sufficient information to make the comparison desired by the researcher? Explain.

c. In a large-sample test of hypothesis to compare the two means, the test statistic was computed to be $z = 1.21$. Compute the two-tailed p-value of the test.

d. Make the appropriate conclusion, using $\alpha = .10$.

MILK

7.24 Detection of rigged school milk prices. Each year, the state of Kentucky invites bids from dairies to supply half-pint containers of fluid milk products for its school districts. In several school districts in northern Kentucky (called the "tricounty" market), two suppliers—Meyer Dairy and Trauth Dairy—were accused of price-fixing—that is, conspiring to allocate the districts so that the winning bidder was predetermined and the price per pint was set above the competitive price. These two dairies were the only two bidders on the milk contracts in the tricounty market between 1983 and 1991. (In contrast, a large number of different dairies won the milk contracts for school districts in the remainder of the northern Kentucky market, called the "surrounding" market.) Did Meyer and Trauth conspire to rig their bids in the tricounty market? Economic theory states that, if so, the mean winning price in the rigged tricounty market will be higher than the mean winning price in the competitive surrounding market. Data on all bids received from the dairies competing for the milk contracts between 1983 and 1991 are saved in the **MILK** file. A MINITAB printout of the comparison of mean prices bid for whole white milk for the two Kentucky milk markets is shown at the bottom of the page. Is there support for the claim that the dairies in the tricounty market participated in collusive practices? Explain in detail.

7.25 Children's use of pronouns. Refer to the *Journal of Communication Disorders* (Mar. 1995) study of specifically language-impaired (SLI) children, presented in Exercise 2.65 (p. 60). The data on deviation intelligence quotient (DIQ) for 10 SLI children and 10 younger, normally developing children are reproduced in the accompanying table. Use the methodology of this section to compare the mean DIQ of the two groups of children. (Take $\alpha = .10$.) What do you conclude?

SLI

SLI Children			YND Children		
86	87	84	110	90	105
94	86	107	92	92	96
89	98	95	86	100	92
110			90		

7.26 Personalities of cocaine abusers. Do cocaine abusers have radically different personalities than nonabusing college students? This was one of the questions researched in *Psychological Assessment* (June 1995). Zuckerman–Kuhlman's Personality Questionnaire (ZKPQ) was administered to a sample of 450 cocaine abusers and a sample of 589 college students. The ZKPQ yields scores (measured on a 20-point scale) on each of five dimensions: impulsive–sensation seeking, sociability, neuroticism–anxiety, aggression–hostility, and activity. The results are summarized in

ZKPQ Dimension	Cocaine Abusers ($n = 450$)		College Students ($n = 589$)	
	Mean	Std. Dev.	Mean	Std. Dev.
Impulsive–sensation seeking	9.4	4.4	9.5	4.4
Sociability	10.4	4.3	12.5	4.0
Neuroticism–anxiety	8.6	5.1	9.1	4.6
Aggression–hostility	8.6	3.9	7.3	4.1
Activity	11.1	3.4	8.0	4.1

Source: Ball, S. A. "The validity of an alternative five-factor measure of personality in cocaine abusers." *Psychological Assessment*, Vol. 7, No. 2, June 1995, p. 150 (Table 1).

MINITAB Output for Exercise 7.24

Two-Sample T-Test and CI: WWBID, Market

```
Two-sample T for WWBID

Market        N    Mean    StDev   SE Mean
SURROUND    254  0.1331   0.0158   0.00099
TRI-COUNTY  100  0.1431   0.0133   0.0013


Difference = mu (SURROUND) - mu (TRI-COUNTY)
Estimate for difference:  -0.009970
95% upper bound for difference:  -0.007232
T-Test of difference = 0 (vs <): T-Value = -6.02  P-Value = 0.000  DF = 213
```

the accompanying table. Compare the mean ZKPQ scores of the two groups on each dimension, using a statistical test of hypothesis. Interpret the results at $\alpha = .01$.

	Blacks	Whites
Sample size	55	159
Mean pain intensity	8.2	6.9

Applying the Concepts—Advanced

7.27 **Ethnicity and pain perception.** An investigation of ethnic differences in reports of pain perception was presented at the annual meeting of the American Psychosomatic Society (March 2001). A sample of 55 blacks and 159 whites participated in the study. Subjects rated (on a 13-point scale) the intensity and unpleasantness of pain felt when a bag of ice was placed on their foreheads for two minutes. (Higher ratings correspond to higher pain intensity.) A summary of the results is provided in the following table:

a. Why is it dangerous to draw a statistical inference from the summarized data? Explain.

b. Give values of the missing sample standard deviations that would lead you to conclude (at $\alpha = .05$) that blacks, on average, have a higher pain intensity rating than whites.

c. Give values of the missing sample standard deviations that would lead you to an inconclusive decision (at $\alpha = .05$) regarding whether blacks or whites have a higher mean intensity rating.

7.2 Comparing Two Population Means: Paired Difference Experiments

In Example 7.4, we compared two methods of teaching reading to "slow learners" by means of a 95% confidence interval. Suppose it is possible to measure the "reading IQs" of the "slow learners" *before* they are subjected to a teaching method. Eight pairs of "slow learners" with similar reading IQs are found, and one member of each pair is randomly assigned to the standard teaching method while the other is assigned to the new method. The data are given in Table 7.3. Do the data support the hypothesis that the population mean reading test score for "slow learners" taught by the new method is greater than the mean reading test score for those taught by the standard method?

 PAIREDSCORES

TABLE 7.3 Reading Test Scores for Eight Pairs of "Slow Learners"

Pair	New Method (1)	Standard Method (2)
1	77	72
2	74	68
3	82	76
4	73	68
5	87	84
6	69	68
7	66	61
8	80	76

We want to test

$$H_0: (\mu_1 - \mu_2) = 0$$

$$H_a: (\mu_1 - \mu_2) > 0$$

Many researchers mistakenly use the t statistic for two independent samples (Section 7.1) to conduct this test. The analysis is shown on the MINITAB printout of Figure 7.9. The test statistic, $t = 1.26$, and the p-value of the test, $p = .115$., are highlighted on the printout. At $\alpha = .10$, the p-value exceeds α. Thus, from *this* analysis, we might conclude that we do not have sufficient evidence to infer a difference in the mean test scores for the two methods.

If you examine the data in Table 7.3 carefully, however, you will find this result difficult to accept. The test score of the new method is larger than the corresponding test score for the standard method *for every one of the eight pairs of "slow learners."* This, in itself, seems to provide strong evidence to indicate that μ_1 exceeds μ_2. Why, then, did the t-test fail to detect the difference? The answer is, *the independent samples t-test is not a valid procedure to use with this set of data.*

```
Two-sample T for NEW vs STANDARD

            N   Mean   StDev   SE Mean
NEW         8   76.00   6.93     2.4
STANDARD    8   71.63   7.01     2.5


Difference = mu (NEW) - mu (STANDARD)
Estimate for difference:   4.37500
95% lower bound for difference:   -1.76200
T-Test of difference = 0 (vs >): T-Value = 1.26   P-Value = 0.115   DF = 14
Both use Pooled StDev = 6.9687
```

FIGURE 7.9

MINITAB analysis of reading test scores in Table 7.3

The *t*-test is inappropriate because the assumption of independent samples is invalid. We have randomly chosen *pairs of test scores;* thus, once we have chosen the sample for the new method, we have *not* independently chosen the sample for the standard method. The dependence between observations within pairs can be seen by examining the pairs of test scores, which tend to rise and fall together as we go from pair to pair. This pattern provides strong visual evidence of a violation of the assumption of independence required for the two-sample *t*-test of Section 9.2. Note also that

$$s_p^2 = \frac{(n_1 - 1)s_1^2 + (n_2 - 1)s_2^2}{n_1 + n_2 - 2} = \frac{(8 - 1)(6.93)^2 + (8 - 1)(7.01)^2}{8 + 8 - 2} = 48.58$$

Hence, there is a *large variation within samples* (reflected by the large value of s_p^2) in comparison to the relatively *small difference between the sample means*. Because s_p^2 is so large, the *t*-test of Section 9.2 is unable to detect a difference between μ_1 and μ_2.

TABLE 7.4 Differences in Reading Test Scores

Pair	New Method	Standard Method	Difference (New Method − Standard Method)
1	77	72	5
2	74	68	6
3	82	76	6
4	73	68	5
5	87	84	3
6	69	68	1
7	66	61	5
8	80	76	4

We now consider a valid method of analyzing the data of Table 7.3. In Table 7.4, we add the column of differences between the test scores of the pairs of "slow learners." We can regard these differences in test scores as a random sample of differences for all pairs (matched on reading IQ) of "slow learners," past and present. Then we can use this sample to make inferences about the mean of the population of differences, μ_d, which is equal to the difference $(\mu_1 - \mu_2)$. That is, the mean of the population (and sample) of differences equals the difference between the population (and sample) means. Thus, our test becomes

$$H_0: \mu_d = 0 \quad (\mu_1 - \mu_2 = 0)$$

$$H_a: \mu_d > 0 \quad (\mu_1 - \mu_2 > 0)$$

The test statistic is a one-sample *t* (Section 6.4), since we are now analyzing a single sample of differences for small *n*. Thus,

$$\text{Test statistic:} \quad t = \frac{\bar{x}_d - 0}{s_d/\sqrt{n_d}}$$

where

$$\bar{x}_d = \text{Sample mean difference}$$

$$s_d = \text{Sample standard deviation of differences}$$

$$n_d = \text{Number of differences} = \text{Number of pairs}$$

Assumptions: The population of differences in test scores is approximately normally distributed. The sample differences are randomly selected from the population differences. [*Note:* We do not need to make the assumption that $\sigma_1^2 = \sigma_2^2$.]

Rejection region: At significance level $\alpha = .05$, we will reject H_0 if $t > t_{.05}$, where $t_{.05}$ is based on $(n_d - 1)$ degrees of freedom.

Referring to Table IV in Appendix A, we find the *t*-value corresponding to $\alpha = .05$ and $n_d - 1 = 8 - 1 = 7$ df to be $t_{.05} = 1.895$. Then we will reject the null hypothesis if $t > 1.895$. (See Figure 7.10.) Note that the number of degrees of freedom decreases from $n_1 + n_2 - 2 = 14$ to 7 when we use the paired difference experiment rather than the two independent random samples design.

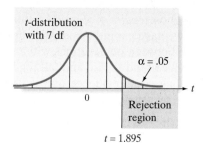

FIGURE 7.10

Rejection region for
Example 7.4

Summary statistics for the $n_d = 8$ differences are shown in the MINITAB printout of Figure 7.11. Note that $\bar{x}_d = 4.375$ and $s_d = 1.685$. Substituting these values into the formula for the test statistic, we have

$$t = \frac{\bar{x}_d - 0}{s_d/\sqrt{n_d}} = \frac{4.375}{1.685/\sqrt{8}} = 7.34$$

Because this value of t falls into the rejection region, we conclude (at $\alpha = .05$) that the population mean test score for "slow learners" taught by the new method exceeds the population mean score for those taught by the standard method. We can reach the same conclusion by noting that the *p*-value of the test, highlighted in Figure 7.11, is much smaller than $\alpha = .05$.

```
Paired T for NEW - STANDARD

             N      Mean     StDev   SE Mean
NEW          8   76.0000    6.9282    2.4495
STANDARD     8   71.6250    7.0089    2.4780
Difference   8    4.37500   1.68502   0.59574


95% lower bound for mean difference: 3.24632
T-Test of mean difference = 0 (vs > 0): T-Value = 7.34   P-Value = 0.000
```

FIGURE 7.11

MINITAB paired difference
analysis of reading test scores

Now Work Exercises 7.33a–b

This kind of experiment, in which observations are paired and the differences are analyzed, is called a **paired difference experiment.** In many cases, a paired difference experiment can provide more information about the difference between population means than an independent samples experiment can. The idea is to compare population means by comparing the differences between pairs of experimental units (objects, people, etc.) that were similar prior to the experiment. The differencing removes sources of variation that tend to inflate σ^2. For example, when two children are taught to read by two different methods, the observed difference in achievement may be due to a difference in the effectiveness of the two teaching methods, *or* it may be due to differences in the initial reading levels and IQs of the two children (random error). To reduce the effect of differences in the children on the observed differences in reading achievement, the two methods of reading are imposed on two children who are more likely to possess similar intellectual capacity, namely, children with nearly equal IQs. The effect of this pairing is to remove the larger source of variation that would be present if children with different abilities were randomly assigned to the two samples. Making comparisons within groups of similar experimental units is called **blocking,** and the paired difference experiment is a simple example of a **randomized block experiment.** In our example, pairs of children with matching IQ scores represent the blocks.

Some other examples for which the paired difference experiment might be appropriate are the following:

1. Suppose you want to estimate the difference $(\mu_1 - \mu_2)$ in mean price per gallon between two major brands of premium gasoline. If you choose two independent random samples of stations for each brand, the variability in price due to geographic location may be large. To eliminate this source of variability, you could choose pairs of stations of similar size, one station for each brand, in close geographic proximity and use the sample of differences between the prices of the brands to make an inference about $(\mu_1 - \mu_2)$.

2. Suppose a college placement center wants to estimate the difference $(\mu_1 - \mu_2)$ in mean starting salaries for men and women graduates who seek jobs through the center. If it independently samples men and women, the starting salaries may vary because of their different college majors and differences in grade point averages. To eliminate these sources of variability, the placement center could match male and female job seekers according to their majors and grade point averages. Then the differences between the starting salaries of each pair in the sample could be used to make an inference about $(\mu_1 - \mu_2)$.

3. Suppose you wish to estimate the difference $(\mu_1 - \mu_2)$ in mean absorption rate into the bloodstream for two drugs that relieve pain. If you independently sample people, the absorption rates might vary because of age, weight, sex, blood pressure, etc. In fact, there are many possible sources of nuisance variability, and pairing individuals who are similar in all the possible sources would be quite difficult. However, it may be possible to obtain two measurements *on the same person*. First, we administer one of the two drugs and record the time until absorption. After a sufficient amount of time, the other drug is administered and a second measurement on absorption time is obtained. The differences between the measurements for each person in the sample could then be used to estimate $(\mu_1 - \mu_2)$. This procedure would be advisable only if the amount of time allotted between drugs is sufficient to guarantee little or no carry-over effect. Otherwise, it would be better to use different people matched as closely as possible on the factors thought to be most important.

| Now Work Exercise 7.31 |

The hypothesis-testing procedures and the method of forming confidence intervals for the difference between two means in a paired difference experiment are summarized in the following boxes for both large and small n:

Paired Difference Confidence Interval for $\mu_d = \mu_1 - \mu_2$

Large Sample

$$\bar{x}_d \pm z_{\alpha/2} \frac{\sigma_d}{\sqrt{n_d}} \approx \bar{x}_d \pm z_{\alpha/2} \frac{s_d}{\sqrt{n_d}}$$

Small Sample

$$\bar{x}_d \pm t_{\alpha/2} \frac{s_d}{\sqrt{n_d}}$$

where $t_{\alpha/2}$ is based on $(n_d - 1)$ degrees of freedom

---

Paired Difference Test of Hypothesis for $\mu_d = \mu_1 - \mu_2$

One-Tailed Test

$H_0: \mu_d = D_0$
$H_a: \mu_d < D_0$
  [or $H_a: \mu_d > D_0$]

Two-Tailed Test

$H_0: \mu_d = D_0$
$H_a: \mu_d \neq D_0$

Large Sample

Test statistic: $z = \dfrac{\bar{x}_d - D_0}{\sigma_d/\sqrt{n_d}} \approx \dfrac{\bar{x}_d - D_0}{s_d/\sqrt{n_d}}$

Rejection region: $z < -z_\alpha$
[or $z > z_\alpha$ when $H_a: \mu_d > D_0$]

Rejection region: $|z| > z_{\alpha/2}$

Small Sample

Test statistic: $t = \dfrac{\bar{x}_d - D_0}{s_d/\sqrt{n_d}}$

Rejection region: $t < -t_\alpha$
  [or $t > t_\alpha$ when $H_a: \mu_d > D_0$]

Rejection region: $|t| > t_{\alpha/2}$

where t_α and $t_{\alpha/2}$ are based on $(n_d - 1)$ degrees of freedom

Conditions Required for Valid Large-Sample Inferences about μ_d

1. A random sample of differences is selected from the target population of differences.
2. The sample size n_d is large (i.e., $n_d \geq 30$). (By the central limit theorem, this condition guarantees that the test statistic will be approximately normal, regardless of the shape of the underlying probability distribution of the population.)

Conditions Required for Valid Small-Sample Inferences about μ_d

1. A random sample of differences is selected from the target population of differences.
2. The population of differences has a distribution that is approximately normal.

EXAMPLE 7.5

CONFIDENCE INTERVAL FOR μ_d: Comparing Mean Salaries of Males and Females

Problem An experiment is conducted to compare the starting salaries of male and female college graduates who find jobs. Pairs are formed by choosing a male and a female with the same major and similar grade point averages (GPAs). Suppose a random sample of 10 pairs is formed in this manner and the starting annual salary of each person is recorded. The results are shown in Table 7.5. Compare the mean starting salary μ_1 for males with the mean starting salary μ_2 for females, using a 95% confidence interval. Interpret the results.

GRADPAIRS

TABLE 7.5 Data on Annual Salaries for Matched Pairs of College Graduates

Pair	Male	Female	Difference Male − Female	Pair	Male	Female	Difference Male − Female
1	$29,300	$28,800	$ 500	6	$37,800	$38,000	$−200
2	41,500	41,600	−100	7	69,500	69,200	300
3	40,400	39,800	600	8	41,200	40,100	1,100
4	38,500	38,500	0	9	38,400	38,200	200
5	43,500	42,600	900	10	59,200	58,500	700

Solution Since the data on annual salary are collected in pairs of males and females matched on GPA and major, a paired difference experiment is performed. To conduct the analysis, we first compute the differences between the salaries, as shown in Table 7.5. Summary statistics for these $n = 10$ differences are displayed at the top of the SAS printout shown in Figure 7.12.

The 95% confidence interval for $\mu_d = (\mu_1 - \mu_2)$ for this small sample is

$$\bar{x}_d \pm t_{\alpha/2}\frac{s_d}{\sqrt{n_d}}$$

where $t_{\alpha/2} = t_{.025} = 2.262$ (obtained from Table IV, Appendix A) is based on $n_d - 1 = 9$ degrees of freedom. Substituting the values of $\bar{x}_d$ and s_d shown on the printout, we obtain

$$\bar{x}_d \pm 2.262\frac{s_d}{\sqrt{n_d}} = 400 \pm 2.262\left(\frac{434.613}{\sqrt{10}}\right)$$
$$= 400 \pm 310.88 \approx 400 \pm 311 = (\$89, \$711)$$

```
                       The MEANS Procedure

                   Analysis Variable : DIFF

      Mean            Std Dev     N        Minimum         Maximum

  400.0000000      434.6134937    10     -200.0000000      1100.00


          Two Sample Paired t-test for the Means of MALE and FEMALE

Sample Statistics

     Group          N       Mean     Std. Dev.    Std. Error
     --------------------------------------------------------
     MALE           10      43930       11665        3688.8
     FEMALE         10      43530       11617        3673.6


Hypothesis Test

     Null hypothesis:      Mean of (MALE - FEMALE) =  0
     Alternative:          Mean of (MALE - FEMALE) ^= 0

         t Statistic       Df        Prob > t
         ----------------------------------------
            2.910           9          0.0173


95% Confidence Interval for the Difference between Two Paired Means

          Lower Limit       Upper Limit
          -----------       -----------
             89.10             710.90
```

FIGURE 7.12

SAS analysis of salary differences

[*Note:* This interval is also shown highlighted at the bottom of the SAS printout of Figure 7.12.] Our interpretation is that the true mean difference between the starting salaries of males and females falls between $89 and $711, with 95% confidence. Since the interval falls above 0, we infer that $\mu_1 - \mu_2 > 0$; that is, the mean salary for males exceeds the mean salary for females.

Look Back Remember that $\mu_d = \mu_1 - \mu_2$. So if $\mu_d > 0$, then $\mu_1 > \mu_2$. Alternatively, if $\mu_d < 0$, then $\mu_1 < \mu_2$.

Now Work Exercise 7.40

To measure the amount of information about $(\mu_1 - \mu_2)$ gained by using a paired difference experiment in Example 7.5 rather than an independent samples experiment, we can compare the relative widths of the confidence intervals obtained by the two methods. A 95% confidence interval for $(\mu_1 - \mu_2)$ obtained from a paired difference experiment is, from Example 7.5, ($89, $711). If we analyzed the same data as though this were an independent samples experiment,* we would first obtain the descriptive statistics shown in the SAS printout of Figure 7.13. Then we substitute the sample means and standard deviations shown on the printout into the formula for a 95% confidence interval for $(\mu_1 - \mu_2)$ using independent samples. The result is

$$(\bar{x}_1 - \bar{x}_2) \pm t_{.025}\sqrt{s_p^2\left(\frac{1}{n_1} + \frac{1}{n_2}\right)}$$

where

$$s_p^2 = \frac{(n_1 - 1)s_1^2 + (n_2 - 1)s_2^2}{n_1 + n_2 - 2}$$

SPSS performed these calculations and obtained the interval ($-10,537.50$, $11,337.50$), highlighted in Figure 7.13.

Notice that the independent samples interval includes 0. Consequently, if we were to use this interval to make an inference about $(\mu_1 - \mu_2)$, we would incorrectly conclude that the mean starting salaries of males and females do not differ! You can see that the confidence interval for the independent sampling experiment is about 35 times wider than for the corresponding paired difference confidence interval. Blocking out the variability due to differences in majors and grade point averages significantly increases the information

Group Statistics

	GENDER	N	Mean	Std. Deviation	Std. Error Mean
SALARY	M	10	43930.00	11665.148	3688.844
	F	10	43530.00	11616.946	3673.601

Independent Samples Test

		Levene's Test for Equality of Variances		t-test for Equality of Means						95% Confidence Interval of the Difference	
		F	Sig.	t	df	Sig. (2-tailed)	Mean Difference	Std. Error Difference	Lower	Upper
SALARY	Equal variances assumed	.000	.991	.077	18	.940	400.00	5206.046	-10537.5	11337.50
	Equal variances not assumed			.077	18.000	.940	400.00	5206.046	-10537.5	11337.51

FIGURE 7.13

SPSS analysis of salaries, assuming independent samples

*This is done only to provide a measure of the increase in the amount of information obtained by a paired design in comparison to an unpaired design. Actually, if an experiment were designed that used pairing, an unpaired analysis would be invalid because the assumption of independent samples would not be satisfied.

about the difference in males' and females' mean starting salaries by providing a much more accurate (a smaller confidence interval for the same confidence coefficient) estimate of $(\mu_1 - \mu_2)$.

You may wonder whether a paired difference experiment is always superior to an independent samples experiment. The answer is, Most of the time, but not always. We sacrifice half the degrees of freedom in the t-statistic when a paired difference design is used instead of an independent samples design. This is a loss of information, and unless that loss is more than compensated for by the reduction in variability obtained by blocking (pairing), the paired difference experiment will result in a net loss of information about $(\mu_1 - \mu_2)$. Thus, we should be convinced that the pairing will significantly reduce variability before performing a paired difference experiment. Most of the time, this will happen.

One final note: The pairing of the observations is determined *before* the experiment is performed (i.e., by the *design* of the experiment). A paired difference experiment is *never* obtained by pairing the sample observations *after* the measurements have been acquired.

What Do You Do When the Assumption of a Normal Distribution for the Population of Differences Is Not Satisfied?

Answer: Use the Wilcoxon signed rank test for the paired difference design (Optional Section 7.5).

ACTIVITY 7.1 *Box Office Receipts:* **Comparing Population Means**

Use the Internet to find the daily box office receipts for two different hit movies during the first eight weeks after their releases. In this activity, you will compare the mean daily box office receipts of these movies in two different ways.

1. Independently select random samples of size $n = 30$ from the data sets for each of the movies' daily box office receipts. Find the mean and standard deviation of each sample. Then find a confidence interval for the difference of the means.

2. Now pair the data for the two movies by day; that is the box office receipts for the day of release of each movie are paired, the box office receipts for each movie's second day are paired, etc. Calculate the difference in box office re-

ceipts for each day, and select a random sample of size $n = 30$ from the daily differences. Then find a confidence interval for the sample mean.

3. Compare the confidence intervals from Exercises 1 and 2. Explain how the sampling for the paired difference experiment is different from the independent sampling. How might the paired difference sampling technique yield a better comparison of the two means in the box office example?

4. Compute the actual means for the daily box office receipts for each of the movies, and then find the difference of the means. Does the difference of the means lie in both confidence intervals you found? Is the exact difference remarkably closer to one of the estimates? Explain.

Confidence Interval for a Paired Difference Mean

Using the TI-84/TI-83 Graphing Calculator

Note: There is no paired difference option on the calculator. The following instructions demonstrate how to calculate the differences and then use the one-sample t-interval:

Step 1 *Enter the data and calculate the differences.*

Press **STAT** and select **1:Edit**.
Note: If the lists already contain data, clear the old data. Use the up **ARROW** to highlight 'L1'.
Press **CLEAR ENTER**.

(continued)

Use the up **ARROW** to highlight '**L2**'.
Press **CLEAR ENTER**.
Use the **ARROW** and **ENTER** keys to enter the first data set into **L1**.
Use the **ARROW** and **ENTER** keys to enter the second data set into **L2**.
The differences will be calculated in **L3.**
Use the up **ARROW** to highlight '**L3**'.
Press **CLEAR.** This will clear any old data, but **L3** will remain highlighted.
To enter the equation
$L3 = L1 - L2$, use the following keystrokes:
Press 2ND '**1**'. (This will enter L1.)
Press the **MINUS** button.
Press 2ND '**2**'. (This will enter L2.)
(Notice the equation at the bottom of the screen.)
Press **ENTER**. (The differences should be calculated in L3.)

L1	L2	L3	3
65	62	3	
58	53	5	
78	36	42	
60	34	26	
68	56	12	
69	50	19	
66	42	24	

$L3 = L_1 - L_2$

Step 2   *Access the Statistical Tests Menu.*

Press **STAT**.
Arrow right to **TESTS**.
Arrow down to **TInterval (even for large sample case)**.
Press **ENTER**.

```
EDIT CALC TESTS
2↑T-Test…
3:2-SampZTest…
4:2-SampTTest…
5:1-PropZTest…
6:2-PropZTest…
7:ZInterval…
8↓TInterval…
```

Step 3   *Choose "Data".*

Press **ENTER**.
Set **List** to **L3**.
Set **Freq** to **1**.
Set **C-Level** to the confidence level.
Arrow down to "**Calculate**".
Press **ENTER**.

```
TInterval
 Inpt:Data Stats
 List:L₃
 Freq:1
 C-Level:.95
 Calculate
```

The confidence interval will be displayed with the mean, standard deviation, and sample size of the differences.

Hypothesis Test for a Paired Difference Mean

Using the TI-84/TI-83 Graphing Calculator

Note: There Is no paired difference option on the calculator. The following instructions demonstrate how to calculate the differences and then use the one-sample *t*-test:

Step 1   *Enter the data and calculate the differences.*

Press **STAT** and select **1:Edit**.
Note: If the lists already contain data, clear the old data. Use the up

ARROW to highlight '**L1**'.
Press **CLEAR ENTER**.
Use the up **ARROW** to highlight '**L2**'.
Press **CLEAR ENTER**.
Use the **ARROW** and **ENTER** keys to enter
the first data set into **L1**.
Use the **ARROW** and **ENTER** keys to enter
the second data set into **L2**.
The differences will be calculated in **L3**.
Use the up **ARROW** to highlight '**L3**'.
Press **CLEAR**. This will clear any old data, but **L3** will remain highlighted.
To enter the equation L3 = L1 − L2, use the following keystrokes:
Press 2ND '**1**'. (This will enter L1.)
Press the **MINUS** button.
Press 2ND '**2**'. (This will enter L2.)
(Notice the equation at the bottom of the screen.)
Press **ENTER**. (The differences should be calculated in L3.)

L1	L2	$\blacksquare$	3
65	62	3	
58	53	5	
78	36	42	
60	34	26	
68	56	12	
69	50	19	
66	42	24	

L3 =L1−L2■

Step 2   *Access the Statistical Tests Menu.*

Press **STAT**.
Arrow right to **TESTS**.
Arrow down to **T-Test (even for a large-sample case).**
Press **ENTER**.

```
EDIT CALC TESTS
1:Z-Test…
2:T-Test…
3:2-SampZTest…
4:2-SampTTest…
5:1-PropZTest…
6:2-PropZTest…
7↓ZInterval…
```

Step 3   *Choose "Data".*

Press **ENTER**.
Enter the values for the hypothesis test, where μ_0 = the value for μ_d in the null
hypothesis.
Set **List** to **L3**.
Set **Freq** to **1**.
Use the **ARROW** to highlight the appropriate alternative hypothesis.
Press **ENTER**.
Arrow down to "**Calculate**".
Press **ENTER**.

The test statistic and the *p*-value will be displayed, as will the sample mean,
standard deviation, and sample size of the differences.

Exercises 7.28–7.47

Understanding the Principles

7.28 What are the advantages of using a paired difference experiment over an independent samples design?

7.29 In a paired difference experiment, when should the observations be paired, before or after the data are collected?

7.30 What conditions are required for valid large-sample inferences about μ_d? small-sample inferences?

Learning the Mechanics

7.31 A paired difference experiment yielded n_d pairs of observa-
NW tions. In each case, what is the rejection region for testing
$H_0: \mu_d = 2$ against $H_a: \mu_d > 2$?
a. $n_d = 10, \alpha = .05$
b. $n_d = 20, \alpha = .10$
c. $n_d = 5, \alpha = .025$
d. $n_d = 9, \alpha = .01$

7.32 A paired difference experiment produced the following data:

$$n_d = 16 \quad \bar{x}_1 = 143 \quad \bar{x}_2 = 150 \quad \bar{x}_d = -7 \quad s_d^2 = 64$$

a. Determine the values of *t* for which the null hypothesis $\mu_1 - \mu_2 = 0$ would be rejected in favor of the alternative hypothesis $\mu_1 - \mu_2 < 0$. Use $\alpha = .10$.

b. Conduct the paired difference test described in part **a**. Draw the appropriate conclusions.

c. What assumptions are necessary so that the paired difference test will be valid?

d. Find a 90% confidence interval for the mean difference μ_d.

e. Which of the two inferential procedures, the confidence interval of part **d** or the test of hypothesis of part **b**, provides more information about the difference between the population means?

7.33 The data for a random sample of six paired observations
NW are shown in the following table:

🌐 **LM7_33**

Pair	Sample from Population 1	Sample from Population 2
1	7	4
2	3	1
3	9	7
4	6	2
5	4	4
6	8	7

a. Calculate the difference between each pair of observations by subtracting observation 2 from observation 1. Use the differences to calculate $\bar{x}_d$ and s_d^2.
b. If μ_1 and μ_2 are the means of populations 1 and 2, respectively, express μ_d in terms of μ_1 and μ_2.
c. Form a 95% confidence interval for μ_d.
d. Test the null hypothesis $H_0: \mu_d = 0$ against the alternative hypothesis $H_a: \mu_d \neq 0$. Use $\alpha = .05$.

7.34 The data for a random sample of 10 paired observations are shown in the following table:

🌐 **LM7_34**

Pair	Population 1	Population 2
1	19	24
2	25	27
3	31	36
4	52	53
5	49	55
6	34	34
7	59	66
8	47	51
9	17	20
10	51	55

a. If you wish to test whether these data are sufficient to indicate that the mean for population 2 is larger than that for population 1, what are the appropriate null and alternative hypotheses? Define any symbols you use.
b. Conduct the test from part **a**, using $\alpha = .10$. What is your decision?
c. Find a 90% confidence interval for μ_d. Interpret this interval.
d. What assumptions are necessary to ensure the validity of the preceding analysis?

7.35 A paired difference experiment yielded the following results:

$$n_d = 40$$
$$\bar{x}_d = 11.7$$
$$s_d = 6.$$

a. Test $H_0: \mu_d = 10$ against $H_a: \mu_d \neq 10$, where $\mu_d = (\mu_1 - \mu_2)$. Use $\alpha = .05$.
b. Report the p-value for the test you conducted in part **a**. Interpret the p-value.

Applying the Concepts—Basic

7.36 **Laughter among deaf signers.** The *Journal of Deaf Studies and Deaf Education* (Fall 2006) published an article on vocalized laughter among deaf users of American Sign Language (ASL). In videotaped ASL conversations among deaf participants, 28 laughed at least once. The researchers wanted to know if they laughed more as speakers (while signing) or as audience members (while listening). For each of the 28 deaf participants, the number of laugh episodes as a speaker and the number of laugh episodes as an audience member were determined. One goal of the research was to compare the mean numbers of laugh episodes of speakers and audience members.
a. Explain why the data should be analyzed as a paired difference experiment.
b. Identify the study's target parameter.
c. The study yielded a sample mean of 3.4 laughter episodes for speakers and a sample mean of 1.3 laughter episodes for audience members. Is this sufficient evidence to conclude that the population means are different? Explain.
d. A paired difference t-test resulted in $t = 3.14$ and p-value $< .01$. Interpret the results in the words of the problem.

7.37 **Animal-assisted therapy for heart patients.** Refer to the *American Heart Association Conference* (Nov. 2005) study to gauge whether animal-assisted therapy can improve the physiological responses of heart failure patients, presented in Exercise 2.102 (p. 74). Recall that a sample of $n = 26$ heart patients was visited by a human volunteer accompanied by a trained dog; the anxiety level of each patient was measured (in points) both before and after the visits. The drop (before minus after) in anxiety level for patients is summarized as follows: $\bar{x}_d = 10.5$, $s_d = 7.6$. Does animal-assisted therapy significantly reduce the mean anxiety level of heart failure patients? Support your answer with a 95% confidence interval.

7.38 **Life expectancy of Oscar winners.** Does winning an Academy of Motion Picture Arts and Sciences award lead to long-term mortality for movie actors? In an article in the *Annals of Internal Medicine* (May 15, 2001), researchers sampled 762 Academy Award winners and matched each one with another actor of the same sex who was in the same winning film and was born in the same era. The life expectancies (ages) of the pairs of actors were compared.
a. Explain why the data should be analyzed as a paired difference experiment.
b. Set up the null hypothesis for a test to compare the mean life expectancies of Academy Award winners and nonwinners.
c. The sample mean life expectancies of Academy Award winners and nonwinners were reported as 79.7 years and 75.8 years, respectively. The p-value for comparing the two population means was reported as $p = .003$. Interpret this value in the context of the problem.

7.39 **The placebo effect and pain.** According to research published in *Science* (Feb. 20, 2004), the mere belief that you are receiving an effective treatment for pain can reduce the pain you actually feel. Researchers from the University of Michigan and Princeton University tested this placebo effect on 24 volunteers as follows: Each volunteer was put inside a magnetic resonance imaging (MRI) machine for two consecutive sessions. During the first session, electric shocks were applied to their arms and the blood oxygen level-dependent (BOLD) signal (a measure related to neural activity in the brain) was recorded during pain. The second session was identical to the first, except that, prior to applying the electric shocks, the researchers smeared a cream on the volunteer's arms. The volunteers were informed that the cream would block the pain when, in fact, it was just a regular skin lotion (i.e., a placebo). If the placebo is effective in reducing the pain experience, the BOLD measurements should be higher, on average, in the first MRI session than in the second.

a. Identify the target parameter for this study.

b. What type of design was used to collect the data?

c. Give the null and alternative hypotheses for testing the placebo effect theory.

d. The differences between the BOLD measurements in the first and second sessions were computed and summarized in the study as follows: $n_d = 24$, $\bar{x}_d = .21$, $s_d = .47$. Use this information to calculate the test statistic.

e. The p-value of the test was reported as p-value $= .02$. Make the appropriate conclusion at $\alpha = .05$.

CRASH

7.40 NHTSA new car crash tests. Refer to the National Highway Traffic Safety Administration (NHTSA) crash test data on new cars, saved in the **CRASH** file. Crash test dummies were placed in the driver's seat and front passenger's seat of a new car model, and the car was steered by remote control into a head-on collision with a fixed barrier while traveling at 35 miles per hour. Two of the variables measured for each of the 98 new cars in the data set are (1) the severity of the driver's chest injury and (2) the severity of the passenger's chest injury. (The more points assigned to the chest injury rating, the more severe the injury is.) Suppose the NHTSA wants to determine whether the true mean driver chest injury rating exceeds the true mean passenger chest injury rating and, if so, by how much.

a. State the parameter of interest to the NHTSA.

b. Explain why the data should be analyzed as matched pairs.

c. Find a 99% confidence interval for the true difference between the mean chest injury ratings of drivers and front-seat passengers.

d. Interpret the interval you found in part **c**. Does the true mean driver chest injury rating exceed the true mean passenger chest injury rating? If so, by how much?

e. What conditions are required for the analysis to be valid? Do these conditions hold for these data?

Applying the Concepts—Intermediate

7.41 Reading tongue twisters. According to *Webster's New World Dictionary*, a tongue twister is "a phrase that is hard to speak rapidly." Do tongue twisters have an effect on the length of time it takes to read silently? To answer this question, 42 undergraduate psychology students participated in a reading experiment. (*Memory & Cognition*, Sept. 1997.) Two lists, each composed of 600 words, were constructed. One list contained a series of tongue twisters, and the other list (called the *control*) did not contain any tongue twisters. Each student read both lists, and the length of time (in minutes) required to complete the lists was recorded. The researchers used a test of hypothesis to compare the mean reading response times for the tongue-twister and control lists.

a. Set up the null hypothesis for the test.

b. Use the information in the accompanying table to find the test statistic and p-value of the test.

c. Give the appropriate conclusion. Use $\alpha = .05$.

List Type	Response Time (minutes)	
	Mean	Standard Deviation
Tongue twister	6.59	1.94
Control	6.34	1.92
Difference	.25	.78

Source: Robinson, D. H., and Katayama, A. D. "At-lexical, articulatory interference in silent reading: The 'upstream' tongue-twister effect," *Memory & Cognition*, Vol. 25, No. 5, Sept. 1997, p. 663.

7.42 Visual search and memory study. In searching for an item (e.g., a roadside traffic sign, a lost earring, or a tumor in a mammogram), common sense dictates that you will not reexamine items previously rejected. However, researchers at Harvard Medical School found that a visual search has no memory. (*Nature,* Aug. 6, 1998.) In their experiment, nine subjects searched for the letter "T" mixed among several letters "L." Each subject conducted the search under two conditions: random and static. In the random condition, the locations of the letters were changed every 111 milliseconds; in the static condition, the locations of the letters remained unchanged. In each trial, the reaction time in milliseconds (i.e., the amount of time it took the subject to locate the target letter) was recorded.

a. One goal of the research was to compare the mean reaction times of subjects in the two experimental conditions. Explain why the data should be analyzed as a paired difference experiment.

b. If a visual search has no memory, then the main reaction times in the two conditions will not differ. Specify H_0 and H_a for testing the "no-memory" theory.

c. The test statistic was calculated as $t = 1.52$ with p-value $= .15$. Draw the appropriate conclusion.

7.43 Linking dementia and leisure activities. Does participation in leisure activities in your youth reduce the risk of Alzheimer's disease and other forms of dementia? To answer this question, a group of university researchers studied a sample of 107 same-sex Swedish pairs of twins. (*Journal of Gerontology: Psychological Sciences and Social Sciences*, Sept. 2003.) Each pair of twins was discordant for dementia; that is, one member of each pair was diagnosed with Alzheimer's disease while the other member (the control) was nondemented for at least five years after the sibling's onset of dementia. The level of overall leisure activity (measured on an 80-point scale, where higher values indicate higher levels of leisure activity) of each twin of each pair 20 years prior to the onset of dementia was obtained from the Swedish Twin Registry database. The leisure activity scores (simulated on the basic of summary information presented in the journal article) are saved in the **DEMENTIA** file. The first five and last five observations are shown in the next table:

a. Explain why the data should be analyzed as a paired difference experiment.

b. Conduct the appropriate analysis, using $\alpha = .05$. Make an inference about which member of the pair, the demented or control (nondemented) twin, had the largest average level of leisure activity.

DEMENTIA (first and last 5 observations)

Pair	Control	Demented
1	27	13
2	57	57
3	23	31
4	39	46
5	37	37
.	.	.
.	.	.
.	.	.
103	22	14
104	32	23
105	33	29
106	36	37
107	24	1

7.44 Testing electronic circuits. Japanese researchers have developed a compression–depression method of testing electronic circuits based on Huffman coding. (*IEICE Transactions on Information & Systems*, Jan. 2005.) The new method is designed to reduce the time required for input decompression and output compression—called the compression ratio. Experimental results were obtained by testing a sample of 11 benchmark circuits (all of different sizes) from a SUN Blade 1000 workstation. Each circuit was tested with the standard compression–depression method and the new Huffman-based coding method and the compression ratio recorded. The data are given in the next table. Compare the two methods with a 95% confidence interval. Which method has the smaller mean compression ratio?

CIRCUITS

Circuit	Standard Method	Huffman Coding Method
1	.80	.78
2	.80	.80
3	.83	.86
4	.53	.53
5	.50	.51
6	.96	.68
7	.99	.82
8	.98	.72
9	.81	.45
10	.95	.79
11	.99	.77

Source: Ichihara, H., Shintani, M., and Inoue, T. "Huffman-based test response coding," *IEICE Transactions on Information & Systems*, Vol. E88-D, No. 1, Jan. 2005 (Table 3).

7.45 Light-to-dark transition of genes. *Synechocystis,* a type of cyanobacterium that can grow and survive under a wide range of conditions, is used by scientists to model DNA behavior. In the *Journal of Bacteriology* (July 2002), scientists isolated genes of the bacterium responsible for photosynthesis and respiration and investigated the sensitivity of the genes to light. Each gene sample was grown to mid-exponential phase in a growth incubator in "full light." The lights were then extinguished, and any growth of the sample was measured after 24 hours in the dark ("full

dark"). The lights were then turned back on for 90 minutes ("transient light"), followed immediately by an additional 90 minutes in the dark ("transient dark"). Standardized growth measurements in each light–dark condition were obtained for 103 genes. The complete data set is saved in the **GENEDARK** file. Data on first 10 genes are shown in the following table:

GENEDARK (first 10 observations shown)

Gene ID	FULL-DARK	TR-LIGHT	TR-DARK
SLR2067	−0.00562	1.40989	−1.28569
SLR1986	−0.68372	1.83097	−0.68723
SSR3383	−0.25468	−0.79794	−0.39719
SLL0928	−0.18712	−1.20901	−1.18618
SLR0335	−0.20620	1.71404	−0.73029
SLR1459	−0.53477	2.14156	−0.33174
SLL1326	−0.06291	1.03623	0.30392
SLR1329	−0.85178	−0.21490	0.44545
SLL1327	0.63588	1.42608	−0.13664
SLL1325	−0.69866	1.93104	−0.24820

Source: Gill, R. T., et al. "Genome-wide dynamic transcriptional profiling of the light to dark transition in *Synechocystis Sp.* PCC6803," *Journal of Bacteriology,* Vol. 184, No. 13, July 2002.

a. Treat the data for the first 10 genes as a random sample collected from the population of 103 genes, and test the hypothesis that there is no difference between the mean standardized growth of genes in the full-dark condition and genes in the transient-light condition. Use $\alpha = .01$.

b. Use a statistical software package to compute the mean difference in standardized growth of the 103 genes in the full-dark condition and the transient-light condition. Did the test you carried out in part **a** detect this difference?

c. Repeat parts **a** and **b** for a comparison of the mean standardized growth of genes in the full-dark condition and genes in the transient-dark condition.

d. Repeat parts **a** and **b** for a comparison of the mean standardized growth of genes in the transient-light condition and genes in the transient-dark condition.

Applying the Concepts—Advanced

7.46 Homophone confusion in Alzheimer's patients. A *homophone* is a word whose pronunciation is the same as that of another word having a different meaning and spelling (e.g., *nun* and *none, doe* and *dough,* etc.). *Brain and Language* (Apr. 1995) reported on a study of homophone spelling in patients with Alzheimer's disease. Twenty Alzheimer's patients were asked to spell 24 homophone pairs given in random order. Then the number of homophone confusions (e.g., spelling *doe* given the context, *bake bread dough*) was recorded for each patient. One year later, the same test was given to the same patients. The data for the study are provided in the table on p. 383. The researchers posed the following question: "Do Alzheimer's patients show a significant increase in mean homophone confusion errors over time?" Perform an analysis of the data to answer the researchers' question. What assumptions are necessary for the procedure used to be valid? Are they satisfied?

⊙ **HOMOPHONE**

Patient	Time 1	Time 2
1	5	5
2	1	3
3	0	0
4	1	1
5	0	1
6	2	1
7	5	6
8	1	2
9	0	9
10	5	8
11	7	10
12	0	3
13	3	9
14	5	8
15	7	12
16	10	16
17	5	5
18	6	3
19	9	6
20	11	8

Source: Neils, J., Roeltgen, D. P., and Constantinidou, F. "Decline in homophone spelling associated with loss of semantic influence on spelling in Alzheimer's disease," *Brain and Language,* Vol. 49, No. 1, Apr. 1995, p. 36 (Table 3).

7.47 Alcoholic fermentation in wines. Determining alcoholic fermentation in wine is critical to the wine-making process. Must/wine density is a good indicator of the fermentation point, since the density value decreases as sugars are converted into alcohol. For decades, winemakers have measured must/wine density with a hydrometer. Although accurate, the hydrometer employs a manual process that is very time consuming. Consequently, large wineries are searching for more rapid measures of density measurement. An alternative method utilizes the hydrostatic balance instrument (similar to the hydrometer, but digital). A winery in Portugal collected must/wine density measurements on white wine samples randomly selected from the fermentation process for a recent harvest. For each sample, the density of the wine at 20°C was measured with both the hydrometer and the hydrostatic balance. The densities for 40 wine samples are saved in the **WINE40** file. The first five and last five observations are shown in the accompanying table. The winery will use the alternative method of measuring wine density only if it can be demonstrated that the mean difference between the density measurements of the two methods does not exceed .002. Perform the analysis for the winery. Provide the winery with a written report of your conclusions.

⊙ **WINE40 (first and last five observations)**

Sample	Hydrometer	Hydrostatic
1	1.08655	1.09103
2	1.00270	1.00272
3	1.01393	1.01274
4	1.09467	1.09634
5	1.10263	1.10518
.	.	.
.	.	.
36	1.08084	1.08097
37	1.09452	1.09431
38	0.99479	0.99498
39	1.00968	1.01063
40	1.00684	1.00526

Source: Cooperative Cellar of Borba *(Adega Cooperativ a de Borba),* Portugal.

7.3 Determining the Sample Size

You can find the appropriate sample size to estimate the difference between a pair of means with a specified sampling error (SE) and degree of reliability by using the method described in Section 5.4. That is, to estimate the difference between a pair of means correct to within SE units with confidence level $(1 - \alpha)$, let $z_{\alpha/2}$ standard deviations of the sampling distribution of the estimator equal SE. Then solve for the sample size. To do this, you have to solve the problem for a specific ratio between n_1 and n_2. Most often, you will want to have equal sample sizes, that is, $n_1 = n_2 = n$. We will illustrate the procedure with two examples.

EXAMPLE 7.6

FINDING THE SAMPLE SIZES FOR ESTIMATING

$\mu_1 - \mu_2$: Comparing Mean Crop Yields

Problem New fertilizer compounds are often advertised with the promise of increased crop yields. Suppose we want to compare the mean yield μ_1 of wheat when a new fertilizer is used to the mean yield μ_2 with a fertilizer in common use. The estimate of the difference in mean yield per acre is to be correct to within .25 bushel with a confidence coefficient of .95. If the sample sizes are to be equal, find $n_1 = n_2 = n$, the number of 1-acre plots of wheat assigned to each fertilizer.

Solution To solve the problem, you need to know something about the variation in the bushels of yield per acre. Suppose from past records you know the yields of wheat possess a range of approximately 10 bushels per acre. You could then approximate $\sigma_1 = \sigma_2 = \sigma$ by letting the range equal 4σ. Thus,

$$4\sigma \approx 10 \text{ bushels}$$
$$\sigma \approx 2.5 \text{ bushels}$$

The next step is to solve the equation

$$z_{\alpha/2}\sigma_{(\bar{x}_1-\bar{x}_2)} = SE \quad \text{or} \quad z_{\alpha/2}\sqrt{\frac{\sigma_1^2}{n_1} + \frac{\sigma_2^2}{n_2}} = SE$$

for n, where $n = n_1 = n_2$. Since we want the estimate to lie within $SE = .25$ of $(\mu_1 - \mu_2)$ with confidence coefficient equal to .95, we have $z_{\alpha/2} = z_{.025} = 1.96$. Then, letting $\sigma_1 = \sigma_2 = 2.5$ and solving for n, we have

$$1.96\sqrt{\frac{(2.5)^2}{n} + \frac{(2.5)^2}{n}} = .25$$

$$1.96\sqrt{\frac{2(2.5)^2}{n}} = .25$$

$$n = 768.32 \approx 769 \text{ (rounding up)}$$

Consequently, you will have to sample 769 acres of wheat for each fertilizer to estimate the difference in mean yield per acre to within .25 bushel.

Look Back Since $n = 769$ would necessitate extensive and costly experimentation, you might decide to allow a larger sampling error (say, $SE = .50$ or $SE = 1$) in order to reduce the sample size, or you might decrease the confidence coefficient. The point is that we can obtain an idea of the experimental effort necessary to achieve a specified precision in our final estimate by determining the approximate sample size *before* the experiment is begun.

Now Work Exercises 7.51

EXAMPLE 7.7

FINDING THE SAMPLE SIZES FOR ESTIMATING μ_d: Comparing Two Measuring Instruments

Problem A laboratory manager wishes to compare the difference in the mean reading of two instruments, A and B, designed to measure the potency (in parts per million) of an antibiotic. To conduct the experiment, the manager plans to select n_d specimens of the antibiotic from a vat and to measure each specimen with both instruments. The difference $(\mu_A - \mu_B)$ will be estimated based on the n_d paired differences $(x_A - x_B)$ obtained in the experiment. If preliminary measurements suggest that the differences will range between plus or minus 10 parts per million, how many differences will be needed to estimate $(\mu_A - \mu_B)$ correct to within 1 part per million with confidence equal to .99?

Solution The estimator for $(\mu_A - \mu_B)$, based on a paired difference experiment, is $\bar{x}_d = (\bar{x}_A - \bar{x}_B)$ and

$$\sigma_{\bar{x}_d} = \frac{\sigma_d}{\sqrt{n_d}}$$

Thus, the number n_d of pairs of measurements needed to estimate $(\mu_A - \mu_B)$ to within 1 part per million can be obtained by solving for n_d in the equation

$$z_{\alpha/2}\left(\frac{\sigma_d}{\sqrt{n_d}}\right) = SE$$

where $z_{.005} = 2.58$ and $SE = 1$. To solve this equation for n_d, we need to have an approximate value for σ_d.

We are given the information that the differences are expected to range from -10 to 10 parts per million. Letting the range equal $4\sigma_d$, we find

$$\text{Range} = 20 \approx 4\sigma_d$$

$$\sigma_d \approx 5$$

Substituting $\sigma_d = 5$, $SE = 1$, and $z_{.005} = 2.58$ into the equation and solving for n_d, we obtain

$$2.58\left(\frac{5}{\sqrt{n_d}}\right) = 1$$

$$n_d = [(2.58)(5)]^2$$

$$= 166.41$$

Therefore, it will require $n_d = 167$ pairs of measurements to estimate $(\mu_A - \mu_B)$ correct to within 1 part per million using the paried difference experiment.

The box summarizes the procedures for determining the sample sizes necessary for estimating $(\mu_1 - \mu_2)$ for the case $n_1 = n_2$ and for estimating μ_d.

Determination of Sample Size for Comparing Two Means

Independent Random Samples

To estimate $(\mu_1 - \mu_2)$ to within a given sampling error SE and with confidence level $(1 - \alpha)$, use the following formula to solve for equal sample sizes that will achieve the desired reliability:

$$n_1 = n_2 = \frac{(z_{\alpha/2})^2(\sigma_1^2 + \sigma_2^2)}{(SE)^2}$$

You will need to substitute estimates for the values of σ_1^2 and σ_2^2 before solving for the sample size. These estimates might be sample variances s_1^2 and s_2^2 from prior sampling (e.g., a pilot study), or from an educated (and conservatively large) guess based on the range—that is, $s \approx R/4$.

Paired Difference Experiment

To estimate μ_d to within a given sampling error SE and with confidence level $(1 - \alpha)$, use the following formula to solve for n_d that will achieve the desired reliability:

$$n_d = \frac{(z_{\alpha/2})^2\sigma_d^2}{(SE)^2}$$

You will need to substitute an estimate of σ_d^2 before solving for the sample size. This estimate might be the sample variance s_d^2 from prior sampling (e.g., a pilot study), or from an educated (and conservatively large) guess based on the range—that is, $s_d = R/4$.

Exercises 7.48–7.59

Understanding the Principles

7.48 When determining the sample sizes for estimating $\mu_1 - \mu_2$, how do you obtain estimates of the population variances $(\sigma_1)^2$ and $(\sigma_2)^2$ used in the calculations?

7.49 When determining the sample size for estimating μ_d, how do you obtain an estimates of the population variance σ_d^2 used in the calculations?

7.50 If the sample size calculation yields a value of n that is too large to be practical, how should you proceed?

Learning the Mechanics

7.51 Find the appropriate values of n_1 and n_2 (assume $n_1 = n_2$)
NW   needed to estimate $(\mu_1 - \mu_2)$ with
   a. A sampling error equal to 3.2 with 95% confidence. From prior experience it is known that $\sigma_1 \approx 15$ and $\sigma_2 \approx 17$.
   b. A sampling error equal to 8 with 99% confidence. The range of each population is 60.
   c. A 90% confidence interval of width 1.0. Assume that $\sigma_1^2 \approx 5.8$ and $\sigma_2^2 \approx 7.5$.

7.52 Suppose you want to estimate the difference between two population means correct to within 2.2 with probability .95. If prior information suggests that the population variances are approximately equal to $\sigma_1^2 = \sigma_2^2 = 15$ and you want to select independent random samples of equal size from the populations, how large should the sample sizes, n_1 and n_2, be?

7.53 Enough money has been budgeted to collect $n = 100$ paired observations from populations 1 and 2 in order to estimate $(\mu_1 - \mu_2)$. Prior information indicates that $\sigma_d = 12$. Have sufficient funds been allocated to construct a 90% confidence interval for $(\mu_1 - \mu_2)$ of width 5 or less? Justify your answer.

Applying the Concepts—Basic

7.54 **Bulimia study.** Refer to *The American Statistician* (May 2001) study comparing the "fear of negative evaluation" (FNE) scores for bulimic and normal female students, Exercise 7.17 (p. 367). Suppose you want to estimate $(\mu_B - \mu_N)$, the difference between the population means of the FNE scores for bulimic and normal female students, using a 95% confidence interval with a sampling error of two points. Find the sample sizes required to obtain such an estimate. Assume equal sample sizes of $\sigma_B^2 = \sigma_N^2 = 25$.

7.55 **Incomes of high school graduates.** A high school counselor wants to estimate the difference in mean income per day between high school graduates who have a college education and those who have not gone on to college. Suppose it is decided to compare the daily incomes of 30-year-olds, and the range of daily incomes for both groups is approximately $200 per day. How many people from each group should be sampled in order to estimate the true difference between mean daily incomes correct to within $10 per day with probability .9? Assume that $n_1 = n_2$.

7.56 **Laughter among deaf signers.** Refer to the *Journal of Deaf Studies and Deaf Education* (Fall 2006) paired difference study on vocalized laughter among deaf users of sign language, presented in Exercise 7.36 (p. 380). Suppose you want to estimate $\mu_d = (\mu_S - \mu_A)$, the difference between the population mean number of laugh episodes of deaf speakers and deaf audience members, using a 90% confidence interval with a sampling error of .75. Find the number of pairs of deaf people required to obtain such an estimate, assuming that the variance of the paired differences is $\sigma_d^2 = 3$.

Applying the Concepts—Intermediate

7.57 **Health hazards of housework.** Is housework hazardous to your health? A study in the *Public Health Reports* (July–Aug. 1992) compares the life expectancies of 25-year-old white women in the labor force to those who are housewives. How large a sample would have to be taken from each group in order to be 95% confident that the estimate of difference in mean life expectancies for the two groups is within 1 year of the true difference? Assume that equal sample sizes will be selected from the two groups, and that the standard deviation for both groups is approximately 15 years.

7.58 **Study of inflamed ear lobes.** In a *Journal of Ethnopharmacology* (June 1995) study, mice with inflamed ear lobes were randomly divided into two groups; one group was treated with bear bile while the other was treated with normal saline. Assuming equal-sized groups, how many mice should be included in each group in order to estimate the difference in the mean degree of swelling to within 2 milligrams with 95% confidence? Use the fact that the degree of swelling standard deviations for the bear bile and normal saline groups are approximately 4 milligrams and 3 milligrams, respectively.

7.59 **Scouting an NFL free agent.** In seeking a free agent NFL running back, a general manager is looking for a player with high mean yards gained per carry and a small standard deviation. Suppose the GM wishes to compare the mean yards gained per carry for two free agents based on independent random samples of their yards gained per carry. Data from last year's pro football season indicate that $\sigma_1 = \sigma_2 \approx 5$ yards. If the GM wants to estimate the difference in means correct to within 1 yard with a confidence level of .90, how many runs would have to be observed for each player? (Assume equal sample sizes.)

7.4 A Nonparametric Test for Comparing Two Populations: Independent Sampling (Optional)

Suppose two independent random samples are to be used to compare two populations and the *t*-test of Section 7.1 is inappropriate for making the comparison. We may be unwilling to make assumptions about the form of the underlying population probability distributions or we may be unable to obtain exact values of the sample measurements. If the data can be ranked in order of magnitude for either of these situations, the nonparametric **Wilcoxon rank sum test** (developed by Frank Wilcoxon) can be used to test the hypothesis that the probability distributions associated with the two populations are equivalent.

Suppose, for example, an experimental psychologist wants to compare reaction times for adult males under the influence of drug A with reaction times for those under the influence of drug B. Experience has shown that the populations of reaction-time measurements often possess probability distributions that are skewed to the right, as shown in Figure 7.14. Consequently, a *t*-test should not be used to compare the mean reaction times for the two drugs, because the normality assumption that is required for the *t*-test may not be valid.

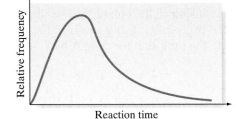

FIGURE 7.14

Typical probability distribution of reaction times

Suppose the psychologist randomly assigns seven subjects to each of two groups, one group to receive drug A and the other to receive drug B. The reaction time for each subject is measured at the completion of the experiment. These data (with the exception of the measurement for one subject in group A who was eliminated from the experiment for personal reasons) are shown in Table 7.6.

DRUGS

TABLE 7.6   Reaction Times of Subjects under the Influence of Drug A or B

Drug A		Drug B	
Reaction Time (seconds)	Rank	Reaction Time (seconds)	Rank
1.96	4	2.11	6
2.24	7	2.43	9
1.71	2	2.07	5
2.41	8	2.71	11
1.62	1	2.50	10
1.93	3	2.84	12
		2.88	13

The population of reaction times for either of the drugs—say, drug A—is that which could conceptually be obtained by giving drug A to all adult males. To compare the probability distributions for populations A and B, *we first rank the sample observations as though they were all drawn from the same population.* That is, we pool the measurements from both samples and then rank all the measurements from the smallest (a rank of 1) to the largest (a rank of 13). The results of this ranking process are also shown in Table 7.6.

If, on the one hand, the two populations were identical, we would expect the ranks to be *randomly mixed* between the two samples. If, on the other hand, one population tends to have longer reaction times than the other, we would expect the larger ranks to be mostly in one sample and the smaller ranks mostly in the other. Thus, the test statistic for the Wilcoxon test is based on the totals of the ranks for each of the two samples— that is, on the **rank sums**. When, for example, the sample sizes are equal, the greater the

TABLE 7.7 **Reproduction of Part of Table V in Appendix A: Critical Values for the Wilcoxon Rank Sum Test**

$\alpha = .025$ one-tailed; $\alpha = .05$ two-tailed

n_2 \ n_1	3		4		5		6		7		8		9		10	
	T_L	T_U	T_L	T_U	T_L	T_U	T_L	T_U	T_L	T_U	T_L	T_U	T_L	T_U	T_L	T_U
3	5	16	6	18	6	21	7	23	7	26	8	28	8	31	9	33
4	6	18	11	25	12	28	12	32	13	35	14	38	15	41	16	44
5	6	21	12	28	18	37	19	41	20	45	21	49	22	53	24	56
6	7	23	12	32	19	41	26	52	28	56	29	61	31	65	32	70
7	7	26	13	35	20	45	28	56	37	68	39	73	41	78	43	83
8	8	28	14	38	21	49	29	61	39	73	49	87	51	93	54	98
9	8	31	15	41	22	53	31	65	41	78	51	93	63	108	66	114
10	9	33	16	44	24	56	32	70	43	83	54	98	66	114	79	131

difference in the rank sums, the greater will be the weight of evidence indicating a difference between the probability distributions of the populations. In the reaction-times example, we denote the rank sum for drug A by T_1 and that for drug B by T_2. Then

$$T_1 = 4 + 7 + 2 + 8 + 1 + 3 = 25$$

$$T_2 = 6 + 9 + 5 + 11 + 10 + 12 + 13 = 66$$

The sum of T_1 and T_2 will always equal $n(n + 1)/2$, where $n = n_1 + n_2$. So, for this example, $n_1 = 6$, $n_2 = 7$, and

$$T_1 + T_2 = \frac{13(13 + 1)}{2} = 91$$

Since $T_1 + T_2$ is fixed, a small value for T_1 implies a large value for T_2 (and vice versa) and a large difference between T_1 and T_2. Therefore, the smaller the value of one of the rank sums, the greater is the evidence indicating that the samples were selected from different populations.

The test statistic for this test is the rank sum for the smaller sample; or, in the case where $n_1 = n_2$, either rank sum can be used. Values that locate the rejection region for this rank sum are given in Table V of Appendix A, a partial reproduction of which is shown in Table 7.7. The columns of the table represent n_1, the first sample size, and the rows represent n_2, the second sample size. *The T_L and T_U entries in the table are the boundaries of the lower and upper regions, respectively, for the rank sum associated with the sample that has fewer measurements.* If the sample sizes n_1 and n_2 are the same, either rank sum may be used as the test statistic. To illustrate, suppose $n_1 = 8$ and $n_2 = 10$. For a two-tailed test with $\alpha = .05$, we consult the table and find that the null hypothesis will be rejected if the rank sum of sample 1 (the sample with fewer measurements), T, is less than or equal to $T_L = 54$ *or* greater than or equal to $T_U = 98$. The Wilcoxon rank sum test is summarized in the next two boxes.

Wilcoxon Rank Sum Test: Independent Samples*

Let D_1 and D_2 represent the probability distributions for populations 1 and 2, respectively.

One-Tailed Test

H_0: D_1 and D_2 are identical
H_a: D_1 is shifted to the right of D_2
[or H_a: D_1 is shifted to the left of D_2]

Two-Tailed Test

H_0: D_1 and D_2 are identical
H_a: D_1 is shifted either to the left or to the right of D_2

(continued)

*Another statistic used to compare two populations on the basis of independent random samples is the *Mann–Whitney U-statistic,* a simple function of the rank sums. It can be shown that the Wilcoxon rank sum test and the Mann–Whitney U-test are equivalent.

Test statistic:

T_1, if $n_1 < n_2$; T_2, if $n_2 < n_1$
(Either rank sum can be used if $n_1 = n_2$.)

Rejection region:
T_1: $T_1 \geq T_U$ [or $T_1 \leq T_L$]
T_2: $T_2 \leq T_L$ [or $T_2 \geq T_U$]

Test statistic:

T_1, if $n_1 < n_2$; T_2, if $n_2 < n_1$
(Either rank sum can be used if $n_1 = n_2$.)
We will denote this rank sum as T.

Rejection region:
$T \leq T_L$ or $T \geq T_U$

where T_L and T_U are obtained from Table V of Appendix A.

Ties: Assign tied measurements the average of the ranks they would receive if they were unequal, but occurred in successive order. For example, if the third-ranked and fourth-ranked measurements are tied, assign each a rank of $(3 + 4)/2 = 3.5$.

Conditions Required for a Valid Rank Sum Test:

1. The two samples are random and independent.

2. The two probability distributions from which the samples are drawn are continuous.

Note that the assumptions necessary for the validity of the Wilcoxon rank sum test do not specify the shape or type of probability distribution. However, the distributions are assumed to be continuous so that the probability of tied measurements is 0 (see Chapter 4) and each measurement can be assigned a unique rank. In practice, however, rounding of continuous measurements will sometimes produce ties. As long as the number of ties is small relative to the sample sizes, the Wilcoxon test procedure will still have an approximate significance level of α. The test is not recommended to compare discrete distributions, for which many ties are expected.

EXAMPLE 7.8

APPLYING THE RANK SUM TEST—Comparing Reaction Times of Two Drugs

Problem Do the data given in Table 7.6 provide sufficient evidence to indicate a shift in the probability distributions for drugs A and B—that is, that the probability distribution corresponding to drug A lies either to the right or to the left of the probability distribution corresponding to drug B? Test at the .05 level of significance.

Solution

H_0: The two populations of reaction times corresponding to drug A and drug B have the same probability distribution.

H_a: The probability distribution for drug A is shifted to the right or left of the probability distribution for drug B.*

Test statistic: Since drug A has fewer subjects than drug B, the test statistic is T_1, the rank sum of drug A's reaction times.

Rejection region: Since the test is two sided, we consult part a of Table XII for the rejection region corresponding to $\alpha = .05$. We will reject H_0 for $T_1 \leq T_L$ or $T_1 \geq T_U$. Thus, we will reject H_0 if $T_1 \leq 28$ or $T_1 \geq 56$.

Since T_1, the rank sum of drug A's reaction times in Table 7.6, is 25, it is in the rejection region. (See Figure 7.15.)† Therefore, there is sufficient evidence to reject H_0. This same conclusion can be reached with a statistical software package. The SAS printout of the analysis is shown in Figure 7.16. Both the test statistic ($T_1 = 25$) and one-tailed p-value ($p = .007$) are highlighted on the printout. The one-tailed p-value is less than $\alpha = .05$, leading us to reject H_0.

---

*The alternative hypotheses in this chapter will be stated in terms of a difference in the *location* of the distributions. However, since the shapes of the distributions may also differ under H_a, some of the figures (e.g., Figure 7.15) depicting the alternative hypothesis will show probability distributions with different shapes.
†Figure 7.15 depicts only one side of the two-sided alternative hypothesis. The other would show the distribution for drug A shifted to the right of the distribution for drug B.

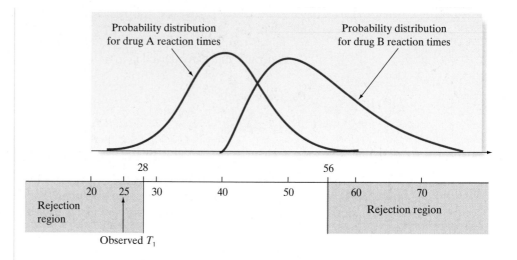

FIGURE 7.15

Alternative hypothesis and rejection region for Example 7.8.

The NPAR1WAY Procedure

Wilcoxon Scores (Rank Sums) for Variable REACTIME
Classified by Variable DRUG

DRUG	N	Sum of Scores	Expected Under H0	Std Dev Under H0	Mean Score
A	6	25.0	42.0	7.0	4.166667
B	7	66.0	49.0	7.0	9.428571

Wilcoxon Two-Sample Test

Statistic (S)	25.0000

Normal Approximation
Z	-2.4286
One-Sided Pr < Z	0.0076
Two-Sided Pr > |Z|	0.0152

t Approximation
| One-Sided Pr < Z | 0.0159 |
| Two-Sided Pr > |Z| | 0.0318 |

Exact Test
| One-Sided Pr <= S | 0.0070 |
| Two-Sided Pr >= |S - Mean| | 0.0140 |

Kruskal-Wallis Test

Chi-Square	5.8980
DF	1
Pr > Chi-Square	0.0152

FIGURE 7.16

SAS printout for Example 7.8

Look Back Our conclusion is that the probability distributions for drugs A and B are not identical. In fact, it appears that drug B tends to be associated with reaction times that are larger than those associated with drug A (because T_1 falls into the lower tail of the rejection region).

Now Work Exercise 7.64

Table V in Appendix A gives values of T_L and T_U for values of n_1 and n_2 less than or equal to 10. When both sample sizes, n_1 and n_2, are 10 or larger, the sampling distribution of T_1 can be approximated by a normal distribution with mean

$$E(T_1) = \frac{n_1(n_1 + n_2 + 1)}{2}$$

and variance

$$\sigma_{T_1}^2 = \frac{n_1 n_2(n_1 + n_2 + 1)}{12}$$

Therefore, for $n_1 \geq 10$ and $n_2 \geq 10$, we can conduct the Wilcoxon rank sum test using the familiar z-test of Section 7.1. The test is summarized in the following box:

The Wilcoxon Rank Sum Test for Large Samples ($n_1 \geq 10$ and $n_2 \geq 10$)

Let D_1 and D_2 represent the probability distributions for populations 1 and 2, respectively.

One-Tailed Test

H_0: D_1 and D_2 are identical
H_a: D_1 is shifted to the right of D_2
(or H_a: D_1 is shifted to the left of D_2)

Two-Tailed Test

H_0: D_1 and D_2 are identical
H_a: D_1 is shifted to the right or to the left of D_2

$$\text{Test statistic: } z = \frac{T_1 - \dfrac{n_1(n_1 + n_2 + 1)}{2}}{\sqrt{\dfrac{n_1 n_2(n_1 + n_2 + 1)}{12}}}$$

Rejection region:

$z > z_\alpha$ (or $z < -z_\alpha$)

Rejection region:

$|z| > z_{\alpha/2}$

STATISTICS IN ACTION REVISITED

Comparing Treatment and Control Roach Trails Using a Nonparametric Test

We return to the study of the trail-following ability of German cockroaches (p. 351) and the entomologist's theory that cockroaches, on average, will tend to follow a feces extract trail closer than they would follow a methanol (control) trail. Recall that four roach groups were utilized in the experiment: adult males, adult females, gravid (pregnant) females, and nymphs (immatures). In this application, we test the theory for adult males only.

Now let μ_1 represent the population mean trail deviation of adult male cockroaches assigned to the control trail and μ_2 represent the population mean trail deviation of adult male cockroaches following the feces extract trail. Again, we want to know whether $\mu_1 > \mu_2$; that is, we want to conduct the one-tailed test, H_0: $(\mu_1 - \mu_2) = 0$ versus H_a: $(\mu_1 - \mu_2) > 0$.

Focusing only on the data for adult males in the **ROACH** file, we find that $n_1 = 10$ and $n_2 = 20$; thus, both samples are small. In order to apply the small-sample t-test, we must assume that both trail deviation distributions are approximately normal. Figure SIA7.2 shows MINITAB histograms for the two samples. Neither appear to come from an approximate normal distribution. Consequently, we will apply the appropriate nonparametric test—the Wilcoxon rank sum test.

FIGURE SIA7.2
MINITAB histograms of trail deviations for two cockroach trails

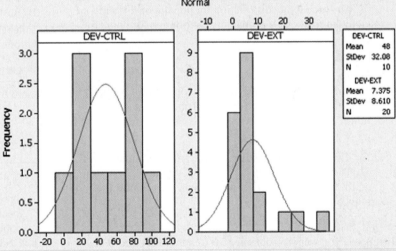

(continued)

We conducted this test using MINITAB. The output is shown in Figure SIA7.3. (*Note:* MINITAB conducts the equivalent Mann-Whitney *U*-test.) The one-tailed *p*-value of the test (.0000) is highlighted on Figure SIA7.3. Therefore, at $\alpha = .05$, there is sufficient evidence to conclude that the trail deviation distribution of adult male cockroaches following the control is shifted significantly above the trail deviation distribution of adult male cockroaches following the feces extract trail. Consequently, the entomologist's theory is supported for adult male cockroaches.

FIGURE SIA7.3
MINITAB nonparametric comparison of two cockroach trails for adult males

```
Mann-Whitney Test and CI: DEV-CTRL, DEV-EXT

               N   Median
DEV-CTRL   10    46.15
DEV-EXT    20     4.50


Point estimate for ETA1-ETA2 is 39.70
95.5 Percent CI for ETA1-ETA2 is (13.11,71.69)
W = 244.0
Test of ETA1 = ETA2 vs ETA1 > ETA2 is significant at 0.0000
The test is significant at 0.0000 (adjusted for ties)
```

Exercises 7.60–7.76

Understanding the Principles

7.60 What is a rank sum?

7.61 *True or False.* If the rank sum for sample 1 is much larger than the rank sum for sample 2 when $n_1 = n_2$, then the distribution of population 1 is likely to be shifted to the right of the distribution of population 2.

7.62 What conditions are required for a valid application of the Wilcoxon rank sum test?

Learning the Mechanics

7.63 Specify the test statistic and the rejection region for the Wilcoxon rank sum test for independent samples in each of the following situations:

a. H_0: Two probability distributions, 1 and 2, are identical
H_a: The probability distribution for population 1 is shifted to the right or left of the probability distribution for population 2
$n_1 = 7, n_2 = 8, \alpha = .10$

b. H_0: Two probability distributions, 1 and 2, are identical
H_a: The probability distribution for population 1 is shifted to the right of the probability distribution for population 2
$n_1 = 6, n_2 = 6, \alpha = .05$

c. H_0: Two probability distributions, 1 and 2, are identical
H_a: The probability distribution for population 1 is shifted to the left of the probability distribution for population 2
$n_1 = 7, n_2 = 10, \alpha = .025$

d. H_0: Two probability distributions, 1 and 2, are identical
H_a: The probability distribution for population 1 is shifted to the right or left of the probability distribution for population 2
$n_1 = 20, n_2 = 20, \alpha = .05$

7.64 Suppose you want to compare two treatments, A and B. In **NW** particular, you wish to determine whether the distribution for population B is shifted to the right of the distribution for population A. You plan to use the Wilcoxon rank sum test.

a. Specify the null and alternative hypotheses you would test.
b. Suppose you obtained the following independent random samples of observations on experimental units subjected to the two treatments:

LM7_64

Sample A	37,	40,	33,	29,	42,	33,	35,	28,	34,
Sample B	65,	35,	47,	52					

Conduct a test of the hypotheses you specified in part **a.** Test, using $\alpha = .05$.

7.65 Suppose you wish to compare two treatments, A and B, on the basis of independent random samples of 15 observations selected from each of the two populations. If $T_1 = 173$, do the data provide sufficient evidence to indicate that distribution A is shifted to the left of distribution B? Test, using $\alpha = .05$.

7.66 Independent random samples are selected from two populations. The data are shown in the following table:

LM7_66

Sample 1		Sample 2		
15	16	5	9	5
10	13	12	8	10
12	8	9	4	

a. Use the Wilcoxon rank sum test to determine whether the data provide sufficient evidence to indicate a shift in the locations of the probability distributions of the sampled populations. Test, using $\alpha = .05$.
b. Do the data provide sufficient evidence to indicate that the probability distribution for population 1 is shifted to

the right of the probability distribution for population 2? Use the Wilcoxon rank sum test with $\alpha = .05$.

Applying the Concepts—Basic

7.67 Research on eating disorders. The "fear of negative evaluation" (FNE) scores for 11 female students known to suffer from the eating disorder bulimia and 14 female students with normal eating habits, first presented in Exercise 2.38 (p. 48), are reproduced in the table at the bottom of the page. (Recall that the higher the score, the greater is the fear of a negative evaluation.) Suppose you want to determine whether the distribution of the FNE scores for bulimic female students is shifted above the corresponding distribution for female students with normal eating habits.

a. Specify H_0 and H_a for the test.
b. Rank all 25 FNE scores in the data set from smallest to largest.
c. Sum the ranks of the 11 FNE scores for bulimic students.
d. Sum the ranks of the 14 FNE scores for students with normal eating habits.
e. Give the rejection region for a nonparametric test of the data if $\alpha = .10$.
f. Conduct the test and give the conclusion in the words of the problem.

7.68 Bursting strength of bottles. Polyethylene terephthalate (PET) bottles are used for carbonated beverages. A critical property of PET bottles is their bursting strength (i.e., the pressure at which bottles filled with water burst when pressurized). In the *Journal of Data Science* (May 2003), researchers measured the bursting strength of PET bottles made from two different designs: an old design and a new design. The data (in pounds per square inch) for 10 bottles of each design are shown in the accompanying table. Suppose you want to compare the distributions of bursting strengths for the two designs.

PET

Old Design	210	212	211	211	190	213	212	211	164	209
New Design	216	217	162	137	219	216	179	153	152	217

a. Rank all 20 observed pressures from smallest to largest, and assign ranks from 1 to 20.
b. Sum the ranks of the observations from the old design.
c. Sum the ranks of the observations from the new design.
d. Compute the Wilcoxon rank sum statistic.
e. Carry out a nonparametric test (at $\alpha = .05$) to compare the distribution of bursting strengths for the two designs.

7.69 Children's recall of TV ads. Refer to the *Journal of Advertising* (Spring 2006) study of children's recall of television advertisements, presented in Exercise 7.11 (p. 366). Two groups of children were shown a 60-second commercial for Sunkist Fun

Fruit Rock-n-Roll Shapes. One group (the A/V group) was shown both the audio and video portions of the ad; the other group (the video-only group) was shown only the video portion of the commercial. The number out of 10 specific items from the ad recalled correctly by each child is shown in the accompanying table. Recall that the researchers theorized that children who receive an audiovisual presentation will have the same level of recall as those who receive only the visual aspects of the ad. Consider testing the researchers' theory, using the Wilcoxon rank sum test.

FUNFRUIT

A/V group:	0 4 6 6 1 2 2 6 6 4 1 2 6 1 3 0 2 5 4 5
Video-only group:	6 3 6 2 2 4 7 6 1 3 6 2 3 1 3 2 5 2 4 6

Source: Maher, J. K., Hu, M. Y., & Kolbe, R. H. "Children's recall of television ad elements," *Journal of Advertising*, Vol. 35, No. 1, Spring 2006 (simulated from summary information in Table 1).

a. Set up the appropriate null and alternative hypotheses for the test.
b. Find the value of the test statistic.
c. Give the rejection region for $\alpha = .10$.
d. Make the appropriate inference. What can you say about the researchers' theory?

7.70 Reading Japanese books. Refer to the *Reading in a Foreign Language* (Apr. 2004) experiment to improve the Japanese reading comprehension levels of University of Hawaii students, presented in Exercise 7.13 (p. 366). Recall that 14 students participated in a 10-week extensive reading program in a second-semester Japanese course. The number of books read by each student and the student's course grade are repeated in the accompanying table. Consider a comparison of the distributions of number of books read by students who earn an "A" grade and those who earn a "B" or "C" grade.

JAPANESE

Number of Books	Course Grade	Number of Books	Course Grade
53	A	30	A
42	A	28	B
40	A	24	A
40	B	22	C
39	A	21	B
34	A	20	B
34	A	16	B

Source: Hitosugi, C. I., and Day, R. R. "Extensive reading in Japanese," *Reading in a Foreign Language,* Vol. 16, No. 1, Apr. 2004 (Table 4).

a. Rank all 14 observations from smallest to largest, and assign ranks from 1 to 14.
b. Sum the ranks of the observations for students with an "A" grade.

BULIMIA

Bulimic students	21	13	10	20	25	19	16	21	24	13	14			
Normal students	13	6	16	13	8	19	23	18	11	19	7	10	15	20

Source: Randles, R. H. "On neutral responses (zeros) in the sign test and ties in the Wilcoxon–Mann–Whitney test," *American Statistician,* Vol. 55, No. 2, May 2001 (Figure 3).

c. Sum the ranks of the observations for students with either a "B" or "C" grade.

d. Compute the Wilcoxon rank sum statistic.

e. Carry out a nonparametric test (at $\alpha = .10$) to compare the distribution of the number of books read by the two populations of students.

7.71 Visual acuity of children. In a comparison of the visual acuity of deaf and hearing children, eye movement rates are taken on 10 deaf and 10 hearing children. The data are shown in the accompanying table. A clinical psychologist believes that deaf children have greater visual acuity than hearing children. (The larger a child's eye movement rate, the more visual acuity the child possesses.)

EYEMOVE

Deaf Children		Hearing Children	
2.75	1.95	1.15	1.23
3.14	2.17	1.65	2.03
3.23	2.45	1.43	1.64
2.30	1.83	1.83	1.96
2.64	2.23	1.75	1.37

a. Use the Wilcoxon rank sum procedure to test the psychologist's claim at $\alpha = .05$.

b. Conduct the test by using the large-sample approximation for the Wilcoxon rank sum test. Compare the results with those found in part **a**.

Applying the Concepts—Intermediate

7.72 Rain in Colorado. The data in the accompanying table, extracted from *Technometrics* (Feb. 1986), represent daily accumulated streamflow and precipitation (in inches) for two U.S. Geological Survey stations in Colorado. Conduct a test to determine whether the distributions of daily accumulated streamflow and precipitation for the two stations differ with location. Use $\alpha = .10$. Why is a nonparametric test appropriate for these data?

COLORAIN

Station 1			Station 2		
127.96	108.91	100.85	114.79	85.54	280.55
210.07	178.21	85.89	109.11	117.64	145.11
203.24	285.37		330.33	302.74	95.36

Source: Gastwirth, J. L., and Mahmoud, H. "An efficient robust nonparametric test for scale change for data from a gamma distribution," *Technometrics*, Vol. 28, No. 1, Feb. 1986, p. 83 (Table 2).

7.73 Computer-mediated communication study. Computer-mediated communication (CMC) is a form of interaction that heavily involves technology (e.g., instant messaging, e-mail). A study was conducted to compare relational intimacy in people interacting via CMC with people meeting face-to-face (FTF). (*Journal of Computer-Mediated Communication*, Apr. 2004.) Participants were 48 undergraduate students, of which half were randomly assigned to the CMC group and half to the FTF group. Each group was given a task that required communication among its group members. Those in the CMC group communicated via the "chat" mode of instant-messaging software; those in the FTF

group met in a conference room. The variable of interest, relational intimacy score, was measured (on a seven-point scale) for each participant after each of three different meetings. Scores for the first meeting are given in the accompanying table. The researchers hypothesized that the relational intimacy scores for participants in the CMC group will tend to be lower than the relational intimacy scores for participants in the FTF group.

INTIMACY

CMC group:	4 3 3 4 3 3 3 3 4 4 3 4 3 3 2 4 2 4 5 4 4 4 5 3
FTF group:	5 4 4 4 3 3 3 4 3 3 3 3 4 4 4 4 4 3 3 3 4 4 2 4

Note: Data simulated from descriptive statistics provided in article.

a. Which nonparametric procedure should be used to test the researchers' hypothesis?

b. Specify the null and alternative hypotheses of the test.

c. Give the rejection region for the test, using $\alpha = .10$.

d. Conduct the test and give the appropriate conclusion in the context of the problem.

7.74 Brood-parasitic birds. The term *brood-parasitic intruder* is used to describe a bird that searches for and lays eggs in a nest built by a bird of another species. For example, the brown-headed cowbird is known to be a brood parasite of the smaller willow flycatcher. Ornithologists theorize that those flycatchers which recognize, but do not vocally react to, cowbird calls are more apt to defend their nests and less likely to be found and parasitized. In a study published in *The Condor* (May 1995), each of 13 active flycatcher nests was categorized as parasitized (if at least one cowbird egg was present) or nonparasitized. Cowbird songs were taped and played back while the flycatcher pairs were sitting in the nest prior to incubation. The vocalization rate (number of calls per minute) of each flycatcher pair was recorded. The data for the two groups of flycatchers are given in the table. Do the data suggest (at $\alpha = .05$) that the vocalization rates of parasitized flycatchers are higher than those of nonparasitized flycatchers?

COWBIRD

Parasitized	Not Parasitized
2.00	1.00
1.25	1.00
8.50	0
1.10	3.25
1.25	1.00
3.75	.25
5.50	

Source: Uyehara, J. C., and Narins, P. M. "Nest defense by Willow Flycatchers to brood-parasitic intruders." *The Condor*, Vol. 97, No. 2, May 1995, p. 364 (Figure 1).

7.75 Patent infringement case. Refer to the *Chance* (Fall 2002) study of a patent infringement case brought against Intel Corp., presented in Exercise 7.18 (p. 367). Recall that the case rested on whether a patent witness' signature was written on top of key text in a patent notebook or under the key text. Using an X-ray beam, zinc measurements were taken at several spots on the notebook page. The zinc measurements for three notebook locations—on a text line, on a witness line, and on the intersection of the witness and text lines—are reproduced in the following table:

PATENT

Text line:	.335	.374	.440			
Witness line:	.210	.262	.188	.329	.439	.397
Intersection:	.393	.353	.285	.295	.319	

a. Why might the Student's *t*-procedure you applied in Exercise 9.20 be inappropriate for analyzing these data?

b. Use a nonparametric test (at $\alpha = .05$) to compare the distribution of zinc measurements for the text line with the distribution for the intersection.

c. Use a nonparametric test (at $\alpha = .05$) to compare the distribution of zinc measurements for the witness line with the distribution for the intersection.

d. From the results you obtained in parts **b** and **c**, what can you infer about the mean zinc measurements at the three notebook locations?

7.76 Family involvement in homework. *The Journal of Educational Research* (July/Aug. 2003) presented a study of the impact of the interactive Teachers Involve Parents in Schoolwork (TIPS) program. In the study, 128 middle school students were assigned to complete TIPS homework assignments, while 98 students were assigned traditional, noninteractive homework assignments (ATIPS). At the end of the study, all students reported on the level of family involvement in their homework on a five-point scale (0 = Never, 1 = Rarely, 2 = Sometimes, 3 = Frequently, 4 = Always). The data for the science,

math, and language arts homework are saved in the **HWSTUDY** file. (The first five and last five observations in the data set are reproduced in the accompanying table.)

a. Why might a nonparametric test be the most appropriate test to apply in order to compare the levels of family involvement in homework assignments of TIPS and ATIPS students?

b. Conduct a nonparametric analysis to compare the involvement in science homework assignments of TIPS and ATIPS students. Use $\alpha = .05$.

c. Repeat part **b** for mathematics homework assignments.

d. Repeat part **a** for language arts homework assignments.

HWSTUDY (selected observations)

Homework Condition	Science	Math	Language
ATIPS	1	0	0
ATIPS	0	1	1
ATIPS	0	1	0
ATIPS	1	2	0
ATIPS	1	1	2
⋮	⋮	⋮	⋮
TIPS	2	3	2
TIPS	1	4	2
TIPS	2	4	2
TIPS	4	0	3
TIPS	2	0	1

Source: Van Voorhis, F. L. "Teachers' use of interactive homework and its effects on family involvement and science achievement of middle grade students," paper presented at the annual meeting of the American Educational Research Association, Seattle, April 2001.

7.5 A Nonparametric Test for Comparing Two Populations: Paired Difference Experiment (Optional)

Nonparametric techniques may also be employed to compare two probability distributions when a paired difference design is used. For example, consumer preferences for two competing products are often compared by having each of a sample of consumers rate both products. Thus, the ratings have been paired on each consumer. Following is an example of this type of experiment.

For some paper products, softness is an important consideration in determining consumer acceptance. One method of determining softness is to have judges give a sample of the products a softness rating. Suppose each of 10 judges is given a sample of two products that a company wants to compare. Each judge rates the softness of each product on a scale from 1 to 10, with higher ratings implying a softer product. The results of the experiment are shown in Table 7.8.

Since this is a paired difference experiment, we analyze the differences between the measurements. (See Section 7.2.) However, a nonparametric approach developed by Wilcoxon requires that we calculate the ranks of the absolute values of the differences between the measurements (i.e., the ranks of the differences after removing any minus signs). *Note that tied absolute differences are assigned the average of the ranks they would receive if they were unequal, but successive, measurements.* After the absolute differences are ranked, the sum of the ranks of the positive differences of the original measurements, T_+, and the sum of the ranks of the negative differences of the original measurements, T_-, are computed.

SOFTPAPER

TABLE 7.8   Softness Ratings of Paper

Judge	Product A	B	Difference $(A - B)$	Absolute Value of Difference	Rank of $(A - B)$ Absolute Value
1	6	4	2	2	5
2	8	5	3	3	7.5
3	4	5	−1	1	2
4	9	8	1	1	2
5	4	1	3	3	7.5
6	7	9	−2	2	5
7	6	2	4	4	9
8	5	3	2	2	5
9	6	7	−1	1	2
10	8	2	6	6	10

$T_+ = $ Sum of positive ranks $= 46$

$T_- = $ Sum of negative ranks $= 9$

We are now prepared to test the nonparametric hypotheses:

H_0: The probability distributions of the ratings for products A and B are identical.

H_a: The probability distributions of the ratings differ (in location) for the two products. (Note that this is a two-sided alternative and that it implies a two-tailed test.)

Test statistic: $T = $ Smaller of the positive and negative rank sums T_+ and T_-

The smaller the value of T, the greater is the evidence indicating that the two probability distributions differ in location. The rejection region for T can be determined by consulting Table VI in Appendix A, part of which is shown in Table 7.9. This table gives a value T_0 for both one-tailed and two-tailed tests for each value of n, the number of matched pairs. For a two-tailed test with $\alpha = .05$, we will reject H_0 if $T \leq T_0$. You can see in Table 7.9 that the value of T_0 which locates the boundary of the rejection region for the judges' ratings for $\alpha = .05$ and $n = 10$ pairs of observations is 8. Thus, the rejection region for the test (see Figure 7.17) is

Rejection region: $T \leq 8$ for $\alpha = .05$

Since the smaller rank sum for the paper data, $T_- = 9$, does not fall within the rejection region, the experiment has not provided sufficient evidence indicating that the two paper products differ with respect to their softness ratings at the $\alpha = .05$ level.

Note that if a significance level of $\alpha = .10$ had been used, the rejection region would have been $T \leq 11$ and we would have rejected H_0. In other words, the samples do provide evidence that the probability distributions of the softness ratings differ at the $\alpha = .10$ significance level.

The **Wilcoxon signed rank test** is summarized in the next box. Note that the difference measurements are assumed to have a continuous probability distribution so that the absolute differences will have unique ranks. Although tied (absolute) differences can be assigned ranks by averaging, in order to ensure the validity of the test, the number of ties should be small relative to the number of observations.

FIGURE 7.17

Rejection region for paired difference experiment

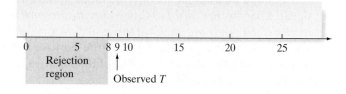

TABLE 7.9   Reproduction of Part of Table VI of Appendix A: Critical Values for the Wilcoxon Paired Difference Signed Rank Test

One-Tailed	Two-Tailed	$n = 5$	$n = 6$	$n = 7$	$n = 8$	$n = 9$	$n = 10$
$\alpha = .05$	$\alpha = .10$	1	2	4	6	8	11
$\alpha = .025$	$\alpha = .05$		1	2	4	6	8
$\alpha = .01$	$\alpha = .02$			0	2	3	5
$\alpha = .005$	$\alpha = .01$				0	2	3
		$n = 11$	$n = 12$	$n = 13$	$n = 14$	$n = 15$	$n = 16$
$\alpha = .05$	$\alpha = .10$	14	17	21	26	30	36
$\alpha = .025$	$\alpha = .05$	11	14	17	21	25	30
$\alpha = .01$	$\alpha = .02$	7	10	13	16	20	24
$\alpha = .005$	$\alpha = .01$	5	7	10	13	16	19
		$n = 17$	$n = 18$	$n = 19$	$n = 20$	$n = 21$	$n = 22$
$\alpha = .05$	$\alpha = .10$	41	47	54	60	68	75
$\alpha = .025$	$\alpha = .05$	35	40	46	52	59	66
$\alpha = .01$	$\alpha = .02$	28	33	38	43	49	56
$\alpha = .005$	$\alpha = .01$	23	28	32	37	43	49
		$n = 23$	$n = 24$	$n = 25$	$n = 26$	$n = 27$	$n = 28$
$\alpha = .05$	$\alpha = .10$	83	92	101	110	120	130
$\alpha = .025$	$\alpha = .05$	73	81	90	98	107	117
$\alpha = .01$	$\alpha = .02$	62	69	77	85	93	102
$\alpha = .005$	$\alpha = .01$	55	61	68	76	84	92

Wilcoxon Signed Rank Test for a Paired Difference Experiment

Let D_1 and D_2 represent the probability distributions for populations 1 and 2, respectively.

One-Tailed Test

H_0: D_1 and D_2 are identical

H_a: D_1 is shifted to the right of D_2 [or H_a: D_1 is shifted to the left of D_2]

Two-Tailed Test

H_0: D_1 and D_2 are identical

H_a: D_1 is shifted either to the left or to the right of D_2

Calculate the difference within each of the n matched pairs of observations. Then rank the absolute value of the n differences from the smallest (rank 1) to the highest (rank n), and calculate the rank sum T_- of the negative differences and the rank sum T_+ of the positive differences. [*Note:* Differences equal to 0 are eliminated, and the number n of differences is reduced accordingly.]

Test statistic:

T_-, the rank sum of the negative differences [or T_+, the rank sum of the positive differences]

Rejection region:

$T_- \leq T_0$ [or $T_+ \leq T_0$]

Test statistic:

T, the smaller of T_+ or T_-

Rejection region:

$T \leq T_0$

where T_0 is given in Table VI in Appendix A.

Ties: Assign tied absolute differences the average of the ranks they would receive if they were unequal, but occurred in successive order. For example, if the third-ranked and fourth-ranked differences are tied, assign both a rank of $(3 + 4)/2 = 3.5$.

Conditions Required for a Valid Signed Rank Test

1. The sample of differences is randomly selected from the population of differences.

2. The probability distribution from which the sample of paired differences is drawn is continuous.

EXAMPLE 7.9

APPLYING THE SIGNED RANK TEST—Comparing Two Crime Prevention Plans

Problem Suppose the police commissioner in a small community must choose between two plans for patrolling the town's streets. Plan A, the less expensive plan, uses voluntary citizen groups to patrol certain high-risk neighborhoods. In contrast, plan B would utilize police patrols. As an aid in reaching a decision, both plans are examined by 10 trained criminologists, each of whom is asked to rate the plans on a scale from 1 to 10. (High ratings imply a more effective crime prevention plan.) The city will adopt plan B (and hire extra police) only if the data provide sufficient evidence that criminologists tend to rate plan B more effective than plan A. The results of the survey are shown in Table 7.10. Do the data provide evidence at the $\alpha = .05$ level that the distribution of ratings for plan B lies above that for plan A?

 CRIMEPLAN

TABLE 7.10   Effectiveness Ratings by 10 Qualified Crime Prevention Experts

Crime Prevention Expert	Plan A	Plan B	Difference (A − B)	Rank of Absolute Difference
1	7	9	−2	4.5
2	4	5	−1	2
3	8	8	0	(Eliminated)
4	9	8	1	2
5	3	6	−3	6
6	6	10	−4	7.5
7	8	9	−1	2
8	10	8	2	4.5
9	9	4	5	9
10	5	9	−4	7.5

Positive rank sum = $T_+ = 15.5$

Solution The null and alternative hypotheses are as follows:

H_0: The two probability distributions of effectiveness ratings are identical

H_a: The effectiveness ratings of the more expensive plan (B) tend to exceed those of plan A

Observe that the alternative hypothesis is one sided (i.e., we only wish to detect a shift in the distribution of the B ratings to the right of the distribution of A ratings); therefore, it implies a one-tailed test of the null hypothesis. (See Figure 7.18.) If the alternative hypothesis is true, the B ratings will tend to be larger than the paired A ratings, more negative differences in pairs will occur, T_- will be large, and T_+ will be small. Because Table VI is constructed to give lower-tail values of T_0, we will use T_+ as the test statistic and reject H_0 for $T_+ \leq T_0$.

The differences in ratings for the pairs (A − B) are shown in Table 7.10. Note that one of the differences equals 0. Consequently, we eliminate this pair from the ranking and reduce the number of pairs to $n = 9$. Looking in Table VI, we have $T_0 = 8$ for a one-tailed test with $\alpha = .05$ and $n = 9$. Therefore, the test statistic and rejection region for the test are

Test statistic: T_+, the positive rank sum

Rejection region: $T_+ \leq 8$

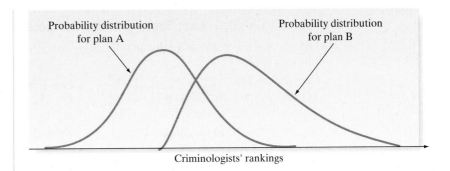

FIGURE 7.18
The alternative hypothesis for Example 7.9

Summing the ranks of the positive differences from Table 7.10, we find that $T_+ = 15.5$. Since this value exceeds the critical value, $T_0 = 8$, we conclude that the sample provides insufficient evidence at the $\alpha = .05$ level to support the alternative hypothesis. The commissioner *cannot* conclude that the plan utilizing police patrols tends to be rated higher than the plan using citizen volunteers. That is, on the basis of this study, extra police will not be hired.

Wilcoxon Signed Ranks Test

Ranks

		N	Mean Rank	Sum of Ranks
A - B	Negative Ranks	6[a]	4.92	29.50
	Positive Ranks	3[b]	5.17	15.50
	Ties	1[c]		
	Total	10		

a. A < B

b. A > B

c. A = B

Test Statistics[b]

	A - B
Z	-.834[a]
Asymp. Sig. (2-tailed)	.404

a. Based on positive ranks.

b. Wilcoxon Signed Ranks Test

FIGURE 7.19
SPSS printout for Example 7.9

An SPSS printout of the analysis, shown in Figure 7.19, confirms the preceding conclusion. Both the test statistic and two-tailed *p*-value are highlighted on the printout. Since the one-tailed *p*-value, $.404/2 = .202$, exceeds $\alpha = .05$, we fail to reject H_0.

Now Work Exercise 7.81

As is the case for the rank sum test for independent samples, the sampling distribution of the signed rank statistic can be approximated by a normal distribution when the number n of paired observations is large (say, $n \geq 25$). The large-sample z-test is summarized in the following box:

Wilcoxon Signed Rank Test for Large Samples ($n \geq 25$)

Let D_1 and D_2 represent the probability distributions for populations 1 and 2, respectively.

One-Tailed Test

H_0: D_1 and D_2 are identical

H_a: D_1 is shifted to the right of D_2 [or

H_a: D_1 is shifted to the left of D_2]

Two-Tailed Test

H_0: D_1 and D_2 are identical

H_a: D_1 is shifted either to the left or to the right of D_2

$$\text{Test statistic: } z = \frac{T_+ - [n(n + 1)/4]}{\sqrt{[n(n + 1)(2n + 1)]/24}}$$

Rejection region:

$z > z_\alpha$ [or $z < -z_\alpha$]

Rejection region:

$|z| > z_{\alpha/2}$

Assumptions: The sample size n is greater than or equal to 25. Differences equal to 0 are eliminated and the number n of differences is reduced accordingly. Tied absolute differences receive ranks equal to the average of the ranks they would have received had they not been tied.

ACTIVITY 7.2: *Box Office Receipts:* Nonparametric Statistics

In this activity, you will refer to your results from Activity 6.2: *Keep the Change: Tests of Hypotheses* and Activity 7.1: *Box Office Receipts: Comparing Population Means.*

1. Referring to Exercise 2 of Activity 7.1, explain why a Wilcoxon signed-rank test might be a better method for testing the difference in mean daily box office receipts of the two movies, especially if the sample is small. Perform the corresponding signed-rank test.

2. Refer to Activity 6.2, in which you investigated bank matching. Consider a study to compare the mean amounts for bank matching in California and bank matching in Florida. Design a Wilcoxon rank sum test to compare the corresponding probability distributions. Be specific about sample sizes, hypotheses, and how a conclusion will be reached. Under what conditions might the rank sum test provide more useful information than the mean comparison test?

Exercises 7.77–7.91

Understanding the Principles

7.77 In order to conduct the Wilcoxon signed rank test, why do we need to assume that the probability distribution of differences is continuous?

7.78 Explain the difference between the one- and two-tailed versions of the Wilcoxon signed rank test for the paired difference experiment.

Learning the Mechanics

7.79 Specify the test statistic and the rejection region for the Wilcoxon signed rank test for the paired difference design in each of the following situations:

a. H_0: Two probability distributions, A and B, are identical

H_a: The probability distribution for population A is shifted to the right or left of the probability distribution for population B

$n = 20, \alpha = .10$

b. H_0: Two probability distributions, A and B, are identical

H_a: The probability distribution for population A is shifted to the right of the probability distribution for population B

$n = 39, \alpha = .05$

c. H_0: Two probability distributions, A and B, are identical

H_a: The probability distribution for population A is shifted to the left of the probability distribution for population B

$n = 7, \alpha = .005$

7.80 A random sample of nine pairs of measurements is shown in the following table:

LM7_80

Pair	Sample Data from Population 1	Sample Data from Population 2
1	8	7
2	10	1
3	6	4
4	10	10
5	7	4
6	8	3
7	4	6
8	9	2
9	8	4

a. Use the Wilcoxon signed rank test to determine whether the data provide sufficient evidence to indicate that the probability distribution for population 1 is shifted to the right of the probability distribution for population 2. Test, using $\alpha = .05$.

b. Use the Wilcoxon signed rank test to determine whether the data provide sufficient evidence to indicate that the probability distribution for population 1 is shifted either to the right or to the left of the probability distribution for population 2. Test, using $\alpha = .05$.

7.81 Suppose you want to test a hypothesis that two treatments, A and B, are equivalent against the alternative hypothesis that the responses for A tend to be larger than those for B. You plan to use a paired difference experiment and to analyze the resulting data with the Wilcoxon signed rank test.

a. Specify the null and alternative hypotheses you would test.

b. Suppose the paired difference experiment yielded the data in the accompanying table. Conduct the test, of part **a**. Test using $\alpha = .025$.

🔵 **LM7_81**

Pair	A	B	Pair	A	B
1	54	45	6	77	75
2	60	45	7	74	63
3	98	87	8	29	30
4	43	31	9	63	59
5	82	71	10	80	82

7.82 Suppose you wish to test a hypothesis that two treatments, A and B, are equivalent against the alternative that the responses for A tend to be larger than those for B.

a. If the number of pairs equals 25, give the rejection region for the large-sample Wilcoxon signed rank test for $\alpha = .05$.

b. Suppose that $T_+ = 273$. State your test conclusions.

c. Find the p-value for the test and interpret it.

Applying the Concepts—Basic

7.83 **Treating psoriasis with the "Doctorfish of Kangal."** Refer to the *Evidence-Based Research in Complementary and Alternative Medicine* (Dec. 2006) study of treating psoriasis with ichthyotherapy, presented in Exercise 2.127 (p. 87). (Recall that the therapy is also known as the "Doctorfish of Kangal," since it uses fish from the hot pools of Kangal, Turkey, to feed on skin scales.) In the study, 67 patients diagnosed with psoriasis underwent three weeks of ichthyotherapy. The Psoriasis Area Severity Index (PASI) of each patient was measured both before and after treatment. (The lower the PASI score, the better is the skin condition.) Before and after-treatment PASI scores were compared with the use of the Wilcoxon signed rank test.

a. Explain why the PASI scores should be analyzed with a test for paired differences.

b. Refer to the box plots shown in Exercise 2.127. Give a reason that the researchers opted to use a nonparametric test to compare the PASI scores.

c. The p-value for the Wilcoxon signed ranks test was reported as $p < .0001$. Interpret this result, and com-

ment on the effectiveness of ichthyotherapy in treating psoriasis.

7.84 **Computer-mediated communication study.** Refer to the *Journal of Computer-Mediated Communication* (Apr. 2004) study comparing people who interact via computer-mediated communication (CMC) with those who meet face-to-face (FTF), presented in Exercise 7.73 (p. 394). Relational intimacy scores (measured on a seven-point scale) were obtained for each participant after each of three different meetings. The researchers hypothesized that relational intimacy scores for participants in the CMC group will tend to be higher at the third meeting than at the first meeting; however, they hypothesize that there are no differences in scores between the first and third meetings for the FTF group.

a. Explain why a nonparametric Wilcoxon signed ranks test is appropriate for analyzing the data.

b. For the CMC group comparison, give the null and alternative hypotheses of interest.

c. Give the rejection region (at $\alpha = .05$) for conducting the test mentioned in part **b**. Recall that there were 14 participants assigned to the CMC group.

d. For the FTF group comparison, give the null and alternative hypotheses of interest.

e. Give the rejection region (at $\alpha = .05$) for conducting the test mentioned in part **d**. Recall that there were 14 participants assigned to the FTF group.

7.85 **Reading comprehension strategies of elementary school children.** An investigation of the reading comprehension strategies employed by good and average elementary school readers was the topic of research published in *The Reading Matrix* (April 2004). Both good and average readers were recruited on the basis of their scores on a midterm language test. Each group was evaluated on how often its members employed each of eight different reading strategies. The accompanying table gives the proportion of times the reading group used each strategy (called the Factor Specificity Index, or FSI score). The researchers conducted a Wilcoxon signed rank test to compare the FSI score distributions of good and average readers.

🔵 **READSTRAT**

STRATEGY	FSI Scores	
	Good Readers	Average Readers
Word meaning	.38	.32
Words in context	.29	.25
Literal comprehension	.42	.25
Draw inference from single string	.60	.26
Draw inference from multiple string	.45	.31
Interpretation of metaphor	.32	.14
Find salient or main idea	.21	.03
From judgment	.73	.80

Source: Ahmed, S., & Asraf, R.M. "Making sense of text: Strategies used by good and average readers," *The Reading Matrix*, Vol. 4, No. 1, April 2004 (Table 2).

a. State H_0 and H_a for the desired test of hypothesis.

b. For each strategy, compute the difference between the FSI scores of good and average readers.

c. Rank the absolute values of the differences.

d. Calculate the value of the signed rank test statistic.

e. Find the rejection region for the test, using $\alpha = .05$.

f. Make the appropriate inference in the words of the problem.

CRASH

7.86 NHTSA new car crash tests. Refer to the National Highway Traffic Safety Administration (NHTSA) new-car crash test data saved in the **CRASH** file. In Exercise 7.40 (p. 381), you compared the chest injury ratings of drivers and front-seat passengers by using the Student's t-procedure for matched pairs. Suppose you want to make the comparison for only those cars which have a driver's rating of five stars (the highest rating). The data for these 18 cars are listed in the accompanying table. Now consider analyzing the data by using the Wilcoxon signed rank test.

a. State the null and alternative hypotheses.

b. Use a statistical software package to find the signed rank test statistic.

c. Give the rejection region for the test, using $\alpha = .01$.

d. State the conclusion in practical terms. Report the p-value of the test.

CRASH

Chest Injury Rating			Chest Injury Rating		
Car	Driver	Passenger	Car	Driver	Passenger
1	42	35	10	36	37
2	42	35	11	36	37
3	34	45	12	43	58
4	34	45	13	40	42
5	45	45	14	43	58
6	40	42	15	37	41
7	42	46	16	37	41
8	43	58	17	44	57
9	45	43	18	42	42

Applying the Concepts—Intermediate

7.87 Concrete-pavement response to temperature. Civil engineers at West Virginia University have developed a three-dimensional model to predict the response of jointed concrete pavement to temperature variations. (*The International Journal of Pavement Engineering*, Sep. 2004.) To validate the model, its predictions were compared with field measurements of key concrete stress variables taken at a newly constructed highway. One variable measured was slab top transverse strain (i.e., change in length per unit length per unit time) at a distance of 1 meter from the longitudinal joint. The 5-hour changes (8:20 P.M. to 1:20 A.M.) in slab top transverse strain for six days are listed in the accompanying table. Analyze the data, using a nonparametric test. Is there a shift in the change in transverse strain distributions between field measurements and the model? Test, using $\alpha = .05$.

SLABSTRAIN

Day	Change in Temperature (°C)	Change in Transverse Strain	
		Field Measurement	3D Model
Oct. 24	−6.3	−58	−52
Dec. 3	13.2	69	59
Dec. 15	3.3	35	32
Feb. 2	−14.8	−32	−24
Mar. 25	1.7	−40	−39
May. 24	−.2	−83	−71

Source: Shoukry, S., William, G., and Riad, M. "Validation of 3DFE model of jointed concrete pavement response to temperature variations," *International Journal of Pavement Engineering,* Vol. 5, No. 3, Sep. 2004 (Table IV).

7.88 Thematic atlas topics. The regional atlas is an important educational resource that is updated on a periodic basis. One of the most critical aspects of a new atlas design is its thematic content. In a survey of atlas users (*Journal of Geography*, May/June 1995), a large sample of high school teachers in British Columbia ranked 12 thematic atlas topics for usefulness. The consensus rankings of the teachers (based on the percentage of teachers who responded that they "would definitely use" the topic) are given in the accompanying table. These teacher rankings were compared with the rankings a group of university geography alumni made three years earlier. Compare the distributions of theme rankings for the two groups with an appropriate nonparametric test. Use $\alpha = .05$. Interpret the results practically.

ATLAS

Theme	Rankings	
	High School Teachers	Geography Alumni
Tourism	10	2
Physical	2	1
Transportation	7	3
People	1	6
History	2	5
Climate	6	4
Forestry	5	8
Agriculture	7	10
Fishing	9	7
Energy	2	8
Mining	10	11
Manufacturing	12	12

Source: Keller, C. P., et al. "Planning the next generation of regional atlases: Input from educators," *Journal of Geography,* Vol. 94, No. 3, May/June 1995, p. 413 (Table 1).

7.89 Treatment for tendon pain. The *British Journal of Sports Medicine* (Feb. 1, 2004) published a study of chronic Achilles tendon pain. Each in a sample of 25 patients with chronic Achilles tendinosis was treated with heavy-load eccentric calf muscle training. Tendon thickness (in millimeters) was measured both before and following the treatment of each patient. The experimental data are listed in the next table. Use a nonparametric test to determine whether the treatment for tendonitis tends to reduce the thickness of tendons. Test, using $\alpha = .10$.

TENDON

Patient	Before Thickness (millimeters)	After Thickness (millimeters)
1	11.0	11.5
2	4.0	6.4
3	6.3	6.1
4	12.0	10.0
5	18.2	14.7
6	9.2	7.3
7	7.5	6.1
8	7.1	6.4
9	7.2	5.7
10	6.7	6.5
11	14.2	13.2
12	7.3	7.5
13	9.7	7.4
14	9.5	7.2
15	5.6	6.3
16	8.7	6.0
17	6.7	7.3
18	10.2	7.0
19	6.6	5.3
20	11.2	9.0
21	8.6	6.6
22	6.1	6.3
23	10.3	7.2
24	7.0	7.2
25	12.0	8.0

Source: Ohberg, L. et al. "Eccentric training in patients with chronic Achilles tendinosis: normalized tendon structure and decreased thickness at follow up." *British Journal of Sports Medicine,* Vol. 38, No. 1, Feb. 1, 2004 (Table 2).

7.90 Neurological impairment of POWs. Eleven prisoners of war during the war in Croatia were evaluated for neurological impairment after their release from a Serbian detention camp. (*Collegium Antropologicum,* June 1997.) All 11 experienced blows to the head and neck and/or loss of consciousness during imprisonment. Neurological impairment was assessed by measuring the amplitude of the visual evoked potential (VEP) in both eyes at two points in time: 157 days and 379 days after their release. (The higher the VEP value, the greater the neurological impairment.) The data on the 11 POWs are shown in the accompanying table. Determine whether the VEP measurements of POWs 379 days after their release tend to be greater than the VEP measurements of POWs 157 days after their release. Test, using $\alpha = .05$.

POWVEP

POW	157 Days after Release	379 Days after Release
1	2.46	3.73
2	4.11	5.46
3	3.93	7.04
4	4.51	4.73
5	4.96	4.71
6	4.42	6.19
7	1.02	1.42
8	4.30	8.70
9	7.56	7.37
10	7.07	8.46
11	8.00	7.16

Source: Vrca, A., et al. "The use of visual evoked potentials to follow-up prisoners of war after release from detention camps." *Collegium Antropologicum,* Vol. 21, No. 1, June 1997, p. 232. (Data simulated from information provided in Table 3.)

Applying the Concepts—Advanced

7.91 Bowlers' hot hand. Is the probability of a bowler rolling a strike higher after he has thrown four consecutive strikes? An investigation into the phenomenon of a "hot hand" in bowling was published in *The American Statistician* (Feb. 2004). Frame-by-frame results were collected on 43 professional bowlers from the 2002–2003 Professional Bowlers Association (PBA) season. For each bowler, the researchers calculated the proportion of strikes rolled after bowling four consecutive strikes and the proportion after bowling four consecutive nonstrikes. The data on 4 of the 43 bowlers, saved in the **HOTBOWLER** file, are shown in the following table:

    a. Do the data on the sample of four bowlers provide support for the "hot hand" theory in bowling? Explain.

    b. When the data on all 43 bowlers are used, the *p*-value for the hypothesis test is approximately 0. Interpret this result.

HOTBOWLER

Bowler	Proportion of Strikes After Four Strikes	Proportion of Strikes After Four Nonstrikes
Paul Fleming	.683	.432
Bryon Smith	.684	.400
Mike DeVaney	.632	.421
Dave D'Entremont	.610	.529

Source: Dorsey-Palmateer, R., & Smith, G. "Bowlers' Hot Hands," *American Statistician,* Vol. 58, No. 1, Feb. 2004 (Table 3).

7.6 Comparing Three or More Population Means: Analysis of Variance (Optional)

Suppose we are interested in comparing the means of three or more populations. For example, we may want to compare the mean SAT scores of seniors at three different high schools. Or, we could compare the mean income per house-hold of residents in four census districts. Since the methods of Sections 7.1–7.5 apply to two populations only, we require an alternative technique. In this optional section, we discuss a method for comparing two or more populations based on independent random samples, called an **analysis of variance (ANOVA)**.

Biography

SIR RONALD A. FISHER (1890–1962)— THE FOUNDER OF MODERN STATISTICS

At a young age, Ronald Fisher demonstrated special abilities in mathematics, astronomy, and biology. (Fisher's biology teacher once divided all his students for "sheer brilliance" into two groups—Fisher and the rest.) Fisher graduated from prestigious Cambridge University in London in 1912 with a B.A. degree in astronomy, and, after several years teaching mathematics, he found work at the Rothamsted Agricultural Experiment station. There, Fisher began his extraordinary career as a statistician. Many consider Fisher to be the leading founder of modern statistics. His contributions to the field include the notion of unbiased statistics, the development of *p*-values for hypothesis tests, the invention of analysis of variance for designed experiments, the maximum likelihood estimation theory, and the mathematical distributions of several well-known statistics. Fisher's book, *Statistical Methods for Research Workers* (written in 1925), revolutionized applied statistics, demonstrating how to analyze data and interpret the results with very readable and practical examples. In 1935, Fisher wrote *The Design of Experiments*, where he first described his famous experiment on the "lady tasting tea." (Fisher showed, through a designed experiment, that the lady really could determine whether tea poured into milk tastes better than milk poured into tea.) Before his death, Fisher was elected a Fellow of the Royal Statistical Society, was awarded numerous medals, and was knighted by the Queen of England.

In the jargon of ANOVA, "treatments" represent the groups or populations of interest. Thus, the primary objective of an analysis of variance is to compare the treatment (or population) means for the quantitative (or dependent) variable of interest. If we denote the true means of the k treatments as $\mu_1, \mu_2, \dots, \mu_k$, then we will test the null hypothesis that the treatment means are all equal against the alternative that at least two of the treatment means differ:

$H_0: \mu_1 = \mu_2 = \cdots = \mu_k$

$H_a:$ At least two of the k treatment means differ

The μ's might represent the means of *all* female and male high school seniors' SAT scores or the means of *all* households' income in each of the four census districts.

To conduct a statistical test of these hypotheses, we will use the means of the independent random samples selected from the treatment populations. That is, we compare the k sample means $\bar{x}_1, \bar{x}_2, \dots, \bar{x}_k$.

For example, suppose you select independent random samples of five female and five male high school seniors and obtain sample mean SAT scores of 550 and 590, respectively. Can we conclude that males score 40 points higher, on average, than females? To answer this question, we must consider the amount of sampling variability among the experimental units (students). If the scores are as depicted in the dot plot shown in Figure 7.20, then the difference between the means is small relative to the sampling variability of the scores within the treatments, namely, Female and Male. We would be inclined not to reject the null hypothesis of equal population means in this case.

In contrast, if the data are as depicted in the dot plot of Figure 7.21, then the sampling variability is small relative to the difference between the two means. In this case, we would be inclined to favor the alternative hypothesis that the population means differ.

FIGURE 7.20

Dot plot of SAT scores: difference between means dominated by sampling variability.

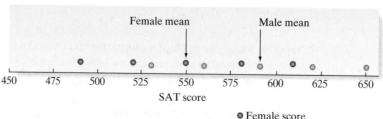

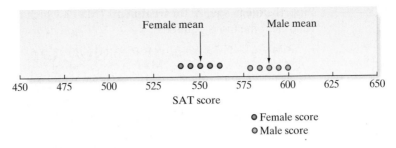

FIGURE 7.21
Dot plot of SAT scores: difference between means large relative to sampling variability

Now Work Exercise 7.97a

You can see that the key is to compare the difference between the treatment means with the amount of sampling variability. To conduct a formal statistical test of the hypothesis requires numerical measures of the difference between the treatment means and the sampling variability within each treatment. The variation between the treatment means is measured by the **sum of squares for treatments** (SST), which is calculated by squaring the distance between each treatment mean and the overall mean of *all* sample measurements, then multiplying each squared distance by the number of sample mea-surements for the treatment, and, finally, adding the results over all treatments:

$$\text{SST} = \sum_{i=1}^{k} n_i(\bar{x}_i - \bar{x})^2 = 5(550 - 570)^2 + 5(590 - 570)^2 = 4{,}000$$

In this equation, we use $\bar{x}$ to represent the overall mean response of all sample measurements—that is, the mean of the combined samples. The symbol n_i is used to denote the sample size for the ith treatment. You can see that the value of SST is 4,000 for the two samples of five female and five male SAT scores depicted in Figures 7.20 and 7.21.

Next, we must measure the sampling variability within the treatments. We call this the **sum of squares for error** (SSE), because it measures the variability around the treatment means that is attributed to sampling error. Suppose the 10 measurements in the first dot plot (Figure 7.20) are 490, 520, 550, 580, and 610 for females and 530, 560, 590, 620, and 650 for males. Then the value of SSE is computed by summing the squared distance between each response measurement and the corresponding treatment mean and then adding the squared differences over all measurements in the entire sample:

$$\text{SSE} = \sum_{j=1}^{n_1}(x_{1j} - \bar{x}_1)^2 + \sum_{j=1}^{n_2}(x_{2j} - \bar{x}_2)^2 + \ldots \sum_{j=1}^{n_k}(x_{kj} - \bar{x}_k)^2$$

Here, the symbol x_{1j} is the jth measurement in sample 1, x_{2j} is the jth measurement in sample 2, and so on. This rather complex-looking formula can be simplified by recalling the formula for the sample variance s^2 given in Chapter 2:

$$s^2 = \sum_{i=1}^{n} \frac{(x_i - \bar{x})^2}{n - 1}$$

Note that each sum in SSE is simply the numerator of s^2 for that particular treatment. Consequently, we can rewrite SSE as

$$\text{SSE} = (n_1 - 1)s_1^2 + (n_2 - 1)s_2^2 + \cdots + (n_k - 1)s_k^2$$

where $s_1^2, s_2^2, \ldots, s_k^2$ are the sample variances for the k treatments. For our samples of SAT scores, we find that $s_1^2 = 2{,}250$ (for females) and $s_2^2 = 2{,}250$ (for males); then we have

$$\text{SSE} = (5 - 1)(2{,}250) + (5 - 1)(2{,}250) = 18{,}000$$

To make the two measurements of variability comparable, we divide each by the number of degrees of freedom in order to convert the sums of squares to mean squares.

First, the **mean square for treatments** (MST), which measures the variability *among* the treatment means, is equal to

$$\text{MST} = \frac{\text{SST}}{k-1} = \frac{4{,}000}{2-1} = 4{,}000$$

where the number of degrees of freedom for the k treatments is $(k-1)$. Next, the **mean square for error** (MSE), which measures the sampling variability *within* the treatments, is

$$\text{MSE} = \frac{\text{SSE}}{n-k} = \frac{18{,}000}{10-2} = 2{,}250$$

Finally, we calculate the ratio of MST to MSE—an **F-statistic**:

$$F = \frac{\text{MST}}{\text{MSE}} = \frac{4{,}000}{2{,}250} = 1.78$$

Values of the F-statistic near 1 indicate that the two sources of variation, between treatment means and within treatments, are approximately equal. In this case, the difference between the treatment means may well be attributable to sampling error, which provides little support for the alternative hypothesis that the population treatment means differ. Values of F well in excess of 1 indicate that the variation among treatment means well exceeds that within means and therefore support the alternative hypothesis that the population treatment means differ.

When does F exceed 1 by enough to reject the null hypothesis that the means are equal? This depends on the sampling distribution of F. An **F-distribution,** which depends on $v_1 = (k-1)$ numerator degrees of freedom and $v_2 = (n-k)$ denominator degrees of freedom, is shown in Figure 7.22. As you can see, the distribution is skewed to the right, since $F = \text{MST/MSE}$ cannot be less than 0, but can increase without bound.

We need to be able to find F values corresponding to the tail areas of this distribution in order to establish the rejection region for our test of hypothesis. The upper-tail F values for $\alpha = .10, .05, .025,$ and $.01$ can be found in Tables VII, VIII, IX, and X of Appendix A. Table VIII is partially reproduced in Table 7.11. It gives F values that correspond to $\alpha = .05$ upper-tail areas for different degrees of freedom v_1 (columns) and v_2 (rows).

Returning to the SAT score example, the F statistic has $v_1 = (2-1) = 1$ numerator degree of freedom and $v_2 = (10-2) = 8$ denominator degrees of freedom. Thus, for $\alpha = .05$, we look in the first column and eighth row of Table 7.11 and find (shaded)

$$F_{.05} = 5.32$$

The implication is that MST would have to be 5.32 times greater than MSE before we could conclude, at the .05 level of significance, that the two population treatment means differ. Since the data yielded $F = 1.78$, our initial impressions of the dot plot in Figure 7.20 are confirmed: There is insufficient information to conclude that the mean SAT scores differ for the populations of female and male high school seniors. The rejection region and the calculated F value are shown in Figure 7.22.

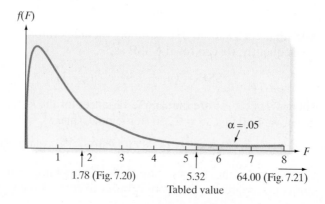

FIGURE 7.22

Rejection region and calculated F-values for SAT score samples

TABLE 7.11   Reproduction of Part of Table VIII in Appendix A: Percentage Points of the F-distribution, $\alpha = .05$

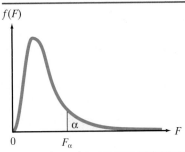

ν_2 \ ν_1	Numerator Degrees of Freedom								
	1	2	3	4	5	6	7	8	9
1	161.4	199.5	215.7	224.6	230.2	234.0	236.8	238.9	240.5
2	18.51	19.00	19.16	19.25	19.30	19.33	19.35	19.37	19.38
3	10.13	9.55	9.28	9.12	9.01	8.94	8.89	8.85	8.81
4	7.71	6.94	6.59	6.39	6.26	6.16	6.09	6.04	6.00
5	6.61	5.79	5.41	5.19	5.05	4.95	4.88	4.82	4.77
6	5.99	5.14	4.76	4.53	4.39	4.28	4.21	4.15	4.10
7	5.59	4.74	4.35	4.12	3.97	3.87	3.79	3.73	3.68
8	5.32	4.46	4.07	3.84	3.69	3.58	3.50	3.44	3.39
9	5.12	4.26	3.86	3.63	3.48	3.37	3.29	3.23	3.18
10	4.96	4.10	3.71	3.48	3.33	3.22	3.14	3.07	3.02
11	4.84	3.98	3.59	3.36	3.20	3.09	3.01	2.95	2.90
12	4.75	3.89	3.49	3.26	3.11	3.00	2.91	2.85	2.80
13	4.67	3.81	3.41	3.18	3.03	2.92	2.83	2.77	2.71
14	4.60	3.74	3.34	3.11	2.96	2.85	2.76	2.70	2.65

Denominator Degrees of Freedom

In contrast, consider the dot plot in Figure 7.21. Since the means are the same as in the first example, 550 and 590, respectively, the variation between the means is the same, MST = 4,000. But the variation within the two treatments appears to be considerably smaller. The observed SAT scores are 540, 545, 550, 555, and 560 for females and 580, 585, 590, 595, and 600 for males. These values yield $s_1^2 = 62.5$ and $s_2^2 = 62.5$. Thus, the variation within the treatments is measured by

$$\text{SSE} = (5 - 1)(62.5) + (5 - 1)(62.5) = 500$$

$$\text{MSE} = \frac{\text{SSE}}{n - k} = \frac{500}{8} = 62.5$$

Then the F-ratio is

$$F = \frac{\text{MST}}{\text{MSE}} = \frac{4,000}{62.5} = 64.0$$

Again, our visual analysis of the dot plot is confirmed statistically: $F = 64.0$ well exceeds the table's F value, 5.32, corresponding to the .05 level of significance. We would therefore reject the null hypothesis at that level and conclude that the SAT mean score of males differs from that of females.

Now Work Exercise 7.97b–h

Recall that we performed a hypothesis test for the difference between two means in Section 7.1, using a two-sample t-statistic for two independent samples. When two independent samples are being compared, the t- and F-tests are equivalent. To see this, recall the formula

$$t = \frac{\bar{x}_1 - \bar{x}_2}{\sqrt{s_p^2\left(\frac{1}{n_1} + \frac{1}{n_2}\right)}} = \frac{590 - 550}{\sqrt{(62.5)\left(\frac{1}{5} + \frac{1}{5}\right)}} = \frac{40}{5} = 8$$

where we used the fact that $s_p^2 = $ MSE, which you can verify by comparing the formulas. Note that the calculated F for these samples ($F = 64$) equals the square of the calculated t for the same samples ($t = 8$). Likewise, the table's F-value (5.32) equals the square of the table's t-value at the two-sided .05 level of significance ($t_{.025} = 2.306$ with 8 df). Since both the rejection region and the calculated values are related in the same way, the tests are equivalent. Moreover, the assumptions that must be met to ensure the validity of the t- and F-tests are the same:

1. The probability distributions of the populations of responses associated with each treatment must all be normal.
2. The probability distributions of the populations of responses associated with each treatment must have equal variances.
3. The samples of experimental units selected for the treatments must be random and independent.

In fact, the only real difference between the tests is that the F-test can be used to compare *more than two* treatment means, whereas the t-test is applicable to two samples only. The **F-test** is summarized in the following box:

ANOVA F-Test to Compare k Treatment Means:
Independent Samples Design

H_0: $\mu_1 = \mu_2 = \cdots = \mu_k$

H_a: At least two treatment means differ.

Test statistic: $F = \dfrac{\text{MST}}{\text{MSE}}$

Rejection region: $F > F_\alpha$, where F_α is based on $v_1 = (k - 1)$ numerator degrees of freedom (associated with MST) and $v_2 = (n - k)$ denominator degrees of freedom (associated with MSE).

Conditions Required for a Valid ANOVA F-Test:
Independent Samples Design

1. The samples are randomly selected in an independent manner from the k treatment populations. (This can be accomplished by randomly assigning the experimental units to the treatments.)
2. All k sampled populations have distributions that are approximately normal.
3. The k population variances are equal (i.e., $\sigma_1^2 = \sigma_2^2 = \sigma_3^2 = \cdots = \sigma_k^2$).

Computational formulas for MST and MSE are given in Appendix B. We will rely on statistical software to compute the F statistic, concentrating on the interpretation of the results rather than their calculation.

EXAMPLE 7.10

CONDUCTING
AN ANOVA
F-TEST—
Comparing Golf
Ball Brands

Problem Suppose the USGA wants to compare the mean distances reached of four different brands of golf balls struck with a driver. An independent-samples design is employed, with Iron Byron, the USGA's robotic golfer, using a driver to hit a random sample of 10 balls of each brand in a random sequence. The distance is recorded for each hit. This quantitative variable represents the *dependent* variable in the study. results are shown in Table 7.12, organized by brand.

a. Set up the test to compare the mean distances for the four brands. Use $\alpha = .10$.
b. Use statistical software to obtain the test statistic and p-value. Interpret the results.

GOLFCRD

TABLE 7.12 Results of Independent-Samples Design: Iron Byron Driver

	Brand A	Brand B	Brand C	Brand D
	251.2	263.2	269.7	251.6
	245.1	262.9	263.2	248.6
	248.0	265.0	277.5	249.4
	251.1	254.5	267.4	242.0
	260.5	264.3	270.5	246.5
	250.0	257.0	265.5	251.3
	253.9	262.8	270.7	261.8
	244.6	264.4	272.9	249.0
	254.6	260.6	275.6	247.1
	248.8	255.9	266.5	245.9
Sample means	250.8	261.1	270.0	249.3

Solution

a. To compare the mean distances of the $k = 4$ brands, we first specify the hypotheses to be tested. Denoting the population mean of the ith brand by μ_i, we test

$$H_0: \mu_1 = \mu_2 = \mu_3 = \mu_4$$
$$H_a: \text{The mean distances differ for at least two of the brands.}$$

The test statistic compares the variation among the four treatment (Brand) means with the sampling variability within each of the treatments:

$$\text{Test statistic:} \quad F = \frac{\text{MST}}{\text{MSE}}$$

$$\text{Rejection region:} \quad F > F_\alpha = F_{.10} \text{ with } v_1 = (k - 1) = 3 \text{ df}$$
$$\text{and } v_2 = (n - k) = 36 \text{ df}$$

From Table VII of Appendix A, we find that $F_{.10} \approx 2.25$ for 3 and 36 df. Thus, we will reject H_0 if $F > 2.25$. (See Figure 7.23.)

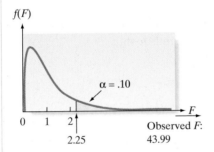

FIGURE 7.23

F-test for completely randomized design: golf ball experiment

The assumptions necessary to ensure the validity of the test are as follows:

1. The samples of 10 golf balls for each brand are selected randomly and independently.
2. The probability distributions of the distances for each brand are normal.
3. The variances of the distance probability distributions for each brand are equal.

b. The MINITAB printout for the data in Table 7.12 resulting from this completely randomized design is given in Figure 7.24. The total sum of squares is designated "Total" and is partitioned into the "Brand" (i.e., treatments) and "Error" sum of squares (SS).

The values of the mean squares, MST and MSE (highlighted on the printout), are 931.5 and 21.2, respectively. The *F*-ratio, 43.99, also highlighted on the printout, exceeds the table value of 2.25. We therefore reject the null hypothesis at the .10 level of significance, concluding that at least two of the brands differ with respect to mean distance traveled when struck by the driver.

Look Back We can also arrive at the appropriate conclusion by noting that the observed significance level of the F-test (highlighted on the printout) is p-value $= .000$. This implies that we would reject the null hypothesis that the means are equal at any α level.

One-way ANOVA: DISTANCE versus BRAND

```
Source   DF      SS       MS       F       P
BRAND     3   2794.4    931.5   43.99   0.000
Error    36    762.3     21.2
Total    39   3556.7

S = 4.602    R-Sq = 78.57%    R-Sq(adj) = 76.78%


                                 Individual 95% CIs For Mean Based on
                                 Pooled StDev
Level    N     Mean   StDev   --------+---------+---------+---------+-
BrandA   10   250.78    4.74   (---*---)
BrandB   10   261.06    3.87                 (---*---)
BrandC   10   269.95    4.50                             (----*---)
BrandD   10   249.32    5.20   (---*---)
                               --------+---------+---------+---------+-
                                 252.0     259.0     266.0     273.0

Pooled StDev = 4.60
```

FIGURE 7.24

MINITAB ANOVA for completely randomized design

Now Work Exercise 7.104

The results of an **analysis of variance (ANOVA)** for independent samples (often called a **one-way ANOVA**) can be summarized in a simple tabular format similar to that obtained from the MINITAB program in Example 7.10. The general form of the table is shown in Table 7.13, where the symbols df, SS, and MS stand for degrees of freedom, sum of squares, and mean square, respectively. Note that the two sources of variation, Treatments and Error, add to the total sum of squares, SS(Total). The ANOVA summary table for Example 7.10 is given in Table 7.14.

TABLE 7.13 General ANOVA Summary Table for a Completely Randomized Design

Source	df	SS	MS	F
Treatments	$k - 1$	SST	$MST = \dfrac{SST}{k-1}$	$\dfrac{MST}{MSE}$
Error	$n - k$	SSE	$MSE = \dfrac{SSE}{n-k}$	
Total	$n - 1$	SS(Total)		

TABLE 7.14 ANOVA Summary Table for Example 7.10

Source	df	SS	MS	F	p-Value
Brands	3	2,794.39	931.46	43.99	.000
Error	36	762.30	21.18		
Total	39	3,556.69			

Suppose the F-test results in a rejection of the null hypothesis that the treatment means are equal. Is the analysis then complete? Usually, the conclusion that at least two of the treatment means differ leads to other conclusions and—for instance, other questions. Which of the means differ and by how much? For example, the *F*-test in Example 7.10 leads to the conclusion that at least two of the brands of golf balls have different mean distances traveled when struck with a driver. Now the questions are, Which of the brands differ? and How are the brands ranked with respect to mean distance?

One way to obtain this information is to construct a confidence interval for the difference between the means of any pair of treatments, using the method of Section 7.1. For example, if a 95% confidence interval for $\mu_A - \mu_C$ in Example 7.10 is found to be $(-24, -13)$, we are confident that the mean distance for Brand C exceeds the mean for Brand A (since all differences in the interval are negative). Constructing these confidence intervals for all possible pairs of brands allows you to rank the brand means. A method for conducting these *multiple comparisons*—one that controls for Type I errors—is beyond the scope of this introductory text. Consult the references to learn about this methodology.

Analysis of Variance

Using the TI-84/TI-83 Graphing Calculator

Computing a One-Way ANOVA

Step 1   *Enter each data set into its own list (i.e., sample 1 into L1, sample 2 into L2, sample 3 into L3, etc.).*

Step 2   *Access the Statistical Test Menu.*

Press **STAT**.
Arrow right to **TESTS**.
Arrow down to **ANOVA**.
Press **ENTER**.
Type in each List name, separated by commas (*e.g., L1, L2, L3, L4*).
Press **ENTER**.

Step 3   *View Display.*

The calculator will display the *F*-test statistic, as well as the *p*-value, the Factor degrees of freedom, the sum of squares, the mean square, and, by arrowing down, the Error degrees of freedom, the sum of squares, the mean square, and the pooled standard deviation.

Example   Shown are four different samples. At the $\alpha = .05$ level of significance, test whether the four population means are equal. The null hypothesis will be $H_0: \mu_1 = \mu_2 = \mu_3 = \mu_4$. The alternative hypothesis H_a. At least one mean is different.

SAMPLE1	SAMPLE2	SAMPLE3	SAMPLE4
60	59	55	58
61	52	55	58
56	51	52	55

The screens for this example are as follows:

As you can see from the screen, the *p*-value is 0.1598, which is **not less than** 0.05; therefore, we should **not reject** H_0. The differences are not significantly different.

EXAMPLE 7.11
CHECKING THE ANOVA ASSUMPTIONS

Problem Refer to the independent samples ANOVA design conducted in Example 7.10. Are the assumptions required for the test approximately satisfied?

Solution The assumptions for the test are repeated as follows:

1. The samples of golf balls for each brand are selected randomly and independently.
2. The probability distributions of the distances for each brand are normal.
3. The variances of the distance probability distributions for each brand are equal.

Since the sample consisted of 10 randomly selected balls of each brand, and since the robotic golfer Iron Byron was used to drive all the balls, the first assumption of independent random samples is satisfied. To check the next two assumptions, we will employ two graphical methods presented in Chapter 2: histograms and box plots. A MINITAB histogram of driving distances for each brand of golf ball is shown in Figure 7.25, and SAS box plots are shown in Figure 7.26.

The normality assumption can be checked by examining the histograms in Figure 7.25. With only 10 sample measurements for each brand, however, the displays are not very informative. More data would need to be collected for each brand before we could assess whether the distances come from normal distributions. Fortunately, analysis of variance has been shown to be a very **robust method** when the assumption of normality is not satisfied exactly. That is, *moderate departures from normality do not*

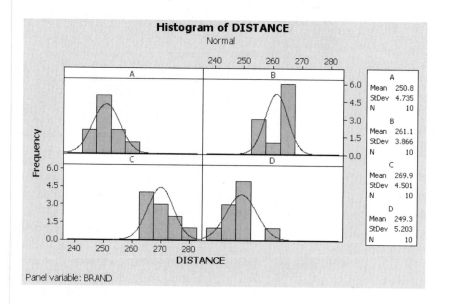

FIGURE 7.25
MINITAB histograms for golf ball driving distances

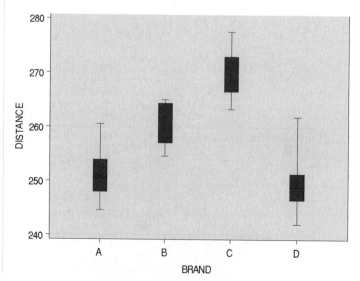

FIGURE 7.26
SAS box plots for golf ball distances

have much effect on the significance level of the ANOVA F-test or on confidence coefficients. Rather than spend the time, energy, or money to collect additional data for this experiment in order to verify the normality assumption, we will rely on the robustness of the ANOVA methodology.

Box plots are a convenient way to obtain a rough check on the assumption of equal variances. With the exception of a possible outlier for Brand D, the box plots in Figure 7.26 show that the spread of the distance measurements is about the same for each brand. Since the sample variances appear to be the same, the assumption of equal population variances for the brands is probably satisfied. Although robust with respect to the normality assumption, ANOVA is *not robust* with respect to the equal-variances assumption. Departures from the assumption of equal population variances can affect the associated measures of reliability (e.g., *p*-values and confidence levels). Fortunately, the effect is slight when the sample sizes are equal, as in this experiment.

> **Now Work Exercise 7.106**

Although graphs can be used to check the ANOVA assumptions as in Example 7.11, no measures of reliability can be attached to these graphs. When you have a plot that is unclear as to whether or not an assumption is satisfied, you can use formal statistical tests that are beyond the scope of this text. Consult the references at the end of the chapter for information on these tests. When the validity of the ANOVA assumptions is in doubt, nonparametric statistical methods are useful.

What Do You Do When the Assumptions Are Not Satisfied for the Analysis of Variance for a Completely Randomized Design?

Answer: Use a nonparametric statistical method such as the Kruskal–Wallis *H*-Test. Consult the references to learn about this method.

STATISTICS IN ACTION REVISITED

Analysis of Variance of the Cockroach Data

We return to the experiment designed to investigate the trail-following ability of German cockroaches (p. 351) and the trail deviation data stored in the **ROACH** file.

For this application, consider only the cockroaches assigned to the fecal extract trail. Four roach groups were utilized in the experiment—adult males, adult females, gravid females, and nymphs—with 20 roaches of each type independently and randomly selected. Is there sufficient evidence to say that the ability to follow the extract trail differs among cockroaches of different age, sex, and reproductive status? In other words, is there evidence to suggest that the mean trail deviation μ differs for the four roach groups?

To answer this question, we conduct a one-way analysis of variance on the **ROACH** data. The quantitative dependent variable of interest is deviation from the extract trail, while the treatments are the four different roach groups. Thus, we want to test the null hypothesis:

$$H_0: \mu_{Male} = \mu_{Female} = \mu_{Gravid} = \mu_{Nymph}$$

A MINITAB printout of the ANOVA is displayed in Figure SIA7.4. The *p*-value of the test (highlighted on the printout) is 0. Since this value is less than, say, $\alpha = .05$, we reject the null hypothesis and conclude (at the .05 level of significance) that the mean deviation from the extract trail differs among the populations of adult male, adult female, gravid, and nymph cockroaches.

The sample means for the four cockroach groups are also highlighted in Figure SIA7.4. Note that gravids have the largest sample mean deviation (44.03). A multiple-comparisons-of-means procedure shows that the gravid mean is statistically larger than the means for the other three groups.

(continued)

FIGURE SIA7.4
MINITAB one-way ANOVA
for deviation from extract trail

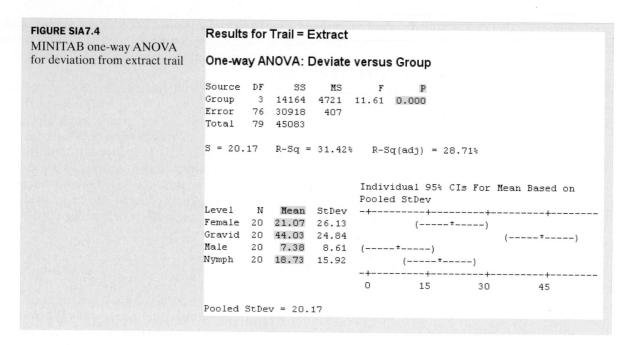

Results for Trail = Extract

One-way ANOVA: Deviate versus Group

```
Source  DF     SS    MS      F      P
Group    3  14164  4721  11.61  0.000
Error   76  30918   407
Total   79  45083

S = 20.17   R-Sq = 31.42%   R-Sq(adj) = 28.71%

                            Individual 95% CIs For Mean Based on
                            Pooled StDev
Level    N   Mean  StDev  -+---------+---------+---------+--------
Female  20  21.07  26.13              (-----*-----)
Gravid  20  44.03  24.84                          (-----*-----)
Male    20   7.38   8.61  (-----*-----)
Nymph   20  18.73  15.92            (-----*-----)
                         -+---------+---------+---------+--------
                          0        15        30        45

Pooled StDev = 20.17
```

Exercises 7.92–7.112

Understanding the Principles

7.92 Explain how to collect the data for a one-way ANOVA.

7.93 What conditions are required for a valid ANOVA F-test?

7.94 *True or False.* The ANOVA method is robust when the assumption of normality is not exactly satisfied.

Learning the Mechanics

7.95 Use Tables VII, VIII, IX, and X of Appendix A to find each of the following F values:
  a. $F_{.05}, v_1 = 3, v_2 = 4$
  b. $F_{.01}, v_1 = 3, v_2 = 4$
  c. $F_{.10}, v_1 = 20, v_2 = 40$
  d. $F_{.025}, v_1 = 12, v_2 = 9$

7.96 Find the following probabilities:
  a. $P(F \leq 3.48)$ for $v_1 = 5, v_2 = 9$
  b. $P(F > 3.09)$ for $v_1 = 15, v_2 = 20$
  c. $P(F > 2.40)$ for $v_1 = 15, v_2 = 15$
  d. $P(F \leq 1.83)$ for $v_1 = 8, v_2 = 40$

7.97 Consider dot plots A and B (shown below). Assume that
NW the two samples represent independent random samples corresponding to two treatments in a completely randomized design.

a. In which dot plot is the difference between the sample means small relative to the variability within the sample observations? Justify your answer.

b. Calculate the treatment means (i.e., the means of samples 1 and 2) for both dot plots.

c. Use the means to calculate the sum of squares for treatments (SST) for each dot plot.

d. Calculate the sample variance for each sample and use these values to obtain the sum of squares for error (SSE) for each dot plot.

e. Calculate the total sum of squares [SS(Total)] for the two dot plots by adding the sums of squares for treatment and error. What percentage of SS(Total) is accounted for by the treatments—that is, what percentage of the total sum of squares is the sum of squares for treatment—in each case?

f. Convert the sums of squares for treatment and error to mean squares by dividing each by the appropriate number of degrees of freedom. Calculate the F-ratio of the mean square for treatment (MST) to the mean square for error (MSE) for each dot plot.

g. Use the F-ratios to test the null hypothesis that the two samples are drawn from populations with equal means. Take $\alpha = .05$.

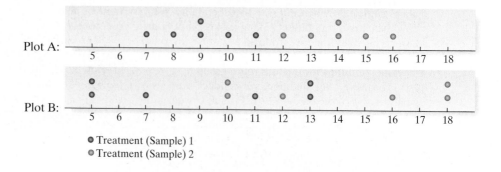

Dot Plots for Exercise 7.97

h. What assumptions must be made about the probability distributions corresponding to the responses for each treatment in order to ensure the validity of the F-tests conducted in part **g**?

7.98 Refer to Exercise 7.97. Conduct a two-sample t-test (Section 7.1) of the null hypothesis that the two treatment means are equal for each dot plot. Use $\alpha = .05$ and two-tailed tests. In the course of the test, compare each of the following with the F-tests in Exercise 7.97:

a. The pooled variances and the MSEs
b. The t- and the F-test statistics
c. The tabled values of t and F that determine the rejection regions
d. The conclusions of the t- and F-tests
e. The assumptions that must be made in order to ensure the validity of the t- and F-tests

7.99 Refer to Exercises 7.97 and 7.98. Complete the following ANOVA table for each of the two dot plots:

Source	df	SS	MS	F
Treatments				
Error				
Total				

7.100 The data in the following table resulted from an experiment that utilized an independent-samples design:

⊚ **LM7_100**

Treatment 1	Treatment 2	Treatment 3
3.9	5.4	1.3
1.4	2.0	.7
4.1	4.8	2.2
5.5	3.8	
2.3	3.5	

a. Use statistical software (or the formulas in Appendix B) to complete the following ANOVA table:

Source	df	SS	MS	F
Treatments				
Error				
Total				

b. Test the null hypothesis that $\mu_1 = \mu_2 = \mu_3$, where μ_i represents the true mean for treatment i, against the alternative that at least two of the means differ. Use $\alpha = .01$.

Applying the Concepts—Basic

7.101 **College tennis recruiting with a team website.** Most university athletic programs now have a website with information on individual sports and a Prospective Student Athlete Form that allows high school athletes to submit information about their academic and sports achievements directly to the college coach. *The Sport Journal* (Winter 2004) published a study of how important team websites are to the recruitment of college tennis players. A survey was conducted of National Collegiate Athletic Association (NCAA) tennis coaches, of which 53 were

from Division I schools, 20 were from Division II schools, and 53 were from Division III schools. Coaches were asked to respond to a series of statements, including "The Prospective Student Athlete Form on the website contributes very little to the recruiting process." Responses were measured on a seven-points scale (where $1 = $ strongly disagree and $7 = $ strongly agree). In order to compare the mean responses of tennis coaches from the three NCAA divisions, the data were analyzed with a one-way ANOVA design.

a. Identify the experimental unit, the dependent variable, and the treatments in this study.
b. Give the null and alternative hypothesis for the ANOVA F-test.
c. The observed significance level of the test was found to be p-value $< .003$. What conclusion can you draw if you want to test at $\alpha = .05$?

7.102 **A new dental bonding agent.** Refer to the *Trends in Biomaterials & Artificial Organs* (Jan. 2003) study of a new bonding adhesive for teeth, presented in Exercise 6.63 (p. 327). Recall that the new adhesive (called "Smartbond") has been developed to eliminate the necessity of a dry field. In one portion of the study, 30 extracted teeth were bonded with Smartbond and each was randomly assigned one of three different bonding times: 1 hour, 24 hours, or 48 hours. At the end of the bonding period, the breaking strength (in Mpa) of each tooth was determined. The data were analyzed with the use of analysis of variance in order to determine whether the true mean breaking strength of the new adhesive varies with the bonding time.

a. Identify the experimental units, treatments, and dependent variable for this study.
b. Set up the null and alternative hypotheses for the ANOVA.
c. Find the rejection region for the test, using $\alpha = .01$.
d. The test results were $F = 61.62$ and p-value ≈ 0. Give the appropriate conclusion for the test.
e. What conditions are required for the test results to be valid?

7.103 **Robots trained to behave like ants.** Robotics researchers investigated whether robots could be trained to behave like ants in an ant colony. (*Nature*, Aug. 2000.) Robots were trained and randomly assigned to "colonies" (i.e., groups) consisting of 3, 6, 9, or 12 robots. The robots were assigned the tasks of foraging for "food" and recruiting another robot when they identified a resource-rich area. One goal of the experiment was to compare the mean energy expended (per robot) of the four different sizes of colonies.

a. What type of experimental design was employed?
b. Identify the treatments and the dependent variable.
c. Set up the null and alternative hypotheses of the test.
d. The following ANOVA results were reported: $F = 7.70$, numerator df $= 3$, denominator df $= 56$, p-value $< .001$. Conduct the test at a significance level of $\alpha = .05$ and interpret the result.

⊚ **TVADRECALL**

7.104 **Study of recall of TV commercials.** Television advertisers
NW seek to promote their products on TV programs that attract the most viewers. Do TV shows with violence and sex impair memory for commercials? To answer this question,

	Sample Size	Youngest Tertile Mean Height	Middle Tertile Mean Height	Oldest Tertile Mean Height	F-Value	p-Value
Boys	1439	0.33	0.33	0.16	4.57	0.01
Girls	1409	0.27	0.18	0.21	0.85	0.43

Source: Wake, M., Coghlan, D., and Hesketh, K. "Does height influence progression through primary school grades?" *The Archives of Disease in Childhood,* Vol. 82, Apr. 2000 (Table 2).

Iowa St. professors B. Bushman and A. Bonacci conducted a designed experiment in which 324 adults were randomly assigned to one of three viewer groups of 108 participants each. (*Journal of Applied Psychology*, June 2002.) One group watched a TV program (e.g., *Tour of Duty*) with a violent content code (V) rating, the second group viewed a show (e.g., *Strip Mall*) with a sex content code (S) rating, and the last group watched a neutral TV program (e.g., *Candid Camera*) with neither a V nor an S rating. Nine commercials were embedded into each TV show. After viewing the program, each participant was scored on his or her recall of the brand names in the commercial messages, with scores ranging from 0 (no brands recalled) to 9 (all brands recalled). The data (simulated from information provided in the article) are saved in the **TVADRECALL** file. The researchers compared the mean recall scores of the three viewing groups with an analysis of variance for a completely randomized design.

a. Identify the experimental units in the study.

b. Identify the dependent variable in the study.

c. Identify the treatments in the study.

d. The sample mean recall scores for the three groups were $\bar{x}_v = 2.08$, $\bar{x}_s = 1.71$, and $\bar{x}_{Neutral} = 3.17$. Explain why one should not draw an inference about differences in the population mean recall scores on the basis of only these summary statistics.

One-way ANOVA: VIOLENT, SEX, NEUTRAL

```
Source    DF      SS      MS      F      P
Factor     2   123.27   61.63   20.45  0.000
Error    321   967.35    3.01
Total    323  1090.62

S = 1.736   R-Sq = 11.30%   R-Sq(adj) = 10.75%
```

e. An ANOVA on the data in the **TVADRECALL** file yielded the results shown in the accompanying MINITAB printout. Locate the test statistic and p-value on the printout.

f. Interpret the results from part **e**, using $\alpha = 0.01$. What can the researchers conclude about the three groups of TV ad viewers?

7.105 **Heights of grade school repeaters.** Refer to the *Archives of Disease in Childhood* (Apr. 2000) study of whether height influences a child's progression through elementary school, presented in Exercise 7.15 (p. 367). Within each grade, Australian schoolchildren were divided into equal thirds (tertiles) based on age (youngest third, middle third, and oldest third). The researchers compared the average heights of the three groups, using an analysis of variance. (All height measurements were standardized with z-scores.) A summary of the results for all grades combined, by gender, is shown in the table at the top of the page:

a. What is the null hypothesis for the ANOVA of the boys' data?

b. Interpret the results of the test, part **a**. Use $\alpha = .05$.

c. Repeat parts **a** and **b** for the girls' data.

d. Summarize the results of the hypothesis tests in the words of the problem.

7.106 **Most powerful American women.** Refer to *Fortune* (Nov. 14, 2002) magazine's study of the most powerful women in America, presented in Exercise 2.58 (p. 59). Recall that the data on age (in years) and title of each of the 50 women in the survey are stored in the **WPOWER50** file. (Some of the data are listed in the accompanying table.) Suppose you want to compare the average ages of the most powerful American women in four groups based on their position (title) within the firm: Group 1 (CEO); Group 2 (Chairman, President CFO, COO, or CRO); Group 3 (EVP, SVP, and Vice Chair); and Group 4 (Founder, Treasurer, or Executive).

ANOVA

AGE

	Sum of Squares	df	Mean Square	F	Sig.
Between Groups	191.140	3	63.713	2.500	.071
Within Groups	1172.140	46	25.481		
Total	1363.280	49			

AGE

GROUP	Mean	N	Std. Deviation
1	51.07	15	5.216
2	48.11	18	5.603
3	49.57	14	3.502
4	56.00	3	7.000
Total	49.88	50	5.275

SPSS output for Exercise 7.106

Rank	Name	Age	Company	Title
1	Meg Whitman	49	eBay	CEO/Chairman
2	Anne Mulcahy	52	Xerox	CEO/Chairman
3	Brenda Barnes	51	Sara Lee	CEO/President
4	Oprah Winfrey	51	Harpo	Chairman
5	Andrea Jung	47	Avon	CEO/Chairman
⋮	⋮	⋮	⋮	⋮
49	Safra Catz	43	Oracle	President
50	Kathy Cassidy	51	General Electric	Treasurer

Source: Fortune, Nov. 14, 2005.

a. Give the null and alternative hypotheses to be tested.

b. An SPSS analysis-of-variance printout for the test you stated in part **a** is shown at the bottom of p. 416. The sample means for the four groups appear at the bottom of the printout. Why is it insufficient to make a decision about the null hypothesis based solely on these sample means?

c. Locate the test statistic and *p*-value on the printout. Use this information to make the appropriate conclusion at $\alpha = .10$.

d. Use the data in the **WPOWER50** file to determine whether the ANOVA assumptions are reasonably satisfied.

Applying the Concepts—Intermediate

7.107 Income and road rage. The phenomenon of road rage has received much media attention in recent years. Is a driver's propensity to engage in road rage related to his or her income? Researchers at Mississippi State University attempted to answer this question by conducting a survey of a representative sample of over 1,000 U.S. adult drivers. (*Accident Analysis and Prevention*, Vol. 34, 2002.) Based on how often each driver engaged in certain road rage behaviors (e.g., making obscene gestures at, tailgating, and thinking about physically hurting another driver), a road rage score was assigned. (Higher scores indicate a greater pattern of road rage behavior.) The drivers were also grouped by annual income: under \$30,000, between \$30,000 and \$60,000, and over \$60,000. The data were subjected to an analysis of variance, with the results summarized in the next table. Interpret the results fully. Is a driver's propensity to engage in road rage related to his or her income?

Income Group	Sample Size	Mean Road Rage Score
Under \$30,000	379	4.60
\$30,000 to \$60,000	392	5.08
Over \$60,000	267	5.15
ANOVA results:	*F*-value = 3.90	*p*-value < .01

7.108 Restoring self-control when intoxicated. Does coffee or some other form of stimulation (e.g., an incentive to stop when seeing a flashing red light on a car) really allow a person suffering from alcohol intoxication to "sober up"? Psychologists from the University of Waterloo investigated the matter in *Experimental and Clinical Psychopharmacology* (February 2005). A sample of 44 healthy male college students participated in the experiment. Each student was asked to memorize a list of 40 words (20 words on a green list and 20 words on a red list). The students were then ran-

domly assigned to one of four different treatment groups (11 students in each group). Students in three of the groups were each given two alcoholic beverages to drink prior to performing a word completion task. Students in Group A received only the alcoholic drinks. Participants in Group AC had caffeine powder dissolved in their drinks. Group AR participants received a monetary award for correct responses on the word completion task. Students in Group P (the placebo group) were told that they would receive alcohol, but instead received two drinks containing a carbonated beverage (with a few drops of alcohol on the surface to provide an alcoholic scent). After consuming their drinks and resting for 25 minutes, the students performed the word completion task. Their scores (simulated on the basis of summary information from the article) are reported in the accompanying table. (*Note:* A task score represents the difference between the proportion of corrects responses on the green list of words and the proportion of incorrect responses on the red list of words.)

a. What type of experimental design is employed in this study?

b. Analyze the data for the researchers, using $\alpha = .05$. Are there differences among the mean task scores for the four groups?

c. What assumptions must be met in order to ensure the validity of the inference you made in part **b**?

◉ DRINKERS

AR	AC	A	P
.51	.50	.16	.58
.58	.30	.10	.12
.52	.47	.20	.62
.47	.36	.29	.43
.61	.39	− .14	.26
.00	.22	.18	.50
.32	.20	− .35	.44
.53	.21	.31	.20
.50	.15	.16	.42
.46	.10	.04	.43
.34	.02	− .25	.40

Adapted from Grattan-Miscio, K.E., and Vogel-Sprott, M. "Alcohol, intentional control, and inappropriate behavior: Regulation by caffeine or an incentive," *Experimental and Clinical Psychopharmacology*, Vol. 13, No. 1, February 2005 (Table 1).

7.109 Effect of scopolamine on memory. The drug scopolamine is often used as a sedative to induce sleep in patients. In *Behavioral Neuroscience* (Feb. 2004), medical researchers examined scopolamine's effects on memory with associated word pairs. A total of 28 human subjects, recruited from a university community, were given a list of related word pairs to memorize. For every word pair in the list (e.g., robber–jail), there was an associated word pair with the same first word, but a different second word (e.g., robber–police). The subjects were then randomly divided into three treatment groups. Group 1 subjects were administered an injection of scopolamine, group 2 subjects were given an injection of glycopyrrolate (an active placebo), and group 3 subjects were not given any drug. Four hours later, subjects were shown 12 word pairs from the list and tested on how many they could recall. The data on number of pairs recalled (simulated on the basis of summary information provided in the research article) are listed in the accompanying table. Prior to the analysis, the researchers theorized that the mean number of word pairs recalled for the scopolamine subjects (group 1) would be less than the corresponding means for the other two groups.

a. Explain why this is an independent-samples design.

b. Identify the treatments and dependent variable.

c. Find the sample means for the three groups. Is this sufficient information to support the researchers' theory? Explain.

d. Conduct an ANOVA *F*-test on the data. Is there sufficient evidence (at $\alpha = .05$) to conclude that the mean number of word pairs recalled differs among the three treatment groups?

SCOPOLAMINE

Group 1 (Scopolamine):	5 8 8 6 6 6 6 8 6 4 5 6
Group 2 (Placebo):	8 10 12 10 9 7 9 10
Group 3 (No drug):	8 9 11 12 11 10 12 12

7.110 The "name game." Psychologists at Lancaster University (United Kingdom) evaluated three methods of name retrieval in a controlled setting. (*Journal of Experimental Psychology—Applied*, June 2000.) A sample of 139 students was randomly divided into three groups, and each group of students used a different method to learn the names of the other students in the group. Group 1 used the "simple name game," in which the first student states his or her full name, the second student announces his or her name and the name of the first student, the third student says his or her name and the names of the first two students, etc. Group 2 used the "elaborate name game," a modification of the simple name game such that the students state not only their names, but also their favorite activity (e.g., sports). Group 3 used "pairwise introductions," according to which students are divided into pairs and each student must introduce the other member of the pair. One year later, all subjects were sent pictures of the students in their group and asked to state the full name of each. The researchers measured the

NAMEGAME

Simple Name Game

24	43	38	65	35	15	44	44	18	27	0	38	50	31
7	46	33	31	0	29	0	0	52	0	29	42	39	26
51	0	42	20	37	51	0	30	43	30	99	39	35	19
24	34	3	60	0	29	40	40						

Elaborate Name Game

39	71	9	86	26	45	0	38	5	53	29	0	62	0
1	35	10	6	33	48	9	26	83	33	12	5	0	0
25	36	39	1	37	2	13	26	7	35	3	8	55	50

Pairwise Introductions

5	21	22	3	32	29	32	0	4	41	0	27	5	9
66	54	1	15	0	26	1	30	2	13	0	2	17	14
5	29	0	45	35	7	11	4	9	23	4	0	8	2
18	0	5	21	14									

Source: Morris, P. E., and Fritz, C. O. "The name game: Using retrieval practice to improve the learning of names," *Journal of Experimental Psychology—Applied*, Vol. 6, No. 2, June 2000 (data simulated from Figure 1).

percentage of names recalled by each student respondent. The data (simulated on the basis of summary statistics provided in the research article) are shown in the next table. Conduct an analysis of variance to determine whether the mean percentages of names recalled differ for the three name-retrieval methods. Use $\alpha = .05$.

7.111 Estimating the age of glacial drifts. Refer to the *American Journal of Science* (Jan. 2005) study of the chemical make-up of buried tills (glacial drifts) in Wisconsin, presented in Exercise 2.130 (p. 88). The ratio of the elements aluminum (AI) and beryllium (Be) in sediment is related to the duration of burial. Recall the AI/Be ratios for a sample of 26 buried till specimens were determined and are saved in the **TILLRATIO** file. The till specimens were obtained from five different boreholes (labeled UMRB-1, UMRB-2, UMRB-3, SWRA, and SD). The data are shown in the accompanying table. Conduct an analysis of variance of the data. Is there sufficient evidence to indicate differences among the mean AI/Be ratios for the five boreholes? Test, using $\alpha = .10$.

TILLRATIO

UMRB-1:	3.75	4.05	3.81	3.23	3.13	3.30	3.21
UMRB-2:	3.32	4.09	3.90	5.06	3.85	3.88	
UMRB-3:	4.06	4.56	3.60	3.27	4.09	3.38	3.37
SWRA:	2.73	2.95	2.25				
SD:	2.73	2.55	3.06				

Source: Adapted from *American Journal of Science*, Vol. 305, No. 1, Jan. 2005, p. 16 (Table 2).

Applying the Concepts—Advanced

7.112 Animal-assisted therapy for heart patients. Refer to the *American Heart Association Conference* (Nov. 2005) study to gauge whether animal-assisted therapy can improve the physiological responses of heart failure patients, presented in Exercise 2.102 (p. 74). Recall that 76 heart patients were randomly assigned to one of three groups. Each patient in group T was visited by a human volunteer accompanied by a trained dog, each patient in group V was visited by a volunteer only, and the patients in group C were not visited at all. The anxiety level of each patient was measured (in points) both before and after the visits. The accompanying table gives summary statistics for the drop in anxiety level for patients in the three groups. The mean drops in anxiety levels of the three groups of patients were compared with the use of an analysis of variance. Although the ANOVA table was not provided in the article, sufficient information is given to reconstruct it.

a. Compute SST for the ANOVA, using the formula (on p. 405)

$$SST = \sum_{i=1}^{3} n_i(\bar{x}_i - \bar{x})^2$$

	Sample Size	Mean Drop	Std. Dev.
Group T: Volunteer + Trained Dog	26	10.5	7.6
Group V: Volunteer only	25	3.9	7.5
Group C: Control group (no visit)	25	1.4	7.5

Source: Cole, K., et al. "Animal-assisted therapy decreases hemodynamics, plasma epinephrine and state anxiety in hospitalized heart failure patients," *American Heart Association Conference*, Dallas, Texas, Nov. 2005.

where $\bar{x}$ is the overall mean drop in anxiety level of all 76 subjects. [Hint: $\bar{x} = (\sum_{i=1}^{3} n_i (\bar{x}_i)/76$.]

b. Recall that SSE for the ANOVA can be written as
$$\text{SSE} = (n_1 - 1)s_1^2 + (n_2 - 1)s_2^2 + (n_3 - 1)s_3^2$$
where s_1^2, s_2^2, and s_3^2 are the sample variances associated with the three treatments. Compute SSE for the ANOVA.

c. Use the results from parts **a** and **b** to construct the ANOVA table.

d. Is there sufficient evidence (at $\alpha = .01$) of differences among the mean drops in anxiety levels by the patients in the three groups?

e. Comment on the validity of the ANOVA assumptions. How might this affect the results of the study?

KEY TERMS

Note: Starred (*) terms are from the optional section in this chapter.

Analysis of variance (ANOVA)*   403
Blocking   372
F-distribution*   406
F-statistic*   406
Mean square for error (MSE)*   406

Mean square for treatment (MST)*   406
One-way ANOVA*   410
Paired difference experiment   372
Pooled sample estimator   356
Rank sum*   387
Randomized block experiment   372

Robust method   412
Standard error   354
Sum of squares for treatment (SST)*   405
Sum of squares for error (SSE)*   405
Wilcoxon rank sum test*   386
Wilcoxon signed rank test*   396

CHAPTER NOTES

Key Words for Identifying the Target Parameter

$\mu_1 - \mu_2$	Difference between Means or Averages
μ_d	Paired Difference in Means or Averages
$\mu_1, \mu_2, \ldots \mu_k$	Comparing three or more means or averages

Key Symbols

$\mu_1 - \mu_2$	Difference between population means
μ_d	Paired difference in population means
D_0	Hypothesized value of difference
$\bar{x}_1 - \bar{x}_2$	Difference between sample means
$\bar{x}_d$	Mean of sample differences
s_p^2	Pooled sample variance
$\sigma_{(\bar{x}_1 - \bar{x}_2)}$	Standard error for $\bar{x}_1 - \bar{x}_2$
$\sigma_{\bar{d}}$	Standard error for $\bar{d}$
SE	Sampling error in estimation
*T_1	Sum of ranks of observations in sample 1
*T_2	Sum of ranks of observations in sample 2
*T_L	Critical lower Wilcoxon rank sum value
*T_U	Critical upper Wilcoxon rank sum value
*T_+	Sum of ranks of positive differences of paired observations
*T_-	Sum of ranks of negative differences of paired observations
*T_0	Critical value of Wilcoxon signed rank test
*F_α	Critical value of F associated with tail area α
*ν_1	Numerator degrees of freedom for F statistic
*ν_2	Denominator degrees of freedom for F statistic
*ANOVA	Analysis of variance
*SST	Sum of squares for treatments
*SSE	Sum of squares for error
*MST	Mean squares for treatments
*MSE	Mean square for error

GUIDE TO COMPARING POPULATION MEANS

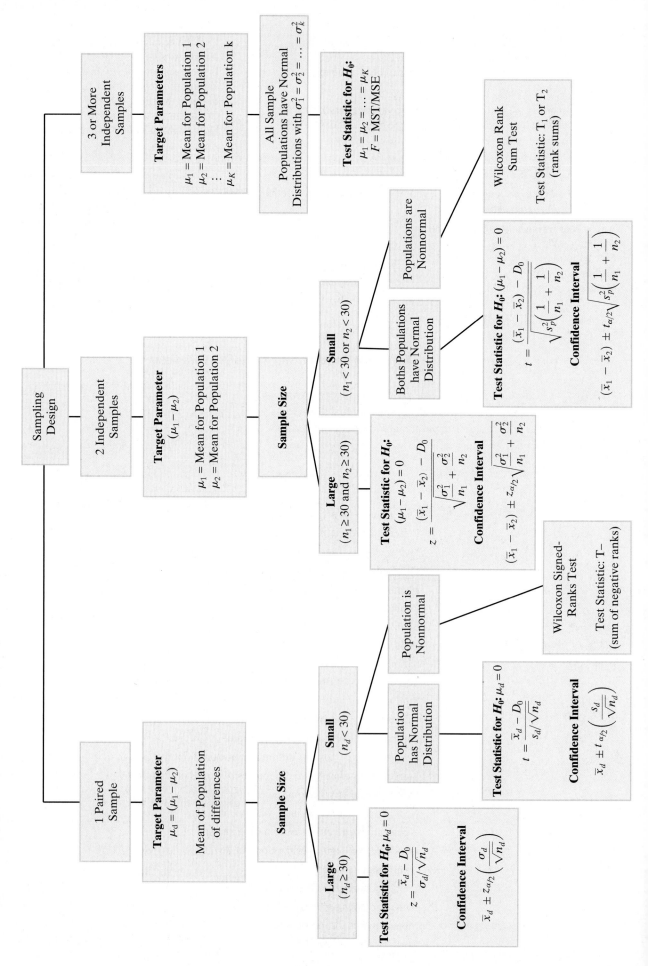

Determining the Sample Size

Estimating $\mu_1 - \mu_2$: $n_1 = n_2 = (z_{\alpha/2})^2(\sigma_1^2 + \sigma_2^2)/(SE)^2$

Estimating μ_d: $n_d = \dfrac{(z_{\alpha/2})^2\sigma_d^2}{(SE)^2}$

Conditions Required for Parametric Inferences about $\mu_1 - \mu_2$

Large samples:
1. Independent random samples
2. $n_1 \geq 30, n_2 \geq 30$

Small samples:
1. Independent random samples
2. Both populations normal
3. $\sigma_1^2 = \sigma_2^2$

Conditions Required for Parametric Inferences about μ_d

Large Samples:
1. Random sample of paired differences
2. $n_d \geq 30$

Small samples:
1. Random sample of paired differences
2. Population of differences is normal

Conditions Required for ANOVA: $H_0: \mu_1 = \mu_2 = \ldots = \mu_k$

1. Independent random samples
2. All k populations normal
3. $\sigma_1^2 = \sigma_2^2 = \ldots = \sigma_k^2$

Using a Confidence Interval for $(\mu_1 - \mu_2)$ to Determine whether a Difference Exists:

1. If the confidence interval includes all *positive* numbers $(+ , +)$: $\rightarrow$ Infer $\mu_1 > \mu_2$
2. If the confidence interval includes all *negative* numbers $(- , -)$ $\rightarrow$ Infer $\mu_1 < \mu_2$
3. If the confidence interval includes 0 $(-, +)$: $\rightarrow$ Infer "no evidence of a difference."

Supplementary Exercises 7.113–7.139

Note: Starred (*) exercises refer to the optional sections in this chapter.

Understanding the Principles

7.113 List the assumptions necessary for each of the following inferential techniques:
  a. Large-sample inferences about the difference $(\mu_1 - \mu_2)$ between population means, using a two-sample z-statistic
  b. Small-sample inferences about $(\mu_1 - \mu_2)$, using an independent samples design and a two-sample t-statistic
  c. Small-sample inferences about $(\mu_1 - \mu_2)$, using a paired difference design and a single-sample t-statistic to analyze the differences
  *d. Inferences about three treatment means, μ_1, μ_2 and μ_3

7.114 For each of the following, identify the target parameter:
  a. Comparison of average SAT scores of males and females matched on IQ
  b. Difference between mean waiting times at two supermarket checkout lanes using independent random samples
  c. Comparison of mean starting salaries of mathematics, psychology, and engineering graduates

7.115 For each of the following, give the appropriate nonparametric test to apply:
  a. Comparing two populations with independent samples
  b. Comparing two populations with matched pairs

Learning the Mechanics

7.116 Independent random samples were selected from two normally distributed populations with means μ_1 and μ_2, respectively. The sample sizes, means, and variances are shown in the following table:

Sample 1	Sample 2
$n_1 = 12$	$n_2 = 14$
$\bar{x}_1 = 17.8$	$\bar{x}_2 = 15.3$
$s_1^2 = 74.2$	$s_2^2 = 60.5$

  a. Test $H_0: (\mu_1 - \mu_2) = 0$ against $H_a: (\mu_1 - \mu_2) > 0$. Use $\alpha = .05$.

b. Form a 99% confidence interval for $(\mu_1 - \mu_2)$.
c. How large must n_1 and n_2 be if you wish to estimate $(\mu_1 - \mu_2)$ to within two units with 99% confidence? Assume that $n_1 = n_2$.

7.117 Two independent random samples are taken from two populations. The results of these samples are summarized in the following table:

Sample 1	Sample 2
$n_1 = 135$	$n_2 = 148$
$\bar{x}_1 = 12.2$	$\bar{x}_2 = 8.3$
$s_1^2 = 2.1$	$s_2^2 = 3.0$

a. Form a 90% confidence interval for $(\mu_1 - \mu_2)$.
b. Test H_0: $(\mu_1 - \mu_2) = 0$ against H_a: $(\mu_1 - \mu_2) \neq 0$. Use $\alpha = .01$.
c. What sample size would be required if you wish to estimate $(\mu_1 - \mu_2)$ to within .2 with 90% confidence? Assume that $n_1 = n_2$.

7.118 A random sample of five pairs of observations were selected, one observation from a population with mean μ_1, the other from a population with mean μ_2. The data are shown in the following table:

LM7_118

Pair	Value from Population 1	Value from Population 2
1	28	22
2	31	27
3	24	20
4	30	27
5	22	20

a. Test the null hypothesis H_0: $\mu_d = 0$ against H_a: $\mu_d \neq 0$, where $\mu_d = \mu_1 - \mu_2$. Use $\alpha = .05$.
b. Form a 95% confidence interval for μ_d.
c. When are the procedures you used in parts **a** and **b** valid?

*7.119 A random sample of nine pairs of observations is recorded on two variables x and y. The data are shown in the following table:

LM7_119

Pair	x	y	Pair	x	y
1	19	12	6	29	10
2	27	19	7	16	16
3	15	7	8	22	10
4	35	25	9	16	18
5	13	11			

Do the data provide sufficient evidence to indicate that the probability distribution for x is shifted to the right of that for y? Test, using $\alpha = .05$.

*7.120 Two independent random samples produced the measurements listed in the next table. Do the data provide sufficient evidence to conclude that there is a difference between the locations of the probability distributions for the sampled populations? Test, using $\alpha = .05$.

LM7_120

Sample 1		Sample 2	
1.2	1.0	1.5	1.9
1.9	1.8	1.3	2.7
.7	1.1	2.9	3.5
2.5			

Learning the Mechanics

*7.121 A completely randomized design is utilized to compare four treatment means. The data are shown in the accompanying table.
a. Given that SST = 36.95 and SS(Total) = 62.55, complete an ANOVA table for this experiment.
b. Is there evidence that the treatment means differ? Use $\alpha = .10$.

LM7_121

Treatment 1	Treatment 2	Treatment 3	Treatment 4
8	6	9	12
10	9	10	13
9	8	8	10
10	8	11	11
11	7	12	11

Applying the Concepts—Basic

EVOS

7.122 Oil spill impact on seabirds. Refer to the *Journal of Agricultural, Biological, and Environmental Statistics* (Sept. 2000) study of the impact of a tanker oil spill on the seabird population in Alaska, presented in Exercise 2.181 (p. 105). Recall that for each of 96 shoreline locations (called transects), the number of seabirds found, the length (in kilometers) of the transect, and whether or not the transect was in an oiled area were recorded. (The data are saved in the **EVOS** file.) *Observed seabird density* is defined as the observed count divided by the length of the transect. A comparison of the mean densities of oiled and unoiled transects is displayed in the MINITAB printout on p. 423. Use this information to make an inference about the difference in the population mean seabird densities of oiled and unoiled transects.

7.123 Effect of altitude on climbers. Dr. Philip Lieberman, a neuroscientist at Brown University, recently conducted a field experiment to gauge the effect of high altitude on a person's ability to think critically. (*New York Times*, Aug. 23, 1995.) The subjects of the experiment were five males who took part in an American expedition climbing Mount Everest. At the base camp, Lieberman read sentences to the climbers while they looked at simple pictures in a book. The length of time (in seconds) it took for each climber to match the picture with a sentence was recorded. Using a radio, Lieberman repeated the task when the climbers reached a camp 5 miles above sea level. At this altitude, he noted that the climbers took 50% longer to complete the task.

Two-Sample T-Test and CI: Density, Oil

```
Two-sample T for Density

Oil   N   Mean   StDev   SE Mean
no    36  3.27   6.70    1.1
yes   60  3.50   5.97    0.77


Difference = mu (no ) - mu (yes)
Estimate for difference:  -0.221165
95% CI for difference:  (-2.927767, 2.485436)
T-Test of difference = 0 (vs not =): T-Value = -0.16  P-Value = 0.871  DF = 67
```

MINITAB output for
Exercise 7.122

a. What is the variable measured in this experiment?
b. What are the experimental units?
c. Discuss how the data should be analyzed.

*7.124 Ordering exam questions. An educational psychologist claims that the order in which test questions are asked affects a student's ability to answer correctly. To investigate this assertion, a professor randomly divides a class of 13 students into two groups: 7 in one group and 6 in the other. The professor prepares one set of test questions, but arranges the questions in two different orders. On test A, the questions are arranged in order of increasing difficulty (i.e., from easiest to most difficult); on test B, the order is reversed. One group of students is given test A, the other test B, and the test score is recorded for each student. The results are shown in the next table:

⚙ **TESTORDER**

Test A	90	71	83	82	75	91	65
Test B	66	78	50	68	80	60	

Use the Wilcoxon rank sum procedure to test for a difference (a shift in location) in the probability distributions of student scores on the two tests. Test using $\alpha = .05$.

7.125 Self-concepts of female students. A study reported in the *Journal of Psychology* (Mar. 1991) measures the change in female students' self-concepts as they move from high school to college. A sample of 133 Boston College female students was selected for the study. Each participant was asked to evaluate several aspects of her life at two points in time: at the end of her senior year of high school and during her sophomore year of college. Each student was asked to evaluate where she believed she stood on a scale that ranged from the top 10% of the class (1) to the lowest 10% of the class (5). The results for three of the traits evaluated are reported in the following table:

Trait	n	Senior Year of High School $\bar{x}_1$	Sophomore Year of College $\bar{x}_2$
Leadership	133	2.09	2.33
Popularity	133	2.48	2.69
Intellectual self-confidence	133	2.29	2.55

a. What null and alternative hypotheses would you test to determine whether the mean self-concept of females decreases between the senior year of high school and the sophomore year of college, as measured by each of the three traits listed?
b. Are these traits more appropriately analyzed with an independent samples test or a paired difference test? Explain.
c. Noting the size of the sample, state what assumptions are necessary to ensure the validity of the tests.
d. The article reports that the leadership test results in a *p*-value greater than .05, while the tests for popularity and intellectual self-confidence result in *p*-values less than .05. Interpret these results.

7.126 Environmental impact study. Some power plants are located near rivers or oceans so that the available water can be used to cool the condensers. Suppose that, as part of an environmental impact study, a power company wants to estimate the difference in mean water temperature between the discharge of its plant and the offshore waters. How many sample measurements must be taken at each site in order to estimate the true difference between means to within .2°C with 95% confidence? Assume that the range in readings will be about 4°C at each site and that the same number of readings will be taken at each site.

*7.127 Children's use of pronouns. Refer to the *Journal of Communication Disorders* (Mar. 1995) study of pronoun misusage by specifically language-impaired (SLI) children, presented in Exercise 2.65 (p. 60). Recall that percentage of pronoun errors was recorded for three groups of children: 10 five-year-old SLI children, 10 younger (three-year-old) normally developing (YND) children, and 10 older (five-year-old) normally developing (OND) children. An ANOVA was conducted on the data, producing the results shown in the accompanying table. Interpret these results.

Source	df	SS	MS	F	*p*-Value
Groups	2	11,292.45	5,646.22	8.295	.0016
Error	27	18,377.65	680.65		
Total	29	29,670.10			

Mean percentage of errors:	0.00	30.17	46.88
Group:	OND	SLI	YND

Source: Moore, M. E. "Error analysis of pronouns by normal and language-impaired children," *Journal of Communication Disorders*, Vol. 28, No. 1, Mar. 1995, p. 67 (Table 6).

7.128 Students' attitudes toward parents. Researchers at the University of South Alabama compared the attitudes of male college students toward their fathers with their attitudes toward their mothers. (*Journal of Genetic Psychology,* March 1998.) Each of a sample of 13 males was asked to complete the following statement about each of his parents: My relationship with my father (mother) can best be described as (1) Awful, (2) Poor, (3) Average, (4) Good, or (5) Great. The following data were obtained:

⊙ **FMATTITUDES**

Student	Attitude toward Father	Attitude toward Mother
1	2	3
2	5	5
3	4	3
4	4	5
5	3	4
6	5	4
7	4	5
8	2	4
9	4	5
10	5	4
11	4	5
12	5	4
13	3	3

Source: Adapted from Vitulli, W. F., and Richardson, D. K. "College students' attitudes toward relationships with parents: A five-year comparative analysis." *Journal of Genetic Psychology,* Vol. 159, No. 1, (March 1998), pp. 45–52.

a. Specify the appropriate hypotheses for testing whether male students' attitudes toward their fathers differ from their attitudes toward their mothers, on average.

b. Conduct the test of part **a** at $\alpha = .05$. Interpret the results in the context of the problem.

****c.** Use the Wilcoxon signed-rank test at $\alpha = .05$ to analyze the data.

Applying the Concepts—Intermediate

7.129 Depo-Provera drug study. Hypersexual behavior caused by traumatic brain injury (TBI) is often treated with the drug Depo-Provera. In one clinical study, eight young male TBI patients who exhibited hypersexual behavior were treated weekly with 400 milligrams of Depo-Provera for six months. (*Journal of Head Trauma Rehabilitation,* June 1995.) The testosterone levels (in nanograms per deciliter) of the patients both prior to treatment and at the end of the six-month treatment period are given in the next table:

a. Construct a 99% confidence interval for the true mean difference between the pretreatment and after-treatment testosterone levels of young male TBI patients.

b. Use the interval you found in part **a** to make an inference about the effectiveness of Depo-Provera in reducing testosterone levels of young male TBI patients.

****c.** Using $\alpha = .05$, apply a nonparametric test to the data. What do you conclude?

⊙ **DEPOPROV**

Patient	Pretreatment	After 6 months on Depo-Provera
1	849	96
2	903	41
3	890	31
4	1,092	124
5	362	46
6	900	53
7	1,006	113
8	672	174

Source: Emory, L. E., Cole, C. M., and Meyer, W. J. "Use of Depo-Provera to control sexual aggression in persons with traumatic brain injury," *Journal of Head Trauma Rehabilitation,* Vol. 10, No. 3, June 1995, p. 52 (Table 2).

7.130 Mating habits of snails. Hermaphrodites are animals that possess the reproductive organs of both sexes. *Genetical Research* (June 1995) published a study of the mating systems of hermaphroditic snail species. The mating habits of the snails were classified into two groups: (1) self-fertilizing (selfing) snails that mate with snails of the same sex and (2) cross-fertilizing (outcrossing) snails that mate with snails of the opposite sex. One variable of interest in the study was the effective population size of the snail species. The means and standard deviations of the effective population size for independent random samples of 17 outcrossing snail species and 5 selfing snail species are given in the accompanying table.

Snail Mating System	Sample Size	Effective Population Size	
		Mean	Standard Deviation
Outcrossing	17	4,894	1,932
Selfing	5	4,133	1,890

Source: Jarne, P. "Mating system, bottlenecks, and genetic polymorphism in hermaphroditic animals." *Genetical Research,* Vol. 65, No. 3, June 1995, p. 197 (Table 4).

Compare the mean effective population sizes of the two types of snail species with a 90% confidence interval. Interpret the result.

7.131 Teaching nursing skills. A traditional approach to teaching basic nursing skills was compared with an innovative approach in the *Journal of Nursing Education* (Jan. 1992). The innovative approach utilizes Vee heuristics and concept maps to link theoretical concepts with practical skills. Forty-two students enrolled in an upper-division nursing course participated in the study. Half (21) were randomly assigned to laboratories that utilized the innovative approach. After completing the course, all students were given short-answer questions about scientific principles underlying each of 10 nursing skills. The objective of the research was to compare the mean scores of the two groups of students.

a. What is the appropriate test to use to compare the two groups?

b. Are any assumptions required for the test?

c. One question dealt with the use of clean or sterile gloves. The mean scores for this question were 3.28 (traditional) and 3.40 (innovative). Is there sufficient information to perform the test?

d. Refer to part **c**. The p-value for the test was reported to be $p = .79$. Interpret this result.

e. Another question concerned the choice of a stethoscope. The mean scores of the two groups were 2.55 (traditional) and 3.60 (innovative), with an associated p-value of .02. Interpret these results.

7.132 Identical twins reared apart. Because they share an identical genotype, twins make ideal subjects for investigating the degree to which various environmental conditions affect personality. The classical method of studying this phenomenon, and the subject of an interesting book by Susan Farber (*Identical Twins Reared Apart*, New York: Basic Books, 1981), is the study of identical twins separated early in life and reared apart. Much of Farber's discussion focuses on a comparison of IQ scores. The data for this analysis appear in the accompanying table. One member (A) of each of the $n = 32$ pairs of twins was reared by a natural parent; the other member (B) was reared by a relative or some other person. Is there a significant difference between the average IQ scores of identical twins when one member of the pair is reared by the natural parents and the other member of the pair is not? Use $\alpha = .05$ to draw your conclusion.

TWINSIQ

Pair ID	Twin A	Twin B	Pair ID	Twin A	Twin B
112	113	109	228	100	88
114	94	100	232	100	104
126	99	86	236	93	84
132	77	80	306	99	95
136	81	95	308	109	98
148	91	106	312	95	100
170	111	117	314	75	86
172	104	107	324	104	103
174	85	85	328	73	78
180	66	84	330	88	99
184	111	125	338	92	111
186	51	66	342	108	110
202	109	108	344	88	83
216	122	121	350	90	82
218	97	98	352	79	76
220	82	94	416	97	98

Source: Adapted from *Identical Twins Reared Apart*, by Susan L. Farber © 1981 by Basic Books, Inc.

*7.133 Facial expression study. What do people infer from facial expressions of emotion? This was the research question of interest in an article published in the *Journal of Nonverbal Behavior* (Fall 1996). A sample of 36 introductory psychology students was randomly divided into six groups. Each group was assigned to view one of six slides showing a person making a facial expression.* The six expressions were (1) angry, (2) disgusted, (3) fearful, (4) happy, (5) sad, and (6) neutral. After viewing the slides, the students rated the degree of dominance they inferred from the facial expression (on a scale ranging from -15 to $+15$). The data (simulated from summary information provided in the article) are listed in the accompanying table. Conduct an analysis of variance to determine whether the mean dominance ratings differ among the six facial expressions. Use $\alpha = .10$.

FACES

Angry	Disgusted	Fearful	Happy	Sad	Neutral
2.10	.40	.82	1.71	.74	1.69
.64	.73	−2.93	−.04	−1.26	−.60
.47	−.07	−.74	1.04	−2.27	−.55
.37	−.25	.79	1.44	−.39	.27
1.62	.89	−.77	1.37	−2.65	−.57
−.08	1.93	−1.60	.59	−.44	−2.16

7.134 Aerobic exercise study. Excess postexercise oxygen consumption (EPOC) describes energy expended during the body's recovery period immediately following aerobic exercise. *The Journal of Sports Medicine and Physical Fitness* (Dec. 1994) published a study designed to investigate the effect of fitness level on the magnitude and duration of EPOC. Ten healthy young adult males volunteered for the study. Five were endurance trained and made up the fit group; the other five were not engaged in any systematic training and constituted the sedentary group. Each volunteer engaged in a weight-supported exercise on a cycle ergometer until 300 kilocalories were expended. The magnitude (in kilocalories) and duration (in minutes) of the EPOC of each exerciser were measured. The study results are summarized in the following table:

Variable		Fit ($n = 5$)	Sedentary ($n = 5$)	p-value
Magnitude (kcal)	Mean	12.2	12.2	.998
	Std. dev.	3.1	4.3	
Duration (min)	Mean	16.6	20.4	.344
	Std. dev.	3.1	7.8	

Source: Sedlock, D. A. "Fitness levels and postexercise energy expenditure," *The Journal of Sports Medicine and Physical Fitness*, Vol. 34, No. 4, Dec. 1994, p. 339 (Table III).

a. Conduct a test of hypothesis to determine whether the true mean magnitude of EPOC differs for fit and sedentary young adult males. Use $\alpha = .10$.

b. The p-value for the test you conducted in part **a** is given in the table. Interpret this value.

c. Conduct a test of hypothesis to determine whether the true mean duration of EPOC differs for fit and sedentary young adult males. Use $\alpha = .10$.

d. The p-value for the test you conducted in part **c** is given in the table. Interpret this value.

7.135 Spending on information systems technology. University of Queensland researchers sampled private-sector and public-sector organizations in Australia to study the planning undertaken by their information systems departments. (*Management Science*, July 1996.) As part of the process, they asked each sample organization how much it had spent on information systems and technology in the previous fiscal year, as a percentage of the organization's total revenues. The results are reported in the next table:

a. Do the two sampled populations have identical probability distributions, or is the distribution for public-sector organizations in Australia located to the right of the distribution for Australia's private-sector firms? Test using $\alpha = .05$.

b. Is the *p*-value for the test less than or greater than .05? Justify your answer.

c. What assumption must be met to ensure the validity of the test you conducted in part **a**?

INFOSYS

Private Sector	Public Sector
2.58%	5.40%
5.05	2.55
.05	9.00
2.10	10.55
4.30	1.02
2.25	5.11
2.50	24.42
1.94	1.67
2.33	3.33

Adapted from Hann, J., and Weber, R. "Information systems planning: A model and empirical tests." *Management Science,* Vol. 42, No. 2 July 1996, pp. 1043–1064.

7.136 Students attitudes towards parents. Refer to the *Journal of Genetic Psychology* (Mar. 1998) study comparing male students' attitudes toward their fathers and mothers, presented in Exercise 7.128 (p. 424). Suppose you want to estimate $\mu_d = (\mu_F - \mu_M)$, the difference between the population mean attitudes toward fathers and mothers, using a 90% confidence interval with a sampling error of .3. Find the number of male students required to obtain such an estimate. Assume that the variance of the paired differences is $\sigma_d^2 = 1$.

Applying the Concepts—Advanced

7.137 Isolation timeouts for students. Researchers at Rochester Institute of Technology investigated the use of isolation time-out as a behavioral management technique (*Exceptional Children,* Feb. 1995.) Subjects for the study were 155 emotionally disturbed students enrolled in a special-education facility. The students were randomly assigned to one of two types of classrooms: Option II classrooms (one teacher, one paraprofessional, and a maximum of 12 students) and Option III classrooms (one teacher, one paraprofessional, and a maximum of 6 students). Over the academic year, the number of behavioral incidents resulting in an isolation time-out was recorded for each student. Summary statistics for the two groups of students are shown in the following table:

	Option II	Option III
Number of students	100	55
Mean number of time-out incidents	78.67	102.87
Standard deviation	59.08	69.33

Source: Costenbader, V., and Reading-Brown, M. "Isolation timeout used with students with emotional disturbance," *Exceptional Children,* Vol. 61, No. 4, Feb. 1995, p. 359 (Table 3).

Do you agree with the following statement: "On average, students in Option III classrooms had significantly more time-out incidents than students in Option II classrooms?"

*7.138 Therapy for binge eaters. Do you experience episodes of excessive eating accompanied by being overweight? If so, you may suffer from binge eating disorder. Cognitive-behavioral therapy (CBT), in which patients are taught how to make changes in specific behavior patterns (e.g., exercise, eat only low-fat foods), can be effective in treating the disorder. A group of Stanford University researchers investigated the effectiveness of interpersonal therapy (IPT) as a second level of treatment for binge eaters. (*Journal of Consulting and Clinical Psychology,* June 1995.) The researchers employed a design that randomly assigned a sample of 41 overweight individuals diagnosed with binge eating disorder to either a treatment group (30 subjects) or a control group (11 subjects). Subjects in the treatment group received 12 weeks of CBT and then were subdivided into two groups. Those who responded successfully to CBT (17 subjects) were assigned to a weight loss therapy (WLT) program for the next 12 weeks. Those CBT subjects who did not respond to treatment (13 subjects) received 12 weeks of IPT. The subjects in the control group received no therapy of any type. Thus, the study ultimately consisted of three groups of overweight binge eaters: the CBT-WLT group, the CBT-IPT group, and the control group. One outcome (response) variable measured for each subject was the number *x* of binge eating episodes per week. Summary statistics for each of the three groups at the end of the 24-week period are shown in the accompanying table. The data were analyzed as a completely randomized design with three treatments (CBT-WLT, CBT-IPT, and Control). Although the ANOVA tables were not provided in the article, sufficient information is given in the table to reconstruct them. [See Exercise 7.112 (p. 418). Is CBT effective in reducing the mean number of binges experienced per work?

	CBT-WLT	CBT-IPT	Control
Sample size	17	13	11
Mean number of binges per week	0.2	1.9	2.9
Standard deviation	0.4	1.7	2.0

Source: Agras, W. S., et al. "Does interpersonal therapy help patients with binge eating disorder who fail to respond to cognitive–behavioral therapy?" *Journal of Consulting and Clinical Psychology,* Vol. 63, No. 3, June 1995, p. 358 (Table 1).

Critical Thinking Challenge

7.139 Self-managed work teams and family life. To improve quality, productivity, and timeliness, more and more American industries are utilizing self-managed work teams (SMWTs). A team typically consists of 5 to 15 workers who are collectively responsible for making decisions and performing all tasks related to a particular project. Researchers L. Stanley-Stevens (Tarleton State University), D. E. Yeatts, and R. R. Seward (both from the University of North Texas) investigated the connection between SMWTs, work characteristics, and workers' perceptions of positive spillover into family life (*Quality Management Journal,* Summer 1995.) Survey data were collected from 114 AT&T [employees who worked on 1 of 15 SMWTs at an AT&T technical division.] The workers were divided into two groups: (1) those who reported a positive spillover of work skills to family life and (2) those who did not report any such positive work spillover. The two groups were compared on a variety of job and demographic characteristics, several of which are shown in the accompanying table. All but the demographic characteristics were measured on a seven-point scale, ranging from 1 = "strongly disagree" to 7 = "strongly agree"; thus, the larger the number, the more the characteristic was indicated. The file named **SPILLOVER** includes the values of the variables listed in the table for each of the 114 survey participants. The researchers' objectives were to compare the two groups of workers on each characteristic. In particular, they wanted to know which job-related characteristics are most highly associated with positive work spillover. Conduct a complete analysis of the data for the researchers.

SPILLOVER

Variables Measured in the SMWT Survey

Characteristic	Variable
Information Flow	Use of creative ideas (seven-point scale)
Information Flow	Utilization of information (seven-point scale)
Decision Making	Participation in decisions regarding personnel matters (seven-point scale)
Job	Good use of skills (seven-point scale)
Job	Task identity (seven-point scale)
Demographic	Age (years)
Demographic	Education (years)
Comparison	Group (positive spillover or no spillover)

SUMMARY ACTIVITY: Paired Vs. Unpaired Experiments

We have now discussed two methods of collecting data to compare two population means. In many experimental situations, a decision must be made either to collect two independent samples or to conduct a paired difference experiment. The importance of this decision cannot be overemphasized, since the amount of information obtained and the cost of the experiment are both directly related to the method of experimentation that is chosen.

Choose two populations (pertinent to your school major) that have unknown means and for which you could both collect two independent samples and collect paired observations. Before conducting the experiment, state which method of sampling you think will provide more information (and why). Compare the two methods, first performing the independent sampling procedure by collecting 10 observations from each population (a total of 20 measurements) and then performing the paired difference experiment by collecting 10 pairs of observations.

Construct two 95% confidence intervals, one for each experiment you conduct. Which method provides the narrower confidence interval and hence more information on this performance of the experiment? Does your result agree with your preliminary expectations?

REFERENCES

Agresti, A., and Agresti, B. F. *Statistical Methods for the Social Sciences*, 2nd ed. San Francisco: Dellen, 1986.

Cochran, W. G., and Cox, G. M. *Experimental Designs*, 2nd ed. New York: Wiley, 1957.

Conover, W. J. *Practical Nonparametric Statistics*, 2nd ed. New York: Wiley, 1980.

Daniel, W. W. *Applied Nonparametric Statistics*, 2nd ed. Boston: PWS-Kent, 1990.

Dunn, O. J. "Multiple comparisons using rank sums." *Technometrics*, Vol. 6, 1964.

Freedman, D., Pisani, R., and Purves, R. *Statistics*. New York: W. W. Norton and Co., 1978.

Friedman, M. "The use of ranks to avoid the assumption of normality implicit in the analysis of variance." *Journal of the American Statistical Association*, Vol. 32, 1937.

Gibbons, J. D. *Nonparametric Statistical Inference*, 2nd ed. New York: McGraw-Hill, 1985.

Hollander, M., and Wolfe, D. A. *Nonparametric Statistical Methods*. New York: Wiley, 1973.

Hsu, J. C. *Multiple Comparisons: Theory and Methods*. London: Chapman & Hall, 1996.

Kramer, C. Y. "Extension of multiple range tests to group means with unequal number of replications." *Biometrics*, Vol. 12, 1956, pp. 307–310.

Mason, R. L., Gunst, R. F., and Hess, J. L. *Statistical Design and Analysis of Experiments*. New York: Wiley, 1989.

Mendenhall, W. *Introduction to Linear Models and the Design and Analysis of Experiments*. Belmont, Calif.: Wadsworth, 1968.

Miller, R. G., Jr. *Simultaneous Statistical Inference*. New York: Springer-Verlag, 1981.

Satterthwaite, F. W. "An approximate distribution of estimates of variance components." *Biometrics Bulletin*, Vol. 2, 1946, pp. 110–114.

Scheffé, H. "A method for judging all contrasts in the Analysis of Variance," *Biometrica*, Vol. 40, 1953, pp. 87–104.

Scheffé, H. *The Analysis of Variance*. New York: Wiley, 1959.

Snedecor, G. W., and Cochran, W. *Statistical Methods*, 7th ed. Ames: Iowa State University Press, 1980.

Steel, R. G. D., and Torrie, J. H. *Principles and Procedures of Statistics*, 2nd ed. New York: McGraw-Hill, 1980.

Tukey, J. "Comparing individual means in the Analysis of Variance," *Biometrics*, Vol. 5, 1949, pp. 99–114.

Using Technology

Comparing Means with MINITAB

MINITAB can be used to make both parametric and nonparametric inferences about $(\mu_1 - \mu_2)$ for independent samples, μ_d for paired samples, and $\mu_1, \mu_2, \ldots, \mu_k$ in an analysis of variance.

Independent Samples $(\mu_1 - \mu_2)$

To carry out an analysis for $(\mu_1 - \mu_2,)$ first access the MINITAB worksheet that contains the sample data. Next, click on the "Basic Statistics" button on the MINITAB menu bar, and then click on "2-Sample t", as shown in Figure 7.M.1. The dialog box shown in Figure 7.M.2 then appears.

If the worksheet contains data for one quantitative variable (on which the means will be computed) and one qualitative variable (which represents the two groups or populations), select "Samples in one column", and then specify the quantitative variable in the "Samples" area and the qualitative variable in the "Subscripts" area. (See Figure 7.M.2.)

If the worksheet contains the data for the first sample in one column and the data for the second sample in another column, select "Samples in different columns" and then specify the "First" and "Second" variables. Alternatively, if you have only summarized data (i.e., sample sizes, sample means, and sample standard deviations), select "Summarized data" and enter these summarized values in the appropriate boxes.

Once you have made the appropriate menu selection, click the "Options" button on the MINITAB "2-Sample T" dialog box. Specify the confidence level for a confidence interval, the null-hypothesized value of the difference $(\mu_1 - \mu_2)$, and the form of the alternative hypothesis (lower tailed, two tailed, or upper tailed) in the resulting dialog box, as shown in Figure 7.M.3. Click "OK" to return to the "2-Sample T" dialog box, and then click "OK" again to generate the MINITAB printout.

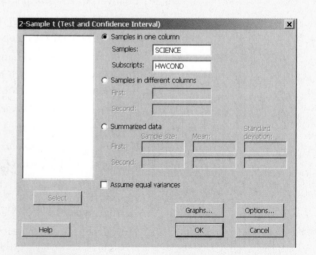

FIGURE 7.M.1

MINITAB menu options for inferences about two means

FIGURE 7.M.2

MINITAB two-sample t dialog box

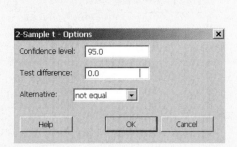

FIGURE 7.M.3
MINITAB two-sample t options

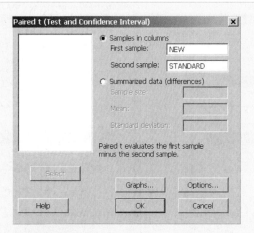

FIGURE 7.M.4
MINITAB paired samples dialog box

[*Important Note:* The MINITAB two-sample *t*-procedure uses the *t*-statistic to conduct the test of hypothesis. When the sample sizes are small, this is the appropriate method. When the sample sizes are large, the *t*-value will be approximately equal to the large-sample *z*-value and the resulting test will still be valid.]

Paired Samples (μ_d)

To carry out an analysis of μ_d for matched pairs, first access the MINITAB worksheet that contains the sample data. The data file should contain two quantitative variables: one with the data values for the first group (or population) and one with the data values for the second group. (*Note:* The sample size should be the same for each group.) Next, click on the "Basic Statistics" button on the MINITAB menu bar, and then click on "Paired t" (see Figure 7.M.1). The dialog box shown in Figure 7.M.4 then appears.

Select the "Samples in columns" option, and specify the two quantitative variables of interest in the "First sample" and "Second sample" boxes, as shown in Figure 7.M.4. [Alternatively, if you have only summarized data of the paired differences, select the "Summarized data (differences)" option and enter the sample size, sample mean, and sample standard deviation in the appropriate boxes.]

Next, click the "Options" button and specify the confidence level for a confidence interval, the null-hypothesized value μ_d, of the difference, and the form of the alternative hypothesis (lower tailed, two tailed, or upper tailed) in the resulting dialog box. (See Figure 7.M.3.) Click "OK" to return to the "Paired t" dialog box, and then click "OK" again to generate the MINITAB printout.

Rank Sum Test

The MINITAB worksheet file with the sample data should contain two quantitative variables—one for each of the two samples being compared. Click on the "Stat" button on the MINITAB menu bar, and then click on "Nonparametrics" and "Mann-Whitney". (See Figure 7.M.5.) The dialog box as shown in Figure 7.M.6 then appears.

Specify the variable for the first sample in the "First Sample" box and the variable for the second sample in the "Second Sample" box. Specify the form of the alternative hypothesis ("not equal", "less than", or "greater than"), and then click "OK" to generate the MINITAB printout.

Signed Rank Test

The MINITAB worksheet file with the matched-pairs data should contain two quantitative variables, one for each of the two groups being compared. You will need to compute the difference between these two variables and save it in a column on the worksheet. (Use the "Calc" button on the MINITAB menu bar.) Next, click on the "Stat" button on the MINITAB menu bar, then click on "Nonparametrics" and "1-Sample Wilcoxon". (See Figure 7.M.5.) The dialog box shown in Figure 7.M.7 then appears.

Enter the variable representing the paired differences in the "Variables" box. Select the "Test median" option and specify the hypothesized value of the median as "0". Select the form of the alternative hypothesis ("not equal", "less than", or "greater than"), and then click "OK" to generate the MINITAB printout.

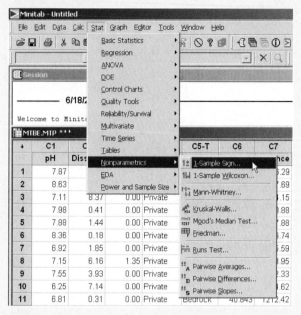

FIGURE 7.M.5
MINITAB nonparametric menu options

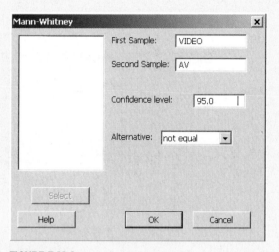

FIGURE 7.M.6
MINITAB Mann–Whitney (rank sum test) dialog box

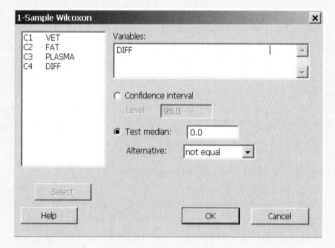

FIGURE 7.M.7
MINITAB 1-sample Wilcoxon dialog box

ANOVA *F*-test

To conduct an ANOVA for an independent-samples design, first access the MINITAB worksheet file that contains the sample data. The data file should contain one quantitative variable (the response, or dependent, variable) and one qualitative factor variable with at least two levels. Next, click on the "Stat" button on the MINITAB menu bar, and then click on "ANOVA" and "One-Way," as shown in Figure 7.M.8.

The resulting dialog box appears as shown in Figure 7.M.9. Specify the response variable in the "Response" box and the factor variable in the "Factor" box. Click "OK" to generate the MINITAB printout.

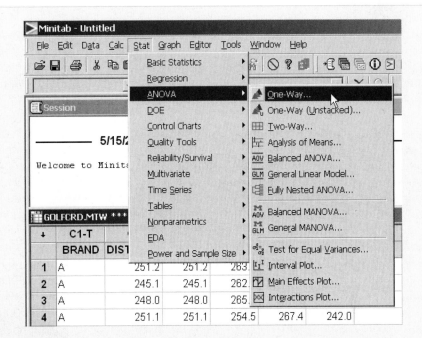

FIGURE 7.M.8
MINITAB menu options for
one-way ANOVA

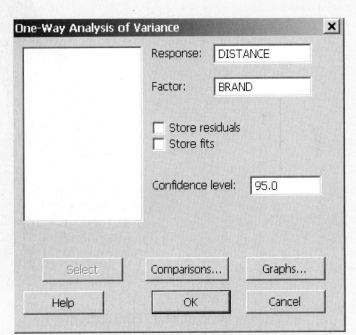

FIGURE 7.M.9
MINITAB one-way ANOVA
dialog box

CHAPTER

8

Comparing Population Proportions

CONTENTS

8.1 Comparing Two Population Proportions: Independent Sampling

8.2 Determining the Sample Size

8.3 Testing Categorical Probabilities: Multinomial Experiment

8.4 Testing Categorical Probabilities: Two-Way (Contingency) Table

STATISTICS IN ACTION

College Students and Alcohol: Is Amount Consumed Related to Drinking Frequency

USING TECHNOLOGY

Categorical Data Analyses with MINITAB

WHERE WE'VE BEEN

■ Presented both parametric and nonparametric methods for comparing two or more population means

WHERE WE'RE GOING

■ Discuss methods for comparing two population proportions.

■ Present a *chi-square* hypothesis test for comparing the category proportions associated with a single qualitative variable—called a *one-way analysis*

■ Present a chi-square hypothesis test for relating two qualitative variables—called a *two-way analysis*.

STATISTICS IN ACTION

College Students and Alcohol: Is Amount Consumed Related to Drinking Frequency?

Traditionally, a common social activity on American college campuses is drinking alcohol. Despite laws on underage drinking, fraternities, sororities, and other campus groups often have alcohol available at their weekend parties. For some students, this activity leads to binge drinking and excessive alcohol use, often resulting in academic failure, physical violence, accidental injury, and even death. In fact, the Journal of Studies on Alcohol *(Vol. 63, 2002) recently reported that about 1,400 alcohol-related deaths occur each year on American college campuses.*

To gain insight into the alcohol consumption behavior of college students, professors Soyeon Shim (University of Arizona) and Jennifer Maggs (Pennsylvania State University) designed a study and reported their results in *Family and Consumer Sciences Research Journal* (Mar. 2005). Among the researchers' main objectives were (1) to segment college students on the basis of their rates of alcohol consumption and (2) to establish a statistical link between the frequency of drinking and the amount of alcohol consumed. They collected survey data from undergraduate students enrolled in a variety of courses at the University of Arizona, a large state university in the Southwest. To increase the likelihood of obtaining a representative sample, the researchers balanced the sample with both lower and upper division students, as well as students with majors in the social sciences, humanities, business, engineering, and the natural sciences. A total of 657 students completed usable surveys.

The survey consisted of a six-page booklet that took approximately 10 minutes to complete. Two of the many questions on the survey (and the subject of this *Statistics in Action*) pertained to the frequency with which the student drank alcohol (beer, wine, or liquor) during the previous one-month period and the average number of drinks the student consumed per occasion. From this information, the researchers categorized students according to Type of drinker. Responses for the three variables of interest were classified qualitatively as shown in Table SIA8.1. The data for the 657 students are saved in the **COLLDRINKS** file.

In an attempt to help the researcher's achieve their objectives, we apply the statistical methodology presented in this chapter to this data set in two *Statistics in Action Revisited* examples.

Statistics in Action Revisited

- Testing Category Proportions for Type of College Drinker (p. 450)

- Testing whether Frequency of Drinking Is Related to Amount of Alcohol Consumed (p. 462)

 COLLDRINKS

TABLE SIA8.1 Qualitative Variables Measured in the Drinking Study

Variable Name	Levels (possible values)
AMOUNT	None, 1 drink, 2–3 drinks, 4–6 drinks, 7–9 drinks, 10 or more drinks
FREQUENCY	None, Once a month, Once or twice per week, More
TYPE	Non/Seldom, Social, Typical binge, Heavy binge

Many experiments are conducted in the biological, physical, and social sciences to compare two or more proportions. Those conducted in business and the social sciences to sample the opinions of people are called **sample surveys**. For example, a state government might wish to estimate the difference between the proportions of people in two regions of the state who would qualify for a new welfare program. Or, after an innovative process change, an engineer might wish to determine whether the proportion of defective items produced by a manufacturing process was less than the proportion of defectives produced before the change. In Section 8.1 we show you how to test hypotheses about the difference between two population proportions based on independent random sampling. We will also show how to find a confidence interval for the difference. Then, in Section 8.3 we will compare more than two population proportions, and in Section 8.4 we will present a related problem.

8.1 Comparing Two Population Proportions: Independent Sampling

Suppose a presidential candidate wants to compare the preferences of registered voters in the northeastern United States with those in the southeastern United States. Such a comparison would help determine where to concentrate campaign efforts. The candidate hires a professional pollster to randomly choose 1,000 registered voters in the northeast and 1,000 in the southeast and interview each to learn her or his voting preference. The objective is to use this sample information to make an inference about the difference $(p_1 - p_2)$ between the proportion p_1 of *all* registered voters in the northeast and the proportion p_2 of *all* registered voters in the southeast who plan to vote for the presidential candidate.

The two samples represent independent binomial experiments. (See Section 4.3 for the characteristics of binomial experiments.) The binomial random variables are the numbers x_1 and x_2 of the 1,000 sampled voters in each area who indicate that they will vote for the candidate. The results are summarized in Table 8.1.

We can now calculate the sample proportions $\hat{p}_1$ and $\hat{p}_2$ of the voters in favor of the candidate in the northeast and southeast, respectively:

TABLE 8.1 Results of Poll

Northeast	Southeast
$n_1 = 1{,}000$	$n_2 = 1{,}000$
$x_1 = 546$	$x_2 = 475$

$$\hat{p} = \frac{x_1}{n_1} = \frac{546}{1{,}000} = .546 \quad \hat{p}_2 = \frac{x_2}{n_2} = \frac{475}{1{,}000} = .475$$

The difference between the sample proportions $(\hat{p}_1 - \hat{p}_2)$ makes an intuitively appealing point estimator of the difference between the population $(p_1 - p_2)$. For this example, the estimate is

$$(\hat{p}_1 - \hat{p}_2) = .546 - .475 = .071$$

To judge the reliability of the estimator $(\hat{p}_1 - \hat{p}_2)$, we must observe its performance in repeated sampling from the two populations. That is, we need to know the sampling distribution of $(\hat{p}_1 - \hat{p}_2)$. The properties of the sampling distribution are given in the next box. Remember that $\hat{p}_1$ and $\hat{p}_2$ can be viewed as means of the number of successes per trial in the respective samples, so the central limit theorem applies when the sample sizes are large.

Properties of the Sampling Distribution of $(\hat{p}_1 - \hat{p}_2)$

1. The mean of the sampling distribution of $(\hat{p}_1 - \hat{p}_2)$ is $(p_1 - p_2)$; that is,

$$E(\hat{p}_1 - \hat{p}_2) = p_1 - p_2$$

Thus, $(\hat{p}_1 - \hat{p}_2)$ is an unbiased estimator of $(p_1 - p_2)$.

2. The standard deviation of the sampling distribution of $(\hat{p}_1 - \hat{p}_2)$ is

$$\sigma_{(\hat{p}_1 - \hat{p}_2)} = \sqrt{\frac{p_1 q_1}{n_1} + \frac{p_2 q_2}{n_2}}$$

3. If the sample sizes n_1 and n_2 are large (see Section 5.3 for a guideline), the sampling distribution of $(\hat{p}_1 - \hat{p}_2)$ is approximately normal.

Since the distribution of $(\hat{p}_1 - \hat{p}_2)$ in repeated sampling is approximately normal, we can use the z-statistic to derive confidence intervals for $(p_1 - p_2)$ or to test a hypothesis about $(p_1 - p_2)$.

For the voter example, a 95% confidence interval for the difference $(p_1 - p_2)$ is

$$(\hat{p}_1 - \hat{p}_2) \pm 1.96\sigma_{(\hat{p}_1 - \hat{p}_2)}, \text{ or } (\hat{p}_1 - \hat{p}_2) \pm 1.96\sqrt{\frac{p_1 q_1}{n_1} + \frac{p_2 q_2}{n_2}}$$

The quantities $p_1 q_1$ and $p_2 q_2$ must be estimated in order to complete the calculation of the standard deviation $\sigma_{(\hat{p}_1 - \hat{p}_2)}$ and, hence, the calculation of the confidence interval. In Section 5.3, we showed that the value of pq is relatively insensitive to the value chosen to approximate p. Therefore, $\hat{p}_1 \hat{q}_1$ and $\hat{p}_2 \hat{q}_2$ will provide satisfactory approximations of $p_1 q_1$ and $p_2 q_2$, respectively. Then

$$\sqrt{\frac{p_1 q_1}{n_1} + \frac{p_2 q_2}{n_2}} \approx \sqrt{\frac{\hat{p}_1 \hat{q}_1}{n_1} + \frac{\hat{p}_2 \hat{q}_2}{n_2}}$$

and we will approximate the 95% confidence interval by

$$(\hat{p}_1 - \hat{p}_2) \pm 1.96\sqrt{\frac{\hat{p}_1 \hat{q}_1}{n_1} + \frac{\hat{p}_2 \hat{q}_2}{n_2}}$$

Substituting the sample quantities yields

$$(.546 - .475) \pm 1.96\sqrt{\frac{(.546)(.454)}{1,000} + \frac{(.475)(.525)}{1,000}}$$

or $.071 \pm .044$. Thus, we are 95% confident that the interval from .027 to .115 contains $(p_1 - p_2)$.

We infer that there are between 2.7% and 11.5% more registered voters in the northeast than in the southeast who plan to vote for the presidential candidate. It seems that the candidate should direct a greater campaign effort in the southeast than in the northeast.

> **Now Work Exercise 8.59**

The general form of a confidence interval for the difference $(p_1 - p_2)$ between population proportions is given in the following box:

Large-Sample $100(1 - \alpha)$% Confidence Interval for $(p_1 - p_2)$

$$(\hat{p}_1 - \hat{p}_2) \pm z_{\alpha/2}\sigma_{(\hat{p}_1 - \hat{p}_2)} = (\hat{p}_1 - \hat{p}_2) \pm z_{\alpha/2}\sqrt{\frac{p_1 q_1}{n_1} + \frac{p_2 q_2}{n_2}}$$

$$\approx (\hat{p}_1 - \hat{p}_2) \pm z_{\alpha/2}\sqrt{\frac{\hat{p}_1 \hat{q}_1}{n_1} + \frac{\hat{p}_2 \hat{q}_2}{n_2}}$$

The z-statistic,

$$z = \frac{(\hat{p}_1 - \hat{p}_2) - (p_1 - p_2)}{\sigma_{(\hat{p}_1 - \hat{p}_2)}}$$

is used to test the null hypothesis that $(p_1 - p_2)$ equals some specified difference, say, D_0. For the special case where $D_0 = 0$—that is, where we want to test the null hypothesis $H_0: (p_1 - p_2) = 0$ (or, equivalently, $H_0: p_1 = p_2$)—the best estimate of $p_1 = p_2 = p$ is obtained by dividing the total number of successes $(x_1 + x_2)$ for the two samples by the total number of observations $(n_1 + n_2)$ that is,

$$\hat{p} = \frac{x_1 + x_2}{n_1 + n_2}, \text{ or } \hat{p} = \frac{n_1 \hat{p}_1 + n_2 \hat{p}_2}{n_1 + n_2}$$

The second equation shows that $\hat{p}$ is a weighted average of $\hat{p}_1$ and $\hat{p}_2$, with the larger sample receiving more weight. If the sample sizes are equal, then $\hat{p}$ is a simple average of the two sample proportions of successes.

We now substitute the weighted average $\hat{p}$ for both p_1 and p_2 in the formula for the standard deviation of $(\hat{p}_1 - \hat{p}_2)$:

$$\sigma_{(\hat{p} - \hat{p}_2)} = \sqrt{\frac{p_1 q_1}{n_1} + \frac{p_2 q_2}{n_2}} \approx \sqrt{\frac{\hat{p}\hat{q}}{n_1} + \frac{\hat{p}\hat{q}}{n_2}} = \sqrt{\hat{p}\hat{q}\left(\frac{1}{n_1} + \frac{1}{n_2}\right)}$$

The test is summarized in the following box:

Large-Sample Test of Hypothesis about $(p_1 - p_2)$

One-Tailed Test

$H_0: (p_1 - p_2) = 0$ *

$H_a: (p_1 - p_2) < 0$

[or $H_a: (p_1 - p_2) > 0$]

Two-Tailed Test

$H_0: (p_1 - p_2) = 0$

$H_a: (p_1 - p_2) \neq 0$

Test statistic:

$$z = \frac{(\hat{p}_1 - \hat{p}_2)}{\sigma_{(\hat{p}_1 - \hat{p}_2)}}$$

Rejection region: $z < -z_\alpha$

[or $z > z_\alpha$ when $H_a: (p_1 - p_2) > 0$]

Rejection region: $|z| > z_{\alpha/2}$

Note: $\sigma_{(\hat{p}_1 - \hat{p}_2)} = \sqrt{\dfrac{p_1 q_1}{n_1} + \dfrac{p_2 q_2}{n_2}} \approx \sqrt{\hat{p}\hat{q}\left(\dfrac{1}{n_1} + \dfrac{1}{n_2}\right)}$ where $\hat{p} = \dfrac{x_1 + x_2}{n_1 + n_2}$

Conditions Required for Valid Large-Sample Inferences about $(p_1 - p_2)$

1.   The two samples are randomly selected in an independent manner from the two target populations.

2.   The sample sizes, n_1 and n_2, are both large, so the sampling distribution of $(\hat{p}_1 - \hat{p}_2)$ will be approximately normal. (This condition will be satisfied if both $n_1\hat{p}_1 \geq 15$, $n_1\hat{q}_1 \geq 15$, and $n_2\hat{p}_2 \geq 15$, $n_2\hat{q}_2 \geq 15$.)

EXAMPLE 8.1

A LARGE-SAMPLE TEST ABOUT $(p_1 - p_2)$: Comparing Fractions of Smokers for Two Years

Problem In the past decade, intensive antismoking campaigns have been sponsored by both federal and private agencies. Suppose the American Cancer Society randomly sampled 1,500 adults in 1997 and then sampled 1,750 adults in 2007 to determine whether there was evidence that the percentage of smokers had decreased. The results of the two sample surveys are shown in Table 8.2, where x_1 and x_2 represent the numbers of smokers in the 1997 and 2007 samples, respectively. Do these data indicate that the fraction of smokers decreased over this 10-year period? Use $\alpha = .05$.

Solution If we define p_1 and p_2 as the true proportions of adult smokers in 1997 and 2007, respectively, then the elements of our test are

$$H_0: (p_1 - p_2) = 0$$
$$H_a: (p_1 - p_2) > 0$$

---

*The test can be adapted to test for a difference $D_0 \neq 0$. Because most applications call for a comparison of p_1 and p_2, implying that $D_0 = 0$, we will confine our attention to this case.

TABLE 8.2 Results of Smoking Survey	
1997	2007
$n_1 = 1,500$	$n_2 = 1,750$
$x_1 = 555$	$x_2 = 578$

(The test is one tailed, since we are interested only in determining whether the proportion of smokers *decreased*.)

$$Test\ statistic:\ z = \frac{(\hat{p}_1 - \hat{p}_2) - 0}{\sigma_{(\hat{p}_1 - \hat{p}_2)}}$$

$$Rejection\ region\ using\ \alpha = .05:$$

$$z > z_\alpha = z_{.05} = 1.645 \qquad \text{(see Figure 8.1)}$$

We now calculate the sample proportions of smokers:

$$\hat{p}_1 = \frac{555}{1,500} = .37 \qquad \hat{p}_2 = \frac{578}{1,750} = .33$$

Then

$$z = \frac{(\hat{p}_1 - \hat{p}_2) - 0}{\sigma_{(\hat{p}_1 - \hat{p}_2)}} \approx \frac{(\hat{p}_1 - \hat{p}_2)}{\sqrt{\hat{p}\hat{q}\left(\frac{1}{n_1} + \frac{1}{n_2}\right)}}$$

where

$$\hat{p} = \frac{x_1 + x_2}{n_1 + n_2} = \frac{555 + 578}{1,500 + 1,750} = .349$$

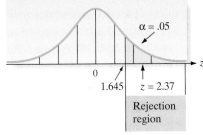

FIGURE 8.1
Rejection region for Example 8.1

Note that $\hat{p}$ is a weighted average of $\hat{p}_1$ and $\hat{p}_2$, with more weight given to the larger (2007) sample.

Thus, the computed value of the test statistic is

$$z = \frac{.37 - .33}{\sqrt{(.349)(.651)\left(\frac{1}{1,500} + \frac{1}{1,750}\right)}} = \frac{.040}{.0168} = 2.38$$

There is sufficient evidence at the $\alpha = .05$ level to conclude that the proportion of adults who smoke has decreased over the 1997–2007 period.

Look Back We could place a confidence interval on $(p_1 - p_2)$ if we were interested in estimating the extent of the decrease.

Now Work Exercise 8.12

EXAMPLE 8.2

FINDING THE OBSERVED SIGNIFICANCE LEVEL OF A TEST FOR $p_1 - p_2$

Problem Use a statistical software package to conduct the test presented in Example 8.1. Find and interpret the *p*-value of the test.

Solution We entered the sample sizes (n_1 and n_2) and numbers of successes (x_1 and x_2) into MINITAB and obtained the printout shown in Figure 8.2. The test statistic for this one-tailed test, $z = 2.37$, as well as the *p*-value of the test, are highlighted on the printout. Note that *p*-value $= .009$ is smaller than $\alpha = .05$. Consequently, we have strong evidence to reject H_0 and conclude that p_1 exceeds p_2.

Test and CI for Two Proportions

```
Sample    X      N   Sample p
1        555   1500   0.370000
2        578   1750   0.330286


Difference = p (1) - p (2)
Estimate for difference:   0.0397143
95% lower bound for difference:   0.0121024
Test for difference = 0 (vs > 0):   Z = 2.37   P-Value = 0.009
```

FIGURE 8.2

MINITAB output for test of two proportions

ACTIVITY 8.1   *Keep the Change:* Inferences Based on Two Samples

In this activity, you will compare the mean amounts transferred for two different Bank of America customers, as well as design some studies that might help the marketing department determine where to allocate more of its advertising budget. You will be working with data sets from Activity 1.1, *Keep the Change: Collecting Data*, (p. 12).

1. You will need to work with another student in the class on this exercise. Each of you should use your data set *Amounts Transferred* from the Activity from Chapter 1 as a random sample from a theoretically larger set of all your amounts ever transferred. Then the means and standard deviations of your data sets will be the sample means and standard deviations. Write a confidence interval for the difference of the two means at the 95% level. Does the interval contain 0? Are your mean amounts transferred significantly different? Explain.

2. Design a study to determine whether there is a significant difference in the mean amount of Bank of America matches for customers in California enrolled in the program and the mean amount of Bank of America matches for customers in Florida enrolled in the program. Be specific about sample sizes, tests used, and how a conclusion will be reached. How might the results of this study help Bank of America estimate costs in the program?

3. Design a study to determine whether there is a significant difference in the percentage of Bank of America customers in California enrolled in the program and the percentage of Bank of America customers in Florida enrolled in the program. Be specific about sample sizes, tests used, and how a conclusion will be reached. How might the results of this study help Bank of America's marketing department?

Keep the results from this activity for use in other activities.

Confidence Interval for $(p_1 - p_2)$
Using the TI-84/TI-83 Graphing Calculator

Step 1   *Access the Statistical Tests Menu.*
Press **STAT**.
Arrow right to **TESTS**.
Arrow down to **2-PropZInt**.
Press **ENTER**.

Step 2   *Enter the values from the sample information and **the confidence level**,*

where x_1 = number of successes in the first sample
n_1 = sample size for the first sample
x_2 = number of successes in the second sample
n_2 = sample size for the second sample

Set **C-Level** to the confidence level.
Arrow down to "**Calculate**".
Press **ENTER**.

Example   Find a 95% confidence interval for the difference in the proportions of two leading automobile models that need major repairs.

Model 1: A sample of 400 owners is contacted. Fifty-three owners report that their cars needed major repairs within the first two years after they purchased them.

Model 2: A sample of 500 owners is contacted. Seventy-eight owners report that their cars needed major repairs within the first two years after they purchased them.

The screens for this example are as follows:

 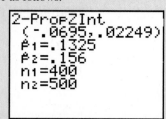

The 95% confidence interval for the difference in the two proportions is $(-.0695, .02249)$.

Hypothesis Test for $(p_1 - p_2)$
Using the TI-84/TI-83 Graphing Calculator

Step 1   *Access the Statistical Tests Menu.*

Press **STAT**.
Arrow right to **TESTS**.
Arrow down to **2-PropZTest**.
Press **ENTER**.

Step 2   *Enter the values from the sample information and* **select the alternative hypothesis,**

where x_1 = number of successes in the first sample
n_1 = sample size for the first sample
x_2 = number of successes in the second sample
n_2 = sample size for the second sample

Use the **ARROW** to highlight the appropriate alternative hypothesis.
Press **ENTER**.
Arrow down to "**Calculate**".
Press **ENTER**.

Example   Test the hypothesis that no difference exists between the proportions of two leading automobile models that need major repairs.

Model 1: A sample of 400 owners is contacted. Fifty-three owners report that their cars needed major repairs within the first two years after they purchased them.

Model 2: A sample of 500 owners is contacted. Seventy-eight owners report that their cars needed major repairs within the first two years after they purchased them.

The screens for this example are as follows:

Since the p-value (.3205) is greater than α (.10). do not reject the null hypothesis. There is insufficient evidence to detect a difference between the population proportions.

Exercises 8.1–8.21

Understanding the Principles

8.1   Consider making an inference about $p_1 - p_2$, where there are x_1 successes in n_1 binomial trials and x_2 successes in n_2 binomial trials.
 a. Describe the distributions of x_1 and x_2.
 b. For large samples, describe the sampling distribution of $(\hat{p}_1 - \hat{p}_2)$.

8.2   What is the problem with using the z-statistic to make inferences about $p_1 - p_2$ when the sample sizes are both small?

8.3   What conditions are required for valid large-sample inferences about $p_1 - p_2$?

Learning the Mechanics

8.4   For each of the following values of α, find the values of z for which H_0: $(p_1 - p_2) = 0$ would be rejected in favor of H_a: $(p_1 - p_2) < 0$.

 a. $\alpha = .01$ **b.** $\alpha = .025$
 c. $\alpha = .05$ **d.** $\alpha = .10$

8.5   In each case, determine whether the sample sizes are large enough to conclude that the sampling distribution of $(\hat{p}_1 - \hat{p}_2)$ is approximately normal.
 a. $n_1 = 10, n_2 = 12, \hat{p}_1 = .50, \hat{p}_2 = .50$
 b. $n_1 = 10, n_2 = 12, \hat{p}_1 = .10, \hat{p}_2 = .08$
 c. $n_1 = n_2 = 30, \hat{p}_1 = .20, \hat{p}_2 = .30$
 d. $n_1 = 100, n_2 = 200, \hat{p}_1 = .05, \hat{p}_2 = .09$
 e. $n_1 = 100, n_2 = 200, \hat{p}_1 = .95, \hat{p}_2 = .91$

8.6   Independent random samples, each containing 800 observations, were selected from two binomial populations. The samples from populations 1 and 2 produced 320 and 400 successes, respectively.
 a. Test H_0: $(p_1 - p_2) = 0$ against H_a: $(p_1 - p_2) \neq 0$. Use $\alpha = .05$.

b. Test H_0: $(p_1 - p_2) = 0$ against H_a: $(p_1 - p_2) \neq 0$. Use $\alpha = .01$.

c. Test H_0: $(p_1 - p_2) = 0$ against H_a: $(p_1 - p_2) < 0$. Use $\alpha = .01$.

d. Form a 90% confidence interval for $(p_1 - p_2)$.

8.7 Construct a 95% confidence interval for $(p_1 - p_2)$ in each of the following situations:

a. $n_1 = 400$, $\hat{p}_1 = .65$; $n_2 = 400$, $\hat{p}_2 = .58$

b. $n_1 = 180$, $\hat{p}_1 = .31$; $n_2 = 250$, $\hat{p}_2 = .25$

c. $n_1 = 100$, $\hat{p}_1 = .46$; $n_2 = 120$, $\hat{p}_2 = .61$

8.8 Sketch the sampling distribution of $(\hat{p}_1 - \hat{p}_2)$ based on independent random samples of $n_1 = 100$ and $n_2 = 200$ observations from two binomial populations with probabilities of success $p_1 = .1$ and $p_2 = .5$, respectively.

Applying the Concepts—Basic

8.9 **Bullying behavior study.** School bullying is a form of aggressive behavior that occurs when a student is exposed repeatedly to negative actions (e.g., name-calling, hitting, kicking, spreading slander) from another student. In order to study the effectiveness of an antibullying policy at Dutch elementary schools, a survey of over 2,000 elementary school children was conducted (*Health Education Research*, Feb. 2005). Each student was asked if he or she ever bullied another student. In a sample of 1,358 boys, 746 claimed they had never bullied another student. In a sample of 1,379 girls, 967 claimed they had never bullied another student.

a. Estimate the true proportion of Dutch boys who have never bullied another student.

b. Estimate the true proportion of Dutch girls who have never bullied another student.

c. Estimate the difference in the proportions with a 90% confidence interval.

d. Make a statement about how likely the interval you used in part **c** contains the true difference in proportions.

e. Which group is more likely to bully another student, Dutch boys or Dutch girls?

8.10 **Executive workout dropouts.** Refer to the *Journal of Sport Behavior* (2001) study of variety in exercise workouts, presented in Exercise 5.58 (p. 283). One group of 40 people varied their exercise routine in workouts, while a second group of 40 exercisers had no set schedule or regulations for their workouts. By the end of the study, 15 people had dropped out of the first exercise group and 23 had dropped out of the second group.

a. Find the dropout rates (i.e., the percentage of exercisers who had dropped out of the exercise group) for each of the two groups of exercisers.

b. Find a 90% confidence interval for the difference between the dropout rates of the two groups of exercisers.

c. Give a practical interpretation of the confidence interval you found in part **c**.

8.11 **Treating depression with St. John's wort.** The *Journal of the American Medical Association* (April 18, 2001) published a study of the effectiveness of using extracts of the herbal medicine St. John's wort in treating major depression. In an eight-week randomized, controlled trial, 200 patients diagnosed with major depression were divided into two groups, one of which ($n_1 = 98$) received St. John's wort extract while the other ($n_2 = 102$) received a placebo (no drug). At the end of

the study period, 14 of the St. John's wort patients were in remission, compared with 5 of the placebo patients.

a. Compute the proportion of the St. John's wort patients who were in remission.

b. Compute the proportion of the placebo patients who were in remission.

c. If St. John's wort is effective in treating major depression, then the proportion of St. John's wort patients in remission will exceed the proportion of placebo patients in remission. At $\alpha = .01$, is St. John's wort effective in treating major depression?

d. Repeat part **c**, but use $\alpha = .10$.

e. Explain why the choice of α is critical for this study.

8.12 **Planning-habits survey.** *American Demographics* (Jan. 2002) reported the results of a survey on the planning habits of men and women. In response to the question "What is your preferred method of planning and keeping track of meetings, appointments, and deadlines?" 56% of the men and 46% of the women answered "I keep them in my head." A nationally representative sample of 1,000 adults participated in the survey; therefore, assume that 500 were men and 500 were women.

a. Set up the null and alternative hypotheses for testing whether the percentage of men who prefer keeping track of appointments in their head is larger than the corresponding percentage of women.

b. Compute the test statistic for the test.

c. Give the rejection region for the test, using $\alpha = .01$.

d. Find the p-value for the test.

e. Draw the appropriate conclusion.

8.13 **Racial profiling by the LAPD.** *Racial profiling* is a term used to describe any police action that relies on ethnicity rather than behavior to target suspects engaged in criminal activities. Does the Los Angeles Police Department (LAPD) invoke racial profiling in stops and searches of Los Angeles drivers? This question was addressed in *Chance* (Spring 2006).

a. Data on stops and searches of both African-Americans and white drivers from January through June 2005 are summarized in the accompanying table. Conduct a test (at $\alpha = .05$) to determine whether there is a disparity in the proportions of African-American and white drivers who are searched by the LAPD after being stopped.

Race	Number Stopped	Number Searched	Number of "Hits"
African-American	61,688	12,016	5,134
White	106,892	5,312	3,006

Source: Khadjavi, L. S. "Driving while black in the City of Angels," *Chance*, Vol. 19, No. 2, Spring 2006 (Tables 1 and 2).

b. The LAPD defines a "hit rate" as the proportion of searches that result in a discovery of criminal activity. Use the data in the table to estimate the disparity in the hit rates for African-American and white drivers under a 95% confidence interval. Interpret the results.

Applying the Concepts—Intermediate

8.14 **Angioplasty's benefits challenged.** Each year, more than 1 million heart patients undergo an angioplasty. The benefits of an angioplasty were challenged in a recent study of 2,287 patients (2007 Annual Conference of the American College of Cardiology, New Orleans). All the patients had substantial blockage of the arteries, but were medically stable. All were treated with medication such as aspirin and beta-blockers. However, half the patients were randomly assigned to get an angioplasty and half were not. After five years, the researchers found that 211 of the 1,145 patients in the angioplasty group had subsequent heart attacks, compared with 202 of 1,142 patients in the medication-only group. Do you agree with the study's conclusion that "There was no significant difference in the rate of heart attacks for the two groups"? Support your answer with a 95% confidence interval.

8.15 Killing insects with low oxygen. Refer to the *Journal of Agricultural, Biological, and Environmental Statistics* (Sep. 2000) study of the mortality of rice weevils exposed to low oxygen, presented in Exercise 8.82 (p. 335). Recall that 31,386 of 31,421 rice weevils were found dead after exposure to nitrogen gas for 4 days. In a second experiment, 23,516 of 23,676 rice weevils were found dead after exposure to nitrogen gas for 3.5 days. Conduct a test of hypothesis to compare the mortality rates of adult rice weevils exposed to nitrogen at the two exposure times. Is there a significant difference (at $\alpha = .10$) in the mortality rates?

8.16 Effectiveness of drug tests of Olympic athletes. Erythropoietin (EPO) is a banned drug used by athletes to increase the oxygen-carrying capacity of their blood. New tests for EPO were first introduced prior to the 2000 Olympic Games held in Sydney, Australia. *Chance* (Spring 2004) reported that of a sample of 830 world-class athletes, 159 did not compete in the 1999 World Championships (a year prior to the introduction of the new EPO test). Similarly, 133 of 825 potential athletes did not compete in the 2000 Olympic games. Was the new test effective in deterring an athlete from participating in the 2000 Olympics? If so, then the proportion of non participating athletes in 2000 will be more than the proportion of nonparticipating athletes in 1999. Conduct the analysis (at $\alpha = .10$) and draw the proper conclusion.

8.17 "Tip-of-the-tongue" study. Trying to think of a word you know, but can't instantly retrieve, is called the "tip-of-the-tongue" phenomenon. *Psychology and Aging* (Sept. 2001) published a study of this phenomenon in senior citizens. The researchers compared 40 people between 60 and 72 years of age with 40 between 73 and 83 years of age. When primed with the initial syllable of a missing word (e.g., seeing the word *include* to help recall the word *incisor*), the younger seniors had a higher recall rate. Suppose 31 of the 40 seniors in the younger group could recall the word when primed with the initial syllable, while only 22 of the 40 seniors could recall the word. Compare the recall rates of the two groups, using $\alpha = .05$. Does one group of elderly people have a significantly higher recall rate than the other?

⊙ MILK

8.18 Detection of rigged school milk prices (cont'd). Refer to the investigation of collusive bidding in the northern Kentucky school milk market, presented in Exercise 7.24 (p. 369). Market allocation is a common form of collusive behavior in bid-rigging conspiracies. Under collusion, the same dairy usually controls the same school districts year after year. The *incumbency rate* for a market is defined as the proportion of school districts that are won by the vendor that won the previous year. Past experience with milk bids in a competitive environment reveals that a typical incumbency rate is .7. That is, 70% of the school districts are expected to purchase their milk from the dairy that won the previous year. Incumbency rates of .9 or higher are strong indicators of collusive bidding. Over the years, when bid collusion was alleged to have occurred in northern Kentucky, there were 51 potential vendor transitions (i.e., changes in milk supplier from one year to the next in a district) in the tricounty market and 134 potential vendor transitions in the surrounding market. These values represent the sample sizes ($n_1 = 134$ and $n_2 = 51$) for calculating incumbency rates. Examining the data saved in the **MILK** file, you'll find that in 50 of the 51 potential vendor transitions for the tricounty market, the winning dairy from the previous year won the bid the next year; similarly, you'll find that in 91 of the 134 potential vendor transitions for the surrounding area, the same dairy won the bid the next year.

a. Estimate the incumbency rates for the tricounty and surrounding milk markets.

b. A MINITAB printout comparing the two incumbency rates is shown below. Give a practical interpretation of the results. Do they show further support for the bid collusion theory?

8.19 Does sleep improve mental performance? Is creativity and problem solving linked to adequate sleep? This question was the subject of research conducted by German scientists at the University of Lübeck (*Nature*, Jan. 22, 2004). One hundred volunteers were divided into two equal-sized groups. Each volunteer took a math test that involved transforming strings of eight digits into a new string that fit a set of given rules, as well as a third, hidden rule. Prior to taking the test, one group received eight hours of sleep, while the other group stayed awake all night. The scientists monitored the volunteers to determine whether and when they figured out the third rule. Of the volunteers who slept, 39 discovered the third rule; of the volunteers who stayed awake all night, 15 discovered the third rule. From the study results, what can you infer about the proportions of volunteers in the two groups who discover the third rule? Support your answer with a 90% confidence interval.

Test and CI for Two Proportions

```
Sample   X    N   Sample p
1       91   134  0.679104
2       50    51  0.980392


Difference = p (1) - p (2)
Estimate for difference:  -0.301288
95% upper bound for difference:  -0.227669
Test for difference = 0 (vs < 0):  Z = -4.30  P-Value = 0.000
```

MINITAB output for Exercise 8.18

Applying the Concepts—Advanced

8.20 Gambling in public high schools. With the rapid growth in legalized gambling in the United States, there is concern that the involvement of youth in gambling activities is also increasing. University of Minnesota professor Randy Stinchfield compared the rates of gambling among Minnesota public school students between 1992 and 1998. (*Journal of Gambling Studies*, Winter 2001.) Based on survey data, the following table shows the percentages of ninth-grade boys who gambled weekly or daily on any game (e.g., cards, sports betting, lotteries) for the two years:

	1992	1998
Number of ninth-grade boys in survey	21,484	23,199
Number who gambled weekly/daily	4,684	5,313

a. Are the percentages of ninth-grade boys who gambled weekly or daily on any game in 1992 and 1998 significantly different? (Use $\alpha = .01$.)

b. Professor Stinchfield states that "because of the large sample sizes, even small differences may achieve statistical significance, so interpretations of the differences should include a judgement regarding the magnitude of the difference and its public health significance." Do you agree with this statement? If not, why not? If so, obtain a measure of the magnitude of the difference between 1992 and 1998 and attach a measure of reliability to the difference.

8.21 Food-craving study. Do you have an insatiable craving for chocolate or some other food? Since many people apparently do, psychologists are designing scientific studies to examine the phenomenon. According to the *New York Times* (Feb. 22, 1995), one of the largest studies of food cravings involved a survey of 1,000 McMaster University (Canada) students. The survey revealed that 97% of the women in the study acknowledged specific food cravings while only 67% of the men did.

a. How large do n_1 and n_2 have to be to conclude that the true proportion of women who acknowledge having food cravings exceeds the corresponding proportion of men? Assume that $\alpha = .01$.

b. Why is it dangerous to conclude from the study that women have a higher incidence of food cravings than men?

8.2 Determining the Sample Size

The sample sizes n_1 and n_2 required to compare two population proportions can be found in a manner similar to the method described in Section 7.3 for comparing two population means. We will assume equal sized (i.e., $n_1 = n_2 = n$) and then choose n so that $(\hat{p}_1 - \hat{p}_2)$ will differ from $(p_1 - p_2)$ by no more than a sampling error SE with a specified level of confidence. We will illustrate the procedure with an example.

EXAMPLE 8.3

FINDING THE SAMPLE SIZES FOR ESTIMATING $(p_1 - p_2)$:

Comparing Defect Rates of Two Machines

Problem A production supervisor suspects that a difference exists between the proportions p_1 and p_2 of defective items produced by two different machines. Experience has shown that the proportion defective for each of the two machines is in the neighborhood of .03. If the supervisor wants to estimate the difference in the proportions to within .005, using a 95% confidence interval, how many items must be randomly sampled from the output produced by each machine? (Assume that the supervisor wants $n_1 = n_2 = n$.)

Solution In this sampling problem, the sampling error $SE = .005$, and for the specified level of reliability, $z_{\alpha/2} = z_{.025} = 1.96$. Then, letting $p_1 = p_2 = .03$ and $n_1 = n_2 = n$, we find the required sample size per machine by solving the following equation for n:

$$z_{\alpha/2}\sigma_{(\hat{p}_1 - \hat{p}_2)} = SE$$

or

$$z_{\alpha/2}\sqrt{\frac{p_1 q_1}{n_1} + \frac{p_2 q_2}{n_2}} = SE$$

$$1.96\sqrt{\frac{(.03)(.97)}{n} + \frac{(.03)(.97)}{n}} = .005$$

$$1.96\sqrt{\frac{2(.03)(.97)}{n}} = .005$$

$$n = 8,943.2$$

Look Back This large n will likely result in a tedious sampling procedure. If the supervisor insists on estimating $(p_1 - p_2)$ correct to within .005 with 95% confidence, approximately 9,000 items will have to be inspected for each machine.

Now Work Exercise 8.24a

You can see from the calculations in Example 8.3 that $\sigma_{(\hat{p}_1 - \hat{p}_2)}$ (and hence the solution, $n_1 = n_2 = n$) depends on the actual (but unknown) values of p_1 and p_2. In fact, the required sample size $n_1 = n_2 = n$ is largest when $p_1 = p_2 = .5$. Therefore, if you have no prior information on the approximate values of p_1 and p_2, use $p_1 = p_2 = .5$ in the formula for $\sigma_{(\hat{p}_1 - \hat{p}_2)}$. If p_1 and p_2 are in fact close to .5, then the values of n_1 and n_2 that you have calculated will be correct. If p_1 and p_2 differ substantially from .5, then your solutions for n_1 and n_2 will be larger than needed. Consequently, using $p_1 = p_2 = .5$ when solving for n_1 and n_2 is a conservative procedure because the sample sizes n_1 and n_2 will be at least as large as (and probably larger than) needed.

The procedure for determining sample sizes necessary for estimating $(p_1 - p_2)$ for the case $n_1 = n_2$ is given in the following box:

Determination of Sample Size for Estimating $p_1 - p_2$

To estimate $(p_1 - p_2)$ to within a given sampling error SE and with confidence level $(1 - \alpha)$, use the following formula to solve for equal sample sizes that will achieve the desired reliability:

$$n_1 = n_2 = \frac{(z_{\alpha/2})^2 (p_1 q_1 + p_2 q_2)}{(SE)^2}$$

You will need to substitute estimates for the values of p_1 and p_2 before solving for the sample size. These estimates might be based on prior samples, obtained from educated guesses or, most conservatively, specified as $p_1 = p_2 = .5$.

Exercises 8.22–8.30

Understanding the Principles

8.22 In determining the sample sizes for estimating $p_1 - p_2$, how do you obtain estimates of the binomial proportions p_1 and p_2 used in the calculations?

8.23 If the sample-size calculation yields a value of n that is too large to be practical, how should you proceed?

Learning the Mechanics

8.24 Assuming that $n_1 = n_2$, find the sample sizes needed to estimate $(p_1 - p_2)$ for each of the following situations:
NW
  a. SE = .01 with 99% confidence. Assume that $p_1 \approx .4$ and $p_2 \approx .7$.
  b. A 90% confidence interval of width .05. Assume there is no prior information available with which to obtain approximate values of p_1 and p_2.
  c. SE = .03 with 90% confidence. Assume that $p_1 \approx .2$ and $p_2 \approx .3$.

8.25 Enough money has been budgeted to collect independent random samples of size $n_1 = n_2 = 100$ from populations 1 and 2 in order to estimate $(p_1 - p_2)$. Prior information indicates that $p_1 = p_2 \approx .6$. Have sufficient funds been allocated to construct a 90% confidence interval for $(p_1 - p_2)$ of width .1 or less? Justify your answer.

Applying the Concepts—Basic

8.26 Size of a political poll. A pollster wants to estimate the difference between the proportions of men and women who favor a particular national candidate using a 90% confidence interval of width .04. Suppose the pollster has no

prior information about the proportions. If equal numbers of men and women are to be polled, how large should the sample sizes be?

8.27 Executive workout dropouts. Refer to the *Journal of Sport Behavior* (2001) study comparing the dropout rates of two groups of exercisers, presented in Exercise 8.10 (p. 440). Suppose you want to reduce the sampling error for estimating $(p_1 - p_2)$ with a 90% confidence interval to .1. Determine the number of exercisers to be sampled from each group in order to obtain such an estimate. Assume equal sample sizes, and assume that $p_1 \approx .4$ and $p_2 \approx .6$.

Applying the Concepts—Intermediate

8.28 Cable-TV home shoppers. All cable television companies carry at least one home-shopping channel. Who uses these home-shopping services? Are the shoppers primarily men or women? Suppose you want to estimate the difference in the percentages of men and women who say they have used or expect to use televised home shopping. You want an 80% confidence interval of width .06 or less.
  a. Approximately how many people should be included in your samples?
  b. Suppose you want to obtain individual estimates for the two percentages of interest. Will the sample size found in part **a** be large enough to provide estimates of each percentage correct to within .02 with probability equal to .90? Justify your response.

8.29 Buyers of TVs. A manufacturer of large-screen televisions wants to compare with a competitor the proportions of its best sets that need repair within 1 year. If it is desired to estimate the difference between proportions to within .05 with 90% confidence, and if the manufacturer plans to sample twice as many buyers (n_1) of its sets as buyers (n_2) of the competitor's sets, how many buyers of each brand must be sampled? Assume that the proportion of sets that need repair will be about .2 for both brands.

8.30 Rat damage in sugarcane. Poisons are used to prevent rat damage in sugarcane fields. The U.S. Department of Agriculture is investigating whether the rat poison should be located in the middle of the field or on the outer perimeter. One way to answer this question is to determine where the greater amount of damage occurs. If damage is measured by the proportion of cane stalks that have been damaged by rats, how many stalks from each section of the field should be sampled in order to estimate the true difference between proportions of stalks damaged in the two sections, to within .02 with 95% confidence?

8.3 Testing Categorical Probabilities: Multinomial Experiment

Recall from Section 1.4 (p. 12) that observations on a qualitative variable can only be categorized. For example, consider the highest level of education attained by a professional hockey player. Level of education is a qualitative variable with several categories, including some high school, high school diploma, some college, college undergraduate degree, and graduate degree. If we were to record education level for all professional hockey players, the result of the categorization would be a count of the numbers of players falling in the respective categories.

When the qualitative variable of interest results in one of two responses (e.g., yes or no, success or failure, favor or do not favor), the data—called *counts*—can be analyzed with the binomial probability distribution discussed in Section 4.3. However, qualitative variables, such as level of education, that allow for more than two categories for a response are much more common, and these must be analyzed by a different method.

Qualitative data with more than two levels often result from a **multinomial experiment**. The characteristics for a multinomial experiment with k outcomes are described in the next box. You can see that the binomial experiment of Chapter 4 is a multinomial experiment with $k = 2$.

Properties of the Multinomial Experiment

1. The experiment consists of n identical trials.

2. There are k possible outcomes to each trial. These outcomes are sometimes called **classes**, **categories**, or **cells**.

3. The probabilities of the k outcomes, denoted by $p_1, p_2, \ldots, p_k$, where $p_1 + p_2 + \cdots + p_k = 1$, remain the same from trial to trial.

4. The trials are independent.

5. The random variables of interest are the **cell counts** $n_1, n_2, \ldots, n_k$ of the number of observations that fall into each of the k categories.

EXAMPLE 8.4

IDENTIFYING A MULTINOMIAL EXPERIMENT

Problem Consider the problem of determining the highest level of education attained by each of a sample of $n = 40$ National Hockey League (NHL) players. Suppose we categorize level of education into one of five categories—some high school, high school diploma, some college, college undergraduate degree, and graduate degree—and count the number of the 40 players that fall into each category. Is this a multinomial experiment, to a reasonable degree of approximation?

Solution Checking the five properties of a multinomial experiment shown in the box, we have the following:

1. The experiment consists of $n = 40$ identical trials, each of which is undertaken to determine the education level of an NHL player.

2. There are $k = 5$ possible outcomes to each trial, corresponding to the five education-level responses.

3. The probabilities of the $k = 5$ outcomes p_1, p_2, p_3, p_4, and p_5, where p_i represents the true probability that an NHL player attains level-of-education category i, remain the same from trial to trial (to a reasonable degree of approximation).
4. The trials are independent; that is, the education level attained by one NHL player does not affect the level attained by any other player.
5. We are interested in the count of the number of hockey players who fall into each of the five education-level categories. These five cell counts are denoted n_1, n_2, n_3, n_4, and n_5.

Thus, the properties of a multinomial experiment are satisfied.

In this section, we consider a multinomial experiment with k outcomes that correspond to the categories of a *single* qualitative variable. The results of such an experiment are summarized in a **one-way table.** The term *one-way* is used because only one variable is classified. Typically, we want to make inferences about the true percentages that occur in the k categories on the basis of the sample information in the one-way table.

To illustrate, suppose three political candidates are running for the same elective position. Prior to the election, we conduct a survey to determine the voting preferences of a random sample of 150 eligible voters. The qualitative variable of interest is *preferred candidate*, which has three possible outcomes: candidate 1, candidate 2, and candidate 3. Suppose the number of voters preferring each candidate is tabulated and the resulting count data appear as in Table 8.3.

TABLE 8.3 Results of Voter Preference Survey

Candidate		
1	**2**	**3**
61 votes	53 votes	36 votes

Note that our voter preference survey satisfies the properties of a multinomial experiment for the qualitative variable, preferred candidate. The experiment consists of randomly sampling $n = 150$ voters from a large population of voters containing an unknown proportion p_1 that favors candidate 1, a proportion p_2 that favors candidate 2, and a proportion p_3 that favors candidate 3. Each voter sampled represents a single trial that can result in one of three outcomes: The voter will favor candidate 1, 2, or 3 with probabilities p_1, p_2, and p_3, respectively. (Assume that all voters will have a preference.) The voting preference of any single voter in the sample does not affect the preference of any other; consequently, the trials are independent. Finally, you can see that the recorded data are the numbers of voters in each of the three preference categories. Thus, the voter preference survey satisfies the five properties of a multinomial experiment.

In this survey, and in most practical applications of the multinomial experiment, the k outcome probabilities p_1, p_2, ..., p_k are unknown and we want to use the survey data to make inferences about their values. The unknown probabilities in the voter preference survey are

$$p_1 = \text{Proportion of all voters who favor candidate 1}$$
$$p_2 = \text{Proportion of all voters who favor candidate 2}$$
$$p_3 = \text{Proportion of all voters who favor candidate 3}$$

To decide whether the voters, in total, have a preference for any one of the candidates, we will test the null hypothesis that the candidates are equally preferred (i.e., $p_1 = p_2 = p_3 = \frac{1}{3}$) against the alternative hypothesis that one candidate is preferred (i.e., at least one of the probabilities p_1, p_2, and p_3 exceeds $\frac{1}{3}$). Thus, we want to test

H_0: $p_1 = p_2 = p_3 = \frac{1}{3}$ (no preference)

H_a: At least one of the proportions exceeds $\frac{1}{3}$ (a preference exists)

If the null hypothesis is true and $p_1 = p_2 = p_3 = \frac{1}{3}$, then the expected value (mean value) of the number of voters who prefer candidate 1 is given by

$$E_1 = np_1 = (n)\frac{1}{3} = (150)\frac{1}{3} = 50$$

Similarly, $E_2 = E_3 = 50$ if the null hypothesis is true and no preference exists.

The **chi-square test** measures the degree of disagreement between the data and the null hypothesis:

$$\chi^2 = \frac{[n_1 - E_1]^2}{E_1} + \frac{[n_2 - E_2]^2}{E_2} + \frac{[n_3 - E_3]^2}{E_3}$$
$$= \frac{(n_1 - 50)^2}{50} + \frac{(n_2 - 50)^2}{50} + \frac{(n_3 - 50)^2}{50}$$

Note that the farther the observed numbers n_1, n_2, and n_3 are from their expected value (50), the larger χ^2 will become. That is, large values of χ^2 imply that the null hypothesis is false.

Biography

KARL PEARSON
(1895–1980)—the Father of Statistics

While attending college, London-born Karl Pearson exhibited a wide range of interests, including mathematics, physics, religion, history, socialism, and Darwinism. After earning a law degree at Cambridge University and a Ph.D. in political science at the University of Heidelberg (Germany), Pearson became a professor of applied mathematics at University College in London. His 1892 book, *The Grammar of Science*, illustrated his convic-

tion that statistical data analysis lies at the foundation of all knowledge; consequently, many consider Pearson to be the "father of statistics." Among Pearson's many contributions to the field are introducing the term *standard deviation* and its associated symbol (σ); developing the distribution of the correlation coefficient; cofounding and editing the prestigious statistics journal *Biometrika*; and (what many consider his greatest achievement) creating the first chi-square "goodness-of-fit" test. Pearson inspired his students (including his son, Egon, and William Gossett) with his wonderful lectures and enthusiasm for statistics.

We have to know the distribution of χ^2 in repeated sampling before we can decide whether the data indicate that a preference exists. When H_0 is true, χ^2 can be shown to have (approximately) a **chi-square (χ^2) distribution.** The shape of the chi-square distribution is nonsymmetric (see Figure 8.3) and for this one-way classification, has $(k - 1)$ degrees of freedom.* Critical values of χ^2 are provided in Table XI of Appendix A, a portion of which is shown in Table 8.4. To illustrate, the rejection region for the voter preference survey for $\alpha = .05$ and $k - 1 = 3 - 1 = 2$ df is

$$\textit{Rejection region: } \chi^2 > \chi^2_{.05}$$

This value of $\chi^2_{.05}$ (found in Table XI) is 5.99147. (See Figure 8.3.) The computed value of the test statistic is

$$\chi^2 = \frac{(n_1 - 50)^2}{50} + \frac{(n_2 - 50)^2}{50} + \frac{(n_3 - 50)^2}{50}$$
$$= \frac{(61 - 50)^2}{50} + \frac{(53 - 50)^2}{50} + \frac{(36 - 50)^2}{50} = 6.52$$

Since the computed $\chi^2 = 6.52$ exceeds the critical value of 5.99147, we conclude at the $\alpha = .05$ level of significance that there does exist a voter preference for one or more of the candidates.

Now that we have evidence to indicate that the proportions p_1, p_2, and p_3 are unequal, we can use the methods of Section 5.4 to make inferences concerning their individ-

*The derivation of the number of degrees of freedom for χ^2 involves the number of linear restrictions imposed on the count data. In the present case, the only constraint is that $\Sigma n_i = n$, where n (the sample size) is fixed in advance. Therefore, df $= k - 1$. For other cases, we will give the number of degrees of freedom for each usage of χ^2 and refer the interested reader to the references for more detail.

TABLE 8.4 Reproduction of Part of Table XI in Appendix A

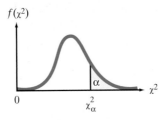

$f(\chi^2)$

Degrees of Freedom	$\chi^2_{.100}$	$\chi^2_{.050}$	$\chi^2_{.025}$	$\chi^2_{.010}$	$\chi^2_{.005}$
1	2.70554	3.84146	5.02389	6.63490	7.87944
2	4.60517	5.99147	7.37776	9.21034	10.5966
3	6.25139	7.81473	9.34840	11.3449	12.8381
4	7.77944	9.48773	11.1433	13.2767	14.8602
5	9.23635	11.0705	12.8325	15.0863	16.7496
6	10.6446	12.5916	14.4494	16.8119	18.5476
7	12.0170	14.0671	16.0128	18.4753	20.2777

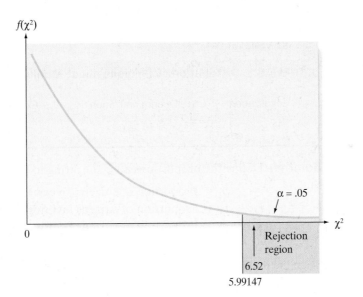

FIGURE 8.3

Rejection region for voter preference survey

ual values. [*Note:* We cannot use the methods of Section 5.4 to compare two proportions, because the cell counts are dependent random variables.] The general form for a test of hypothesis concerning multinomial probabilities is shown in the following box:

A Test of a Hypothesis about Multinomial Probabilities: One-Way Table

H_0: $p_1 = p_{1,0}$, $p_2 = p_{2,0}, \ldots, p_k = p_{k,0}$

where $p_{1,0}, p_{2,0}, \ldots, p_{k,0}$ represent the hypothesized values of the multinomial probabilities

H_a: At least one of the multinomial probabilities does not equal its hypothesized value

Test statistic: $\chi^2 = \sum \dfrac{[n_i - E_i]^2}{E_i}$

where $E_i = np_{i,0}$ is the **expected cell count**—that is, the expected number of outcomes of type i, assuming that H_0 is true. The total sample size is n.

Rejection region: $\chi^2 > \chi^2_\alpha$, where χ^2_α has $(k - 1)$ df

> **Conditions Required for a Valid χ^2 Test: One-Way Table**
>
> 1. A multinomial experiment has been conducted. This is generally satisfied by taking a random sample from the population of interest.
>
> 2. The sample size n will be large enough so that, for every cell, the expected cell count $E(n_i)$ will be equal to 5 or more.*

EXAMPLE 8.5
A ONE-WAY χ^2 TEST—Effectiveness of a TV Program on Marijuana

Problem Suppose an educational television station has broadcast a series of programs on the physiological and psychological effects of smoking marijuana. Now that the series is finished, the station wants to see whether the citizens within the viewing area have changed their minds about how the possession of marijuana should be considered legally. Before the series was shown, it was determined that 7% of the citizens favored legalization, 18% favored decriminalization, 65% favored the existing law (an offender could be fined or imprisoned), and 10% had no opinion.

A summary of the opinions (after the series was shown) of a random sample of 500 people in the viewing area is given in Table 8.5. Test at the $\alpha = .01$ level to see whether these data indicate that the distribution of opinions differs significantly from the proportions that existed before the educational series was aired.

 MARIJUANA

TABLE 8.5 Distribution of Opinions about Marijuana Possession

Legalization	Decriminalization	Existing Laws	No Opinion
39	99	336	26

Solution Define the proportions after the airing to be

$$p_1 = \text{Proportion of citizens favoring legalization}$$
$$p_2 = \text{Proportion of citizens favoring decriminalization}$$
$$p_3 = \text{Proportion of citizens favoring existing laws}$$
$$p_4 = \text{Proportion of citizens with no opinion}$$

Then the null hypothesis representing no change in the distribution of percentages is

$$H_0: p_1 = .07, p_2 = .18, p_3 = .65, p_4 = .10$$

and the alternative is

H_a: At least one of the proportions differs from its null hypothesized value

Thus, we have

$$\text{Test statistic: } \chi^2 = \sum \frac{[n_i - E_i]^2}{E_i}$$

where

$$E_1 = np_{1,0} = 500(.07) = 35$$
$$E_2 = np_{2,0} = 500(.18) = 90$$
$$E_3 = np_{3,0} = 500(.65) = 325$$
$$E_4 = np_{4,0} = 500(.10) = 50$$

Since all these values are larger than 5, the χ^2 approximation is appropriate. Also, if the citizens in the sample were randomly selected, then the properties of the multinomial probability distribution are satisfied.

---

*The assumption that all expected cell counts are at least 5 is necessary in order to ensure that the χ^2 approximation is appropriate. Exact methods for conducting the test of hypothesis exist and may be used for small expected cell counts, but these methods are beyond the scope of this text.

OPINION

	Observed N	Expected N	Residual
LEGAL	39	35.0	4.0
DECRIM	99	90.0	9.0
EXISTLAW	336	325.0	11.0
NONE	26	50.0	-24.0
Total	500		

Test Statistics

	OPINION
Chi-Square[a]	13.249
df	3
Asymp. Sig.	.004

a. 0 cells (.0%) have expected frequencies less than 5. The minimum expected cell frequency is 35.0.

FIGURE 8.4

SPSS analysis of data in Table 8.5

Rejection region: For $\alpha = .01$ and df $= k - 1 = 3$, reject H_0 if $\chi^2 > \chi^2_{.01}$, where (from Table XI in Appendix A) $\chi^2_{.01} = 11.3449$.

We now calculate the test statistic:

$$\chi^2 = \frac{(39-35)^2}{35} + \frac{(99-90)^2}{90} + \frac{(336-325)^2}{325} + \frac{(26-50)^2}{50} = 13.249$$

Since this value exceeds the table value of χ^2 (11.3449), the data provide sufficient evidence ($\alpha = .01$) that the opinions on the legalization of marijuana have changed since the series was aired.

The χ^2 test can also be conducted with the use of an available statistical software package. Figure 8.4 is an SPSS printout of the analysis of the data in Table 8.5. The test statistic and p-value of the test are highlighted on the printout. Since $\alpha = .01$ exceeds $p = .004$, there is sufficient evidence to reject H_0.

Look Back If the conclusion for the χ^2 test is "fail to reject H_0," then there is insufficient evidence to conclude that the distribution of opinions differs from the proportions stated in H_0. Be careful not to "accept H_0" and conclude that $p_1 = .07$, $p_2 = .18$, $p_3 = .65$, and $p_4 = .10$. The probability (β) of a Type II error is unknown.

Now Work Exercise 8.39

If we focus on one particular outcome of a multinomial experiment, we can use the methods developed in Section 5.4 for a binomial proportion to establish a confidence interval for any one of the multinomial probabilities.* For example, if we want a 95% confidence interval for the proportion of citizens in the viewing area who have no opinion about the issue, we calculate

$$\hat{p}_4 \pm 1.96\sigma_{\hat{p}_4}$$

where

$$\hat{p}_4 = \frac{n_4}{n} = \frac{26}{500} = .052 \quad \text{and} \quad \sigma_{\hat{p}_4} \approx \sqrt{\frac{\hat{p}_4(1-\hat{p}_4)}{n}}$$

Thus, we get

$$.052 \pm 1.96\sqrt{\frac{(.052)(.948)}{500}} = .052 \pm .019$$

or (.033, .071). Consequently, we estimate that between 3.3% and 7.1% of the citizens now have no opinion on the issue of the legalization of marijuana. The series of programs may have helped citizens who formerly had no opinion on the issue to form an opinion, since it appears that the proportion of "no opinions" is now less than 10%.

*Note that focusing on one outcome has the effect of lumping the other ($k - 1$) outcomes into a single group. Thus, we obtain, in effect, two outcomes—or a binomial experiment.

ACTIVITY 8.2: **Binomial versus Multinomial Experiments**

In this activity, you will study the difference between binomial and multinomial experiments.

1. A television station has hired an independent research group to determine whether television viewers in the area prefer its local news program to the news programs of two other stations in the same city. Explain why a multinomial experiment would be appropriate, and design a poll that satisfies the five properties of a multinomial experiment. State the null and alternative hypotheses for the corresponding χ^2 test.

2. Suppose the television station believes that a majority of local viewers prefers its news program to those of its two competitors. Explain why a binomial experiment would be appropriate to support this claim, and design a poll that satisfies the five properties of a binomial experiment. State the null and alternative hypotheses for the corresponding test.

3. Generalize the situations in Exercises 1 and 2 in order to describe conditions under which a multinomial experiment can be rephrased as a binomial experiment. Is there any advantage in doing so? Explain.

STATISTICS IN ACTION REVISITED

Testing Category Proportions for Type of College Drinker

In the *Family and Consumer Sciences Research Journal* (Mar. 2005) study of college students and drinking (p. 433), one of the researchers' main objectives was to segment college students according to their rates of alcohol consumption. A segmentation was developed on the basis of the students' responses to the questions on frequency of drinking and average number of drinks per occasion. Four types, or groups, of college drinkers emerged: non/seldom drinkers, social drinkers, typical binge drinkers, and heavy binge drinkers. What are the proportions of students in each of these groups, and are these proportions statistically different?

FIGURE SIA8.1

SPSS descriptive statistics and graph for type of drinker

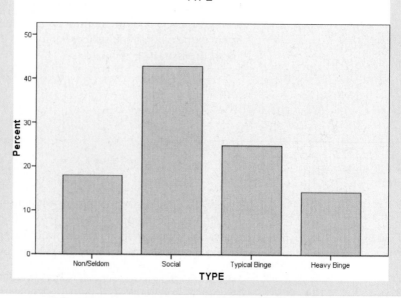

TYPE

		Frequency	Percent	Valid Percent	Cumulative Percent
Valid	Non/Seldom	118	18.0	18.0	18.0
	Social	282	42.9	42.9	60.9
	Typical Binge	163	24.8	24.8	85.7
	Heavy Binge	94	14.3	14.3	100.0
	Total	657	100.0	100.0	

(continued)

To answer these questions, we used SPSS to analyze the type-of-drinker variable in the **COLLDRINKS** file. Figure SIA8.1 shows summary statistics and a graph describing the four categories. From the summary table at the top of the printout, you can see that 118 (or 18%) of the students are non/seldom drinkers, 282 (or 43%) are social drinkers, 163 (or 25%) are typical binge drinkers, and 94 (or 14%) are heavy binge drinkers. These sample percentages are illustrated in the bar graph in Figure SIA8.1. In this sample of students, the largest percentage (43%) consists of social drinkers.

Is this sufficient evidence to indicate that the true proportions in the population of college students are different? Letting p_1, p_2, p_3, and p_4 represent the true proportions for non/seldom, social, typical binge, and heavy binge drinkers, respectively, we tested H_0: $p_1 = p_2 = p_3 = p_4 = .25$, using the chi-square test in SPSS. The printout is displayed in Figure SIA8.2. The cell frequencies and expected numbers are shown in the top table of the figure, while the chi-square test statistic (127.5) and p-value (.000) are shown in the bottom table. At any reasonably selected α-level (say, $\alpha = .01$), the small p-value indicates that there is sufficient evidence to reject the null hypothesis; thus, we conclude that the true proportions associated with the four type-of-drinker categories are indeed statistically different.

Chi-Square Test

Frequencies

TYPE

	Observed N	Expected N	Residual
Non/Seldom	118	164.3	-46.3
Social	282	164.3	117.8
Typical Binge	163	164.3	-1.3
Heavy Binge	94	164.3	-70.3
Total	657		

Test Statistics

	TYPE
Chi-Square[a]	127.493
df	3
Asymp. Sig.	.000

a. 0 cells (.0%) have expected frequencies less than 5. The minimum expected cell frequency is 164.3.

FIGURE SIA8.2
SPSS chi-square test for type-of-drinker categories

Exercises 8.31–8.48

Understanding the Principles

8.31 What are the characteristics of a multinomial experiment? Compare the characteristics with those of a binomial experiment.

8.32 What conditions must n satisfy to make the χ^2 test for a one-way table valid?

Learning the Mechanics

8.33 Use Table IX of Appendix A to find each of the following χ^2 values:
a. $\chi^2_{.05}$ for df = 10
b. $\chi^2_{.990}$ for df = 50
c. $\chi^2_{.10}$ for df = 16
d. $\chi^2_{.005}$ for df = 50

8.34 Use Table IX of Appendix A to find the following probabilities:
a. $P(\chi^2 \le 1.063623)$ for df = 4
b. $P(\chi^2 > 30.5779)$ for df = 15
c. $P(\chi^2 \ge 82.3581)$ for df = 100
d. $P(\chi^2 < 18.4926)$ for df = 30

8.35 Find the rejection region for a one-dimensional χ^2-test of a null hypothesis concerning $p_1, p_2, \ldots, p_k$ if
a. $k = 3$; $\alpha = .05$
b. $k = 5$; $\alpha = .10$
c. $k = 4$; $\alpha = .01$

8.36 A multinomial experiment with $k = 3$ cells and $n = 320$ produced the data shown in the accompanying table. Do these data provide sufficient evidence to contradict the null hypothesis that $p_1 = .25$, $p_2 = .25$, and $p_3 = .50$? Test, using $\alpha = .05$.

	Cell		
	1	2	3
n_i	78	60	182

8.37 A multinomial experiment with $k = 4$ cells and $n = 205$ produced the data shown in the following table:

	Cell			
	1	2	3	4
n_i	43	56	59	47

a. Do these data provide sufficient evidence to conclude that the multinomial probabilities differ? Test, using $\alpha = .05$.
b. What are the Type I and Type II errors associated with the test of part **a**?
c. Construct a 95% confidence interval for the multinomial probability associated with cell 3.

8.38 A multinomial experiment with $k = 4$ cells and $n = 400$ produced the data shown in the accompanying table. Do these data provide sufficient evidence to contradict the null hypothesis that $p_1 = .2$, $p_2 = .4$, $p_3 = .1$, and $p_4 = .3$? Test, using $\alpha = .05$.

	Cell			
	1	2	3	4
n_i	70	196	46	88

Applying the Concepts—Basic

8.39 **Jaw dysfunction study.** A report on dental patients with [NW] temporomandibular (jaw) joint dysfunction (TMD) was published in *General Dentistry* (Jan/Feb. 2004). A random sample of 60 patients was selected for an experimental treatment of TMD. Prior to treatment, the patients filled out a survey on two nonfunctional jaw habits—bruxism (teeth grinding) and teeth clenching—that have been linked to TMD. Of the 60 patients, 3 admitted to bruxism, 11 admitted to teeth clenching, 30 admitted to both habits, and 16 claimed they had neither habit.

a. Describe the qualitative variable of interest in the study. Give the levels (categories) associated with the variable.

b. Construct a one-way table for the sample data.

c. Give the null and alternative hypotheses for testing whether the percentages associated with the admitted habits are the same.

d. Calculate the expected numbers for each cell of the one-way table.

e. Calculate the appropriate test statistic.

f. Give the rejection region for the test at $\alpha = .05$

g. Give the appropriate conclusion in the words of the problem.

h. Find and interpret a 95% confidence interval for the true proportion of dental patients who admit to both habits.

8.40 **Location of major sports venues.** There has been a recent trend for professional sports franchises in Major League Baseball (MLB), the National Football League (NFL), the National Basketball Association (NBA), and the National Hockey League (NHL) to build new stadiums and ballparks in urban, downtown venues. An article in *Professional Geographer* (Feb. 2000) investigated whether there has been a significant suburban-to-urban shift in the location of major sport facilities. In 1985, 40% of all major sport facilities were located downtown, 30% in central cities, and 30% in suburban areas. In contrast, of the 113 major sports franchises that existed in 1997, 58 were built downtown, 26 in central cities, and 29 in suburban areas.

a. Describe the qualitative variable of interest in the study. Give the levels (categories) associated with the variable.

b. Give the null hypothesis for a test to determine whether the proportions of major sports facilities in downtown, central city, and suburban areas in 1997 are the same as in 1985.

c. If the null hypothesis of part **b** is true, how many of the 113 sports facilities in 1997 would you expect to be located in downtown, central city, and suburban areas, respectively?

d. Find the value of the chi-square statistic for testing the null hypothesis of part **b**.

e. Find the (approximate) p-value of the test, and give the appropriate conclusion in the words of the problem. Assume that $\alpha = .05$.

8.41 **Excavating ancient pottery.** Refer to the *Chance* (Fall 2000) study of ancient Greek pottery, presented in Exercise 2.12 (p. 35). Recall that 837 pottery pieces were uncovered at the excavation site. The table describing the types of pottery found is reproduced here:

Pot Category	Number Found
Burnished	133
Monochrome	460
Painted	183
Other	61
Total	**837**

Source: Berg, I., and Bliedon, S. "The Pots of Phyiakopi: Applying statistical Techniques to Archaeology." *Chance,* Vol. 13, No. 4, Fall 2000.

a. Describe the qualitative variable of interest in the study. Give the levels (categories) associated with the variable.

b. Assume that the four types of pottery occur with equal probability at the excavation site. What are the values of p_1, p_2, p_3, and p_4, the probabilities associated with the four pottery types?

c. Give the null and alternative hypotheses for testing whether one type of pottery is more likely to occur at the site than any of the other types.

d. Find the test statistic for testing the hypotheses stated in part **c**.

e. Find and interpret the p-value of the test. State the conclusion in the words of the problem if you use $\alpha = .10$.

8.42 **"Made in the USA" survey.** Refer to the *Journal of Global Business* (Spring 2002) study of what "Made in the USA" on product labels means to the typical consumer, presented in Exercise 2.13 (p. 36). Recall that 106 shoppers participated in the survey. Their responses, given as a percentage of U.S. labor and materials in four categories, are summarized in the accompanying table. Suppose a consumer advocate group claims that half of all consumers believe that "Made in the USA" means "100%" of labor and materials are produced in the United States, one-fourth believe that "75 to 99%" are produced in the United States, one-fifth believe that "50 to 74%" are produced in the United States, and 5 percent believe that "less than 50%" are produced in the United States.

Response to "Made in the USA"	Number of Shoppers
100%	64
75 to 99%	20
50 to 74%	18
Less than 50%	4

Source: "'Made in the USA': Consumer perceptions, deception and policy alternatives," *Journal of Global Business*, Vol. 13, No. 24, Spring 2002 (Table 3).

a. Describe the qualitative variable of interest in the study. Give the levels (categories) associated with the variable.

b. What are the values of p_1, p_2, p_3, and p_4, the probabilities associated with the four response categories hypothesized by the consumer advocate group?

c. Give the null and alternative hypotheses for testing the consumer advocate group's claim.

d. Compute the test statistic for testing the hypotheses stated in part **c**.

e. Find the rejection region of the test at $\alpha = .10$.

f. State the conclusion in the words of the problem.

g. Find and interpret a 90% confidence interval for the true proportion of consumers who believe that "Made in the USA" means that "100%" of labor and materials are produced in the United States.

Applying the Concepts—Intermediate

8.43 Top Internet search engines. Nielsen/NetRatings is a global leader in Internet media and market research. In May 2006, the firm reported on the "search" shares (i.e., the percentage of all Internet searches) for the most popular search engines available on the Web. Google Search accounted for 50% of all searches, Yahoo! Search for 22%, MSN Search for 11%, and all other search engines for 17%. Suppose that, in a random sample of 1,000 recent Internet searches, 487 used Google Search, 245 used Yahoo! Search, 121 used MSN Search, and 147 used another search engine.

a. Do the sample data disagree with the percentages reported by Nielsen/NetRatings? Test, using $\alpha = .05$.

b. Find and interpret a 95% confidence interval for the percentage of all Internet searches that use the Google Search engine.

8.44 Sociology fieldwork methods. Refer to the *Teaching Sociology* (July 2006) study of the fieldwork methods used by qualitative sociologists, presented in Exercise 2.14 (p. 36). Recall that fieldwork methods can be categorized as follows: Interview, Observation plus Participation, Observation Only, and Grounded Theory. The accompanying table shows the number of papers published over the past seven years in each category. Suppose a sociologist claims that 70%, 15%, 10%, and 5% of the fieldwork methods involve interview, observation plus participation, observation only, and grounded theory, respectively. Do the data support or refute the claim? Explain.

FIELDWORK

Fieldwork Method	Number of Papers
Interview	5,079
Observation + Participation	1,042
Observation Only	848
Grounded Theory	537

Source: Hood, J.C. "Teaching against the Text: The Case of Qualitative Methods," *Teaching Sociology,* Vol. 34, July 2006 (Exhibit 2).

BIBLE

8.45 Do you believe in the Bible? Refer to the General Social Survey (GSS) and the question pertaining to a person's belief in the Bible, presented in Exercise 2.17 (p. 36). Recall that approximately 2,800 Americans selected from one of the following answers: (1) The Bible is the actual word of God and is to be taken literally; (2) the Bible is the inspired word of God, but not everything is to be taken literally; (3) the Bible is an ancient book of fables; or (4) the Bible has some other origin, but is recorded by men. The variable "Bible1" in the **BIBLE** file contains the responses.

a. Summarize the responses in a one-way table.

b. State the null and alternative hypotheses for testing whether the true proportions in each category are equal.

c. Find the expected number of responses in each answer category for the test mentioned in part **b**.

d. Compute the chi-square statistic for the test.

e. Give the appropriate conclusion for the test if $\alpha = .10$.

f. A more realistic null hypothesis is that 30% of Americans believe that the Bible is the actual word of God; 50% believe that it is inspired by God, but not to be taken literally; 15% believe that it is an ancient book of fables; and 5% believe that the Bible has some other origin. Repeat parts **b–e** for this hypothesis.

PONDICE

8.46 Characteristics of ice-melt ponds. The National Snow and Ice Data Center (NSIDC) collected data on 504 ice-melt ponds in the Canadian Arctic. The data are saved in the **PONDICE** file. One variable of interest to environmental engineers studying the ponds is the type of ice observed in each. Ice type is classified as first-year ice, multiyear ice, or landfast ice. The SAS summary table for the types of ice of the 504 ice-melt ponds is reproduced at the bottom of the page.

a. Use a 90% confidence interval to estimate the proportion of ice-melt ponds in the Canadian Arctic that have first-year ice.

b. Suppose environmental engineers hypothesize that 15% of Canadian Arctic ice-melt ponds have first-year ice, 40% have landfast ice, and 45% have multiyear ice. Test the engineers' theory, using $\alpha = .01$.

Applying the Concepts—Advanced

8.47 Analysis of a Scrabble game. In the board game Scrabble™, a player initially draws a "hand" of seven tiles at random from 100 tiles. Each tile has a letter of the alphabet, and the player attempts to form a word from the letters in his or her hand. In *Chance* (Winter 2002), scientist C. J. Robinove investigated whether a handheld electronic version of the game, called ScrabbleExpress™, produces too

The FREQ Procedure

ICETYPE	Frequency	Percent	Cumulative Frequency	Cumulative Percent
First-year	88	17.46	88	17.46
Landfast	196	38.89	284	56.35
Multi-year	220	43.65	504	100.00

SAS output for Exercise 8.46

SCRABBLE

Letter	Relative Frequency in Board Game	Frequency in Electronic Game
A	.09	39
B	.02	18
C	.02	30
D	.04	30
E	.12	31
F	.02	21
G	.03	35
H	.02	21
I	.09	25
J	.01	17
K	.01	27
L	.04	18
M	.02	31
N	.06	36
O	.08	20
P	.02	27
Q	.01	13
R	.06	27
S	.04	29
T	.06	27
U	.04	21
V	.02	33
W	.02	29
X	.01	15
Y	.02	32
Z	.01	14
# (blank)	.02	34
Total		700

Source: Robinove, C. J. "Letter-frequency Bias in an Electronic Scrabble Game," *Chance,* Vol. 15, No. 1, Winter 2002, p. 31 (Table 3).

few vowels in the 7-letter draws. For each of the 26 letters (and "blank" for any letter), the accompanying table gives the true relative frequency of the letter in the board game, as well as the frequency of occurrence of the letter in a sample of 700 tiles (i.e., 100 "hands") randomly drawn in the electronic game.

a. Do the data support the scientist's contention that ScrabbleExpress™ "presents the player with unfair word selection opportunities" that are not the same as the Scrabble™ board game? Test, using $\alpha = .05$.

b. Use a 95% confidence interval to estimate the true proportion of letters drawn in the electronic game that are vowels. Compare the results with the true relative frequency of a vowel in the board game.

8.48 **Political representation of religious groups.** Do those elected to the U.S. House of Representatives really "represent" their constituents demographically? This was a question of interest in *Chance* (Summer 2002). One of several demographics studied was religious affiliation. The accompanying table gives the proportion of the U.S. population for several religions, as well as the number of the 435 seats in the House of Representatives which are affiliated with that religion. Give your opinion on whether or not the members of the House of Representatives are statistically representative of the religious affiliation of their constituents in the United States.

USHOUSE

Religion	Proportion of U.S. Population	Number of Seats in House
Catholic	.28	117
Methodist	.04	61
Jewish	.02	30
Other	.66	227
TOTALS	1.00	435

8.4 Testing Categorical Probabilities: Two-Way (Contingency) Table

In Section 8.3, we introduced the multinomial probability distribution and considered data classified according to a single criterion. We now consider multinomial experiments in which the data are classified according to two criteria—that is, *classification with respect to two qualitative factors.*

Consider a study in the *Journal of Marketing* (Fall 1992) on the impact of using celebrities in television advertisements. The researchers investigated the relationship between the gender of a viewer and the viewer's brand awareness. Three hundred TV viewers were asked to identify products advertised by male celebrity spokespersons. The data are summarized in the **two-way table** shown in Table 8.6. This table, called a **contingency table**, presents multinomial count data classified on two scales, or **dimensions, of classification**: gender of viewer and brand awareness.

TABLE 8.6 Contingency Table for Marketing Example

		Gender		
		Male	Female	Totals
Brand Awareness	**Could Identify Product**	95	41	136
	Could Not Identify Product	55	109	164
	Totals	150	150	300

The symbols representing the cell counts for the multinomial experiment in Table 8.6 are shown in Table 8.7a, and the corresponding cell, row, and column probabilities are shown in Table 8.7b. Thus, n_{11} represents the number of viewers who are male and could identify the brand, and p_{11} represents the corresponding cell probability. Note the symbols for the row and column totals and also the symbols for the probability totals. The latter are called **marginal probabilities** for each row and column. The marginal probability p_{r1} is the probability that a TV viewer identifies the product; the marginal probability p_{c1} is the probability that a TV viewer is male. Thus,

$$p_{r1} = p_{11} + p_{12} \text{ and } p_{c1} = p_{11} + p_{21}$$

TABLE 8.7a   Observed Counts for Contingency Table 8.6

		Gender		
		Male	Female	Totals
Brand Awareness	**Could Identify Product**	n_{11}	n_{12}	R_1
	Could Not Identify Product	n_{21}	n_{22}	R_2
	Totals	C_1	C_2	n

TABLE 8.7b   Probabilities for Contingency Table 8.6

		Gender		
		Male	Female	Totals
Brand Awareness	**Could Identify Product**	p_{11}	p_{12}	p_{r1}
	Could Not Identify Product	p_{21}	p_{22}	p_{r2}
	Totals	p_{c1}	p_{c2}	1

We can see, then, that this really is a multinomial experiment with a total of 300 trials, $(2)(2) = 4$ cells or possible outcomes, and probabilities for each cell as shown in Table 8.7b. If the 300 TV viewers are randomly chosen, the trials are considered independent and the probabilities are viewed as remaining constant from trial to trial.

Suppose we want to know whether the two classifications of gender and brand awareness are dependent. That is, if we know the gender of the TV viewer, does that information give us a clue about the viewer's brand awareness? In a probabilistic sense, we know (Chapter 3) that the independence of events A and B implies that $P(AB) = P(A)P(B)$. Similarly, in the contingency table analysis, if the **two classifications are independent**, the probability that an item is classified into any particular cell of the table is the product of the corresponding marginal probabilities. Thus, under the hypothesis of independence, in Table 8.7b we must have

$$p_{11} = p_{r1}p_{c1} \quad p_{12} = p_{r1}p_{c2}$$
$$p_{21} = p_{r2}p_{c1} \quad p_{22} = p_{r2}p_{c2}$$

To test the hypothesis of independence, we use the same reasoning employed in the one-dimensional tests of Section 8.3. First, we calculate the *expected*, or *mean, count in each cell*, assuming that the null hypothesis of independence is true. We do this by noting that the expected count in a cell of the table is just the total number of multinomial trials, n, times the cell probability. Recall that n_{ij} represents the **observed count** in the cell located in the ith row and jth column. Then the expected cell count for the upper left-hand cell (first row, first column) is

$$E_{11} = np_{11}$$

or, when the null hypothesis (the classifications are independent) is true,

$$E_{11} = n p_{r1} p_{c1}$$

Since these true probabilities are not known, we estimate p_{r1} and p_{c1} by the same proportions $\hat{p}_{r1} = R_1/n$ and $\hat{p}_{c1} = C_1/n$. Thus, the estimate of the expected value $E(n_{11})$ is

$$E_{11} = n \left(\frac{R_1}{n} \right) \left(\frac{C_1}{n} \right) = \frac{R_1 C_1}{n}$$

Similarly, for each i, j,

$$E_{ij} = \frac{(\text{Row total})(\text{Column total})}{\text{Total sample size}}$$

Hence,

$$E_{12} = \frac{R_1 C_2}{n}$$

$$E_{21} = \frac{R_2 C_1}{n}$$

$$E_{22} = \frac{R_2 C_2}{n}$$

Finding Expected Cell Counts for a Two-Way Contingency Table

The estimate of the expected number of observations falling into the cell in row i and column j is given by

$$E_{ij} = \frac{R_i C_j}{n}$$

where R_i = total for row i, C_j = total for column j, and n = sample size.

Using the data in Table 8.6, we find that

$$E_{11} = \frac{R_1 C_1}{n} = \frac{(136)(150)}{300} = 68$$

$$E_{12} = \frac{R_1 C_2}{n} = \frac{(136)(150)}{300} = 68$$

$$E_{21} = \frac{R_2 C_1}{n} = \frac{(164)(150)}{300} = 82$$

$$E_{22} = \frac{R_2 C_2}{n} = \frac{(164)(150)}{300} = 82$$

The observed data and the estimated expected values (in parentheses) are shown in Table 8.8.

We now use the χ^2 statistic to compare the observed and expected (estimated) counts in each cell of the contingency table:

$$\chi^2 = \frac{[n_{11} - E_{11}]^2}{E_{11}} + \frac{[n_{12} - E_{12}]^2}{E_{12}} + \frac{[n_{21} - E_{21}]^2}{E_{21}} + \frac{[n_{22} - E_{22}]^2}{E_{22}}$$

$$= \sum \frac{[n_{ij} - E_{ij}]^2}{E_{ij}}$$

(*Note:* The use of $\sum$ in the context of a contingency table analysis refers to a sum over all cells in the table.)

TABLE 8.8 Observed and Estimated Expected (in Parentheses) Counts

| | | Gender | | |
		Male	Female	Totals
Brand Awareness	**Could Identify Product**	95	41	136
		(68)	(68)	
	Could Not Identify Product	55	109	164
		(82)	(82)	
	Totals	150	150	300

Substituting the data of Table 8.8 into this expression, we get

$$\chi^2 = \frac{(95-68)^2}{68} + \frac{(41-68)^2}{68} + \frac{(55-82)^2}{82} + \frac{(109-82)^2}{82} = 39.22$$

Large values of χ^2 imply that the observed counts do not closely agree and hence that the hypothesis of independence is false. To determine how large χ^2 must be before it is too large to be attributed to chance, we make use of the fact that the sampling distribution of χ^2 is approximately a χ^2 probability distribution when the classifications are independent.

In testing the null hypothesis of independence in a two-way contingency table, the appropriate degrees of freedom will be $(r-1)(c-1)$, where r is the number of rows and c is the number of columns in the table.

For the brand awareness example, the number of degrees of freedom for χ^2 is $(r-1)(c-1) = (2-1)(2-1) = 1$. Then, for $\alpha = .05$, we reject the hypothesis of independence when

$$\chi^2 > \chi^2_{.05} = 3.84146$$

Since the computed $\chi^2 = 39.22$ exceeds the value 3.84146, we conclude that viewer gender and brand awareness are dependent events.

The pattern of **dependence** can be seen more clearly by expressing the data as percentages. We first select one of the two classifications to be used as the base variable. In the preceding example, suppose we select gender of the TV viewer as the classificatory variable to be the base. Next, we represent the responses for each level of the second categorical variable (brand awareness here) as a percentage of the subtotal for the base variable. For example, from Table 8.8, we convert the response for males who identify the brand (95) to a percentage of the total number of male viewers (150). That is,

$$\left(^{95}\!/_{150}\right)100\% = 63.3\%$$

All of the entries in Table 8.8 are similarly converted, and the values are shown in Table 8.9. The value shown at the right of each row is the row's total, expressed as a percentage of the

TABLE 8.9 Percentage of TV Viewers Who Identify Brand, by Gender

| | | Gender | | |
		Male	Female	Totals
Brand Awareness	**Could Identify Product**	63.3	27.3	45.3
	Could Not Identify Product	36.7	72.7	54.7
	Totals	100	100	100

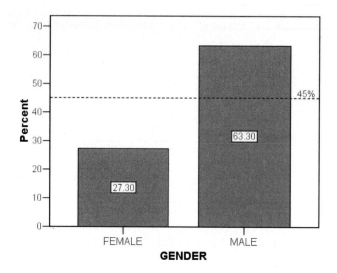

FIGURE 8.5

SPSS bar graph showing percentage of viewers who identified the TV product

total number of responses in the entire table. Thus, the percentage of TV view-ers who iden-tify the product is $(\frac{136}{300})100\% = 45.3\%$ (rounded to the nearest percent).

If the gender and brand awareness variables are independent, then the percent-ages in the cells of the table are expected to be approximately equal to the correspond-ing row percentages. Thus, we would expect the percentage of viewers who identify the brand for each gender to be approximately 45% if the two variables are independent. The extent to which each gender's percentage departs from this value determines the dependence of the two classifications, with greater variability of the row percentages meaning a greater degree of dependence. A plot of the percentages helps summarize the observed pattern. In the SPSS bar graph in Figure 8.5, we show the gender of the viewer (the base variable) on the horizontal axis and the percentage of TV viewers who identi-fy the brand on the vertical axis. The "expected" percentage under the assumption of in-dependence is shown as a dotted horizontal line.

Figure 8.5 clearly indicates the reason that the test resulted in the conclusion that the two classifications in the contingency table are dependent. The percentage of male TV viewers who identify the brand promoted by a male celebrity is more than twice as high as the percentage of female TV viewers who identify the brand. Statistical measures of the degree of dependence and procedures for making comparisons of pairs of levels for classifications are beyond the scope of this text, but can be found in the references. We will utilize descriptive summaries such as Figure 8.5 to examine the degree of de-pendence exhibited by the sample data.

The general form of a two-way contingency table containing r rows and c columns (called an $r \times c$ contingency table) is shown in Table 8.10. Note that the observed count in the ijth cell is denoted by n_{ij}, the ith row total is r_i, the jth column total is c_j, and the total sample size is n. Using this notation, we give the general form of the contingency table test for independent classifications in the following box:

General Form of a Two-way (Contingency) Table Analysis: A Test for Independence

H_0: The two classifications are independent

H_a: The two classifications are dependent

Test statistic: $\chi^2 = \Sigma \dfrac{[n_{ij} - E_{ij}]^2}{E_{ij}}$

where $E_{ij} = \dfrac{R_i C_j}{n}$

Rejection region: $\chi^2 > \chi_\alpha^2$, where χ_α^2 has $(r - 1)(c - 1)$ df

Conditions Required for a Valid χ^2 Test: Contingency Tables

1. The n observed counts are a random sample from the population of interest. We may then consider this to be a multinomial experiment with $r \times c$ possible outcomes.

2. The sample size n will be large enough so that, for every cell, the expected count $E(n_{ij})$ will be equal to 5 or more.

TABLE 8.10 General $r \times c$ Contingency Table

		Column				
		1	2	...	c	Row Totals
Row	1	n_{11}	n_{12}	...	n_{1c}	R_1
	2	n_{21}	n_{22}	...	n_{2c}	R_2
	⋮	⋮	⋮		⋮	⋮
	r	n_{r1}	n_{r2}	...	n_{rc}	R_r
Column Totals		C_1	C_2	...	C_c	n

EXAMPLE 8.6

CONDUCTING A TWO-WAY ANALYSIS: Marital Status and Religion

Problem A social scientist wants to determine whether the marital status (divorced or not divorced) of U.S. men is independent of their religious affiliation (or lack thereof). A sample of 500 U.S. men is surveyed, and the results are tabulated as shown in Table 8.11.

a. Test to see whether there is sufficient evidence to indicate that the marital status of men who have been or are currently married is dependent on religious affiliation. Take $\alpha = .01$.

b. Graph the data and describe the patterns revealed. Is the result of the test supported by the graph?

TABLE 8.11 Survey Results (Observed Counts), Example 8.6

		Religious Affiliation					
		A	B	C	D	None	Totals
Marital Status	**Divorced**	39	19	12	28	18	116
	Married, never divorced	172	61	44	70	37	384
	Totals	211	80	56	98	55	500

Solution

a. The first step is to calculate estimated expected cell frequencies under the assumption that the classifications are independent. Rather than compute these values by hand, we resort to a computer. The SAS printout of the analysis of Table 8.11 is displayed in Figure 8.6, each cell of which contains the observed (top) and expected (bottom) frequency in that cell. Note that E_{11}, the estimated expected count for the Divorced, A cell, is 48.952. Similarly, the estimated expected count for the Divorced, B cell, is $E_{12} = 18.56$. Since all the estimated expected cell frequencies are greater than 5, the χ^2 approximation for the test statistic is appropriate. Assuming that the men chosen were randomly selected from all married or previously married American men, the characteristics of the multinomial probability distribution are satisfied.

The FREQ Procedure

Table of MARITAL by RELIGION

MARITAL   RELIGION

Frequency Expected	A	B	C	D	NONE	Total
DIVORCED	39 48.952	19 18.56	12 12.992	28 22.736	18 12.76	116
NEVER	172 162.05	61 61.44	44 43.008	70 75.264	37 42.24	384
Total	211	80	56	98	55	500

Statistics for Table of MARITAL by RELIGION

Statistic	DF	Value	Prob
Chi-Square	4	7.1355	0.1289
Likelihood Ratio Chi-Square	4	6.9854	0.1367
Mantel-Haenszel Chi-Square	1	6.4943	0.0108
Phi Coefficient		0.1195	
Contingency Coefficient		0.1186	
Cramer's V		0.1195	

Fisher's Exact Test

Table Probability (P)	6.936E-06
Pr <= P	0.1251

Sample Size = 500

The null and alternative hypotheses we want to test are

H_0: The marital status of U.S. men and their religious affiliation are independent
H_a: The marital status of U.S. men and their religious affiliation are dependent

The test statistic, $\chi^2 = 7.135$, is highlighted at the bottom of the printout, as is the observed significance level (*p*-value) of the test. Since $\alpha = .01$ is less than $p = .129$, we fail to reject H_0; that is, we cannot conclude that the marital status of U.S. men depends on their religious affiliation. (Note that we could not reject H_0 even with $\alpha = .10$.)

b. The marital status frequencies can be expressed as percentages of the number of men in each religious affiliation category. The expected percentage of divorced men under the assumption of independence is $(^{116}/_{500})100\% = 23\%$. An SAS graph of the percentages is shown in Figure 8.7. Note that the percentages of divorced

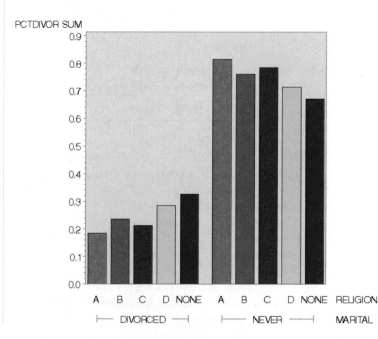

FIGURE 8.7

SAS side-by-side bar graphs showing percentage of divorced and never divorced males by religion

men (see the bars in the "DIVORCED" block of the SAS graph) deviate only slightly from that expected under the assumption of independence, supporting the result of the test in part **a**. That is, neither the descriptive bar graph nor the statistical test provides evidence that the male divorce rate depends on (varies with) religious affiliation.

Now Work Exercise 8.57

ACTIVITY 8.3: Contingency Tables

In this Activity, you will revisit Activity 3.1, *Exit Polls* (p. 142). For convenience, the table shown in that activity is repeated here.

2004 Presidential Election, Vote by Gender

	Bush	Kerry	Other
Male (46%)	55%	44%	1%
Female (54%)	48%	51%	1%

Source: CNN.com

1. Determine whether this table and the similar tables that you found in the activity on page 142 are contingency tables. If not, do you have enough information to create a contingency

table for the data? If you need more information, state specifically what information you need.

2. Choose one of your examples from Activity 8.2 (p. 450) if it contains a contingency table or enough information to create one, or use the Internet or some other source to find a new example with a contingency table given. Determine whether the conditions for a valid χ^2 test are met. If not, choose a different example in which the conditions are met.

3. Perform a χ^2 test for independence for the example chosen in Exercise 2. Are the results what you would expect in the given situation? Explain.

Contingency Tables

Using the TI-84/TI-83 Graphing Calculator

Finding *p*-values for Contingency Tables

Step 1   *Access the Matrix menu to enter the observed values.*

Press **2nd x⁻¹** for **MATRX**. (*Note*: On the TI-83, press **MATRX**.)
Arrow right to **EDIT**.
Press **ENTER**.
Use the **ARROW** key to enter the row and column dimensions of your observed Matrix.
Use the **ARROW** key to enter your observed values into Matrix [A].

Step 2   *Access the Matrix menu to enter the expected values.*

Press **2nd x⁻¹** for **MATRX**. (*Note*: On the TI-83, press **MATRX**.)
Arrow right to **EDIT**.
Arrow down to **2:[B]**.
Press **ENTER**.
Use the **ARROW** key to enter the row and column dimensions of your expected matrix. (The dimensions will be the same as in Matrix A.)
Use the **ARROW** key to enter your expected values into Matrix [B].

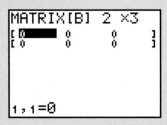

continued

Step 3 *Access the Statistical Tests menu and perform the Chi-square test.*

Press **STAT**.
Arrow right to **TESTS**.
Arrow down to χ^2 **Test**.
Press **ENTER**.
Arrow down to **Calculate**.
Press **ENTER**.

```
X²-Test
 Observed: [A]
 Expected: [B]
 Calculate Draw
```

Step 4 *Reject H_0 if the p-value $< \alpha$*

Example

Our observed Matrix is [A] = 39 19 12 28 18
                             172 61 44 70 37

Our Expected Matrix is [B] = 48.952 18.56 12.992 22.736 12.76
                             162.05 61.44 43.008 75.264 42.24

Use $\alpha = .05$ to test the following hypotheses:

H_0: The Matrix entries represent independent events.
H_a: The Matrix entries represent events that **are not** independent.

The screens for this example are as follows:

```
MATRIX[A]  2 ×5
[ 39      19      12    _
[ 172     61      44    _
```

```
MATRIX[B]  2 ×5
[ 48.952  18.56   12.992 _
[ 162.05  61.44   43.008 _




2,1=162.05
```

```
X²-Test
 X²=7.135464193
 P=.128900399
 df=4
```

As you can see from the last screen, the p-value is 0.1289. Since the p-value is **greater** than $\alpha = .05$, we **do not** reject H_0.

STATISTICS IN ACTION REVISITED

Testing whether Frequency of Drinking Is Related to Amount of Alcohol Consumed

Refer again to the *Family and Consumer Sciences Research Journal* (Mar. 2005) study of college students and drinking (p. 433). A second objective of the researchers was to establish a statistical link between the frequency of drinking and the amount of alcohol consumed. That is, the researchers sought a link between frequency of drinking alcohol over the previous one-month period and average number of drinks consumed per occasion. Since both of these variables (FREQUENCY and AMOUNT) measured on the sample of 657 students in the **COLLDRINKS** file are qualitative, a contingency table analysis is appropriate.

Figure SIA8.3 shows the SPSS contingency table analyses relating frequency of drinking to average amount of alcohol consumed. The null hypothesis for the test is H_0: Frequency and Amount are independent.

The chi-square test statistic (756.6) and the p-value of the test (.000) are highlighted on the printout. If we conduct the test at $\alpha = .01$, there is sufficient evidence to reject H_0. That is, the data provide evidence indicating that, for college students, the average amount of alcohol consumed per occasion is associated with the frequency of drinking.

The row percentages highlighted in the contingency table of Figure SIA8.3 reveal the differences in drinking amounts for the different levels of drinking frequency. For frequency of drinking "None" and "Once a month", 0% drink heavily (7–9 or 10 or more drinks per occasion). However, for frequency of drinking "Twice a week" and "More," 12.7% and 17.1%, respectively, have 7–9 drinks per occasion, while 4.0% and 11.2%, respectively, have 10 or more drinks per occasion. These results led the researchers to report that "The frequent drinkers were more likely to consume more [alcohol] on each occasion, a tendency that clearly makes them heavy drinkers."

FREQNCY * AMOUNT Crosstabulation

			None	1 drink	2-3 drinks	4-6 drinks	7-9 drinks	More	Total
					AMOUNT				
FREQNCY	None	Count	118	0	0	0	0	0	118
		Expected Count	21.2	11.0	51.2	17.8	11.5	5.4	118.0
		% within FREQNCY	100.0%	.0%	.0%	.0%	.0%	.0%	100.0%
	OnceMonth	Count	0	30	43	20	0	0	93
		Expected Count	16.7	8.6	40.3	14.0	9.1	4.2	93.0
		% within FREQNCY	.0%	32.3%	46.2%	21.5%	.0%	.0%	100.0%
	TwiceWeek	Count	0	26	158	46	35	11	276
		Expected Count	49.6	25.6	119.7	41.6	26.9	12.6	276.0
		% within FREQNCY	.0%	9.4%	57.2%	16.7%	12.7%	4.0%	100.0%
	More	Count	0	5	84	33	29	19	170
		Expected Count	30.5	15.8	73.7	25.6	16.6	7.8	170.0
		% within FREQNCY	.0%	2.9%	49.4%	19.4%	17.1%	11.2%	100.0%
Total		Count	118	61	285	99	64	30	657
		Expected Count	118.0	61.0	285.0	99.0	64.0	30.0	657.0
		% within FREQNCY	18.0%	9.3%	43.4%	15.1%	9.7%	4.6%	100.0%

Chi-Square Tests

	Value	df	Asymp. Sig. (2-sided)
Pearson Chi-Square	756.606[a]	15	.000
Likelihood Ratio	706.412	15	.000
Linear-by-Linear Association	322.813	1	.000
N of Valid Cases	657		

a. 1 cells (4.2%) have expected count less than 5. The minimum expected count is 4.25.

FIGURE SIA8.3
SPSS contingency table analysis: frequency of drinking vs. average amount

Exercises 8.49–8.70

Understanding the Principles

8.49 What is a two-way (contingency) table?

8.50 *True* or *False*. One goal of a contingency table analysis is to determine whether the two classifications are dependent.

8.51 What conditions are required for a valid chi-square test of data from a contingency table?

Learning the Mechanics

8.52 Find the rejection region for a test of independence of two classifications for which the contingency table contains r rows and c columns and
  a. $r = 5$, $c = 5$, $\alpha = .05$
  b. $r = 3$, $c = 6$, $\alpha = .10$
  c. $r = 2$, $c = 3$, $\alpha = .01$

8.53 Consider the following 2×3 (i.e., $r = 2$ and $c = 3$) contingency table:

		Column		
		1	2	3
Row	**1**	9	34	53
	2	16	30	25

a. Specify the null and alternative hypotheses that should be used in testing the independence of the row and column classifications.
b. Specify the test statistic and the rejection region that should be used in conducting the hypothesis test of part **a.** Use $\alpha = .01$.
c. Assuming that the row classification and the column classification are independent, find estimates for the expected cell counts.
d. Conduct the hypothesis test of part **a.** Interpret your result.

8.54 Refer to Exercise 8.53.
a. Convert the frequency responses to percentages by calculating the percentage of each column total falling in each row. Also, convert the row totals to percentages of the total number of responses. Display the percentages in a table.
b. Create a bar graph with row 1 percentage on the vertical axis and column number on the horizontal axis. Show the row 1 total percentage as a horizontal line on the graph.
c. What pattern do you expect to see if the rows and columns are independent? Does the plot support the result of the test of independence in Exercise 8.53?

8.55 Test the null hypothesis of independence of the two classifications A and B of the 3×3 contingency table shown here. Use $\alpha = .05$.

		B		
		B_1	B_2	B_3
	A_1	40	72	42
A	A_2	63	53	70
	A_3	31	38	30

8.56 Refer to Exercise 8.55. Convert the responses to percentages by calculating the percentage of each B class total falling into each A classification. Also, calculate the percentage of the total number of responses that constitute each of the A classification totals.
  a. Create a bar graph with row A_1 percentage on the vertical axis and B classification on the horizontal axis. Does the graph support the result of the test of hypothesis in Exercise 13.25? Explain.
  b. Repeat part **a** for the row A_2 percentages.
  c. Repeat part **a** for the row A_3 percentages.

Applying the Concepts—Basic

8.57 Children's perceptions of their neighborhood. In *Health* **NW** *Education Research* (Feb. 2005), nutrition scientists at Deakin University (Australia) investigated children's perceptions of their environments. Each in a sample of 147 ten-year-old children drew maps of their home and neighborhood environment. The researchers examined the maps for certain themes (e.g., presence of a pet, television in the bedroom, opportunities for physical activity). The results, broken down by gender, for two themes (presence of a dog and TV in the bedroom) are shown in the tables below.
  a. Find the sample proportion of boys who drew a dog on their maps.

  b. Find the sample proportion of girls who drew a dog on their maps.
  c. Compare the proportions you found in parts **a** and **b**. Does it appear that the likelihood of drawing a dog on the neighborhood map depends on gender?
  d. Give the null hypothesis for testing whether the likelihood of a drawing a dog on the neighborhood map depends on gender.
  e. Use the accompanying MINITAB printout to conduct the test mentioned in part **d** at $\alpha = .05$.
  f. Conduct a test to determine whether the likelihood of drawing a TV in the bedroom is different for boys and girls. Use $\alpha = .05$.

⊙ MAPDOG

Presence of a Dog	Number of Boys	Number of Girls
Yes	6	11
No	71	59
TOTAL	77	70

⊙ MAPTV

Presence of TV in Bedroom	Number of Boys	Number of Girls
Yes	11	9
No	66	61
TOTAL	77	70

Source: Hume, C., Salmon, J., and Ball, K. "Children's perceptions of their home and neighborhood environments, and their association with objectively measured physical activity: A qualitative and quantitative study," *Health Education Research*, Vol. 20, No. 1, February 2005 (Table III).

8.58 Late-emerging reading disabilities. Studies of children with reading disabilities typically focus on "early-emerging" difficulties identified prior to the fourth grade. Psychologists at

Tabulated statistics: DOG, GENDER

```
Using frequencies in NUMBER


Rows: DOG    Columns: GENDER

          Boy    Girl    All

No         71      59     130
        68.10   61.90  130.00

Yes         6      11      17
         8.90    8.10   17.00

All        77      70     147
        77.00   70.00  147.00

Cell Contents:      Count
                    Expected count
```

MINITAB output for Exercise 8.57

```
Pearson Chi-Square = 2.250, DF = 1, P-Value = 0.134
Likelihood Ratio Chi-Square = 2.268, DF = 1, P-Value = 0.132
```

Haskins Laboratories recently studied children with "late-emerging" reading difficulties (i.e., children who appeared to undergo a fourth-grade "slump" in reading achievement) and published their findings in the *Journal of Educational Psychology* (June 2003). A sample of 161 children was selected from fourth and fifth graders at elementary schools in Philadelphia. In addition to recording the grade level, the researchers determined whether each child had a previously undetected reading disability. Sixty-six children were diagnosed with a reading disability. Of these children, 32 were fourth graders and 34 were fifth graders. Similarly, of the 95 children with normal reading achievement, 55 were fourth graders and 40 were fifth graders.

a. Identify the two qualitative variables (and corresponding levels) measured in the study.

b. From the information provided, form a contingency table.

c. Assuming that the two variables are independent, calculate the expected cell counts.

d. Find the test statistic for determining whether the proportions of fourth and fifth graders with reading disabilities differs from the proportions of fourth and fifth graders with normal reading skills.

e. Find the rejection region for the test if $\alpha = .10$.

f. Is there a link between reading disability and grade level? Give the appropriate conclusion of the test.

8.59 **Masculinity and crime.** Refer to the *Journal of Sociology* (July 2003) study on the link between the level of masculinity and criminal behavior in men, presented in Exercise 7.23 (p. 369). The researcher identified events that a sample of newly incarcerated men were involved in and classified each event as "violent" (involving the use of a weapon, the throwing of objects, punching, choking, or kicking) or "avoided-violent" (involving pushing, shoving, grabbing, or threats of violence that did not escalate into a violent event). Each man (and corresponding event) was also classified as possessing "high-risk masculinity" (scored high on the Masculinity–Femininity Scale test and low on the Traditional Outlets of Masculinity Scale test) or "low-risk masculinity." The data on 1,507 events are summarized in the following table:

HRM

	Violent Events	Avoided-Violent Events	Totals
High-Risk Masculinity	236	143	379
Low-Risk Masculinity	801	327	1,128
Totals	1,037	470	1,507

Source: Krienert, J. L. "Masculinity and crime: A quantitative exploration of Messerschmidt's hypothesis," *Journal of Sociology,* Vol. 7, No. 2, July 2003 (Table 4).

a. Identify the two categorical variables measured (and their levels) in the study.

b. Identify the experimental units.

c. If the type of event (violent or avoided-violent) is independent of high- low-risk masculinity, how many of the 1,507 events would you expect to be violent and involve a high-risk-masculine man?

d. Repeat part **c** for the other combinations of event type and high- low-risk masculinity.

e. Calculate the χ^2 statistic for testing whether event type depends on high- low-risk masculinity.

f. Give the appropriate conclusion of the test mentioned in part **e**, using $\alpha = .05$.

8.60 **Healing heart patients with music, imagery, touch, and prayer.** "Frontier medicine" is a term used to describe medical therapies (e.g., energy healing, therapeutic prayer, spiritual healing) for which there is no plausible explanation. *The Lancet* (July 16, 2005) published the results of a study designed to test the effectiveness of two types of frontier medicine—music, imagery, and touch (MIT) therapy and therapeutic prayer—in healing cardiac care patients. Patients were randomly assigned to receive one of four types of treatment: (1) prayer, (2) MIT, (3) prayer and MIT, and (4) standard care (no prayer and no MIT). Six months after therapy, the patients were evaluated for a major adverse cardiovascular event (e.g., a heart attack). The results of the study are summarized in the accompanying table.

a. Identify the two qualitative variables (and associated levels) measured in the study.

b. State H_o and H_a for testing whether a major adverse cardiovascular event depends on type of therapy.

c. Use the MINITAB printout on p. 466 to conduct the test mentioned in part **b** at $\alpha = .10$. On the basis of this test, what can the researchers infer about the effectiveness of music, imagery, and touch therapy and the effectiveness of healing prayer in heart patients?

HEALING

Therapy	Number of Patients with Major Cardiovascular Events	Number of Patients with No Events	TOTAL
Prayer	43	139	182
MIT	47	138	185
Prayer and MIT	39	150	189
Standard	50	142	192

Source: Krucoff, M. W., et al. "Music, imagery, touch, and prayer as adjuncts to interventional cardiac care: The Monitoring and Actualization of Noetic Trainings (MANTRA) II randomized study," *The Lancet,* Vol. 366, July 16, 2005 (Table 4).

Applying the Concepts—Intermediate

8.61 **IQ and mental retardation.** A person is diagnosed with mental retardation if, before the age of 18, his or her score on a standard IQ test is no higher than 70 (two standard deviations below the mean of 100). Researchers at Cornell and West Virginia Universities examined the impact of rising IQ scores on diagnoses of mental retardation (MR). (*American Psychologist,* October, 2003.) IQ data were collected from different school districts across the United States, and the students were tested with either the Wechsler Intelligence Scale for Children—Revised (WISC-R) or the Wechsler Intelligence Scale for Children—Third Revision (WISC-III) IQ tests. The researchers focused on those students with IQs just above the mental retardation cutoff (between 70 and 85), based on the original IQ test. These "borderline" MR students were then retested one year later with one of the IQ tests. The accompanying table gives the number of students diagnosed with mental retardation on the basis of the retest.

Tabulated statistics: THERAPY, EVENT

```
Using frequencies in NUMBER


Rows: THERAPY    Columns: EVENT

                   No     Yes     All

MIT                138     47     185
                 140.7    44.3   185.0

Prayer             139     43     182
                 138.4    43.6   182.0

Prayer&MIT         150     39     189
                 143.8    45.2   189.0

Standard           142     50     192
                 146.1    45.9   192.0

All                569    179     748
                 569.0   179.0   748.0


Cell Contents:       Count
                     Expected count
```

MINITAB output for
Exercise 8.60

```
Pearson Chi-Square = 1.828, DF = 3, P-Value = 0.609
Likelihood Ratio Chi-Square = 1.855, DF = 3, P-Value = 0.603
```

Conduct a chi-square test for independence to determine whether the proportion of students diagnosed with MR depends on the IQ test/retest method. Use $\alpha = .01$.

MRIQ

Test/Retest	Diagnosed with MR	Above MR Cutoff IQ	TOTAL
WISC-R / WISC-R	25	167	192
WISC-R / WISC-III	54	103	157
WISC-III / WISC-III	36	141	177

Source: Kanaya, T., Scullin, M. H., and Ceci, S. J. "The Flynn effect and U.S. Policies," *American Psychologist,* Vol. 58, No. 10, Oct. 2003 (Figure 1).

8.62 Creating menus to influence others. Refer to the *Journal of Consumer Research* (Mar. 2003) study on influencing the choices of others by offering undesirable alternatives, presented in Exercise 6.131 (p. 346). In another experiment conducted by the researcher, 96 subjects were asked to imagine that they had just moved to an apartment with two others and that they were shopping for a new appliance (e.g., a television, a microwave oven). Each subject was asked to create a menu of three brand choices for his or her roommates; then subjects were randomly assigned (in equal numbers) to one of three different "goal" conditions: (1) Create the menu in order to influence roommates to buy a preselected brand, (2) create the menu in order to influence roommates to buy a brand of your choice, and (3) create the menu with no intent to influence roommates. The researcher theorized that the menus created to influence others would likely include undesirable alternative brands. Consequently, the number of

menus in each goal condition that was consistent with the theory was determined. The data are summarized in the accompanying table. Analyze the data for the purpose of determining whether the proportion of subjects who select menus consistent with the theory depends on the goal condition. Use $\alpha = .01$.

MENU3

Goal Condition	Number Consistent with Theory	Number Not Consistent with Theory	Totals
Influence/preselected brand	15	17	32
Influence/own brand	14	18	32
No influence	3	29	32

Source: Hamilton, R. W. "Why do people suggest what they do not want? Using context effects to influence others' choices," *Journal of Consumer Research,* Vol. 29, March 2003 (Table 2).

8.63 Politics and religion. University of Maryland professor Ted R. Gurr examined the political strategies used by ethnic groups worldwide in their fight for minority rights. (*Political Science & Politics,* June 2000.) Each in a sample of 275 ethnic groups was classified according to world region and highest level of political action reported. The data are summarized in the contingency table on p. 467. Conduct a test at $\alpha = .10$ to determine whether political strategy of ethnic groups depends on world region. Support your answer with a graph.

ETHNIC

		Political Strategy		
		No Political Action	Mobilization, Mass Action	Terrorism, Rebellion, Civil War
World Region	**Latin American**	24	31	7
	Post-Communist	32	23	4
	South, Southeast, East Asia	11	22	26
	Africa/Middle East	39	36	20

Source: Gurr, T. R. "Nonviolence in ethnopolitics: Strategies for the attainment of group rights and autonomy." *Political Science & Politics,* Vol. 33, No. 2, June 2000 (Table 1).

8.64 Trapping grain moths. In an experiment described in the *Journal of Agricultural, Biological, and Environmental Statistics* (Dec. 2000), bins of corn were stocked with various parasites (e.g., grain moths) in late winter. In early summer (June), three bowl-shaped traps were placed on the surface of the grain in order to capture the moths. All three traps were baited with a sex pheromone lure; however, one trap used an unmarked sticky adhesive, one was marked with a fluorescent red powder, and one was marked with a fluorescent blue powder. The traps were set on a Wednesday, and the catch was collected the following Thursday and Friday. The accompanying table shows the number of moths captured in each trap on each day. Conduct a test (at $\alpha = .10$) to determine whether the percentages of moths caught by the three traps depends on the day of the week.

MOTHTRAP

	Adhesive, No Mark	Red Mark	Blue Mark
Thursday	136	41	17
Friday	101	50	18

Source: Wileyto, E. P. et al. "Self-marking recapture models for estimating closed insect populations," *Journal of Agricultural, Biological, and Environmental Statistics,* Vol. 5, No. 4, December 2000 (Table 5A).

8.65 Classifying air threats with heuristics. The *Journal of Behavioral Decision Making* (Jan. 2007) published a study on the use of heuristics to classify the threat level of approaching aircraft. Of special interest was the use of a fast and frugal heuristic—a computationally simple procedure for making judgments with limited information—named "Take-the-Best-for-Classification" (TTB-C). The subjects were 48 men and women, some from a Canadian Forces reserve unit, others university students. Each subject was presented with a radar screen on which simulated approaching aircraft were identified with asterisks. By using the computer mouse to click on the asterisk, one could receive further information about the aircraft. The goal was to identify the aircraft as "friend" or "foe" as fast as possible. Half the subjects were given cue-based instructions for determining the type of aircraft, while the other half were given pattern-based instructions. The researcher also classified the heuristic strategy used by the subject as TTB-C, Guess, or Other. Data on the two variables Instruction type and Strategy, measured for each of the 48 subjects, are saved in the **AIRTHREAT** file. (Data on the first and last five subjects are shown in the accompanying table.) Do the data provide sufficient evidence at $\alpha = .05$ to indicate that choice of heuristic strategy depends on type of instruction provided? How about at $\alpha = .01$?

AIRTHREAT (Selected observations)

Instruction	Strategy
Pattern	Other
Pattern	Other
Pattern	Other
Cue	TTBC
Cue	TTBC
⋮	⋮
Pattern	TTBC
Cue	Guess
Cue	TTBC
Cue	Guess
Pattern	Guess

Source: Bryant, D. J. "Classifying simulated air threats with fast and frugal heuristics," *Journal of Behavioral Decision Making,* Vol. 20, January 2007 (Appendix C).

SEEDLING

8.66 Subarctic plant study. The traits of seed-bearing plants indigenous to subarctic Finland were studied in *Arctic, Antarctic, and Alpine Research* (May 2004). Plants were categorized according to *type* (dwarf shrub, herb, or grass), *abundance of seedlings* (no seedlings, rare seedlings, or abundant seedlings), *regenerative group* (no vegetative reproduction, vegetative reproduction possible, vegetative reproduction ineffective, or vegetative reproduction effective), *seed weight class* (0–.1, .1–.5, .5–1.0, 1.0–5.0, and > 5.0 milligrams), and *diaspore morphology* (no structure, pappus, wings, fleshy fruits, or awns/hooks). The data on a sample of 73 plants are saved in the **SEEDLING** file.

Tabulated statistics: Abundance, Type

```
Rows: Abundance   Columns: Type

          DwarfShrub   Grasses   Herbs   All

NS             3          1         1      5
SA             5         14        32     51
SR             5          2        10     17
All           13         17        43     73

Cell Contents:        Count
```

MINITAB output for Exercise 8.66

HYPNOSIS

		CAHS Level				
		Low	Medium	High	Very High	Totals
SHSS: C Level	**Low**	32	14	2	0	48
	Medium	11	14	6	0	31
	High	6	14	19	3	42
	Very High	0	2	4	3	9
	Totals	49	44	31	6	130

Source: Grant, C. D., and Nash, M. R. "The Computer-Assisted Hypnosis Scale: Standardization and Norming of a Computer-administered Measure of Hypnotic Ability," *Psychological Assessment*, Vol. 7, No. 1, Mar. 1995, p. 53 (Table 4).

a. A contingency table for plant type and seedling abundance, produced by MINITAB, is shown on p. 467. (*Note:* NS = no seedlings, SA = seedlings abundant, and SR = seedlings rare.) Suppose you want to perform a chi-square test of independence to determine whether seedling abundance depends on plant type. Find the expected cell counts for the contingency table. Are the assumptions required for the test satisfied?

b. Reformulate the contingency table by combining the NS and SR categories of seedling abundance. Find the expected cell counts for this new contingency table. Are the assumptions required for the test satisfied?

c. Reformulate the contingency table of part **b** by combining the dwarf shrub and grasses categories of plant type. Find the expected cell counts for this contingency table. Are the assumptions required for the test satisfied?

d. Carry out the chi-square test for independence on the contingency table you came up with in part **c**, using $\alpha = .10$. What do you conclude?

8.67 **Susceptibility to hypnosis.** A standardized procedure for determining a person's susceptibility to hypnosis is the Stanford Hypnotic Susceptibility Scale, Form C (SHSS:C). Recently, a new method called the Computer-Assisted Hypnosis Scale (CAHS), which uses a computer as a facilitator of hypnosis, has been developed. Each scale classifies a person's hypnotic susceptibility as low, medium, high, or very high. Researchers at the University of Tennessee compared the two scales by administering both tests to each of 130 undergraduate volunteers. (*Psychological Assessment,* Mar. 1995.) The hypnotic classifications are summarized in the table at the top of the page. A contingency table analysis will be performed to determine whether CAHS level and SHSS level are independent.

a. Check to see if the assumption of expected cell counts of 5 or more is satisfied. Should you proceed with the analysis? Explain.

b. One way to satisfy the assumption of part **a** is to combine the data for two or more categories (e.g., high and very high) in the contingency table. Form a new contingency table by combining the data for the high and very high categories in both the rows and the columns.

c. Calculate the expected cell counts in the new contingency table you formed in part **c**. Is the assumption now satisfied?

d. Perform the chi-square test on the new contingency table. Use $\alpha = .05$. Interpret the results.

8.68 **Gangs and homemade weapons.** The National Gang Crime Research Center (NGCRC) has developed a six-level gang classification system for both adults and juveniles. The six categories are shown in the accompanying table. The classification system was developed as a potential predictor of a gang member's propensity for violence in prison, jail, or a correctional facility. To test the system, the NGCRC collected data on approximately 10,000 confined offenders and assigned each a score from the gang classification system. (*Journal of Gang Research,* Winter 1997.) One of several other variables measured by the NGCRC was whether or not the offender had ever carried a homemade weapon (e.g., knife) while in custody. The data on gang score and homemade weapon are summarized in the table below. Conduct a test to determine whether carrying a homemade weapon in custody depends on gang classification score. (Use $\alpha = .01$.) Support your conclusion with a graph.

GANGS

	Homemade Weapon Carried	
Gang Classification Score	Yes	No
0 (Never joined a gang, no close friends in a gang)	255	2,551
1 (Never joined a gang, 1–4 close friends in a gang)	110	560
2 (Never joined a gang, 5 or more friends in a gang)	151	636
3 (Inactive gang member)	271	959
4 (Active gang member, no position of rank)	175	513
5 (Active gang member, holds position of rank)	476	831

Source: Knox, G. W., et al. "A gang classification system for corrections," *Journal of Gang Research,* Vol. 4, No. 2, Winter 1997, p. 54 (Table 4).

Applying the Concepts—Advanced

8.69 Efficacy of an HIV vaccine. New, effective AIDS vaccines are now being developed through the process of "sieving"—that is, sifting out infections with some strains of HIV. Harvard School of Public Health statistician Peter Gilbert demonstrated how to test the efficacy of an HIV vaccine in *Chance* (Fall 2000). As an example, using the 2×2 table shown below, Gilbert reported the results of VaxGen's preliminary HIV vaccine trial. The vaccine was designed to eliminate a particular strain of the virus called the "MN strain." The trial consisted of 7 AIDS patients vaccinated with the new drug and 31 AIDS patients who were treated with a placebo (no vaccination). The table shows the number of patients who tested positive and negative for the MN strain in the trial follow-up period.

HIVVAC1

		MN Strain		
		Positive	Negative	Totals
Patient Group	**Unvaccinated**	22	9	31
	Vaccinated	2	5	7
	Totals	24	14	38

Source: Gilbert, P. "Developing an AIDS vaccine by sieving," *Chance,* Vol. 13, No. 4, Fall 2000.

HIVVAC2

		MN Strain		
		Positive	Negative	Totals
	Unvaccinated	23	8	31
Patient Group	**Vaccinated**	1	6	7
	Totals	24	14	38

HIVVAC3

		MN Strain		
		Positive	Negative	Totals
	Unvaccinated	24	7	31
Patient Group	**Vaccinated**	0	7	7
	Totals	24	14	38

a. Conduct a test to determine whether the vaccine is effective in treating the MN strain of HIV. Use $\alpha = .05$.

b. Are the assumptions for the test you carried out in part **a,** satisfied? What are the consequences if the assumptions are violated?

c. In the case of a 2×2 contingency table, R. A. Fisher (1935) developed a procedure for computing the exact *p*-value for the test (called *Fisher's exact test*). The method utilizes the *hypergeometric probability distribution*. Consider the hypergeometric probability

$$\frac{\binom{7}{2}\binom{31}{22}}{\binom{38}{24}}$$

SAS output for Exercise 8.69

The FREQ Procedure

Table of GROUP by MNSTRAIN

```
GROUP       MNSTRAIN

Frequency
Expected  NEG        POS        Total

UN           9         22          31
          11.421     19.579

V            5          2           7
          2.5789     4.4211

Total       14         24          38
```

Statistics for Table of GROUP by MNSTRAIN

Statistic	DF	Value	Prob
Chi-Square	1	4.4112	0.0357
Likelihood Ratio Chi-Square	1	4.2893	0.0384
Continuity Adj. Chi-Square	1	2.7773	0.0956
Mantel-Haenszel Chi-Square	1	4.2952	0.0382
Phi Coefficient		-0.3407	
Contingency Coefficient		0.3225	
Cramer's V		-0.3407	

WARNING: 50% of the cells have expected counts less than 5. Chi-Square may not be a valid test.

```
          Fisher's Exact Test

Cell (1,1) Frequency (F)          9
Left-sided Pr <= F            0.0498
Right-sided Pr >= F          0.9940

Table Probability (P)        0.0438
Two-sided Pr <= P            0.0772

        Sample Size = 38
```

which represents the probability that 2 out of 7 vaccinated AIDS patients test positive and 22 out of 31 unvaccinated patients test positive—that is, the probability of the result shown in table, given that the null hypothesis of independence is true. Compute this probability (called the *probability of the contingency table*).

d. Refer to part **c.** Two contingency tables (with the same marginal totals as the original table) that are more unsupportive of the null hypothesis of independence than the observed table are shown below. First, explain why these tables provide more evidence to reject H_0 than the original table does. Then compute the probability of each table, using the hypergeometric formula.

e. The *p*-value of Fisher's exact test is the probability of observing a result at least as unsupportive of the null hypothesis as is the observed contingency table, given the same marginal totals. Sum the probabilities of parts **c** and **d** to obtain the *p*-value of Fisher's exact test. (To verify your calculations, check the *p*-value labeled **Left-sided Pr < = F**) at the bottom of the SAS printout shown above.) Interpret this value in the context of the vaccine trial.

8.70 Examining the "Monty Hall Dilemma." In Exercise 3.182 (p. 177) you solved the game show problem of whether or not to switch your choice of three doors, one of which hides a prize, after the host reveals what is behind a door that is not chosen. (Despite the natural inclination of many to keep one's first choice, the correct answer is that you should switch your choice of doors.) This problem is sometimes called the "Monty Hall Dilemma," named for Monty Hall, the host of the popular TV game show *Let's Make a Deal.* In *Thinking & Reasoning*

(Oct. 2006), Wichita State University professors set up an experiment designed to influence subjects to switch their original choice of doors. Each subject participated in 23 trials. In trial 1, three (boxes) representing doors were presented on a computer screen; only one box hid a prize. In each subsequent trial, an additional box was presented, so that in trial 23, twenty-five boxes were presented. In each trial, after a box was selected, all of the remaining boxes except for one either (1) were shown to be empty (*Empty* condition), (2) disappeared (*Vanish* condition), (3) disappeared, and the chosen box was enlarged (*Steroids* condition), or (4) disappeared, and the remaining box not chosen was enlarged (*Steroids2* condition). Twenty-seven subjects were assigned to each condition. The number of subjects who ultimately switched boxes is tallied, by condition, in the following table for both the first trial and the last trial:

a. For a selected trial, does the likelihood of switching boxes depend on condition?

⊚ **MONTYHALL**

	First Trial (1)		Last Trial (23)	
Condition	Switch Boxes	No Switch	Switch Boxes	No Switch
Empty	10	17	23	4
Vanish	3	24	12	15
Steroids	5	22	21	6
Steroids2	8	19	19	8

Source: Howard, J. N., Lambdin, C. G., and Datteri, D. L. "Let's Make a Deal: Quality and Availability of Second-stage Information as a Catalyst for Change," *Thinking & Reasoning,* October 2006 (Table 2).

b. For a given condition, does the likelihood of switching boxes depend on trial number?

c. On the basis of the results you obtained in parts **a** and **b**, what factors influence a subject to switch choices?

KEY TERMS

Categories   444
Cell   444
Cell counts   444
Chi-square distribution   446
Chi-square test   446
Classes   444
Contingency table   454

Dependence   457
Dimensions of classification   454
Expected cell count   447
Independence of two classifications   455
Marginal probabilities   455
Multinomial experiment   444

Observed cell count   455
One-way table   445
Sample survey   433
Sampling distribution of $(\hat{p}_1 - \hat{p}_2)$ 434
Two-way table   454

CHAPTER NOTES

Multinomial Data

Qualitative data that fall into more than two categories (or classes)

Properties of a Multinomial Experiment

(1) n identical trials

(2) k possible outcomes to each trial

(3) probabilities of the k outcomes $(p_1, p_2, \ldots, p_k)$ where $p_1 + p_2 + \ldots + p_k = 1$, remain the same from trial to trial

(4) trials are independent

(5) variables of interest: *cell counts* (i.e., number of observations falling into each outcome category), denoted $n_1, n_2, \ldots, n_k$

Key Symbols/Notation

$p_1 - p_2$	Difference between population proportions
$\hat{p}_1 - \hat{p}_2$	Difference between sample proportions
SE	Error in estimation sampling
$p_{i,0}$	Value of multinomial probability p_1 hypothesized in H_0
χ^2	Chi-square test statistic used in analysis of categorical data
n_i	Number of observed outcomes in cell i of a one-way table
E_i	Expected number of outcomes in cell i of a one-way table
p_{ij}	Probability of an outcome in row i and column j of a two-way table
n_{ij}	Number of observed outcomes in row i and column j of a two-way table
E_{ij}	Expected number of outcomes in row i and column j of a two-way table
R_i	Total number of outcomes in row i of a two-way table
C_j	Total number of outcomes in column j of a two-way table

CATEGORICAL DATA ANALYSIS GUIDE

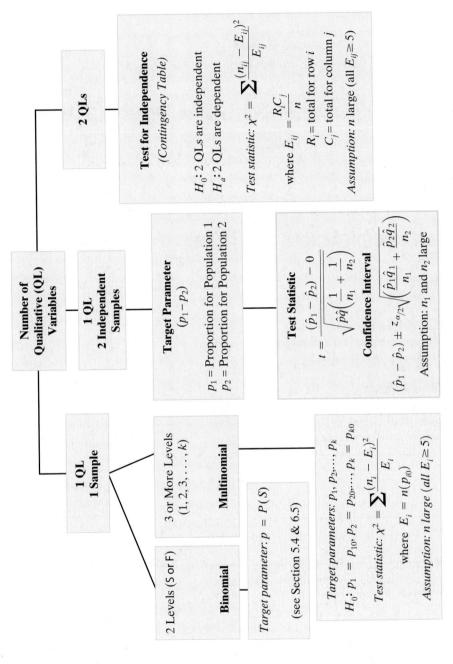

Number of Qualitative (QL) Variables

2 QLs

Test for Independence
(Contingency Table)

H_0: 2 QLs are independent
H_a: 2 QLs are dependent

Test statistic: $\chi^2 = \sum \dfrac{(n_{ij} - E_{ij})^2}{E_{ij}}$

where $E_{ij} = \dfrac{R_i C_j}{n}$

R_i = total for row i
C_j = total for column j

Assumption: n large (all $E_{ij} \geq 5$)

1 QL 2 Independent Samples

Target Parameter
$(p_1 - p_2)$

p_1 = Proportion for Population 1
p_2 = Proportion for Population 2

Test Statistic

$t = \dfrac{(\hat{p}_1 - \hat{p}_2) - 0}{\sqrt{\hat{p}\hat{q}\left(\dfrac{1}{n_1} + \dfrac{1}{n_2}\right)}}$

Confidence Interval

$(\hat{p}_1 - \hat{p}_2) \pm z_{\alpha/2} \sqrt{\dfrac{\hat{p}_1 \hat{q}_1}{n_1} + \dfrac{\hat{p}_2 \hat{q}_2}{n_2}}$

Assumption: n_1 and n_2 large

1 QL 1 Sample

Binomial
2 Levels (S or F)

Target parameter: $p = P(S)$

(see Section 5.4 & 6.5)

Multinomial
3 or More Levels
$(1, 2, 3, \ldots, k)$

Target parameters: $p_1, p_2, \ldots, p_k$

H_0: $p_1 = p_{10}, p_2 = p_{20}, \ldots, p_k = p_{k0}$

Test statistic: $\chi^2 = \sum \dfrac{(n_i - E_i)^2}{E_i}$

where $E_i = n(p_{i0})$

Assumption: n large (all $E_i \geq 5$)

471

One-Way Table	Two-Way (Contingency) Table
Summary table for a *single* qualitative variable	Summary table for *two* qualitative variables

Chi-Square (χ^2) Statistic	Chi-square tests for independence
used to test category probabilities in one-way and two-way tables	should **not** be used to *infer a causal relationship between two QLs*

Conditions Required for Valid χ^2 Tests

(1) multinomial experiment

(2) sample size n is large (expected cell counts are all greater than or equal to 5)

Conditions Required for Inferences about $p_1 - p_2$

Large samples: (1) Independent random samples

(2) $n_1p_1 \geq 15, n, q, \geq 15$

(3) $n_2p_2 \geq 15, n, q, \geq 15$

Determining the Sample Size for Estimating ($p_1 - p_2$)

$$n_1 = n_2 = (z_{\alpha/2})^2 (p_1q_1 + p_2q_2)/(SE)^2$$

SUPPLEMENTARY EXERCISES 8.71–8.98

Understanding the Principles

8.71 What is the difference between a one-way chi-square analysis and a two-way chi-square analysis?

8.72 *True or False*. Rejecting the null hypothesis in a chi-square test for independence implies that a causal relationship exists between the two categorical variables.

Learning the Mechanics

8.73 Independent random samples were selected from two binomial populations. The size and number of observed successes for each sample are shown in the table.

Sample 1	Sample 2
$n_1 = 200$	$n_2 = 200$
$x_1 = 110$	$x_2 = 130$

a. Test $H_0: (p_1 - p_2) = 0$ against $H_a: (p_1 - p_2) < 0$. Use $\alpha = .10$.

b. Form a 95% confidence interval for $(p_1 - p_2)$.

c. What sample sizes would be required if we wish to use a 95% confidence interval of width .01 to estimate $(p_1 - p_2)$?

8.74 A random sample of 150 observations was classified into the categories shown in the table.

	Category				
	1	2	3	4	5
n_i	28	35	33	25	29

a. Do the data provide sufficient evidence that the categories are not equally likely? Use $\alpha = .10$.

b. Form a 90% confidence interval for p_2, the probability that an observation will fall into category 2.

8.75 A random sample of 250 observations was classified according to the row and column categories shown in the table.

		Column		
		1	2	3
	1	20	20	10
Row	2	10	20	70
	3	20	50	30

a. Do the data provide sufficient evidence to conclude that the rows and columns are dependent? Test using $\alpha = .05$.

b. Would the analysis change if the row totals were fixed before the data were collected?

c. Do the assumptions required for the analysis to be valid differ according to whether the row (or column) totals are fixed? Explain.

d. Convert the table entries to percentages by using each column total as a base and calculating each row response as a percentage of the corresponding column total. In addition, calculate the row totals and convert them to percentages of all 250 observations.

e. Create a bar graph with row 1 percentages on the vertical axis and the column number on the horizontal axis. Draw a horizontal line corresponding to the row 1 total percentage. Does the graph support the result of the test conducted in part **a**?

f. Repeat part **e** for the row 2 percentages.

g. Repeat part **e** for the row 3 percentages.

Applying the Concepts—Basic

8.76 Tapeworms in brill fish. The *Journal of Fish Biology* (Aug. 1990) reported on a study to compare the incidence of parasites (tapeworms) in species of Mediterranean and Atlantic fish. In the Mediterranean Sea, 588 brill were captured and dissected, and 211 were found to be infected by the parasite. In the Atlantic Ocean, 123 brill were captured and dissected, and 26 were found to be infected. Compare the proportions of infected brill at the two capture sites using a 90% confidence interval. Interpret the interval.

8.77 Dosing errors at hospitals. Each year, approximately 1.3 million people in the United States suffer adverse drug effects (ADEs)—that is, unintended injuries caused by prescribed medication. A study in the *Journal of the American Medical Association* (July 5, 1995) identified the cause of 247 ADEs that occurred at two Boston hospitals. The researchers found that dosing errors (i.e., wrong dosage prescribed or dispensed) were the most common. The table summarizes the proximate cause of 95 ADEs that resulted from a dosing error. Conduct a test (at $\alpha = .10$) to determine whether the true percentages of ADEs in the five "cause" categories are different.

Cause of Wrong Dosage	Number of ADEs
(1) Lack of knowledge of drug	29
(2) Rule violation	17
(3) Faulty dose checking	13
(4) Slips	9
(5) Other	27

8.78 Scanning Internet messages. *Inc. Technology* (Mar. 18, 1997) reported the results of an Equifax/Harris Consumer Privacy Survey in which 328 Internet users indicated their level of agreement with the following statement: "The government needs to be able to scan Internet messages and user communications to prevent fraud and other crimes." The number of users in each response category is summarized as follows:

Agree Strongly	Agree Somewhat	Disagree Somewhat	Disagree Strongly
59	108	82	79

a. Specify the null and alternative hypotheses you would use to determine whether the opinions of Internet users are evenly divided among the four categories.
b. Conduct the test of part **a**, using $\alpha = .05$.
c. In the context of this exercise, what is a Type I error? A Type II error?
d. What assumptions must hold in order to ensure the validity of the test you conducted in part **b**?

8.79 Risk factor for lumbar disease. One of the most common musculoskeletal disorders is lumbar disk disease (LDD). Medical researchers reported finding a common genetic risk factor for LDD (*Journal of the American Medical Association*, Apr. 11, 2001). The study included 171 Finnish patients diagnosed with LDD (the patient group) and 321 without LDD (the control group). Of the 171 LDD patients, 21 were discovered to have the genetic trait. Of the 321 people in the control group, 15 had the genetic trait.
a. Consider the two categorical variables group and presence/absence of genetic trait. Form a 2 × 2 contingency table for these variables.
b. Conduct a χ^2 test to determine whether the genetic trait occurs at a higher rate in LDD patients than in the controls. Use $\alpha = .01$.

c. Construct a bar graph that will visually support your conclusion in part **b**.
d. Conduct the test, part b, using the large-sample test for $(p_1 - p_2)$.

8.80 Seat-belt use study. The *American Journal of Public Health* (July 1995) reported on a population-based study of trauma in Hispanic children. One of the objectives of the study was to compare the use of protective devices in motor vehicles that transported Hispanic children and non-Hispanic white children. On the basis of data collected from the San Diego County Regionalized Trauma System, 792 children treated for injuries sustained in vehicular accidents were classified according to ethnic status (Hispanic or non-Hispanic white) and seat-belt usage (worn or not worn) during the accident. The data are summarized in the following table:

TRAUMA

	Hispanic	Non-Hispanic White	Totals
Seat belts worn	31	148	179
Seat belts not worn	283	330	613
Totals	314	478	792

Source: Matteneci, R. M., et al. "Trauma among Hispanic children: A population-based study in a regionalized system of trauma care," *American Journal of Public Health,* Vol. 85, No. 7, July 1995, p. 1007 (Table 2).

a. Calculate the sample proportion of injured Hispanic children who were not wearing seat belts during the accident.
b. Calculate the sample proportion of injured non-Hispanic white children who were not wearing seatbelts during the accident.
c. Compare the sample proportions from parts **a** and **b**. Do you think the true population proportions differ?
d. Conduct a test to determine whether seat-belt usage in motor vehicle accidents depends on ethnic status in the San Diego County Regionalized Trauma System. Use $\alpha = .01$.
e. Construct a 99% confidence interval for the difference between the proportions you found in parts **a** and **b**. Interpret the interval.

8.81 High-fiber food and cancer. According to research reported in the *Journal of the National Cancer Institute* (Apr. 1991), eating foods high in fiber may help protect against breast cancer. The researchers randomly divided 120 laboratory rats into four groups of 30 each. All of the rats were injected with a drug that causes breast cancer. Then each rat was fed a diet of fat and fiber for 15 weeks. However, the levels of fat and fiber varied from group to group. At the end of the feeding period, the number of rats with cancer tumors was determined for each group. The data are summarized in the table on p. 474.
a. Does the sampling appear to satisfy the assumptions for a multinomial experiment? Explain.
b. Calculate the expected cell counts for the contingency table.
c. Calculate the χ^2 statistic.
d. Is there evidence to indicate that diet and presence/absence of cancer are independent? Test, using $\alpha = 05$.
e. Compare the percentage of rats on a high-fat, no-fiber diet with cancer with the percentage of rats on a high-fat,

TUMORS

		Diet				
		High Fat/No Fiber	High Fat/Fiber	Low Fat/No Fiber	Low Fat/Fiber	Totals
Cancer Tumors	**Yes**	27	20	19	14	80
	No	3	10	11	16	40
	Totals	30	30	30	30	120

Source: Tampa Tribune, Apr. 3, 1991.

fiber diet with cancer, using a 95% confidence interval. Interpret the result.

8.82 Travel habits of retirees. A study in the *Annals of Tourism Research* (Vol. 19, 1992) investigates the relationship of retirement status (pre- and postretirement) to various items related to the travel industry. One part of the study investigated the differences in the length of stay of a trip for pre- and postretirees. A sample of 703 travelers were asked how long they stayed on a typical trip. The results are shown in the accompanying table. Use the information in the table to determine whether the retirement status of a traveler and the duration of a typical trip are dependent. Test, using $\alpha = .05$.

TRAVEL

Number of Nights	Preretirement	Postretirement
4–7	247	172
8–13	82	67
14–21	35	52
22 or more	16	32
Total	380	323

8.83 The "winner's curse" in auction bidding. In auction bidding, the "winner's curse" is the phenomenon of the winning (or highest) bid price being above the expected value of the item being auctioned. *The Review of Economics and Statistics* (Aug. 2001) published a study on whether experience in bidding affects the likelihood of the winner's curse occurring. Two groups of bidders in a sealed-bid auction were compared: (1) superexperienced bidders and (2) less experienced bidders. In the superexperienced group, 29 of 189 winning bids were above the item's expected value; in the less experienced group, 32 of 149 winning bids were above the item's expected value.

a. Find an estimate of p_1, the true proportion of super-experienced bidders who fall prey to the winner's curse.

b. Find an estimate of p_2, the true proportion of less experienced bidders who fall prey to the winner's curse.

c. Construct a 90% confidence interval for $p_1 - p_2$.

d. Give a practical interpretation of the confidence interval you constructed in part **c.** Make a statement about whether experience in bidding affects the likelihood of the winner's curse occurring.

8.84 Butterfly hot spots. *Nature* (Sept. 1993) reported on a study of animal and plant species "hot spots" in Great Britain. A hot spot is defined as a 10-km^2 area that is species rich—that is, heavily populated by a species of interest. Analogously, a cold spot is a 10-km^2 area that is species poor. The accompanying table gives the number of butterfly hot spots and the number of butterfly cold spots in a sample of 2,588 10-km^2 areas. In theory, 5% of the areas should be butterfly hot spots and 5% should be butterfly cold spots, while the remaining areas (90%) are neutral. Test the theory, using $\alpha = .01$.

Butterfly hot spots	123
Butterfly cold spots	147
Neutral areas	2,318
Total	**2,588**

Source: Prendergast, J. R., et al. "Rare species, the coincidence of diversity hotspots and conservation strategies." *Nature,* Vol. 365, No. 6444, Sept. 23, 1993, p. 335 (Table 1).

Applying the Concepts—Intermediate

8.85 Switching majors in college. When female undergraduates switch from science, mathematics, and engineering (SME) majors into disciplines that are not based on science, are their reasons different from those of their male counterparts? This question was investigated in *Science Education* (July 1995). A sample of 335 junior/senior undergraduates—172 females and 163 males—at two large research universities were identified as "switchers"; that is, they left a declared SME major for a non-SME major. Each student listed one or more factors that contributed to the switching decision.

a. Of the 172 females in the sample, 74 listed lack or loss of interest in SME (i.e., they were "turned off" by science) as a major factor, compared with 72 of the 163 males. Conduct a test (at $\alpha = .10$) to determine whether the proportion of female switchers who give "lack of interest in SME" as a major reason for switching differs from the corresponding proportion of males.

b. Thirty-three of the 172 females in the sample indicated that they because discouraged or lost confidence because of low grades in SME during their early years, compared with 44 of 163 males. Construct a 90% confidence interval for the difference between the proportions of female and male switchers who lost confidence due to low grades in SME. Interpret the result.

8.86 Rating music teachers. Students enrolled in music classes at the University of Texas (Austin) participated in a study to compare the observations and teacher evaluations of music education majors and nonmusic majors. (*Journal of Research in Music Education,* Winter 1991.) Independent random samples of 100 music majors and 100 nonmajors rated the overall performance of their teacher, using a six-point scale, where 1 was the lowest rating and 6 the highest. Use the information in the next table to compare the mean teacher ratings of the two groups of music students with a 95% confidence interval. Interpret the result.

	Music Majors	Nonmusic Majors
Sample size	100	100
Mean "overall" rating	4.26	4.59
Standard deviation	.81	.78

Source: Duke, R. A., and Blackman, M. D. "The relationship between observers' recorded teacher behavior and evaluation of music instruction," *Journal of Research in Music Education*, Vol. 39, No. 4, Winter 1991 (Table 2).

8.87 Iraq War survey. The Pew Internet & American Life Project commissioned Princeton Survey Research Associates to develop and carry out a survey of what Americans think about the recent War in Iraq. Some of the results of the March 2003 survey of over 1,400 American adults are saved in the **IRAQWAR** file. Responses to the following questions were recorded:
1. Do you support or oppose the Iraq War? (1 = Support, 2 = Oppose)
2. Do you ever go online to access the Internet or World Wide Web? (1 = Yes, 2 = No)
3. Do you consider yourself a Republican, Democrat, or Independent? (1 = Rep., 2 = Dem., 3 = Ind.)
4. Have you or anyone in your household served in the U.S. military? (1 = Yes, I have; 2 = Yes, other; 3 = Yes, both; 4 = No)
5. In general, would you describe your political views as very conservative, conservative, moderate, liberal, or very liberal? (1 = Very conservative, 2 = Conservative, 3 = Moderate, 4 = Liberal, 5 = Very liberal)
6. What is your race? (1 = White, 2 = African-American, 3 = Asian, 4 = Mixed, 5 = Native American, 6 = Other)
7. What is your income range? (1 = < 10K, 2 = 10–20K, 3 = 20–30K, 4 = 30–40K, 5 = 40–50K, 6 = 50–75K, 7 = 75–100K, 8 = > 100K)
8. Do you live in a suburban, rural, or urban community? (1 = urban, 2 = suburban, 3 = rural)
Conduct a series of contingency table analyses to determine whether support for the Iraq War depends on one or more of the other categorical variables measured in the March 2003 survey.

8.88 Pig farmer study. An article in *Sociological Methods & Research* (May 2001) analyzed the data presented in the accompanying table. A sample of 262 Kansas pig farmers were classified according to their education level (college or not) and size of their pig farm (number of pigs). Conduct a test to determine whether a pig farmer's education level has an impact on the size of the pig farm. Use $\alpha = .05$ and support your answer with a graph.

PIGFARM

		Education Level		
		No College	College	Totals
Farm Size	**<1,000 pigs**	42	53	95
	1,000–2,000 pigs	27	42	69
	2,001–5,000 pigs	22	20	42
	>5,000 pigs	27	29	56
	Totals	118	144	262

Source: Agresti, A., and Liu, I. "Strategies for modeling a categorical variable allowing multiple category choices." *Sociological Methods & Research*, Vol. 29, No. 4, May 2001 (Table I).

8.89 Multiple-sclerosis drug. Interferons are proteins produced naturally by the human body that help fight infections and regulate the immune system. A drug developed from interferons, called Avonex, is now available for treating patients with multiple sclerosis (MS). In a clinical study, 85 MS patients received weekly injections of Avonex over a two-year period. The number of exacerbations (i.e., flare-ups of symptoms) was recorded for each patient and is summarized in the accompanying table. For MS patients who take a placebo (no drug) over a similar two-year period, it is known from previous studies that 26% will experience no exacerbations, 30% one exacerbation, 11% two exacerbations, 14% three exacerbations, and 19% four or more exacerbations.

Number of Exacerbations	Number of Patients
0	32
1	26
2	15
3	6
4 or more	6

Source: Biogen, Inc.

a. Conduct a test to determine whether the exacerbation distribution of MS patients who take Avonex differs from the percentages reported for placebo patients. Use $\alpha = .05$.
b. Find a 95% confidence interval for the true percentage of Avonex MS patients who remain free of exacerbations during a two-year period.
c. Refer to part **b**. Is there evidence that Avonex patients are more likely to have no exacerbations than placebo patients? Explain.

8.90 Flight response of geese to helicopter traffic. Offshore oil drilling near an Alaskan estuary has led to increased air traffic—mostly large helicopters—in the area. The U.S. Fish and Wildlife Service commissioned a study to investigate the impact these helicopters have on the flocks of Pacific brant geese that inhabit the estuary in the fall before migrating. (*Statistical Case Studies: A Collaboration between Academe and Industry*, 1998.) Two large helicopters were flown repeatedly over the estuary at different altitudes and lateral distances from the flock. The flight responses of the geese (recorded as "low" or "high"), the altitude (in hundreds of meters), and the lateral distance (also in hundreds of meters) for each of 464 helicopter overflights were recorded and are saved in the **PACGEESE** file. The data for the first 10 overflights are shown in the table on p. 476:
a. The researchers categorized altitude as follows: less than 300 meters, 300–600 meters, and 600 or more meters. Summarize the data in the **PACGEESE** file by creating a contingency table for altitude category and flight response.
b. Conduct a test to determine whether flight response of the geese depends on altitude of the helicopter. Test, using $\alpha = .01$.
c. The researchers categorized lateral distance as follows: less than 1,000 meters, 1,000–2,000 meters, 2,000–3,000 meters, and 3,000 or more meters. Summarize the data in the **PACGEESE** file by creating a contingency table for lateral distance category and flight response.

PACGEESE (First 10 observations shown)

Overflight	Altitude	Lateral Distance	Flight Response
1	0.91	4.99	HIGH
2	0.91	8.21	HIGH
3	0.91	3.38	HIGH
4	9.14	21.08	LOW
5	1.52	6.60	HIGH
6	0.91	3.38	HIGH
7	3.05	0.16	HIGH
8	6.10	3.38	HIGH
9	3.05	6.60	HIGH
10	12.19	6.60	HIGH

Source: Erickson, W., Nick, T., and Ward, D. "Investigating Flight Response of Pacific Brant to Helicopters at Izembek Lagoon, Alaska by Using Logistic Regression," *Statistical Case Studies: A Collaboration between Academe and Industry*, ASA-SIAM Series on Statistics and Applied Probability, 1998.

d. Conduct a test to determine whether flight response of the geese depends on lateral distance of helicopter from the flock. Test, using $\alpha = .01$.

e. The current Federal Aviation Authority (FAA) minimum altitude standard for flying over the estuary is 2,000 feet (approximately 610 meters). On the basis of the results obtained in parts **a–d**, what changes to the FAA regulations do you recommend in order to minimize the effects to Pacific brant geese.

8.91 Birds feeding on gypsy moths. A field study was conducted to identify the natural predators of the gypsy moth. (*Environmental Entomology*, June 1995.) For one part of the study, 24 black-capped chickadees (common wintering birds) were captured in mist nets and individually caged. Each bird was offered a mass of gypsy moth eggs attached to a piece of bark. Half the birds were offered no other food (no choice), and half were offered a variety of other naturally occurring foods such as spruce and pine seeds (choice). The numbers of birds that did and did not feed on the gypsy moth egg mass are given in the accompanying table. Analyze the data in the table to determine whether a relationship exists between food choice and feeding or not feeding on gypsy moth eggs. Use $\alpha = .10$.

MOTH

	Fed on Egg Mass	
	Yes	No
Choice of foods	2	10
No choice	8	4

8.92 Social play of children. The *American Journal on Mental Retardation* (Jan. 1992) published a study of the social interactions of two groups of children. Independent random samples of 15 children with and 15 children without developmental delays (i.e., mild mental retardation) comprised the subjects of the experiment. After observing the children during "free play," researchers recorded the number of children who exhibited disruptive behavior (e.g., ignoring or rejecting other children, taking toys from another child) for each group. The data are summarized in the two-way table shown at the bottom of the page. Analyze the data and interpret the results.

8.93 Kicking the cigarette habit. Can taking an antidepressant drug help cigarette smokers kick their habit? The *New England Journal of Medicine* (Oct. 23, 1997) published a study in which 615 smokers (all of whom wanted to give up smoking) were randomly assigned to receive either Zyban (an antidepressant) or a placebo (a dummy pill) for six weeks. Of the 309 patients who received Zyban, 71 were not smoking one year later. Of the 306 patients who received a placebo, 37 were not smoking one year later. Conduct a test of hypothesis (at $\alpha = .05$) to answer the research question posed in the first sentence of this exercise.

8.94 Battle simulation trials. In order to evaluate their situational awareness, fighter aircraft pilots participate in battle simulations. At a random point in the trial, the simulator is frozen and data on situational awareness are immediately collected. The simulation is then continued until, ultimately, performance (e.g., number of kills) is measured. A study reported in *Human Factors* (Mar. 1995) investigated whether temporarily stopping the simulation results in any change in pilot performance. Trials were designed so that some simulations were stopped to collect situational awareness data while others were not. Each trial was then classified according to the number of kills made by the pilot. The data for 180 trials are summarized in the accompanying contingency table. Conduct a contingency table analysis and interpret the results fully.

SIMKILLS

	Number of Kills					
	0	1	2	3	4	Totals
Stops	32	33	19	5	2	91
No Stops	24	36	18	8	3	89
Totals	56	69	37	13	5	180

DISRUPT

	Disruptive Behavior	Nondisruptive Behavior	Totals
With Developmental Delays	12	3	15
Without Developmental Delays	5	10	15
Total	17	13	30

Source: Kopp, C. B., Baker, B., and Brown, K. W. "Social skills and their correlates: Preschoolers with developmental delays," *American Journal on Mental Retardation,* Vol. 96, No. 4, Jan. 1992.

Applying the Concepts—Advanced

8.95 Goodness-of-fit test. A statistical analysis is to be done on a set of data consisting of 1,000 monthly salaries. The analysis requires the assumption that the sample was drawn from a normal distribution. A preliminary test, called the χ^2 *goodness-of-fit test*, can be used to help determine whether it is reasonable to assume that the sample is from a normal distribution. Suppose the mean and standard deviation of the 1,000 salaries are hypothesized to be $1,200 and $200, respectively. Using the standard normal table, we can approximate the probability of a salary being in the intervals listed in the accompanying table. The third column represents the expected number of the 1,000 salaries to be found in each interval if the sample was drawn from a normal distribution with $\mu = \$1,200$ and $\sigma = \$200$. Suppose the last column contains the actual observed frequencies in the sample. Large differences between the observed and expected frequencies cast doubt on the normality assumption.

Interval	Probability	Expected Frequency	Observed Frequency
Less than $800	.023	23	26
$800 < $1,000	.136	136	146
$1,000 < $1,200	.341	341	361
$1,200 < $1,400	.341	341	311
$1,400 < $1,600	.136	136	143
$1,600 or above	.023	23	13

   a. Compute the χ^2 statistic on the basis of the observed and expected frequencies.

   b. Find the tabulated χ^2 value when $\alpha = .05$ and there are five degrees of freedom. (There are $k - 1 = 5$ df associated with this χ^2 statistic.)

   c. On the basis of the χ^2 statistic and the tabulated χ^2 value, is there evidence that the salary distribution is nonnormal?

   d. Find the approximate observed significance level for the test in part **c.**

8.96 Testing normality. Suppose a random variable is hypothesized to be normally distributed with a mean of 0 and a standard deviation of 1. A random sample of 200 observations of the variable yields frequencies in the intervals listed in the table shown below. Do the data provide sufficient evidence to contradict the hypothesis that x is normally distributed with $\mu = 0$ and $\sigma = 1$? Use the technique developed in Exercise 13.63.

8.97 Coupon usage study. A hot topic in marketing research is the exploration of a technology-based self-service (TBSS) encounter, in which various technologies (e.g., ATMs, online banking, self-scanning at retail stores) allow the customer to perform all or part of the service. Marketing professor Dan Ladik of the University of Suffolk investigated whether there were differences in customer characteristics and customer satisfaction between users of discount coupons distributed through the mail (nontechnology users) and users of coupons distributed via the Internet (TBSS users). A questionnaire measured several qualitative variables (defined in the accompanying table) for each of 440 coupon users. The data are saved in the **COUPONS** file.

COUPONS

Variable Name	Levels (Possible Values)
Coupon User Type	Mail, Internet, or Both
Gender	Male or Female
Education	High School, Vo-Tech/College, 4-year College Degree, or Graduate School
Work Status	Full Time, Part Time, Not Working, Retired
Coupon Satisfaction	Satisfied, Unsatisfied, Indifferent

   a. Consider the variable Coupon User Type. Conduct a test (at $\alpha = .05$) to determine whether the proportions of mail-only users, Internet-only users, and users of both media are statistically different. Illustrate the results with a graph.

   b. The researcher wants to know whether there are differences in customer characteristics (i.e., Gender, Education, Work Status, and Coupon Satisfaction) among the three types of coupon users. For each characteristic, conduct a contingency table analysis (at $\alpha = .05$) to determine whether Coupon User Type is related to that characteristic. Illustrate your results with graphs.

Critical Thinking Challenge

8.98 A "rigged" election? *Chance* (Spring 2004) presented data from a recent election held to determine the board of directors of a local community. There were 27 candidates for the board, and each of 5,553 voters was allowed to choose 6 candidates. The claim was that "a fixed vote with fixed percentages [was] assigned to each and every candidate, making it impossible to participate in an honest election." Votes were tallied in six time slots: after 600 total votes were in, after 1200, after 2,444, after 3,444, after 4,444, and, finally, after 5,553 votes. The data on three of the candidates (Smith, Coppin, and Montes) are shown in the accompanying table. A residential organization believes that "there was nothing random about the count and tallies for each time slot, and specific unnatural or rigged percentages were being assigned to each and every candidate." Give your opinion. Is the probability of a candidate receiving votes independent of the time slot, and if so, does this imply a rigged election?

RIGVOTE

Time Slot	1	2	3	4	5	6
Votes for Smith	208	208	451	392	351	410
Votes for Coppin	55	51	109	98	88	104
Votes for Montes	133	117	255	211	186	227
Total Votes	600	600	1,244	1,000	1,000	1,109

Source: Gelman, A. "55,000 residents desperately need your help!" *Chance*, Vol. 17, No. 2, Spring 2004 (Figures 1 and 5).

Table for Exercise 8.96

Interval	$x < -2$	$-2 \leq x < -1$	$-1 \leq x < 0$	$0 \leq x < 1$	$1 \leq x < 2$	$x \geq 2$
Frequency	7	20	61	77	26	9

SUMMARY ACTIVITY: "Guesstimates" of Population Proportions

Many researchers rely on surveys to estimate the proportions of experimental units in populations that possess certain specified characteristics. A political scientist may want to estimate the proportion of an electorate in favor of a certain legislative bill. A social scientist may be interested in the proportions of people in a geographical region who fall into certain socioeconomic classifications. A psychologist might want to compare the proportions of patients who have different psychological disorders.

Choose a specific topic, similar to those described in the previous paragraph, that interests you. Define clearly the population of interest, identify data categories of specific interest, and identify the proportions associated with them. Now "*guesstimate*" the pro-

portions of the population that you think fall into each of the categories. For example, you might guess that all the proportions are equal, or that the first proportion is twice as large as the second but equal to the third, etc.

You are now ready to collect the data by obtaining a random sample from your population of interest. Select a sample size so that all expected cell counts are at least 5 (preferably larger), and then collect the data.

Use the count data you have obtained to test the null hypothesis that the true proportions in the population are equal to your presampling guesstimates. Would failure to reject this null hypothesis imply that your guesstimates are correct?

REFERENCES

Agresti, A. *Categorical Data Analysis.* New York: Wiley, 1990.

Cochran, W. G. "The χ^2 test of goodness of fit." *Annals of Mathematical Statistics*, 1952, 23.

Conover, W. J. *Practical Nonparametric Statistics*, 2d ed. New York: Wiley, 1980.

Fisher, R. A. "The logic of inductive inference (with discussion)." *Journal of the Royal Statistical Society*, Vol. 98, 1935, pp. 39–82.

Hollander, M., and Wolfe, D. A. *Nonparametric Statistical Methods.* New York: Wiley, 1973.

Savage, I. R. "Bibliography of nonparametric statistics and related topics." *Journal of the American Statistical Association*, 1953, 48.

Using Technology

Categorical Data Analyses with MINITAB

MINITAB can analyze the difference between two proportions, and conduct chi-square tests on both one-way and two-way (contingency) tables.

Inference on $(p_1 - p_2)$: To analyze the difference between two proportions $(p_1 - p_2)$, first access the MINITAB worksheet that contains the sample data. Next, click on the "Basic Statistics" button on the MINITAB menu bar, and then and click on "2 Proportions", as shown in Figure 8.M.1. The dialog box shown in Figure 8.M.2 then appears. Select the data option ("Samples in one column", "Samples in different columns", or "Summarized data"), and make the appropriate menu choices. (Figure 8.M.2 shows the menu options when you select "Summarized data".)

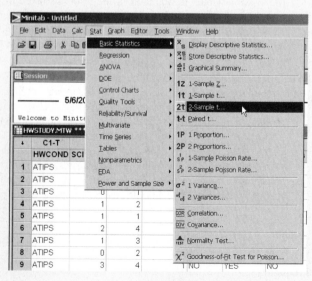

FIGURE 8.M.1

MINITAB menu options for inferences about two means

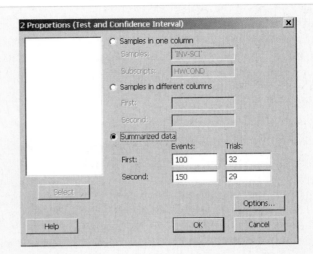

FIGURE 8.M.2
MINITAB two-proportions
dialog box

Next, click the "Options" button and specify the confidence level for a confidence interval, the null-hypothesized value of the difference, and the form of the alternative hypothesis (lower tailed, two tailed, or upper tailed) in the resulting dialog box, as shown in Figure 8.M.3. (If you desire a pooled estimate of p for the test, be sure to check the appropriate box.) Click "OK" to return to the "2 Proportions" dialog box, and then click "OK" again to generate the MINITAB printout.

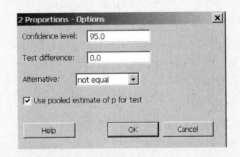

FIGURE 8.M.3
MINITAB two-proportions
options

One-Way Table: To conduct a chi-square test on a one-way table, first access the MINITAB worksheet file that contains the sample data for the qualitative variable of interest. [*Note:* The data file can have actual values (levels) of the variable for each observation or, alternatively, two columns, one listing the levels of the qualitative variable and the other with the observed counts for each level.] Next, click on the "Stat" button on the MINITAB menu bar, and then click on "Tables" and "Chi-Square Goodness-of-Fit Test (One Variable)", as shown in Figure 8.M.4.

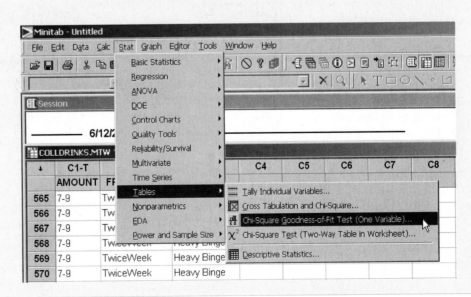

FIGURE 8.M.4
MINITAB menu options for a
one-way chi-square analysis

The dialog box shown in Figure 8.M.5 then appears. If you have one column of data for your qualitative variable, select "Categorical data" and specify the variable name (or column) in the box. If, instead, you have summary information in two columns (see note in previous paragraph), select "Observed counts" and specify the column with the counts and the column with the variable names in the respective boxes. Select "Equal proportions" for a test of equal proportions, or select "Specific proportions" and enter the hypothesized proportion next to each level in the resulting box. Click "OK" to generate the MINITAB printout.

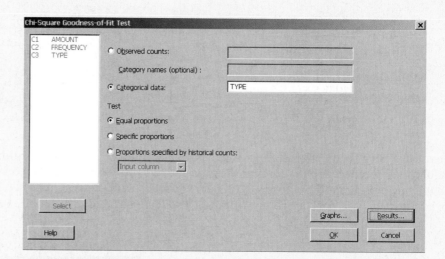

FIGURE 8.M.5

MINITAB one-way chi-square dialog box

Two-way Table: To conduct a chi-square test for a two-way table, first access the MINITAB worksheet file that contains the sample data. The data file should contain two qualitative variables, with category values for each of the *n* observations in the data set. Alternatively, the worksheet can contain the cell counts for each of the categories of the two qualitative variables. Next, click on the "Stat" button on the MINITAB menu bar, and then click on "Tables" and "Cross Tabulation and Chi-Square," as shown in Figure 8.M.6.

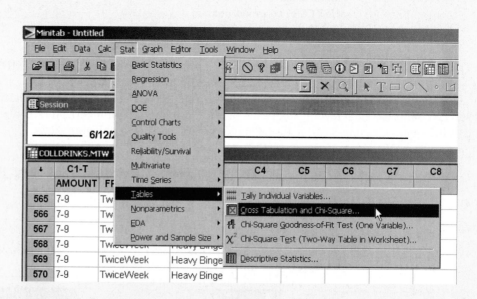

FIGURE 8.M.6

MINITAB menu options for a two-way chi-square analysis

The dialog box shown in Figure 8.M.7 then appears. Specify one qualitative variable in the "For rows" box and the other qualitative variable in the "For columns" box. [*Note:* If your worksheet contains cell counts for the categories, enter the variable with the cell counts in the "Frequencies are in" box.] Next, select the summary statistics (e.g., counts, percentages) you want to display in the contingency table. Then click the "Chi-square" button. The dialog box shown in Figure 8.M.8 then appears. Select "Chi-Square analysis" and "Expected cell counts" and click "OK." When you return to the "Cross Tabulation" menu screen, click "OK" to generate the MINITAB printout.

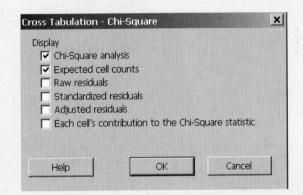

FIGURE 8.M.7
MINITAB cross-tabulation
dialog box

FIGURE 8.M.8
MINITAB cross-tabulation
chi-square options

[*Note:* If your MINITAB worksheet contains only the cell counts for the contingency table in columns, click the "Chi-Square Test (Two-way Table in Worksheet)" menu option (see Figure 8.M.6) and specify the columns in the "Columns containing the table" box. Click "OK" to produce the MINITAB printout.]

Simple Linear Regression

CONTENTS

9.1 Probabilistic Models

9.2 Fitting the Model: The Least Squares Approach

9.3 Model Assumptions

9.4 Assessing the Utility of the Model: Making Inferences about the Slope β_1

9.5 The Coefficients of Correlation and Determination

9.6 Using the Model for Estimation and Prediction

9.7 A Complete Example

9.8 A Nonparametric Test for Correlation (Optional)

STATISTICS IN ACTION
Can Dowsers Really Detect Water?

USING TECHNOLOGY
Simple Linear Regression with MINITAB

WHERE WE'VE BEEN

- Presented methods for estimating and testing population parameters (e.g., the mean, proportion, and variance) for a single sample

- Extended these methods to allow for a comparison of population parameters for multiple samples

WHERE WE'RE GOING

- Introduce the straight-line (*simple linear regression*) model as a means of relating one quantitative variable to another quantitative variable

- Introduce the *correlation coefficient* as a means of relating one quantitative variable to another quantitative variable

- Assess how well the simple linear regression model fits the sample data

- Utilize the simple linear regression model to predict the value of one variable from a specified value of another variable

STATISTICS IN ACTION

Can Dowsers Really Detect Water?

The act of searching for and finding underground supplies of water with the use of nothing more than a divining rod is commonly known as "dowsing." Although widely regarded among scientists as no more than a superstitious relic from medieval times, dowsing remains popular in folklore, and to this day, there are individuals who claim to have this mysterious skill.

Many dowsers in Germany claim that they respond to "earthrays" which emanate from the water source. Earthrays, say the dowsers, are a subtle form of radiation that is potentially hazardous to human health. As a result of these claims, in the mid-1980s the German government conducted a two-year experiment to investigate the possibility that dowsing is a genuine skill. If such a skill could be demonstrated, reasoned government officials, then dangerous levels of radiation in Germany could be detected, avoided, and disposed of.

A group of university physicists in Munich, Germany, was provided a grant of 400,000 marks (about 250,000) to conduct the study. Approximately 500 candidate dowsers were recruited to participate in preliminary tests of their skill. To avoid fraudulent claims, the 43 individuals who seemed to be the most successful in the preliminary tests were selected for the final, carefully controlled, experiment.

The researchers set up a 10-meter-long line on the ground floor of a vacant barn, along which a small wagon could be moved. Attached to the wagon was a short length of pipe, perpendicular to the test line, that was connected by hoses to a pump with running water. The location of the pipe along the line for each trial of the experiment was assigned by a computer-generated random number. On the upper floor of the barn, directly above the experimental line, a 10-meter test line was painted. In each trial, a dowser was admitted to this upper level and required, with his or her rod, stick, or other tool of choice, to ascertain where the pipe with running water on the ground floor was located.

Each dowser participated in at least one test series constituting a sequence of from 5 to 15 trials (typically, 10), with the pipe randomly repositioned after each trial. (Some dowsers undertook only 1 test series, whereas selected others underwent more than 10 test series.) Over the two-year experimental period, the 43 dowsers participated in a total of 843 tests. The experiment was "double blind" in that neither the observer (researcher) on the top floor nor the dowser knew the pipe's location, even after a guess was made. [*Note:* Before the experiment began, a professional magician inspected the entire arrangement for potential deception or cheating by the dowsers.]

For each trial, two variables were recorded: the actual location of the pipe (in decimeters from the beginning of the line) and the dowser's guess (also measured in decimeters). On the basis of an examination of these data, the German physicists concluded in their final report that although most dowsers did not do particularly well in the experiments, "some few dowsers, in particular tests, showed an extraordinarily high rate of success, which can scarcely if at all be explained as due to chance...a real core of dowser-phenomena can be regarded as empirically proven..." (Wagner, Betz, and König, 1990. Final Report 01 KB8602, Federal Ministry for Research and Technology.).

This conclusion was critically assessed by Professor J. T. Enright of the University of California at San Diego. (*Skeptical Inquirer*, Jan./Feb. 1999.) In the Statistics in Action Revisited sections of this chapter, we demonstrate how Enright concluded the exact opposite of the German physicists.

Statistics in Action Revisited

- Estimating a Straight-Line Regression Model for the Dowsing Data (p. 495)

- Assessing How Well the Straight-Line Model Fits the Dowsing Data (p. 511)

- Using the Coefficients of Correlation and Determination to Assess the Dowsing Data (p. 522)

- Using the Straight-Line Model to Predict Pipe Location for the Dowsing Data (p. 530)

In Chapters 5–7, we described methods for making inferences about population means. The mean of a population has been treated as a *constant*, and we have shown how to use sample data to estimate or to test hypotheses about this constant mean. In many applications, the mean of a population is not viewed as a constant, but rather as a variable. For example, the mean sale price of residences in a large city might be treated as a variable that depends on the number of square feet of living space in the residence. The relationship might be

$$\text{Mean sale price} = \$30,000 + \$60 \text{ (Square feet)}$$

This formula implies that the mean sale price of 1,000-square-foot homes is $90,000, the mean sale price of 2,000-square-foot homes is $150,000, and the mean sale price of 3,000-square-foot homes is $210,000.

In this chapter, we discuss situations in which the mean of the population is treated as a variable, dependent on the value of another variable. The dependence of the residential sale price on the number of square feet of living space is one illustration. Other examples include the dependence of the mean reaction time on the amount of a drug in the bloodstream, the dependence of the mean starting salary of a college graduate on the student's GPA, and the dependence of the mean number of years to which a criminal is sentenced on the number of previous convictions.

Here, we present the simplest of all models relating a populating mean to another variable: the *straight-line model*. We show how to use the sample data to estimate the straight-line relationship between the mean value of one variable, *y*, as it relates to a second variable, *x*. The methodology of estimating and using a straight-line relationship is referred to as *simple linear regression analysis*.

9.1 Probabilistic Models

An important consideration in taking a drug is how it may affect one's perception or general awareness. Suppose you want to model the length of time it takes to respond to a stimulus (a measure of awareness) as a function of the percentage of a certain drug in the bloodstream. The first question to be answered is this: "Do you think that an exact relationship exists between these two variables?" That is, do you think that it is possible to state the exact length of time it takes an individual (subject) to respond if the amount of the drug in the bloodstream is known? We think that you will agree with us that this is *not* possible, for several reasons: The reaction time depends on many variables other than the percentage of the drug in the bloodstream—for example, the time of day, the amount of sleep the subject had the night before, the subject's visual acuity, the subject's general reaction time without the drug, and the subject's age. Even if many variables are included in a model, it is still unlikely that we would be able to predict the subject's reaction time *exactly*. There will almost certainly be some variation in response times due strictly to *random phenomena* that cannot be modeled or explained.

If we were to construct a model that hypothesized an exact relationship between variables, it would be called a **deterministic model**. For example, if we believe that *y*, the reaction time (in seconds), will be exactly one-and-one-half times *x*, the amount of drug in the blood, we write

$$y = 1.5x$$

This represents a **deterministic relationship** between the variables *y* and *x*. It implies that *y* can always be determined exactly when the value of *x* is known. *There is no allowance for error in this prediction.*

If, however, we believe that there will be unexplained variation in reaction times—perhaps caused by important, but unincluded, variables or by random phenomena—we discard the deterministic model and use a model that accounts for this **random error**. Our **probabilistic model** will include both a deterministic component and a random-error component. For example, if we hypothesize that the response time *y* is related to the percentage *x* of drug by

$$y = 1.5x + \text{Random error}$$

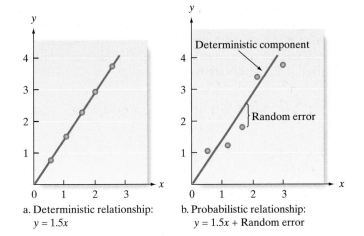

FIGURE 9.1

Possible reaction times y for five different drug percentages x

a. Deterministic relationship:
$y = 1.5x$

b. Probabilistic relationship:
$y = 1.5x$ + Random error

we are hypothesizing a **probabilistic relationship** between y and x. Note that the deterministic component of this probabilistic model is $1.5x$.

Figure 9.1a shows the possible responses for five different values of x, the percentage of drug in the blood, when the model is deterministic. All the responses must fall exactly on the line, because a deterministic model leaves no room for error.

Figure 9.1b shows a possible set of responses for the same values of x when we are using a probabilistic model. Note that the deterministic part of the model (the straight line itself) is the same. Now, however, the inclusion of a random-error component allows the response times to vary from this line. Since we know that the response time does vary randomly for a given value of x, the probabilistic model for y is more realistic than the deterministic model.

General Form of Probabilistic Models

$$y = \text{Deterministic component} + \text{Random error}$$

where y is the variable of interest. We always assume that the mean value of the random error equals 0. This is equivalent to assuming that the mean value of y, $E(y)$, equals the deterministic component of the model; that is,

$$E(y) = \text{Deterministic component}$$

Biography

FRANCIS GALTON (1822–1911)—
The Law of Universal Regression

Francis Galton was the youngest of seven children born to a middle-class English family of Quaker faith. A cousin of Charles Darwin, Galton attended Trinity College (Cambridge, England) to study medicine. Due to the death of his father, Galton was unable to obtain his degree. His competence in both medicine and mathematics, however, led Galton to pursue a career as a scientist. He made major contributions to the fields of genetics, psychology, meteorology, and anthro-

pology. Some consider Galton to be the first social scientist for his applications of the novel statistical concepts of the time—in particular, regression and correlation. While studying natural inheritance in 1886, Galton collected data on heights of parents and adult children. He noticed the tendency for tall (or short) parents to have tall (or short) children, but that the children were not as tall (or short), on average as their parents. Galton called this phenomenon the "law of universal regression," for the average heights of adult children tended to "regress" to the mean of the population. With the help of his friend and disciple, Karl Pearson, Galton applied the straight-line model to the height data, and the term *regression model* was coined.

In this chapter, we present the simplest of probabilistic models—the **straight-line model**—which gets its name from the fact that the deterministic portion of the model graphs as a straight line. Fitting this model to a set of data is an example of **regression analysis**, or **regression modeling**. The elements of the straight-line model are summarized in the following box:

A First-Order (Straight-Line) Probabilistic Model

$$y = \beta_0 + \beta_1 x + \varepsilon$$

where

$y =$ **Dependent** *or* **response variable** (variable to be modeled)

$x =$ **Independent** *or* **predictor variable** (variable used as a predictor of y)*

$\beta_0 + \beta_1 x = E(y) =$ Deterministic component

ε (epsilon) $=$ Random error component

β_0 (beta zero) $=$ **y-intercept of the line** — that is, the point at which the line intersects, or cuts through, the y-axis (see Figure 9.2)

β_1 (beta one) $=$ **Slope of the line** — that is, the amount of increase (or decrease) in the deterministic component of y for every one-unit increase in x.

[*Note:* A *positive* slope implies that *$E(y)$ increases* by the amount β_1. (See Figure 9.2.) A *negative* slope implies that *$E(y)$ decreases* by the amount β_1.]

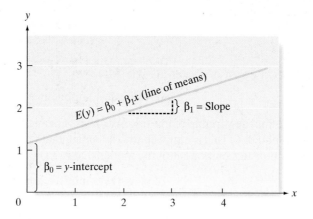

FIGURE 9.2

The straight-line model

In the probabilistic model, the deterministic component is referred to as the **line of means**, because the mean of y, $E(y)$, is equal to the straight-line component of the model. That is,

$$E(y) = \beta_0 + \beta_1 x$$

Note that the Greek symbols β_0 and β_1 respectively represent the y-intercept and slope of the model. They are population parameters that will be known only if we have access to the entire population of (x, y) measurements. Together with a specific value of the independent variable x, they determine the mean value of y, which is just a specific point on the line of means (Figure 9.2).

The values of β_0 and β_1 will be unknown in almost all practical applications of regression analysis. The process of developing a model, estimating the unknown parameters, and using the model can be viewed as the five-step procedure shown in the following box:

---

*The word *independent* should not be interpreted in a probabilistic sense as defined in Chapter 3. The phrase *independent variable* is used in regression analysis to refer to a predictor variable for the response y.

Step 1   Hypothesize the deterministic component of the model that relates the mean $E(y)$ to the independent variable x (Section 9.2).

Step 2   Use the sample data to estimate unknown parameters in the model (Section 9.2).

Step 3   Specify the probability distribution of the random-error term and estimate the standard deviation of this distribution (Section 9.3).

Step 4   Statistically evaluate the usefulness of the model (Sections 9.4 and 9.5).

Step 5   When satisfied that the model is useful, use it for prediction, estimation, and other purposes (Section 9.6).

Exercises 9.1–9.10

Understanding the Principles

9.1 Why do we generally prefer a probabilistic model to a deterministic model? Give examples for which the two types of models might be appropriate.

9.2 What is the difference between a dependent variable and an independent variable in a probabilistic model?

9.3 What is the line of means?

9.4 If a straight-line probabilistic relationship relates the mean $E(y)$ to an independent variable x, does it imply that every value of the variable y will always fall exactly on the line of means? Why or why not?

Learning the Mechanics

9.5 In each case, graph the line that passes through the given points.
   a. $(1, 1)$ and $(5, 5)$
   b. $(0, 3)$ and $(3, 0)$
   c. $(-1, 1)$ and $(4, 2)$
   d. $(-6, -3)$ and $(2, 6)$

9.6 Give the slope and y-intercept for each of the lines graphed in Exercise 9.5.

9.7 The equation (deterministic) for a straight line is

$$y = \beta_0 + \beta_1 x$$

If the line passes through the point $(-2, 4)$, then $x = -2$, $y = 4$ must satisfy the equation; that is,

$$4 = \beta_0 + \beta_1(-2)$$

Similarly, if the line passes through the point $(4, 6)$, then $x = 4$, $y = 6$ must satisfy the equation; that is,

$$6 = \beta_0 + \beta_1(4)$$

Use these two equations to solve for β_0 and β_1; then find the equation of the line that passes through the points $(-2, 4)$ and $(4, 6)$. $\beta_1 = \frac{1}{3}$; $\beta_0 = \frac{14}{3}$

9.8 Refer to Exercise 9.7. Find the equations of the lines that pass through the points listed in Exercise 9.5.

9.9 Plot the following lines:
   a. $y = 4 + x$ **b.** $y = 5 - 2x$
   c. $y = -4 + 3x$ **d.** $y = -2x$
   e. $y = x$ **f.** $y = .50 + 1.5x$

9.10 Give the slope and y-intercept for each of the lines defined in Exercise 9.9.

9.2 Fitting the Model: The Least Squares Approach

After the straight-line model has been hypothesized to relate the mean $E(y)$ to the independent variable x, the next step is to collect data and to estimate the (unknown) population parameters, the y-intercept β_0 and the slope β_1.

To begin with a simple example, suppose an experiment involving five subjects is conducted to determine the relationship between the percentage of a certain drug in the bloodstream and the length of time it takes to react to a stimulus. The results are shown in Table 9.1. (The number of measurements and the measurements themselves are unrealistically simple in order to avoid arithmetic confusion in this introductory example.) This set of data will be used to demonstrate the five-step procedure of regression modeling given in the previous section. In the current section, we hypothesize the deterministic component of the model and estimate its unknown parameters (steps 1 and 2). The model's assumptions and the random-error component (step 3) are the subjects of Section 9.3, whereas Sections 9.4 and 9.5 assess the utility of the model (step 4). Finally, we use the model for prediction and estimation (step 5) in Section 9.6.

STIMULUS

TABLE 9.1 Reaction Time versus Drug Percentage

Subject	Percent x of Drug	Reaction Time y (seconds)
1	1	1
2	2	1
3	3	2
4	4	2
5	5	4

Step 1 *Hypothesize the deterministic component of the probabilistic model.* As stated before, we will consider only straight-line models in this chapter. Thus, the complete model relating mean response time $E(y)$ to drug percentage x is given by

$$E(y) = \beta_0 + \beta_1 x$$

Step 2 *Use sample data to estimate unknown parameters in the model.* This step is the subject of this section—namely, how can we best use the information in the sample of five observations in Table 9.1 to estimate the unknown y-intercept β_0 and slope β_1?

To determine whether a linear relationship between y and x is plausible, it is helpful to plot the sample data in a **scattergram**. Recall (Section 2.9) that a scattergram locates each data point on a graph, as shown in Figure 9.3 for the five data points of Table 9.1. Note that the scattergram suggests a general tendency for y to increase as x increases. If you place a ruler on the scattergram, you will see that a line may be drawn through three of the five points, as shown in Figure 9.4. To obtain the equation of this visually fitted line, note that the line intersects the y-axis at $y = -1$, so the y-intercept is -1. Also, y increases exactly one unit for every one-unit increase in x, indicating that the slope is $+1$. Therefore, the equation is

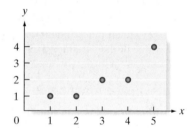

FIGURE 9.3

Scattergram for data in Table 9.1

$$\tilde{y} = -1 + 1(x) = -1 + x$$

where $\tilde{y}$ is used to denote the y that is predicted from the visual model.

One way to decide quantitatively how well a straight line fits a set of data is to note the extent to which the data points deviate from the line. For example, to evaluate the model in Figure 9.4, we calculate the magnitude of the *deviations* (i.e., the differences between the observed and the predicted values of y). These deviations, or **errors of prediction**, are the vertical distances between observed and predicted values (see Figure 9.4). The observed and predicted values of y, their differences, and their squared differences are shown in Table 9.2. Note that the *sum of errors* equals 0 and the *sum of squares of the errors* (SSE), which places a greater emphasis on large deviations of the points from the line, is equal to 2.

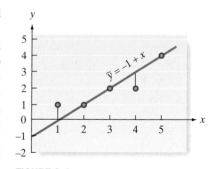

FIGURE 9.4

Visual straight line fitted to the data in Figure 9.3

TABLE 9.2 Comparing Observed and Predicted Values for the Visual Model

x	y	$\tilde{y} = -1 + x$	$(y - \tilde{y})$	$(y - \tilde{y})^2$
1	1	0	$(1 - 0) = 1$	1
2	1	1	$(1 - 1) = 0$	0
3	2	2	$(2 - 2) = 0$	0
4	2	3	$(2 - 3) = -1$	1
5	4	4	$(4 - 4) = 0$	0
			Sum of errors = 0	Sum of squared errors (SSE) = 2

You can see by shifting the ruler around the graph that it is possible to find many lines for which the sum of errors is equal to 0, but it can be shown that there is one (and only one) line for which the SSE is a *minimum*. This line is called the **least squares line**, the **regression line**, or the **least squares prediction equation**. The methodology used to obtain that line is called the **method of least squares**.

Now Work Exercise 9.16a–d

To find the least squares prediction equation for a set of data, assume that we have a sample of n data points consisting of pairs of values of x and y, say, $(x_1, y_1), (x_2, y_2), \ldots, (x_n, y_n)$. For example, the $n = 5$ data points shown in Table 9.2 are $(1, 1), (2, 1), (3, 2), (4, 2)$, and $(5, 4)$. The fitted line, which we will calculate on the basis of the five data points, is written as

$$\hat{y} = \hat{\beta}_0 + \hat{\beta}_1 x$$

The "hats" indicate that the symbols below them are estimates: $\hat{y}$ (y-hat) is an estimator of the mean value of y, $E(y)$, and is a predictor of some future value of y; and $\hat{\beta}_0$ and $\hat{\beta}_1$ are estimators of β_0 and β_1, respectively.

For a given data point—say, the point (x_i, y_i),—the observed value of y is y_i and the predicted value of y would be obtained by substituting x_i into the prediction equation:

$$\hat{y}_i = \hat{\beta}_0 + \hat{\beta}_1 x_i$$

The deviation of the ith value of y from its predicted value is

$$(y_i - \hat{y}_i) = [y_i - (\hat{\beta}_0 + \hat{\beta}_1 x_i)]$$

Then the sum of the squares of the deviations of the y-values about their predicted values for all the n data points is

$$\text{SSE} = \sum [y_i - (\hat{\beta}_0 + \hat{\beta}_1 x_i)]^2$$

The quantities $\hat{\beta}_0$ and $\hat{\beta}_1$ that make the SSE a minimum are called the **least squares estimates** of the population parameters β_0 and β_1, and the prediction equation $\hat{y} = \hat{\beta}_0 + \hat{\beta}_1 x$ is called the *least squares line*.

Definition 9.1

The **least squares line** $\hat{y} = \hat{\beta}_0 + \hat{\beta}_1 x$ is the line that has the following two properties:

1. The sum of the errors (SE) equals 0.
2. The sum of squared errors (SSE) is smaller than that for any other straight-line model

The values of $\hat{\beta}_0$ and $\hat{\beta}_1$ that minimize the SSE are given by the formulas in the following box (proof omitted):*

Formulas for the Least Squares Estimates

Slope: $\hat{\beta}_1 = \dfrac{SS_{xy}}{SS_{xx}}$

y-intercept: $\hat{\beta}_0 = \bar{y} - \hat{\beta}_1 \bar{x}$

where

$$SS_{xy} = \sum(x_i - \bar{x})(y_i - \bar{y}) = \sum x_i y_i - \frac{\left(\sum x_i\right)\left(\sum y_i\right)}{n}$$

$$SS_{xx} = \sum(x_i - \bar{x})^2 = \sum x_i^2 - \frac{\left(\sum x_i\right)^2}{n}$$

n = Sample size

EXAMPLE 9.1

APPLYING THE
METHOD OF
LEAST SQUARES—
Drug Reaction Data

Problem Refer to the reaction data presented in Table 9.1. Consider the straight-line model $E(y) = \beta_0 + \beta_1 x$, where y = reaction time (in seconds) and x = percent of drug received.

a. Use the method of least squares to estimate the values of β_0 and β_1.

b. Predict the reaction time when $x = 2\%$.

c. Find SSE for the analysis.

d. Give practical interpretations of $\hat{\beta}_0$ and $\hat{\beta}_1$.

Solution

a. Preliminary computations for finding the least squares line for the drug reaction example are presented in Table 9.3. We can now calculate

$$SS_{xy} = \sum x_i y_i - \frac{\left(\sum x_i\right)\left(\sum y_i\right)}{5} = 37 - \frac{(15)(10)}{5} = 37 - 30 = 7$$

$$SS_{xx} = \sum x_i^2 - \frac{\left(\sum x_i\right)^2}{5} = 55 - \frac{(15)^2}{5} = 55 - 45 = 10$$

TABLE 9.3 Preliminary Computations for the Drug Reaction Example

	x_i	y_i	x_i^2	$x_i y_i$
	1	1	1	1
	2	1	4	2
	3	2	9	6
	4	2	16	8
	5	4	25	20
Totals	$\sum x_i = 15$	$\sum y_i = 10$	$\sum x_i^2 = 55$	$\sum x_i y_i = 37$

*Students who are familiar with calculus should note that the values of β_0 and β_1 that minimize
SSE $= \Sigma(y_i - \hat{y}_i)^2$ are obtained by setting the two partial derivatives $\partial SSE/\partial\beta_0$ and $\partial SSE/\partial\beta_1$ equal
to 0. The solutions of these two equations yield the formulas shown in the box. Furthermore, we denote the
sample solutions of the equations by $\hat{\beta}_0$ and $\hat{\beta}_1$, where the "hat" denotes that these are sample estimates of
the true population intercept β_0 and slope β_1.

Then the slope of the least squares line is

$$\hat{\beta}_1 = \frac{SS_{xy}}{SS_{xx}} = \frac{7}{10} = .7$$

and the y-intercept is

$$\hat{\beta}_0 = \bar{y} - \hat{\beta}_1 \bar{x} = \frac{\sum y_i}{5} - \hat{\beta}_1 \frac{\sum x_i}{5}$$

$$= \frac{10}{5} - (.7)\left(\frac{15}{5}\right) = 2 - (.7)(3) = 2 - 2.1 = -.1$$

The least squares line is thus

$$\hat{y} = \hat{\beta}_0 + \hat{\beta}_1 x = -.1 + .7x$$

The graph of this line is shown in Figure 9.5.

b. The predicted value of y for a given value of x can be obtained by substituting into the formula for the least squares line. Thus, when $x = 2$, we predict y to be

$$\hat{y} = -.1 + .7x = -.1 + .7(2) = 1.3$$

We show how to find a prediction interval for y in Section 9.6.

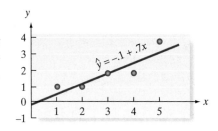

FIGURE 9.5
The line $\hat{y} = -.1 + .7x$ fitted to the data

c. The observed and predicted values of y, the deviations of the y values about their predicted values, and the squares of these deviations are shown in Table 9.4. Note that the sum of the squares of the deviations, SSE, is 1.10 and (as we would expect) this is less than the SSE = 2.0 obtained in Table 9.2 for the visually fitted line.

d. The estimated y-intercept, $\hat{\beta}_0 = -.1$, appears to imply that the estimated mean reaction time is equal to $-.1$ second when the percent x of drug is equal to 0%. Since negative reaction times are not possible, this seems to make the model nonsensical. However, *the model parameters should be interpreted only within the sampled range of the independent variable*—in this case, for amounts of drug in the bloodstream between 1% and 5%. Thus, the y-intercept—which is, by definition, at $x = 0$ (0% drug)—is not within the range of the sampled values of x and is not subject to meaningful interpretation.

The slope of the least squares line, $\hat{\beta} = .7$, implies that for every unit increase in x, the mean value of y is estimated to increase by .7 unit. In terms of this example, for every 1% increase in the amount of drug in the bloodstream, the mean reaction time is estimated to increase by .7 second *over the sampled range of drug amounts from 1% to 5%*. Thus, the model does not imply that increasing the drug amount from 5% to 10% will result in an increase in mean reaction time of 3.5 seconds, because the range of x in the sample does not extend to 10% ($x = 10$). In fact, 10% might be such a high concentration that the drug would kill the subject! Be careful to interpret the estimated parameters only within the sampled range of x.

TABLE 9.4 Comparing Observed and Predicted Values for the Least Squares Prediction Equation

x	y	$\hat{y} = -.1 + .7x$	$(y - \hat{y})$	$(y - \hat{y})^2$
1	1	.6	$(1 - .6) = .4$	.16
2	1	1.3	$(1 - 1.3) = -.3$	.09
3	2	2.0	$(2 - 2.0) = 0$	.00
4	2	2.7	$(2 - 2.7) = -.7$	.49
5	4	3.4	$(4 - 3.4) = .6$	.36
			Sum of errors = 0	SSE = 1.10

Look Back The calculations required to obtain $\hat{\beta}_0$, $\hat{\beta}_1$, and SSE in simple linear regression, although straightforward, can become rather tedious. Even with the use of a pocket calculator, the process is laborious and susceptible to error, especially when the sample size is large. Fortunately, the use of statistical computer software can significantly reduce the labor involved in regression calculations. The SAS, SPSS, and MINITAB outputs for the simple linear regression of the data in Table 9.1 are displayed in Figure 9.6a–c. The val-

Dependent Variable: TIME_Y

Number of Observations Read	5
Number of Observations Used	5

Analysis of Variance

Source	DF	Sum of Squares	Mean Square	F Value	Pr > F
Model	1	4.90000	4.90000	13.36	0.0354
Error	3	1.10000	0.36667		
Corrected Total	4	6.00000			

Root MSE	0.60553	R-Square	0.8167
Dependent Mean	2.00000	Adj R-Sq	0.7556
Coeff Var	30.27650		

Parameter Estimates

Variable	DF	Parameter Estimate	Standard Error	t Value	Pr > |t|
Intercept	1	-0.10000	0.63509	-0.16	0.8849
DRUG_X	1	0.70000	0.19149	3.66	0.0354

FIGURE 9.6a

SAS printout for the time–drug regression

Model Summary

Model	R	R Square	Adjusted R Square	Std. Error of the Estimate
1	.904[a]	.817	.756	.606

a. Predictors: (Constant), DRUG_X

ANOVA[b]

Model		Sum of Squares	df	Mean Square	F	Sig.
1	Regression	4.900	1	4.900	13.364	.035[a]
	Residual	1.100	3	.367		
	Total	6.000	4			

a. Predictors: (Constant), DRUG_X

b. Dependent Variable: TIME_Y

Coefficients[a]

Model		Unstandardized Coefficients		Standardized Coefficients		
		B	Std. Error	Beta	t	Sig.
1	(Constant)	-.100	.635		-.157	.885
	DRUG_X	.700	.191	.904	3.656	.035

a. Dependent Variable: TIME_Y

FIGURE 9.6b

SPSS printout for the time–drug regression

Regression Analysis: TIME_Y versus DRUG_X

```
The regression equation is
TIME_Y = - 0.100 + 0.700 DRUG_X


Predictor       Coef   SE Coef       T      P
Constant     -0.1000    0.6351   -0.16  0.885
DRUG_X        0.7000    0.1915    3.66  0.035


S = 0.605530   R-Sq = 81.7%   R-Sq(adj) = 75.6%


Analysis of Variance

Source        DF      SS      MS      F      P
Regression     1  4.9000  4.9000  13.36  0.035
Residual Error 3  1.1000  0.3667
Total          4  6.0000
```

FIGURE 9.6c

MINITAB printout for the time–drug regression

ues of $\hat{\beta}_0$ and $\hat{\beta}_1$ are highlighted on the printouts. These values, $\hat{\beta}_0 = -.1$ and $\hat{\beta}_1 = .7$, agree exactly with our hand-calculated values. The value of SSE $= 1.10$ is also highlighted on the printouts.

> **Now Work Exercise 9.23**

Simple Linear Regression
Using the TI-84/TI-83 Graphing Calculator

I. Finding the least squares regression equation

Step 1   *Enter the data.*
Press **STAT** and select **1:Edit**.
(*Note*: If a list already contains data, clear the old data.)
Use the up arrow to highlight the list name, "**L1**" or "**L2**".
Press **CLEAR ENTER**.
Enter your *x*-data into **L1** and your *y*-data into **L2**.

Step 2   *Find the equation.*

Press **STAT** and highlight **CALC**.
Press **4** for **LinReg(ax + b)**.
Press **ENTER**.
The screen will show the values for *a* and *b* in the equation $y = ax + b$.

Example   The following figures show a table of data entered on the TI-84/TI-83 and the regression equation obtained by following the preceding steps.

```
L1      L2      L3     2        LinReg
2       67      ------            y=ax+b
1.9     68                        a=2.495967742
2.3     70                        b=63.3391129
3.9     74
6.4     78
7.4     81
6.8     82

L2(7) =82
```

II. Graphing the least squares line with the scatterplot

Step 1   *Enter the data as shown in part I.*

Step 2   *Set up the data plot.*

Press **Y =** and **CLEAR** all functions from the Y register.
Press **2nd Y =** for **STAT PLOT**.

(continued)

Press **1** for **Plot 1**.
Set the cursor so that **ON** is flashing, and press **ENTER**.
For **Type**, use the **ARROW** and **ENTER** keys to highlight and select the scatterplot (first icon in the first row).
For **Xlist**, choose the column containing the *x*-data.
For **Ylist**, choose the column containing the *y*-data.

Step 3 *Find the regression equation and store the equation in Y1.*

Press **STAT** and highlight **CALC**.
Press **4** for **LinReg(ax + b)** (*Note:* Don't press ENTER here, because you want to store the regression equation in Y1.)
Press **VARS**.
Use the right arrow to highlight **Y-VARS**.
Press **ENTER** to select **1:Function**.
Press **ENTER** to select **1:Y1**.
Press **ENTER**.

Step 4 *View the scatterplot and regression line.*

Press **ZOOM**, and then press **9** to select **9:ZoomStat**.
You should see the data graphed along with the regression line.

Example The figure shows a graph of the scatterplot and least squares line obtained by following the preceding steps.

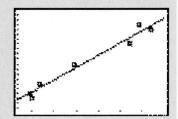

ACTIVITY 9.1: *Keep the Change:* **Least Squares Models**

In this activity, you will once again use data collected in Activity 1.1, *Keep the Change: Collecting Data* (p. 12). For each student in your class, collect the sum of the data set Purchase Total, the sum of the data set. *Amounts Transferred*, and the number of purchases (the number of data items in the set *Purchase Totals*).

1. For each student in your class, form the ordered pair

(*Sum of Purchase Totals, Sum of Amounts Transferred*)

Use these ordered pairs to create a scattergram. Then use the ordered pairs to find the values of $\hat{\beta}_0$ and $\hat{\beta}_1$ in the least squares line for the data.

2. Suppose that three customers have made only one debit-card purchase each, as shown in the following table:

Customer	Purchases	Actual Amount Transferred	Estimated Amount Transferred	Difference
A	$ 3.49			
B	$ 30.49			
C	$300.49			

Complete the table by first finding the actual amount that will be transferred for each customer. Then use the least squares line $y = \hat{\beta}_0 + \hat{\beta}_1 x$ from Exercise 1 to estimate the amount transferred. Finally, calculate the difference between the actual and estimated amounts for each customer.

3. On the basis of your results in Exercise 2, comment on the utility of using your model from Exercise 1 to estimate the transferred amounts. Do you believe that the least squares model is an appropriate model in this situation? Explain.

4. For each student in your class, form the ordered pair

(*Number of Purchases, Sum of Amounts Transferred*)

Use these ordered pairs to create a scattergram. Then use the ordered pairs to find the values of $\hat{\beta}_0$ and $\hat{\beta}_1$ in the least squares line for the data. Create your own hypothetical data as in Exercise 2 to test your model. Do you believe that this model is more or less useful than the model of Exercise 1? Explain.

Even when the interpretations of the estimated parameters in a simple linear regression are meaningful, we need to remember that they are only estimates based on the sample. As such, their values will typically change in repeated sampling. How much confidence do we have that the estimated slope $\hat{\beta}_1$ accurately approximates the true slope

β_1? Determining this requires statistical inference, in the form of confidence intervals and tests of hypotheses, which we address in Section 9.5.

To summarize, we defined the best-fitting straight line to be the line that minimizes the sum of squared errors around it, and we called it the least squares line. We should interpret the least squares line only within the sampled range of the independent variable. In subsequent sections, we show how to make statistical inferences about the model.

STATISTICS IN ACTION REVISITED

Estimating a Straight-Line Regression Model for the Dowsing Data

After conducting a series of experiments in a Munich barn, a group of German physicists concluded that dowsing (i.e., the ability to find underground water with a divining rod) "can be regarded as empirically proven." This observation was based on the data col- lected on 3 (of the participating 500) dowsers who had particularly impressive results. All of these "best" dowsers (numbered 99, 18, and 108) performed the experiment multiple times, and the best test series (se- quence of trials) for each of them was identified. These data, saved in the **DOWSING** file, are listed in Table SIA9.1.

DOWSING

TABLE SIA9.1 Dowsing Trial Results: Best Series for the Three Best Dowsers

Trial	Dowser Number	Pipe Location	Dowser's Guess
1	99	4	4
2	99	5	87
3	99	30	95
4	99	35	74
5	99	36	78
6	99	58	65
7	99	40	39
8	99	70	75
9	99	74	32
10	99	98	100
11	18	7	10
12	18	38	40
13	18	40	30
14	18	49	47
15	18	75	9
16	18	82	95
17	108	5	52
18	108	18	16
19	108	33	37
20	108	45	40
21	108	38	66
22	108	50	58
23	108	52	74
24	108	63	65
25	108	72	60
26	108	95	49

Source: Enright, J. T. "Testing dowsing: The failure of the Munich experiments," *Skeptical Inquirer*, Jan./Feb. 1999, p. 45 (Figure 6a).

Recall (p. 483) that for various hidden pipe loca- tions, each dowser guessed where the pipe with run- ning water was located. Let x = dowser's guess (in meters) and y = pipe location (in meters) for each trial. One way to determine whether the "best" dowsers are effective is to fit the straight-line model $E(y) = \beta_0 + \beta_1 x$ to the data in Table SIA9.1.

A MINITAB scatterplot of the data is shown in Figure SIA9.1. The least squares line, obtained from the MINITAB regression printout shown in Figure SIA9.2, is also displayed on the scatterplot. Although the least squares line has a slight upward trend, the variation of the data points around the line is large. It does not appear that a dowser's guess (x) will be a very

(continued)

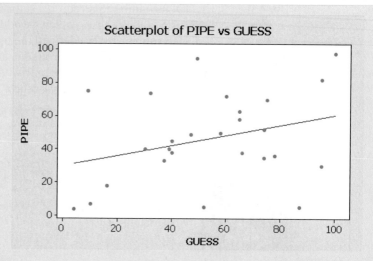

FIGURE SIA9.1

MINITAB scatterplot of dowsing data

good predictor of the actual pipe location (y). In fact, the estimated slope (obtained from Figure SIA9.2) is $\hat{\beta}_1 = .31$. Thus, for every 1-meter increase in a dowser's guess, we estimate that the actual pipe location will increase only .31 meter. In the Statistics in Action Revisited sections that follow, we will provide a measure of reliability for this inference and investigate the phenomenon of dowsing further.

Regression Analysis: PIPE versus GUESS

```
The regression equation is
PIPE = 30.1 + 0.308 GUESS


Predictor      Coef   SE Coef      T       P
Constant      30.07     11.41   2.63   0.015
GUESS        0.3079    0.1900   1.62   0.118


S = 26.0298    R-Sq = 9.9%    R-Sq(adj) = 6.1%


Analysis of Variance

Source          DF        SS       MS      F       P
Regression       1    1778.9   1778.9   2.63   0.118
Residual Error  24   16261.2    677.6
Total           25   18040.2
```

FIGURE SIA9.2

MINITAB simple linear regression for dowsing data

Exercises 9.11–9.31

Understanding the Principles

9.11 In regression, what is an error of prediction?

9.12 Give two properties of the line estimated with the method of least squares.

9.13 *True or False*. The estimates of β_0 and β_1 should be interpreted only within the sampled range of the independent variable, x.

Learning the Mechanics

9.14 The accompanying table is similar to Table 9.3. It is used to make the preliminary computations for finding the least squares line for the given pairs of x and y values.
  a. Complete the table. **b.** Find SS_{xy}.

c. Find SS_{xx}. **d.** Find $\hat{\beta}_1$.
e. Find $\bar{x}$ and $\bar{y}$. **f.** Find $\hat{\beta}_0$.
g. Find the least squares line.

x_i	y_i	x_i^2	$x_i y_i$
7	2	—	—
4	4	—	—
6	2	—	—
2	5	—	—
1	7	—	—
1	6	—	—
3	5	—	—
Totals $\sum x_i =$	$\sum y_i =$	$\sum x_i^2 =$	$\sum x_i y_i =$

9.15 Refer to Exercise 9.14. After the least squares line has been obtained, the following table (which is similar to Table 9.4) can be used (1) to compare the observed and the predicted values of y and (2) to compute SSE.

x	y	$\hat{y}$	$(y - \hat{y})$	$(y - \hat{y})^2$
7	2	—	—	—
4	4	—	—	—
6	2	—	—	—
2	5	—	—	—
1	7	—	—	—
1	6	—	—	—
3	5	—	—	—
			$\sum(y - \hat{y}) =$	SSE $= \sum(y - \hat{y})^2 =$

a. Complete the table.

b. Plot the least squares line on a scattergram of the data. Plot the following line on the same graph:
$$\hat{y} = 14 - 2.5x$$

c. Show that SSE is larger for the line in part **b** than it is for the least squares line.

9.16 Construct a scattergram of the data in the following table.
NW

x	.5	1	1.5
y	2	1	3

a. Plot the following two lines on your scattergram:
$$y = 3 - x \quad \text{and} \quad y = 1 + x$$

b. Which of these lines would you choose to characterize the relationship between x and y? Explain.

c. Show that the sum of errors for both of these lines equals 0.

d. Which of these lines has the smaller SSE?

e. Find the least squares line for the data, and compare it with the two lines described in part **a**.

9.17 Consider the following pairs of measurements:

x	5	3	−1	2	7	6	4
y	4	3	0	1	8	5	3

a. Construct a scattergram of these data.

b. What does the scattergram suggest about the relationship between x and y?

c. Given that $SS_{xx} = 43.4286$, $SS_{xy} = 39.8571$, $\bar{y} = 3.4286$, and $\bar{x} = 3.7143$, calculate the least squares estimates of β_0 and β_1.

d. Plot the least squares line on your scattergram. Does the line appear to fit the data well? Explain.

e. Interpret the y-intercept and slope of the least squares line. Over what range of x are these interpretations meaningful?

APPLET Applet Exercise 9.1

Use the applet entitled *Regression by Eye* to explore the relationship between the pattern of data in a scattergram and the corresponding least squares model.

a. Run the applet several times. For each time, attempt to move the green line into a position that appears to minimize the vertical distances of the points from the line. Then click *Show regression line* to see the actual regression line. How close is your line to the actual line? Click *New data* to reset the applet.

b. Click the trash can to clear the graph. Use the mouse to place five points on the scattergram that are approximately in a straight line. Then move the green line to approximate the regression line. Click *Show regression line* to see the actual regression line. How close were you this time?

c. Continue to clear the graph, and plot sets of five points with different patterns among the points. Use the green line to approximate the regression line. How close do you come to the actual regression line each time?

d. On the basis of your experiences with the applet, explain why we need to use more reliable methods of finding the regression line than just "eyeing" it.

Applying the Concepts—Basic

9.18 Quantitative models of music. Writing in *Chance* (Fall 2004), University of Konstanz (Germany) statistics professor Jan Beran demonstrated that certain aspects of music can be described by quantitative models. For example, the information content of a musical composition (called *entropy*) can be quantified by determining how many times a certain pitch occurs. In a sample of 147 famous compositions ranging from the 13th to the 20th century, Beran computed the Z12-note entropy (y) and plotted it against the year of birth (x) of the composer. The graph is reproduced here.

a. Do you observe a trend, especially since the year 1400?

b. The least squares line for the data since 1400 is shown on the graph. Is the slope of the line positive or negative? What does this imply?

c. Explain why the line shown is not the true line of means.

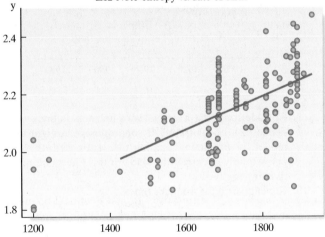

Z12-Note-entropy vs. date of birth

9.19 Wind turbine blade stress. Mechanical engineers at the University of Newcastle (Australia) investigated the use of timber in high-efficiency small wind turbine blades (*Wind Engineering*, Jan. 2004). The strengths of two types of timber—radiata pine and hoop pine—were compared.

Twenty specimens (called "coupons") of each timber blade were fatigue tested by measuring the stress (in MPa) on the blade after various numbers of blade cycles. A simple linear regression analysis of the data, one conducted for each type of timber, yielded the following results (where y = stress and x = natural logarithm of number of cycles):

$$\text{Radiata Pine: } \hat{y} = 97.37 - 2.50x$$
$$\text{Hoop Pine: } \hat{y} = 122.03 - 2.36x$$

a. Interpret the estimated slope of each line.
b. Interpret the estimated y-intercept of each line.
c. On the basis of these results, which type of timber blade appears to be stronger and more fatigue resistant? Explain.

9.20 Winning marathon times. In *Chance* (Winter 2000), statistician Howard Wainer and two students compared men's and women's winning times in the Boston Marathon. One of the graphs used to illustrate gender differences is reproduced here. The scattergram plots the winning times (in minutes) against the year in which the race was run. Men's times are represented by purple dots and women's times by red dots.

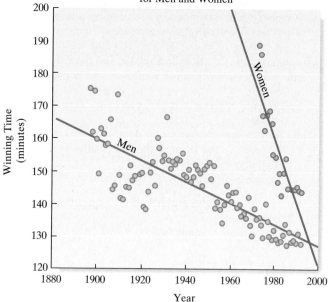

Winning Times in the Boston Marathon for Men and Women

a. Consider only the winning times for men. Is there evidence of a linear trend? If so, propose a straight-line model for predicting winning time (y) based on year (x). Would you expect the slope of this line to be positive or negative?
b. Repeat part **b** for women's times.

c. Which slope, the men's or the women's, will be greater in absolute value?
d. Would you recommend using the straight-line models to predict the winning time in the 2020 Boston Marathon? Why or why not?

9.21 College protests of labor exploitation. Refer to the *Journal of World-Systems Research* (Winter 2004) study of student "sit-ins" for a "sweat-free campus" at universities, presented in Exercise 2.145 (p. 93). Recall that the **SITIN** file contains data on the duration (in days) of each sit-in, as well as the number of student arrests. The data for 5 sit-ins in which there was at least one arrest are shown in the accompanying table. Let y = number of arrests and x = duration.

a. Give the equation of a straight-line model relating y to x.

b. SPSS was used to fit the model to the data for the 5 sit-ins. The SPSS printout is shown at the bottom of the page. Give the least squares prediction equation.

c. Interpret the estimates of β_0 and β_1 in the context of the problem.

SITIN2 (Selected observations)

Sit-In	University	Duration (days)	Number of Arrests
12	Wisconsin	4	54
14	SUNY Albany	1	11
15	Oregon	3	14
17	Iowa	4	16
18	Kentucky	1	12

Source: Ross, R. J. S. "From antisweatshop to global justice to antiwar: How the new new left is the same and different from the old new left," *Journal of Word-Systems Research*, Vol. X, No. 1, Winter 2004 (Tables 1 and 3).

GOBIANTS

9.22 Mongolian desert ants. Refer to the *Journal of Biogeography* (Dec. 2003) study of ants in Mongolia, presented in Exercise 2.147 (p. 93). Data on annual rainfall, maximum daily temperature, and number of ant species recorded at each of 11 study sites are listed in the next table.
a. Consider a straight-line model relating annual rainfall (y) and maximum daily temperature (x). A MINITAB printout of the simple linear regression is shown (p. 575). Give the least squares prediction equation.
b. Construct a scatterplot for the analysis you performed in part **a**. Include the least square line on the plot. Does the line appear to be a good predictor of annual rainfall?
c. Now consider a straight-line model relating number of ant species (y) to annual rainfall (x). On the basis of the MINITAB printout (p. 499), repeat parts **a** and **b**.

Coefficients[a]

Model		Unstandardized Coefficients		Standardized Coefficients	t	Sig.
		B	Std. Error	Beta		
1	(Constant)	2.522	16.352		.154	.887
	DURATION	7.261	5.576	.601	1.302	.284

SPSS output for Exercise 9.21   a. Dependent Variable: ARRESTS

Regression Analysis: Rain versus Temp

```
The regression equation is
Rain = 295 - 16.4 Temp

Predictor       Coef    SE Coef      T       P
Constant      295.25      22.41   13.18   0.000
Temp          -16.364      2.346   -6.97   0.000

S = 17.5111    R-Sq = 84.4%    R-Sq(adj) = 82.7%

Analysis of Variance
Source             DF       SS       MS       F       P
Regression          1    14915    14915   48.64   0.000
Residual Error      9     2760      307
Total              10    17675
```

Regression Analysis: AntSpecies versus Rain

```
The regression equation is
AntSpecies = 10.5 + 0.016 Rain

Predictor       Coef    SE Coef      T       P
Constant       10.52      22.03    0.48   0.644
Rain          0.0160      0.1480   0.11   0.916

S = 19.6726    R-Sq = 0.1%    R-Sq(adj) = 0.0%

Analysis of Variance
Source             DF       SS       MS       F       P
Regression          1      4.5      4.5    0.01   0.916
Residual Error      9   3483.1    387.0
Total              10   3487.6
```

MINITAB output for
Exercise 9.22

GOBIANTS

Site	Region	Annual Rainfall (mm)	Max. Daily Temp. (°C)	Number of Ant Species
1	Dry Steppe	196	5.7	3
2	Dry Steppe	196	5.7	3
3	Dry Steppe	179	7.0	52
4	Dry Steppe	197	8.0	7
5	Dry Steppe	149	8.5	5
6	Gobi Desert	112	10.7	49
7	Gobi Desert	125	11.4	5
8	Gobi Desert	99	10.9	4
9	Gobi Desert	125	11.4	4
10	Gobi Desert	84	11.4	5
11	Gobi Desert	115	11.4	4

Source: Pfeiffer, M., et al. "Community organization and species richness of ants in Mongolia along an ecological gradient from steppe to Gobi desert." *Journal of Biogeography*, Vol. 30, No. 12, Dec. 2003 (Tables 1 and 2).

9.23 Redshifts of Quasi Stellar Objects. Astronomers call a shift in the spectrum of galaxies a "redshift." A correlation between redshift level and apparent magnitude (i.e., brightness on a logarithmic scale) of a quasi-stellar object was discovered and reported in the *Journal of Astrophysics & Astronomy* (Mar./Jun. 2003). Physicist D. Basu (Carleton University, Ottawa) applied simple linear regression to data collected for a sample of over 6,000 quasi-stellar objects with confirmed redshifts. The analysis yielded the following results for a specific range of magnitudes: $\hat{y} = 18.13 + 6.21x$, where y = magnitude and x = redshift level.

a. Graph the least squares line. Is the slope of the line positive or negative?

b. Interpret the estimate of the y-intercept in the words of the problem.

c. Interpret the estimate of the slope in the words of the problem.

Applying the Concepts—Intermediate

9.24 Extending the life of an aluminum smelter pot. An investigation of the properties of bricks used to line aluminum smelter pots was published in *The American Ceramic Society Bulletin* (Feb. 2005). Six different commercial bricks were evaluated. The life span of a smelter pot depends on the porosity of the brick lining (the less porosity, the longer is the life); consequently, the researchers measured the apparent porosity of each brick specimen, as well as the mean pore diameter of each brick. The data are given in the next table:

a. Find the least squares line relating porosity (y) to mean pore diameter (x).

b. Interpret the y-intercept of the line.

c. Interpret the slope of the line.

d. Predict the apparent percentage of porosity for a brick with a mean pore diameter of 10 micrometers.

SMELTPOT

Brick	Apparent Porosity (%)	Mean Pore Diameter (micrometers)
A	18.8	12.0
B	18.3	9.7
C	16.3	7.3
D	6.9	5.3
E	17.1	10.9
F	20.4	16.8

Source: Bonadia, P., et al. "Aluminosilicate refractories for aluminum cell linings," *The American Ceramic Society Bulletin*, Vol. 84, No. 2, Feb. 2005 (Table II).

9.25 Ranking driving performance of professional golfers. Refer to *The Sport Journal* (Winter 2007) study of a new method for ranking the total driving performance of golfers on the Professional Golf Association (PGA) tour, presented in Exercise 2.60 (p. 59). Recall that the method computes a driving performance index based on a golfer's average driving distance (yards) and driving accuracy (percent of drives that land in the fairway). Data for the top 40 PGA golfers (as ranked by the new method) are saved in the **PGADRIVER** file. (The first five and last five observations are listed in the accompanying table.)
a. Write the equation of a straight-line model relating driving accuracy (y) to diving distance (x).
b. Use simple linear regression to fit the model you found in part **a** to the data. Give the least squares prediction equation.
c. Interpret the estimated y-intercept of the line.
d. Interpret the estimated slope of the line.
e. In Exercise 2.149 (p. 93), you were informed that a professional golfer practicing a new swing to increase his average driving distance is concerned that his driving accuracy will be lower. Which of the two estimates, y-intercept or slope, will help you determine whether the golfer's concern is a valid one? Explain.

PGADRIVER (Selected observations shown)

Rank	Player	Driving Distance (yards)	Driving Accuracy (%)	Driving Performance Index
1	Woods	316.1	54.6	3.58
2	Perry	304.7	63.4	3.48
3	Gutschewski	310.5	57.9	3.27
4	Wetterich	311.7	56.6	3.18
5	Hearn	295.2	68.5	2.82
⋮	⋮	⋮	⋮	⋮
36	Senden	291	66	1.31
37	Mickelson	300	58.7	1.30
38	Watney	298.9	59.4	1.26
39	Trahan	295.8	61.8	1.23
40	Pappas	309.4	50.6	1.17

Source: Wiseman, F. et al. "A New Method for Ranking Total Driving Performance on the PGA Tour," *The Sport Journal*, Vol. 10, No. 1, Winter 2007 (Table 2).

9.26 FCAT scores and poverty. In the state of Florida, elementary school performance is based on the average score obtained by students on a standardized exam, called the Florida Comprehensive Assessment Test (FCAT). An analysis of the link between FCAT scores and sociodemo-graphic factors was published in the *Journal of Educational and Behavioral Statistics* (Spring 2004). Data on average math and reading FCAT scores of third graders, as well as the percentage of students below the poverty level, for a sample of 22 Florida elementary schools are listed in the accompanying table.
a. Propose a straight-line model relating math score (y) to percentage (x) of students below the poverty level.
b. Use the method of least squares to fit the model to the data in the FCAT file.
c. Graph the least squares line on a scattergram of the data. Is there visual evidence of a relationship between the two variables? Is the relationship positive or negative?
d. Interpret the estimates of the y-intercept and slope in the words of the problem.
e. Now consider a model relating reading score (y) to percentage (x) of students below the poverty level. Repeat parts **a–d** for this model.

FCAT

Elementary School	FCAT—Math	FCAT—Reading	% Below Poverty
1	166.4	165.0	91.7
2	159.6	157.2	90.2
3	159.1	164.4	86.0
4	155.5	162.4	83.9
5	164.3	162.5	80.4
6	169.8	164.9	76.5
7	155.7	162.0	76.0
8	165.2	165.0	75.8
9	175.4	173.7	75.6
10	178.1	171.0	75.0
11	167.1	169.4	74.7
12	177.0	172.9	63.2
13	174.2	172.7	52.9
14	175.6	174.9	48.5
15	170.8	174.8	39.1
16	175.1	170.1	38.4
17	182.8	181.4	34.3
18	180.3	180.6	30.3
19	178.8	178.0	30.3
20	181.4	175.9	29.6
21	182.8	181.6	26.5
22	186.1	183.8	13.8

Source: Tekwe, C. D., et al. "An empirical comparison of statistical models for value-added assessment of school performance," *Journal of Educational and Behavioral Statistics*, Vol. 29, No. 1, Spring 2004 (Table 2).

9.27 New method of estimating rainfall. Accurate measurements of rainfall are critical for many hydrological and meteorological projects. Two standard methods of monitoring rainfall use rain gauges and weather radar. Both, however, can be contaminated by human and environmental interference. In the *Journal of Data Science* (Apr. 2004), researchers employed artificial neural networks (i.e., computer-based mathematical models) to estimate rainfall at a meteorological station in Montreal. Rainfall estimates were made every 5 minutes over a 70-minute period by each of the three methods. The data (in millimeters) are listed in the next table.

RAINFALL

Time	Radar	Rain Gauge	Neural Network
8:00 A.M.	3.6	0	1.8
8:05	2.0	1.2	1.8
8:10	1.1	1.2	1.4
8:15	1.3	1.3	1.9
8:20	1.8	1.4	1.7
8:25	2.1	1.4	1.5
8:30	3.2	2.0	2.1
8:35	2.7	2.1	1.0
8:40	2.5	2.5	2.6
8:45	3.5	2.9	2.6
8:50	3.9	4.0	4.0
8:55	3.5	4.9	3.4
9:00 A.M.	6.5	6.2	6.2
9:05	7.3	6.6	7.5
9:10	6.4	7.8	7.2

Source: Hessami, M. et al. "Selection of an artificial neural network model for the post-calibration of weather radar rainfall estimation," *Journal of Data Science,* Vol. 2, No. 2, Apr. 2004. (Adapted from Figures 2 and 4.)

a. Propose a straight-line model relating rain gauge amount (y) to weather radar rain estimate (x).

b. Use the method of least squares to fit the model to the data in the **RAINFALL** file.

c. Graph the least squares line on a scattergram of the data. Is there visual evidence of a relationship between the two variables? Is the relationship positive or negative?

d. Interpret the estimates of the y-intercept and slope in the words of the problem.

e. Now consider a model relating rain gauge amount (y) to the artificial neural network rain estimate (x). Repeat parts **a–d** for this model.

9.28 Sweetness of orange juice The quality of the orange juice produced by a manufacturer is constantly monitored. There are numerous sensory and chemical components that combine to make the best-tasting orange juice. For example, one manufacturer has developed a quantitative index of the "sweetness" of orange juice. (The higher the index, the sweeter is the juice.) Is there a relationship between the sweetness index and a chemical measure such as the amount of water-soluble pectin (parts per million) in the orange juice? Data collected on these two variables during 24 production runs at a juice-manufacturing plant are shown in the table (top, right). Suppose a manufacturer wants to use simple linear regression to predict the sweetness (y) from the amount of pectin (x).

a. Find the least squares line for the data.

b. Interpret $\hat{\beta}_0$ and $\hat{\beta}_1$ in the words of the problem.

c. Predict the sweetness index if the amount of pectin in the orange juice is 300 ppm. [*Note:* A measure of reliability of such a prediction is discussed in Section 9.6.]

9.29 Are geography journals worth their cost? Refer to the *Geoforum* (Vol. 37, 2006) study of whether the price of a geography journal is correlated with quality, presented in Exercise 2.144 (p. 92). Several quantitative variables were recorded for each in a sample of 28 geography journals: cost of a one-year subscription (dollars); journal impact factor (JIF), the average number of times articles from the journal

OJUICE

Run	Sweetness Index	Pectin (ppm)
1	5.2	220
2	5.5	227
3	6.0	259
4	5.9	210
5	5.8	224
6	6.0	215
7	5.8	231
8	5.6	268
9	5.6	239
10	5.9	212
11	5.4	410
12	5.6	256
13	5.8	306
14	5.5	259
15	5.3	284
16	5.3	383
17	5.7	271
18	5.5	264
19	5.7	227
20	5.3	263
21	5.9	232
22	5.8	220
23	5.8	246
24	5.9	241

Note: The data in the table are authentic. For reasons of confidentiality, the name of the manufacturer cannot be disclosed.

have been cited; number of citations for the journals over the past five years; and relative price index (RPI). The data for the 28 journals are saved in the **GEOJRNL** file. Selected observations are listed in the accompanying table.

a. Fit a straight-line model relating cost (y) to JIF (x). Give a practical interpretation of the estimated slope of the line.

b. Fit a straight-line model relating cost (y) to number of citations (x). Give a practical interpretation of the estimated slope of the line.

c. Fit a straight-line model relating cost (y) to RPI (x). Give a practical interpretation of the estimated slope of the line.

GEOJRNL (selected observations)

Journal	Cost ($)	JIF	Citations	RPI
J. Econ. Geogr	468	3.139	207	1.16
Prog. Hum. Geog	624	2.943	544	0.77
T. I. Brit. Geogr	499	2.388	249	1.11
Econ. Geogr.	90	2.325	173	0.30
A. A. A. Geogr.	698	2.115	377	0.93
⋮	⋮	⋮	⋮	⋮
Geogr. Anal.	213	0.902	106	0.88
Geogr. J.	223	0.857	81	0.94
Appl. Geogr	646	0.853	74	3.38

Source: Blomley, N. "Is this journal worth US$1118?" *Geoforum,* Vol. 27, 2006.

9.30 The "name game." Refer to the *Journal of Experimental Psychology—Applied* (June 2000) study in which the "name game" was used to help groups of students learn the names of other students in the group, presented in

Exercise 7.110 (p. 418). Recall that the "name game" requires the first student in the group to state his or her full name, the second student to say his or her name and the name of the first student, the third student to say his or her name and the names of the first two students, etc. After making their introductions, the students listened to a seminar speaker for 30 minutes. At the end of the seminar, all students were asked to remember the full name of each of the other students in their group, and the researchers measured the proportion of names recalled for each. One goal of the study was to investigate the linear trend between y = proportion of names recalled and x = position (order) of the student during the game. The data (simulated on the basis of summary statistics provided in the research article) for 144 students in the first eight positions are saved in the **NAMEGAME2** file. The first five and last five observations in the data set are listed in the accompanying table. [*Note:* Since the student in position 1 actually must recall the names of all the other students, he or she is assigned position number 9 in the data set.] Use the method of least squares to estimate the line $E(y) = \beta_0 + \beta_1 x$. Interpret the β estimates in the words of the problem.

NAMEGAME2

Position	Recall
2	0.04
2	0.37
2	1.00
2	0.99
2	0.79
⋮	⋮
9	0.72
9	0.88
9	0.46
9	0.54
9	0.99

Source: Morris, P.E., and Fritz, C.O. "The name game: Using retrieval practice to improve the learning of names," *Journal of Experimental Psychology—Applied*, Vol. 6, No. 2, June 2000 (data simulated from Figure 2).

Applying the Concepts—Advanced

9.31 Spreading rate of spilled liquid. Refer to the *Chemical Engineering Progress* (Jan. 2005) study of the rate which a spilled volatile liquid will spread across a surface, presented in Exercise 2.150 (p. 94). Recall that a DuPont Corp. engineer calculated the mass (in pounds) of a 50-gallon methanol spill after a period ranging from 0 to 60 minutes. Do the data shown in the accompanying table indicate that the mass of the spill tends to diminish as time increases? If so, how much will the mass diminish each minute?

LIQUIDSPILL

Time (minutes)	Mass (pounds)
0	6.64
1	6.34
2	6.04
4	5.47
6	4.94
8	4.44
10	3.98
12	3.55
14	3.15
16	2.79
18	2.45
20	2.14
22	1.86
24	1.60
26	1.37
28	1.17
30	0.98
35	0.60
40	0.34
45	0.17
50	0.06
55	0.02
60	0.00

Source: Barry, J. "Estimating rates of spreading and evaporation of volatile liquids," *Chemical Engineering Progress*, Vol. 101, No. 1, Jan. 2005.

9.3 Model Assumptions

In Section 9.2, we assumed that the probabilistic model relating the drug reaction time y to the percentage x of drug in the bloodstream is

$$y = \beta_0 + \beta_1 x + \varepsilon$$

We also recall that the least squares estimate of the deterministic component of the model, $\beta_0 + \beta_1 x$, is

$$\hat{y} = \hat{\beta}_0 + \hat{\beta}_1 x = -.1 + .7x$$

Now we turn our attention to the random component ε of the probabilistic model and its relation to the errors in estimating β_0 and β_1. We will use a probability distribution to characterize the behavior of ε. We will see how the probability distribution of ε determines how well the model describes the relationship between the dependent variable y and the independent variable x.

Step 3 in a regression analysis requires us to specify the probability distribution of the random error ε. We will make four basic assumptions about the general form of this probability distribution:

Assumption 1 The mean of the probability distribution of ε is 0. That is, the average of the values of ε over an infinitely long series of experiments is 0 for each setting of the independent variable x. This assumption implies that the mean value of y, for a given value of x is $E(y) = \beta_0 + \beta_1 x$.

Assumption 2 The variance of the probability distribution of ε is constant for all settings of the independent variable x. For our straight-line model, this assumption means that the variance of ε is equal to a constant—say, σ^2—for all values of x.

Assumption 3 The probability distribution of ε is normal.

Assumption 4 The values of ε associated with any two observed values of y are independent. That is, the value of ε associated with one value of y has no effect on any of the values of ε associated with any other y values.

The implications of the first three assumptions can be seen in Figure 9.7, which shows distributions of errors for three values of x, namely, 5, 10, and 15. Note that the relative frequency distributions of the errors are normal with a mean of 0 and a constant variance σ^2. (All of the distributions shown have the same amount of spread or variability.) The straight line shown in Figure 9.7 is the line of means; it indicates the mean value of y for a given value of x. We denote this mean value as $E(y)$. Then the line of means is given by the equation

$$E(y) = \beta_0 + \beta_1 x$$

These assumptions make it possible for us to develop measures of reliability for the least squares estimators and to devise hypothesis tests for examining the usefulness of the least squares line. We have various techniques for checking the validity of these assumptions, and we have remedies to apply when they appear to be invalid. These topics are beyond the scope of this text, but they are discussed in some of the chapter references. Fortunately, the assumptions need not hold exactly in order for least squares estimators to be useful. The assumptions will be satisfied adequately for many applications encountered in practice.

It seems reasonable to assume that the greater the variability of the random error ε (which is measured by its variance σ^2), the greater will be the errors in the estimation of the model parameters β_0 and β_1 and in the error of prediction when $\hat{y}$ is used to predict y for some value of x. Consequently, you should not be surprised, as we proceed through this chapter, to find that σ^2 appears in the formulas for all confidence intervals and test statistics that we will be using.

In most practical situations, σ^2 is unknown and we must use our data to estimate its value. The best estimate of σ^2, denoted by s^2, is obtained by dividing the sum of the squares of the deviations of the y values from the prediction line, or

$$\text{SSE} = \sum (y_i - \hat{y}_i)^2$$

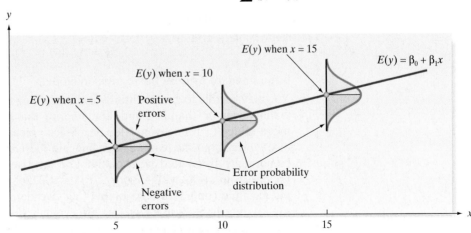

FIGURE 9.7

The probability distribution of ε

by the number of degrees of freedom associated with this quantity. We use 2 df to estimate the two parameters β_0 and β_1 in the straight-line model, leaving $(n - 2)$ df for the estimation of the error variance.

Estimation of σ^2 for a (First-Order) Straight-Line Model

$$s^2 = \frac{\text{SSE}}{\text{Degrees of freedom for error}} = \frac{\text{SSE}}{n - 2}$$

where $\text{SSE} = \sum (y_i - \hat{y}_i)^2 = \text{SS}_{yy} - \hat{\beta}_1 \text{SS}_{xy}$

in which

$$\text{SS}_{yy} = \sum (y_i - \bar{y})^2 = \sum y_i^2 - \frac{\left(\sum y_i \right)^2}{n}$$

To estimate the standard deviation σ of ε, we calculate

$$s = \sqrt{s^2} = \sqrt{\frac{\text{SSE}}{n - 2}}$$

We will refer to s as the **estimated standard error of the regression model**.

Warning

When performing these calculations, you may be tempted to round the calculated values of SS_{yy}, $\hat{\beta}_1$, and SS_{xy}. Be certain to carry at least six significant figures for each of these quantities, to avoid substantial errors in calculating the SSE.

EXAMPLE 9.2

ESTIMATING σ IN REGRESSION—Drug Reaction Data

Problem   Refer to Example 9.1 and the simple linear regression of the drug reaction data in Table 9.3.

a. Compute an estimate of σ.
b. Give a practical interpretation of the estimate.

Solution

a. We previously calculated SSE = 1.10 for the least squares line $\hat{y} = -.1 + .7x$. Recalling that there were $n = 5$ data points, we have $n - 2 = 5 - 2 = 3$ df for estimating σ^2. Thus,

$$s^2 = \frac{\text{SSE}}{n - 2} = \frac{1.10}{3} = .367$$

is the estimated variance, and

$$s = \sqrt{.367} = .61$$

is the standard error of the regression model.

b. You may be able to grasp s intuitively by recalling the interpretation of a standard deviation given in Chapter 2 and remembering that the least squares line estimates the mean value of y for a given value of x. Since s measures the spread of the distribution of y values about the least squares line and these errors of prediction are assumed to be normally distributed, we should not be surprised to find that most (about 95%) of the observations lie within $2s$, or $2(.61) = 1.22$, of the least squares line. For this simple example (only five data points), all five data points fall within $2s$ of the least squares line. In Section 9.6, we use s to evaluate the error of prediction when the least squares line is used to predict a value of y to be observed for a given value of x.

Dependent Variable: TIME_Y

Number of Observations Read		5
Number of Observations Used		5

Analysis of Variance

Source	DF	Sum of Squares	Mean Square	F Value	Pr > F
Model	1	4.90000	4.90000	13.36	0.0354
Error	3	1.10000	0.36667		
Corrected Total	4	6.00000			

Root MSE	0.60553	R-Square	0.8167	
Dependent Mean	2.00000	Adj R-Sq	0.7556	
Coeff Var	30.27650			

Parameter Estimates

Variable	DF	Parameter Estimate	Standard Error	t Value	Pr > |t|
Intercept	1	-0.10000	0.63509	-0.16	0.8849
DRUG_X	1	0.70000	0.19149	3.66	0.0354

FIGURE 9.8

SAS printout for the time–drug regression

Look Back The values of s^2 and s can also be obtained from a simple linear regression printout. The SAS printout for the drug reaction example is reproduced in Figure 9.8. The value of s^2 is highlighted on the printout (in the **Mean Square** column in the row labeled **Error**). The value $s^2 = .36667$, rounded to three decimal places, agrees with the one calculated by hand. The value of s is also highlighted in Figure 9.8 (next to the heading **Root MSE**). This value, $s = .60553$, agrees (except for rounding) with our hand-calculated value.

> **Now Work Exercise 9.36a-b**

Interpretation of s, the Estimated Standard Deviation of ε

We expect most ($\approx 95\%$) of the observed y values to lie within $2s$ of their respective least squares predicted values, $\hat{y}$.

Exercises 9.32–9.46

Understanding the Principles

9.32 What are the four assumptions made about the probability distribution of ε in regression?

9.33 Illustrate the assumptions of Exercise 9.32 with a graph.

9.34 Visually compare the scattergrams shown at right. If a least squares line were determined for each data set, which do you think would have the smallest variance s^2? Explain.

Learning the Mechanics

9.35 Calculate SSE and s^2 for each of the following cases:
    a. $n = 20$, $SS_{yy} = 95$, $SS_{xy} = 50$, $\hat{\beta}_1 = .75$

    b. $n = 40$, $\sum y^2 = 860$, $\sum y = 50$,

       $SS_{xy} = 2,700$, $\hat{\beta}_1 = .2$

    c. $n = 10$, $\sum (y_i - \bar{y})^2 = 58$,

       $SS_{xy} = 91$, $SS_{xx} = 170$

9.36 Suppose you fit a least squares line to 12 data points and
NW the calculated value of SSE is .429.

a.
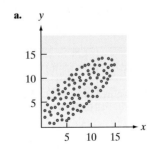

b.
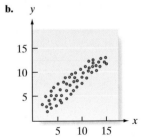

c.
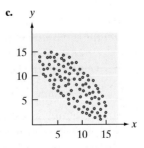

a. Find s^2, the estimator of σ^2 (the variance of the random error term ε).
b. Find s, the estimate of σ.
c. What is the largest deviation that you might expect between any one of the 12 points and the least squares line?

9.37 Refer to Exercises 9.14 and 9.17 (pp. 496-497). Calculate SSE, s^2, and s for the least squares lines obtained in those exercises. Interpret the standard errors of the regression model for each.

Applying the Concepts—Basic

9.38 Quantitative models of music. Refer to the *Chance* (Fall 2004) study on modeling a certain pitch of a musical composition, presented in Exercise 9.18 (p. 497). Recall that the number of times (y) a certain pitch occurs—called entropy—was modeled as a straight-line function of year of birth (x) of the composer. On the basis of the scatterplot of the data (p. 573), the standard deviation σ of the model is estimated to be $s = .1$. For a given year (x), about 95% of the actual entropy values (y) will fall within d units of their predicted values. Find the value of d.

9.39 Winning marathon times. Refer to Exercise 9.20 (p. 498) and the *Chance* (Winter 2000) study on predicting the winning time (y) in the Boston Marathon. On the basis of the variation of the data around the scatterplots (p. 498), which simple linear regression model, the men's or the women's, will have the smallest estimate of σ? Explain.

SITIN

9.40 College protests of labor exploitation. Refer to the *Journal of World-Systems Research* (Winter 2004) study of college student "sit-ins," presented in Exercise 9.21 (p. 498). The data in the **SITIN** file was used to estimate the straight-line model relating number of arrests (y) to duration of sit-in (x).
a. Give the values of SSE, s^2, and s, shown on the SPSS printout below.
b. Give a practical interpretation of the value of s.

GOBIANTS

9.41 Mongolian desert ants. Refer to the *Journal of Biogeography* (Dec. 2003) study of ant sites in Mongolia, presented in Exercise 9.22 (p. 498). The data in the **GOBIANTS** file was used to estimate the straight-line model relating annual rainfall (y) to maximum daily temperature (x).
a. Give the values of SSE, s^2, and s, shown on the MINITAB printout (p. 575).
b. Give a practical interpretation of the value of s.

Applying the Concepts—Intermediate

SMELTPOT

9.42 Extending the life of an aluminum smelter pot. Refer to *The American Ceramic Society Bulletin* (Feb. 2005) study of bricks that line aluminum smelter pots, presented in Exercise 9.24 (p. 499). You fit the simple linear regression model relating brick porosity (y) to mean pore diameter (x) to the data in the **SMELTPOT** file.
a. Find an estimate of the standard deviation σ of the model.
b. In Exercise 9.24d, you predicted brick porosity percentage when $x = 10$ micrometers. Use the result of part **a** to estimate the error of prediction.

FCAT

9.43 FCAT scores and poverty. Refer to the *Journal of Educational and Behavioral Statistics* (Spring 2004) study of scores on the Florida Comprehensive Assessment Test (FCAT), presented in Exercise 9.26 (p. 500).
a. Consider the simple linear regression relating math score (y) to percentage (x) of students below the poverty level. Find and interpret the value of s for this regression.
b. Consider the simple linear regression relating reading score (y) to percentage (x) of students below the poverty level. Find and interpret the value of s for this regression.
c. Which dependent variable, math score or reading score, can be more accurately predicted by percentage (x) of students below the poverty level? Explain.

RAINFALL

9.44 New method of estimating rainfall. Refer to the *Journal of Data Science* (Apr. 2004) comparison of methods for estimating rainfall, presented in Exercise 9.27 (p. 500).

Model Summary

Model	R	R Square	Adjusted R Square	Std. Error of the Estimate
1	.601[a]	.361	.148	16.913

a. Predictors: (Constant), DURATION

ANOVA[b]

Model		Sum of Squares	df	Mean Square	F	Sig.
1	Regression	485.026	1	485.026	1.696	.284[a]
	Residual	858.174	3	286.058		
	Total	1343.200	4			

a. Predictors: (Constant), DURATION
b. Dependent Variable: ARRESTS

SPSS output for Exercise 9.40

a. Consider the simple linear regression relating rain gauge amount (y) to weather radar rain estimate (x). Find and interpret the value of s for this regression.

b. Consider the simple linear regression relating rain gauge amount (y) to the artificial neural network rain estimate (x). Find and interpret the value of s for this regression.

c. Which independent variable, radar estimate or neural network estimate, is a more accurate predictor of rain gauge amount (y)? Explain.

GEOJRNL

9.45 Are geography journals worth their cost? Refer to the *Geoforum* (Vol. 37, 2006) study of whether the price of a geography journal is correlated with quality, presented in Exercise 9.29 (p. 501). The data are saved in the **GEOJRNL** file.

a. In Exercise 9.29a, you fit a straight-line model relating cost (y) to JIF(x). Within how many dollars can you expect to predict cost with this model?

b. In Exercise 9.29b, you fit a straight-line model relating cost (y) to number of citation(x). Within how many dollars can you expect to predict cost with this model?

c. In Exercise 9.29c, you fit a straight-line model relating cost (y) to RPI(x). Within how many dollars can you expect to predict cost with this model?

Applying the Concepts—Advanced

9.46 Life tests of cutting tools. To improve the quality of the output of any production process, it is necessary first to understand the capabilities of the process (Gitlow, et al., *Quality*

Management: Tools and Methods for Improvement, 1995). In a particular manufacturing process, the useful life of a cutting tool is linearly related to the speed at which the tool is operated. The data in the accompanying table were derived from life tests for the two different brands of cutting tools currently used in the production process. For which brand would you feel more confident using the least squares line to predict useful life for a given cutting speed? Explain.

CUTTOOL

Cutting Speed (meters per minute)	Useful Life (hours)	
	Brand A	Brand B
30	4.5	6.0
30	3.5	6.5
30	5.2	5.0
40	5.2	6.0
40	4.0	4.5
40	2.5	5.0
50	4.4	4.5
50	2.8	4.0
50	1.0	3.7
60	4.0	3.8
60	2.0	3.0
60	1.1	2.4
70	1.1	1.5
70	.5	2.0
70	3.0	1.0

9.4 Assessing the Utility of the Model: Making Inferences about the Slope β_1

Now that we have specified the probability distribution of ε and found an estimate of the variance σ^2, we are ready to make statistical inferences about the linear model's usefulness in predicting the response y. This is step 4 in our regression modeling procedure.

Refer again to the data of Table 9.1, and suppose the reaction times are *completely unrelated* to the percentage of drug in the bloodstream. What could then be said about the values of β_0 and β_1 in the hypothesized probabilistic model

$$y = \beta_0 + \beta_1 x + \varepsilon$$

if x contributes no information for the prediction of y? The implication is that the mean of y—that is, the deterministic part of the model $E(y) = \beta_0 + \beta_1 x$—does not change as x changes. In the straight-line model, this means that the true slope, β_1, is equal to 0. (See Figure 9.9.) Therefore, to test the null hypothesis that the linear model contributes no

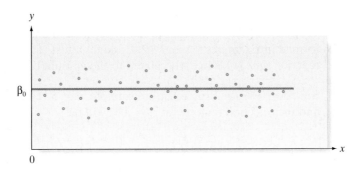

FIGURE 9.9

Graph of the straight-line model when the slope is zero, i.e., $y = \beta_0 + \varepsilon$

information for the prediction of y against the alternative hypothesis that the linear model is useful in predicting y, we test

$$H_0: \beta_1 = 0$$
$$H_a: \beta_1 \neq 0$$

If the data support the alternative hypothesis, we will conclude that x does contribute information for the prediction of y with the straight-line model [although the true relationship between $E(y)$ and x could be more complicated than a straight line]. In effect, then, this is a test of the usefulness of the hypothesized model.

The appropriate test statistic is found by considering the sampling distribution of $\hat{\beta}_1$, the least squares estimator of the slope β_1, as shown in the following box:

Sampling Distribution of $\hat{\beta}_1$

If we make the four assumptions about ε (see Section 9.3), the sampling distribution of the least squares estimator $\hat{\beta}_1$ of the slope will be normal with mean β_1 (the true slope) and standard deviation

$$\sigma_{\hat{\beta}_1} = \frac{\sigma}{\sqrt{SS_{xx}}} \quad \text{(see Figure 9.10)}$$

We estimate $\sigma_{\hat{\beta}_1}$ by $s_{\hat{\beta}_1} = \dfrac{s}{\sqrt{SS_{xx}}}$ and refer to $S_{\hat{\beta}_1}$ as the **estimated standard error of the least squares slope $\hat{\beta}_1$.**

FIGURE 9.10
Sampling distribution of $\hat{\beta}_1$

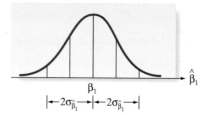

Since σ is usually unknown, the appropriate test statistic is a t-statistic, formed as:

$$t = \frac{\hat{\beta}_1 - \text{Hypothesized value of } \beta_1}{s_{\hat{\beta}_1}} \quad \text{where} \quad s_{\hat{\beta}_1} = \frac{s}{\sqrt{SS_{xx}}}$$

Thus,

$$t = \frac{\hat{\beta}_1 - 0}{s/\sqrt{SS_{xx}}}$$

Note that we have substituted the estimator s for σ and then formed the estimated standard error $s_{\hat{\beta}_1}$ by dividing s by $\sqrt{SS_{xx}}$. The number of degrees of freedom associated with this t statistic is the same as the number of degrees of freedom associated with s. Recall that this number is $(n-2)$ df when the hypothesized model is a straight line. (See Section 9.3.) The setup of our test of the usefulness of the straight-line model is summarized in the two boxes on p. 509.

EXAMPLE 9.3

TESTING THE REGRESSION SLOPE, β_1—Drug Reaction Model

Problem Refer to the simple linear regression analysis of the drug reaction data performed in Examples 9.1 and 9.2. Conduct a test (at $\alpha = .05$) to determine whether the reaction time (y) is linearly related to the amount of drug (x).

Solution For the drug reaction example, $n = 5$. Thus, t will be based on $n - 2 = 3$ df, and the rejection region t (at $\alpha = .05$) will be

$$|t| > t_{.025} = 3.182$$

A Test of Model Utility: Simple Linear Regression

One-Tailed Test
$H_0: \beta_1 = 0$
$H_a: \beta_1 < 0$ (or $H_a: \beta_1 > 0$)

Two-Tailed Test
$H_0: \beta_1 = 0$
$H_a: \beta_1 \neq 0$

Test statistic: $t = \dfrac{\hat{\beta}_1}{s_{\hat{\beta}_1}} = \dfrac{\hat{\beta}_1}{s/\sqrt{SS_{xx}}}$

Rejection region: $t < -t_\alpha$
(or $t > t_\alpha$ when $H_a: \beta_1 > 0$)
where t_α and $t_{\alpha/2}$ are based on $(n-2)$ degrees of freedom

Rejection region: $|t| > t_{\alpha/2}$

Conditions Required for a Valid Test: Simple Linear Regression
The four assumptions about ε listed in Section 9.3.

We previously calculated $\hat{\beta}_1 = .7$, $s = .61$, and $SS_{xx} = 10$. Thus,

$$t = \frac{\hat{\beta}_1}{s/\sqrt{SS_{xx}}} = \frac{.7}{.61/\sqrt{10}} = \frac{.7}{.19} = 3.7$$

Since this calculated t-value falls into the upper-tail rejection region (see Figure 9.11), we reject the null hypothesis and conclude that the slope β_1 is not 0. The sample evidence indicates that the percentage x of drug in the bloodstream contributes information for the prediction of the reaction time y when a linear model is used.

   [*Note:* We can reach the same conclusion by using the observed significance level (p-value) of the test from a computer printout. The MINITAB printout for the drug reaction example is reproduced in Figure 9.12. The test statistic and the two-tailed

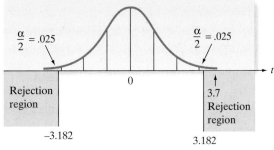

FIGURE 9.11
Rejection region and calculated t value for testing $H_0: \beta_1 = 0$ versus $H_a: \beta_1 \neq 0$

Regression Analysis: TIME_Y versus DRUG_X

```
The regression equation is
TIME_Y = - 0.100 + 0.700 DRUG_X


Predictor        Coef    SE Coef        T       P
Constant      -0.1000     0.6351    -0.16   0.885
DRUG_X         0.7000     0.1915     3.66   0.035


S = 0.605530    R-Sq = 81.7%    R-Sq(adj) = 75.6%


Analysis of Variance

Source           DF        SS        MS       F       P
Regression        1    4.9000    4.9000   13.36   0.035
Residual Error    3    1.1000    0.3667
Total             4    6.0000
```

FIGURE 9.12
MINITAB printout for the time–drug regression

p-value are highlighted on the printout. Since the *p*-value = .035 is smaller than $\alpha = .05$, we will reject H_0.]

Look Back What conclusion can be drawn if the calculated *t*-value does not fall into the rejection region or if the observed significance level of the test exceeds α? We know from previous discussions of the philosophy of hypothesis testing that such a *t*-value does *not* lead us to accept the null hypothesis. That is, we do not conclude that $\beta_1 = 0$. Additional data might indicate that β_1 differs from 0, or a more complicated relationship may exist between *x* and *y*, requiring the fitting of a model other than the straight-line model.

Now Work Exercise 9.54

Interpreting *p*-Values for β Coefficients in Regression

Almost all statistical computer software packages report a *two-tailed p*-value for each of the β parameters in the regression model. For example, in simple linear regression, the *p*-value for the two-tailed test $H_0: \beta_1 = 0$ versus $H_a: \beta_1 \neq 0$ is given on the printout. If you want to conduct a *one-tailed* test of hypothesis, you will need to adjust the *p*-value reported on the printout as follows:

Upper-tailed test ($H_a: \beta_1 > 0$):   *p*-value = $\begin{cases} p/2 & \text{if } t > 0 \\ 1 - p/2 & \text{if } t < 0 \end{cases}$

Lower-tailed test ($H_a: \beta_1 < 0$):   *p*-value = $\begin{cases} p/2 & \text{if } t < 0 \\ 1 - p/2 & \text{if } t > 0 \end{cases}$

Here, *p* is the *p*-value reported on the printout and *t* is the value of the test statistic.

Another way to make inferences about the slope β_1 is to estimate it with a confidence interval, formed as shown in the following box:

A $100(1 - \alpha)\%$ Confidence Interval for the Simple Linear Regression Slope β_1

$$\hat{\beta}_1 \pm t_{\alpha/2} s_{\hat{\beta}_1}$$

where the estimated standard error of $\hat{\beta}_1$ is calculated by

$$s_{\hat{\beta}_1} = \frac{s}{\sqrt{SS_{xx}}}$$

and $t_{\alpha/2}$ is based on $(n - 2)$ degrees of freedom.

Conditions Required for a Valid Confidence Interval: Simple Linear Regression

The four assumptions about ε listed in Section 9.3.

For the simple linear regression for the drug reaction (Examples 9.1–9.3), $t_{\alpha/2}$ is based on $(n - 2) = 3$ degrees of freedom. Therefore, a 95% confidence interval for the slope β_1, the expected change in reaction time for a 1% increase in the amount of drug in the bloodstream, is

$$\hat{\beta}_1 \pm t_{.025} s_{\hat{\beta}_1} = .7 \pm 3.182\left(\frac{s}{\sqrt{SS_{xx}}}\right) = .7 \pm 3.182\left(\frac{.61}{\sqrt{10}}\right) = .7 \pm .61$$

Coefficients[a]

Model		Unstandardized Coefficients B	Unstandardized Coefficients Std. Error	Standardized Coefficients Beta	t	Sig.	95% Confidence Interval for B Lower Bound	95% Confidence Interval for B Upper Bound
1	(Constant)	-.100	.635		-.157	.885	-2.121	1.921
	TIME_X	.700	.191	.904	3.656	.035	.091	1.309

a. Dependent Variable: DRUG_Y

FIGURE 9.13

SPSS printout with 95% confidence intervals for the time–drug regression betas

Thus, the estimate of the interval for the slope parameter β_1 is from .09 to 1.31. [*Note:* This interval can also be obtained with statistical software and is highlighted on the SPSS printout shown in Figure 9.13.] In terms of this example, the implication is that we can be 95% confident that the *true* mean increase in reaction time per additional 1% of the drug is between .09 and 1.31 seconds. This inference is meaningful only over the sampled range of x—that is, from 1% to 5% of the drug in the bloodstream.

Now Work Exercise 9.59

Since all the values in this interval are positive, it appears that β_1 is positive and that the mean of y, $E(y)$, increases as x increases. However, the rather large width of the confidence interval reflects the small number of data points (and, consequently, a lack of information) used in the experiment. We would expect a narrower interval if the sample size were increased.

STATISTICS IN ACTION REVISITED

Assessing How Well the Straight-Line Model Fits the Dowsing Data

In the previous Statistics in Action Revisited, we fit the straight-line model $E(y) = \beta_0 + \beta_1 x$, where x = dowser's guess (in meters) and y = pipe location (in meters) for each trial. The MINITAB regression printout is reproduced in Figure SIA9.3. The two-tailed p-value for testing the null hypothesis $H_0: \beta_1 = 0$ (highlighted on the printout) is p-value = .118. Even for an α-level as high as $\alpha = .10$, there is insufficient evidence to reject H_0. Consequently, the dowsing data in Table SIA9.1 provide no statistical support for the German researchers' claim that the three best dowsers have an ability to find underground water with a divining rod.

This lack of support for the dowsing theory is made clearer with a confidence interval for the slope of the line. When $n = 26$, df $= (n - 2) = 24$ and $t_{.025} = 2.064$. Substituting the latter value and the relevant values shown on the MINITAB printout, we find that a 95% confidence interval for β_1 is

$$\hat{\beta}_1 \pm t_{.025}(s_{\hat{\beta}_1}) = .31 \pm (2.064)(.19)$$
$$= .31 \pm .39, \text{ or } (-.08, .70)$$

Regression Analysis: PIPE versus GUESS

```
The regression equation is
PIPE = 30.1 + 0.308 GUESS


Predictor    Coef    SE Coef    T       P
Constant     30.07   11.41      2.63    0.015
GUESS        0.3079  0.1900     1.62    0.118


S = 26.0298   R-Sq = 9.9%   R-Sq(adj) = 6.1%


Analysis of Variance

Source           DF      SS        MS       F      P
Regression       1       1778.9    1778.9   2.63   0.118
Residual Error   24      16261.2   677.6
Total            25      18040.2
```

FIGURE SIA9.3

MINITAB simple linear regression for dowsing data

(continued)

Thus, for every 1-meter increase in a dowser's guess, we estimate (with 95% confidence) that the change in the actual pipe location will range anywhere from a decrease of .08 meter to an increase of .70 meter. In other words, we're not sure whether the pipe location will increase or decrease along the 10-meter pipeline!

Keep in mind also that the data in Table SIA9.1 represent the "best" performances of the three dowsers (i.e., the outcome of the dowsing experiment in its most favorable light). When the data for all trials are considered and plotted, there is not even a hint of a trend.

Exercises 9.47–9.65

Understanding the Principles

9.47 In the equation $E(y) = \beta_0 + \beta_1 x$, what is the value of β_1 if x has no linear relationship to y?

9.48 What conditions are required for valid inferences about the β's in simple linear regression?

9.49 How do you adjust the p-value obtained from a computer printout when you perform a one-tailed test of β_1 in simple linear regression?

9.50 For each of the following 95% confidence intervals for β_1 in simple linear regression, decide whether there is evidence of a positive or negative linear relationship between y and x:
a. $(22, 58)$ **b.** $(-30, 111)$ **c.** $(-45, -7)$

Learning the Mechanics

9.51 Construct both a 95% and a 90% confidence interval for β_1 for each of the following cases:
a. $\hat{\beta}_1 = 31, s = 3, SS_{xx} = 35, n = 12$
b. $\hat{\beta}_1 = 64, SSE = 1,960, SS_{xx} = 30, n = 18$
c. $\hat{\beta}_1 = -8.4, SSE = 146, SS_{xx} = 64, n = 24$

9.52 Consider the following pairs of observations:

x	1	5	3	2	6	6	0
y	1	3	3	1	4	5	1

a. Construct a scattergram of the data.
b. Use the method of least squares to fit a straight line to the seven data points in the table.
c. Plot the least squares line on your scattergram of part **a**.
d. Specify the null and alternative hypotheses you would use to test whether the data provide sufficient evidence to indicate that x contributes information for the (linear) prediction of y.
e. What is the test statistic that should be used in conducting the hypothesis test of part **d**? Specify the number of degrees of freedom associated with the test statistic.
f. Conduct the hypothesis test of part **d**, using $\alpha = .05$.
g. Construct a 95% confidence interval for β_1.

9.53 Consider the following pairs of observations:

y	4	2	5	3	2	4
x	1	4	5	3	2	4

a. Construct a scattergram of the data.
b. Use the method of least squares to fit a straight line to the six data points.

c. Plot the least squares line on the scattergram of part **a**.
d. Compute the test statistic for determining whether x and y are linearly related.
e. Carry out the test you set up in part **d**, using $\alpha = .01$.

f. Find a 99% confidence interval for β_1.

Applying the Concepts—Basic

9.54 **English as a second language reading ability.** What are the factors that allow a native Spanish-speaking person to understand and read English? A study published in the *Bilingual Research Journal* (Summer 2006) investigated the relationship of Spanish (first-language) grammatical knowledge to English (second-language) reading. The study involved a sample of $n = 55$ native Spanish-speaking adults who were students in an English as a second language (ESL) college class. Each student took four standardized exams: Spanish grammar (SG), Spanish reading (SR), English grammar (EG), and English reading (ESLR). Simple linear regression was used to model the ESLR score (y) as a function of each of the other exam scores (x). The results are summarized in the following table:

Independent variable (x)	p-value for testing H_0: $\beta_1 = 0$
SG score	.739
SR score	.012
ER score	.022

a. At $\alpha = .05$, is there sufficient evidence to indicate that ESLR score is linearly related to SG score?

b. At $\alpha = .05$, is there sufficient evidence to indicate that ESLR score is linearly related to SR score?
c. At $\alpha = .05$, is there sufficient evidence to indicate that ESLR score is linearly related to ER score?

9.55 **Ranking driving performance of professional golfers.** Refer to *The Sport Journal* (Winter 2007) study of a new method for ranking the total driving performance of golfers on the Professional Golf Association (PGA) tour, presented in Exercise 9.25 (p. 500). You fit a straight-line model relating driving accuracy (y) to driving distance (x) to the data saved in the **PGADRIVER** file.
a. Give the null and alternative hypotheses for testing whether driving accuracy (y) decreases linearly as driving distance (x) increases.
b. Find the test statistic and p-value of the test you set up in part **a**.
c. Make the appropriate conclusion at $\alpha = .01$.

FCAT

9.56 FCAT scores and poverty. Refer to the *Journal of Educational and Behavioral Statistics* (Spring 2004) study of scores on the Florida Comprehensive Assessment Test (FCAT), first presented in Exercise 9.26 (p. 500). Consider the simple linear regression relating math score (y) to percentage (x) of students below the poverty level.
a. Test whether y is negatively related to x. Use $\alpha = .01$.
b. Construct a 99% confidence interval for β_1. Interpret the result practically.

RAINFALL

9.57 New method of estimating rainfall. Refer to the *Journal of Data Science* (Apr. 2004) comparison of methods for estimating rainfall, first presented in Exercise 9.27 (p. 500). Consider the simple linear regression relating the rain gauge amount (y) to the artificial neural network rain estimate (x).
a. Test whether y is positively related to x. Use $\alpha = .10$.
b. Construct a 90% confidence interval for β_1. Interpret the result practically.

GEOJRNL

9.58 Are geography journals worth their cost? Refer to the *Geoforum* (Vol. 37, 2006) study of whether the price of a geography journal is correlated with the quality of the journal, first presented in Exercise 9.29 (p. 501). The data are saved in the **GEOJRNL** file.
a. In Exercise 9.29a, you fit a straight-line model relating cost (y) to JIF(x). Find and interpret a 95% confidence interval for the slope of the line.
b. In Exercise 9.29b, you fit a straight-line model relating cost (y) to number of citations (x). Find and interpret a 95% confidence interval for the slope of the line.
c. In Exercise 9.29c, you fit a straight-line model relating cost (y) to RPI(x). Find and interpret a 95% confidence interval for the slope of the line.

OJUICE

9.59 Sweetness of orange juice. Refer to Exercise 9.28
NW (p. 501) and the simple linear regression relating the sweetness index (y) of an orange juice sample to the amount of water-soluble pectin (x) in the juice. Find a 90% confidence interval for the true slope of the line. Interpret the result.

Applying the Concepts—Intermediate

9.60 Effect of massage on boxers. The *British Journal of Sports Medicine* (Apr. 2000) published a study of the effect of massage on boxing performance. Two variables measured on the boxers were blood lactate concentration (in mM) and the boxer's perceived recovery (on a 28-point scale). On the basis of information provided in the article, the data shown in the accompanying table were obtained for 16 five-round boxing performances in which a massage was given to the boxer between rounds. Conduct a test to determine whether blood lactate level (y) is linearly related to perceived recovery (x). Use $\alpha = .10$.

BOXING2

Blood Lactate Level	Perceived Recovery
3.8	7
4.2	7
4.8	11
4.1	12
5.0	12
5.3	12
4.2	13
2.4	17
3.7	17
5.3	17
5.8	18
6.0	18
5.9	21
6.3	21
5.5	20
6.5	24

Source: Hemmings, B., Smith, M., Graydon, J., and Dyson, R. "Effects of massage on physiological restoration, perceived recovery, and repeated sports performance," *British Journal of Sports Medicine*, Vol. 34, No. 2, Apr. 2000 (data adapted from Figure 3).

9.61 Forest fragmentation study. Refer to the *Conservation Ecology* (Dec. 2003) study on the causes of fragmentation of 54 South American forests, presented in Exercise 2.148 (p. 93). Recall that researchers developed two fragmentation indexes for each forest—one index for anthropogenic (human development activities) fragmentation and one for fragmentation from natural causes. Data on 5 of the 54 forests saved in the **FORFRAG** file are listed in the following table:

FORFRAG (First 5 observations listed)

Ecoregion (forest)	Anthropogenic Index, y	Natural Origin Index, x
Araucaria moist forests	34.09	30.08
Atlantic Coast *restingas*	40.87	27.60
Bahia coastal forests	44.75	28.16
Bahia interior forests	37.58	27.44
Bolivian *Yungas*	12.40	16.75

Source: Wade, T. G., et al. "Distribution and causes of global forest fragmentation," *Conservation Ecology*, Vol. 72, No. 2, Dec. 2003 (Table 6).

a. Ecologists theorize that a linear relationship exists between the two fragmentation indexes. Write the model relating y to x.
b. Fit the model to the data in the **FORFRAG** file, using the method of least squares. Give the equation of the least squares prediction equation.
c. Interpret the estimates of β_0 and β_1 in the context of the problem.
d. Is there sufficient evidence to indicate that the natural origin index (x) and the anthropogenic index (y) are positively linearly related? Test, using $\alpha = .05$.
e. Find and interpret a 95% confidence interval for the change in the anthropogenic index (y) for every 1-point increase in the natural origin index (x).

9.62 Pain empathy and brain activity. Empathy refers to being able to understand and vicariously feel what others actually feel. Neuroscientists at University College of London investigated the relationship between brain activity and pain-related empathy in persons who watch others in pain. (*Science*, Feb. 20, 2004.) Sixteen couples participated in the experiment. The female partner watched while painful stimulation was applied to the finger of her male partner. Two variables were measured for each female: y = pain-related brain activity (measured on a scale ranging from −2 to 2) and x = score on the Empathic Concern Scale (0 to 25 points). The data are listed in the accompanying table. The research question of interest was "Do people scoring higher in empathy show higher pain-related brain activity?" Use simple linear regression analysis to answer this question.

BRAINPAIN

Couple	Brain Activity (y)	Empathic Concern (x)
1	.05	12
2	−.03	13
3	.12	14
4	.20	16
5	.35	16
6	0	17
7	.26	17
8	.50	18
9	.20	18
10	.21	18
11	.45	19
12	.30	20
13	.20	21
14	.22	22
15	.76	23
16	.35	24

Source: Singer, T. et al. "Empathy for pain involves the affective but not sensory components of pain," *Science*, Vol. 303, Feb. 20, 2004. (Adapted from Figure 4.)

9.63 Relation of eye and head movements. How do eye and head movements relate to body movements when a person reacts to a visual stimulus? Scientists at the California Institute of Technology designed an experiment to answer this question and reported their results in *Nature* (Aug. 1998). Adult male rhesus monkeys were exposed to a visual stimulus (i.e., a panel of light-emitting diodes), and their eye, head, and body movements were electronically recorded. In one variation of the experiment, two variables were measured: active head movement (x, percent per degree) and body-plus-head rotation (y, percent per degree). The data for n = 39 trials were subjected to a simple linear regression analysis, with the following results: $\hat{\beta}_1 = .88, s_{\hat{\beta}_1} = .14$

a. Conduct a test to determine whether the two variables, active head movement x and body-plus-head rotation y are positively linearly related. Use $\alpha = .05$.

b. Construct and interpret a 90% confidence interval for β_1.

c. The scientists want to know whether the true slope of the line differs significantly from 1. On the basis of your answer to part **b**, make the appropriate inference.

NAMEGAME2

9.64 The "name game." Refer to the *Journal of Experimental Psychology—Applied* (June 2000) name-retrieval study,

presented in Exercise 9.30 (p. 502). Recall that the goal of the study was to investigate the linear trend between proportion of names recalled (y) and position (order) of the student (x) during the "name game." Is there sufficient evidence (at $\alpha = .01$) of a linear trend? Answer the question by analyzing the data for 144 students saved in the **NAMEGAME2** file. $t = 2.858$

Applying the Concepts—Advanced

9.65 Does elevation affect hitting performance in baseball? The Colorado Rockies play their major league home baseball games in Coors Field, Denver. Each year, the Rockies are among the leaders in team batting statistics (e.g., home runs, batting average, and slugging percentage). Many baseball experts attribute this phenomenon to the "thin air" of Denver—called the "Mile-High City" due to its elevation. *Chance* (Winter 2006) investigated the effects of elevation on slugging percentage in Major League Baseball. Data were compiled on players' composite slugging percentages at each of 29 cities for the 2003 season, as well as on each city's elevation (feet above sea level.) The data are saved in the **MLBPARKS** file. (Selected observations are shown in the accompanying table.) Consider a straight-line model relating slugging percentage (y) to elevation (x).

a. The model was fit to the data with the use of MINITAB, with the results shown in the printout (p. 591). Locate the estimates of the model parameters on the printout.

b. Is there sufficient evidence (at $\alpha = .01$) of a positive linear relationship between elevation (x) and slugging percentage (y)? Use the p-value shown on the printout to make the inference.

c. Construct a scatterplot of the data and draw the least squared line on the graph. Locate the data point for Denver on the graph. What do you observe?

d. Remove the data point for Denver from the data set and refit the straight-line model to the remaining data. Repeat parts **a** and **b**. What conclusions can you draw about the "thin air" theory from this analysis?

MLBPARKS (Selected observations)

City	Slug Pct.	Elevation
Anaheim	.480	160
Arlington	.605	616
Atlanta	.530	1,050
Baltimore	.505	130
Boston	.505	20
⋮	⋮	⋮
Denver	.625	5,277
⋮	⋮	⋮
Seattle	.550	350
San Francisco	.510	63
St. Louis	.570	465
Tampa	.500	10
Toronto	.535	566

Source: Schaffer, J., & Heiny, E.L. "The effects of elevation on slugging percentage in Major League Baseball," *Chance*, Vol. 19, No. 1, Winter 2006 (adapted from Figure 2).

Regression Analysis: SLUGPCT versus ELEVATION

```
The regression equation is
SLUGPCT = 0.515 + 0.000021 ELEVATION

Predictor          Coef      SE Coef       T       P
Constant       0.515140     0.007954   64.76   0.000
ELEVATION     0.00002074   0.00000719   2.89   0.008

S = 0.0369803    R-Sq = 23.6%    R-Sq(adj) = 20.7%

Analysis of Variance

Source            DF        SS         MS       F       P
Regression         1   0.011390   0.011390    8.33   0.008
Residual Error    27   0.036924   0.001368
Total             28   0.048314
```

MINITAB output for
Exercise 9.65

9.5 The Coefficients of Correlation and Determination

In this section, we present two statistics that describe the adequacy of a model: the *coefficient of correlation* and the *coefficient of determination*.

Coefficient of Correlation

Recall (from optional Section 2.9) that a **bivariate relationship** describes a relationship—or correlation—between two variables x and y. Scattergrams are used to describe a bivariate relationship graphically. In this section, we will discuss the concept of **correlation** and how it can be used to measure the linear relationship between two variables x and y. A numerical descriptive measure of correlation is provided by the *Pearson product moment coefficient of correlation*, r.

Definition 9.2

The **coefficient of correlation**,* r, is a measure of the strength of the *linear* relationship between two variables x and y. It is computed (for a sample of n measurements on x and y) as follows:

$$r = \frac{SS_{xy}}{\sqrt{SS_{xx}\,SS_{yy}}}$$

Note that the computational formula for the correlation coefficient r given in Definition 9.2 involves the same quantities that were used in computing the least squares prediction equation. In fact, since the numerators of the expressions for $\hat{\beta}_1$ and r are identical, it is clear that $r = 0$ when $\hat{\beta}_1 = 0$ (the case where x contributes no information for the prediction of y) and that r is positive when the slope is positive and negative when the slope is negative. Unlike $\hat{\beta}_1$, the correlation coefficient r is *scaleless* and assumes a value between -1 and $+1$, regardless of the units of x and y.

A value of r near or equal to 0 implies little or no linear relationship between y and x. In contrast, the closer r comes to 1 or -1, the stronger is the linear relationship between y and x. And if $r = 1$ or $r = -1$, all the sample points fall exactly on the least squares line. Positive values of r imply a positive linear relationship between y and x;

*The value of r is often called the *Pearson correlation coefficient* to honor its developer, Karl Pearson. (See Biography, p. 446.)

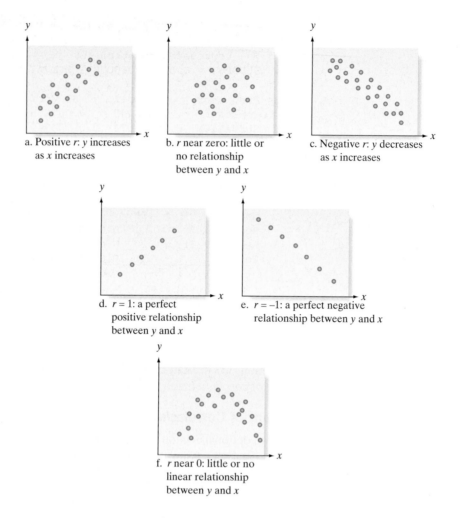

FIGURE 9.14
Values of r and their
implications

that is, y increases as x increases. Negative values of r imply a negative linear relationship between y and x; that is, y decreases as x increases. Each of these situations is portrayed in Figure 9.14.

Now Work Exercise 9.69

We use the data in Table 9.1 for the drug reaction example to demonstrate how to calculate the coefficient of correlation, r. The quantities needed to calculate r are SS_{xy}, SS_{xx}, and SS_{yy}. The first two quantities have been calculated previously and are repeated here for convenience:

$$SS_{xy} = 7, \quad SS_{xx} = 10, \quad SS_{yy} = \sum y^2 - \frac{\left(\sum y\right)^2}{n}$$

$$= 26 - \frac{(10)^2}{5} = 26 - 20 = 6$$

We now find the coefficient of correlation:

$$r = \frac{SS_{xy}}{\sqrt{SS_{xx}\,SS_{yy}}} = \frac{7}{\sqrt{(10)(6)}} = \frac{7}{\sqrt{60}} = .904$$

The fact that r is positive and near 1 indicates that the reaction time tends to increase as the amount of drug in the bloodstream increases—*for the given sample of five subjects*. This is the same conclusion we reached when we found the calculated value of the least squares slope to be positive.

EXAMPLE 9.4

USING THE
CORRELATION
COEFFICIENT—
Relating Crime
Rate and Casino
Employment

Problem Legalized gambling is available on several riverboat casinos operated by a city in Mississippi. The mayor of the city wants to know the correlation between the number of casino employees and the yearly crime rate. The records for the past 10 years are examined, and the results listed in Table 9.5 are obtained. Calculate the coefficient of correlation, r, for the data. Interpret the result.

 CASINO

TABLE 9.5 Data on Casino Employees and Crime Rate, Example 9.4

Year	Number x of Casino Employees (thousands)	Crime Rate y (number of crimes per 1,000 population)
1998	15	1.35
1999	18	1.63
2000	24	2.33
2001	22	2.41
2002	25	2.63
2003	29	2.93
2004	30	3.41
2005	32	3.26
2006	35	3.63
2007	38	4.15

Solution Rather than use the computing formula given in Definition 9.2, we resort to a statistical software package. The data of Table 9.5 were entered into a computer and MINITAB was used to compute r. The MINITAB printout is shown in Figure 9.15.

The coefficient of correlation, highlighted at the top of the printout, is $r = .987$. Thus, the size of the casino workforce and crime rate in this city are very highly correlated—at least over the past 10 years. The implication is that a strong positive linear relationship exists between these variables. (See Figure 9.15.) We must be careful, however, not to jump to

FIGURE 9.15

MINITAB correlation printout and scattergram for Example 9.4

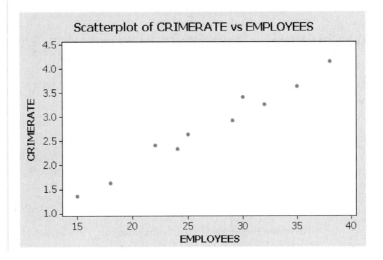

Correlations: EMPLOYEES, CRIMERAT

Pearson correlation of EMPLOYEES and CRIMERAT = 0.987
P-Value = 0.000

any unwarranted conclusions. For instance, the mayor may be tempted to conclude that hiring more casino workers next year will increase the crime rate—that is, that there is a *causal relationship* between the two variables. However, high correlation does not imply causality. The fact is, many things have probably contributed both to the increase in the casino workforce and to the increase in crime rate. The city's tourist trade has undoubtedly grown since riverboat casinos were legalized, and it is likely that the casinos have expanded both in services offered and in number. *We cannot infer a causal relationship on the basis of high sample correlation. When a high correlation is observed in the sample data, the only safe conclusion is that a linear trend may exist between x and y.*

Look Back Another variable, such as the increase in tourism, may be the underlying cause of the high correlation between *x* and *y*.

> **Now Work Exercise 9.75a**

Warning

Two caveats apply in using the sample correlation coefficient *r* to infer the nature of the relationship between *x* and *y*: (1) A *high correlation* does not necessarily imply that a causal relationship exists between *x* and *y*—only that a linear trend may exist; (2) a *low correlation* does not necessarily imply that *x* and *y* are unrelated—only that *x* and *y* are not strongly linearly related.

Keep in mind that the correlation coefficient *r* measures the linear correlation between *x* values and *y* values in the sample, and a similar linear coefficient of correlation exists for the population from which the data points were selected. The **population correlation coefficient** is denoted by the symbol ρ (rho). As you might expect, ρ is estimated by the corresponding sample statistic *r*. Or, instead of estimating ρ, we might want to test the null hypothesis $H_0: \rho = 0$ against $H_a: \rho \neq 0$; that is, we can test the hypothesis that *x* contributes no information for the prediction of *y* by using the straight-line model against the alternative that the two variables are at least linearly related.

However, we already performed this *identical* test in Section 9.5 when we tested $H_0: \beta_1 = 0$ against $H_a: \beta_1 \neq 0$. That is, the null hypothesis $H_0: \rho = 0$ is equivalent to the hypothesis $H_0: \beta_1 = 0$.* When we tested the null hypothesis $H_0: \beta_1 = 0$ in connection with the drug reaction example, the data led to a rejection of the null hypothesis at the $\alpha = .05$ level. This rejection implies that the null hypothesis of a 0 linear correlation between the two variables (drug and reaction time) can also be rejected at the $\alpha = .05$ level. The only real difference between the least squares slope $\hat{\beta}_1$ and the coefficient of correlation, *r*, is the measurement scale. Therefore, the information they provide about the usefulness of the least squares model is to some extent redundant. For this reason, we will use the slope to make inferences about the existence of a positive or negative linear relationship between two variables.

ACTIVITY 9.2: *Keep the Change:* **Correlation Coefficients**

In this activity, you will use data collected in Activity 1.1 *Keep the Change: Collecting Data* (p. 12), and the results from Activity 9.1, *Keep the Change: Least Squares Models* (p. 494), to study the strength of a linear relationship.

1. For each of the least squares models in Exercises 1 and 4 in Activity 9.1, calculate the corresponding correlation coefficient. Do these values support your conclusion about which model is more useful? Explain?

2. For each purchase in your original data set *Purchase Totals*, form the ordered pair.

    (Purchase Total, Amount Transferred)

    Calculate the correlation coefficient and discuss the strength of the linear relationship between the two variables. Explain why you might expect the slope of the corresponding least squares line to be close to zero. Find the least squares model and comment on its usefulness.

---

*The correlation test statistic that is equivalent to $t = \hat{\beta}_1/s_{\hat{\beta}_1}$ is $t = \dfrac{r}{\sqrt{(1 - r^2)/(n - 2)}}$.

Does the model produce estimated values that are meaningless for large purchase totals? Explain.

3. For each student in your class, form the ordered pair

(Sum of Amounts Transferred, Bank Matching)

Calculate the correlation coefficient. How strong is the linear relationship between the two variables? Find the least squares model. Is the slope $\hat{\beta}_1$ approximately 13 and the y-intercept $\hat{\beta}_0$ approximately 0? Explain why the line should have that slope and y-intercept. How good are the estimates given by the model in this situation?

Coefficient of Determination

Another way to measure the usefulness of a linear model is to measure the contribution of x in predicting y. To accomplish this, we calculate how much the errors of prediction of y were reduced by using the information provided by x. To illustrate, consider the sample shown in the scattergram of Figure 9.16a. If we assume that x contributes no information for the prediction of y, the best prediction for a value of y is the sample mean $\bar{y}$, which is shown as the horizontal line in Figure 9.16b. The vertical line segments in Figure 9.16b are the deviations of the points about the mean $\bar{y}$. Note that the sum of the squares of the deviations for the prediction equation $\hat{y} = \bar{y}$ is

$$SS_{yy} = \sum (y_i - \bar{y})^2$$

Now suppose you fit a least squares line to the same set of data and locate the deviations of the points about the line, as shown in Figure 9.16c. Compare the deviations about the prediction lines in Figures 9.16b and 9.16c. You can see that

1. If x contributes little or no information for the prediction of y, the sums of the squares of the deviations for the two lines

$$SS_{yy} = \sum (y_i - \bar{y})^2 \quad \text{and} \quad SSE = \sum (y_i - \hat{y}_i)^2$$

will be nearly equal.

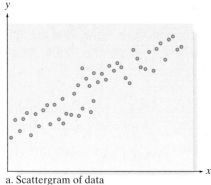

a. Scattergram of data

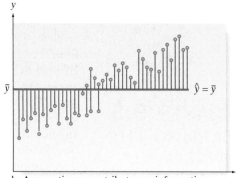

b. Assumption: x contributes no information for predicting y, $\hat{y} = \bar{y}$

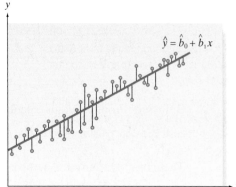

c. Assumption: x contributes information for predicting y, $\hat{y} = \hat{\beta}_0 + \hat{\beta}_1 x$

FIGURE 9.16

A comparison of the sum of squares of deviations for two models

2. If x does contribute information for the prediction of y, the SSE will be smaller than SS_{yy}. In fact, if all the points fall on the least squares line, then SSE = 0.

Consequently, the reduction in the sum of the squares of the deviations that can be attributed to x, expressed as a proportion of SS_{yy}, is

$$\frac{SS_{yy} - SSE}{SS_{yy}}$$

Note that SS_{yy} is the "total sample variability" of the observations around the mean $\bar{y}$ and that SSE is the remaining "unexplained sample variability" after fitting the line $\hat{y}$. Thus, the difference $(SS_{yy} - SSE)$ is the "explained sample variability" attributable to the linear relationship with x. Thus, a verbal description of the proportion is

$$\frac{SS_{yy} - SSE}{SS_{yy}} = \frac{\text{Explained sample variability}}{\text{Total sample variability}}$$

$$= \text{Proportion of total sample variability explained by the linear relationship}$$

In simple linear regression, it can be shown that this proportion—called the *coefficient of determination*—is equal to the square of the simple linear coefficient of correlation, r.

Definition 9.3

The **coefficient of determination** is

$$r^2 = \frac{SS_{yy} - SSE}{SS_{yy}} = 1 - \frac{SSE}{SS_{yy}}$$

and represents the proportion of the total sample variability around $\bar{y}$ that is explained by the linear relationship between y and x. (In simple linear regression, it may also be computed as the square of the coefficient of correlation, r.)

Note that r^2 is always between 0 and 1, because r is between -1 and $+1$. Thus, an r^2 of .60 means that the sum of the squares of the deviations of the y values about their predicted values has been reduced 60% by the use of the least squares equation $\hat{y}$, instead of $\bar{y}$, to predict y.

EXAMPLE 9.5

OBTAINING THE VALUE OF r^2—

Drug Reaction Regression

Problem Calculate the coefficient of determination for the drug reaction example. The data are repeated in Table 9.6 for convenience. Interpret the result.

 STIMULUS

TABLE 9.6

Percent x of Drug	Reaction Time y (seconds)
1	1
2	1
3	2
4	2
5	4

Solution From previous calculations,

$$SS_{yy} = 6 \quad \text{and} \quad SSE = \sum (y - \hat{y})^2 = 1.10$$

Then, from Definition 9.3, the coefficient of determination is

$$r^2 = \frac{SS_{yy} - SSE}{SS_{yy}} = \frac{6.0 - 1.1}{6.0} = \frac{4.9}{6.0} = .817$$

Model Summary

Model	R	R Square	Adjusted R Square	Std. Error of the Estimate
1	.904[a]	.817	.756	.606

a. Predictors: (Constant), DRUG_X

FIGURE 9.17
Portion of SPSS printout for time-drug regression

Another way to compute r^2 is to recall from Section 9.5 that $r = .904$. Then we have $r^2 = (.904)^2 = .817$. A third way to obtain r^2 is from a computer printout. Its value is highlighted on the SPSS printout in Figure 9.17. Our interpretation is as follows: We know that using the percent x of drug in the blood to predict y with the least squares line

$$\hat{y} = -.1 + .7x$$

accounts for nearly 82% of the total sum of the squares of the deviations of the five sample y values about their mean. Or, stated another way, 82% of the sample variation in reaction time (y) can be "explained" by using the percent x of drug in a straight-line model.

Now Work Exercise 9.75b

Practical Interpretation of the Coefficient of Determination, r^2

$100(r^2)$% of the sample variation in y (measured by the total sum of the squares of the deviations of the sample y values about their mean $\bar{y}$) can be explained by (or attributed to) using x to predict y in the straight-line model.

Simple Linear Regression

Using the TI-84/TI-83 Graphing Calculator

Finding r and r^2

Step 1 *Enter the data.*

Press **STAT** and select **1:Edit**.
Note: If a list already contains data, clear the old data. Use the up arrow to highlight the list name, **L1** or **L.2'**.
Press **CLEAR ENTER**.
Enter your x-data into **L1** and your y-data into **L2**.

Step 2 *Turn the diagnostics feature on.*

Press **2nd 0** for **CATALOG**.
Press the key for **D**.
Press the down arrow key until **Diagnostics On** is highlighted.
Press **ENTER** twice.

Step 3 *Find the equation.*

Press **START** and highlight **CALC**.
Press 4 for LinReg (**ax + b**).
Press **ENTER**.
The screen will show the values for a and b in the equation $y = ax + b$.
The values for r and r^2 will appear on the screen as well.

```
LinReg
y=ax+b
a=2.495967742
b=63.3391129
r²=.9620199454
r=.9808261545
```

Example The figure shows the form of the output with the **DiagnosticsOn**.

STATISTICS IN ACTION REVISED

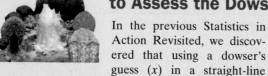

Using the Coefficients of Correlation and Determination to Assess the Dowsing Data

In the previous Statistics in Action Revisited, we discovered that using a dowser's guess (x) in a straight-line model was not statistically useful in predicting actual pipe location (y). Both the coefficient of correlation and the coefficient of determination (highlighted on the MINITAB printouts in Figure SIA9.4) also support this conclusion. The value of the correlation coefficient, $r = .314$, indicates a fairly weak positive linear relationship between the variables. This value, however, is not statistically significant (p-value $= .118$). In other words, there is no evidence to indicate that the population correlation coefficient is different from 0. The coefficient of determination, $r^2 = .099$, implies that only about 10% of the sample variation in pipe location values can be explained by the simple linear model.

Regression Analysis: PIPE versus GUESS

```
The regression equation is
PIPE = 30.1 + 0.308 GUESS

Predictor     Coef    SE Coef     T       P
Constant     30.07      11.41   2.63    0.015
GUESS       0.3079     0.1900   1.62    0.118

S = 26.0298    R-Sq = 9.9%    R-Sq(adj) = 6.1%

Analysis of Variance

Source           DF       SS       MS      F       P
Regression        1   1778.9   1778.9   2.63   0.118
Residual Error   24  16261.2    677.6
Total            25  18040.2
```

FIGURE SIA9.4

MINITAB printouts with coefficients of correlation and determination for the dowsing data

Correlations: PIPE, GUESS

```
Pearson correlation of PIPE and GUESS = 0.314
P-Value = 0.118
```

Exercises 9.66–9.86

Understanding the Principles

9.66 Describe the slope of the least squares line if
  a. $r = .7$ **b.** $r = -.7$ **c.** $r = 0$ **d.** $r^2 = .64$

9.67 *True or False.* The correlation coefficient is a measure of the strength of the linear relationship between x and y.

9.68 *True or False.* A value of the correlation coefficient near 1 or near -1 implies a causal relationship between x and y.

9.69 Explain what each of the following sample correlation coef-
NW ficients tells you about the relationship between the x and y values in the sample:
  a. $r = 1$ **b.** $r = -1$
  c. $r = 0$ **d.** $r = .90$
  e. $r = .10$ **f.** $r = -.88$

Learning the Mechanics

9.70 Calculate r^2 for the least squares line in Exercise 9.14 (p. 496).

9.71 Calculate r^2 for the least squares line in Exercise 9.17 (p. 497).

9.72 Construct a scattergram for each data set. Then calculate r and r^2 for each data set. Interpret their values.

a.

x	-2	-1	0	1	2
y	-2	1	2	5	6

b.

x	-2	-1	0	1	2
y	6	5	3	2	0

c.

x	1	2	2	3	3	3	4
y	2	1	3	1	2	3	2

d.

x	0	1	3	5	6
y	0	1	2	1	0

Applet Exercise 9.2

Use the applet entitled *Correlation by the Eye* to explore the relationship between the pattern of data in a scattergram and the corresponding correlation coefficient.

    a. Run the applet several times. Each times, guess the value of the correlation coefficient. Then click *Show r* to see the actual correlation coefficient. How close is your value to the actual value of *r*? Click *New data* to reset the applet.

    b. Click the trash can to clear the graph. Use the mouse to place five points on the scattergram that are approximately in a straight line. Then guess the value of the correlation coefficient. Click *Show r* to see the actual correlation coefficient. How close were you this time?

    c. Continue to clear the graph and plot sets of five points with different patterns among the points. Guess the value of *r*. How close do you come to the actual value of *r* each time?

    d. On the basis of your experiences with the applet, explain why we need to use more reliable methods of finding the correlation coefficient than just "eyeing" it.

Applying the Concepts—Basic

9.73 Physical activity of obese young adults. Refer to the *International Journal of Obesity* (January 2007) study of the physical activity of obese young adults, presented in Exercise 4.151 (p. 240). For two groups of young adults— 13 obese and 15 of normal weight—researchers recorded the total number of registered movements (counts) of each your adult over a period of time. *Baseline* physical activity was then computed as the number of counts per minute (cpm). Four years later, physical activity measurements were taken again—called physical activity *at follow-up*.

    a. For the 13 obese young adults, the researchers reported a correlation of $r = .50$ between baseline and follow-up physical activity, with an associated *p*-value of .07. Give a practical interpretation of this correlation coefficient and *p*-value.

    b. Refer to part **a**. Construct a scatterplot of the 13 data points that would yield a value or $r = .50$.

    c. For the 15 young adults of normal weight, the researchers reported a correlation of $r = -.12$ between baseline and follow-up physical activity, with an associated *p*-value of .66. Give a practical interpretation of this correlation coefficient and *p*-value.

    d. Refer to part **c**. Construct a scatterplot of the 15 data points that would yield a value of $r = -12$.

9.74 Wind turbine blade stress. Refer to the *Wind Engineering* (Jan. 2004) study of two types of timber—radiata pine and hoop pine—used in high-efficiency small wind turbine blades, presented in Exercise 9.19 (p. 497). Data on stress (*y*) and the natural logarithm of the number of blade cycles (*x*) for each type of timber were analyzed by means of simple linear regression. The results are as follows, with additional information on the coefficient of determination:

Radiata Pine: $\hat{y} = 97.37 - 2.50x$, $r^2 = .84$
Hoop Pine: $\hat{y} = 122.03 - 2.36x$, $r^2 = .90$

    Interpret the value of r^2 for each type of timber.

9.75 Sports news on local TV broadcasts. *The Sports Journal* (Winter 2004) published the results of a study conducted to assess the factors that affect the time allotted to sports news on local television news broadcasts. Information on total time (in minutes) allotted to sports and on audience ratings of the TV news broadcast (measured on a 100-point scale) was obtained from a national sample of 163 news directors. A correlation analysis of the data yielded $r = .43$.

    a. Interpret the value of the correlation coefficient *r*.

    b. Find and interpret the value of the coefficient of determination r^2.

9.76 English as a second language reading ability. Refer to the *Bilingual Research Journal* (Summer 2006) study of the relationship of Spanish (first-language) grammatical knowledge to English (second-language) reading, presented in Exercise 9.54 (p. 512). Recall that each in a sample of $n = 55$ native Spanish-speaking adults took four standardized exams: Spanish grammar (SG), Spanish reading, (SR), English grammar (EG), and English reading (ESLR). Simple linear regressions were used to model the ESLR score (*y*) as a function of each of the other exam scores (*x*). The coefficient of determination, r^2, for each model is listed in the accompanying table. Give a practical interpretation of each of these values.

Independent Variable (x)	r^2
SG score	.002
SR score	.099
ER score	.078

9.77 Redshifts of quasi-stellar objects. Refer to the *Journal of Astrophysics & Astronomy* (Mar./Jun. 2003) study of redshifts in quasi-stellar objects presented in Exercise 9.23 (p. 499). Recall that simple linear regression was used to model the magnitude (*y*) of a quasu-stellar object as a function of the redshift level (*x*). In addition to the least squares line, $\hat{y} = 18.13 + 6.21x$, the coefficient of correlation was determined to be $r = .84$.

    a. Interpret the value of *r* in the words of the problem.

    b. What is the relationship between *r* and the estimated slope of the line?

    c. Find and interpret the value of r^2.

9.78 Removing metal from water. In the *Electronic Journal of Biotechnology* (Apr. 15, 2004), Egyptian scientists studied a new method for removing heavy metals from water. Metal solutions were prepared in glass vessels, and then biosorption was used to remove the metal ions. Two variables were measured for each test vessel: $y =$ metal uptake (milligrams of metal per gram of biosorbent) and $x =$ final concentration of metal in the solution (milligrams per liter).

    a. Write a simple linear regression model relating *y* to *x*.

    b. For one metal, a simple linear regression analysis yielded $r^2 = .92$. Interpret this result.

Applying the Concepts—Intermediate

9.79 Performance in online courses. Florida State University information scientists assessed the impact of online courses on student performance (*Educational Technology & Society*, Jan. 2005). Each in a sample of 24 graduate students enrolled in an online advanced Web application course was asked, "How many courses per semester (on average) do you take online?"

Each student's performance on weekly quizzes was also recorded. The information scientists found that the number of online courses and the weekly quiz grade were negatively correlated at $r = -.726$.

a. Give a practical interpretation of r.

b. The researchers concluded that there was a "significant negative correlation" between the number of online courses and the weekly quiz grade. Do you agree?

9.80 Salary linked to height. Are short people shortchanged when it comes to salary? According to business professors T. A. Judge (University of Florida) and D. M. Cable (University of North Carolina), tall people tend to earn more money over their career than short people earn. (*Journal of Applied Psychology*, June 2004.) Using data collected from participants in the National Longitudinal Surveys begun in 1979, the researchers computed the correlation between average earnings (in dollars) from 1985 to 2000 and height (in inches) for several occupations. The results are given in the following table:

Occupation	Correlation, r	Sample Size, n
Sales	.41	117
Managers	.35	455
Blue Collar	.32	349
Service Workers	.31	265
Professional/Technical	.30	453
Clerical	.25	358
Crafts/Forepersons	.24	250

Source: Judge, T. A., & Cable, D. M. "The effect of physical height on workplace success and income: Preliminary test of a theoretical model," *Journal of Applied Psychology*, Vol. 89, No. 3, June 2004 (Table 5).

a. Interpret the value of r for people in sales occupations.

b. Compute r^2 for people in sales occupations. Interpret the result.

c. Give H_0 and H_a for testing whether average earnings and height are positively correlated.

d. The test statistic for testing H_0 and H_a in part **c** is $t = \dfrac{r\sqrt{n-2}}{\sqrt{1-r^2}}$. Compute the value of t for people in sales occupations.

e. Use the result you obtained in part **d** to conduct the test at $\alpha = .01$. State the appropriate conclusion.

f. Select another occupation and repeat parts **a–e**.

9.81 View of rotated objects. *Perception & Psychophysics* (July 1998) reported on a study of how people view three-dimensional objects projected onto a rotating two-dimensional image. Each in a sample of 25 university students viewed various depth-rotated objects (e.g., a hairbrush, a duck, and a shoe) until they recognized the object. The recognition exposure time—that is, the minimum time (in milliseconds) required for the subject to recognize the object—was recorded for each object. In addition, each subject rated the "goodness of view" of the object on a numerical scale, with lower scale values corresponding to better views. The following table gives the correlation coefficient r between recognition exposure time and goodness of view for several different rotated objects:

Object	r	t
Piano	.447	2.40
Bench	−.057	.27
Motorbike	.619	3.78
Armchair	.294	1.47
Teapot	.949	14.50

a. Interpret the value of r for each object.

b. Calculate and interpret the value of r^2 for each object.

c. The table also includes the t-value for testing the null hypothesis of no correlation (i.e., for testing $H_0: \beta_1 = 0$). Interpret these results.

9.82 Snow geese feeding trial. Botanists at the University of Toronto conducted a series of experiments to investigate the feeding habits of baby snow geese. (*Journal of Applied Ecology*, Vol. 32, 1995.) Goslings were deprived of food until their guts were empty and then were allowed to feed for 6 hours on a diet of plants or Purina® Duck Chow®. For each feeding trial, the change in the weight of the gosling after 2.5 hours was recorded as a percentage of the bird's initial weight. Two other variables recorded were digestion efficiency (measured as a percentage) and amount of acid-detergent fiber in the digestive tract (also measured as a

SNOWGEESE (First and last 5 observations listed)

Feeding Trial	Diet	Weight Change (%)	Digestion Efficiency (%)	Acid-Detergent Fiber (%)
1	Plants	−6	0	28.5
2	Plants	−5	2.5	27.5
3	Plants	−4.5	5	27.5
4	Plants	0	0	32.5
5	Plants	2	0	32
⋮	⋮	⋮	⋮	⋮
38	Duck Chow	9	59	8.5
39	Duck Chow	12	52.5	8
40	Duck Chow	8.5	75	6
41	Duck Chow	10.5	72.5	6.5
42	Duck Chow	14	69	7

Source: Gadallah, F. L., and Jefferies, R. L. "Forage quality in brood rearing areas of the lesser snow goose and the growth of captive goslings," *Journal of Applied Biology*, Vol. 32, No. 2, 1995, pp. 281–282 (adapted from Figures 2 and 3).

percentage). Data on 42 feeding trials are saved in the **SNOWGEESE** file. The first and last 5 observations are listed in the table (bottom, p. 524).

a. The botanists were interested in the correlation between weight change (y) and digestion efficiency (x). Plot the data for these two variables in a scattergram. Do you observe a trend?

b. Find the coefficient of correlation relating weight change y to digestion efficiency x. Interpret this value.

c. Conduct a test to determine whether weight change y is correlated with digestion efficiency x. Use $\alpha = .01$.

d. Repeat parts **b** and **c**, but exclude the data for trials that used duck chow. What do you conclude?

e. The botanists were also interested in the correlation between digestion efficiency y and acid-detergent fibre x. Repeat parts **a–d** for these two variables.

9.83 Dance/movement therapy. In cotherapy, two or more therapists lead a group. An article in the *American Journal of Dance Therapy* (Spring/Summer 1995) examined the use of cotherapy in dance/movement therapy. Two of several variables measured on each of a sample of 136 professional dance/movement therapists were years x of formal training and reported success rate y (measured as a percentage) of coleading dance/movement therapy groups.

a. Propose a linear model relating y to x.

b. The researcher hypothesized that dance/movement therapists with more years in formal dance training will report higher perceived success rates in cotherapy relationships. State the hypothesis in terms of the parameter of the model you proposed in part **a**.

c. The correlation coefficient for the sample data was reported as $r = -.26$. Interpret this result.

d. Does the value of r in part **c** support the hypothesis in part **b**? Test, using $\alpha = .05$. [*Hint:* See the last footnote at the bottom of page 518.]

⊙ **NAMEGAME2**

9.84 The "name game." Refer to the *Journal of Experimental Psychology—Applied* (June 2000) name-retrieval study, first presented in Exercise 9.30 (p. 501). Find and interpret the values of r and r^2 for the simple linear regression relating the proportion of names recalled (y) and the position (order) of the student (x) during the "name game."

⊙ **BOXING2**

9.85 Effect of massage on boxing. Refer to the *British Journal of Sports Medicine* (April 2000) study of the effect of massage on boxing performance, presented in Exercise 9.60 (p. 513). Find and interpret the values of r and r^2 for the simple linear regression relating the blood lac-tate concentration and the boxer's perceived recovery.

Applying the Concepts—Advanced

9.86 Pain tolerance study. A study published in *Psychosomatic Medicine* (Mar./Apr. 2001) explored the relationship between reported severity of pain and actual pain tolerance in 337 patients who suffer from chronic pain. Each patient reported his or her severity of chronic pain on a seven-point scale (1 = no pain, 7 = extreme pain). To obtain a pain tolerance level, a tourniquet was applied to the arm of each patient and twisted. The maximum pain level tolerated was measured on a quantitative scale.

a. According to the researchers, "Correlational analysis revealed a small but significant inverse relationship between [actual] pain tolerance and the reported severity of chronic pain." On the basis of this statement, is the value of r for the 337 patients positive or negative?

b. Suppose that the result reported in part **a** is significant at $\alpha = .05$. Find the approximate value of r for the sample of 337 patients. [*Hint:* Use the formula $t = r\sqrt{(n-2)}/\sqrt{(1-r^2)}$.]

9.6 Using the Model for Estimation and Prediction

If we are satisfied that a useful model has been found to describe the relationship between reaction time and percent of drug in the bloodstream, we are ready for step 5 in our regression modeling procedure: using the model for estimation and prediction.

> *The most common uses of a probabilistic model for making inferences can be divided into two categories. The first is the use of the model for estimating the mean value of y, E(y), for a specific value of x.*

For our drug reaction example, we may want to estimate the mean response time for all people whose blood contains 4% of the drug.

> *The second use of the model entails predicting a new individual y value for a given x.*

That is, we may want to predict the reaction time for a specific person who possesses 4% of the drug in the bloodstream.

In the first case, we are attempting to estimate the mean value of y for a very large number of experiments at the given x value. In the second case, we are trying to predict the outcome of a single experiment at the given x value. Which of these uses of the model—estimating the mean value of y or predicting an individual new value of y (for the same value of x)—can be accomplished with the greater accuracy?

Before answering this question, we first consider the problem of choosing an estimator (or predictor) of the mean (or a new individual) y value. We will use the least squares prediction equation

$$\hat{y} = \hat{\beta}_0 + \hat{\beta}_1 x$$

both to estimate the mean value of y and to predict a specific new value of y for a given value of x. For our example, we found that

$$\hat{y} = -.1 + .7x$$

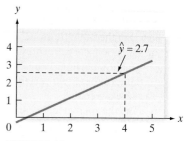

FIGURE 9.18

Estimated mean value and predicted individual value of reaction time y for $x = 4$

so the estimated mean reaction time for all people when $x = 4$ (the drug is 4% of the blood content) is

$$\hat{y} = -.1 + .7(4) = 2.7 \text{ seconds}$$

The same value is used to predict a new y value when $x = 4$. That is, both the estimated mean and the predicted value of y are $\hat{y} = 2.7$ when $x = 4$, as shown in Figure 9.18.

The difference between these two uses of the model lies in the accuracies of the estimate and the prediction, best measured by the sampling errors of the least squares line when it is used as an estimator and as a predictor, respectively. These errors are reflected in the standard deviations given in the following box:

Sampling Errors for the Estimator of the Mean of y and the Predictor of an Individual New Value of y

1. The standard deviation of the sampling distribution of the estimator $\hat{y}$ of the mean value of y at a specific value of x, say x_p, is

    $$\sigma_{\hat{y}} = \sigma \sqrt{\frac{1}{n} + \frac{(x_p - \bar{x})^2}{SS_{xx}}}$$

    where σ is the standard deviation of the random error ε. We refer to $\sigma_{\hat{y}}$ as the standard error of $\hat{y}$.

2. The standard deviation of the prediction error for the predictor $\hat{y}$ of an individual new y value at a specific value of x is

    $$\sigma_{(y-\hat{y})} = \sigma \sqrt{1 + \frac{1}{n} + \frac{(x_p - \bar{x})^2}{SS_{xx}}}$$

    where σ is the standard deviation of the random error ε. We refer to $\sigma_{(y-\hat{y})}$ as the standard error of prediction.

The true value of σ is rarely known, so we estimate σ by s and calculate the estimation and prediction intervals as shown in the next two boxes:

A 100$(1 - \alpha)$% Confidence Interval for the Mean Value of y at $x = x_p$

$$\hat{y} + t_{\alpha/2}(\text{Estimated standard error of } \hat{y})$$

or

$$\hat{y} \pm t_{\alpha/2} s \sqrt{\frac{1}{n} + \frac{(x_p - \bar{x})^2}{SS_{xx}}}$$

where $t_{\alpha/2}$ is based on $(n - 2)$ degrees of freedom.

A 100$(1 - \alpha)$% Prediction Interval* for an Individual New Value of y at $x = x_p$

$$\hat{y} \pm t_{\alpha/2}(\text{Estimated standard error of prediction})$$

or

$$\hat{y} \pm t_{\alpha/2}s\sqrt{1 + \frac{1}{n} + \frac{(x_p - \bar{x})^2}{\text{SS}_{xx}}}$$

where $t_{\alpha/2}$ is based on $(n - 2)$ degrees of freedom.

EXAMPLE 9.6

ESTIMATING THE MEAN OF y—Drug Reaction Regression

Problem Refer to the simple linear regression on drug reaction. Find a 95% confidence interval for the mean reaction time when the concentration of the drug in the bloodstream is 4%.

Solution For a 4% concentration, $x = 4$ and the confidence interval for the mean value of y is

$$\hat{y} \pm t_{\alpha/2}s\sqrt{\frac{1}{n} + \frac{(x_p - \bar{x})^2}{\text{SS}_{xx}}} = \hat{y} \pm t_{.025}s\sqrt{\frac{1}{5} + \frac{(4 - \bar{x})^2}{\text{SS}_{xx}}}$$

where $t_{.025}$ is based on $n - 2 = 5 - 2 = 3$ degrees of freedom. Recall that $\hat{y} = 2.7$, $s = .61$, $\bar{x} = 3$, and $\text{SS}_{xx} = 10$. From Table VI in Appendix A, $t_{.025} = 3.182$. Thus, we have

$$2.7 \pm (3.182)(.61)\sqrt{\frac{1}{5} + \frac{(4 - 3)^2}{10}} = 2.7 \pm (3.182)(.61)(.55)$$

$$= 2.7 \pm (3.182)(.34)$$

$$= 2.7 \pm 1.1$$

Therefore, when the percentage of drug in the bloodstream is 4%, we can be 95% confident that the mean reaction time for all possible subjects will range from 1.6 to 3.8 seconds.

Look Back Note that we used a small amount of data (a small sample size) for purposes of illustration in fitting the least squares line. The interval would probably be narrower if more information had been obtained from a larger sample.

Now Work Exercise 9.92a–d

EXAMPLE 9.7

PREDICTING AN INDIVIDUAL VALUE OF y—Drug Reaction Regression

Problem Refer again to the drug reaction regression. Predict the reaction time for the next performance of the experiment for a subject with a drug concentration of 4%. Use a 95% prediction interval.

Solution To predict the response time for an individual new subject for whom $x = 4$, we calculate the 95% prediction interval as

$$\hat{y} \pm t_{\alpha/2}s\sqrt{1 + \frac{1}{n} + \frac{(x_p - \bar{x})^2}{\text{SS}_{xx}}} = 2.7 \pm (3.182)(.61)\sqrt{1 + \frac{1}{5} + \frac{(4 - 3)^2}{10}}$$

$$= 2.7 \pm (3.182)(.61)(1.14)$$

$$= 2.7 \pm (3.182)(.70)$$

$$= 2.7 \pm 2.2$$

*The term *prediction interval* is used when the interval formed is intended to enclose the value of a random variable. The term *confidence interval* is reserved for the estimation of population parameters (such as the mean).

Therefore, when the drug concentration for an individual is 4%, we predict with 95% confidence that the reaction time for this new individual will fall into the interval from .5 to 4.9 seconds.

Look Back Like the confidence interval for the mean value of y, the prediction interval for y is quite large. This is because we have chosen a simple example (one with only five data points) to fit the least squares line. The width of the prediction interval could be reduced by using a larger number of data points.

Now Work Exercise 9.92e

Both the confidence interval for $E(y)$ and the prediction interval for y can be obtained from a statistical software package. Figure 9.19 is a MINITAB printout showing the confidence interval and prediction interval, respectively, for the data in the drug example.

The 95% confidence interval for $E(y)$ when $x = 4$, highlighted under "95% CI" in Figure 9.19, is (1.645, 3.755). The 95% prediction interval for y when $x = 4$, highlighted in Figure 9.19 under "95% PI", is (.503, 4.897). These agree with the ones computed in Examples 9.6 and 9.7.

Note that the prediction interval for an individual new value of y is *always* wider than the corresponding confidence interval for the mean value of y. Will this always be true? The answer is "Yes." The error in estimating the mean value of y, $E(y)$, for a given value of x, say, x_p, is the distance between the least squares line and the true line of means, $E(y) = \beta_0 + \beta_1 x$. This error, $[\hat{y} - E(y)]$, is shown in Figure 9.20. In contrast, *the error $(y_p - \hat{y})$ in predicting some future value of y is the sum of two errors*: the error in estimating the mean of y, $E(y)$, shown in Figure 9.20, plus the random error that is a component of the value of y that is to be predicted. (See Figure 9.21.) Consequently, the

```
Predicted Values for New Observations

New
Obs    Fit   SE Fit      95% CI            95% PI
 1    2.700  0.332   (1.645, 3.755)   (0.503, 4.897)

Values of Predictors for New Observations

New
Obs   DRUG_X
 1     4.00
```

FIGURE 9.19

MINITAB printout giving 95% confidence interval for $E(y)$ and 95% prediction interval for y

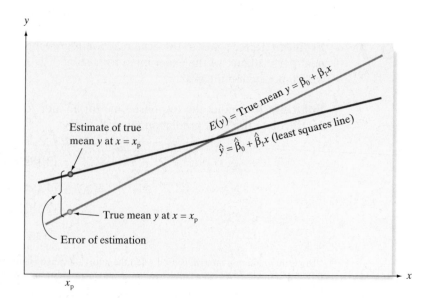

FIGURE 9.20

Error in estimating the mean value of y for a given value of x

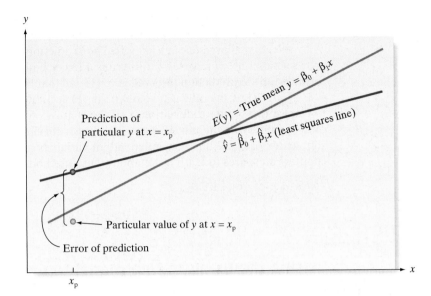

FIGURE 9.21

Error in predicting a future value of y for a given value of x

error in predicting a particular value of y will be larger than the error in estimating the mean value of y for a particular value of x. Note from their formulas that both the error of estimation and the error of prediction take their smallest values when $x_p = \bar{x}$. The farther x_p lies from $\bar{x}$, the larger will be the errors of estimation and prediction. You can see why this is true by noting the deviations for different values of x_p between the actual line of means $E(y) = \beta_0 + \beta_1 x$ and the predicted line of means $\hat{y} = \hat{\beta}_0 + \hat{\beta}_1 x$ shown in Figure 9.21. The deviation is larger at the extremes of the interval, where the largest and smallest values of x in the data set occur.

Both the confidence intervals for mean values and the prediction intervals for new values are depicted over the entire range of the regression line in Figure 9.22. You can see that the confidence interval is always narrower than the prediction interval and that they are both narrowest at the mean $\bar{x}$, increasing steadily as the distance $|x - \bar{x}|$ increases. In fact, when x is selected far enough away from $\bar{x}$ so that it falls outside the range of the sample data, it is dangerous to make any inferences about $E(y)$ or y.

Caution

Using the least squares prediction equation to estimate the mean value of y or to predict a particular value of y for values of x that fall outside the range of the values of x contained in your sample data may lead to errors of estimation or prediction that are much larger than expected. Although the least squares model may provide a very good fit to the data over the range of x values contained in the sample, it could give a poor representation of the true model for values of x outside that region.

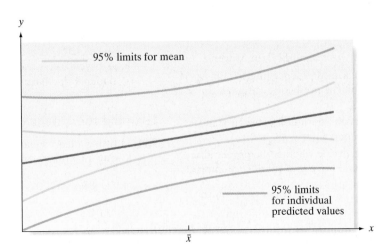

FIGURE 9.22

Confidence intervals for mean values and prediction intervals for new values

The width of the confidence interval grows smaller as n is increased; thus, in theory, you can obtain as precise an estimate of the mean value of y as desired (at any given x) by selecting a large enough sample. The prediction interval for a new value of y also grows smaller as n increases, but there is a lower limit on its width. If you examine the formula for the prediction interval, you will see that the interval can get no smaller than $\hat{y} \pm z_{\alpha/2}\sigma$.* Thus, the only way to obtain more accurate predictions for new values of y is to reduce the standard deviation σ of the regression model. This can be accomplished only by improving the model, either by using a curvilinear (rather than linear) relationship with x or by adding new independent variables to the model (or both). Consult the chapter references to learn more about these methods of improving the model.

Now Work Exercise 9.92f

STATISTICS IN ACTION REVISITED

Using the Straight-Line Model to Predict Pipe Location for the Dowsing Data

The group of German physicists who conducted the dowsing experiments stated that the data for the three "best" dowsers empirically support the dowsing theory. If so, then the straight-line model relating a dowser's guess (x) to actual pipe location (y) should yield accurate predictions. The MINITAB printout shown in Figure SIA9.5 gives a 95% prediction interval for y when a dowser guesses $x = 50$ meters (the middle of the 100-meter-long waterpipe). The highlighted interval is $(-9.3, 100.23)$. Thus, we can be 95% confident that the actual pipe location will fall between -9.3 meters and 100.23 meters for this guess. Since the pipe is only 100 meters long, the interval in effect ranges from 0 to 100 meters—the entire length of the pipe! This result, of course, is due to the fact that the straight-line model is not a statistically useful predictor of pipe location, a fact we discovered in the previous Statistics in Action Revisited sections.

```
Predicted Values for New Observations

New
Obs     Fit   SE Fit       95% CI              95% PI
  1   45.47     5.15  (34.83, 56.10)   (-9.30, 100.23)


Values of Predictors for New Observations

New
Obs   GUESS
  1    50.0
```

FIGURE SIA9.5

MINITAB prediction interval for dowsing data

Exercises 9.87–9.103

Understanding the Principles

9.87 Explain the difference between y and $E(y)$ for a given x.

9.88 *True or False*. For a given x, a confidence interval for $E(y)$ will always be wider than a prediction interval for y.

9.89 *True or False*. The greater the deviation between x and $\bar{x}$, the wider the prediction interval for y will be.

9.90 For each of the following, decide whether the proper inference is a prediction interval for y or a confidence interval for $E(y)$:

a. A jeweler wants to predict the selling price of a diamond stone on the basis of its size (number of carats).

b. A psychologist wants to estimate the average IQ of all patients who have a certain income level.

Learning the Mechanics

9.91 In fitting a least squares line to $n = 10$ data points, the following quantities were computed:

$$SS_{xx} = 32, \bar{x} = 3, SS_{yy} = 26, \bar{y} = 4, SS_{xy} = 28$$

---

*The result follows from the facts that, for large n, $t_{\alpha/2} \approx z_{\alpha/2}$, $s \approx \sigma$, and the last two terms under the radical in the standard error of the predictor are approximately 0.

a. Find the least squares line.

b. Graph the least squares line.

c. Calculate SSE.

d. Calculate s^2.

e. Find a 95% confidence interval for the mean value of y when $x_p = 2.5$.

f. Find a 95% prediction interval for y when $x_p = 4$.

9.92 Consider the following pairs of measurements:
NW

🔘 **LM9_92**

x	1	2	3	4	5	6	7
y	3	5	4	6	7	7	10

a. Construct a scattergram of these data.

b. Find the least squares line, and plot it on your scattergram.

c. Find s^2.

d. Find a 90% confidence interval for the mean value of y when $x = 4$. Plot the upper and lower bounds of the confidence interval on your scattergram.

e. Find a 90% prediction interval for a new value of y when $x = 4$. Plot the upper and lower bounds of the prediction interval on your scattergram.

f. Compare the widths of the intervals you constructed in parts **d** and **e**. Which is wider and why?

9.93 Consider the pairs of measurements shown in the following
NW   table:

🔘 **LM9_93**

x	4	6	0	5	2	3	2	6	2	1
y	3	5	−1	4	3	2	0	4	1	1

For these data, $SS_{xx} = 38.900, SS_{yy} = 33.600, SS_{xy} = 32.8$, and $\hat{y} = -.414 + .843x$.

a. Construct a scattergram of the data.

b. Plot the least squares line on your scattergram.

c. Use a 95% confidence interval to estimate the mean value of y when $x_p = 6$. Plot the upper and lower bounds of the interval on your scattergram.

d. Repeat part **c** for $x_p = 3.2$ and $x_p = 0$.

e. Compare the widths of the three confidence intervals you constructed in parts **c** and **d**, and explain why they differ.

9.94 Refer to Exercise 9.93.

a. Using no information about x, estimate and calculate a 95% confidence interval for the mean value of y. [*Hint:* Use the one-sample t methodology of Section 5.3.]

b. Plot the estimated mean value and the confidence interval as horizontal lines on your scattergram.

c. Compare the confidence intervals you calculated in parts **c** and **d** of Exercise 9.93 with the one you calculated in part **a** of this exercise. Does x appear to contribute information about the mean value of y?

d. Check the answer you gave in part **c** with a statistical test of the null hypothesis $H_0: \beta_1 = 0$ against $H_a: \beta_1 \neq 0$. Use $\alpha = .05$.

Applying the Concepts—Basic

9.95 **English as a second language reading ability.** Refer to the *Bilingual Research Journal* (Summer 2006) study of the relationship of Spanish (first-language) grammatical knowledge to English (second-language) reading, presented in Exercise 9.54 (p. 512). Recall that three simple linear regressions were used to model the English reading (ESLR) score (y) as a function of Spanish grammar (SG), Spanish reading (SR), and English grammar (EG), respectively.

a. If the researchers want to predict the ELSR score (y) of a native Spanish-speaking adult who scored 50% in Spanish grammar (x), how should they proceed?

b. If the researchers want to estimate the mean ELSR score $E(y)$ of all native Spanish-speaking adults who scored 70% in Spanish grammar (x), how should they proceed?

🔘 **PGADRIVER**

9.96 **Ranking driving performance of professional golfers.** Refer to *The Sport Journal* (Winter 2007) study of a new method for ranking the total driving performance of golfers on the Professional Golf Association (PGA) tour, presented in Exercise 9.25 (p. 500). You fit a straight-line model relating driving accuracy (y) to driving distance (x) to the data saved in the **PGADRIVER** file. A MINITAB printout with prediction and confidence intervals for a driving distance of $x = 300$ yards is shown below.

a. Locate the 95% prediction interval for driving accuracy (y) on the printout, and give a practical interpretation of the result.

b. Locate the 95% prediction interval for mean driving accuracy (y) on the printout, and give a practical interpretation of the result.

c. If you are interested in knowing the average driving accuracy of all PGA golfers who have a driving distance of 300 yards, which of the intervals is relevant? Explain.

🔘 **GOBIANTS**

9.97 **Mongolian desert ants.** Refer to the *Journal of Biogeography* (Dec. 2003) study of ant sites in Mongolia, presented in Exercise 9.22 (p. 498). You applied the method of least squares to

```
Predicted Values for New Observations

New
Obs      Fit   SE Fit        95% CI              95% PI
  1   61.309    0.357   (60.586, 62.032)   (56.724, 65.894)


Values of Predictors for New Observations

New
Obs   DISTANCE
  1        300
```

MINITAB output for
Exercise 9.96

Predictions

Obs	Region	Rain	Temp	Predicted Rain	Lower prediction limit of Rain	Upper prediction limit of Rain
1	DryStepp	196	5.7	201.977	179.525	224.430
2	DryStepp	196	5.7	201.977	179.525	224.430
3	DryStepp	179	7.0	180.704	163.694	197.714
4	DryStepp	197	8.0	164.340	150.594	178.085
5	DryStepp	149	8.5	156.157	143.513	168.802
6	GobiDese	112	10.7	120.156	106.038	134.274
7	GobiDese	125	11.4	108.701	92.298	125.104
8	GobiDese	99	10.9	116.883	102.172	131.595
9	GobiDese	125	11.4	108.701	92.298	125.104
10	GobiDese	84	11.4	108.701	92.298	125.104
11	GobiDese	115	11.4	108.701	92.298	125.104

SAS output for Exercise 9.97

the data in the **GOBIANTS** file to estimate the straight-line model relating annual rainfall (y) and maximum daily temperature (x). An SAS printout giving 95% prediction intervals for the amount of rainfall at each of the 11 sites is shown above. Select the interval associated with site (observation) 7 and interpret it practically.

9.98 Sweetness of orange juice. Refer to the simple linear, regression of sweetness index y and amount of pectin, x, for $n = 24$ orange juice samples, presented in Exercise 9.28 (p. 501). The SPSS printout of the analysis is shown below. A 90% confidence interval for the mean sweetness index $E(y)$ for each value of x is shown on the SPSS spreadsheet. Select an observation and interpret this interval.

Applying the Concepts—Intermediate

9.99 Sports participation survey. The Sasakawa Sports Foundation conducted a national survey to assess the physical activity patterns of Japanese adults. The accompanying table lists the frequency (average number of days in the past year) and the amount of time (average number of minutes per single activity) Japanese adults spent participating in a sample of 11 sports activities.

JAPANSPORTS

Activity	Frequency x (days/years)	Amount of Time y (minutes)
Jogging	135	43
Cycling	68	99
Aerobics	44	61
Swimming	39	60
Volleyball	30	80
Tennis	21	100
Softball	16	91
Baseball	19	127
Skating	7	115
Skiing	10	249
Golf	5	262

Source: J. Bennett, ed. *Statistics in Sport.* London: Arnold, 1998 (adapted from Figure 11.6).

	run	sweet	pectin	lower90m	upper90m
1	1	5.2	220	5.64898	5.83848
2	2	5.5	227	5.63898	5.81613
3	3	6.0	259	5.57819	5.72904
4	4	5.9	210	5.66194	5.87173
5	5	5.8	224	5.64337	5.82560
6	6	6.0	215	5.65564	5.85493
7	7	5.8	231	5.63284	5.80379
8	8	5.6	268	5.55553	5.71011
9	9	5.6	239	5.61947	5.78019
10	10	5.9	212	5.65946	5.86497
11	11	5.4	410	5.05526	5.55416
12	12	5.6	256	5.58517	5.73592
13	13	5.8	306	5.43785	5.65219
14	14	5.5	259	5.57819	5.72904
15	15	5.3	284	5.50957	5.68213
16	16	5.3	383	5.15725	5.57694
17	17	5.7	271	5.54743	5.70434
18	18	5.5	264	5.56591	5.71821
19	19	5.7	227	5.63898	5.81613
20	20	5.3	263	5.56843	5.72031
21	21	5.9	232	5.63125	5.80075
22	22	5.8	220	5.64898	5.83848
23	23	5.8	246	5.60640	5.76091
24	24	5.9	241	5.61587	5.77454

SPSS output for Exercise 9.98

a. Write the equation of a straight-line model relating duration (y) to frequency (x).

b. Find the least squares prediction equation.

c. Is there evidence of a linear relationship between y and x? Test, using $\alpha = .05$.

d. Use the least squares line to predict the amount of time Japanese adults participate in a sport that they play 25 times a year. Form a 95% confidence interval around the prediction and interpret the result.

NAMEGAME2

9.100 **The "name game."** Refer to the *Journal of Experimental Psychology—Applied* (June 2000) name-retrieval study, presented in Exercise 9.30 (p. 501).

a. Find a 99% confidence interval for the mean recall proportion for students in the fifth position during the "name game." Interpret the result.

b. Find a 99% prediction interval for the recall proportion of a particular student in the fifth position during the "name game." Interpret the result.

c. Compare the intervals you found in parts **a** and **b**. Which interval is wider? Will this always be the case? Explain.

LIQUIDSPILL

9.101 **Spreading rate of spilled liquid.** Refer to the *Chemicial Engineering Progress* (Jan. 2005) study of the rate at which a spilled volatile liquid will spread across a surface, presented in Exercise 9.31 (p. 502). Recall that simple linear regression was used to model y = mass of the spill as a function of y = elapsed time of the spill.

a. Find a 99% confidence interval for the mean mass of all spills with an elapsed time of 15 minutes. Interpret the result.

b. Find a 99% prediction interval for the mass of a single spill with an elapsed time of 15 minutes. Interpret the result.

c. Compare the intervals you found in parts **a** and **b**. Which interval is wider? Will this always be the case? Explain.

SNOWGEESE

9.102 **Feeding habits of snow geese.** Refer to the *Journal of Applied Ecology* feeding study of the relationship between the weight change y of baby snow geese and their digestion efficiency x, presented in Exercise 9.82 (p. 524).

a. Fit the simple linear regression model to the data.

b. Do you recommend using the model to predict weight change y? Explain.

c. Use the model to form a 95% confidence interval for the mean weight change of all baby snow geese with a digestion efficiency of $x = 15\%$. Interpret the interval.

Applying the Concepts—Advanced

CUTTOOL

9.103 **Life tests of cutting tools.** Refer to the data saved in the **CUTTOOL** file of Exercise 9.46 (p. 507).

a. Use a 90% confidence interval to estimate the mean useful life of a brand-A cutting tool when the cutting speed is 45 meters per minute. Repeat for brand B. Compare the widths of the two intervals and comment on the reasons for any difference.

b. Use a 90% prediction interval to predict the useful life of a brand-A cutting tool when the cutting speed is 45 meters per minute. Repeat for brand B. Compare the widths of the two intervals with each other and with the two intervals you calculated in part **a**. Comment on the reasons for any differences.

c. Note that the estimation and prediction you performed in parts **a** and **b** were for a value of x that was not included in the original sample. That is, the value $x = 45$ was not part of the sample. However, the value is within the range of x values in the sample, so that the regression model spans the x value for which the estimation and prediction were made. In such situations, estimation and prediction represent **interpolations**.

Suppose you were asked to predict the useful life of a brand-A cutting tool for a cutting speed of $x = 100$ meters per minute. Since the given value of x is outside the range of the sample x values, the prediction is an example of **extrapolation**. Predict the useful life of a brand-A cutting tool that is operated at 100 meters per minute, and construct a 95% confidence interval for the actual useful life of the tool. What additional assumption do you have to make in order to ensure the validity of an extrapolation?

9.7 A Complete Example

In the previous sections, we presented the basic elements necessary to fit and use a straight-line regression model. In this section, we will assemble these elements by applying them in an example with the aid of computer software.

Suppose a fire insurance company wants to relate the amount of fire damage in major residential fires to the distance between the burning house and the nearest fire station. The study is to be conducted in a large suburb of a major city; a sample of 15 recent fires in this suburb is selected. The amount of damage, y, and the distance between the fire and the nearest fire station, x, are recorded for each fire. The results are given in Table 9.7.

Step 1 First, we hypothesize a model to relate fire damage, y, to the distance from the nearest fire station, x. We hypothesize a straight-line probabilistic model:

$$y = \beta_0 + \beta_1 x + \varepsilon$$

 **FIREDAM**

TABLE 9.7 Fire Damage Data

Distance from Fire Station, x (miles)	Fire Damage y (thousands of dollars)
3.4	26.2
1.8	17.8
4.6	31.3
2.3	23.1
3.1	27.5
5.5	36.0
.7	14.1
3.0	22.3
2.6	19.6
4.3	31.3
2.1	24.0
1.1	17.3
6.1	43.2
4.8	36.4
3.8	26.1

Step 2 Next, we enter the data of Table 9.7 into a computer and use statistical software to estimate the unknown parameters in the deterministic component of the hypothesized model. The SAS printout for the simple linear regression analysis is shown in Figure 9.23. The least squares estimate of the slope β_1 and intercept β_0, highlighted on the printout, are

$$\hat{\beta}_1 = 4.91933$$
$$\hat{\beta}_0 = 10.27793$$

Dependent Variable: DAMAGE

Analysis of Variance

Source	DF	Sum of Squares	Mean Square	F Value	Pr > F
Model	1	841.76636	841.76636	156.89	<.0001
Error	13	69.75098	5.36546		
Corrected Total	14	911.51733			

Root MSE	2.31635	R-Square	0.9235	
Dependent Mean	26.41333	Adj R-Sq	0.9176	
Coeff Var	8.76961			

Parameter Estimates

Variable	DF	Parameter Estimate	Standard Error	t Value	Pr > |t|	95% Confidence Limits	
Intercept	1	10.27793	1.42028	7.24	<.0001	7.20960	13.34625
DISTANCE	1	4.91933	0.39275	12.53	<.0001	4.07085	5.76781

Output Statistics

Obs	DISTANCE	Dep Var DAMAGE	Predicted Value	Std Error Mean Predict	95% CL Predict		Residual
1	3.4	26.2000	27.0037	0.5999	21.8344	32.1729	-0.8037
2	1.8	17.8000	19.1327	0.8340	13.8141	24.4514	-1.3327
3	4.6	31.3000	32.9068	0.7915	27.6186	38.1951	-1.6068
4	2.3	23.1000	21.5924	0.7112	16.3577	26.8271	1.5076
5	3.1	27.5000	25.5279	0.6022	20.3573	30.6984	1.9721
6	5.5	36.0000	37.3342	1.0573	31.8334	42.8351	-1.3342
7	0.7	14.1000	13.7215	1.1766	8.1087	19.3342	0.3785
8	3	22.3000	25.0359	0.6081	19.8622	30.2097	-2.7359
9	2.6	19.6000	23.0682	0.6550	17.8678	28.2686	-3.4682
10	4.3	31.3000	31.4311	0.7198	26.1908	36.6713	-0.1311
11	2.1	24.0000	20.6085	0.7566	15.3442	25.8729	3.3915
12	1.1	17.3000	15.6892	1.0444	10.1999	21.1785	1.6108
13	6.1	43.2000	40.2858	1.2587	34.5906	45.9811	2.9142
14	4.8	36.4000	33.8907	0.8450	28.5640	39.2175	2.5093
15	3.8	26.1000	28.9714	0.6320	23.7843	34.1585	-2.8714
16	3.5	.	27.4956	0.6043	22.3239	32.6672	.

FIGURE 9.23

SAS printout for fire damage regression

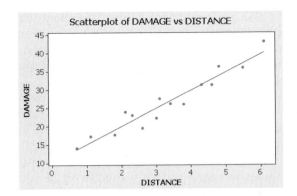

Scatterplot of DAMAGE vs DISTANCE

FIGURE 9.24

MINITAB scatterplot with least squares line for fire damage regression analysis

and the least squares equation is (rounded)

$$\hat{y} = 10.28 + 4.92x$$

This prediction equation is graphed by MINITAB in Figure 9.24, along with a plot of the data points.

The least squares estimate of the slope, $\hat{\beta}_1 = 4.92$, implies that the estimated mean damage increases by \$4,920 for each additional mile from the fire station. This interpretation is valid over the range of x, or from .7 to 6.1 miles from the station. The estimated y-intercept, $\hat{\beta}_0 = 10.28$, has the interpretation that a fire 0 miles from the fire station has an estimated mean damage of \$10,280. Although this would seem to apply to the fire station itself, remember that the y-intercept is meaningfully interpretable only if $x = 0$ is within the sampled range of the independent variable. Since $x = 0$ is outside the range, β_0 has no practical interpretation.

Step 3   Now we specify the probability distribution of the random-error component ε. The assumptions about the distribution are identical to those listed in Section 9.3. Although we know that these assumptions are not completely satisfied (they rarely are for practical problems), we are willing to assume that they are approximately satisfied for this example. The estimate of the standard deviation σ of ε, highlighted on the SAS printout, is

$$s = 2.31635$$

This implies that most of the observed fire damage (y) values will fall within approximately $2s = 4.64$ thousand dollars of their respective predicted values when the least squares line is used.

Step 4   We can now check the usefulness of the hypothesized model—in other words, whether x really contributes information for the prediction of y by the straight-line model. First, test the null hypothesis that the slope β_1 is 0—that is, that there is no linear relationship between fire damage and the distance from the nearest fire station—against the alternative hypothesis that fire damage increases as the distance increases. We test

$$H_0: \beta_1 = 0$$
$$H_a: \beta_1 > 0$$

The two-tailed observed significance level for testing $H_a: \beta_1 \neq 0$, highlighted on the SAS printout, is less than .0001. Thus, the p-value for our one-tailed test is less than half of this value (.00005). This small p-value leaves little doubt that mean fire damage and distance between the fire and the fire station are at least linearly related, with mean fire damage increasing as the distance increases.

We gain additional information about the relationship by forming a confidence interval for the slope β_1. A 95% confidence interval, highlighted on the SAS printout, is (4.071, 5.768). Thus, with 95% confidence, we estimate that the interval from $4,071 to $5,768 encloses the mean increase (β_1) in fire damage per additional mile in distance from the fire station.

Another measure of the utility of the model is the coefficient of determination, r^2. The value (also highlighted on the printout) is $r^2 = .9235$, which implies that about 92% of the sample variation in fire damage (y) is explained by the distance (x) between the fire and the fire station.

The coefficient of correlation, r, that measures the strength of the linear relationship between y and x is not shown on the SAS printout and must be calculated. Using the facts that $r = \sqrt{r^2}$ in simple linear regression and that r and $\hat{\beta}_1$ have the same sign, we calculate

$$r = +\sqrt{r^2} = \sqrt{.9235} = .96$$

The high correlation confirms our conclusion that β_1 is greater than 0; it appears that fire damage and distance from the fire station are positively correlated. All signs point to a strong linear relationship between y and x.

Step 5 We are now prepared to use the least squares model. Suppose the insurance company wants to predict the fire damage if a major residential fire were to occur 3.5 miles from the nearest fire station. The predicted value (highlighted at the bottom of the SAS printout) is $\hat{y} = 27.496$, while the 95% prediction interval (also highlighted) is (22.324, 32.667). Therefore, with 95% confidence, we predict fire damage in a major residential fire 3.5 miles from the nearest station to be between $22,324 and $32,667.

Caution

We would not use this model to make predictions for homes less than .7 mile or more than 6.1 miles from the nearest fire station. A look at the data in Table 9.7 reveals that all the x-values fall between .7 and 6.1. It is dangerous to use the model to make predictions outside the region in which the sample data fall. A straight line might not provide a good model for the relationship between the mean value of y and the value of x when stretched over a wider range of x-values.

9.8 A Nonparametric Test for Correlation (Optional)

When the simple linear regression assumptions (Section 9.3) are violated (e.g., the random error ϵ has a highly skewed distribution), an alternative method of analysis may be required. One approach is to apply a nonparametric test for correlation based on ranks.

To illustrate, suppose 10 new paintings are shown to two art critics and each critic ranks the paintings from 1 (best) to 10 (worst). We want to determine whether the critics' ranks are related. Does a correspondence exist between their ratings? If a painting is ranked high by critic 1, is it likely to be ranked high by critic 2? Or do high rankings by one critic correspond to low rankings by the other? That is, are the rankings of the critics *correlated*?

If the rankings are as shown in the "Perfect Agreement" columns of Table 9.8, we immediately notice that the critics agree on the rank of every painting. High ranks correspond to high ranks and low ranks to low ranks. This is an example of a *perfect positive correlation* between the ranks. In contrast, if the rankings appear as shown in the "Perfect Disagreement" columns of Table 9.9, then high ranks for one critic correspond to low ranks for the other. This is an example of *perfect negative correlation*.

In practice, you will rarely see perfect positive or perfect negative correlation between the ranks. In fact, it is quite possible for the critics' ranks to appear as shown in Table 9.9. Note that these rankings indicate some agreement between the critics, but not perfect agreement, thus pointing up a need for a measure of rank correlation.

TABLE 9.8   Rankings of 10 Paintings by Two Critics

Painting	Perfect Agreement		Perfect Disagreement	
	Critic 1	Critic 2	Critic 1	Critic 2
1	4	4	9	2
2	1	1	3	8
3	7	7	5	6
4	5	5	1	10
5	2	2	2	9
6	6	6	10	1
7	8	8	6	5
8	3	3	4	7
9	10	10	8	3
10	9	9	7	4

Spearman's rank correlation coefficient, r_s, provides a measure of correlation between ranks. The formula for this measure of correlation is given in the box (p. 538). We also give a formula that is identical to r_s when there are no ties in rankings; this formula provides a good approximation to r_s when the number of ties is small relative to the number of pairs.

TABLE 9.9   Rankings of Paintings: Less-than-perfect Agreement

Painting	Critic		Difference between Rank 1 and Rank 2	
	1	2	d	d^2
1	4	5	−1	1
2	1	2	−1	1
3	9	10	−1	1
4	5	6	−1	1
5	2	1	1	1
6	10	9	1	1
7	7	7	0	0
8	3	3	0	0
9	6	4	2	4
10	8	8	0	0
				$\Sigma d^2 = 10$

Biography

CHARLES E. SPEARMAN (1863–1945)—
Spearman's Correlation

London-born Charles Spearman was educated at Leamington College before joining the British Army. After 20 years as a highly decorated officer, Spearman retired from the army and moved to Germany to begin his study of experimental psychology at the University of Liepzig. At the age of 41, he earned his Ph.D. and ultimately became one of the most influential figures in the field of psychology. Spearman was the originator of the classical theory of mental tests and developed the "two-factor" theory of intelligence. These theories were used to develop and support the "Plus-Elevens" tests in England: exams administered to British 11-year-olds that predict whether they should attend a university or a technical school. Spearman was greatly influenced by the works of Francis Galton (p. 561); consequently, he developed a strong statistical background. While conducting his research on intelligence, he proposed the rank-order correlation coefficient—now called "Spearman's correlation coefficient." During his career, Spearman spent time at various universities, including University College (London), Columbia University, Catholic University, and the University of Cairo (Egypt).

Note that if the ranks for the two critics are identical, as in the second and third columns of Table 9.8, the differences between the ranks will all be 0. Thus,

$$r_s = 1 - \frac{6 \sum d^2}{n(n^2 - 1)} = 1 - \frac{6(0)}{10(99)} = 1$$

That is, *perfect positive correlation* between the pairs of ranks is characterized by a Spearman correlation coefficient of $r_s = 1$. When the ranks indicate perfect disagreement, as in the fourth and fifth columns of Table 14.9, $\sum d_i^2 = 330$ and

$$r_s = 1 - \frac{6(330)}{10(99)} = -1.$$

Thus, *perfect negative correlation* is indicated by $r_s = -1$.

Spearman's Rank Correlation Coefficient

$$r_s = \frac{SS_{uv}}{\sqrt{SS_{uu}SS_{vv}}}$$

where

$$SS_{uv} = \sum (u_i - \bar{u})(v_i - \bar{v}) = \sum u_i v_i - \frac{\left(\sum u_i \right)\left(\sum v_i \right)}{n}$$

$$SS_{uu} = \sum (u_i - \bar{u})^2 = \sum u_i^2 - \frac{\left(\sum u_i \right)^2}{n}$$

$$SS_{vv} = \sum (v_i - \bar{v})^2 = \sum v_i^2 - \frac{\left(\sum v_i \right)^2}{n}$$

u_i = Rank of the *i*th observation in sample 1

v_i = Rank of the *i*th observation in sample 2

n = Number of pairs of observations (number of observations in each sample)

Shortcut Formula for r_s*

$$r_s = 1 - \frac{6 \sum d_i^2}{n(n^2 - 1)}$$

where

$d_i = u_i - v_i$ (difference in the ranks of the *i* th observations for samples 1 and 2)

For the data of Table 9.9,

$$r_s = 1 - \frac{6 \sum d^2}{n(n^2 - 1)} = 1 - \frac{6(10)}{10(99)} = 1 - \frac{6}{99} = .94$$

The fact that r_s is *close* to 1 indicates that the critics tend to agree, but the agreement is not perfect.

The value of r_s always falls between -1 and $+1$, with $+1$ indicating perfect positive correlation and -1 indicating a perfect negative correlation. The closer r_s falls to $+1$ or -1, the greater the correlation between the ranks. Conversely, the nearer r_s is to 0, the less is the correlation.

---

*The shortcut formula is not exact when there are tied measurements, but it is a good approximation when the total number of ties is not large relative to *n*.

Note that the concept of correlation implies that two responses are obtained for each experimental unit. In the art critics example, each painting received two ranks (one from each critic) and the objective of the study was to determine the degree of positive correlation between the two rankings. Rank correlation methods can be used to measure the correlation between any pair of variables. If two variables are measured on each of n experimental units, we rank the measurements associated with each variable separately. Ties receive the average of the ranks of the tied observations. Then we calculate the value of r_s for the two rankings. This value measures the rank correlation between the two variables. We illustrate the procedure in Example 9.8.

EXAMPLE 9.8

SPEARMAN'S RANK CORRELATION— Smoking Versus Babies' Weights

Problem A study is conducted to investigate the relationship between cigarette smoking during pregnancy and the weights of newborn infants. The 15 women smokers who make up the sample kept accurate records of the number of cigarettes smoked during their pregnancies, and the weights of their children were recorded at birth. The data are given in Table 9.10.

a. Calculate and interpret Spearman's rank correlation coefficient for the data.

b. Use a nonparametric test to determine whether level of cigarette smoking and weights of newborns are negatively correlated for all smoking mothers. Use $\alpha = .05$.

Solution

a. We first rank the number of cigarettes smoked per day, assigning a 1 to the smallest number (12) and a 15 to the largest (46). Note that the two ties receive the averages of their respective ranks. Similarly, we assign ranks to the 15 babies' weights. Since the number of ties is relatively small, we will use the shortcut formula to calculate r_s. The differences d between the ranks of the babies' weights and the ranks of the number of cigarettes smoked per day are shown in Table 9.10. The squares of the differences, d^2, are also given. Thus,

$$r_s = 1 - \frac{6\sum d_i^2}{n(n^2 - 1)} = 1 - \frac{6(795)}{15(15^2 - 1)} = 1 - 1.42 = -.42$$

The value of r_s can also be obtained by computer. A SAS printout of the analysis is shown in Figure 9.25. The value of r_s, highlighted on the printout, agrees (except for rounding) with our hand-calculated value, $-.42$.

The negative correlation coefficient indicates that in this sample an increase in the number of cigarettes smoked per day is *associated with* (but is not necessarily the *cause of*) a decrease in the weight of the newborn infant.

NEWBORN

TABLE 9.10 Data and Calculations for Example 9.8

Woman	Cigarettes per Day	Rank	Baby's Weight (pounds)	Rank	d	d^2
1	12	1	7.7	5	−4	16
2	15	2	8.1	9	−7	49
3	35	13	6.9	4	9	81
4	21	7	8.2	10	−3	9
5	20	5.5	8.6	13.5	−8	64
6	17	3	8.3	11.5	−8.5	72.25
7	19	4	9.4	15	−11	121
8	46	15	7.8	6	9	81
9	20	5.5	8.3	11.5	−6	36
10	25	8.5	5.2	1	7.5	56.25
11	39	14	6.4	3	11	121
12	25	8.5	7.9	7	1.5	2.25
13	30	12	8.0	8	4	16
14	27	10	6.1	2	8	64
15	29	11	8.6	13.5	−2.5	6.25
					Total =	795

```
                    The CORR Procedure

        2  Variables:    CIGARETTES WEIGHT


        Spearman Correlation Coefficients, N = 15
               Prob > |r| under HO: Rho=0

                         CIGARETTES        WEIGHT

        CIGARETTES        1.00000        -0.42473
                                          0.1145

        WEIGHT           -0.42473         1.00000
                          0.1145
```

FIGURE 9.25

SAS Spearman correlation printout for Example 9.8

b. If we define ρ as the **population rank correlation coefficient** [i.e., the rank correlation coefficient that could be calculated from all (x, y) values in the population], we can determine whether level of cigarette smoking and weights of newborns are negatively correlated by conducting the following test:

H_0: $\rho = 0$ (no population correlation between ranks)

H_a: $\rho < 0$ (negative population correlation between ranks)

Test statistic: r_s (the *sample* Spearman rank correlation coefficient)

To determine a rejection region, we consult Table XII in Appendix A, which is partially reproduced in Table 9.11. Note that the left-hand column gives values of n, the number of pairs of observations. The entries in the table are values for an upper-tail rejection region, since only positive values are given. Thus, for $n = 15$ and $\alpha = .05$, the value .441 is the boundary of the upper-tailed rejection region, so $P(r_s > .441) = .05$ if H_0: $\rho = 0$ is true. Similarly, for negative values of r_s, we have $P(r_s < -.441) = .05$ if $\rho = 0$. That is, we expect to see $r_s < -.441$ only 5% of the time if there is really no relationship between the ranks of the variables.

The lower-tailed rejection region is therefore

Rejection region $(\alpha = .05)$: $r_s < -.441$

TABLE 9.11 Reproduction of Part of Table XII in Appendix A: Critical Values of Spearman's Rank Correlation Coefficient

n	$\alpha = .05$	$\alpha = .025$	$\alpha = .01$	$\alpha = .005$
5	.900	—	—	—
6	.829	.886	.943	—
7	.714	.786	.893	—
8	.643	.738	.833	.881
9	.600	.683	.783	.833
10	.564	.648	.745	.794
11	.523	.623	.736	.818
12	.497	.591	.703	.780
13	.475	.566	.673	.745
14	.457	.545	.646	.716
15	.441	.525	.623	.689
16	.425	.507	.601	.666
17	.412	.490	.582	.645
18	.399	.476	.564	.625
19	.388	.462	.549	.608
20	.377	.450	.534	.591

Since the calculated $r_s = -.42$ is not less than $-.441$, we cannot reject H_0 at the $\alpha = .05$ level of significance. That is, this sample of 15 smoking mothers provides insufficient evidence to conclude that a negative correlation exists between the number of cigarettes smoked and the weight of newborns for the populations of measurements corresponding to all smoking mothers. This does not, of course, mean that no relationship exists. A study using a larger sample of smokers and taking other factors into account (father's weight, sex of newborn child, etc.) would be more likely to reveal whether smoking and the weight of a newborn child are related.

Look Back The two-tailed p-value of the test (.1145) is highlighted on the SAS printout, shown in Figure 9.25. Since the lower-tailed p-value, $.1145/2 = .05725$, exceeds $\alpha = .05$, our conclusion is the same: Do not reject H_0.

Now Work Exercise 9.109

A summary of Spearman's nonparametric test for correlation is given in the following box:

Spearman's Nonparametric Test for Rank Correlation

One-Tailed Test

$H_0: \rho = 0$

$H_a: \rho > 0 \ [\text{or } H_a: \rho < 0]$

Test statistics: r_s, the sample rank correlation (see the formulas for calculating r_s)

Rejection region: $r_s > r_{s,\alpha}$

$[\text{or } r_s < -r_{s,\alpha} \text{ when } H_a: \rho < 0]$

where $r_{s,\alpha}$ is the value from Table XII corresponding to the upper-tail area α and n pairs of observations

Two-Tailed Test

$H_0: \rho = 0$

$H_a: \rho \neq 0$

Rejection region: $|r_s| > r_{s,\alpha/2}$

where $r_{s,\alpha/2}$ is the value from Table XII corresponding to the upper-tail area $\alpha/2$ and n pairs of observations

Ties: Assign tied measurements the average of the ranks they would receive if they were unequal, but occurred in successive order. For example, if the third-ranked and fourth-ranked measurements are tied, assign each a rank of $(3 + 4)/2 = 3.5$. The number of ties should be small relative to the total number of observations.

Conditions Required for a Valid Spearman's Test

1. The sample of experimental units on which the two variables are measured is randomly selected.
2. The probability distributions of the two variables are continuous.

Exercises 9.104–9.120

Understanding the Principles

9.104 What is the value of r_s when there is perfect negative rank correlation between two variables? Perfect positive rank correlation?

9.105 What conditions are required for a valid Spearman's test?

Learning the Mechanics

9.106 Use Table XII of Appendix A to find each of the following probabilities:
   a. $P(r_s > .508)$ when $n = 22$
   b. $P(r_s > .448)$ when $n = 28$
   c. $P(r_s \leq .648)$ when $n = 10$
   d. $P(r_s < -.738 \text{ or } r_s > .738)$ when $n = 8$

9.107 Specify the rejection region for Spearman's nonparametric test for rank correlation in each of the following situations:

   a. $H_0: \rho = 0, H_a: \rho \neq 0, n = 10, \alpha = .05$
   b. $H_0: \rho = 0, H_a: \rho > 0, n = 20, \alpha = .025$
   c. $H_0: \rho = 0, H_a: \rho < 0, n = 30, \alpha = .01$

9.108 Compute Spearman's rank correlation coefficient for each of the following pairs of sample observations:

a.

x	33	61	20	19	40
y	26	36	65	25	35

b.

x	89	102	120	137	41
y	81	94	75	52	136

c.

x	2	15	4	10
y	11	2	15	21

d.

x	5	20	15	10	3
y	80	83	91	82	87

9.109 The following sample data were collected on variables x
NW and y:

LM9_109

x	0	3	0	−4	3	0	4
y	0	2	2	0	3	1	2

a. Specify the null and alternative hypotheses that should be used in conducting a hypothesis test to determine whether the variables x and y are correlated.
b. Conduct the test of part **a**, using $\alpha = .05$.
c. What is the approximate p-value of the test of part **b**?
d. What assumptions are necessary to ensure the validity of the test of part **b**?

Applying the Concepts—Basic

9.110 **Mongolian desert ants.** Refer to the *Journal of Biogeography* (Dec. 2003) study of ants in Mongolia, presented in Exercise 9.22 (p. 498). Data on annual rainfall, maximum daily temperature, and number of ant species recorded at each of 11 study sites are reproduced in the table below.
a. Consider the data for the five sites in the Dry Steppe region only. Rank the five annual rainfall amounts. Then rank the five maximum daily temperature values.
b. Use the ranks, from part **a** to find and interpret the rank correlation between annual rainfall (y) and maximum daily temperature (x).
c. Repeat parts **a** and **b** for the six sites in the Gobi Desert region.
d. Now consider the rank correlation between the number of ant species (y) and annual rainfall (x). Using all the data, compute and interpret Spearman's rank correlation statistic.

GOBIANTS

Site	Region	Annual Rainfall (mm)	Max. Daily Temp. (°C)	Number of Ant Species
1	Dry Steppe	196	5.7	3
2	Dry Steppe	196	5.7	3
3	Dry Steppe	179	7.0	52
4	Dry Steppe	197	8.0	7
5	Dry Steppe	149	8.5	5
6	Gobi Desert	112	10.7	49
7	Gobi Desert	125	11.4	5
8	Gobi Desert	99	10.9	4
9	Gobi Desert	125	11.4	4
10	Gobi Desert	84	11.4	5
11	Gobi Desert	115	11.4	−4

Source: Pfeiffer, M., et al. "Community organization and species richness of ants in Mongolia along an ecological gradient from steppe to Gobi desert," *Journal of Biogeography*, Vol. 30, No. 12, Dec. 2003 (Tables 1 and 2).

9.111 **Extending the life of an aluminum smelter pot.** Refer to the *American Ceramic Society Bulletin* (Feb. 2005) study of the lifetime of an aluminum smelter pot, presented in Exercise 9.24 (p. 499). Since the life of a smelter pot de-

pends on the porosity of the brick lining, the researchers measured the apparent porosity and the mean pore diameter of each of six bricks. The data are reproduced in the following table:

SMELTPOT

Brick	Apparent Porosity (%)	Mean Pore Diameter (micrometers)
A	18.8	12.0
B	18.3	9.7
C	16.3	7.3
D	6.9	5.3
E	17.1	10.9
F	20.4	16.8

Source: Bonadia, P., et al. "Aluminosilicate refractories for aluminum cell linings," *American Ceramic Society Bulletin*, Vol. 84, No. 2, Feb. 2005 (Table II).

a. Rank the apparent porosity values for the six bricks. Then rank the six pore diameter values.
b. Use the ranks from part **a** to find the rank correlation between apparent porosity (y) and mean pore diameter (x). Interpret the result.
c. Conduct a test for positive rank correlation. Use $\alpha = .01$.

9.112 **Organizational use of the Internet.** Researchers from the United Kingdom and Germany attempted to develop a theoretically grounded measure of organizational Internet use (OIU) and published their results in *Internet Research* (Vol. 15, 2005). Using data collected from a sample of 77 websites, they investigated the link between OIU level (measured on a seven-point scale) and several observation-based indicators. Spearman's rank correlation coefficient (and associated p-values) for several indicators are shown in the following table:

Indicator	Correlation with OIU Level	
	r_s	p-value
Navigability	.179	.148
Transactions	.334	.023
Locatability	.590	.000
Information Richness	−.115	.252
Number of files	.114	.255

Source: Brock, J. K., and Zhou, Y. "Organizational use of the internet," *Internet Research,* Vol. 15, No. 1, 2005 (Table IV).

a. Interpret each of the values of r_s given in the table.
b. Interpret each of the p-values given in the table. (Use $\alpha = .10$ to conduct each test.)

9.113 **Effect of massage on boxers.** Refer to the *British Journal of Sports Medicine* (Apr. 2000) study of the effect of massaging boxers between rounds, presented in Exercise 9.60 (p. 513). Two variables measured on the boxers were blood lactate level (y) and the boxer's perceived recovery (x). The data for 16 five-round boxing performances are reproduced in the next table.
a. Rank the values of the 16 blood lactate levels.
b. Rank the values of the 16 perceived recovery values.
c. Use the ranks from parts **a** and **b** to compute Spearman's rank correlation coefficient. Give a practical interpretation of the result.

d. Find the rejection region for a test to determine whether *y* and *x* are rank correlated. Use $\alpha = .10$.

e. What is the conclusion of the test you conducted in part **d**? State your answer in the words of the problem.

BOXING2

Blood Lactate Level	Perceived Recovery
3.8	7
4.2	7
4.8	11
4.1	12
5.0	12
5.3	12
4.2	13
2.4	17
3.7	17
5.3	17
5.8	18
6.0	18
5.9	21
6.3	21
5.5	20
6.5	24

Source: Hemmings, B., Smith, M., Graydon, J., and Dyson, R. "Effects of massage on physiological restoration, perceived recovery, and repeated sports performance," *British Journal of Sports Medicine*, Vol. 34, No. 2, Apr. 2000 (data adapted from Figure 3).

9.114 Assessment of biometric recognition methods. Biometric technologies have been developed to detect or verify an individual's identity. These methods are based on physiological characteristics (called *biometric signatures*), such as facial features, the iris of the eye, fingerprints, the voice, the shape of the hand, and the gait. In *Chance* (Winter 2004), four biometric recognition algorithms were compared. All four were applied to 1,196 biometric signatures, and "match" scores were obtained. The Spearman correlation between match scores for each possible pair of algorithms was determined. The rank correlation matrix is as follows:

Method	I	II	III	IV
I	1	.189	.592	.340
II		1	.205	.324
III			1	.314
IV				1

a. Locate the largest rank correlation and interpret its value.

b. Locate the smallest rank correlation and interpret its value.

Applying the Concepts—Intermediate

9.115 Media coverage of the 9–11 attacks and public opinion. The terrorist attacks of September 11, 2001, and related events (e.g., the war in Iraq) have, and continue to receive, much media coverage. How has this coverage influenced the American public's concern about terrorism? This was the topic of research conducted by journalism professors at the University of Missouri (*International Journal of Public Opinion*, Winter 2004). Using random-digit dialing, they conducted a telephone survey of 235 Americans. Each person was asked to rate, on a scale of 1 to 5, his or her level of concern about each of eight topics: a long war, future terrorist attacks, the effect on the economy, the Israel–Palestine conflict, biological threats, air travel safety, war protests, and Afghan civilian deaths. The eight scores were summed to obtain a "public agenda" score. The respondents were also asked how many days per week they read the newspaper, watch the local television news, and watch national television news. The responses to these three questions were also summed to obtain a "media agenda" score. The researchers hypothesized that the public agenda score would be positively related to the media agenda score.

a. Spearman's rank correlation between the two scores was computed to be $r_s = .643$. Give a practical interpretation of this value.

b. The researchers removed the "length of war" question from the data and recomputed the "public agenda" score. Spearman's rank correlation between the public agenda and media agenda scores was then calculated as $r_s = .714$. Interpret this result.

c. Refer to part b. Conduct Spearman's test for positive rank correlation at $\alpha = .01$.

9.116 The "name game" Refer to the *Journal of Experimental Psychology—Applied* (June 2000) study in which the "name game" was used to help groups of students learn the names of other students in the group, presented in Exercise 9.30 (pp. 501–502). Recall that one goal of the study was to investigate the relationship between proportion *y* of names recalled by a student and position (order *x*), of the student during the game. The data for 144 students in the first eight positions are saved in the **NAMEGAME2** file. (The first five and last five observations in the data set are listed in the next table.)

a. To properly apply the parametric test for correlation on the basis of the Pearson coefficient of correlation, *r* (Section 9.5), both the *x* and *y* variables must be normally distributed. Demonstrate that this assumption is violated for these data. What are the consequences of the violation?

b. Find Spearman's rank correlation coefficient on the accompanying SAS printout and interpret its value.

c. Find the observed significance level for testing for zero rank correlation on the SAS printout, and interpret its value.

SAS output for Exercise 9.116

```
            The CORR Procedure

  2  Variables:    POSITION RECALL


Spearman Correlation Coefficients, N = 144
        Prob > |r| under H0: Rho=0

                 POSITION            RECALL

  POSITION       1.00000             0.20652
                                     0.0130

  RECALL         0.20652             1.00000
                 0.0130
```

d. At $\alpha = .05$, is there sufficient evidence of rank correlation between proportion y of names recalled by a student and position (order x), of the student during the game?

NAMEGAME2 (selected observations)

Position	Recall
2	0.04
2	0.37
2	1.00
2	0.99
2	0.79
⋮	⋮
9	0.72
9	0.88
9	0.46
9	0.54
9	0.99

Source: Morris, P. E., and Fritz, C. O. "The name game: Using retrieval practice to improve the learning of names," *Journal of Experimental Psychology—Applied,* Vol. 6, No. 2, June 2000 (data simulated from Figure 2).

FCAT

9.117 FCAT scores and poverty. Refer to the *Journal of Educational and Behavioral Statistics* (Spring 2004) analysis of the link between Florida Comprehensive Assessment Test (FCAT) scores and sociodemographic factors, presented in Exercise 9.26 (p. 500). Data on average math and reading FCAT scores of third graders, as well as the percentage of students below the poverty level, for a sample of 22 Florida elementary schools are saved in the **FCAT** file.

a. Compute and interpret Spearman's rank correlation between FCAT math score (y) and percentage (x) of students below the poverty level.

b. Compute and interpret Spearman's rank correlation between FCAT reading score (y) and percentage (x) of students below the poverty level.

c. Determine whether the value of r_s in part **a** would lead you to conclude that FCAT math score and percent below poverty level are negatively rank correlated in the population of all Florida elementary schools. Use $\alpha = .01$ to make your decision.

d. Determine whether the value of r_s in part **b** would lead you to conclude that FCAT reading score and percent below poverty level are negatively rank correlated in the population of all Florida elementary schools. Use $\alpha = .01$ to make your decision.

9.118 Pain empathy and brain activity. Refer to the *Science* (Feb. 20, 2004) study on the relationship between brain activity and pain-related empathy in persons who watch others in pain, presented in Exercise 9.62 (pp. 513–514). Recall that 16 female partners watched while painful stimulation was applied to the finger of their respective male partners. The two variables of interest were y = female's pain-related brain activity (measured on a scale ranging from -2 to 2) and x = female's score on the Empathic Concern Scale (0 to 25 points). The data are reproduced in the accompanying table. Use Spearman's rank correlation test to answer the research question, "Do

people scoring higher in empathy show higher pain-related brain activity?"

BRAINPAIN

Couple	Brain Activity (y)	Empathic Concern (x)
1	.05	12
2	−.03	13
3	.12	14
4	.20	16
5	.35	16
6	0	17
7	.26	17
8	.50	18
9	.20	18
10	.21	18
11	.45	19
12	.30	20
13	.20	21
14	.22	22
15	.76	23
16	.35	24

Source: Singer, T. et al. "Empathy for pain involves the affective but not sensory components of pain," *Science,* Vol. 303, Feb. 20, 2004. (Adapted from Figure 4.)

9.119 Study of child bipolar disorders. Psychiatric researchers at the University of Pittsburgh Medical Center have developed a new test for measuring manic symptoms in pediatric bipolar patients. (*Journal of Child and Adolescent Psychopharmacology*, Dec., 2003.) The new test is called the Kiddie Schedule for Affective Disorders and Schizophrenia-Mania Rating Scale (KSADS-MRS). The new test was compared with the standard test, the Clinical Global Impressions—Bipolar Scale (CGI-BP). Both tests were administered to a

MANIA

Patient	Change in KSADS-MRS (%)	Improvement in CGI-BP
1	80	6
2	65	5
3	20	4
4	−15	4
5	−50	4
6	20	3
7	−30	3
8	−70	3
9	−10	2
10	−25	2
11	−35	2
12	−65	2
13	−65	2
14	−70	2
15	−80	2
16	−90	2
17	−95	2
18	−90	1

Source: Axelson, D. et al. "A preliminary study of the Kiddie Schedule for Affective Disorders and Schizophrenia for School-Age Children Mania Rating Scale for children and adolescents," *Journal of Child and Adolescent Psychopharmacology,* Vol. 13, No. 4, Dec. 2003 (adapted from Figure 2).

sample of 18 pediatric patients before and after they were treated for manic symptoms. The changes in the test scores are recorded in the table on p. 544.

a. The researchers used Spearman's statistic to measure the correlation between the changes in the two test scores. Compute the value of r_s.

b. Is there sufficient evidence (at $\alpha = .05$) of positive rank correlation between the two test score changes in the population of all pediatric patients with manic symptoms?

9.120 Public perceptions of health risks. The *Journal of Experimental Psychology: Learning, Memory, and Cognition* (July 2005) published a study of the ability of people to judge the risk of an infectious disease. The researchers asked German college students to estimate the number of people infected with a certain disease in a typical year. The median estimates, as well as the actual incidence for each in a sample of 24 infections, are reproduced in the accompanying table.

a. Use graphs to demonstrate that the variables actual incidence and estimated incidence are not normally distributed.

b. Recall that the researchers used regression to model the relationship between actual incidence and estimated incidence. How does the result you found in part **a** affect this analysis?

c. Find Spearman's correlation coefficient for the two variables. Interpret this value.

d. Refer to part **c.** At $\alpha = .01$, is there a positive association between actual incidence and estimated incidence?

INFECTION

Infection	Actual Incidence	Estimated Incidence
Polio	0.25	300
Diphtheria	1	1000
Trachoma	1.75	691
Rabbit Fever	2	200
Cholera	3	17.5
Leprosy	5	0.8
Tetanus	9	1000
Hemorrhagic Fever	10	150
Trichinosis	22	326.5
Undulant Fever	23	146.5
Well's Disease	39	370
Gas Gangrene	98	400
Parrot Fever	119	225
Typhoid	152	200
Q Fever	179	200
Malaria	936	400
Syphilis	1514	1500
Dysentery	1627	1000
Gonorrhea	2926	6000
Meningitis	4019	5000
Tuberculosis	12619	1500
Hepatitis	14889	1000
Gastroenteritis	203864	37000
Botulism	15	37500

Source: Hertwig, R., Pachur, T., & Kurzenhauser, S. "Judgments of risk frequencies: Tests of possible cognitive mechanisms," *Journal of Experimental Psychology: Learning, Memory, and Cognition*, Vol. 31, No. 4, July 2005 (Table 1).

KEY TERMS

[Note: Starred (*) items are from the optional section in this chapter.]

Bivariate relationship 515
Coefficient of correlation 515
Coefficient of determination 520
Confidence interval for mean
  of *y* 526
Dependent variable 486
Deterministic model 484
Deterministic relationship 484
Errors of prediction 488
Estimated standard error of the
  regression model 504

Estimated standard error of
  the least squares slope 508
Extrapolation 533
Independent variable 486
Interpolation 533
Least squares estimates 489
Least squares line (or regression
  line or prediction equation) 489
Line of means 486
Method of least squares 489
Population correlation
  coefficient 518
*Population rank correlation
  coefficient 540

Prediction interval for *y* 527
Predictor variable 486
Probabilistic model 484
Probabilistic relationship 485
Random error 484
Regression analysis (modeling) 486
Response variable 486
Scattergram 488
Slope 486
*Spearman's rank correlation 537
Straight-line (first-order) model 486
y-intercept 486

GUIDE TO SIMPLE LINEAR REGRESSION

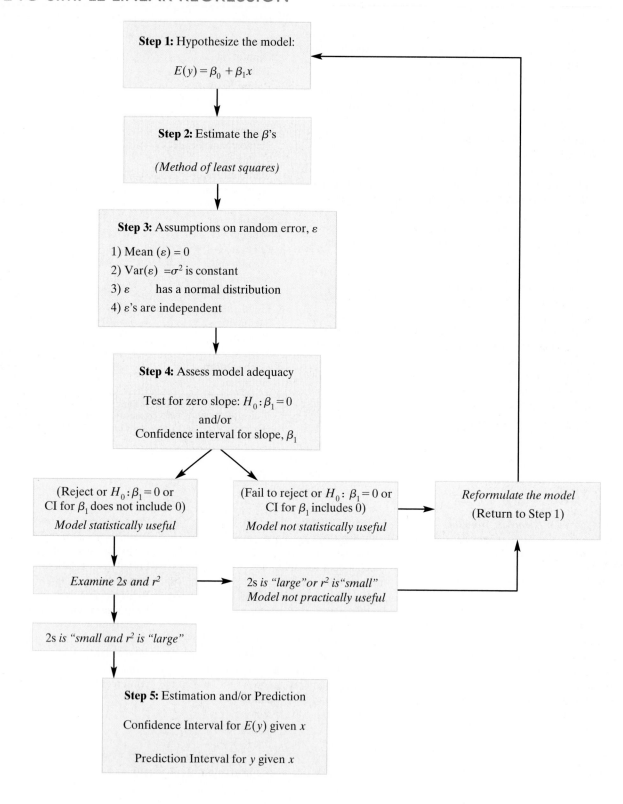

Step 1: Hypothesize the model:

$$E(y) = \beta_0 + \beta_1 x$$

Step 2: Estimate the β's

(Method of least squares)

Step 3: Assumptions on random error, ε

1) Mean $(\varepsilon) = 0$
2) Var$(\varepsilon) = \sigma^2$ is constant
3) ε has a normal distribution
4) ε's are independent

Step 4: Assess model adequacy

Test for zero slope: $H_0 : \beta_1 = 0$
and/or
Confidence interval for slope, β_1

(Reject or $H_0 : \beta_1 = 0$ or
CI for β_1 does not include 0)
Model statistically useful

(Fail to reject or $H_0 : \beta_1 = 0$ or
CI for β_1 includes 0)
Model not statistically useful

Reformulate the model
(Return to Step 1)

Examine 2s and r^2

2s is *"large"* or r^2 is *"small"*
Model not practically useful

2s is *"small and r^2 is "large"*

Step 5: Estimation and/or Prediction

Confidence Interval for $E(y)$ given x

Prediction Interval for y given x

CHAPTER NOTES

Simple linear regression variables:

y = **Dependent** variable (quantitative)

x = **Independent** variable (quantitative)

Method of least squared properties:

1. average error of prediction = 0
2. sum of squared errors is minimum

Practical interpretation of y-interpret:

Predicted y-value when $x = 0$

(no practical interpretation if $x = 0$ is either nonsensical or outside range of sample data)

Practical interpretation of slope:

Increase (or decrease) in y for every one-unit increase in x.

Practical interpretation of model standard deviation s:

Ninety-five percent of y-values fall within $2s$ of their respective predicted values

Width of *confidence interval for E(y)* will always be **narrower** than width of *prediction interval for* y

First-order (straight-line) model:

$$E(y) = \beta_0 + \beta_1 x$$

where $E(y)$ = mean of y

β_0 = *y-intercept* of line (point where line intercepts y-axis)

β_1 = **slope** of line (change in y for every one-unit change in x)

Coefficient of correlation, r:

1. ranges between -1 and $+1$
2. measures strength of *linear relationship* between y and x

Coefficient of determination, r^2:

1. ranges between 0 and 1
2. measures proportion of sample variation in y "explained" by the model.

Nonparametric test for *rank correlation:*

Spearman's test

Key Symbols/Notation

y	Dependent variable (variable to be predicted)
x	Independent variable (variable used to predict)
$E(y)$	Expected (mean) of y
β_0	y-intercept of true line
β_1	slope of true line.
$\hat{\beta}_0$	Least squared estimate of y-intercept
$\hat{\beta}_1$	Least squares estimate of slope
ϵ	Random error
$\hat{y}$	Predicted value of y for a given x-value
$(y - \hat{y})$	Estimated error of prediction
SSE	Sum of squared errors of prediction
r	Coefficient of correlation
r^2	Coefficient of determination
x_p	Value of x used to predict y
$r^2 = \dfrac{SS_{yy} - SSE}{SS_{yy}}$	Coefficient of determination
$\hat{y} \pm t_{\alpha/2}s\sqrt{\dfrac{1}{n} + \dfrac{(x_p - \bar{x})^2}{SS_{xx}}}$	$(1 - \alpha)100\%$ confidence interval for $E(y)$ when $x = x_p$
$\hat{y} \pm t_{\alpha/2}s\sqrt{1 + \dfrac{1}{n} + \dfrac{(x_p - \bar{x})^2}{SS_{xx}}}$	$(1 - \alpha)100\%$ prediction interval for y when $x = x_p$
$*r_s = 1 - \dfrac{6\sum d^2}{n(n^2 - 1)},$	Spearman's rank correlation coefficient

where d_i = difference between ranks of ith observations for x and y

SUPPLEMENTARY EXERCISES 9.121–9.144

Understanding the Principles

[Note: Starred (*) exercises are from the optional section in this chapter.]

9.121 Give the general form of a straight-line model for $E(y)$.

9.122 Explain the difference between a probabilistic model and a deterministic model.

9.123 *True or False.* In simple linear regression, about 95% of the y-values in the sample will fall within $2s$ of their respective predicted values.

9.124 Outline the five steps in a simple linear regression analysis.

Learning the Mechanics

9.125 Consider the following sample data:

y	5	1	3
x	5	1	3

a. Construct a scattergram for the data.
b. It is possible to find many lines for which $\Sigma(y - \hat{y}) = 0$. For this reason, the criterion $\Sigma(y - \hat{y}) = 0$ is not used to identify the "best-fitting" straight line. Find two lines that have $\Sigma(y - \hat{y}) = 0$.
c. Find the least squares line.
d. Compare the value of SSE for the least squares line with that of the two lines you found in part **b.** What principle of least squares is demonstrated by this comparison?

9.126 Consider the following 10 data points:

LM9_126

x	3	5	6	4	3	7	6	5	4	7
y	4	3	2	1	2	3	3	5	4	2

a. Plot the data on a scattergram.
b. Calculate the values of r and r^2.
c. Is there sufficient evidence to indicate that x and y are linearly correlated? Test at the $\alpha = .10$ level of significance.
*d. Calculate Spearman's rank correlation coefficient for the data. Is there evidence that x and y are rank correlated? Test using $\alpha = .10$.

Applying the Concepts—Basic

9.127 Arsenic in soil. In Denver, Colorado, environmentalists have discovered a link between high arsenic levels in soil and a crabgrass killer used in the 1950s and 1960s. (*Environmental Science & Technology*, Sept. 1, 2000.) The recent discovery was based, in part, on the accompanying scattergrams. The graphs plot the level of the metals cadmium and arsenic, respectively, against the distance from a former smelter plant for samples of soil taken from Denver residential properties.

a. Normally, the metal level in soil decreases as distance from the source (e.g., a smelter plant) increases. Pro-

Scattergrams for
Exercise 9.127

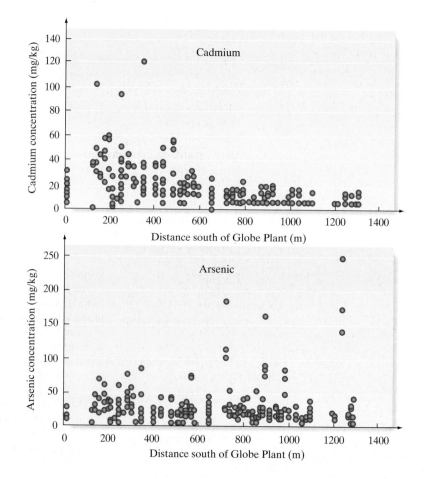

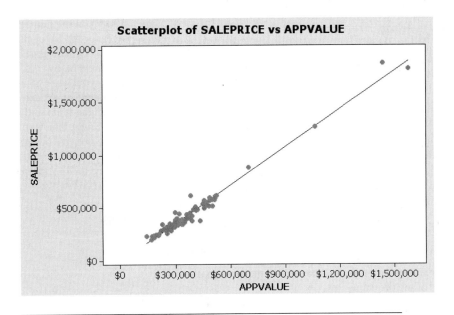

Scatterplot of SALEPRICE vs APPVALUE

Correlations: SALEPRICE, APPVALUE

```
Pearson correlation of SALEPRICE and APPVALUE = 0.987
P-Value = 0.000
```

Regression Analysis: SALEPRICE versus APPVALUE

```
The regression equation is
SALEPRICE = 184 + 1.20 APPVALUE


Predictor      Coef   SE Coef      T      P
Constant        184      9834   0.02  0.985
APPVALUE    1.19956   0.02234  53.70  0.000


S = 44859.7   R-Sq = 97.4%   R-Sq(adj) = 97.4%

Analysis of Variance

Source          DF           SS           MS        F      P
Regression       1  5.80234E+12  5.80234E+12  2883.31  0.000
Residual Error  76  1.52942E+11   2012390103
Total           77  5.95528E+12

Predicted Values for New Observations

New
Obs     Fit  SE Fit        95% CI             95% PI
  1  480007    5105  (469839, 490175)  (390085, 569930)

Values of Predictors for New Observations

New
Obs  APPVALUE
  1    400000
```

MINITAB output for Exercise 9.128

---

pose a straight-line model relating metal level y to distance x from the plant. On the basis of the theory, would you expect the slope of the line to be positive or negative?

b. Examine the scatterplot for cadmium. Does the plot support the theory you set forth in part **a**?

c. Examine the scatterplot for arsenic. Does the plot support the theory of part **a**? (*Note:* This finding led investigators to discover the link between high arsenic levels and the use of the crabgrass killer.)

9.128 Predicting sale prices of homes. Real-estate investors, home buyers, and homeowners often use the appraised value of property as a basis for predicting the sale of that

property. Data on sale prices and total appraised value of 78 residential properties sold in 2006 in an upscale Tampa, Florida, neighborhood named Hunter's Green are saved in the **HUNGREEN** file. Selected observations are listed in the next table.

a. Propose a straight-line model to relate the appraised property value (x) to the sale price (y) for residential properties in this neighborhood.

b. A MINITAB scatterplot of the data with the least squared line is shown above. Does it appear that a straight-line model will be an appropriate fit to the data?

c. A MINITAB simple linear regression printout is also shown. Find the equation of the least squared line.

HUNGREEN (selected observations)

Property	Sale Price	Appraised Value
1	$489,900	$418,601
2	1,825,000	1,577,919
3	890,000	687,836
4	250,00	191,620
5	1,275,000	1,063,901
⋮	⋮	⋮
74	325,000	292,702
75	516,000	407,449
76	309,300	272,275
77	370,000	347,320
78	580,000	511,359

Source: Hillsborough Country (Florida) Property Appraiser's Officer.

Interpret the estimated slope and y-interpret in the words of the problem.

d. Locate the test statistic and p-value for testing $H_0: \beta_1 = 0$ against $H_a: \beta_1 > 0$. Is there sufficient evidence (at $\alpha = .01$) of a positive linear relationship between apprised property value (x) and sale price (y)?

e. Locate and interpret practically the values of r and r^2 on the printout.

f. Locate and interpret practically the 95% prediction interval for sale price (y) on the printout.

9.129 **Baseball batting averages versus wins.** Is the number of games won by a major league baseball team in a season related to the team's batting average? In Exercise 2.141 (p. 92), you examined data from the *Baseball Almanac*

(2007) on the number of games won and the batting averages for the 14 teams in the American League for the 2006 Major League Baseball season. The data are repeated in the table below:

a. If you were to model the relationship between the mean (or expected) number of games won by a major league team and the team's batting average x, using a straight line, would you expect the slope of the line to be positive or negative? Explain.

b. Construct a scattergram of the data. Does the pattern revealed by the scattergram agree with your answer to part **a?**

ALWINS

Team	Games Won	Batting Ave.
New York	97	.285
Toronto	87	.284
Baltimore	70	.277
Boston	86	.269
Tampa Bay	61	.255
Cleveland	78	.280
Detroit	95	.274
Chicago	90	.280
Kansas City	62	.271
Minnesota	96	.287
Los Angeles	89	.274
Texas	80	.278
Seattle	78	.272
Oakland	93	.260

Source: Baseball Almanac, 2007; www.mlb.com

Dependent Variable: WINS

Number of Observations Read 14
Number of Observations Used 14

Analysis of Variance

Source	DF	Sum of Squares	Mean Square	F Value	Pr > F
Model	1	405.26316	405.26316	3.32	0.0936
Error	12	1466.73684	122.22807		
Corrected Total	13	1872.00000			

Root MSE	11.05568	R-Square	0.2165	
Dependent Mean	83.00000	Adj R-Sq	0.1512	
Coeff Var	13.32010			

Parameter Estimates

Variable	DF	Parameter Estimate	Standard Error	t Value	Pr > |t|
Intercept	1	-85.68421	92.68557	-0.92	0.3735
BATAVG	1	614.03509	337.21750	1.82	0.0936

Obs	BATAVG	Dependent Variable	Predicted Value	Std Error Mean Predict	95% CL Predict		Residual
1	0.285	97.0000	89.3158	4.5564	63.2620	115.3696	7.6842
2	0.284	87.0000	88.7018	4.3053	62.8515	114.5520	-1.7018
3	0.277	70.0000	84.4035	3.0536	59.4133	109.3937	-14.4035
4	0.269	86.0000	79.4912	3.5276	54.2065	104.7760	6.5088
5	0.255	61.0000	70.8947	7.2751	42.0590	99.7304	-9.8947
6	0.28	78.0000	86.2456	3.4507	61.0113	111.4800	-8.2456
7	0.274	95.0000	82.5614	2.9646	57.6222	107.5006	12.4386
8	0.28	90.0000	86.2456	3.4507	61.0113	111.4800	3.7544
9	0.271	62.0000	80.7193	3.2093	55.6367	105.8019	-18.7193
10	0.287	96.0000	90.5439	5.0887	64.0265	117.0613	5.4561
11	0.274	89.0000	82.5614	2.9646	57.6222	107.5006	6.4386
12	0.278	80.0000	85.0175	3.1557	59.9672	110.0679	-5.0175
13	0.272	78.0000	81.3333	3.0933	56.3200	106.3467	-3.3333
14	0.26	93.0000	73.9649	5.7750	46.7883	101.1416	19.0351

SAS output for Exercise 9.129

c. A SAS printout of the simple linear regression is shown on p. 550. Find the estimates of the β's on the printout and write the equation of the least squares line.

d. Graph the least squares line on your scattergram. Does your least squares line seem to fit the points on your scattergram?

e. Interpret the estimates of β_0 and β_1 in the words of the problem.

f. Conduct a test (at $\alpha = .05$) to determine whether the mean (or expected) number of games won by a major league baseball team is positively linearly related to the team's batting average.

g. Find the coefficient of determination, r^2, and interpret its value.

h. Predict the number of games won by a team with a .285 batting average.

i. Find a 95% prediction interval for the number of games won by a team with a .285 batting average. Interpret the interval.

9.130 Australian seagrass study. The abundance of tropical sea-grasses in Australia was the subject of research published in *Aquatic Botany* (Mar. 1995). Simple linear regression was used to relate the standing crop (y, measured in grams per meter squared) of a seagrass species to the percentage (x) of land covered by the plants. Data collected for $n = 12$ sites at Rowes Bay, Australia, yielded the following results:

$$\hat{y} = .031 + .089x \qquad t = 2.34 \qquad p\text{-value} = .042$$

a. Give the practical interpretation, if possible, of the estimated y-intercept of the line.

b. Give the practical interpretation, if possible, of the estimated slope of the line.

c. At what α-level is there sufficient evidence of a linear relationship between standing crop and percentage cover of a seagrass species? Explain.

9.131 Feeding habits of fish. Refer to the *Brain and Behavior Evolution* (Apr. 2000) study of the feeding behavior of black-bream fish, presented in Exercise 2.142 (p. 92). Recall that the zoologists recorded the number of aggressive strikes of two black-bream fish feeding at the bottom of an aquarium in the 10-minute period following the addition of food. The table listing the weekly number of strikes and the age of the fish (in days) is reproduced in the next column:

a. Write the equation of a straight-line model relating number of strikes (y) to age of fish (x).

b. Fit the model to the data by the method of least squares and give the least squares prediction equation.

c. Give a practical interpretation of the value of $\hat{\beta}_0$ if possible.

d. Give a practical interpretation of the value of $\hat{\beta}_1$ if possible.

e. Test $H_0: \beta_1 = 0$ versus $H_a: \beta_1 < 0$, using $\alpha = .10$. Interpret the result.

*f. Find Spearman's rank correlation coefficient relating number of strikes (y) to age (x).

*g. Test whether number of strikes (y) and age (x) are negatively correlated. Use $\alpha = .01$.

BLACKBREAM

Week	Number of Strikes	Age of Fish (days)
1	85	120
2	63	136
3	34	150
4	39	155
5	58	162
6	35	169
7	57	178
8	12	184
9	15	190

Source: Shand, J., et al. "Variability in the location of the retinal ganglion cell area centralis is correlated with ontogenetic changes in feeding behavior in the Blackbream, Acanthopagrus 'butcher'." *Brain and Behavior*, Vol. 55, No. 4, Apr. 2000 (Figure H).

9.132 Math anxiety study. Many high school students experience "math anxiety." Does such an attitude carry over to learning computer skills? A researcher at Duquesne University investigated this question and published her results in *Educational Technology* (May–June 1995). A sample of high school students—902 boys and 828 girls—from public schools in Pittsburgh, Pennsylvania, participated in the study. Using five-point Likert scales, where 1 = "strongly disagree" and 5 = "strongly agree," the researcher measured the students' interest and confidence in both mathematics and computers.

a. For boys, math confidence and computer interest were correlated at $r = .14$. Interpret this result fully.

b. For girls, math confidence and computer interest were correlated at $r = .33$. Interpret this result fully.

9.133 "Metaskills" and career management. In today's business environment, effective management of one's own career requires a skill set that includes adaptability, tolerance for ambiguity, self-awareness, and identity change. Management professors at Pace University (New York) used correlation coefficients to investigate the relationship between these "metaskills" and effective career management. (*International Journal of Manpower*, Aug. 2000.) Data were collected on 446 business graduates who had all completed a management "metaskills" course. Two of the many variables measured were self-knowledge skill level (x) and goal-setting ability (y). The correlation coefficient for these two variables was $r = .70$.

a. Give a practical interpretation of the value of r.

b. The p-value for a test of no correlation between the two variables was reported as $p\text{-value} = .001$. Interpret this result.

c. Find the coefficient of determination, r^2, and interpret the result.

9.134 Walking study. Refer to the *American Scientist* (July–Aug. 1998) study of the relationship between self-avoiding and unrooted walks, presented in Exercise 2.143 (p. 92). Recall that in a self-avoiding walk you never retrace or cross your own path, while an unrooted walk is a path in which the starting and ending points are impossible to distinguish. The possible number of walks of each type of various lengths are reproduced in the next table. Consider the straight-line model $y = \beta_0 + \beta_1 x + \varepsilon$, where x is walk length (number of steps).

WALK

Walk Length (number of steps)	Unrooted Walks	Self-Avoiding Walks
1	1	4
2	2	12
3	4	36
4	9	100
5	22	284
6	56	780
7	147	2,172
8	388	5,916

Source: Hayes, B. "How to avoid yourself," *American Scientist*, Vol. 86, No. 4, July–Aug. 1988, p. 317 (Figure 5).

a. Use the method of least squares to fit the model to the data if y is the possible number of unrooted walks.
b. Interpret $\hat{\beta}_0$ and $\hat{\beta}_1$ in the estimated model of part **a**.
c. Repeat parts **a** and **b** if y is the possible number of self-avoiding walks.
d. Find a 99% confidence interval for the number of unrooted walks that are possible when the walk length is four steps.
e. Would you recommend using simple linear regression to predict the number of walks that are possible when walk length is 15 steps? Explain.

Applying the Concepts—Intermediate

9.135 Beanie Babies. Refer to Exercise 2.175 (p. 104) and the data on 50 Beanie Babies collector's items, published in *Beanie World Magazine*. Can the age of a Beanie Baby be used to predict its market value? Answer this question by conducting a complete simple linear regression analysis on the data saved in the **BEANIE** file. (The first and last 5 entries are shown in the table below.)

9.136 Organic chemistry experiment. Chemists at Kyushu University (Japan) examined the linear relationship between the maximum absorption rate y (in nanomoles) and the Hammett substituent constant x for metacyclophane compounds (*Journal of Organic Chemistry*, July 1995.) The data for variants of two compounds are given in the accompanying table. The variants of compound 1 are labeled 1a, 1b, 1d, 1e, 1f, 1g, and 1h; the variants of compound 2 are 2a, 2b, 2c, and 2d.

a. Plot the data in a scattergram. Use two different plotting symbols for the two compounds. What do you observe?
b. Using only the data for compound 1, fit the model $E(y) = \beta_0 + \beta_1 x$.
c. Assess the adequacy of the model you fit in part **b**. Use $\alpha = .01$.
d. Repeat parts **b** and **c**, using only the data for compound 2.
*e. Conduct Spearman's test for rank correlation between x and y. Use $\alpha = .01$.

ORGCHEM

Compound	Maximum Absorption y	Hammett Constant x
1a	298	0.00
1b	346	.75
1d	303	.06
1e	314	−.26
1f	302	.18
1g	332	.42
1h	302	−.19
2a	343	.52
2b	367	1.01
2c	325	.37
2d	331	.53

Source: Adapted from Tsuge, A., et al. "Preparation and spectral properties of disubstituted [2-2] metacyclophanes," *Journal of Organic Chemistry*, Vol. 60, No. 15, July 1995, pp. 4390–4391 (Table 1 and Figure 1).

9.137 Mortality of predatory birds. Two species of predatory birds—collard flycatchers and tits—compete for nest holes during breeding season on the island of Gotland, Sweden. Frequently, dead flycatchers are found in nest boxes occupied by tits. A field study examined whether the risk of mortality to flycatchers is related to the degree of competition between the two bird species for nest sites. (*The Condor*, May 1995.) The table (p. 553) gives data on the number y of flycatchers killed at each of 14 discrete locations (plots) on the island, as well as on the nest box tit occupancy x (i.e., the percentage of nest boxes occupied by tits) at each plot. Consider the simple linear regression model $E(y) = \beta_0 + \beta_1 x$.

a. Plot the data in a scattergram. Does the frequency of flycatcher casualties per plot appear to increase lin-

BEANIE (First and last 5 observations shown.)

Name	Age (months) as of Sept. 1998	Retired (R)/ Current (C)	Value ($)
1. Ally the Alligator	52	R	55.00
2. Batty the Bat	12	C	12.00
3. Bongo the Brown Monkey	28	R	40.00
4. Blackie the Bear	52	C	10.00
5. Bucky the Beaver	40	R	45.00
⋮	⋮	⋮	⋮
46. Stripes the Tiger (Gold/Black)	40	R	400.00
47. Teddy the 1997 Holiday Bear	12	R	50.00
48. Tuffy the Terrier	17	C	10.00
49. Tracker the Basset Hound	5	C	15.00
50. Zip the Black Cat	28	R	40.00

Source: Beanie World Magazine, Sept. 1998.

early with increasing proportion of nest boxes occupied by tits?

b. Use the method of least squares to find the estimates of β_0 and β_1. Interpret their values.

c. Test the utility of the model, using $\alpha = .05$.

d. Find r and r^2 and interpret their values.

e. Find s and interpret the result.

f. Do you recommend using the model to predict the number of flycatchers killed? Explain.

CONDOR2

Plot	Number of Flycatchers Killed y	Nest Box Tit Occupancy x (%)
1	0	24
2	0	33
3	0	34
4	0	43
5	0	50
6	1	35
7	1	35
8	1	38
9	1	40
10	2	31
11	2	43
12	3	55
13	4	57
14	5	64

Source: Merila, J., and Wiggins, D. A. "Interspecific competition for nest holes causes adult mortality in the collard flycatcher." *The Condor*, Vol. 97, No. 2, May 1995, p. 449 (Figure 2), Cooper Ornithological Society.

9.138 Quantum tunneling. At temperatures approaching absolute zero ($-273°C$), helium exhibits traits that seem to defy many laws of Newtonian physics. An experiment has been conducted with helium in solid form at various temperatures near absolute zero. The solid helium is placed in a dilution refrigerator along with a solid impure substance, and the fraction (in weight) of the impurity passing through the solid helium is recorded. (This phenomenon of solids passing directly through solids is known as *quantum tunneling*.) The data are given in the following table:

HELIUM

Temperature x(°C)	Proportion of Impurity
−262.0	.315
−265.0	.202
−256.0	.204
−267.0	.620
−270.0	.715
−272.0	.935
−272.4	.957
−272.7	.906
−272.8	.985
−272.9	.987

a. Find the least squares estimates of the intercept and slope. Interpret them.

b. Use a 95% confidence interval to estimate the slope β_1. Interpret the interval in terms of this application.

Does the interval support the hypothesis that temperature contributes information about the proportion of impurity passing through helium?

c. Interpret the coefficient of determination for this model.

d. Find a 95% prediction interval for the percentage of impurity passing through solid helium at $-273°C$. Interpret the result.

e. Note that the value of x in part **d** is outside the experimental region. Why might this lead to an unreliable prediction?

9.139 Conversing with the hearing impaired. A study was conducted to investigate how people with a hearing impairment communicate with their conversational partners. (*Journal of the Academy of Rehabilitative Audiology*, Vol. 27, 1994.) Each of 13 hearing-impaired subjects, all fitted with a cochlear implant, participated in a structured communication interaction with a familiar conversational partner (a family member) and with an unfamiliar conversational partner (who was instructed not to take the initiative to repair breakdowns in communication). The total number of words used by the subject in each of the two conversations is given in the accompanying table.

HEARAID

Subject	Words with Familiar Partner x	Words with Unfamiliar Partner y
1	65	47
2	160	78
3	55	90
4	83	75
5	0	6
6	140	101
7	49	40
8	164	215
9	62	29
10	56	75
11	207	121
12	207	139
13	93	83

Source: Tye-Murray, N., et al., "Communication breakdowns: Partner contingencies and partner reactions," *Journal of the Academy of Rehabilitative Audiology*, Vol. 27, 1994, pp. 116–117 (Tables 6, 7).

a. Plot the data in a scattergram. Is there visual evidence of a linear relationship between x and y? If so, is it positive or negative?

b. Propose a straight-line model relating y to x.

c. Use the method of least squares to find the estimates of β_0 and β_1.

d. Interpret the values of $\hat{\beta}_0$ and $\hat{\beta}_1$.

9.140 Loneliness in families. Is there a link between the loneliness of parents and that of their offspring? This question was examined in the *Journal of Marriage and Family* (Aug. 1986). The participants in the study were 130 female college undergraduates and their parents. Each triad of daughter, mother, and father completed the UCLA Loneliness Scale, a 20-item questionnaire designed to assess loneliness and several variables theoretically related to loneliness, such as social accessibility to others, difficulty in making friends, and depression.

a. The correlation between daughter's loneliness score y and mother's loneliness score x was determined to be $r = .26$. Interpret this value.

b. The correlation between daughter's loneliness score y and father's loneliness score x was determined to be $r = .19$. Interpret this value.

c. The correlation between daughter's loneliness score y and mother's self-esteem score x was determined to be $r = .14$. Interpret this value.

d. The correlation between daughter's loneliness score y and father's assertiveness score x was determined to be $r = .01$. Interpret this value.

e. Calculate the coefficient of determination r^2 for parts **a–d**. Interpret the results.

Applying the Concepts—Advanced

9.141 Regression through the origin. Sometimes it is known from theoretical considerations that the straight-line relationship between two variables x and y passes through the origin of the xy-plane. Consider the relationship between the total weight y of a shipment of 50-pound bags of flour and the number x of bags in the shipment. Since a shipment containing $x = 0$ bags (i.e., no shipment at all) has a total weight of $y = 0$, a straight-line model of the relationship between x and y should pass through the point $x = 0$, $y = 0$. In such a case, you could assume that $\beta_0 = 0$ and characterize the relationship between x and y with the following model:

$$y = \beta_1 x + \varepsilon$$

The least squares estimate of β_1 for this model is

$$\hat{\beta}_1 = \frac{\Sigma x_i y_i}{\Sigma x_i^2}$$

From the records of past flour shipments, 15 shipments were randomly chosen and the data shown in the following table were recorded:

FLOUR

Weight of Shipment	Number of 50-Pound Bags in Shipment
5,050	100
10,249	205
20,000	450
7,420	150
24,685	500
10,206	200
7,325	150
4,958	100
7,162	150
24,000	500
4,900	100
14,501	300
28,000	600
17,002	400
16,100	400

a. Find the least squares line for the given data under the assumption that $\beta_0 = 0$. Plot the least squares line on a scattergram of the data.

b. Find the least squares line for the given data, using the model

$$y = \beta_0 + \beta_1 x + \varepsilon$$

(i.e., do not restrict β_0 to equal 0). Plot this line on the same scatterplot you constructed in part **a**.

c. Refer to part **b**. Why might $\hat{\beta}_0$ be different from 0 even though the true value of β_0 is known to be 0?

d. The estimated standard error of $\hat{\beta}_0$ is equal to

$$s\sqrt{\frac{1}{n} + \frac{\bar{x}^2}{SS_{xx}}}$$

Use the t-statistic

$$t = \frac{\hat{\beta}_0 - 0}{s\sqrt{(1/n) + (\bar{x}^2/SS_{xx})}}$$

to test the null hypothesis $H_0: \beta_0 = 0$ against the alternative $H_a: \beta_0 \neq 0$. Take $\alpha = .10$. Should you include β_0 in your model?

9.142 Long-jump "takeoff error." The long jump is a track-and-field event in which a competitor attempts to jump a maximum distance into a sandpit after a running start. At the edge of the pit is a takeoff board. Jumpers usually try to plant their toes at the front edge of this board to maximize their jumping distance. The absolute distance between the front edge of the takeoff board and the spot where the toe actually lands on the board prior to jumping is called "takeoff error." Is takeoff error in the long jump linearly related to best jumping distance? To answer this question, kinesiology researchers videotaped the performances of 18 novice long jumpers at a high school track meet. (*Journal of Applied Biomechanics*, May 1995.) The average takeoff error x and the best jumping distance y (out of three jumps) for each jumper are recorded in the table. If a jumper can reduce his or her average takeoff error by .1 meter, how much would you estimate the jumper's best jumping distance to change? On the basis of your answer, comment on the usefulness of the model for predicting best jumping distance.

LONGJUMP

Jumper	Best Jumping Distance y (meters)	Average Takeoff Error x (meters)
1	5.30	.09
2	5.55	.17
3	5.47	.19
4	5.45	.24
5	5.07	.16
6	5.32	.22
7	6.15	.09
8	4.70	.12
9	5.22	.09
10	5.77	.09
11	5.12	.13
12	5.77	.16
13	6.22	.03
14	5.82	.50
15	5.15	.13
16	4.92	.04
17	5.20	.07
18	5.42	.04

Source: Berg, W. P., and Greer, N. L. "A kinematic profile of the approach run of novice long jumpers," *Journal of Applied Biomechanics*, Vol. 11, No. 2, May 1995, p. 147 (Table 1).

Critical Thinking Challenges

9.143 Study of fertility rates. The fertility rate of a country is defined as the number of children a woman citizen bears, on average, in her lifetime. *Scientific American* (Dec. 1993) reported on the declining fertility rate in developing countries. The researchers found that family planning can have a great effect on fertility rate. The accompanying table gives the fertility rate y and contraceptive prevalence x (measured as the percentage of married women who use contraception) for each of 27 developing countries.

a. According to the researchers, "The data reveal that differences in contraceptive prevalence explain about 90% of the variation in fertility rates." Do you concur?

b. The researchers also concluded that "if contraceptive use increases by 18 percent, women bear, on average, one fewer child." Is this statement supported by the data? Explain.

FERTRATE

Country	Contraceptive Prevalence x	Fertility Rate y
Mauritius	76	2.2
Thailand	69	2.3
Colombia	66	2.9
Costa Rica	71	3.5
Sri Lanka	63	2.7
Turkey	62	3.4
Peru	60	3.5
Mexico	55	4.0
Jamaica	55	2.9
Indonesia	50	3.1
Tunisia	51	4.3
El Salvador	48	4.5
Morocco	42	4.0
Zimbabwe	46	5.4
Egypt	40	4.5
Bangladesh	40	5.5
Botswana	35	4.8
Jordan	35	5.5
Kenya	28	6.5
Guatemala	24	5.5
Cameroon	16	5.8
Ghana	14	6.0
Pakistan	13	5.0
Senegal	13	6.5
Sudan	10	4.8
Yemen	9	7.0
Nigeria	7	5.7

Source: Robey, B., et al. "The fertility decline in developing countries," *Scientific American*, Dec. 1993, p. 62. [*Note:* The data values are estimated from a scatterplot.]

9.144 Spall damage in bricks. A recent civil suit revolved around a five-building brick apartment complex located in the Bronx, New York, which began to suffer *spalling* damage (i.e., a separation of some portion of the face of a brick from its body). The owner of the complex alleged that the bricks were manufactured defectively. The brick manufacturer countered that poor design and shoddy management led to the damage. To settle the suit, an estimate of the rate of damage per 1,000 bricks, called the spall rate, was required. (*Chance*, Summer 1994.) The owner estimated the spall rate by using several *scaffold-drop* surveys. (With this method, an engineer lowers a scaffold down at selected places on building walls and counts the number of visible spalls for every 1,000 bricks in the observation area.) The brick manufacturer conducted its own survey by dividing the walls of the complex into 83 wall segments and taking a photograph of each one. (The number of spalled bricks that could be made out from each photo was recorded, and the sum over all 83 wall segments was used as an estimate of total spall damage.) In this court case, the jury was faced with the following dilemma: On the one hand, the scaffold-drop survey provided the most accurate estimate of spall rates in a given wall segment. Unfortunately, however, the drop areas were not selected at random from the entire complex; rather, drops were made at areas with high spall concentrations, leading to an overestimate of the total damage. On the other hand, the photo survey was complete in that all 83 wall segments in the complex were checked for spall damage. But the spall rate estimated by the photos, at least in areas of high spall concentration, was biased low (spalling damage cannot always be seen from a photo), leading to an underestimate of the total damage.

The data in the table are the spall rates obtained from the two methods at 11 drop locations. Use the data, as did expert statisticians who testified in the case, to help the jury estimate the true spall rate at a given wall segment. Then explain how this information, coupled with the data (not given here) on all 83 wall segments, can provide a reasonable estimate of the total spall damage (i.e., total number of damaged bricks).

BRICKS

Drop Location	Drop Spall Rate (per 1,000 bricks)	Photo Spall Rate (per 1,000 bricks)
1	0	0
2	5.1	0
3	6.6	0
4	1.1	.8
5	1.8	1.0
6	3.9	1.0
7	11.5	1.9
8	22.1	7.7
9	39.3	14.9
10	39.9	13.9
11	43.0	11.8

Source: Fairley, W. B., et al. "Bricks, buildings, and the Bronx: Estimating masonry deterioration," *Chance*, Vol. 7. No. 3, Summer 1994, p. 36 (Figure 3). [*Note:* The data points are estimated from the points shown on a scatterplot.]

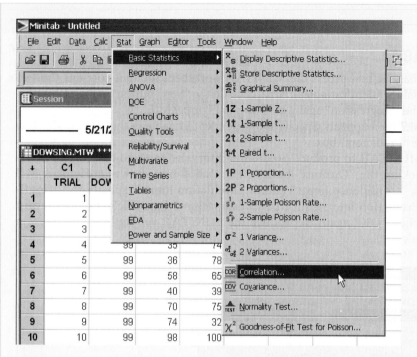

FIGURE 9.M.4

MINITAB menu options for correlation

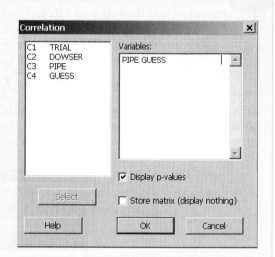

FIGURE 9.M.5

MINITAB correlation dialog box

Rank Correlation

To obtain Spearman's rank correlation coefficient in MINITAB, you must first rank the values of the two quantitative variables of interest. Click the "Calc" button on the MINITAB menu bar, and create two additional columns; one for the ranks of the x-variable and one for the ranks of the y-variable. (Use the "Rank" function on the MINITAB calculator, as shown in Figure 9.M.6.) Next, click on the "Stat" button on the main menu bar, and then click on "Basic Statistics" and "Correlation." The dialog box shown in Figure 9.M.7 then appears. Enter the ranked variables in the "Variables" box and unselect the "Display p-values" option. Click "OK" to obtain the MINITAB printout. (You will need to look up the critical value of Spearman's rank correlation to conduct the test.)

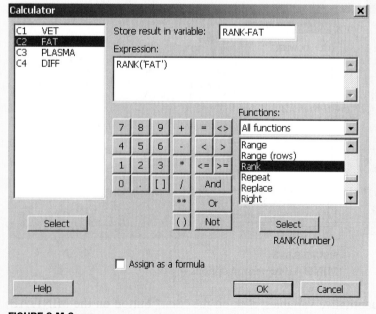

FIGURE 9.M.6

MINITAB calculator options for creating ranks

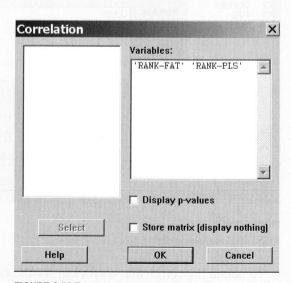

FIGURE 9.M.7

MINITAB correlation dialog box

Tables

Table I	Random Numbers	560
Table II	Binomial Probabilities	563
Table III	Normal Curve Areas	567
Table IV	Critical Values of t	568
Table V	Critical Values of T_L and T_U for the Wilcoxon Rank Sum Test: Independent Samples	569
Table VI	Critical Values of T_0 in the Wilcoxon Paired Difference Signed Ranks Test	570
Table VII	Percentage Points of the F-Distribution, $\alpha = .10$	571
Table VIII	Percentage Points of the F-Distribution, $\alpha = .05$	573
Table IX	Percentage Points of the F-Distribution, $\alpha = .025$	575
Table X	Percentage Points of the F-Distribution, $\alpha = .01$	577
Table XI	Critical Values of χ^2	579
Table XII	Critical Values of Spearman's Rank Correlation Coefficient	581

TABLE I   Random Numbers

Row \ Column	1	2	3	4	5	6	7	8	9	10	11	12	13	14
1	10480	15011	01536	02011	81647	91646	69179	14194	62590	36207	20969	99570	91291	90700
2	22368	46573	25595	85393	30995	89198	27982	53402	93965	34095	52666	19174	39615	99505
3	24130	48360	22527	97265	76393	64809	15179	24830	49340	32081	30680	19655	63348	58629
4	42167	93093	06243	61680	07856	16376	39440	53537	71341	57004	00849	74917	97758	16379
5	37570	39975	81837	16656	06121	91782	60468	81305	49684	60672	14110	06927	01263	54613
6	77921	06907	11008	42751	27756	53498	18602	70659	90655	15053	21916	81825	44394	42880
7	99562	72905	56420	69994	98872	31016	71194	18738	44013	48840	63213	21069	10634	12952
8	96301	91977	05463	07972	18876	20922	94595	56869	69014	60045	18425	84903	42508	32307
9	89579	14342	63661	10281	17453	18103	57740	84378	25331	12566	58678	44947	05585	56941
10	85475	36857	53342	53988	53060	59533	38867	62300	08158	17983	16439	11458	18593	64952
11	28918	69578	88231	33276	70997	79936	56865	05859	90106	31595	01547	85590	91610	78188
12	63553	40961	48235	03427	49626	69445	18663	72695	52180	20847	12234	90511	33703	90322
13	09429	93969	52636	92737	88974	33488	36320	17617	30015	08272	84115	27156	30613	74952
14	10365	61129	87529	85689	48237	52267	67689	93394	01511	26358	85104	20285	29975	89868
15	07119	97336	71048	08178	77233	13916	47564	81056	97735	85977	29372	74461	28551	90707
16	51085	12765	51821	51259	77452	16308	60756	92144	49442	53900	70960	63990	75601	40719
17	02368	21382	52404	60268	89368	19885	55322	44819	01188	65255	64835	44919	05944	55157
18	01011	54092	33362	94904	31273	04146	18594	29852	71585	85030	51132	01915	92747	64951
19	52162	53916	46369	58586	23216	14513	83149	98736	23495	64350	94738	17752	35156	35749
20	07056	97628	33787	09998	42698	06691	76988	13602	51851	46104	88916	19509	25625	58104
21	48663	91245	85828	14346	09172	30168	90229	04734	59193	22178	30421	61666	99904	32812
22	54164	58492	22421	74103	47070	25306	76468	26384	58151	06646	21524	15227	96909	44592
23	32639	32363	05597	24200	13363	38005	94342	28728	35806	06912	17012	64161	18296	22851
24	29334	27001	87637	87308	58731	00256	45834	15398	46557	41135	10367	07684	36188	18510
25	02488	33062	28834	07351	19731	92420	60952	61280	50001	67658	32586	86679	50720	94953
26	81525	72295	04839	96423	24878	82651	66566	14778	76797	14780	13300	87074	79666	95725
27	29676	20591	68086	26432	46901	20849	89768	81536	86645	12659	92259	57102	80428	25280
28	00742	57392	39064	66432	84673	40027	32832	61362	98947	96067	64760	64584	96096	98253
29	05366	04213	25669	26422	44407	44048	37937	63904	45766	66134	75470	66520	34693	90449
30	91921	26418	64117	94305	26766	25940	39972	22209	71500	64568	91402	42416	07844	69618
31	00582	04711	87917	77341	42206	35126	74087	99547	81817	42607	43808	76655	62028	76630
32	00725	69884	62797	56170	86324	88072	76222	36086	84637	93161	76038	65855	77919	88006
33	69011	65795	95876	55293	18988	27354	26575	08625	40801	59920	29841	80150	12777	48501
34	25976	57948	29888	88604	67917	48708	18912	82271	65424	69774	33611	54262	85963	03547
35	09763	83473	73577	12908	30883	18317	28290	35797	05998	41688	34952	37888	38917	88050

(continued)

TABLE I Continued

Row\Column	1	2	3	4	5	6	7	8	9	10	11	12	13	14
36	91576	42595	27958	30134	04024	86385	29880	99730	55536	84855	29080	09250	79656	73211
37	17955	56349	90999	49127	20044	59931	06115	20542	18059	02008	73708	83517	36103	42791
38	46503	18584	18845	49618	02304	51038	20655	58727	28168	15475	56942	53389	20562	87338
39	92157	89634	94824	78171	84610	82834	09922	25417	44137	48413	25555	21246	35509	20468
40	14577	62765	35605	81263	39667	47358	56873	56307	61607	49518	89656	20103	77490	18062
41	98427	07523	33362	64270	01638	92477	66969	98420	04880	45585	46565	04102	46880	45709
42	34914	63976	88720	82765	34476	17032	87589	40836	32427	70002	70663	88863	77775	69348
43	70060	28277	39475	46473	23219	53416	94970	25832	69975	94884	19661	72828	00102	66794
44	53976	54914	06990	67245	68350	82948	11398	42878	80287	88267	47363	46634	06541	97809
45	76072	29515	40980	07391	58745	25774	22987	80059	39911	96189	41151	14222	60697	59583
46	90725	52210	83974	29992	65831	38857	50490	83765	55657	14361	31720	57375	56228	41546
47	64364	67412	33339	31926	14883	24413	59744	92351	97473	89286	35931	04110	23726	51900
48	08962	00358	31662	25388	61642	34072	81249	35648	56891	69352	48373	45578	78547	81788
49	95012	68379	93526	70765	10592	04542	76463	54328	02349	17247	28865	14777	62730	92277
50	15664	10493	20492	38391	91132	21999	59516	81652	27195	48223	46751	22923	32261	85653
51	16408	81899	04153	53381	79401	21438	83035	92350	36693	31238	59649	91754	72772	02338
52	18629	81953	05520	91962	04739	13092	97662	24822	94730	06496	35090	04822	86774	98289
53	73115	35101	47498	87637	99016	71060	88824	71013	18735	20286	23153	72924	35165	43040
54	57491	16703	23167	49323	45021	33132	12544	41035	80780	45393	44812	12512	98931	91202
55	30405	83946	23792	14422	15059	45799	22716	19792	09983	74353	68668	30429	70735	25499
56	16631	35006	85900	98275	32388	52390	16815	69290	82732	38480	73817	32523	41961	44437
57	96773	20206	42559	78985	05300	22164	24369	54224	35083	19687	11052	91491	60383	19746
58	38935	64202	14349	82674	66523	44133	00697	35552	35970	19124	63318	29686	03387	59846
59	31624	76384	17403	53363	44167	64486	64758	75366	76554	31601	12614	33072	60332	92325
60	78919	19474	23632	27889	47914	02584	37680	20801	72152	39339	34806	08930	85001	87820
61	03931	33309	57047	74211	63445	17361	62825	39908	05607	91284	68833	25570	38818	46920
62	74426	33278	43972	10110	89917	15665	52872	73823	73144	88662	88970	74492	51805	99378
63	09066	00903	20795	95452	92648	45454	09552	88815	16553	51125	79375	97596	16296	66092
64	42238	12426	87025	14267	20979	04508	64535	31355	86064	29472	47689	05974	52468	16834
65	16153	08002	26504	41744	81959	65642	74240	56302	00033	67107	77510	70625	28725	34191
66	21457	40742	29820	96783	29400	21840	15035	34537	33310	06116	95240	15957	16572	06004
67	21581	57802	02050	89728	17937	37621	47075	42080	97403	48626	68995	43805	33386	21597
68	55612	78095	83197	33732	05810	24813	86902	60397	16489	03264	88525	42786	05269	92532
69	44657	66999	99324	51281	84463	60563	79312	93454	68876	25471	93911	25650	12682	73572
70	91340	84979	46949	81973	37949	61023	43997	15263	80644	43942	89203	71795	99533	50501

(continued)

TABLE I   Continued

Column / Row	1	2	3	4	5	6	7	8	9	10	11	12	13	14
71	91227	21199	31935	27022	84067	05462	35216	14486	29891	68607	41867	14951	91696	85065
72	50001	38140	66321	19924	72163	09538	12151	06878	91903	18749	34405	56087	82790	70925
73	65390	05224	72958	28609	81406	39147	25549	48542	42627	45233	57202	94617	23772	07896
74	27504	96131	83944	41575	10573	08619	64482	73923	36152	05184	94142	25299	84387	34925
75	37169	94851	39117	89632	00959	16487	65536	49071	39782	17095	02330	74301	00275	48280
76	11508	70225	51111	38351	19444	66499	71945	05422	13442	78675	84081	66938	93654	59894
77	37449	30362	06694	54690	04052	53115	62757	95348	78662	11163	81651	50245	34971	52924
78	46515	70331	85922	38329	57015	15765	97161	17869	45349	61796	66345	81073	49106	79860
79	30986	81223	42416	58353	21532	30502	32305	86482	05174	07901	54339	58861	74818	46942
80	63798	64995	46583	09785	44160	78128	83991	42865	92520	83531	80377	35909	81250	54238
81	82486	84846	99254	67632	43218	50076	21361	64816	51202	88124	41870	52689	51275	83556
82	21885	32906	92431	09060	64297	51674	64126	62570	26123	05155	59194	52799	28225	85762
83	60336	98782	07408	53458	13564	59089	26445	29789	85205	41001	12535	12133	14645	23541
84	43937	46891	24010	25560	86355	33941	25786	54990	71899	15475	95434	98227	21824	19585
85	97656	63175	89303	16275	07100	92063	21942	18611	47348	20203	18534	03862	78095	50136
86	03299	01221	05418	38982	55758	92237	26759	86367	21216	98442	08303	56613	91511	75928
87	79626	06486	03574	17668	07785	76020	79924	25651	83325	88428	85076	72811	22717	50585
88	85636	68335	47539	03129	65651	11977	02510	26113	99447	68645	34327	15152	55230	93448
89	18039	14367	64337	06177	12143	46609	32989	74014	64708	00533	35398	58408	13261	47908
90	08362	15656	60627	36478	65648	16764	53412	09013	07832	41574	17639	82163	60859	75567
91	79556	29068	04142	16268	15387	12856	66227	38358	22478	73373	88732	09443	82558	05250
92	92608	82674	27072	32534	17075	27698	98204	63863	11951	34648	88022	56148	34925	57031
93	23982	25835	40055	67006	12293	02753	14827	23235	35071	99704	37543	11601	35503	85171
94	09915	96306	05908	97901	28395	14186	00821	80703	70426	75647	76310	88717	37890	40129
95	59037	33300	26695	62247	69927	76123	50842	43834	86654	70959	79725	93872	28117	19233
96	42488	78077	69882	61657	34136	79180	97526	43092	04098	73571	80799	76536	71255	64239
97	46764	86273	63003	93017	31204	36692	40202	35275	57306	55543	53203	18098	47625	88684
98	03237	45430	55417	63282	90816	17349	88298	90183	36600	78406	06216	95787	42579	90730
99	86591	81482	52667	61582	14972	90053	89534	76036	49199	43716	97548	04379	46370	28672
100	38534	01715	94964	87288	65680	43772	39560	12918	86537	62738	19636	51132	25739	56947

Source: Abridged from W. H. Beyer (ed.). *CRC Standard Mathematical Tables*, 24th edition. (Cleveland: The Chemical Rubber Company), 1976. Reproduced by permission of the publisher.

TABLE II   Binomial Probabilities

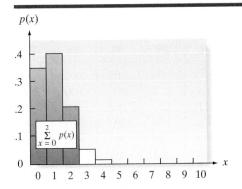

Tabulated values are $\sum_{x=0}^{k} p(x)$. (Computations are rounded at the third decimal place.)

a. $n = 5$

k \ p	.01	.05	.10	.20	.30	.40	.50	.60	.70	.80	.90	.95	.99
0	.951	.774	.590	.328	.168	.078	.031	.010	.002	.000	.000	.000	.000
1	.999	.977	.919	.737	.528	.337	.188	.087	.031	.007	.000	.000	.000
2	1.000	.999	.991	.942	.837	.683	.500	.317	.163	.058	.009	.001	.000
3	1.000	1.000	1.000	.993	.969	.913	.812	.663	.472	.263	.081	.023	.001
4	1.000	1.000	1.000	1.000	.998	.990	.969	.922	.832	.672	.410	.226	.049

b. $n = 6$

k \ p	.01	.05	.10	.20	.30	.40	.50	.60	.70	.80	.90	.95	.99
0	.941	.735	.531	.262	.118	.047	.016	.004	.001	.000	.000	.000	.000
1	.999	.967	.886	.655	.420	.233	.109	.041	.011	.002	.000	.000	.000
2	1.000	.998	.984	.901	.744	.544	.344	.179	.070	.017	.001	.000	.000
3	1.000	1.000	.999	.983	.930	.821	.656	.456	.256	.099	.016	.002	.000
4	1.000	1.000	1.000	.998	.989	.959	.891	.767	.580	.345	.114	.033	.001
5	1.000	1.000	1.000	1.000	.999	.996	.984	.953	.882	.738	.469	.265	.059

c. $n = 7$

k \ p	.01	.05	.10	.20	.30	.40	.50	.60	.70	.80	.90	.95	.99
0	.932	.698	.478	.210	.082	.028	.008	.002	.000	.000	.000	.000	.000
1	.998	.956	.850	.577	.329	.159	.063	.019	.004	.000	.000	.000	.000
2	1.000	.996	.974	.852	.647	.420	.227	.096	.029	.005	.000	.000	.000
3	1.000	1.000	.997	.967	.874	.710	.500	.290	.126	.033	.003	.000	.000
4	1.000	1.000	1.000	.995	.971	.904	.773	.580	.353	.148	.026	.004	.000
5	1.000	1.000	1.000	1.000	.996	.981	.937	.841	.671	.423	.150	.044	.002
6	1.000	1.000	1.000	1.000	1.000	.998	.992	.972	.918	.790	.522	.302	.068

(continued)

TABLE II Continued

d. $n = 8$

k \ p	.01	.05	.10	.20	.30	.40	.50	.60	.70	.80	.90	.95	.99
0	.923	.663	.430	.168	.058	.017	.004	.001	.000	.000	.000	.000	.000
1	.997	.943	.813	.503	.255	.106	.035	.009	.001	.000	.000	.000	.000
2	1.000	.994	.962	.797	.552	.315	.145	.050	.011	.001	.000	.000	.000
3	1.000	1.000	.995	.944	.806	.594	.363	.174	.058	.010	.000	.000	.000
4	1.000	1.000	1.000	.990	.942	.826	.637	.406	.194	.056	.005	.000	.000
5	1.000	1.000	1.000	.999	.989	.950	.855	.685	.448	.203	.038	.006	.000
6	1.000	1.000	1.000	1.000	.999	.991	.965	.894	.745	.497	.187	.057	.003
7	1.000	1.000	1.000	1.000	1.000	.999	.996	.983	.942	.832	.570	.337	.077

e. $n = 9$

k \ p	.01	.05	.10	.20	.30	.40	.50	.60	.70	.80	.90	.95	.99
0	.914	.630	.387	.134	.040	.010	.002	.000	.000	.000	.000	.000	.000
1	.997	.929	.775	.436	.196	.071	.020	.004	.000	.000	.000	.000	.000
2	1.000	.992	.947	.738	.463	.232	.090	.025	.004	.000	.000	.000	.000
3	1.000	.999	.992	.914	.730	.483	.254	.099	.025	.003	.000	.000	.000
4	1.000	1.000	.999	.980	.901	.733	.500	.267	.099	.020	.001	.000	.000
5	1.000	1.000	1.000	.997	.975	.901	.746	.517	.270	.086	.008	.001	.000
6	1.000	1.000	1.000	1.000	.996	.975	.910	.768	.537	.262	.053	.008	.000
7	1.000	1.000	1.000	1.000	1.000	.996	.980	.929	.804	.564	.225	.071	.003
8	1.000	1.000	1.000	1.000	1.000	1.000	.998	.990	.960	.866	.613	.370	.086

f. $n = 10$

k \ p	.01	.05	.10	.20	.30	.40	.50	.60	.70	.80	.90	.95	.99
0	.904	.599	.349	.107	.028	.006	.001	.000	.000	.000	.000	.000	.000
1	.996	.914	.736	.376	.149	.046	.011	.002	.000	.000	.000	.000	.000
2	1.000	.988	.930	.678	.383	.167	.055	.012	.002	.000	.000	.000	.000
3	1.000	.999	.987	.879	.650	.382	.172	.055	.011	.001	.000	.000	.000
4	1.000	1.000	.998	.967	.850	.633	.377	.166	.047	.006	.000	.000	.000
5	1.000	1.000	1.000	.994	.953	.834	.623	.367	.150	.033	.002	.000	.000
6	1.000	1.000	1.000	.999	.989	.945	.828	.618	.350	.121	.013	.001	.000
7	1.000	1.000	1.000	1.000	.998	.988	.945	.833	.617	.322	.070	.012	.000
8	1.000	1.000	1.000	1.000	1.000	.998	.989	.954	.851	.624	.264	.086	.004
9	1.000	1.000	1.000	1.000	1.000	1.000	.999	.994	.972	.893	.651	.401	.096

(continued)

TABLE II Continued

g. n = 15

p k	.01	.05	.10	.20	.30	.40	.50	.60	.70	.80	.90	.95	.99
0	.860	.463	.206	.035	.005	.000	.000	.000	.000	.000	.000	.000	.000
1	.990	.829	.549	.167	.035	.005	.000	.000	.000	.000	.000	.000	.000
2	1.000	.964	.816	.398	.127	.027	.004	.000	.000	.000	.000	.000	.000
3	1.000	.995	.944	.648	.297	.091	.018	.002	.000	.000	.000	.000	.000
4	1.000	.999	.987	.838	.515	.217	.059	.009	.001	.000	.000	.000	.000
5	1.000	1.000	.998	.939	.722	.403	.151	.034	.004	.000	.000	.000	.000
6	1.000	1.000	1.000	.982	.869	.610	.304	.095	.015	.001	.000	.000	.000
7	1.000	1.000	1.000	.996	.950	.787	.500	.213	.050	.004	.000	.000	.000
8	1.000	1.000	1.000	.999	.985	.905	.696	.390	.131	.018	.000	.000	.000
9	1.000	1.000	1.000	1.000	.996	.966	.849	.597	.278	.061	.002	.000	.000
10	1.000	1.000	1.000	1.000	.999	.991	.941	.783	.485	.164	.013	.001	.000
11	1.000	1.000	1.000	1.000	1.000	.998	.982	.909	.703	.352	.056	.005	.000
12	1.000	1.000	1.000	1.000	1.000	1.000	.996	.973	.873	.602	.184	.036	.000
13	1.000	1.000	1.000	1.000	1.000	1.000	1.000	.995	.965	.833	.451	.171	.010
14	1.000	1.000	1.000	1.000	1.000	1.000	1.000	1.000	.995	.965	.794	.537	.140

h. n = 20

p k	.01	.05	.10	.20	.30	.40	.50	.60	.70	.80	.90	.95	.99
0	.818	.358	.122	.012	.001	.000	.000	.000	.000	.000	.000	.000	.000
1	.983	.736	.392	.069	.008	.001	.000	.000	.000	.000	.000	.000	.000
2	.999	.925	.677	.206	.035	.004	.000	.000	.000	.000	.000	.000	.000
3	1.000	.984	.867	.411	.107	.016	.001	.000	.000	.000	.000	.000	.000
4	1.000	.997	.957	.630	.238	.051	.006	.000	.000	.000	.000	.000	.000
5	1.000	1.000	.989	.804	.416	.126	.021	.002	.000	.000	.000	.000	.000
6	1.000	1.000	.998	.913	.608	.250	.058	.006	.000	.000	.000	.000	.000
7	1.000	1.000	1.000	.968	.772	.416	.132	.021	.001	.000	.000	.000	.000
8	1.000	1.000	1.000	.990	.887	.596	.252	.057	.005	.000	.000	.000	.000
9	1.000	1.000	1.000	.997	.952	.755	.412	.128	.017	.001	.000	.000	.000
10	1.000	1.000	1.000	.999	.983	.872	.588	.245	.048	.003	.000	.000	.000
11	1.000	1.000	1.000	1.000	.995	.943	.748	.404	.113	.010	.000	.000	.000
12	1.000	1.000	1.000	1.000	.999	.979	.868	.584	.228	.032	.000	.000	.000
13	1.000	1.000	1.000	1.000	1.000	.994	.942	.750	.392	.087	.002	.000	.000
14	1.000	1.000	1.000	1.000	1.000	.998	.979	.874	.584	.196	.011	.000	.000
15	1.000	1.000	1.000	1.000	1.000	1.000	.994	.949	.762	.370	.043	.003	.000
16	1.000	1.000	1.000	1.000	1.000	1.000	.999	.984	.893	.589	.133	.016	.000
17	1.000	1.000	1.000	1.000	1.000	1.000	1.000	.996	.965	.794	.323	.075	.001
18	1.000	1.000	1.000	1.000	1.000	1.000	1.000	.999	.992	.931	.608	.264	.017
19	1.000	1.000	1.000	1.000	1.000	1.000	1.000	1.000	.999	.988	.878	.642	.182

(continued)

TABLE II   Continued

i. $n = 25$

k \ p	.01	.05	.10	.20	.30	.40	.50	.60	.70	.80	.90	.95	.99
0	.778	.277	.072	.004	.000	.000	.000	.000	.000	.000	.000	.000	.000
1	.974	.642	.271	.027	.002	.000	.000	.000	.000	.000	.000	.000	.000
2	.998	.873	.537	.098	.009	.000	.000	.000	.000	.000	.000	.000	.000
3	1.000	.966	.764	.234	.033	.002	.000	.000	.000	.000	.000	.000	.000
4	1.000	.993	.902	.421	.090	.009	.000	.000	.000	.000	.000	.000	.000
5	1.000	.999	.967	.617	.193	.029	.002	.000	.000	.000	.000	.000	.000
6	1.000	1.000	.991	.780	.341	.074	.007	.000	.000	.000	.000	.000	.000
7	1.000	1.000	.998	.891	.512	.154	.022	.001	.000	.000	.000	.000	.000
8	1.000	1.000	1.000	.953	.677	.274	.054	.004	.000	.000	.000	.000	.000
9	1.000	1.000	1.000	.983	.811	.425	.115	.013	.000	.000	.000	.000	.000
10	1.000	1.000	1.000	.994	.902	.586	.212	.034	.002	.000	.000	.000	.000
11	1.000	1.000	1.000	.998	.956	.732	.345	.078	.006	.000	.000	.000	.000
12	1.000	1.000	1.000	1.000	.983	.846	.500	.154	.017	.000	.000	.000	.000
13	1.000	1.000	1.000	1.000	.994	.922	.655	.268	.044	.002	.000	.000	.000
14	1.000	1.000	1.000	1.000	.998	.966	.788	.414	.098	.006	.000	.000	.000
15	1.000	1.000	1.000	1.000	1.000	.987	.885	.575	.189	.017	.000	.000	.000
16	1.000	1.000	1.000	1.000	1.000	.996	.946	.726	.323	.047	.000	.000	.000
17	1.000	1.000	1.000	1.000	1.000	.999	.978	.846	.488	.109	.002	.000	.000
18	1.000	1.000	1.000	1.000	1.000	1.000	.993	.926	.659	.220	.009	.000	.000
19	1.000	1.000	1.000	1.000	1.000	1.000	.998	.971	.807	.383	.033	.001	.000
20	1.000	1.000	1.000	1.000	1.000	1.000	1.000	.991	.910	.579	.098	.007	.000
21	1.000	1.000	1.000	1.000	1.000	1.000	1.000	.998	.967	.766	.236	.034	.000
22	1.000	1.000	1.000	1.000	1.000	1.000	1.000	1.000	.991	.902	.463	.127	.002
23	1.000	1.000	1.000	1.000	1.000	1.000	1.000	1.000	.998	.973	.729	.358	.026
24	1.000	1.000	1.000	1.000	1.000	1.000	1.000	1.000	1.000	.996	.928	.723	.222

TABLE III   Normal Curve Areas

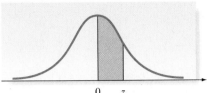

z	.00	.01	.02	.03	.04	.05	.06	.07	.08	.09
.0	.0000	.0040	.0080	.0120	.0160	.0199	.0239	.0279	.0319	.0359
.1	.0398	.0438	.0478	.0517	.0557	.0596	.0636	.0675	.0714	.0753
.2	.0793	.0832	.0871	.0910	.0948	.0987	.1026	.1064	.1103	.1141
.3	.1179	.1217	.1255	.1293	.1331	.1368	.1406	.1443	.1480	.1517
.4	.1554	.1591	.1628	.1664	.1700	.1736	.1772	.1808	.1844	.1879
.5	.1915	.1950	.1985	.2019	.2054	.2088	.2123	.2157	.2190	.2224
.6	.2257	.2291	.2324	.2357	.2389	.2422	.2454	.2486	.2517	.2549
.7	.2580	.2611	.2642	.2673	.2704	.2734	.2764	.2794	.2823	.2852
.8	.2881	.2910	.2939	.2967	.2995	.3023	.3051	.3078	.3106	.3133
.9	.3159	.3186	.3212	.3238	.3264	.3289	.3315	.3340	.3365	.3389
1.0	.3413	.3438	.3461	.3485	.3508	.3531	.3554	.3577	.3599	.3621
1.1	.3643	.3665	.3686	.3708	.3729	.3749	.3770	.3790	.3810	.3830
1.2	.3849	.3869	.3888	.3907	.3925	.3944	.3962	.3980	.3997	.4015
1.3	.4032	.4049	.4066	.4082	.4099	.4115	.4131	.4147	.4162	.4177
1.4	.4192	.4207	.4222	.4236	.4251	.4265	.4279	.4292	.4306	.4319
1.5	.4332	.4345	.4357	.4370	.4382	.4394	.4406	.4418	.4429	.4441
1.6	.4452	.4463	.4474	.4484	.4495	.4505	.4515	.4525	.4535	.4545
1.7	.4554	.4564	.4573	.4582	.4591	.4599	.4608	.4616	.4625	.4633
1.8	.4641	.4649	.4656	.4664	.4671	.4678	.4686	.4693	.4699	.4706
1.9	.4713	.4719	.4726	.4732	.4738	.4744	.4750	.4756	.4761	.4767
2.0	.4772	.4778	.4783	.4788	.4793	.4798	.4803	.4808	.4812	.4817
2.1	.4821	.4826	.4830	.4834	.4838	.4842	.4846	.4850	.4854	.4857
2.2	.4861	.4864	.4868	.4871	.4875	.4878	.4881	.4884	.4887	.4890
2.3	.4893	.4896	.4898	.4901	.4904	.4906	.4909	.4911	.4913	.4916
2.4	.4918	.4920	.4922	.4925	.4927	.4929	.4931	.4932	.4934	.4936
2.5	.4938	.4940	.4941	.4943	.4945	.4946	.4948	.4949	.4951	.4952
2.6	.4953	.4955	.4956	.4957	.4959	.4960	.4961	.4962	.4963	.4964
2.7	.4965	.4966	.4967	.4968	.4969	.4970	.4971	.4972	.4973	.4974
2.8	.4974	.4975	.4976	.4977	.4977	.4978	.4979	.4979	.4980	.4981
2.9	.4981	.4982	.4982	.4983	.4984	.4984	.4985	.4985	.4986	.4986
3.0	.4987	.4987	.4987	.4988	.4988	.4989	.4989	.4989	.4990	.4990

Source: Abridged from Table I of A. Hald, *Statistical Tables and Formulas* (New York: Wiley), 1952. Reproduced by permission of A. Hald.

TABLE IV   Critical Values of t

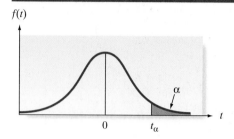

Degrees of Freedom	$t_{.100}$	$t_{.050}$	$t_{.025}$	$t_{.010}$	$t_{.005}$	$t_{.001}$	$t_{.0005}$
1	3.078	6.314	12.706	31.821	63.657	318.31	636.62
2	1.886	2.920	4.303	6.965	9.925	22.326	31.598
3	1.638	2.353	3.182	4.541	5.841	10.213	12.924
4	1.533	2.132	2.776	3.747	4.604	7.173	8.610
5	1.476	2.015	2.571	3.365	4.032	5.893	6.869
6	1.440	1.943	2.447	3.143	3.707	5.208	5.959
7	1.415	1.895	2.365	2.998	3.499	4.785	5.408
8	1.397	1.860	2.306	2.896	3.355	4.501	5.041
9	1.383	1.833	2.262	2.821	3.250	4.297	4.781
10	1.372	1.812	2.228	2.764	3.169	4.144	4.587
11	1.363	1.796	2.201	2.718	3.106	4.025	4.437
12	1.356	1.782	2.179	2.681	3.055	3.930	4.318
13	1.350	1.771	2.160	2.650	3.012	3.852	4.221
14	1.345	1.761	2.145	2.624	2.977	3.787	4.140
15	1.341	1.753	2.131	2.602	2.947	3.733	4.073
16	1.337	1.746	2.120	2.583	2.921	3.686	4.015
17	1.333	1.740	2.110	2.567	2.898	3.646	3.965
18	1.330	1.734	2.101	2.552	2.878	3.610	3.922
19	1.328	1.729	2.093	2.539	2.861	3.579	3.883
20	1.325	1.725	2.086	2.528	2.845	3.552	3.850
21	1.323	1.721	2.080	2.518	2.831	3.527	3.819
22	1.321	1.717	2.074	2.508	2.819	3.505	3.792
23	1.319	1.714	2.069	2.500	2.807	3.485	3.767
24	1.318	1.711	2.064	2.492	2.797	3.467	3.745
25	1.316	1.708	2.060	2.485	2.787	3.450	3.725
26	1.315	1.706	2.056	2.479	2.779	3.435	3.707
27	1.314	1.703	2.052	2.473	2.771	3.421	3.690
28	1.313	1.701	2.048	2.467	2.763	3.408	3.674
29	1.311	1.699	2.045	2.462	2.756	3.396	3.659
30	1.310	1.697	2.042	2.457	2.750	3.385	3.646
40	1.303	1.684	2.021	2.423	2.704	3.307	3.551
60	1.296	1.671	2.000	2.390	2.660	3.232	3.460
120	1.289	1.658	1.980	2.358	2.617	3.160	3.373
∞	1.282	1.645	1.960	2.326	2.576	3.090	3.291

TABLE V Critical Values of T_L and T_U for the Wilcoxon Rank Sum Test: Independent Samples

Test statistic is the rank sum associated with the smaller sample (if equal sample sizes, either rank sum can be used).

a. $\alpha = .025$ one-tailed; $\alpha = .05$ two-tailed

n_2 \ n_1	3 T_L	3 T_U	4 T_L	4 T_U	5 T_L	5 T_U	6 T_L	6 T_U	7 T_L	7 T_U	8 T_L	8 T_U	9 T_L	9 T_U	10 T_L	10 T_U
3	5	16	6	18	6	21	7	23	7	26	8	28	8	31	9	33
4	6	18	11	25	12	28	12	32	13	35	14	38	15	41	16	44
5	6	21	12	28	18	37	19	41	20	45	21	49	22	53	24	56
6	7	23	12	32	19	41	26	52	28	56	29	61	31	65	32	70
7	7	26	13	35	20	45	28	56	37	68	39	73	41	78	43	83
8	8	28	14	38	21	49	29	61	39	73	49	87	51	93	54	98
9	8	31	15	41	22	53	31	65	41	78	51	93	63	108	66	114
10	9	33	16	44	24	56	32	70	43	83	54	98	66	114	79	131

b. $\alpha = .05$ one-tailed; $\alpha = .10$ two-tailed

n_2 \ n_1	3 T_L	3 T_U	4 T_L	4 T_U	5 T_L	5 T_U	6 T_L	6 T_U	7 T_L	7 T_U	8 T_L	8 T_U	9 T_L	9 T_U	10 T_L	10 T_U
3	6	15	7	17	7	20	8	22	9	24	9	27	10	29	11	31
4	7	17	12	24	13	27	14	30	15	33	16	36	17	39	18	42
5	7	20	13	27	19	36	20	40	22	43	24	46	25	50	26	54
6	8	22	14	30	20	40	28	50	30	54	32	58	33	63	35	67
7	9	24	15	33	22	43	30	54	39	66	41	71	43	76	46	80
8	9	27	16	36	24	46	32	58	41	71	52	84	54	90	57	95
9	10	29	17	39	25	50	33	63	43	76	54	90	66	105	69	111
10	11	31	18	42	26	54	35	67	46	80	57	95	69	111	83	127

Source: From F. Wilcoxon and R. A. Wilcox, "Some Rapid Approximate Statistical Procedures," 1964, 20–23.

TABLE VI Critical Values of T_0 in the Wilcoxon Paired Difference Signed Rank Test

One-Tailed	Two-Tailed	n = 5	n = 6	n = 7	n = 8	n = 9	n = 10
$\alpha = .05$	$\alpha = .10$	1	2	4	6	8	11
$\alpha = .025$	$\alpha = .05$		1	2	4	6	8
$\alpha = .01$	$\alpha = .02$			0	2	3	5
$\alpha = .005$	$\alpha = .01$				0	2	3
		n = 11	n = 12	n = 13	n = 14	n = 15	n = 16
$\alpha = .05$	$\alpha = .10$	14	17	21	26	30	36
$\alpha = .025$	$\alpha = .05$	11	14	17	21	25	30
$\alpha = .01$	$\alpha = .02$	7	10	13	16	20	24
$\alpha = .005$	$\alpha = .01$	5	7	10	13	16	19
		n = 17	n = 18	n = 19	n = 20	n = 21	n = 22
$\alpha = .05$	$\alpha = .10$	41	47	54	60	68	75
$\alpha = .025$	$\alpha = .05$	35	40	46	52	59	66
$\alpha = .01$	$\alpha = .02$	28	33	38	43	49	56
$\alpha = .005$	$\alpha = .01$	23	28	32	37	43	49
		n = 23	n = 24	n = 25	n = 26	n = 27	n = 28
$\alpha = .05$	$\alpha = .10$	83	92	101	110	120	130
$\alpha = .025$	$\alpha = .05$	73	81	90	98	107	117
$\alpha = .01$	$\alpha = .02$	62	69	77	85	93	102
$\alpha = .005$	$\alpha = .01$	55	61	68	76	84	92
		n = 29	n = 30	n = 31	n = 32	n = 33	n = 34
$\alpha = .05$	$\alpha = .10$	141	152	163	175	188	201
$\alpha = .025$	$\alpha = .05$	127	137	148	159	171	183
$\alpha = .01$	$\alpha = .02$	111	120	130	141	151	162
$\alpha = .005$	$\alpha = .01$	100	109	118	128	138	149
		n = 35	n = 36	n = 37	n = 38	n = 39	
$\alpha = .05$	$\alpha = .10$	214	228	242	256	271	
$\alpha = .025$	$\alpha = .05$	195	208	222	235	250	
$\alpha = .01$	$\alpha = .02$	174	186	198	211	224	
$\alpha = .005$	$\alpha = .01$	160	171	183	195	208	
		n = 40	n = 41	n = 42	n = 43	n = 44	n = 45
$\alpha = .05$	$\alpha = .10$	287	303	319	336	353	371
$\alpha = .025$	$\alpha = .05$	264	279	295	311	327	344
$\alpha = .01$	$\alpha = .02$	238	252	267	281	297	313
$\alpha = .005$	$\alpha = .01$	221	234	248	262	277	292
		n = 46	n = 47	n = 48	n = 49	n = 50	
$\alpha = .05$	$\alpha = .10$	389	408	427	446	466	
$\alpha = .025$	$\alpha = .05$	361	379	397	415	434	
$\alpha = .01$	$\alpha = .02$	329	345	362	380	398	
$\alpha = .005$	$\alpha = .01$	307	323	339	356	373	

Source: From F. Wilcoxon and R. A. Wilcox, "Some Rapid Approximate Statistical Procedures," 1964, p. 28.

TABLE VII   Percentage Points of the *F*-distribution, $\alpha = .10$

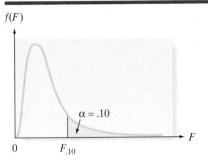

| | **NUMERATOR DEGREES OF FREEDOM** | | | | | | | | |
v_2 \ v_1	**1**	**2**	**3**	**4**	**5**	**6**	**7**	**8**	**9**
1	39.86	49.50	53.59	55.83	57.24	58.20	58.91	59.44	59.86
2	8.53	9.00	9.16	9.24	9.29	9.33	9.35	9.37	9.38
3	5.54	5.46	5.39	5.34	5.31	5.28	5.27	5.25	5.24
4	4.54	4.32	4.19	4.11	4.05	4.01	3.98	3.95	3.94
5	4.06	3.78	3.62	3.52	3.45	3.40	3.37	3.34	3.32
6	3.78	3.46	3.29	3.18	3.11	3.05	3.01	2.98	2.96
7	3.59	3.26	3.07	2.96	2.88	2.83	2.78	2.75	2.72
8	3.46	3.11	2.92	2.81	2.73	2.67	2.62	2.59	2.56
9	3.36	3.01	2.81	2.69	2.61	2.55	2.51	2.47	2.44
10	3.29	2.92	2.73	2.61	2.52	2.46	2.41	2.38	2.35
11	3.23	2.86	2.66	2.54	2.45	2.39	2.34	2.30	2.27
12	3.18	2.81	2.61	2.48	2.39	2.33	2.28	2.24	2.21
13	3.14	2.76	2.56	2.43	2.35	2.28	2.23	2.20	2.16
14	3.10	2.73	2.52	2.39	2.31	2.24	2.19	2.15	2.12
15	3.07	2.70	2.49	2.36	2.27	2.21	2.16	2.12	2.09
16	3.05	2.67	2.46	2.33	2.24	2.18	2.13	2.09	2.06
17	3.03	2.64	2.44	2.31	2.22	2.15	2.10	2.06	2.03
18	3.01	2.62	2.42	2.29	2.20	2.13	2.08	2.04	2.00
19	2.99	2.61	2.40	2.27	2.18	2.11	2.06	2.02	1.98
20	2.97	2.59	2.38	2.25	2.16	2.09	2.04	2.00	1.96
21	2.96	2.57	2.36	2.23	2.14	2.08	2.02	1.98	1.95
22	2.95	2.56	2.35	2.22	2.13	2.06	2.01	1.97	1.93
23	2.94	2.55	2.34	2.21	2.11	2.05	1.99	1.95	1.92
24	2.93	2.54	2.33	2.19	2.10	2.04	1.98	1.94	1.91
25	2.92	2.53	2.32	2.18	2.09	2.02	1.97	1.93	1.89
26	2.91	2.52	2.31	2.17	2.08	2.01	1.96	1.92	1.88
27	2.90	2.51	2.30	2.17	2.07	2.00	1.95	1.91	1.87
28	2.89	2.50	2.29	2.16	2.06	2.00	1.94	1.90	1.87
29	2.89	2.50	2.28	2.15	2.06	1.99	1.93	1.89	1.86
30	2.88	2.49	2.28	2.14	2.05	1.98	1.93	1.88	1.85
40	2.84	2.44	2.23	2.09	2.00	1.93	1.87	1.83	1.79
60	2.79	2.39	2.18	2.04	1.95	1.87	1.82	1.77	1.74
120	2.75	2.35	2.13	1.99	1.90	1.82	1.77	1.72	1.68
∞	2.71	2.30	2.08	1.94	1.85	1.77	1.72	1.67	1.63

(left axis label: **DENOMINATOR DEGREES OF FREEDOM**)

Source: From M. Merrington and C. M. Thompson, "Tables of Percentage Points of the Inverted Beta (*F*)-Distribution," *Biometrika,* 1943, 33, 73–88.

(continued)

TABLE VII   Continued

ν_2	\multicolumn{10}{c}{NUMERATOR DEGREES OF FREEDOM ν_1}
	10	**12**	**15**	**20**	**24**	**30**	**40**	**60**	**120**	**∞**
1	60.19	60.71	61.22	61.74	62.00	62.26	62.53	62.79	63.06	63.33
2	9.39	9.41	9.42	9.44	9.45	9.46	9.47	9.47	9.48	9.49
3	5.23	5.22	5.20	5.18	5.18	5.17	5.16	5.15	5.14	5.13
4	3.92	3.90	3.87	3.84	3.83	3.82	3.80	3.79	3.78	3.76
5	3.30	3.27	3.24	3.21	3.19	3.17	3.16	3.14	3.12	3.10
6	2.94	2.90	2.87	2.84	2.82	2.80	2.78	2.76	2.74	2.72
7	2.70	2.67	2.63	2.59	2.58	2.56	2.54	2.51	2.49	2.47
8	2.54	2.50	2.46	2.42	2.40	2.38	2.36	2.34	2.32	2.29
9	2.42	2.38	2.34	2.30	2.28	2.25	2.23	2.21	2.18	2.16
10	2.32	2.28	2.24	2.20	2.18	2.16	2.13	2.11	2.08	2.06
11	2.25	2.21	2.17	2.12	2.10	2.08	2.05	2.03	2.00	1.97
12	2.19	2.15	2.10	2.06	2.04	2.01	1.99	1.96	1.93	1.90
13	2.14	2.10	2.05	2.01	1.98	1.96	1.93	1.90	1.88	1.85
14	2.10	2.05	2.01	1.96	1.94	1.91	1.89	1.86	1.83	1.80
15	2.06	2.02	1.97	1.92	1.90	1.87	1.85	1.82	1.79	1.76
16	2.03	1.99	1.94	1.89	1.87	1.84	1.81	1.78	1.75	1.72
17	2.00	1.96	1.91	1.86	1.84	1.81	1.78	1.75	1.72	1.69
18	1.98	1.93	1.89	1.84	1.81	1.78	1.75	1.72	1.69	1.66
19	1.96	1.91	1.86	1.81	1.79	1.76	1.73	1.70	1.67	1.63
20	1.94	1.89	1.84	1.79	1.77	1.74	1.71	1.68	1.64	1.61
21	1.92	1.87	1.83	1.78	1.75	1.72	1.69	1.66	1.62	1.59
22	1.90	1.86	1.81	1.76	1.73	1.70	1.67	1.64	1.60	1.57
23	1.89	1.84	1.80	1.74	1.72	1.69	1.66	1.62	1.59	1.55
24	1.88	1.83	1.78	1.73	1.70	1.67	1.64	1.61	1.57	1.53
25	1.87	1.82	1.77	1.72	1.69	1.66	1.63	1.59	1.56	1.52
26	1.86	1.81	1.76	1.71	1.68	1.65	1.61	1.58	1.54	1.50
27	1.85	1.80	1.75	1.70	1.67	1.64	1.60	1.57	1.53	1.49
28	1.84	1.79	1.74	1.69	1.66	1.63	1.59	1.56	1.52	1.48
29	1.83	1.78	1.73	1.68	1.65	1.62	1.58	1.55	1.51	1.47
30	1.82	1.77	1.72	1.67	1.64	1.61	1.57	1.54	1.50	1.46
40	1.76	1.71	1.66	1.61	1.57	1.54	1.51	1.47	1.42	1.38
60	1.71	1.66	1.60	1.54	1.51	1.48	1.44	1.40	1.35	1.29
120	1.65	1.60	1.55	1.48	1.45	1.41	1.37	1.32	1.26	1.19
∞	1.60	1.55	1.49	1.42	1.38	1.34	1.30	1.24	1.17	1.00

(Left vertical label: DENOMINATOR DEGREES OF FREEDOM ν_2)

TABLE VIII   Percentage Points of the F-distribution, $\alpha = .05$

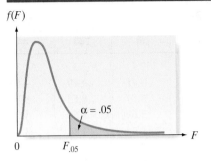

$f(F)$

$\alpha = .05$

0 $F_{.05}$ F

ν_1	**NUMERATOR DEGREES OF FREEDOM**								
ν_2	**1**	**2**	**3**	**4**	**5**	**6**	**7**	**8**	**9**
1	161.4	199.5	215.7	224.6	230.2	234.0	236.8	238.9	240.5
2	18.51	19.00	19.16	19.25	19.30	19.33	19.35	19.37	19.38
3	10.13	9.55	9.28	9.12	9.01	8.94	8.89	8.85	8.81
4	7.71	6.94	6.59	6.39	6.26	6.16	6.09	6.04	6.00
5	6.61	5.79	5.41	5.19	5.05	4.95	4.88	4.82	4.77
6	5.99	5.14	4.76	4.53	4.39	4.28	4.21	4.15	4.10
7	5.59	4.74	4.35	4.12	3.97	3.87	3.79	3.73	3.68
8	5.32	4.46	4.07	3.84	3.69	3.58	3.50	3.44	3.39
9	5.12	4.26	3.86	3.63	3.48	3.37	3.29	3.23	3.18
10	4.96	4.10	3.71	3.48	3.33	3.22	3.14	3.07	3.02
11	4.84	3.98	3.59	3.36	3.20	3.09	3.01	2.95	2.90
12	4.75	3.89	3.49	3.26	3.11	3.00	2.91	2.85	2.80
13	4.67	3.81	3.41	3.18	3.03	2.92	2.83	2.77	2.71
14	4.60	3.74	3.34	3.11	2.96	2.85	2.76	2.70	2.65
15	4.54	3.68	3.29	3.06	2.90	2.79	2.71	2.64	2.59
16	4.49	3.63	3.24	3.01	2.85	2.74	2.66	2.59	2.54
17	4.45	3.59	3.20	2.96	2.81	2.70	2.61	2.55	2.49
18	4.41	3.55	3.16	2.93	2.77	2.66	2.58	2.51	2.46
19	4.38	3.52	3.13	2.90	2.74	2.63	2.54	2.48	2.42
20	4.35	3.49	3.10	2.87	2.71	2.60	2.51	2.45	2.39
21	4.32	3.47	3.07	2.84	2.68	2.57	2.49	2.42	2.37
22	4.30	3.44	3.05	2.82	2.66	2.55	2.46	2.40	2.34
23	4.28	3.42	3.03	2.80	2.64	2.53	2.44	2.37	2.32
24	4.26	3.40	3.01	2.78	2.62	2.51	2.42	2.36	2.30
25	4.24	3.39	2.99	2.76	2.60	2.49	2.40	2.34	2.28
26	4.23	3.37	2.98	2.74	2.59	2.47	2.39	2.32	2.77
27	4.21	3.35	2.96	2.73	2.57	2.46	2.37	2.31	2.25
28	4.20	3.34	2.95	2.71	2.56	2.45	2.36	2.29	2.24
29	4.18	3.33	2.93	2.70	2.55	2.43	2.35	2.28	2.22
30	4.17	3.32	2.92	2.69	2.53	2.42	2.33	2.27	2.21
40	4.08	3.23	2.84	2.61	2.45	2.34	2.25	2.18	2.12
60	4.00	3.15	2.76	2.53	2.37	2.25	2.17	2.10	2.04
120	3.92	3.07	2.68	2.45	2.29	2.17	2.09	2.02	1.96
∞	3.84	3.00	2.60	2.37	2.21	2.10	2.01	1.94	1.88

DENOMINATOR DEGREES OF FREEDOM

Source: From M. Merrington and C. M. Thompson, "Tables of Percentage Points of the Inverted Beta (F)-Distribution," *Biometrika,* 1943, 33, 73–88.

(continued)

TABLE VIII   Continued

ν_1 ν_2	NUMERATOR DEGREES OF FREEDOM									
	10	**12**	**15**	**20**	**24**	**30**	**40**	**60**	**120**	**∞**
1	241.9	243.9	245.9	248.0	249.1	250.1	251.1	252.2	253.3	254.3
2	19.40	19.41	19.43	19.45	19.45	19.46	19.47	19.48	19.49	19.50
3	8.79	8.74	8.70	8.66	8.64	8.62	8.59	8.57	8.55	8.53
4	5.96	5.91	5.86	5.80	5.77	5.75	5.72	5.69	5.66	5.63
5	4.74	4.68	4.62	4.56	4.53	4.50	4.46	4.43	4.40	4.36
6	4.06	4.00	3.94	3.87	3.84	3.81	3.77	3.74	3.70	3.67
7	3.64	3.57	3.51	3.44	3.41	3.38	3.34	3.30	3.27	3.23
8	3.35	3.28	3.22	3.15	3.12	3.08	3.04	3.01	2.97	2.93
9	3.14	3.07	3.01	2.94	2.90	2.86	2.83	2.79	2.75	2.71
10	2.98	2.91	2.85	2.77	2.74	2.70	2.66	2.62	2.58	2.54
11	2.85	2.79	2.72	2.65	2.61	2.57	2.53	2.49	2.45	2.40
12	2.75	2.69	2.62	2.54	2.51	2.47	2.43	2.38	2.34	2.30
13	2.67	2.60	2.53	2.46	2.42	2.38	2.34	2.30	2.25	2.21
14	2.60	2.53	2.46	2.39	2.35	2.31	2.27	2.22	2.18	2.13
15	2.54	2.48	2.40	2.33	2.29	2.25	2.20	2.16	2.11	2.07
16	2.49	2.42	2.35	2.28	2.24	2.19	2.15	2.11	2.06	2.01
17	2.45	2.38	2.31	2.23	2.19	2.15	2.10	2.06	2.01	1.96
18	2.41	2.34	2.27	2.19	2.15	2.11	2.06	2.02	1.97	1.92
19	2.38	2.31	2.23	2.16	2.11	2.07	2.03	1.98	1.93	1.88
20	2.35	2.28	2.20	2.12	2.08	2.04	1.99	1.95	1.90	1.84
21	2.32	2.25	2.18	2.10	2.05	2.01	1.96	1.92	1.87	1.81
22	2.30	2.23	2.15	2.07	2.03	1.98	1.94	1.89	1.84	1.78
23	2.27	2.20	2.13	2.05	2.01	1.96	1.91	1.86	1.81	1.76
24	2.25	2.18	2.11	2.03	1.98	1.94	1.89	1.84	1.79	1.73
25	2.24	2.16	2.09	2.01	1.96	1.92	1.87	1.82	1.77	1.71
26	2.22	2.15	2.07	1.99	1.95	1.90	1.85	1.80	1.75	1.69
27	2.20	2.13	2.06	1.97	1.93	1.88	1.84	1.79	1.73	1.67
28	2.19	2.12	2.04	1.96	1.91	1.87	1.82	1.77	1.71	1.65
29	2.18	2.10	2.03	1.94	1.90	1.85	1.81	1.75	1.70	1.64
30	2.16	2.09	2.01	1.93	1.89	1.84	1.79	1.74	1.68	1.62
40	2.08	2.00	1.92	1.84	1.79	1.74	1.69	1.64	1.58	1.51
60	1.99	1.92	1.84	1.75	1.70	1.65	1.59	1.53	1.47	1.39
120	1.91	1.83	1.75	1.66	1.61	1.55	1.50	1.43	1.35	1.25
∞	1.83	1.75	1.67	1.57	1.52	1.46	1.39	1.32	1.22	1.00

DENOMINATOR DEGREES OF FREEDOM

TABLE IX   Percentage Points of the *F*-distribution, $\alpha = .025$

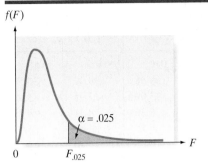

$f(F)$

$\alpha = .025$

F

0   $F_{.025}$

ν_2 \ ν_1	NUMERATOR DEGREES OF FREEDOM								
	1	2	3	4	5	6	7	8	9
1	647.8	799.5	864.2	899.6	921.8	937.1	948.2	956.7	963.3
2	38.51	39.00	39.17	39.25	39.30	39.33	39.36	39.37	39.39
3	17.44	16.04	15.44	15.10	14.88	14.73	14.62	14.54	14.47
4	12.22	10.65	9.98	9.60	9.36	9.20	9.07	8.98	8.90
5	10.01	8.43	7.76	7.39	7.15	6.98	6.85	6.76	6.68
6	8.81	7.26	6.60	6.23	5.99	5.82	5.70	5.60	5.52
7	8.07	6.54	5.89	5.52	5.29	5.12	4.99	4.90	4.82
8	7.57	6.06	5.42	5.05	4.82	4.65	4.53	4.43	4.36
9	7.21	5.71	5.08	4.72	4.48	4.32	4.20	4.10	4.03
10	6.94	5.46	4.83	4.47	4.24	4.07	3.95	3.85	3.78
11	6.72	5.26	4.63	4.28	4.04	3.88	3.76	3.66	3.59
12	6.55	5.10	4.47	4.12	3.89	3.73	3.61	3.51	3.44
13	6.41	4.97	4.35	4.00	3.77	3.60	3.48	3.39	3.31
14	6.30	4.86	4.24	3.89	3.66	3.50	3.38	3.29	3.21
15	6.20	4.77	4.15	3.80	3.58	3.41	3.29	3.20	3.12
16	6.12	4.69	4.08	3.73	3.50	3.34	3.22	3.12	3.05
17	6.04	4.62	4.01	3.66	3.44	3.28	3.16	3.06	2.98
18	5.98	4.56	3.95	3.61	3.38	3.22	3.10	3.01	2.93
19	5.92	4.51	3.90	3.56	3.33	3.17	3.05	2.96	2.88
20	5.87	4.46	3.86	3.51	3.29	3.13	3.01	2.91	2.84
21	5.83	4.42	3.82	3.48	3.25	3.09	2.97	2.87	2.80
22	5.79	4.38	3.78	3.44	3.22	3.05	2.93	2.84	2.76
23	5.75	4.35	3.75	3.41	3.18	3.02	2.90	2.81	2.73
24	5.72	4.32	3.72	3.38	3.15	2.99	2.87	2.78	2.70
25	5.69	4.29	3.69	3.35	3.13	2.97	2.85	2.75	2.68
26	5.66	4.27	3.67	3.33	3.10	2.94	2.82	2.73	2.65
27	5.63	4.24	3.65	3.31	3.08	2.92	2.80	2.71	2.63
28	5.61	4.22	3.63	3.29	3.06	2.90	2.78	2.69	2.61
29	5.59	4.20	3.61	3.27	3.04	2.88	2.76	2.67	2.59
30	5.57	4.18	3.59	3.25	3.03	2.87	2.75	2.65	2.57
40	5.42	4.05	3.46	3.13	2.90	2.74	2.62	2.53	2.45
60	5.29	3.93	3.34	3.01	2.79	2.63	2.51	2.41	2.33
120	5.15	3.80	3.23	2.89	2.67	2.52	2.39	2.30	2.22
∞	5.02	3.69	3.12	2.79	2.57	2.41	2.29	2.19	2.11

DENOMINATOR DEGREES OF FREEDOM

Source: From M. Merrington and C. M. Thompson, "Tables of Percentage Points of the Inverted Beta (*F*)-Distribution," *Biometrika*, 1943, 33, 73–88.

(continued)

TABLE IX   Continued

ν_1 → ν_2 ↓	NUMERATOR DEGREES OF FREEDOM									
	10	12	15	20	24	30	40	60	120	∞
1	968.6	976.7	984.9	993.1	997.2	1,001	1,006	1,010	1,014	1,018
2	39.40	39.41	39.43	39.45	39.46	39.46	39.47	39.48	39.49	39.50
3	14.42	14.34	14.25	14.17	14.12	14.08	14.04	13.99	13.95	13.90
4	8.84	8.75	8.66	8.56	8.51	8.46	8.41	8.36	8.31	8.26
5	6.62	6.52	6.43	6.33	6.28	6.23	6.18	6.12	6.07	6.02
6	5.46	5.37	5.27	5.17	5.12	5.07	5.01	4.96	4.90	4.85
7	4.76	4.67	4.57	4.47	4.42	4.36	4.31	4.25	4.20	4.14
8	4.30	4.20	4.10	4.00	3.95	3.89	3.84	3.78	3.73	3.67
9	3.96	3.87	3.77	3.67	3.61	3.56	3.51	3.45	3.39	3.33
10	3.72	3.62	3.52	3.42	3.37	3.31	3.26	3.20	3.14	3.08
11	3.53	3.43	3.33	3.23	3.17	3.12	3.06	3.00	2.94	2.88
12	3.37	3.28	3.18	3.07	3.02	2.96	2.91	2.85	2.79	2.72
13	3.25	3.15	3.05	2.95	2.89	2.84	2.78	2.72	2.66	2.60
14	3.15	3.05	2.95	2.84	2.79	2.73	2.67	2.61	2.55	2.49
15	3.06	2.96	2.86	2.76	2.70	2.64	2.59	2.52	2.46	2.40
16	2.99	2.89	2.79	2.68	2.63	2.57	2.51	2.45	2.38	2.32
17	2.92	2.82	2.72	2.62	2.56	2.50	2.44	2.38	2.32	2.25
18	2.87	2.77	2.67	2.56	2.50	2.44	2.38	2.32	2.26	2.19
19	2.82	2.72	2.62	2.51	2.45	2.39	2.33	2.27	2.20	2.13
20	2.77	2.68	2.57	2.46	2.41	2.35	2.29	2.22	2.16	2.09
21	2.73	2.64	2.53	2.42	2.37	2.31	2.25	2.18	2.11	2.04
22	2.70	2.60	2.50	2.39	2.33	2.27	2.21	2.14	2.08	2.00
23	2.67	2.57	2.47	2.36	2.30	2.24	2.18	2.11	2.04	1.97
24	2.64	2.54	2.44	2.33	2.27	2.21	2.15	2.08	2.01	1.94
25	2.61	2.51	2.41	2.30	2.24	2.18	2.12	2.05	1.98	1.91
26	2.59	2.49	2.39	2.28	2.22	2.16	2.09	2.03	1.95	1.88
27	2.57	2.47	2.36	2.25	2.19	2.13	2.07	2.00	1.93	1.85
28	2.55	2.45	2.34	2.23	2.17	2.11	2.05	1.98	1.91	1.83
29	2.53	2.43	2.32	2.21	2.15	2.09	2.03	1.96	1.89	1.81
30	2.51	2.41	2.31	2.20	2.14	2.07	2.01	1.94	1.87	1.79
40	2.39	2.29	2.18	2.07	2.01	1.94	1.88	1.80	1.72	1.64
60	2.27	2.17	2.06	1.94	1.88	1.82	1.74	1.67	1.58	1.48
120	2.16	2.05	1.94	1.82	1.76	1.69	1.61	1.53	1.43	1.31
∞	2.05	1.94	1.83	1.71	1.64	1.57	1.48	1.39	1.27	1.00

DENOMINATOR DEGREES OF FREEDOM

TABLE X   Percentage Points of the F-distribution, $\alpha = .01$

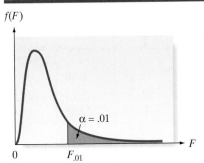

ν_1					NUMERATOR DEGREES OF FREEDOM				
ν_2	**1**	**2**	**3**	**4**	**5**	**6**	**7**	**8**	**9**
1	4,052	4,999.5	5,403	5,625	5,764	5,859	5,928	5,982	6,022
2	98.50	99.00	99.17	99.25	99.30	99.33	99.36	99.37	99.39
3	34.12	30.82	29.46	28.71	28.24	27.91	27.67	27.49	27.35
4	21.20	18.00	16.69	15.98	15.52	15.21	14.98	14.80	14.66
5	16.26	13.27	12.06	11.39	10.97	10.67	10.46	10.29	10.16
6	13.75	10.92	9.78	9.15	8.75	8.47	8.26	8.10	7.98
7	12.25	9.55	8.45	7.85	7.46	7.19	6.99	6.84	6.72
8	11.26	8.65	7.59	7.01	6.63	6.37	6.18	6.03	5.91
9	10.56	8.02	6.99	6.42	6.06	5.80	5.61	5.47	5.35
10	10.04	7.56	6.55	5.99	5.64	5.39	5.20	5.06	4.94
11	9.65	7.21	6.22	5.67	5.32	5.07	4.89	4.74	4.63
12	9.33	6.93	5.95	5.41	5.06	4.82	4.64	4.50	4.39
13	9.07	6.70	5.74	5.21	4.86	4.62	4.44	4.30	4.19
14	8.86	6.51	5.56	5.04	4.69	4.46	4.28	4.14	4.03
15	8.68	6.36	5.42	4.89	4.56	4.32	4.14	4.00	3.89
16	8.53	6.23	5.29	4.77	4.44	4.20	4.03	3.89	3.78
17	8.40	6.11	5.18	4.67	4.34	4.10	3.93	3.79	3.68
18	8.29	6.01	5.09	4.58	4.25	4.01	3.84	3.71	3.60
19	8.18	5.93	5.01	4.50	4.17	3.94	3.77	3.63	3.52
20	8.10	5.85	4.94	4.43	4.10	3.87	3.70	3.56	3.46
21	8.02	5.78	4.87	4.37	4.04	3.81	3.64	3.51	3.40
22	7.95	5.72	4.82	4.31	3.99	3.76	3.59	3.45	3.35
23	7.88	5.66	4.76	4.26	3.94	3.71	3.54	3.41	3.30
24	7.82	5.61	4.72	4.22	3.90	3.67	3.50	3.36	3.26
25	7.77	5.57	4.68	4.18	3.85	3.63	3.46	3.32	3.22
26	7.72	5.53	4.64	4.14	3.82	3.59	3.42	3.29	3.18
27	7.68	5.49	4.60	4.11	3.78	3.56	3.39	3.26	3.15
28	7.64	5.45	4.57	4.07	3.75	3.53	3.36	3.23	3.12
29	7.60	5.42	4.54	4.04	3.73	3.50	3.33	3.20	3.09
30	7.56	5.39	4.51	4.02	3.70	3.47	3.30	3.17	3.07
40	7.31	5.18	4.31	3.83	3.51	3.29	3.12	2.99	2.89
60	7.08	4.98	4.13	3.65	3.34	3.12	2.95	2.82	2.72
120	6.85	4.79	3.95	3.48	3.17	2.96	2.79	2.66	2.56
∞	6.63	4.61	3.78	3.32	3.02	2.80	2.64	2.51	2.41

Source: From M. Merrington and C. M. Thompson, "Tables of Percentage Points of the Inverted Beta (F)-Distribution," *Biometrika,* 1943, 33, 73–88.

(continued)

TABLE X   Continued

ν_2 \ ν_1	NUMERATOR DEGREES OF FREEDOM									
	10	**12**	**15**	**20**	**24**	**30**	**40**	**60**	**120**	**∞**
1	6,056	6,106	6,157	6,209	6,235	6,261	6,287	6,313	6,339	6,366
2	99.40	99.42	99.43	99.45	99.46	99.47	99.47	99.48	99.49	99.50
3	27.23	27.05	26.87	26.69	26.60	26.50	26.41	26.32	26.22	26.13
4	14.55	14.37	14.20	14.02	13.93	13.84	13.75	13.65	13.56	13.46
5	10.05	9.89	9.72	9.55	9.47	9.38	9.29	9.20	9.11	9.02
6	7.87	7.72	7.56	7.40	7.31	7.23	7.14	7.06	6.97	6.88
7	6.62	6.47	6.31	6.16	6.07	5.99	5.91	5.82	5.74	5.65
8	5.81	5.67	5.52	5.36	5.28	5.20	5.12	5.03	4.95	4.86
9	5.26	5.11	4.96	4.81	4.73	4.65	4.57	4.48	4.40	4.31
10	4.85	4.71	4.56	4.41	4.33	4.25	4.17	4.08	4.00	3.91
11	4.54	4.40	4.25	4.10	4.02	3.94	3.86	3.78	3.69	3.60
12	4.30	4.16	4.01	3.86	3.78	3.70	3.62	3.54	3.45	3.36
13	4.10	3.96	3.82	3.66	3.59	3.51	3.43	3.34	3.25	3.17
14	3.94	3.80	3.66	3.51	3.43	3.35	3.27	3.18	3.09	3.00
15	3.80	3.67	3.52	3.37	3.29	3.21	3.13	3.05	2.96	2.87
16	3.69	3.55	3.41	3.26	3.18	3.10	3.02	2.93	2.84	2.75
17	3.59	3.46	3.31	3.16	3.08	3.00	2.92	2.83	2.75	2.65
18	3.51	3.37	3.23	3.08	3.00	2.92	2.84	2.75	2.66	2.57
19	3.43	3.30	3.15	3.00	2.92	2.84	2.76	2.67	2.58	2.49
20	3.37	3.23	3.09	2.94	2.86	2.78	2.69	2.61	2.52	2.42
21	3.31	3.17	3.03	2.88	2.80	2.72	2.64	2.55	2.46	2.36
22	3.26	3.12	2.98	2.83	2.75	2.67	2.58	2.50	2.40	2.31
23	3.21	3.07	2.93	2.78	2.70	2.62	2.54	2.45	2.35	2.26
24	3.17	3.03	2.89	2.74	2.66	2.58	2.49	2.40	2.31	2.21
25	3.13	2.99	2.85	2.70	2.62	2.54	2.45	2.36	2.27	2.17
26	3.09	2.96	2.81	2.66	2.58	2.50	2.42	2.33	2.23	2.13
27	3.06	2.93	2.78	2.63	2.55	2.47	2.38	2.29	2.20	2.10
28	3.03	2.90	2.75	2.60	2.52	2.44	2.35	2.26	2.17	2.06
29	3.00	2.87	2.73	2.57	2.49	2.41	2.33	2.23	2.14	2.03
30	2.98	2.84	2.70	2.55	2.47	2.39	2.30	2.21	2.11	2.01
40	2.80	2.66	2.52	2.37	2.29	2.20	2.11	2.02	1.92	1.80
60	2.63	2.50	2.35	2.20	2.12	2.03	1.94	1.84	1.73	1.60
120	2.47	2.34	2.19	2.03	1.95	1.86	1.76	1.66	1.53	1.38
∞	2.32	2.18	2.04	1.88	1.79	1.70	1.59	1.47	1.32	1.00

DENOMINATOR DEGREES OF FREEDOM

TABLE XI   Critical Values of χ^2

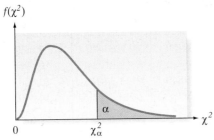

Degrees of Freedom	$\chi^2_{.995}$	$\chi^2_{.990}$	$\chi^2_{.975}$	$\chi^2_{.950}$	$\chi^2_{.900}$
1	.0000393	.0001571	.0009821	.0039321	.0157908
2	.0100251	.0201007	.0506356	.102587	.210720
3	.0717212	.114832	.215795	.351846	.584375
4	.206990	.297110	.484419	.710721	1.063623
5	.411740	.554300	.831211	1.145476	1.61031
6	.675727	.872085	1.237347	1.63539	2.20413
7	.989265	1.239043	1.68987	2.16735	2.83311
8	1.344419	1.646482	2.17973	2.73264	3.48954
9	1.734926	2.087912	2.70039	3.32511	4.16816
10	2.15585	2.55821	3.24697	3.94030	4.86518
11	2.60321	3.05347	3.81575	4.57481	5.57779
12	3.07382	3.57056	4.40379	5.22603	6.30380
13	3.56503	4.10691	5.00874	5.89186	7.04150
14	4.07468	4.66043	5.62872	6.57063	7.78953
15	4.60094	5.22935	6.26214	7.26094	8.54675
16	5.14224	5.81221	6.90766	7.96164	9.31223
17	5.69724	6.40776	7.56418	8.67176	10.0852
18	6.26481	7.01491	8.23075	9.39046	10.8649
19	6.84398	7.63273	8.90655	10.1170	11.6509
20	7.43386	8.26040	9.59083	10.8508	12.4426
21	8.03366	8.89720	10.28293	11.5913	13.2396
22	8.64272	9.54249	10.9823	12.3380	14.0415
23	9.26042	10.19567	11.6885	13.0905	14.8479
24	9.88623	10.8564	12.4011	13.8484	15.6587
25	10.5197	11.5240	13.1197	14.6114	16.4734
26	11.1603	12.1981	13.8439	15.3791	17.2919
27	11.8076	12.8786	14.5733	16.1513	18.1138
28	12.4613	13.5648	15.3079	16.9279	18.9392
29	13.1211	14.2565	16.0471	17.7083	19.7677
30	13.7867	14.9535	16.7908	18.4926	20.5992
40	20.7065	22.1643	24.4331	26.5093	29.0505
50	27.9907	29.7067	32.3574	34.7642	37.6886
60	35.5346	37.4848	40.4817	43.1879	46.4589
70	43.2752	45.4418	48.7576	51.7393	55.3290
80	51.1720	53.5400	57.1532	60.3915	64.2778
90	59.1963	61.7541	65.6466	69.1260	73.2912
100	67.3276	70.0648	74.2219	77.9295	82.3581

Source: From C. M. Thompson, "Tables of the Percentage Points of the χ^2-*Distribution*," *Biometrika*, 1941, 32, 188–189. Reproduced by permission of the *Biometrika* Trustees.

(continued)

TABLE XI   Continued

Degrees of Freedom	$\chi^2_{.100}$	$\chi^2_{.050}$	$\chi^2_{.025}$	$\chi^2_{.010}$	$\chi^2_{.005}$
1	2.70554	3.84146	5.02389	6.63490	7.87944
2	4.60517	5.99147	7.37776	9.21034	10.5966
3	6.25139	7.81473	9.34840	11.3449	12.8381
4	7.77944	9.48773	11.1433	13.2767	14.8602
5	9.23635	11.0705	12.8325	15.0863	16.7496
6	10.6446	12.5916	14.4494	16.8119	18.5476
7	12.0170	14.0671	16.0128	18.4753	20.2777
8	13.3616	15.5073	17.5346	20.0902	21.9550
9	14.6837	16.9190	19.0228	21.6660	23.5893
10	15.9871	18.3070	20.4831	23.2093	25.1882
11	17.2750	19.6751	21.9200	24.7250	26.7569
12	18.5494	21.0261	23.3367	26.2170	28.2995
13	19.8119	22.3621	24.7356	27.6883	29.8194
14	21.0642	23.6848	26.1190	29.1413	31.3193
15	22.3072	24.9958	27.4884	30.5779	32.8013
16	23.5418	26.2962	28.8454	31.9999	34.2672
17	24.7690	27.5871	30.1910	33.4087	35.7185
18	25.9894	28.8693	31.5264	34.8053	37.1564
19	27.2036	30.1435	32.8523	36.1908	38.5822
20	28.4120	31.4104	34.1696	37.5662	39.9968
21	29.6151	32.6705	35.4789	38.9321	41.4010
22	30.8133	33.9244	36.7807	40.2894	42.7956
23	32.0069	35.1725	38.0757	41.6384	44.1813
24	33.1963	36.4151	39.3641	42.9798	45.5585
25	34.3816	37.6525	40.6465	44.3141	46.9278
26	35.5631	38.8852	41.9232	45.6417	48.2899
27	36.7412	40.1133	43.1944	46.9630	49.6449
28	37.9159	41.3372	44.4607	48.2782	50.9933
29	39.0875	42.5569	45.7222	49.5879	52.3356
30	40.2560	43.7729	46.9792	50.8922	53.6720
40	51.8050	55.7585	59.3417	63.6907	66.7659
50	63.1671	67.5048	71.4202	76.1539	79.4900
60	74.3970	79.0819	83.2976	88.3794	91.9517
70	85.5271	90.5312	95.0231	100.425	104.215
80	96.5782	101.879	106.629	112.329	116.321
90	107.565	113.145	118.136	124.116	128.299
100	118.498	124.342	129.561	135.807	140.169

TABLE XII Critical Values of Spearman's Rank Correlation Coefficient

The α values correspond to a one-tailed test of $H_0: \rho = 0$. The value should be doubled for two-tailed tests.

n	$\alpha = .05$	$\alpha = .025$	$\alpha = .01$	$\alpha = .005$	n	$\alpha = .05$	$\alpha = .025$	$\alpha = .01$	$\alpha = .005$
5	.900	—	—	—	18	.399	.476	.564	.625
6	.829	.886	.943	—	19	.388	.462	.549	.608
7	.714	.786	.893	—	20	.377	.450	.534	.591
8	.643	.738	.833	.881	21	.368	.438	.521	.576
9	.600	.683	.783	.833	22	.359	.428	.508	.562
10	.564	.648	.745	.794	23	.351	.418	.496	.549
11	.523	.623	.736	.818	24	.343	.409	.485	.537
12	.497	.591	.703	.780	25	.336	.400	.475	.526
13	.475	.566	.673	.745	26	.329	.392	.465	.515
14	.457	.545	.646	.716	27	.323	.385	.456	.505
15	.441	.525	.623	.689	28	.317	.377	.448	.496
16	.425	.507	.601	.666	29	.311	.370	.440	.487
17	.412	.490	.582	.645	30	.305	.364	.432	.478

Source: From E. G. Olds, "Distribution of Sums of Squares of Rank Differences for Small Samples," *Annals of Mathematical Statistics,* 1938, 9.

Calculation Formulas for Analysis of Variance (Independent Sampling)

$$\text{CM} = \text{Correction for mean}$$

$$= \frac{(\text{Total of all observations})^2}{\text{Total number of abservations}} = \frac{(\Sigma y_i)^2}{n}$$

$$\text{SS(Total)} = \text{Total sum of squares}$$

$$= (\text{Sum of squares of all observations}) - \text{CM} = \Sigma y_i^2 - \text{CM}$$

$$\text{SST} = \text{Sum of squares for treatments}$$

$$= \left(\begin{array}{c} \text{Sum of squares of treatments totals with} \\ \text{each square divided by the number of} \\ \text{observations for that treatment} \end{array} \right) - \text{CM}$$

$$= \frac{T_1^2}{n_1} + \frac{T_2^2}{n_2} + \ldots + \frac{T_k^2}{n_k} - \text{CM}$$

$$\text{SSE} = \text{Sum of squares for error} = \text{SS(Total)} - \text{SST}$$

$$\text{MST} = \text{Mean square for treatments} = \frac{\text{SST}}{k - 1}$$

$$\text{MSE} = \text{Mean square for error} = \frac{\text{SSE}}{n - k}$$

$$F = \text{Test statistic} = \frac{\text{MST}}{\text{MSE}}$$

where

$$n = \text{Total number of observations}$$
$$k = \text{Number of treatments}$$
$$T_i = \text{Total for treatment } i \ (i = 1, 2, \ldots, k)$$

Short Answers to Selected Odd Exercises

Chapter 1

1.3 population; variable(s); sample; inference; measure of reliability **1.11** qualitative; qualitative **1.13 a.** earthquake sites **b.** sample **c.** ground motion (qualitative); magnitude (quantitative); ground acceleration (quantitative) **1.15 b.** qualitative **e.** survey data **1.17 a.** sample **b.** industry (qualitative); compensation (quantitative); stock shares (quantitative); age (quantitative); efficiency rating (quantitative) **1.19** Town, Type of water supply, Presence of hydrogen sulphide are qualitative; all others are quantitative **1.21 a.** quantitative **b.** quantitative **c.** qualitative **d.** quantitative **e.** qualitative **f.** quantitative **g.** qualitative **1.23 a.** designed experiment **b.** smokers **c.** quantitative **d.** population: all smokers in the U.S.; sample: 50,000 smokers in trial **e.** the difference in mean age at which each of the scanning methods first detects a tumor. **1.25 a.** designed experiment **b.** amateur boxers **c.** heart rate (quantitative); blood lactate level (quantitative) **d.** no difference between the two groups of boxers **e.** no **1.27 a.** designed experiment **b.** inferential statistics **c.** all possible burn patients **1.29 b.** number of headers per game and IQ **c.** both quantitative **1.31** ambiguous question

Chapter 2

2.5 a. $X - 8, Y - 9, Z - 3$ **b.** $X - .40, Y - .45, Z - .15$ **2.7 a.** $39/266 = .147$ **b.** level 2 $-$ .286, level 3 $-$ .188, level 4 $-$ .327, level 5 $-$ .041, level 6 $-$ .011 **e.** level 4 **2.9 a.** relative frequencies: Black $-$ .203; White $-$ .637; Sumatran $-$ .017; Javan $-$ .003; Indian $-$ .140 **c.** .839; .161 **2.11 a.** .389 **b.** yes **c.** multi-year ice is most common **2.13 a.** survey **b.** quantitative **c.** about 60% **2.15** 75% of sampled CEOs had advanced degrees **2.17 a.** relative frequencies: 1 $-$ .338; 2 $-$ .470; 3 $-$ .164; 4 $-$ .017; 8 $-$ .011 **2.25 a.** 23 **2.27** frequencies: 50, 75, 125, 100, 25, 50, 50, 25 **2.29 a.** 70% **b.** 10%

2.31 a.

Stem	Leaf
1	6
2	0 1 2 4 8
3	0 4 4 9
4	0 0 2
5	3

b. A students tend to read the most books

2.35 a.

Stem	Leaf
1	0000000
2	0
3	00
4	000
5	
6	
7	
8	
9	0
10	0
11	0
12	00

b. Yes

2.37 most of the PMI's range from 3 to 7.5

2.39 a.

Stem	Leaf
0	1 1 2 3 4 5 5 9
1	1 2 3
2	0 0
3	9
4	6
5	6
6	1 5
7	
8	
9	
10	0

c. eclipses   **2.41 a.** Response rate of the familiarity group is higher.   **b.** Familiarity group has highest rate; control group has lowest rate   **2.43 a.** 33   **b.** 175   **c.** 20   **d.** 71   **e.** 1,089   **2.45 a.** 6   **b.** 50   **c.** 42.8   **2.47** mean, median, mode   **2.49** sample size and variability of the data   **2.51 a.** mean < median   **b.** mean > median   **c.** mean = median   **2.53** mode = 15; mean = 14.545; median = 15   **2.55 a.** 8.5   **b.** 25   **c.** .78   **d.** 13.44   **2.57 a.** mean = 31.6; median = 32; mode = 34 and 40   **b.** little or no skewness   **2.59 a.** 1.47   **b.** 1.49   **2.61 a.** mean = -4.86; median = -4.85; mode = -5.00 **2.63 b.** probably none   **2.65 a.** qualitative, qualitative, quantitative, quantitative   **c.** mean = 93.6; median = 91.5; mode = 86   **d.** mean = 95.3; median = 92; mode = 92   **e.** mean = 101.9; median = 103; mode = 113   **f.** three centers   **g.** YND: mean = 46.88, median = 43.7, mode = 0; SLI: mean = 30.17, median = 32.5, mode = 0; OND: mean = 0, median = 0, mode = 0   **2.67 a.** mean = $-.15$; median = $-.11$; 4 modes   **c.** mean = $-.12$; median = $-.105$; 4 modes   **2.69** largest value minus smallest value   **2.73** more variable   **2.75 a.** 5, 3.7, 1.92   **b.** 99, 1949.25, 44.15   **c.** 98, 1307.84, 36.16   **2.77** data set 1: 1, 1, 2, 2, 3, 3, 4, 4, 5, 5; data set 2: 1, 1, 1, 1, 1, 5, 5, 5, 5, 5   **2.79 a.** 3, 1.3, 1.14   **b.** 3, 1.3, 1.14   **c.** 3, 1.3, 1.14   **2.81 a.** 29, 75.7, 8.70   **b.** 24, 72.7, 8.53   **c.** A students   **2.83 a.** .18   **b.** .0041   **c.** .064   **d.** Morning   **2.85 a.** 12, 9.37, 3.06   **b.** 8, 5.15, 2.27; data less variable   **c.** 8, 5.06, 2.25; data less variable   **2.87 a.** dollars; quantitative   **b.** at least 3/4; at least 8/9; nothing   **2.89 a.** $\approx 68\%$ **b.** $\approx 95\%$ **c.** $\approx$ all   **2.91** range/6 = 104.17, range/4 = 156.25; no   **2.93 a.** $\bar{x} = 94.91$, $s = 4.83$ **b.** (90.08, 99.74); (85.25, 104.57); (80.42, 109.40)   **c.** 81.1%, 97.6%, 98.2%; yes   **2.95 a.** unknown   **b.** $\approx 84\%$ **2.97 a.** (0, 300.5)   **b.** (0, 546)   **c.** handrubbing appears to be more effective   **2.99 a.** at least 8/9 of the velocities will fall within $936 \pm 3(10)$ **b.** No   **2.101 a.** 19 ± 195 **b.** 7 ± 147 **c.** SAT-Math   **2.103 a.** 25%; 75%   **b.** 50%; 50%   **c.** 80%; 20%   **d.** 16%; 84%   **2.105** $\approx 95\%\%$ **2.107 a.** $z = 2$ **b.** $z = -3$ **c.** $z = -2$ **d.** $z = 1.67$ **2.109** 26th percentile   **2.111 a.** $z = -3.50$ **b.** $z = .64$ **2.113 a.** 23 **b.** $z = 3.28$ **2.115 a.** -0.2 **b.** -0.06 **c.** $z = -4.65$ **2.117 a.** $z = 2.0$: 3.7; $z = -1.0$: 2.2; $z = .5$: 2.95; $z = -2.5$: 1.45   **b.** 1.9   **c.** $z = 1.0$ and 2.0; GPA = 3.2 and 3.7; mound-shaped, symmetric   **2.123 a.** no, $z = .73$ **b.** yes, $z = -3.27$ **c.** no, $z = 1.36$ **d.** yes, $z = 3.73$ **2.125 a.** 4   **b.** $Q_U \approx 6$, $Q_L \approx 3$ **c.** 3   **d.** skewed right   **e.** 50%; 75%   **f.** 12, 13, and 16   **2.127 a.** 10, 15, 27.5   **b.** 3.5, 5, 7.5   **c.** effective   **2.129 a.** -1.26 **b.** No   **2.131 b.** medians: Familiar ≈ 45, Treatment ≈ 40, Control ≈ 30 **c.** IQR for Treatment is the largest   **d.** yes   **e.** no   **2.133 a.** 62, 72, 78, and 84   **b.** 62, 72, and 78   **2.135 a.** $z = -3$ **b.** yes   **c.** no   **d.** $z = -1.5$; yes   **2.139** slight positive linear trend   **2.141** yes, positive linear trend   **2.143 a.** nonlinear increasing   **b.** nonlinear increasing   **2.145 a.** no   **b.** yes, slight positive   **c.** reliability is suspect (only 5 data points)   **2.147 a.** negative trend   **b.** positive trend between plant coverage and diversity   **2.149** Yes; accuracy decreases as driving distance increases   **2.157 a.** $-1, 1, 2$ **b.** $-2, 2, 4$ **c.** 1, 3, 4   **d.** .1, .3, .4   **2.159 a.** 3.1234   **b.** 9.0233   **c.** 9.7857   **2.161 a.** $\bar{x} = 5.67$, $s^2 = 1.07$, $s = 1.03$ **b.** $\bar{x} = -\$1.5$, $s^2 = 11.5$, $s = \$3.39$ **c.** $\bar{x} = .413\%$, $s^2 = .088$, $s = .277\%$ **d.** 3; \$10; .7375%   **2.163** yes, positive   **2.165 a.** pie chart   **b.** breast cancer   **c.** 19%   **2.167 b.** $z = -1.06$ **2.169 b.** 5.4%   **c.** average player rating is 1068   **2.171 a.** relative frequencies: violence $- .664$; sympathetic $- .181$; harm $- .126$; comic $- .021$; criticism $- .007$ **2.173 a.** "favorable/recommended"; .635   **b.** yes   **2.175 a.** current: 42.0%; retired: 58.0%   **b.** Most (40 of 50) Beanie babies have values less than \$100.   **c.** yes, positive   **d.** unknown; at least 84%; at least 93.7%   **e.** 44%; 100%; 100%   **f.** 90%; 98%; 98%   **2.177** Over half of the whistle types were "Type a"   **2.179 a.** no outliers   **2.181 a.** Seabirds—quantitative; length—quantitative; oil—qualitative   **b.** transect   **c.** oiled: 38%; unoiled: 62%   **e.** distributions are similar   **f.** 3.27 ± 13.4 **g.** 3.495 ± 11.936 **h.** unoiled   **2.183 b.** yes   **c.** A1775A: $\bar{x} = 19,462.2$, $s = 532.29$; A1775B: $\bar{x} = 22,838.5$, $s = 560.98$ **d.** cluster A1775A   **2.185** yes; $z = -2.5$ **2.187 a.** quantitative: days and year   **b.** lesser crimes   **c.** yes   **d.** mean = 41.4 days; median = 15 days   **e.** no; $z = 2.38$ **f.** no   **2.189** a. median   **b.** mean   **2.191** survey results biased, not reliable

Chapter 3

3.9 a. .5   **b.** .3   **c.** .6   **3.11** $P(A) = .55$; $P(B) = .50$; $P(C) = .70$ **3.13 a.** 10   **b.** 20   **c.** 15,504   **3.15 a.** (R_1R_2), (R_1R_3), (R_2R_3), (R_1B_1), (R_1B_2), (R_2B_1), (R_2B_2), (R_3B_1), (R_3B_2), (B_1B_2) **b.** 1/10 for each sample point   **c.** $P(A) = 1/10$, $P(B) = 6/10$, $P(C) = 3/10$ **3.17** .05   **3.19** a. .01 **b.** yes   **3.21 a.** .261   **b.** Trunk $- .85$, Leaves $- .10$, Branch $- .05$ **3.23 a.** Interview; Observation + Participation; Observation Only; Grounded Theory   **c.** .677, .139, .113, .072   **d.** .748   **3.25 a.** 6   **b.** .282, .065, .339, .032, .008, .274   **c.** .686   **3.27 a.** 28   **b.** 1/28   **3.29** 693/1,686,366 = .000411   **3.31 a.** 15   **b.** 20   **c.** 15   **d.** 6   **3.41 b.** $P(A) = 7/8$, $P(B) = 1/2$, $P(A \cup B) = 7/8$, $P(A^c) = 1/8$, $P(A \cap B) = 1/2$ **c.** 7/8   **d.** no   **3.43 a.** 3/4   **b.** 13/20   **c.** 1   **d.** 2/5   **e.** 1/4   **f.** 7/20   **g.** 1   **h.** 1/4   **3.45 a.** .65   **b.** .72   **c.** .25   **d.** .08   **e.** 35   **f.** .72   **g.** .65   **g.** A and C, B and C, C and D **3.47 a.** School laboratory, In Transit, Chemical plant, Non-chemical plant, Other   **b.** .06, .26, .21, .35, .12   **c.** .06   **d.** .56   **e.** .74   **3.49 b.** .43   **c.** .57   **3.51 a.** {11, 13, 15, 17, 29, 31, 33, 35}   **b.** {2, 4, 6, 8, 10, 11, 13, 15, 17, 20, 22, 24, 26, 28, 29, 31, 33, 35, 1, 3, 5, 7, 9, 19, 21, 23, 25, 27}   **c.** $P(A) = 9/19$, $P(B) = 9/19$, $P(A \cap B) = 4/19$, $P(A \cup B) = 14/19$, $P(C) = 9/19$ **d.** {11, 13, 15, 17}   **e.** 14/19, no   **f.** 2/19   **g.** 1, 2, 3, . . . , 29, 31, 33, 35}   **h.** 16/19   **3.53 b.** .156   **c.** .617   **d.** .210   **e.** .449   **f.** yes   **3.55 a.** (PTW-R, Jury), (PTW-R, Judge), (PTW-A/D, Jury), (PTW-A/D, Judge), (DTW-R, Jury), (DTW-R, Judge), (DTW-A/D, Jury), (DTW-A/D, Judge)   **b.** .684   **c.** .124   **d.** no   **e.** .316   **f.** .717   **g.** .091   **3.57** $P(9) = 25/216$ is less than $P(10) = 27/216$ **3.59** $P(A \cap B) = P(A) \cdot P(B)$ **3.61** $P(A \cap B) = P(A) P(B|A)$ **3.63 a.** .5   **b.** .25   **c.** no   **3.65 a.** .08   **b.** .40   **c.** .52   **3.67 a.** $P(A) = .4$; $P(B) = .4$; $P(A \cap B) = .3$ **b.** $P(E_1|A) = .25$, $P(E_2|A) = .25$, $P(E_3|A) = .5$ **c.** .75   **3.69** no   **3.71** 1/3, 0, 1/14, 1/7, 1   **3.73** .40   **3.75 a.** .1   **b.** .522   **c.** The person is guessing   **3.77 a.** .7   **b.** .5   **c.** .5   **3.79 a.** .23   **b.** .729   **3.81 a.** 38/132 = .2879   **b.** 29/123 = .236   **3.83 a.** $P(A \mid I) = .9$, $P(B \mid I) = .95$, $P(A \mid N) = .2$, $P(B \mid N) = .1$ **b.** .855   **c.** .02   **d.** .995   **3.85 a.** .333   **b.** .120   **c.** No   **d.** probability of playing Center depends on race   **3.87** $P(A \mid B) = 0$ **3.89 a.** (A defeats B, C defeats D, A defeats C), (A defeats B, C defeats D, C defeats A), (A defeats B, D defeats C, A defeats D), (A defeats B, D defeats C, D defeats A), (B defeats A, C defeats D, B defeats C), (B defeats A, C defeats D, C defeats B), (B defeats A, D defeats C, D defeats B), (B defeats A, D defeats C, B defeats D)   **b.** $P(A$ wins$) = 2/8 = .25$ **c.** .576   **3.91 a.** .553   **b.** $P(W \mid CA) = 1$; $P(W \mid CB) = .873$; $P(W \mid CC) = .559$; $P(W \mid BA) = .741$; $P(W \mid BB) = .547$; $P(W \mid BC) = .342$; $P(W \mid AA) = .536$; $P(W \mid AB) = .216$; $P(W \mid AC) = .128$ **c.** .856   **d.** .552   **3.93 b.** Worst $- .250$, 2nd worst $- .200$, 3rd worst $- .157$, 4th worst $- .120$, 5th worst $- .089$, 6th worst $- .064$, 7th worst $- .044$, 8th worst $- .029$, 9th worst $- .018$, 10th worst $- .011$, 11th worst $- .007$, 12th worst $- .006$, 13th worst $- .005$ **c.** 250/800 = .313   **d.** .297   **e.** .288   **3.99 a.** 35,820,200   **b.** 1/35,820,200   **c.** highly unlikely   **3.111 a.** 4   **b.** 8

c. 32   **d.** 2^n **3.113 a.** 35   **b.** 15   **c.** 435   **d.** 45   **e.** $\binom{q}{r} = \dfrac{q!}{r!(q-r)!}$ **3.115 a.** 56   **b.** 1,680   **c.** 6,720   **3.117 a.** 24   **b.** 12   **3.119 a.** 30
b. 1,050   **3.121 a.** 6   **b.** 1/3   **3.123 a.** 18   **b.** 4/18   **3.125 a.** 21/252   **b.** 21/252   **c.** 105/252   **3.127 a.** 2,598,960   **b.** .002   **c.** .00394
d. .0000154   **3.131 a.** .225   **b.** .125   **c.** .35   **d.** .643   **e.** .357   **3.133** .2   **3.135 a.** .5   **b.** .99   **c.** .847   **3.137 a.** .158   **b.** .316   **c.** .526
d. #3   **3.139** .966   **3.141 a.** $A \cup B$ **b.** B^c **c.** $A \cap B$ **d.** $A^c | B$ **3.143 a.** 0   **b.** no   **3.145** .5   **3.147 c.** $P(A) = 1/4$; $P(B) = \frac{1}{2}$
e. $P(A^c) = 3/4$; $P(B^c) = 1/2$; $P(A \cap B) = 1/4$; $P(A \cup B) = 1/2$; $P(A|B) = 1/2$; $P(B|A) = 1$ **f.** no, no   **3.149 a.** no   **b.** .3, .1   **c.** .37
3.151 a. 720   **b.** 10   **c.** 10   **d.** 30   **e.** 20   **f.** 1   **g.** 5,040   **h.** 2,450   **3.153 a.** false   **b.** true   **c.** true   **d.** false   **3.155** $A = \{$8th-grader
scores above 655$\}$; $P(A^c) = .95$ **3.157 b.** .147   **c.** .85   **d.** .65   **e.** yes   **f.** African Americans are stopped for speeding more often than
expected   **3.159 a.** {Single, shore parallel; Other; Planar}   **b.** .1/3, 1/3, 1/3   **c.** 2/3   **d.** {No dunes/flat; Bluff/scarp; Single dune;
Not observed}   **e.** 1/3, 1/6, 1/3, 1/6 **f.** 2/3   **3.161 a.** Total population, Agricultural change, Presence of industry, Growth, Population
concentration   **b.** .18, .05, .27, .05, .45 **c.** .63   **3.163 a.** .385   **b.** .378   **c.** .183   **3.165 a.** .64, .32, .04   **b.** .72, .22, .06   **c.** dependent
3.167 a. .116   **b.** .728   **3.169 a.** .7127 **b.** .2873   **c.** .9639   **d.** .3078   **e.** .0361   **f.** at least 3   **3.171 a.** #4   **b.** #4 or #6   **3.173 a.** 1 to 2
b. .5   **c.** .4   **3.175 a.** .25   **b.** .0156   **c.** .4219   **d.** .000000001, .9970   **3.177** .993   **3.179** 4.4739×10^{-28} **3.181** Probabilities are the
same   **3.183** yes

Chapter 4

4.3 a. discrete   **b.** continuous   **c.** continuous   **d.** discrete   **e.** continuous   **f.** continuous   **4.5 a.** continuous   **b.** discrete   **c.** discrete
d. discrete   **e.** discrete   **f.** continuous   **4.7** $0, 1, 2, 3, \ldots$; discrete   **4.9** gender; IQ score   **4.11** allergy to penicillin; blood pressure
4.13 table, graph, formula   **4.15** no   **4.17 a.** .25   **b.** .40   **c.** .75   **4.19 a.** .7   **b.** .3   **c.** 1   **d.** .2   **e.** .8   **4.21 a.** 3.8   **b.** 10.56   **c.** 3.2496
e. no   **f.** yes   **4.23 a.** $\mu_x = 1$, $\mu_y = 1$ **b.** distribution of x **c.** $\mu_x = 1$, $\sigma^2 = .6$; $\mu_y = 1$, $\sigma^2 = .2$ **4.25 a.** 1   **b.** .24   **c.** .39   **d.** 1.8
e. .99   **f.** .96   **4.27 a.** $P(6) = .282$, $P(7) = .065$, $P(8) = .339$, $P(9) = .032$, $P(10) = .008$, $P(11) = .274$ **b.** .274   **4.29 a.** .9985, .0015,
0, 0 **c.** 1   **4.31 a.** MC, MS, MB, MO, ML, CS, CB, CO, CL, SB, SO, SL, BO, BL, OL   **b.** equally likely, with p = 1/15   **d.** . $P(0) = 6/15$,
$P(1) = 8/15$, $P(2) = 1/15$ **e.** .667   **4.33 a.** .23   **b.** .08   **c.** .77   **4.35** $P(-3) = .1$, $P(-1) = .105$, $P(5) = .795$ **4.37** $-\$0.263$
4.39 a. 0, 1, 2   **b.** $P(0) = .625$, $P(1) = .250$, $P(2) = .125$ **4.43 a.** 5   **b.** .7   **4.45 a.** .4096   **b.** .3456   **c.** .027   **d.** .0081   **e.** .3456   **f.** .027
4.47 a. $\mu = 12.5$, $\sigma^2 = 6.25$, $\sigma = 2.5$ **b.** $\mu = 16$, $\sigma^2 = 12.8$, $\sigma = 3.578$ **c.** $\mu = 60$, $\sigma^2 = 24$, $\sigma = 4.899$ **d.** $\mu = 63$, $\sigma^2 = 6.3$,
$\sigma = 2.510$ **e.** $\mu = 48$, $\sigma^2 = 9.6$, $\sigma = 3.098$ **f.** $\mu = 40$, $\sigma^2 = 38.4$, $\sigma = 6.197$ **4.49 a.** 0   **b.** .998   **c.** .137   **4.51 b.** $n = 20$, $p = .8$
c. .174 **d.** .804   **e.** 16   **4.53 a.** .001   **b.** .322   **c.** .994   **4.55 a.** .064   **b.** .936   **c.** 1.8; .85   **4.57 a.** .986   **b.** ≈ 0 **c.** 5   **4.59 a.** .1   **b.** .7
c. .4783 **d.** .0078   **e.** yes   **4.61** 15   **4.63 b.** $\mu = 2.4$, $\sigma = 1.47$ **c.** $p = .9$, $q = .1$, $n = 24$, $\mu = 21.6$, $\sigma = 1.47$ **4.65** standard normal
4.67 a. .4772   **b.** .3413   **c.** .4545   **d.** .2190   **4.69 a.** 0   **b.** .8413   **c.** .8413   **d.** .1587   **e.** .6826   **f.** .9544   **g.** .6934   **h.** .6378   **4.71 a.** 0
b. 1   **c.** 2.5   **d.** -3 **e.** 5   **f.** 1.4   **4.73 a.** 1.645   **b.** 1.96   **c.** -1.96 **d.** 1.28   **e.** 1.28   **4.75 a.** .3830   **b.** .3023   **c.** .1525   **d.** .7333
e. .1314   **f.** .9545   **4.77** 182   **4.79 a.** .1442   **b.** .2960   **c.** .0023   **d.** .9706   **4.81 a.** .8413   **b.** .7528   **4.83 a.** .4107   **b.** .1508   **c.** .9066
d. .0162   **e.** 841.8   **4.85 f.** (62.26, 65.74); (61.01, 66.99); (59.72, 68.28); (58.90, 69.10); (57.30, 70.70)   **4.87 a.** .0735   **b.** .3651   **c.** 7.29
4.89 $d = 56.24$ **4.91 a.** $z_L = -.675$, $z_U = .675$ **b.** $-2.68, 2.68$ **c.** $-4.69, 4.69$ **d.** .0074, 0   **4.95 a.** .68   **b.** .95   **c.** .997   **4.97** plot c
4.99 a. not normal (skewed right)   **b.** $Q_L = 37.5$, $Q_U = 152.8$, s = 95.8   **c.** IQR/s = 1.204   **4.101 a.** 6   **b.** 5.275   **c.** IQR/s = 1.14
4.103 no; IQR/s = .82   **4.105 a.** histogram too peaked   **b.** more than 95%   **4.107** both distributions approx. normal
4.109 IQR/s = 1.3, histogram approx. normal   **4.111** lowest score of 0 is less than one standard deviation below the mean
4.115 a. yes   **b.** $\mu = 10$, $\sigma^2 = 6$ **c.** .726   **d.** .7291   **4.117 a.** .1788   **b.** .5236   **c.** .6950   **4.119 a.** 16.25   **b.** 3.49   **c.** 1.07   **d.** .1762
4.121 a. $\mu = 1,500$, $\sigma^2 = 1,275$ **b.** .0026   **c.** no   **4.123** ≈ 0 **4.125 a.** no   **b.** yes   **c.** yes   **4.127** ≈ 0 **4.129 a.** 300 **b.** 800 **c.** 1
4.133 c. 1/16   **4.135 c.** .05   **d.** no   **4.141** equal   **4.145 a.** $\mu = 100$, $\sigma = 5$ **b.** $\mu = 100$, $\sigma = 2$ **c.** $\mu = 100$, $\sigma = 1$ **d.** $\mu = 100$,
$\sigma = 1.414$ **e.** $\mu = 100$, $\sigma = .447$ **f.** $\mu = 100$, $\sigma = .316$ **4.147 a.** $\mu = 2.9$, $\sigma^2 = 3.29$, $\sigma = 1.814$ **c.** $\mu_{\bar{x}} = 2.9$, $\sigma_{\bar{x}} = 1.283$
4.149 a. $\mu_{\bar{x}} = 30$, $\sigma_{\bar{x}} = 1.6$ **b.** approx. normal   **c.** .8944   **d.** .0228   **e.** .1303   **f.** .9699   **4.151 a.** $\mu_{\bar{x}} = 320$, $\sigma_{\bar{x}} = 10$, approx. normal
b. .1359   **c.** 0   **4.153 a.** 79   **b.** 2.3   **c.** approx. normal   **d.** .43   **e.** .3336   **4.155 a.** .10; .0002   **b.** Central Limit Theorem   **c.** .0170
4.157 a. $\mu_{\bar{x}} = .53$, $\sigma_{\bar{x}} = .0273$, approx. normal   **b.** .0336   **c.** after   **4.159 a.** .0034   **b.** $\mu > 6$ **4.161** handrubbing: $P(\bar{x} < 30) = .2743$;
handwashing: $P(\bar{x} < 30) = .0047$; sample used handrubbing   **4.163 a.** yes   **b.** no   **c.** no   **d.** yes   **4.165** false   **4.167 a.** .192   **b.** .228
c. .772   **d.** .987   **e.** .960   **f.** 14; 4.2; 2.05   **g.** .975   **4.169 a.** .6915   **b.** .0228   **c.** .5328   **d.** .3085   **e.** 0   **f.** .9938   **4.171 a.** .3821   **b.** .5389
c. 0   **d.** .1395   **e.** .0045   **f.** .4602   **4.173 b.** .05   **c.** 10   **4.175 a.** .9406   **b.** .9406   **c.** .1140   **4.177** binomial   **4.179 a.** yes   **b.** .051
c. .757   **d.** .192   **e.** 3.678   **4.181 a.** 5.1; .498   **b.** yes   **c.** .2119   **d.** .4071   **4.183 a.** .3745   **b.** .7553   **4.185** .642   **4.187** .0154; very
unlikely   **4.189 a.** .3264   **b.** 1.881   **c.** valid   **4.191** approx. normal   **4.193** 5.068   **4.195 a.** .05, .20, .50, .20, .05   **b.** z-scores **c.** identical

Chapter 5

5.5 yes   **5.7 a.** 1.645   **b.** 2.58   **c.** 1.96   **d.** 1.28   **5.9 a.** $28 \pm .784$ **b.** $102 \pm .65$ **c.** $15 \pm .0588$ **d.** $4.05 \pm .163$ **e.** no   **5.11 a.** $83.2 \pm$
1.25   **c.** 83.2 ± 1.65 **d.** increases   **e.** yes   **5.13 a.** 19.3   **b.** 19.3 ± 3.44 **d.** random sample, large n **5.15 a.** $.36 \pm .0099$ **c.** first-year:
$.303 \pm .0305$; landfast: $.362 \pm .0177$; multi-year: $.381 \pm .0101$ **5.17 a.** μ **b.** no; apply Central Limit Theorem   **c.** (0.4786, 0.8167)
d. yes   **5.19 a.** $1.13 \pm .67$ **b.** yes   **5.21 a.** 19 ± 7.826 **b.** 7 ± 5.90 **c.** SAT-Math   **5.23 a.** males: 16.79 ± 2.35; females: 10.79 ± 1.67
b. .0975   **c.** males   **5.25** Central Limit Theorem no longer applies; σ unknown   **5.27 a.** large: normal; small: t-distribution   **b.** large:
normal; small: unknown   **5.29 a.** 2.228 **b.** 2.567 **c.** -3.707 **d.** -1.771 **5.31 a.** 5 ± 1.876 **b.** 5 ± 2.394 **c.** 5 ± 3.754 **d.** (a) $5 \pm .780$,
(b) $5 \pm .941$, (c) 5 ± 1.276; width decreases   **5.33 a.** $\bar{x} = 1.86$, s = 1.195   **b.** skewed right   **c.** $1.86 \pm .45$ **e.** 90%   **5.35 a.** .009
b. 2.306   **c.** $.009 \pm .0037$ **5.37 a.** $1.07 \pm .24$ **c.** normal   **5.39 a.** both untreated: 20.9 ± 1.06; male treated: 20.3 ± 1.25;
female treated: 22.9 ± 1.79; both treated: $18.6 \pm .79$ **b.** female treated   **5.41 a.** 7.3 ± 1.41 **c.** PMI values normally distributed
5.43 a. 37.3 ± 7.70 **c.** One hour before: $\mu > 25.5$ **5.47 a.** yes   **b.** $.64 \pm .067$ **5.49 a.** yes   **b.** no   **c.** no   **d.** no   **5.51 a.** set of all
gun ownership status (yes/no) values for all U.S. adults   **b.** true percentage of all adults who own a gun   **c.** .26   **d.** $.26 \pm .02$

5.53 a. $.63 \pm .03$ **b.** yes **5.55** $.219 \pm .024$ **5.57 a.** $.338 \pm .025$ **5.59 a.** $.524 \pm .095$ **5.61** 95% confident that proportion of health care workers with latex allergy who suspect he/she has allergy is between .327 and .541 **5.63** true **5.65** 519 **5.67 a.** 482 **b.** 214
5.69 34 **5.71 a.** $n = 16$: $W = .98$; $n = 25$: $W = .784$; $n = 49$: $W = .56$; $n = 100$: $W = .392$; $n = 400$: $W = .196$ **5.73 a.** small sample
b. 125 **5.75** 21 **5.77** 14,735 **5.79** 129 **5.81** 271 **5.83 a.** μ **b.** μ **c.** p **d.** p **e.** p **5.85 a.** $t = 2.086$ **b.** $z = 1.96$ **c.** $z = 1.96$
d. $z = 1.96$ **e.** neither t nor z **5.87 a.** $.57 \pm .049$ **b.** 2,358 **5.89** (1) p; (2) μ; (3) μ; (4) μ **5.91 a.** $.03$ **b.** $.03 \pm .010$ **5.93 a.** $.90$
b. $.05$ **c.** 259 **5.95** $1.37 \pm .76$ **5.97** 95% confident that true proportion of all adults who are homeless is between .045 and .091
5.99 a. $.660 \pm .029$ **b.** $.301 \pm .035$ **5.101 a.** $.044 \pm .162$ **b.** no evidence of species inbreeding **5.103 a.** $.81 \pm .41$ **b.** 140
5.105 a. $.378 \pm .156$ **b.** $.703 \pm .147$ **c.** 41.405 ± 21.493 **d.** $.378 \pm .156$ **5.107 a.** $.094$ **b.** yes **c.** $.094 \pm .037$ **5.109 a.** 49.3 ± 8.60
c. Population is normal. **d.** 60 **5.111 a.** 781 **5.113 a.** yes **b.** missing measure of reliability **c.** 95% CI for μ: $.932 \pm .037$

Chapter 6

6.1 null hypothesis **6.3** α **6.7** no **6.9** H_0: $p = .75$, H_a: $p \ne .75$ **6.11 a.** H_0: $p = .45$ **b.** H_0: $\mu = 2.5$ **6.13** H_0: $p = .05$, H_a:
$p < .05$ **6.15 a.** H_0: $\mu = 15$, H_a: $\mu < 15$ **b.** conclude mean mercury level is less than 15 ppm when mean equals 15 ppm
c. conclude mean mercury level equals 15 ppm when mean is less than 15 ppm **6.17 a.** H_0: No intrusion occurs **b.** H_a: Intrusion occurs
c. $\alpha = .001$, $\beta = .5$ **6.21 g.** (a) .025, (b) .05, (c) $\approx$.005, (d) $\approx$.10, (e) .10, (f) $\approx$.01 **6.23 a.** $z = 1.67$, reject H_0 **b.** $z = 1.67$,
fail to reject H_0 **6.25 a.** Type I: conclude mean response for all New York City public school children is not 3 when mean equals 3;
Type II: conclude mean response for all New York City public school children equals 3 when mean is not equal to 3 **b.** $z = -85.52$,
reject H_0 **c.** $z = -85.52$, reject H_0 **6.27** yes, $z = 3.95$ **6.29 a.** H_0: $\mu = 16$, H_a: $\mu < 16$ **b.** $z = -4.31$, reject H_0 **6.31 a.** $z = 4.03$,
reject H_0 **6.33 a.** no **b.** $z = .61$, do not reject H_0 **d.** no **e.** $z = -.83$, do not reject H_0 **6.35 a.** $z = -3.72$, reject H_0 **6.37** small
p-values **6.39 a.** fail to reject H_0 **b.** reject H_0 **c.** reject H_0 **d.** fail to reject H_0 **e.** fail to reject H_0 **6.41** $.0150$ **6.43** p-value $=$
$.9279$, fail to reject H_0 **6.45 a.** fail to reject H_0 **b.** fail to reject H_0 **c.** reject H_0 **d.** fail to reject H_0 **6.47 a.** $.3446$ **b.** do not reject
H_0 **6.49** p-value $< .0001$; reject H_0 at $\alpha = .05$ **6.51 b.** reject H_0 at $\alpha = .05$ **c.** reject H_0 at $\alpha = .05$ **6.53 a.** $.1970$ **b.** $.0985$
c. $.3788$ **d.** (a) .20; (b) .10; (c) .40 **e.** $s \le 3.95$ **6.55** small n, normal population **6.57** .10; .05; .05 **6.59 a.** $t = -2.064$; fail to
reject H_0 **b.** $t = -2.064$; fail to reject H_0 **c.** (a) $.05 < p$-value $< .10$; (b) $.10 < p$-value $< .20$ **6.61 b.** reject H_0 at $\alpha = .05$
c. p-value $= .076$; do not reject H_0 at $\alpha = .05$ **6.63 a.** H_0: $\mu = 5.70$, H_a: $\mu < 5.70$ **b.** $t < -2.821$ **c.** $t = -4.33$ **d.** reject H_0
6.65 $t = -.43$; fail to reject H_0 **6.67** $t = 3.725$, p-value $= .0058$, reject H_0 **6.69** yes, $t = -2.53$ **6.71** yes, $t = 1.95$ **6.73** qualitative
6.75 a. yes **b.** no **c.** no **d.** no **e.** no **6.77 a.** $z = -2.00$ **c.** reject H_0 **d.** $.0228$ **6.79 a.** $z = 1.13$, fail to reject H_0 **b.** $.1292$
6.81 a. p **b.** H_0: $p = .02$, H_a: $p > .02$ **c.** $z = 14.23$; $z > 1.645$ **d.** reject H_0 **e.** large n; yes **6.83** yes, $z = 1.74$ **6.85** yes, $z = -2.71$
6.87 no, $z = 1.52$ **6.89 a.** no, $z = 1.49$ **b.** p-value $\approx .07$ **6.91** $z = 33.47$, reject H_0: $p = .5$ **6.93 a.** $.5$ **b.** $.5$ **c.** unknown **d.** $.5$
6.95 a. do not reject H_0 **b.** do not reject H_0 **c.** reject H_0 **d.** reject H_0 **6.97** $S = 8$, p-value $= .0391$, reject H_0 **6.99 a.** H_0:
$\eta = 1.5$, H_a: $\eta > 1.5$ **b.** 3 **c.** $.773$ **d.** do not reject H_0 **6.101 a.** H_0: $\eta = 70$, H_a: $\eta < 70$ **b.** $S = 7$ **c.** $.172$ **d.** do not reject H_0
6.103 a. H_0: $\eta = 5$ **b.** $z = 6.07$, p-value $= 0$ **c.** reject H_0 **6.105** no; $S = 10$, p-value $= .3145$ **6.107** alternative **6.109** H_0, H_a, α
6.111 null **6.113 a.** $z = -1.78$, reject H_0 **b.** $z = -1.78$, fail to reject H_0 **c.** $.29 \pm .063$ **d.** $.29 \pm .083$ **e.** 549 **6.115 a.** $.172$
b. do not reject H_0 **6.117 a.** H_0: Drug is unsafe, H_a: Drug is safe **c.** α **6.119 a.** $t = -3.46$, reject H_0 **b.** normal population
c. p-value $= .194$, do not reject H_0 **6.121 a.** H_0: $p = .5$, H_a: $p \ne .5$ **b.** $z = 5.99$ **c.** yes **d.** reject H_0 **6.123** yes, $z = 12.36$
6.125 a. $t = 1.44$, p-value $= .169$, fail to reject H_0 **6.127 a.** $z = .70$, fail to reject H_0 **b.** yes **6.129 a.** no, $z = 1.41$
b. small α **6.131** $z = 15.46$, reject H_0: $p = .167$ **6.133 b.** $.37, .24$ **6.135 a.** no at $\alpha = .01$, $z = 1.85$ **b.** yes at $\alpha = .01$, $z = 3.09$

Chapter 7

7.1 normally distributed with mean $\mu_1 - \mu_2$ and standard deviation $\sqrt{\dfrac{\sigma_1^2}{n_1} + \dfrac{\sigma_2^2}{n_2}}$ **7.3 a.** no **b.** no **c.** no **d.** yes **e.** no **7.5** b

7.7 a. $14; .4$ **b.** $10; .3$ **c.** $4; .5$ **d.** yes **7.9 a.** $.5989$ **b.** $t = -2.39$, reject H_0 **c.** $-1.24 \pm .98$ **7.11 a.** H_0: $\mu_1 = \mu_2$, H_a: $\mu_1 \ne \mu_2$
b. $t = .62$ **c.** $|t| > 1.684$ **d.** fail to reject H_0; supports theory **7.13 a.** $\mu_1 - \mu_2$ **b.** 12.5 ± 10.2 **7.15 b.** $z = -2.64$, reject H_0
c. $z = .27$, fail to reject H_0 **7.17 a.** $(-.60, 7.96)$ **b.** independent random samples from normal populations with equal variances
7.19 $t = 1.08$, fail to reject H_0 **7.21** no, $t = .18$ **7.23 a.** $\mu_1 - \mu_2$ **b.** no **c.** $.2262$ **d.** fail to reject H_0 **7.25** $t = -.46$, fail to reject
H_0 **7.27 a.** no standard deviations reported **b.** $s_1 = s_2 = 5$ **c.** $s_1 = s_2 = 6$ **7.29** before **7.31 a.** $t > 1.833$ **b.** $t > 1.328$
c. $t > 2.776$ **d.** $t > 2.896$ **7.33 a.** $\bar{x}_d = 2$, $s_d^2 = 2$ **b.** $\mu_d = \mu_1 - \mu_2$ **c.** 2 ± 1.484 **d.** $t = 3.46$, reject H_0 **7.35 a.** $z = 1.79$, fail to
reject H_0 **b.** $.0734$ **7.37** yes; $(7.43, 13.57)$ **7.39 a.** μ_d **b.** paired difference **c.** H_0: $\mu_d = 0$, H_a: $\mu_d > 0$ **d.** $t = 2.19$ **e.** reject H_0
7.41 a. H_0: $\mu_d = 0$, H_a: $\mu_d \ne 0$ **b.** $z = 2.08$, p-value $= .0376$ **c.** reject H_0 **7.43 b.** 95% CI for μ_d: 1.95 ± 1.91; control group has
larger mean **7.45 a.** $t = -2.97$, fail to reject H_0 **b.** $-.42$; no **c.** $t = .57$, fail to reject H_0; $-.23$, no **d.** $t = 3.23$, fail to reject H_0;
19, no **7.47** 95% CI for μ_d: $-.000523 \pm .000400$; use alternative method **7.51 a.** $n_1 = n_2 = 193$ **b.** $n_1 = n_2 = 47$ **c.** $n_1 = n_2 = 144$
7.53 yes **7.55** $n_1 = n_2 = 130$ **7.57** $n_1 = n_2 = 1,729$ **7.59** $n_1 = n_2 = 136$ **7.61** true **7.63 a.** $T_1 \le 41$, $T_1 \ge 71$ **b.** $T_1 \ge 50$
c. $T_1 \le 43$ **d.** $|z| > 1.96$ **7.65** yes, $z = -2.47$ **7.67 a.** H_0: $D_{bulimic} = D_{normal}$, H_a: $D_{bulimic} > D_{normal}$ **c.** 174.5 **d.** 150.5 **e.** $z > 1.28$
f. $z = 1.72$, reject H_0 **7.69 a.** H_0: $D_{AV} = D_{Video}$, H_a: $D_{AV} \ne D_{Video}$ **b.** $z = .67$ **c.** $|z| > 1.645$ **d.** do not reject H_0 **7.71 a.** $T_1 =$
150.5, reject H_0 **b.** $z = 3.44$, reject H_0 **7.73 a.** rank sum test **b.** H_0: $D_{CMC} = D_{FTF}$, H_a: $D_{CMC} > D_{FTF}$ **c.** $z < -1.28$ **d.** $z =$
$-.23$, fail to reject H_0 **7.75 a.** zinc measurements non-normal **b.** $T_1 = 18$, do not reject H_0 **c.** $T_2 = 32$, do not reject H_0
7.81 a. H_a: $D_A > D_B$ **b.** $T_- = 3.5$, reject H_0 **7.83 a.** before and after measurements not independent **b.** scores not normal
c. reject H_0, ichthyotherapy is effective **7.85 a.** H_0: $D_{good} = D_{average}$, H_a: $D_{good} \ne D_{average}$ **d.** $T = 3$ **e.** $T < 4$ **f.** reject H_0 **7.87** no,
$T = 7$ **7.89** yes, $T_- = 50.5$ **7.91 a.** no, $T_- = 0$ **b.** reject H_0; supports the "hot hand" theory **7.95 a.** 6.59 **b.** 16.69 **c.** 1.61 **d.**
3.87 **7.97 a.** plot b **b.** $9; 14$ **c.** $75; 75$ **d.** $20; 144$ **e.** 95 (78.95%); 219 (34.25%) **f.** MST $= 75$, MSE $= 2$, F $= 37.5$; MST $= 75$,
MSE $= 14.4$, F $= 5.21$ **g.** reject H_0; reject H_0 **h.** both populations normal with equal variances

7.99 plot a:

Source	df	SS	MS	F
Treatment	1	75	75	37.5
Error	10	20	2	
Total	11	95		

plot b:

Source	df	SS	MS	F
Treatment	1	75	75	5.21
Error	10	144	14.4	
Total	11	219		

7.101 a. exp. units: coaches; dep. variable: 7-point rating; factor: division; treatments: I, II, III   **b.** $H_o: \mu_I = \mu_{II} = \mu_{III}$ **c.** reject H_o
7.103 a. completely randomized   **b.** treatments: 3, 6, 9, 12 robots; dep. variable.: energy expanded   **c.** $H_o: \mu_3 = \mu_6 = \mu_9 = \mu_{12}$,
H_a: At least 2 μ's differ   **d.** reject H_o **7.105 a.** $H_o: \mu_{young} = \mu_{middle} = \mu_{old}$ **b.** reject H_o **c.** $H_o: \mu_{young} = \mu_{middle} = \mu_{old}$; fail to
reject H_o **7.107** yes, $F = 3.90$ **7.109 b.** treatments: scopolamine, glycopyrrolate, no drug; response: number of pairs recalled
c. 6.17; 9.38; 10.6   **d.** $F = 27.07$, reject H_o **7.111** yes, $F = 7.25$ **7.115 a.** rank sum test   **b.** signed rank test   **7.117 a.** $3.9 \pm .31$
b. $z = 20.60$, reject H_o **c.** $n_1 = n_2 = 346$ **7.119** yes, $T_- = 1.5$ **7.121 b.** yes, $F = 7.70$ **7.123 a.** time needed   **b.** climbers   **c.** paired
experiment   **7.125 a.** $H_o: \mu_d = 0$, $H_a: \mu_d > 0$ **b.** paired difference test   **d.** leadership: fail to reject H_o at $\alpha = .05$; popularity:
reject H_o at $\alpha = .05$; intellectual self-confidence: reject H_o at $\alpha = .05$ **7.127** $F = 8.3$, reject H_o **7.129 a.** 749.5 ± 278.1
7.131 a. t test **b.** yes   **c.** no   **d.** fail to reject H_o **e.** reject H_o at $\alpha = .05$ **7.133 a.** $F = 3.96$, reject H_o **b.** $\mu_{Sad} < \mu_{Happy}$, $\mu_{Sad} <$
μ_{Angry} **7.135 a.** yes, $T_2 = 105$ **b.** less than   **7.137** yes, $t = -2.19$ **7.139** use of creative ideas ($z = 8.85$); good use of job skills
($z = 4.76$)

Chapter 8
8.3 large, independent samples   **8.5 a.** no   **b.** no   **c.** no   **d.** no   **e.** no   **8.7 a.** $.07 \pm .067$ **b.** $.06 \pm .086$ **c.** $-.15 \pm .131$ **8.9 a.** .55
b. .70   **c.** $-.15 \pm .03$ **d.** 90% confidence   **e.** Dutch boys   **8.11 a.** .143   **b.** .049   **c.** $z = 2.27$, fail to reject H_o **d.** reject H_o
8.13 a. $z = 94.35$, reject H_o **b.** $-.139 \pm .016$ **8.15** yes, $z = 11.05$ **8.17** yes; $z = 2.13$, reject H_o at $\alpha = .05$ **8.19** $.48 \pm .144$;
proportion greater for those who slept   **8.21 a.** at least 18   **b.** sample may not be representative of the population   **8.25** $n_1 =$
$n_2 = 520$ **8.27** $n_1 = n_2 = 130$ **8.29** $n_1 = 520$, $n_2 = 260$ **8.33 a.** 18.3070   **b.** 29.7067   **c.** 23.5418   **d.** 79.4900   **8.35 a.** $\chi^2 >$
5.99147   **b.** $\chi^2 > 7.77944$ **c.** $\chi^2 > 11.3449$ **8.37 a.** no, $\chi^2 = 3.293$ **8.39 a.** jaw habits; bruxism, clenching, bruxism and clenching,
neither **c.** $H_o: p_1 = p_2 = p_3 = p_4 = .25$, H_a: At least one p_i differs from .25   **d.** 15   **e.** $\chi^2 = 25.73$ **f.** $\chi^2 > 7.81473$ **g.** reject H_o
h. (.37, .63)   **8.41 a.** Pottery type; burnished, monochrome, painted, other   **b.** $p_1 = p_2 = p_3 = p_4 = .25$ **c.** $H_o: p_1 = p_2 = p_3 = p_4 = .25$
d. $\chi^2 = 436.59$ **e.** p-value ≈ 0, reject H_o **8.43 a.** no, $\chi^2 = 7.39$ **b.** (.456, .518)
8.45 a.

1	2	3	4
450	627	219	23

b. $H_o: p_1 = p_2 = p_3 = p_4 = .25$ **c.** 329.75 for each category   **d.** 634.36   **e.** reject H_o **f.** $\chi^2 = 39.29$, reject H_o **8.47 a.** yes,
$\chi^2 = 360.48$ **b.** (.165, .223)   **8.53 a.** H_o: Row & Column are independent, H_a: Row & Column are dependent **b.** $\chi^2 > 9.201034$
c.

	Column 1	Column 2	Column 3
Row 1	14.37	36.79	44.84
Row 2	10.63	27.21	33.16

d. $\chi^2 = 8.71$, fail to reject H_o **8.55** $\chi^2 = 12.33$, reject H_o **8.57 a.** .078   **b.** .157   **c.** possibly   **d.** H_o: Presence of Dog & Gender are
independent   **e.** $\chi^2 = 2.25$, fail to reject H_o **f.** $\chi^2 = .064$, fail to reject H_o **8.59 a.** masculinity risk (high and low); event (violent
and avoided-violent)   **b.** 1,507 newly incarcerated men   **c.** 260.8   **d.** 118.2, 776.2, 351.8   **e.** $\chi^2 = 10.1$ **f.** reject H_o **8.61** $\chi^2 =$
23.46, reject H_o **8.63** $\chi^2 = 35.41$, reject H_o **8.65** yes, $\chi^2 = 7.38$, p-value $= .025$; no   **8.67 a.** no
b.

SHSS:C	CAHS		
	Low	Medium	High
Low	32	14	2
Medium	11	14	6
High	6	16	29

c. yes   **d.** $\chi^2 = 46.70$, reject H_o **8.69 a.** $\chi^2 = 4.407$, reject H_o **b.** no   **c.** .0438   **d.** .0057; .0003   **e.** p-value $= .0498$, reject H_o **8.73 a.** $z = -2.04$, reject H_o **b.** $-.1 \pm .096$ **c.** $n_1 = n_2 = 72{,}991$ **8.75 a.** yes, $\chi^2 = 54.14$ **b.** no   **c.** yes
d.

	Col. 1	Col. 2	Col. 3	Totals
Row 1	.400	.222	.091	.200
Row 2	.200	.222	.636	.400
Row 3	.400	.556	.273	.400

8.77 $\chi^2 = 16$, reject H_o
8.79 a.

Genetic Trait	LLD		
	Yes	No	Total
Yes	21	15	36
No	150	306	456
Total	171	321	492

b. $\chi^2 = 9.52$, reject H_o **d.** $z = 2.74$, reject H_o **8.81 a.** possibly not   **c.** $\chi^2 = 12.9$ **d.** yes, reject H_o **e.** $.233 \pm .200$
8.83 a. .153   **b.** .215   **c.** $-.062 \pm .070$ **8.85 a.** $z = -.21$, do not reject H_o **b.** $-.078 \pm .076$ **8.87** Internet: $\chi^2 = .512$, do not reject H_o; party: $\chi^2 = 164.76$, reject H_o; military: $\chi^2 = 8.3$, reject H_o; views: $\chi^2 = 174.39$, reject H_o; race: $\chi^2 = 69.18$, reject H_o; income: $\chi^2 = 16.39$, reject H_o; community: $\chi^2 = 17.31$, reject H_o **8.89 a.** $\chi^2 = 17.16$, reject H_o **b.** $.376 \pm .103$ **c.** yes
8.91 $\chi^2 = 6.17$, reject H_o **8.93** $z = 3.55$, reject H_o **8.95 a.** $\chi^2 = 9.65$ **b.** 11.0705   **c.** no   **d.** $.05 < p$-value $< .10$
8.97 a. $\chi^2 = 164.90$, reject H_o **b.** Gender: $\chi^2 = 6.80$, reject H_o; Education: $\chi^2 = 6.59$, fail to reject H_o; Work: $\chi^2 = 11.69$, fail to reject H_o; Satisfaction: $\chi^2 = 30.42$, reject H_o

Chapter 9

9.7 $\beta_1 = 1/3$, $\beta_0 = 14/3$, $y = 14/3 + (1/3)x$ **9.11** difference between the observed and predicted   **9.13** true   **9.15 b.** $\hat{y} = 7.10 - .78x$
9.17 c. $\hat{\beta} = 1/3$, $\beta_0 = 14/3$, $y = 14/3 + (1/3)x$ **e.** -1 to 7   **9.19 c.** hoop pine   **9.21 a.** $y = \beta_0 + \beta_1 x + \epsilon$ **b.** $\hat{y} = 2.522 + 7.261x$
9.23 a. positive   **9.25 a.** $y = \beta_0 + \beta_1 x + \epsilon$ **b.** $\hat{y} = 250.14 - .6294x$ **e.** slope   **9.27 a.** $y = \beta_0 + \beta_1 x + \epsilon$ **b.** $\hat{y} = -.607 + 1.062x$
c. positive **e.** $\hat{y} = -.148 + 1.022x$ **9.29 a.** $\hat{y} = 560.1 + 63.3x$ **b.** $\hat{y} = 326.5 + 1.48x$ **c.** $\hat{y} = 338.9 + 197.21x$ **9.31** yes, $\hat{y} =$ $5.2207 - .11402x$; decrease by .114 pound   **9.35 a.** 57.5; 3.19444   **b.** 257.5; 6.7763   **c.** 9.288; 1.1610   **9.37** 11.14: SSE $= 1.22$, $s^2 =$ $.244$, $s = .494$; 11.17: SSE $= 5.134$, $s^2 = 1.03$, $s = 1.01$ **9.39** women's   **9.41 a.** SSE $= 2760$, $s^2 = 306.6$, and s $= 17.51$ **9.43 a.** 6.36
b. 3.42   **c.** reading score   **9.45 a.** \$908.50   **b.** \$736.66   **c.** \$846.73   **9.47** 0   **9.49** divide the value in half   **9.51 a.** 95%: $.31 \pm 1.13$; 90%: $.31 \pm .92$ **b.** 95%: 64 ± 4.28; 90%: 64 ± 3.53 **c.** 95%: $-8.4 \pm .67$; 90%: $-8.4 \pm .55$ **9.53 b.** $\hat{y} = 2.554 + .246x$ **d.** $t = .627$
e. fail to reject H_o **f.** $.246 \pm 1.81$ **9.55 a.** $H_o: \beta_1 = 0$, $H_a: \beta_1 < 0$ **b.** $t = -13.23$, p-value $= .000$ **c.** reject H_o **9.57 a.** $t = 10.05$, reject H_o **b.** $1.022 \pm .180$ **9.59** $-.0023 \pm .0016$ **9.61 a.** $y = \beta_0 + \beta_1 x + \epsilon$ **b.** $\hat{y} = -8.524 + 1.665x$ **d.** yes, $t = 7.25$ **e.** $1.67 \pm$ .46   **9.63 a.** $t = 6.286$, reject H_o **b.** $.88 \pm .236$ **c.** no evidence that slope differs from 1   **9.65 a.** $\hat{\beta}_0 = .515$, $\hat{\beta}_1 = .000021$ **b.** yes
c. very influential   **d.** $\hat{\beta} = .515$. $\hat{\beta}_1 = .000020$, p-value $= .332$, fail to reject H_o **9.67** true   **9.69 a.** perfect positive linear   **b.** perfect negative linear   **c.** no linear   **d.** strong positive linear   **e.** weak positive linear   **f.** strong negative linear   **9.71** .877   **9.73 a.** moderate positive linear relationship; not significantly different from 0 at $\alpha = .05$ **c.** weak negative linear relationship; not significantly different from 0 at $\alpha = .10$ **9.75 b.** .185   **9.77 b.** both positive   **c.** .706   **9.79 a.** moderately strong negative linear relationship between the number of online courses taken and weekly quiz grade   **b.** yes, $t = -4.95$ **9.81 b.** piano: $r^2 = .1998$; bench: $r^2 = .0032$; motorbike: $r^2 = .3832$, armchair: $r^2 = .0864$; teapot: $r^2 = .9006$ **c.** Reject H_o for all objects except bench and armchair
9.83 a. $y = \beta_0 + \beta_1 x + \epsilon$ **b.** $\beta_1 > 0$ **d.** no, $t = -3.12$ **9.85** $r = .570$, $r^2 = .325$ **9.89** true   **9.91 a.** $\hat{y} = 1.375 + .875x$ **c.** 1.5
d. .1875   **e.** $3.56 \pm .33$ **f.** 4.88 ± 1.06 **9.93 c.** 4.65 ± 1.12 **d.** $2.28 \pm .63$; $-.414 \pm 1.717$ **9.95 a.** prediction interval for y when $x = 50$ **b.** confidence interval for $E(y)$ when $x = 70$ **9.97** $(92.298, 125.104)$ **9.99 a.** $y = \beta_0 + \beta_1 x + \epsilon$ **b.** $\hat{y} = 155.912 - 1.086x$
c. no, $t = -2.05$ **d.** $(-21.56, 279.09)$ **9.101 a.** $(2.955, 4.066)$ **b.** $(1.020, 6.000)$ **c.** prediction interval; yes   **9.103 a.** Brand A: 3.349 $\pm$ .587; Brand B: $4.464 \pm .296$ **b.** Brand A: 3.349 ± 2.224; Brand B: 4.464 ± 1.120 **c.** $-.65 \pm 3.606$ **9.107** $|r_s| > .648$
b. $r_s > .450$ **c.** $r_s < -.432$ **9.109 b.** $r_s = .745$, do not reject H_o **c.** $.05 < p$-value $< .10$ **9.111 b.** $r_s = .943$ **c.** reject H_o
9.113 c. .713   **d.** $|r_s| > .425$ **e.** reject H_o **9.115 a.** moderate positive association between the two scores   **c.** reject H_o
9.117 a. $-.877$ **b.** $-.907$ **c.** reject H_o **d.** reject H_o **9.119 a.** .714   **b.** reject H_o **9.121** $E(y) = \beta_0 + \beta_1 x$ **9.123** true
9.125 b. $\hat{y} = x$; $\hat{y} = 3$ **c.** $\hat{y} = x$ **d.** least squares line has the smallest SSE   **9.127 a.** $y = \beta_0 + \beta_1 x + \epsilon$; negative   **b.** yes   **c.** no
9.129 a. positive   **b.** yes   **c.** $\hat{y} = -85.68 + 614.04x$ **f.** $t = 1.82$, p-value $= .0936$, fail to reject H_o **g.** .2165   **i.** 89.32   **j.** (63.26, 115.37)   **9.131 a.** $y = \beta_0 + \beta_1 x + \epsilon$ **b.** $\hat{y} = 175.7033 - .8195x$ **e.** $t = -3.43$, reject H_o **f.** $r_s = -.733$ **g.** do not reject H_o
9.133 b. reject H_o **c.** .49   **9.135** $\hat{y} = -92.458 + 8.347x$; $t = 3.248$, reject H_o; $r = .42$ **9.137 a.** yes **b.** $\hat{\beta}_0 = -3.05$, $\hat{\beta}_1 = .108$
c. $t = 4.00$, reject H_o **d.** $r = .756$, $r^2 = .572$ **e.** 1.09   **f.** yes   **9.139 a.** yes; positive   **b.** $y = \beta_0 + \beta_1 x + \epsilon$ **c.** $\hat{\beta}_0 = 20.1275$ $\hat{\beta} = .62442$
9.141 a. $\hat{y} = 46.4x$ **b.** $\hat{y} = 478.44 + 45.15x$ **d.** no, $t = .91$ **9.143 a.** no; $r^2 = .748$ **b.** yes, $18(\beta_1) = -.98$

Subject Index

A

Additive rule of probability, 132
Alcohol consumption on college campuses, 433, 450–451, 462–463
Alternative hypothesis, 302
  selecting, 309
Analysis of variance (ANOVA), 403–419
  assumptions, checking, 412–413
  calculation formulas for, 582
  defined, 403
  for independent samples, results of, 410
  one-way ANOVA, 410
  using the TI-84/TI-83 graphing calculator, 411
Analysis of variance (ANOVA) F-test:
  to compare k treatment means, 408
  conducting, 408–410
  valid, conditions required for, 408
Arithmetic mean, 51

B

Bar graph, 29, 31
Bell curve, 197
Bell-shaped probability distribution, 197
Benford's law, 38
Bernoulli distribution, 184
Bernoulli, Jacob, 184
Bias, 14
Binomial distribution, 183–194
  applying, 188
  approximating with a normal distribution, 221–226
  cumulative binomial probabilities, 190
  defined, 187
Binomial experiments, 184
Binomial probabilities:
  calculating with the TI-84/TI-83 graphing calculator, 193–194
  table, 563–566
Binomial probability distribution, deriving, 185–187
Binomial random variables, 184
  characteristics, 184
  exploring, 189
  mean of, 190
  standard deviation of, 190
  variance of, 190
Binomial table, using, 191–193
Biometric signatures, 543
Bivariate data, graphing, 89–90
Bivariate relationships:
  defined, 89
  graphing, 89–94
Blind study, 156
Blocking, 373
Bound on the estimation error, 9
Box office receipts, nonparametric statistics, 400
Box plots, 79
  aids to the interpretation of, 81
  comparing, 82–83
  elements of, 81
  generating with a computer, 82
  stimulus reaction experiment, 82–83
  using the TI-84/TI-83 graphing calculator, 85
  and variation in the data set, 82

C

Categories, 444
Cell counts, 444
Cells, 444
Census, 7
Central Limit Theorem, 234–239
  applications of, 238
  demonstration of, 249
  statement of, 235
  using to find a probability, 236–237
Central tendency, 336
  defined, 51
  measures of, 65
  numerical measures of, 51–61
Chebyshev, Pafnuty L., 68
Chebyshev's rule, 68, 176–177
Chi-square (χ^2) distribution, 446
Chi-square (χ^2) goodness-of-fit test, 477
Chi-square test, 446
Class frequency, 28–29
Class intervals, 40
Class percentage, 29
Class relative frequency, 28–29
Classes, 444
  defined, 28
Cockroach trails (study), 351, 364–365, 391–392
  analysis of variance (ANOVA), 413–414
Combinations rule, 122
  using, 122–123
Combinatorial mathematics, 122
Complementary events, 130–131
Complements:
  coin tosses, 131
  coin-tossing experiment, 130–131
  rule of, 130
    applying, 131
Compound events, 127
Conclusion, 304–306
Conditional probability, 138–141
  defined, 138
  exit polls, 142
  formula for, 138–139
    applying in a two-way table, 139–140
  in a two-way table, 141
Confidence coefficient, 258
Confidence intervals, 10, 258
  using the TI-84/TI-83 graphing calculator, 262
    for ($\mu_1 - \mu_2$), 361–362
    for ($p1 - p2$), 438
    for a paired difference mean, 377–378
    for a population mean, 272–273
    for a population proportion, 281
Confidence Intervals for a Mean applet, 263, 273
Confidence Intervals for a Proportion applet, 281
Confidence level, 258

Contingency table, 454–470
  for marketing example, 454
  using the TI-84/TI-83 graphing calculator, 461–462
Continuous random variables:
  continuous probability distributions for, 196–197
  defined, 171
  values of, 171
Control group, 13, 156
Correlation by the Eye applet, 523
Critical values:
  of χ^2 (table), 579–580
  of Spearman's rank correlation coefficient (table), 581
  of t (table), 568
  of T_0, in the Wilcoxon paired difference signed rank test
    (table), 570
  of T_L and T_U, for Wilcoxon rank sum test (table), 569
Cumulative binomial probabilities, 190

D
Data, 4
  methods for describing sets of, 26–111
  from popular sources, 108
  qualitative, 11
  quantitative, 11
  types of, 11–12
Data collection, 12, 13–15
  designed experiments, 13
  observational studies, 13
  published source, 13
  scientific phenomena/business operations/government
    activities, growth in, 15–16
  surveys, 13
de Moivre, Abraham, 223
Degrees of freedom (df), 267, 357
Descriptive statistical problems, elements of, 10
Descriptive statistics, 4
  defined, 5
  distorting the truth with, 94–97
Designed experiments, 13
Destructive sampling, 270–271
Dichotomous responses, 183
Dimensions of classification, 454
Discrete probability distributions, properties of, 243
Discrete random variables, 170
  defined, 171
  examples of, 171–172
  probability distributions for, 173–179
  standard deviation of, 176
  using the TI-84/TI-83 graphing calculator, 178–179
  values of, 171
Distribution-free tests, 336
Dot plots, 38–39, 43
Dowsers (study), 483, 511–512, 522, 530
  straight-line regression model, estimating, 495

E
Efficient market theory, testing, 347
Efron, Bradley, 357
Empirical rule, 68, 176–177
Erosional hot spots, 161
Error of estimation, 9
Errors of prediction, 488
Estimated standard error:
  of the least squares slope $\beta 1$, 508
  of the regression model, 504
Events:
  complementary, 130–131
  compound, 127

  defined, 119
  probability of:
    calculating, 120
    coin-tossing experiment, 119–120
    defined, 120
Expected cell count, 447
Expected value:
  defined, 175
  finding, 175–176
Experimental unit, defined, 6
Experiments, 114–115
  defined, 115
  sample spaces, 116
Extrapolations, 533
"Eye Cue" test, 27
  interpreting descriptive statistics for, 72
  interpreting pie charts for test data, 33

F
F-distribution, 406
F-ratio, 407
F-statistic, 406
First-order probabilistic model, 486, *See also* Probabilistic
    models
Fisher, Ronald A., 404
Frequency, 40
Frequency function, 196

G
Galton, Francis, 485
Gauss, Carl F., 197
Gosset, William S., 267
Graphical distortions, of descriptive statistics, 94–97
Guassian distribution, 197

H
Hinges, 80–81
Histograms, 40–43
  defined, 43
  determining the number of classes in, 41
  interpreting histograms for the "eye cue" test data, 45
  using the TI-84/TI-83 graphing calculator, 43–44
Hypergeometric random variable model, 185fn
Hypotheses Test for a Mean applet, 307, 313
Hypotheses Test for a Proportion applet, 334
Hypothesis testing:
  for $(p_1 - p_2)$, using the TI-84/TI-83 graphing calculator, 439
  for $(\mu 1 - \mu 2)$, using the TI-84/TI-83 graphing calculator,
    362–363
  Bank of America *Keep the Change* program, 326
  one-tailed (one-sided) statistical test, 309
  population mean:
    carrying out, 311–312
    large-sample test of hypothesis about, 308–313
    setting up, 310
    using the TI-84/TI-83 graphing calculator, 319–320
  possible conclusions for, 312–313
  two-tailed (two-sided) hypothesis, 309
Hypothesis tests, 300–349, 304–305
  alternative hypothesis, 302, 305
  caution about, 305
  challenging a claim, 306
  conclusions and consequences for, 305
  elements of, 301–306
  null hypothesis, 302, 305
  rejection region, 303, 306
  test statistic, 302, 306

I

In-person surveys, 13
Incumbency rate, 441
Independence, checking for, 144–146
Independent classifications, 455
Independent events:
 Bank of America *Keep the Change* program, 148
 consumer complaint study, 145
 defined, 144
 die-tossing experiment, 144–145
 and multiplicative rule of probability, 141–147
 probability of intersection of, 146
 simultaneous occurrence of, 147
Independent sampling, 386–395
 large samples, 352–356
 small samples, 356–364
Independent variables, 486fn
Inferences, 5
 based on a single sample, 254–299
 hypothesis tests, 300–349
 based on two samples, 438
 using z-scores, 84
Inferential statistical problems, elements of, 10
Inferential statistics, 4
 defined, 5
Inner fences, 80–81
Internet (World Wide Web), 13
Interpolations, 533
Interquartile range (IQR), 79–80
Intersections, 127–130
 defined, 127
 die-tossing experiment, 127–128
 probabilities of, 127–128
Interval data, 11fn
Interval estimator, 258

K

KLEENEX® survey, 301, 307, 320, 333
Kruskal–Wallis H-Test, 413

L

LaPlace, Pierre-Simon, 236
Large-sample confidence interval:
 for *p*, 277–279
  conditions required for, 277
 for population mean, 256–261
  conditions required for, 260
 for $(\mu 1 - \mu 2)$, 354
Large-sample inferences about $(\mu 1 - \mu 2)$, 355
 conditions required for, 354
Large-sample inferences about μ_d, conditions required for, 374
Large-sample sign test, for a population median, 338
Large-sample test of hypothesis:
 about a population mean, 308–313
  conditions required for, 312
 for $(\mu 1 - \mu 2)$, 354
Least squares estimates, 489
 formulas for, 490
Least squares line, 489
Least squares method, 489
 Bank of America *Keep the Change* program, 495–496
Least squares prediction equation, 489
Level of significance, 306
Levy, Paul, 236
Line of means, 486
Lot acceptance sampling, 223–224
Lottery Buster, 113–114, 133

Lower quartile, 79
Lower-tailed test, 309, 325

M

Mail surveys, 13
Marginal probabilities, 455
Mean, 51, 175
 of binomial random variables, 190
 compared to median, 54
Mean of a population, 52
Mean of a sample, 51
Mean square for error (MSE), 406
Mean square for treatments (MST), 406
Mean versus Median applet, 58
Measure of reliability, 9
Measurement, 7, 114
Measurement error, 16–17
 defined, 17
Measures of central tendency, 51–61
Median, 52–53
 compared to mean, 54
 computing, 53
 defined, 52
Method of least squares, 489
Middle quartile, 79
MINITAB:
 binomial probabilities, normal probabilities, 249–250
 categorical data analyses with, 478–481
  inference on $(p_1 - p_2)$, 478–479
  one-way table, 479–480
  two-way table, 480–481
 comparing means with, 428–431
  ANOVA F-test, 430–431
  independent samples, 428–429
  paired samples, 429
  rank sum test, 429–430
  signed rank test, 429–430
 confidence intervals with, 297–299
 creating and listing data with, 23–25
 describing data with, 109–111
 generating a random sample with, 166–167
 hypothesis testing with, 348–349
 normal probabilities, 250–251
 sampling distributions for, 252–253
 simple linear regression with, 556–557
  rank correlation, 558
Misleading descriptive statistics, 96–97
Misleading statistics, identifying, 17
Modal class, 55
Mode:
 defined, 55
 finding, 55
Mound-shaped, symmetric distributions, 68
Multinomial experiments, 444–454
 binomial vs., 450
 identifying, 444–445
 properties of, 444
Multinomial probabilities, test of hypothesis about, 447–448
Multiplicative rule of probability:
 applying, 142–143
 defined, 141
 famous psychological experiment, 142
 and independent events, 141–147
Mutually exclusive events, 132–133
 defined, 133
 probability of union of two, 133

N

Neyman, Jerzy, 259
Nielsen survey, 13
Nightingale, Florence, 4
Nominal data, 11fn
Nonparametric statistics, box office receipts, 400
Nonparametric test about a population median, 336–339
Nonparametrics, defined, 336
Nonrandom samples, 16
Nonresponse bias, 16–17
    defined, 17
Nonstandard normal probability, using the TI-84/TI-83 graphing
        calculator, 205–206
Normal curve areas (table), 567
Normal data, checking for, 214–216
Normal distributions, 197–213, 336
    approximating a binomial distribution with, 221–226
    approximating binomial probability with, 223–225
    defined, 197
    property of, 202
    standard, 198
Normal probability:
    finding, 202–203
    using to make an inference, 204–205
Normal probability plot:
    defined, 214
    using the TI-84/TI-83 graphing calculator, 216–217
Normal random variable, 197
    probability distribution for, 198
    standard, defined, 199
    steps for finding a probability corresponding to, 203
Normal table, using in reverse, 207–209
Normality, descriptive methods for assessing, 213–218
Null hypothesis, 302, 305
    selecting, 309
Numerical descriptive measures, 51

O

Observation, 114
Observational studies, 13
Observed count, 455
Observed significance levels, 315–320
One-tailed (one-sided) statistical test, 309
    rejection region, 309
One-variable descriptive statistics, using the TI-84/TI-83
        graphing calculator, 57–58
One-way ANOVA, 410
One-way table, 445, 447–448
Ordinal data, 11fn
Outer fences, 80–81
Outliers:
    defined, 79
    detecting in the "Eye Cue" test data, 86
    methods for detecting, 79–88
    potential, 80
    rules of thumb for detecting, 85

P

p-values, 315–320
    calculating, for a hypothesis test, 316–317
    reporting test results as, 318
    for small-sample test of population mean, 325–326
    for a test about a population proportion, 332–333
    for a test of $(\mu_1 - \mu_2)$, 355–356
    using, 318–319
Paired difference confidence interval, 373
Paired difference experiments, 370–383, 385, 395–403
    blocking, 373
    defined, 372
    examples of, 373
    randomized block experiment, 373
Paired difference mean, hypothesis test for, using the TI-84/TI-
        83 graphing calculator, 378–379
Paired difference test of hypothesis, 374
Parameters, defined, 227
Pareto diagram, 30–31
Pareto, Vilfredo, 30
Pascal, Blaise, 145
Pearson, Egon S., 304
Pearson, Karl, 446
Pearson product moment coefficient of correlation, 515
Percentage points, F-distributions (tables), 571–578
Percentile rankings, and large data sets, 75
Percentiles, finding/interpreting, 75–76
Perfect negative correlation, 536, 538
Perfect positive correlation, 536, 538
Picasso, Pablo, 40
Pie chart, 30–31
Pilot study, conducting, 296
Placebo effect, 336
Political polls, 13
Pooled sample estimator $\sigma 2$, 356
Population, 8fn
    comparing, nonparametric test for, 395–403
    defined, 6
Population correlation coefficient, 518
Population data sets, use of term, 8fn
Population mean(s), 336
    comparing, 350–431
        analysis of variance (ANOVA), 403–419
        box office receipts (activity), 377
        independent sampling, 352–365
        nonparametric test for, 386–395
        paired difference experiments, 370–383
    estimating, 283–285
    interpretation of a confidence interval for, 260–261
    large-sample confidence interval for, 256–261
        conditions required for, 260
    large-sample test of hypothesis about, 308–313
    sample size, determining, 383–386
    small-sample confidence interval for, 266–272
    small-sample test of hypothesis about, 322–326
    symbol for, 52
Population median:
    nonparametric test about, 336–339
    sign test for, 338
Population proportions:
    adjusted confidence interval procedure, 279
        using, 279–280
    categorical probabilities, testing:
        multinomial experiment, 444–454
        two-way (contingency) table, 454–470
    comparing, 432–481
        independent sampling, 434–442
    confidence interval for, 276–281
    estimating, 276–277, 285–287
    "guesstimates" of, 478
    independent sampling:
        conditions required for valid large-sample inferences, 436
        large-sample $100(1 - \alpha)\%$ confidence interval, 435
        large-sample test of hypothesis, 436
        sampling distribution, properties of, 434
    large-sample test of hypothesis about, 329–333
    observed significance level of a test for $p_1 - p_2$, 437
    sample size determination, 442–444
        for estimating $(p_1 - p_2)$, 443
Population rank correlation coefficient, 540
Population variance, 63, 176

Population *z*-score, 76
Potential outliers, 80
Primary source, 13fn
Probabilistic models:
　defined, 484
　deterministic model, 484
　deterministic relationship, 484
　general form of, 485
　probabilistic relationship, 485
　random error, 484–485
　straight-line model, 486
Probability, 112–167
　additive rule of, 132
　calculating, 120–121
　of a complementary event, 130–131
　finding from a two-way table, 128–129
　and inference, 114
　multiplicative rule of, 141–147
　simulating with a deck of cards, 165
Probability density function, 196
Probability distributions, 196–197
　defined, 174
　for discrete random variables, 173–179
　　requirements for, 174
　finding, coin-tossing experiment, 173–174
　and random variables, 168–253
　using a formula, 174–175
*p*th percentile, 75
Published source, 13

Q
Qualitative data, 11
　describing, 28–38
Quantitative data, 11
　dot plots, 38–39, 43
　graphical methods for describing, 38–49
　histograms, 40–43
　stem-and-leaf display, 39, 43
Quartiles, 79

R
Random error, 484–485
Random-number generators, 154
Random numbers (table), 560–562
Random Numbers applet, 18–19, 159, 180
Random sample:
　defined, 14, 154
　selecting, 154–155
Random sampling, 154–157
Random variables:
　Chebyshev's rule, 176–177
　continuous, values of, 171
　defined, 169–170
　discrete, values of, 171
　empirical rule, 176–177
　and probability distributions, 168–253
　types of, 170–172
　variance, 176
Randomization in a designed experiment, 156–157
Randomized block experiment, 373
Range, 62
Rank sums, 387–388
Rare-event approach, 76
Ratio data, 11fn
Regression analysis (modeling), 486
Regression by Eye applet, 497
Regression line, 489
Rejection region, 303
　one- and two-tailed hypothesis, 309

Relative frequency distribution, departure from normality, 271
Relative standing, numerical measures of, 75–79
Reliability, 9
　of an inference, 10
　measure of, defined, 9
Representative sample, 13–15
Research hypothesis, 302
Resource constraints, 9

S
Sample, 8fn
　defined, 7
Sample data sets, use of term, 8fn
Sample from a Population applet, 194, 211
Sample mean:
　computing, 51
　formula for, 51
　symbol for, 52
Sample median, calculating, 52
Sample points:
　coin-tossing experiment, 115
　defined, 115
　probability of, 116–117
　　collections of sample points, 118–119
　　die-tossing experiment, 119
　　hotel guest room survey, 118
　probability rules for, 118
Sample size, 52
　for comparing two means, 385
　determining, 283–288
　　for $100(1-\alpha)\%$ confidence intervals of μ, 285
　　for $100(1-\alpha)\%$ confidence intervals of p, 286
　for estimating μ, 285–286
　for estimating p, 287
Sample spaces, 115
　defined, 116
　experiments, 116
Sample standard deviation, defined, 63
Sample statistic, defined, 228
Sample surveys, 433
Sample variance:
　defined, 63
　formula for, 63
Sample *z*-score, 76
Sampling, 4
Sampling distribution, 354
Sampling Distribution applet, 240
Sampling distributions, 227–232
　defined, 228
　finding, 229–231
　parameters, 227
　simulating, 229, 231–232
　standard error of the mean, 235
Sampling error, 284
Scaffold-drop surveys, 555
Scattergram (scatterplot), 89, 488
　using the TI-84/TI-83 graphing calculator, 91–92
Scholastic Aptitude Test (SAT) scores, 28
Secondary source, 13fn
Selection bias, 16–17
　defined, 17
Sign test, 339
　for population median, 338
　valid application of, conditions required for, 338
Significance level, 306
　observed, 315–320
Simple event, use of term, 115fn
Simple linear regression, 482–558
　applying to your favorite data, 556

Simple linear regression, *continued*
  coefficient of correlation, 515–519, 522–525
    Bank of America *Keep the Change* program, 518–519
    population, 518
    using, 517–518
  coefficient of determination, 519–525
    practical interpretation of, 521
  complete example, 533–536
  conditions required for a valid confidence interval, 510
  fitting the model:, 487–502
  inferences about slope β_1, 507–512
  interpreting *p*-values for β coefficients in regression, 510
  least squares approach, 487–502
  model assumptions, 502–507
  model utility, assessing, 507–515
  nonparametric test for correlation, 536–545
  probabilistic models, 484–487
    defined, 484
    deterministic model, 484
    deterministic relationship, 484
    general form of, 485
    probabilistic relationship, 485
    random error, 484–485
    straight-line model, 486
  Spearman's rank correlation coefficient, applying, 539–540
  testing the regression slope β_1, 508–510
  using the model for estimation and prediction, 525–533
  using the TI-84/TI-83 graphing calculator, 493–494
Simple linear regression analysis, 484
Simulating the Probability of a Head with an Unfair Coin applet, 194
Simulating the Probability of a Head with a Fair Coin applet, 124–125, 180
Simulating the Probability of Rolling a 3 or 4 applet, 135
Simulating the Probability of Rolling a 6 applet, 124, 135, 149
Simulating the Stock Market applet, 194–195
Skewed data set, defined, 54
Skewness, 54
Small-sample confidence interval:
  for population mean, 266–272
  for μ, 270
    valid, conditions required for, 270
  for ($\mu_1 - \mu_2$), 358–360
    independent samples, 357
Small-sample inferences:
  about ($\mu_1 - \mu_2$), conditions required for, 358
  about μd, conditions required for, 374
Small-sample test of hypothesis:
  for ($\mu_1 - \mu_2$), independent samples, 357–358
  about a population mean, 322–326
Smple random sample, 154fn
Spearman, Charles, 537
Spearman's rank correlation coefficient, 537–538
  applying, 539–540
  summary of, 541
Sporting News, The, 13
Spread, 52
  measure of, 62
Sprint time (study), 255
  athletes required to participate in the training program, 287–288
  improvement after speed training, 280
Standard deviation, 446
  of binomial random variables, 190
  Chebyshev's rule, 68
  checking the calculation of, 70
  of discrete random variables, 176
  empirical rule, 68

  interpreting, 67–75, 68
    rat-in-maze experiment, 69–70
  symbol for, 63
Standard Deviation applet, 66, 87
Standard error, 354
Standard error of the mean, 235
Standard normal table, 199–201
Standard normal distribution, 198–199
  defined, 199
Standard normal probabilities, using the TI-84/TI-83 graphing calculator, 203–204
Standard normal random variable, defined, 199
Statistical Abstract of the United States, 13
Statistical applications, types of, 4–5
Statistical experiments, 114
Statistical inference:
  defined, 8
  making, 71–72
  rare-event approach, 76
Statistical thinking, 15–16
  defined, 16
Statistics:
  defined, 4
  fundamental elements of, 6–10
  misleading, identifying, 17
  role in critical thinking, 15–17
  science of, 4
Stem-and-leaf display, 39–40, 43
Straight-line model, 484
Straight-line Probabilistic Model, 486
Subset, 7
Sum of errors, 488–489
Sum of squares for treatments (SST), 405
Sum of squares of the errors (SSE), 405, 488
Summation notation, 49–61
  meaning of, 50
Super weapons development, 169, 209–210, 217–218
Survey of Current Business, 13
Surveys, 13

T
t-statistic, 267
Target parameter, 255–256
  defined, 256
  determining, 256, 306
Telephone surveys, 13
Test of hypothesis, *See* Hypothesis tests
Test statistic, 302, 306
TI-84/TI-83 graphing calculator:
  analysis of variance (ANOVA), 411
  binomial probabilities, 193–194
  box plots, 85
  confidence intervals, 262
    for ($p_1 - p_2$), 438
    for a paired difference mean, 377–378
    for ($\mu_1 - \mu_2$), 361–362
  contingency table, 461–462
  discrete random variables, 178–179
  histograms, 43–44
  hypothesis test:
    for ($p_1 - p_2$), 439
    for ($\mu_1 - \mu_2$), 362–363
  hypothesis testing, 319–320
  nonstandard normal probability, 205–206
  normal probability plot, 216–217
  one-variable descriptive statistics, 57–58
  paired difference mean, hypothesis test for, 378–379
  population mean, 319–320
  scattergram (scatterplot), 91

simple linear regression, 493–494
   standard normal probabilities, 203–204
Treatment group, 13, 156
Tree diagrams, 143–144
Tukey, John, 40
Two-tailed (two-sided) hypothesis, 309
   rejection region, 309
Two-way (contingency) table, 454–470
   dependence, pattern of, 457
   finding expected cell counts for, 456
   for marketing example, 455
   test for independence, 458
   using the TI-84/TI-83 graphing calculator, 461–462
Type I decision error, 302
Type II error, 304–305

U

Uncertainty, 9
Unconditional probabilities, 138
Unethical statistical practice, 16
Unions, 127–130
   defined, 127
   die-tossing experiment, 127–128
   probabilities of, 127–128
   of two mutually exclusive events, 133
Upper quartile, 79
Upper-tailed test, 309, 316, 325, 332
USA WEEKEND Teen Survey (Are Boys Really from Mars and
      Girls from Venus?), 3
   critically assessing the results of, 17
   identifying the data collection method and data type for, 15
   identifying the population, sample, and inference for, 10

V

Variability, 51, 52, 62–67
   measures of, 62, 65
Variables, defined, 6

Variance:
   of binomial random variables, 190
   random variables, 176
   symbol for, 63
Variation:
   computing measures of, 64
   measures of, finding on a printout, 64–65
Venn diagrams, 116, 127–128, 169
Venn, John, 117

W

Wall Street Journal, The, 13
Weighted average, 357
Wells, H. G., 15–16
Whiskers, 80–81
Wilcoxon, Frank, 386–387
Wilcoxon rank sum test, 386–388
   applying, 389–390
   independent samples, 388–389
   for large samples, 391
   valid, conditions required for, 389
Wilcoxon signed rank test:
   applying, 398–399
   defined, 396
   for large samples, 400
   for a paired difference experiment, 397
   valid, conditions required for, 398
Wilson's adjustment for estimating p, 279

Z

z-score:
   defined, 76
   finding, 76–77
   inference using, 84
   interpreting for mound-shaped distributions of data, 77
   population, 76
   sample, 76

Credits

Using Technology box image: Spohn Matthieu
Calculator box image: Courtesy of Texas Instruments

Chapter 1 p. 3 © Abimelec Olan/Courtesy of www.istock.com; p. 8, 10, 15, 17 Steve Gorton © Dorling Kindersley; p. 12, 15 Courtesy of www.istock.com; p. 14 © Justin Horrocks/Courtesy of www.istock.com

Chapter 2 p. 26 Getty Images- Stockbyte; pp. 27, 33, 45, 72, 86 © Skip O'Donnell/Courtesy of www.istockphoto.com; p. 31 Courtesy of www.istockphoto.com; p. 54 © Diane Labombarbe/Courtesy of www.istockphoto.com; p. 55 (top) Paul Bricknell © Dorling Kindersley; p. 55 (bottom) © Alexey Gostev/Courtesy of www.istockphoto.com; p. 69 © Emilia Stasiak/Courtesy of www.istockphoto.com; p. 97 © Bonnie Jacobs/Courtesy of www.istockphoto.com

Chapter 3 p. 113, 123, 133, 147 Simon Askham/Courtesy of www.istock.com; p. 113 © Damon Higgins/The Palm Beach Post; p. 115 Clive Streeter © Dorling Kindersley; p 118 © Christopher O'Driscoll/Courtesy of www.istock.com; p. 119, 156, 159, 160 Courtesy of www.istock.com; p. 120, 147 © Tim Starkey/Courtesy of www.istock.com; p. 127 Pekka Jaakkola/Courtesy of www.istock.com; p. 128 © Yvonne Chamberlain/Courtesy of www.istock.com; p. 139 © Kenny Haner/Courtesy of www.istock.com; p. 162 © Dar Yang Yan/Courtesy of www.istock.com; p. 166 © Neil Sullivan/Courtesy of www.istock.com

Chapter 4 p. 180 Photodisc/Getty Images; p. 181, 207, 218 © Christoph Ermel/Courtesy of www.istock.com; p. 182 (top) Dave King © Dorling Kindersley; p. 182 (bottom) Courtesy of www.istock.com; p. 198 © Peter Miekuz/Courtesy of www.istock.com; p. 204 © Danny Bailey/Courtesy of www.istock.com; p. 217 © Steve Luker/Courtesy of www.istock.com

Chapter 5 p. 226 Todd Pearson/Image Bank/Getty Images; p. 227, 245, 253 © Craig DeBourbon/Courtesy of www.istock.com; p. 230, 240, 244, 250 Courtesy of www.istock.com; p. 238 © Nathan Gutshall-Kresge/Courtesy of www.istock.com; p. 264 © Matt Matthews/Courtesy of www.istock.com

Chapter 6 p. 277, 294 © Stephanie Horrocks/Courtesy of www.istock.com; p. 280 © Doris Kindersley; p. 291 © Aldo Murillo/Courtesy of www.istock.com

Chapter 7 p. 304 Ed Honowitz/Taxi/Getty Images; p. 305, 311, 330, 337 (bottom) © Jason Lugo/Courtesy of www.istock.com; p. 320 © Jaroslaw Wojcik/Courtesy of www.istock.com; p. 326 © Uyen Le/Courtesy of www.istock.com; p. 329 © Leah-Anne Thompson/Courtesy of www.istock.com; p. 335 © Dan Thomberg/Courtesy of www.istock.com; p. 337 (top) © Andrew Dernie/Courtesy of www.istock.com

Chapter 8 p. 350 Andrew Errington/Photographer's Choice/Getty Images; p. 351, 357, 370, 383 Andy Crawford © Dorling Kindersley; p. 360, 367, 389 © Russel Gough/Courtesy of www.istock.com; p. 368 © Lisa F. Young/Courtesy of www.istock.com; p. 374 © Robery Byron/Courtesy of www.istock.com; p. 381 Courtesy of www.istock.com; p. 395 Andrew Johnson/Courtesy of www.istock.com

Chapter 9 p. 409, 423, 449 © Maartje van Caspel/Courtesy of www.istock.com; p. 410 © Chris Hutchison/Courtesy of www.istock.com; p. 417 © Andrzej Tokarski/Courtesy of www.istock.com; p. 434, 446 Courtesy of www.istock.com; p. 458 © Russel Gough/Courtesy of www.istock.com

Chapter 10 p. 481, 498, 508, 537 © Lee Pettet/Courtesy of www.istock.com; p. 484, 493 Matthew Ward © Dorling Kindersley; p. 488 Susanna Price © Dorling Kindersley; p. 518 Alex Balako/Courtesy of www.istock.com

Chapter 11 p. 559, p. 571, 587, 598, 606 © Jim DeLillo/Courtesy of www.istock.com; p. 593 © Wojtek Wojtowicz /Courtesy of www.istock.com

Chapter 12 p. 626 Bruce Laurance/Image Bank/Getty Images p. 627, 650, 692, 716 © Sean Locke/Courtesy of www.istock.com; p. 631, 649, 655, 666, 720 Courtesy of www.istock.com; p. 661 © Jim Jurica/Courtesy of www.istock.com; p. 673 © Cole Vineyard/Courtesy of www.istock.com; p. 688 © Thomas Mounsey/Courtesy of www.istock.com; p. 699 © Kutay Tanir/Courtesy of www.istock.com

Chapter 13 p. 743, 749, 762 © Lise Gagne/Courtesy of www.istock.com; p. 744 Courtesy of www.istock.com

Chapter 14 (CD-only) 14-3, 14-8, 14-16, 14-31, 14-46 © Nathan Reighard/Courtesy of www.istock.com

Selected Formulas

CHAPTER 2

Relative Frequency = (frequency)/n

$$\bar{x} = \frac{\Sigma x}{n}$$

$$s^2 = \frac{\Sigma(x - \bar{x})^2}{n - 1} = \frac{\Sigma x^2 - \frac{(\Sigma x)^2}{n}}{n - 1}$$

$$s = \sqrt{s^2}$$

$$z = \frac{x - \mu}{\sigma} = \frac{x - \bar{x}}{s}$$

Chebyshev = At least $\left(1 - \dfrac{1}{k^2}\right)100\%$

$$\text{IQR} = Q_U - Q_L$$

CHAPTER 3

$$P(A^c) = 1 - P(A)$$

$$P(A \cup B) = P(A) + P(B) - P(A \cap B)$$
$$= P(A) + P(B) \text{ if } A \text{ and } B \text{ mutually exclusive}$$

$$P(A \cap B) = P(A|B) \cdot P(B) = P(B|A) \cdot P(A)$$
$$= P(A) \cdot P(B) \text{ if } A \text{ and } B \text{ independent}$$

$$P(A|B) = \frac{P(A \cap B)}{P(B)}$$

$$\binom{N}{n} = \frac{N!}{n!(N - n)!}$$

CHAPTER 4

KEY FORMULAS

Random Variable	Prob. Dist'n	Mean	Variance
General Discrete:	Table, formula, or graph for $p(x)$	$\displaystyle\sum_{\text{all } x} x \cdot p(x)$	$\displaystyle\sum_{\text{all } x} (x - \mu)^2 \cdot p(x)$
Binomial:	$p(x) = \dbinom{n}{x} p^x q^{n-x}$ $x = 0, 1, 2, \ldots, n$	np	npq
Normal:	$f(x) = \dfrac{1}{\sigma\sqrt{2\pi}} e^{-\frac{1}{2}[(x-\mu)/\sigma]^2}$	μ	σ^2
Standard Normal:	$f(z) = \dfrac{1}{\sqrt{2\pi}} e^{-\frac{1}{2}(z)^2}$ $z = (x - \mu)/\sigma$	$\mu = 0$	$\sigma^2 = 1$
Sample Mean: (large n)	$f(\bar{x}) = \dfrac{1}{\sigma_{\bar{x}}\sqrt{2\pi}} e^{-\frac{1}{2}[(\bar{x}-\mu)/\sigma_{\bar{x}}]^2}$	$\mu_{\bar{x}} = \mu$	$\sigma_{\bar{x}}^2 = \sigma^2/n$

CHAPTER 5

CI for μ: $\bar{x} \pm (z_{\alpha/2})\sigma/\sqrt{n}$ (large n)

$\bar{x} \pm (t_{\alpha/2})s/\sqrt{n}$ (small n, σ unknown)

CI for p: $\hat{p} \pm z_{\alpha/2}\sqrt{\dfrac{\hat{p}\hat{q}}{n}}$

Estimating μ: $n = (z_{\alpha/2})^2(\sigma^2)/(SE)^2$

Estimating p: $n = (z_{\alpha/2})^2(pq)/(SE)^2$

CHAPTER 6

Test for μ: $z = \dfrac{\bar{x} - \mu}{\sigma/\sqrt{n}}$ (large n)

$t = \dfrac{\bar{x} - \mu}{s/\sqrt{n}}$ (small n, σ unknown)

Test for p: $z = \dfrac{\hat{p} - p_0}{\sqrt{p_0 q_0/n}}$

Nonparametric test for median (η):

S = # measurements greater than η (small n)

$z = \dfrac{(S - .5) - .5n}{.5\sqrt{n}}$ (large n)

CHAPTER 7

CI for $\mu_1 - \mu_2$:

$(\bar{x}_1 - \bar{x}_2) \pm z_{\alpha/2}\sqrt{\dfrac{\sigma_1^2}{n_1} + \dfrac{\sigma_2^2}{n_2}}$ (large n_1 and n_2)

Test for $\mu_1 - \mu_2$:

$z = \dfrac{(\bar{x}_1 - \bar{x}_2) + (\mu_1 - \mu_2)}{\sqrt{\dfrac{\sigma_1^2}{n_1} + \dfrac{\sigma_2^2}{n_2}}}$ (large n_1 and n_2)

$s_p^2 = \dfrac{(n_1 - 1)s_1^2 + (n_2 - 1)s_2^2}{n_1 + n_2 - 2}$

CI for $\mu_1 - \mu_2$:

$(\bar{x}_1 - \bar{x}_2) \pm t_{\alpha/2}\sqrt{s_p^2\left(\dfrac{1}{n_1} + \dfrac{1}{n_2}\right)}$ (small n_1 and/or n_2)

Test for $\mu_1 - \mu_2$:

$t = \dfrac{(\bar{x}_1 - \bar{x}_2) - (\mu_1 - \mu_2)}{\sqrt{s_p^2\left(\dfrac{1}{n_1} + \dfrac{1}{n_2}\right)}}$ (small n_1 and/or n_2)

Estimating $\mu_1 - \mu_2$: $n_1 = n_2 = (z_{\alpha/2})^2 (\sigma_1^2 + \sigma_2^2)/(SE)^2$

CI for μ_d: $\bar{x}_d \pm t_{\alpha/2}\dfrac{s_d}{\sqrt{n}}$

Test for μ_d: $t = \dfrac{\bar{x}_d - \mu_d}{s_d/\sqrt{n}}$

ANOVA Test for independent samples design:
$F = MST/MSE$

CHAPTER 8

CI for $p_1 - p_2$: $(\hat{p}_1 - \hat{p}_2) \pm z_{\alpha/2}\sqrt{\dfrac{\hat{p}_1\hat{q}_1}{n_1} + \dfrac{\hat{p}_2\hat{q}_2}{n_2}}$

Test for $p_1 - p_2$: $z = \dfrac{(\hat{p}_1 - \hat{p}_2) - (p_1 - p_2)}{\sqrt{\hat{p}\hat{q}\left(\dfrac{1}{n_1} + \dfrac{1}{n_2}\right)}}$

$\hat{p} = \dfrac{x_1 + x_2}{n_1 + n_2}$

Estimating $p_1 - p_2$: $n_1 = n_2 = (z_{\alpha/2})^2 (p_1 q_1 + p_2 q_2)/(SE)^2$

Multinomial test: $\chi^2 = \sum\dfrac{(n_i - E_i)^2}{E_i}$

$E_i = n(p_{i0})$

Contingency table test: $\chi^2 = \sum\dfrac{(n_{ij} - E_{ij})^2}{E_{ij}}$

$E_{ij} = \dfrac{R_i C_j}{n}$

CHAPTER 9

$SS_{xx} = \Sigma x^2 - \dfrac{(\Sigma x)^2}{n}$

$SS_{yy} = \Sigma y^2 - \dfrac{(\Sigma y)^2}{n}$

$SS_{xy} = \Sigma xy - \dfrac{(\Sigma x)(\Sigma y)}{n}$

$\hat{y} = \hat{\beta}_0 + \hat{\beta}_1 x$

CHAPTER 9 (cont'd)

$$\hat{\beta}_1 = \frac{SS_{xy}}{SS_{xx}}$$

$$\hat{\beta}_0 = \bar{y} - \hat{\beta}_1\bar{x}$$

$$r = \frac{SS_{xy}}{\sqrt{SS_{xx}}\sqrt{SS_{yy}}}$$

$$s^2 = \frac{SSE}{n-2}$$

$$s = \sqrt{s^2}$$

$$r^2 = \frac{SS_{yy} - SSE}{SS_{yy}}$$

CI for β_1: $\hat{\beta}_1 \pm (t_{\alpha/2})s/\sqrt{SS_{xx}}$

Test for β_1: $t = \dfrac{\hat{\beta}_1 - 0}{s/\sqrt{SS_{xx}}}$

CI for $E(y)$ when $x = x_{\hat{p}}$: $\hat{y} \pm t_{\alpha/2}s\sqrt{\dfrac{1}{n} + \dfrac{(x_p - \bar{x})^2}{SS_{xx}}}$

CI for y when $x = x_{\hat{p}}$: $\hat{y} \pm t_{\alpha/2}s\sqrt{1 + \dfrac{1}{n} + \dfrac{(x_p - \bar{x})^2}{SS_{xx}}}$

Rank correlation: $r_s = 1 - \dfrac{6\Sigma d_i^2}{n(n^2 - 1)}$

where d_i = difference in x and y ranks for ith observation

SINGLE PC LICENSE AGREEMENT AND LIMITED WARRANTY

READ THIS LICENSE CAREFULLY BEFORE OPENING THIS PACKAGE. BY OPENING THIS PACKAGE, YOU ARE AGREEING TO THE TERMS AND CONDITIONS OF THIS LICENSE. IF YOU DO NOT AGREE, DO NOT OPEN THE PACKAGE. PROMPTLY RETURN THE UNOPENED PACKAGE AND ALL ACCOMPANYING ITEMS TO THE PLACE YOU OBTAINED THEM [[FOR A FULL REFUND OF ANY SUMS YOU HAVE PAID FOR THE SOFTWARE]]. *THESE TERMS APPLY TO ALL LICENSED SOFTWARE ON THE DISK EXCEPT THAT THE TERMS FOR USE OF ANY SHAREWARE OR FREEWARE ON THE DISKETTES ARE AS SET FORTH IN THE ELECTRONIC LICENSE LOCATED ON THE DISK:*

1. GRANT OF LICENSE and OWNERSHIP: The enclosed computer programs and data ("Software") are licensed, not sold, to you by Pearson Education, Inc. publishing as Prentice-Hall, Inc. ("We" or the "Company") in consideration of your purchase or adoption of the accompanying Company textbooks and/or other materials, and your agreement to these terms. We reserve any rights not granted to you. You own only the disk(s) but we and/or our licensors own the Software itself. This license allows individuals who have purchased the accompanying Company textbook to use and display their copy of the Software on a single computer (i.e., with a single CPU) at a single location for academic use only, so long as you comply with the terms of this Agreement. You may make one copy for back up, or transfer your copy to another CPU, provided that the Software is usable on only one computer.

2. RESTRICTIONS: You may not transfer or distribute the Software or documentation to anyone else. Except for backup, you may not copy the documentation or the Software. You may not network the Software or otherwise use it on more than one computer or computer terminal at the same time. You may not reverse engineer, disassemble, decompile, modify, adapt, translate, or create derivative works based on the Software or the Documentation. You may be held legally responsible for any copying or copyright infringement that is caused by your failure to abide by the terms of these restrictions.

3. TERMINATION: This license is effective until terminated. This license will terminate automatically without notice from the Company if you fail to comply with any provisions or limitations of this license. Upon termination, you shall destroy the Documentation and all copies of the Software. All provisions of this Agreement as to limitation and disclaimer of warranties, limitation of liability, remedies or damages, and our ownership rights shall survive termination.

4. LIMITED WARRANTY AND DISCLAIMER OF WARRANTY: Company warrants that for a period of 60 days from the date you purchase this SOFTWARE (or purchase or adopt the accompanying textbook), the Software, when properly installed and used in accordance with the Documentation, will operate in substantial conformity with the description of the Software set forth in the Documentation, and that for a period of 30 days the disk(s) on which the Software is delivered shall be free from defects in materials and workmanship under normal use. The Company does not warrant that the Software will meet your requirements or that the operation of the Software will be uninterrupted or error-free. Your only remedy and the Company's only obligation under these limited warranties is, at the Company's option, return of the disk for a refund of any amounts paid for it by you or replacement of the disk. THIS LIMITED WARRANTY IS THE ONLY WARRANTY PROVIDED BY THE COMPANY AND ITS LICENSORS, AND THE COMPANY AND ITS LICENSORS DISCLAIM ALL OTHER WARRANTIES, EXPRESS OR IMPLIED, INCLUDING WITHOUT LIMITATION, THE IMPLIED WARRANTIES OF MERCHANTABILITY AND FITNESS FOR A PARTICULAR PURPOSE. THE COMPANY DOES NOT WARRANT, GUARANTEE OR MAKE ANY REPRESENTATION REGARDING THE ACCURACY, RELIABILITY, CURRENTNESS, USE, OR RESULTS OF USE, OF THE SOFTWARE.

5. LIMITATION OF REMEDIES AND DAMAGES: IN NO EVENT, SHALL THE COMPANY OR ITS EMPLOYEES, AGENTS, LICENSORS, OR CONTRACTORS BE LIABLE FOR ANY INCIDENTAL, INDIRECT, SPECIAL, OR CONSEQUENTIAL DAMAGES ARISING OUT OF OR IN CONNECTION WITH THIS LICENSE OR THE SOFTWARE, INCLUDING FOR LOSS OF USE, LOSS OF DATA, LOSS OF INCOME OR PROFIT, OR OTHER LOSSES, SUSTAINED AS A RESULT OF INJURY TO ANY PERSON, OR LOSS OF OR DAMAGE TO PROPERTY, OR CLAIMS OF THIRD PARTIES, EVEN IF THE COMPANY OR AN AUTHORIZED REPRESENTATIVE OF THE COMPANY HAS BEEN ADVISED OF THE POSSIBILITY OF SUCH DAMAGES. IN NO EVENT SHALL THE LIABILITY OF THE COMPANY FOR DAMAGES WITH RESPECT TO THE SOFTWARE EXCEED THE AMOUNTS ACTUALLY PAID BY YOU, IF ANY, FOR THE SOFTWARE OR THE ACCOMPANYING TEXTBOOK. BECAUSE SOME JURISDICTIONS DO NOT ALLOW THE LIMITATION OF LIABILITY IN CERTAIN CIRCUMSTANCES, THE ABOVE LIMITATIONS MAY NOT ALWAYS APPLY TO YOU.

6. GENERAL: THIS AGREEMENT SHALL BE CONSTRUED IN ACCORDANCE WITH THE LAWS OF THE UNITED STATES OF AMERICA AND THE STATE OF NEW YORK, APPLICABLE TO CONTRACTS MADE IN NEW YORK, AND SHALL BENEFIT THE COMPANY, ITS AFFILIATES AND ASSIGNEES. HIS AGREEMENT IS THE COMPLETE AND EXCLUSIVE STATEMENT OF THE AGREEMENT BETWEEN YOU AND THE COMPANY AND SUPERSEDES ALL PROPOSALS OR PRIOR AGREEMENTS, ORAL, OR WRITTEN, AND ANY OTHER COMMUNICATIONS BETWEEN YOU AND THE COMPANY OR ANY REPRESENTATIVE OF THE COMPANY RELATING TO THE SUBJECT MATTER OF THIS AGREEMENT. If you are a U.S. Government user, this Software is licensed with "restricted rights" as set forth in subparagraphs (a)-(d) of the Commercial Computer-Restricted Rights clause at FAR 52.227-19 or in subparagraphs (c)(1)(ii) of the Rights in Technical Data and Computer Software clause at DFARS 252.227-7013, and similar clauses, as applicable.

Should you have any questions concerning this agreement or if you wish to contact the Company for any reason, please contact in writing:

Director, Media Production
Pearson Education
1 Lake Street
Upper Saddle River, NJ 07458